Campbell / Spalding

Analyzing and Troubleshooting Single-Screw Extruders

Gregory A. Campbell
Mark A. Spalding

Analyzing and Troubleshooting Single-Screw Extruders

2nd Edition

HANSER

Hanser Publishers, Munich
Hanser Publications, Cincinnati

The Authors:

Gregory A. Campbell, Ph. D.
Chief Technical Officer, Castle Associates, 1145 Mason Bay Road, Jonesport, ME 04649, USA

Mark A. Spalding, Ph. D.
Fellow, Dow, Inc., 1702 Building, Midland, MI 48667, USA

Distributed by:
Carl Hanser Verlag
Postfach 86 04 20, 81631 Munich, Germany
eMail: info@hanser.de
www.hanser-fachbuch.de

Library of Congress Control Number: 2020946059

Editor: Dr. Mark Smith
Production Management: Der Buchmacher, Arthur Lenner, Windach
Coverconcept: Marc Müller-Bremer, www.rebranding.de, Munich
Coverdesign: Max Kostopoulos
Typesetting: Kösel Media GmbH, Krugzell, Germany

ISBN: 978-1-56990-784-9
E-Book ISBN: 978-1-56990-785-6

Preface

Classically, all prior extrusion books are based on barrel rotation physics. Literature developed over the past 15 years has led to this first book to be published based on the actual physics of the process–screw rotation physics. After the theories and the math models are developed in the first nine chapters, the models are then used to solve actual commercial problems in the remainder of the book. Realistic case studies are unique in that they describe the problem as viewed by the plant engineers and provide the actual dimensions of the screws. Knowledge is developed using a series of hypotheses that are developed and then tested, which allows a series of technical solutions. Several actual solutions are proposed with the final results that solve the problem then clearly presented. Overall, there is not a book on the market with this level of detail and disclosure. New knowledge in this book will be highly useful for production engineers, technical service engineers working with customers, consultants specializing in troubleshooting and process design, and process researchers and designers that are responsible for processes that run at maximum rates and maximum profitability.

Debugging and troubleshooting single-screw extruders is an important skill set for plant engineers since all machines will eventually have a deterioration in their performance or a catastrophic failure. Original design performance must be restored as quickly as possible to mitigate production losses. With troubleshooting knowledge and a fundamental understanding of the process, the performance of the extruder can be restored in a relatively short time, minimizing the economic loss to the plant. Common root causes and their detection are provided. Hypothesis testing is outlined in Chapter 10 and is used throughout the troubleshooting chapters to identify the root causes. Elimination of the root cause is provided by offering the equipment owner several technical solutions, allowing the owner to choose the level of risk associated with the process modification. Mechanical failures are also common with single-screw extruders, and the common problems are identified. Illustrations are provided with the problems along with many numerical simulations of the case studies. Collectively, these instruct the reader on how to determine and solve many common extrusion problems. About 100 case studies and defects are identified in the book with acceptable technical solutions. Lastly, we hope that this book provides the information and technology that is required for the understanding, operation, and troubleshooting of single-screw extruders.

We have focused on two things as we developed the second edition of this treatise on single-screw fundamentals and application of those fundamentals to the engineering art of troubleshooting, research/development, and production single screw extruders. The stoichiometry of several important chemical reactions relating to the number of important commercial polymer monomers and their polymers was addressed in Chapter 2. Also, in Chapter 2 a table of 10 commercial polymers and their properties and structure was added so the reader can relate to some important polymers finished properties after extrusion. With the constraint of few new monomers and the cost and risk of building new monomer and polymer plants, the polymer industry has been focusing on developing rigid and fiber-based composite materials. A major addition to Chapter 3 is the development of an energy-based model for the shear thinning power law. The technique of using this new concept on polymer particulate composites is then developed and the utility of this concept is applied to the limiting extrusion-injection rate for injection molding. The difficulties of extruding a sound-deadening composite due to a filler change is related to the importance of understanding that filled system rheology is dependent on the volume and not the weight fraction of the filler. Other additions include an expansion of the design of Maddock mixers, transfer line designs, economic evaluations, and new case studies.

Gregory A. Campbell

Mark A. Spalding

The views and opinions expressed in this book are soley those of the authors and contributors. These views and opinions do not necessarily reflect the views and opinions of any affiliated individuals, companies, or trade associations.

Acknowledgements

My interest in fundamental polymer research began in 1964 when I began my graduate career. My research efforts were strongly influenced by my mentor Professor Edward G. Bobalek, one of finest gentleman and innovative research minds I have ever met. My research philosophy was strongly influenced by many encounters with Ed before and after I defended my dissertation. One particularly important encounter occurred when I was lamenting that my dissertation research did not appear to be a really important breakthrough. He took a long draw on his ever present pipe and said "Greg, that is why we call it research and not search." From that time on I have always looked at my efforts as learning from the previous researchers that have laid the technical foundation in the area that is now being addressed. My role is thus to continue to build on that foundation when looking for a solution to the research challenge that I am currently addressing.

After leaving the University of Maine, I worked with wonderful groups of exceptional researchers at General Motors research, Mobil Chemical research, and Clarkson University. Many of these individuals spent their valuable time to help me hone my research skills. Probably the most influential individual was Dr. William Meluch; a true genius that I had the pleasure of working with for 13 years. Another good friend that had a major influence on my manner of approaching engineering research was Professor Art Fricke whom I collaborated with at the University of Maine and the University of Florida. My colleague Dr. Don Rassmussen at Clarkson University provided important guidance in all things thermodynamic.

My extrusion experience started when I directed process research at Mobil Chemical Research in the early 1980s. We developed and analyzed data on a 24 to 1 single-screw extruder with 12 infrared probes and 12 pressure probes using high-speed data acquisition. I then changed career paths and accepted a position at Clarkson University teaching chemical engineering while developing the Clarkson Polymer Processing Laboratory. The new concepts developed in this book were first recognized by Dr. Paul Sweeney when he was a graduate student in about 1988. I have to admit that it took considerable effort on Paul's part to convince me to even address these new concepts. Once we became convinced that it was important to complete the solution of the single-screw extruder analysis and bring the solution back to the laboratory frame, it has taken 25 years to reach our current incomplete understanding. I would not have been able to acquire this understanding without the dedication and efforts of my colleagues and students that led the extrusion research in my lab: Paul Sweeney, Jeff Felton, Douglas Small, ChiCheng Wang, Dontula Narasimharao, Diana Hunt, Hongying Cheng, Zirong Tang, Mary Ann te-Riele, Jason C. Baird, Sirisha Bomma, and Sam St. John. An academic without excellent students is severely handicapped and I can truly say that I was not handicapped.

The development of this book has been an interesting and exhausting "trip" which in all likelihood would not have been completed without the encouragement and understanding of Sue, my wife for the past 50 years.

Gregory A. Campbell, Castle Research, Jonesport, Maine

My extrusion career started as one of the founding members of the Polymer Processing Technology Team of The Dow Chemical Company in 1987. The team was built and led by Dr. Kun Sup Hyun and consisted of four members (along with Joseph Dooley and Thomas McCullough). During the early years, the team researched many aspects of polymer processing including single-screw extrusion, twin-screw extrusion, and die technologies. These early years allowed the team to develop strong skills in process fundamentals, design, and troubleshooting. I am grateful to have this experience and the opportunity to develop this skill set. I am also grateful for the many mentors that I have had through my life including my father, Robert Bean, Gene Kratzman, Prof. Lyle F. Albright, and Dr. Hyun.

A book like this would not be possible without the help and contributions from coworkers, industry experts, and family. Many of the figures were contributed by industry experts and their names are provided with the figure. Photographs, content, and assistance were provided by Timothy W. Womer (consultant), Jeffery Kuhlman (Glycon), Jeff Myers (Robert Barr, Inc.), James Fogharty (Plastics Engineering Associates Licensing, Inc.), John Christiano (Davis-Standard), William Kramer (American Kuhne), and many others. Numerous diagrams were made and enhanced by my sons Stephen W. Spalding and Aaron F. Spalding. I also thank those who reviewed the original chapter drafts.

My wife Pamela has been a source of inspiration and motivation during this project. I thank her and my sons for their continued support through the writing of this book. My parents William and Joan provided me with a loving environment while growing up, and they provided the foundation for success.

Mark A. Spalding, The Dow Chemical Company, Midland, MI

Contents

1 Single-Screw Extrusion: Introduction and Troubleshooting

This book was written to provide the extrusion process engineer with a resource for assessing and fixing process problems associated with the use of single-screw extruders. The authors have drawn on their complementary backgrounds; both have worked with industrial extruder design, analysis, and fundamental research in the mechanism, operation, and troubleshooting of the single-screw extrusion process. The use of single-screw extruders in production processes has progressed significantly over the past several decades. As a result, the number of single-screw extruders in use has increased dramatically as has the diameter and length of the machine, especially for melt-fed extruders used in large resin production plants. In addition, resin manufacturers have developed many new resins for final products such as extruded sheet, film, pipe, fibers, coatings, and profiles. The extruder is still the process unit of choice for producing pellets in the production of polymer materials. Two types of extruders are generally used in polymer production: single-screw extruders and twin-screw extruders. The material in this book will be confined to the analysis and troubleshooting of single-screw extruders. The rapid expansion of this part of the polymer industry has been accompanied by the need for many new extrusion engineers. Many of these engineers have not had formal training in the analysis of the extruder and screw design nor have they had extensive education in polymer materials, which would help in troubleshooting problems on production equipment.

All single-screw extruders have several common characteristics, as shown in Figure 1.1 and Figure 1.2. The main sections of the extruder include the barrel, a screw that fits inside the barrel, a motor-drive system for rotating the screw, and a control system for the barrel heaters and motor speed. Many innovations on the construction of these components have been developed by machine suppliers over the years. A hopper is attached to the barrel at the entrance end of the screw and the resin is either gravity-fed (flood-fed) into the feed section of the screw or metered (starve-fed) through the hopper to the screw flights. The resin can be in either a solid particle form or molten. If the resin feedstock is in the solid form, typically pellets (or powders), the extruder screw must first convey the pellets away from the feed opening, melt the resin, and then pump and pressurize it for a downstream process operation. This type of machine is referred to as a plasticating single-screw extruder. The barrel is usually heated with a minimum of three temperature zones. These different temperature zones are consistent with the three utilitarian functions of the screw: solids conveying, melting, and pumping or metering of the polymer.

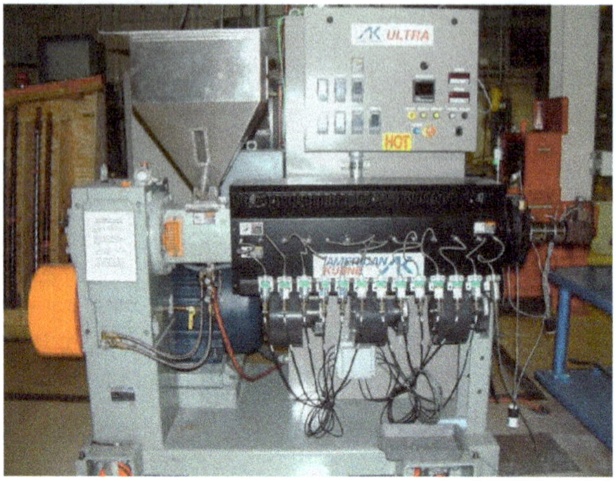

Figure 1.1 Photograph of a highly instrumented 63.5 mm diameter extruder built by American Kuhne

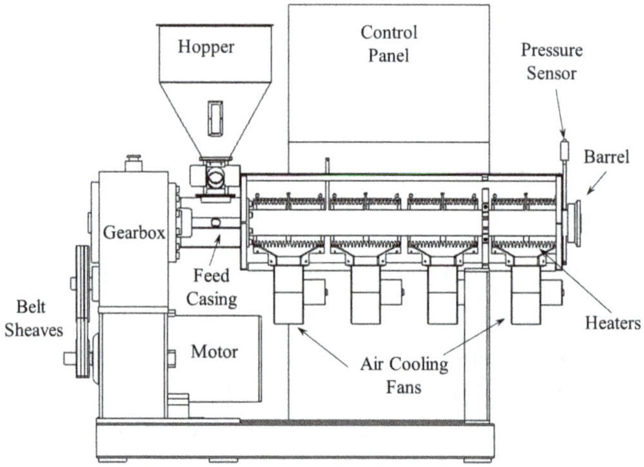

Figure 1.2 Schematic of a typical plasticating single-screw extruder. The extruder is equipped with four barrel heating and cooling zones and a combination belt sheave gearbox speed reduction drivetrain (courtesy of William Kramer of American Kuhne)

The single-screw plasticating process starts with the mixing of the feedstock materials. Typically, several different feedstocks are added to the hopper, such as fresh resin pellets, recycle material, additives, and a color concentrate. The recycle material typically comes from the grinding of edge trim, web material from thermoforming processes, or off-specification film and sheet. Often these components need to be dried and blended prior to adding them to the hopper. Next, the feedstock flows via gravity from the hopper through the feed throat of the feed casing and into the solids-conveying section of the screw. Typically this feed casing is cooled using water. The feed section of the screw is typically designed with a constant depth and is about 4 to 8 barrel diameters in axial length. Directly after the solids-conveying section is a section where the channel depth tapers to a shallow depth-metering section. The tapered-depth section is commonly referred to as the transition or melting section. In general, the metering section is also a constant depth, but many variations exist where the channels oscillate in depth. The metering section pumps and pressurizes the material for the downstream unit operations, including static mixers, screen filtering devices, gear pumps, secondary extruders, and dies. The total length of the extruder screw and barrel is typically measured in barrel diameters or as a length-to-diameter (L/D) ratio. Section lengths are often specified in barrel diameters or simply diameters.

The plasticator on an injection-molding machine is a specialized plasticating single-screw extruder. The plasticator has two main differences: there is a nonreturn valve on the tip of the screw, and the screw retracts as molten material accumulates between the nonreturn valve and the end of the barrel. Pressure is maintained on the accumulated material by a constant force applied to the shank of the screw via the drive system. This force is typically measured as a pressure applied to the shank and is referred to as the "back pressure." During the injection step of the process, the screw is forced forward, the nonreturn valve closes, and the material is injected into the mold. Additional information on the injection-molding process can be obtained elsewhere [1].

1.1 Organization of this Book

This book has been organized so that the information is helpful in troubleshooting extruders and extrusion processes, and it is presented in a manner that is of maximum utility to extrusion engineers. Appendices have been provided that present the theoretical analysis and assumptions in developing the design equations used throughout this text. In order to assess extruder production problems, it is necessary to understand the nature of the polymer that is being extruded, the design of the extruder and screw, and the interaction of these as the extruder is being operated. Numerous case studies are presented that demonstrate these interactions.

Knowledge of the geometry and mathematical description of a screw is required to understand the analysis of the functional sections of the screw and the troubleshooting of case studies. In Chapter 1 the geometry and mathematical descriptions are presented. Also in this chapter, the calculation of the rotational flow (also known as drag flow) and pressure flow rates for a metering channel is introduced. Simple calculation problems are presented and solved so that the reader can understand the value of the calculations.

Resin manufacturers go to extreme measures to produce a reproducible, high-quality, and useful polymer that is ready for final conversion to a product. Every time these polymers are passed through an extruder, however, the polymer has the potential to degrade, changing the chemical and physical properties of the resin. Degradation processes can often be the cause of extrusion problems. Chapter 2 begins with an introduction to how polymers are produced from the perspective of the type of chemical bonds that are important in different polymer families. It is beyond the scope of this book to discuss polymer production processes in detail. The discussion of polymerization is intended to aid the reader with a basic understanding on how the polymer is formed from its monomer. Knowing how the polymer was produced from its monomers will provide the engineer with the knowledge of how the extrusion process interacts with the polymer. This basic understanding will help in troubleshooting situations where the problem is the effect of the extrusion process on the stability of the polymer being extruded.

The physical properties that are important to polymer processing are presented in Chapters 3 and 4. Chapter 3 provides a basic understanding of the viscoelastic characteristics of polymers. In this chapter the fundamental concepts of polymer rheology are developed, and then there is a discussion of Newtonian and Power Law rheological responses of polymeric fluids, followed by a short introduction to the elastic nature of polymer melts. Chapter 4 presents the remaining physical properties, including friction coefficients (or stress at an interface), densities, melting fluxes, and thermal properties. These properties impact the performance of a resin during the extrusion process.

The fundamental processes and mechanisms that control single-screw extrusion are presented in Chapters 5 through 8. These processes include solids conveying, melting, polymer fluid flow, and mixing. The analyses presented in these chapters focus on easily utilized functions needed to assess the operation of the single-screw extruder. The derivation of these relationships will be presented in detail in the appendices for those who desire to explore the theory of extrusion in more detail.

The remaining Chapters 9 through 15 are devoted to different types of extrusion troubleshooting analyses. These chapters include presentations on scale-up techniques, general troubleshooting, screw fabrication, contamination in finished products, flow surging, and rate limitations. The chapters are presented with actual case studies of extrusion troubleshooting problems with the detailed analytical approach that was used to address these problems. As part of this troubleshooting presentation, high-performance screws and their benefits are presented. Lastly, melt-fed extruders will be discussed. Melt-fed extruders are a special class of machines that are rarely discussed in the open literature.

Appendix A1 has a listing of the polymer abbreviations used in this book.

1.2 Troubleshooting Extrusion Processes

All extrusion and injection-molding plastication processes will eventually operate at a performance level less than the designed level. This reduction in performance can be caused by many factors, including but not limited to control failures, a worn screw or barrel, or a process change such as processing a different resin. Moreover, an improper screw design or process operation can limit the performance of the machine and reduce the profitability of the plant. Other processes may be operating properly at the designed rate, but a higher rate may be required to meet market demands. In this case, the rate-limiting step of the process needs to be identified and a strategy developed to remove the limitation.

Troubleshooting is a process for systematically and quickly determining the root cause of the process defect. The troubleshooting process is built on a series of hypotheses, and then experiments are developed to prove or disprove a hypothesis. The ability to build a series of plausible hypotheses is directly related to the knowledge of the engineer troubleshooting the process. Our focus is on providing the knowledge for the proper operation of an extrusion process, helping determine typical root causes that decrease the performance of the machine, and offering methods of removing the root cause defect from the process.

The economic impact of a properly designed troubleshooting process can be significant, especially if the defect is causing very high scrap rates or production requirements are not being met. Returning the process to full production in a timely manner will often require subject matter experts from several disciplines or companies. An excellent example of a troubleshooting process is described next for a processing problem at the Saturn Corporation.

1.2.1 The Injection Molding Problem at Saturn

During the startup of Saturn Corporation's Spring Hill, Tennessee, plant in September of 1990, a serious splay problem was encountered for the injection molding of door panels from a PC/ABS resin [2]. Splay is a common term used to describe surface defects on injection-molded parts. The splay on the surface of the door panels created parts with unacceptable appearances after the painting process. The part rejection rate was higher than 25%, high enough to nearly shut down the entire plant. Teams were formed from the companies involved to determine quickly the root cause for the splay. After a detailed analysis was performed, it was determined that the plasticating screw in the injection molder was not operating properly, causing some of the resin to degrade in the channels of the screw. The splay was created by the volatile components from the degradation of the resin. A high-performance Energy Transfer (ET) screw [3] was designed and built, eliminating the splay. A detailed discussion of the troubleshooting process at Saturn is presented in Section 11.12.5.

The troubleshooting project at Saturn is an excellent example of combining strengths from different companies to diagnose and eliminate a costly defect from a process.

1.3 Introduction to Screw Geometry

In order to simulate an extrusion process or design a screw, the mathematical description of the screw geometry must be understood. This section provides the basic details that describe a screw and the complex mathematics that describe the channels.

The single-screw extruder screw can be single flighted or multiple flighted. A conventional single-flighted screw is shown in Figure 1.3. This screw has a single helix wound around the screw root or core. Multiple-flighted screws with two or more helixes started on the core are very common on high-performance screws and on large-diameter melt-fed machines. For example, barrier melting sections have a secondary barrier flight that is located a fraction of a turn downstream from the primary flight, creating two flow channels: a solids melting channel and a melt-conveying channel. Moreover, many high-performance screws have two or more flights in the metering section of the screw. Barrier screws and other high-performance screws will be presented in Chapter 14. Multiple flights are very common on larger-diameter extruder screws, because this creates a narrower channel for the polymer melt to flow through, leading to less pressure variation due to the rotation of the screw. In addition, the multiple flights spread the bearing forces between the flight tip and the barrel wall. Melt-fed extrusion processes will be discussed in detail in Chapter 15. The screw is rotated by the shank using either specially designed splines or by keys with rectangular cross sections. The mathematical zero position of the screw is set at the pocket where the screw helix starts. Most extruder manufacturers rotate the screw in a counterclockwise direction for viewers positioned on the shank and looking towards the tip. This rotation convention, however, is not standard.

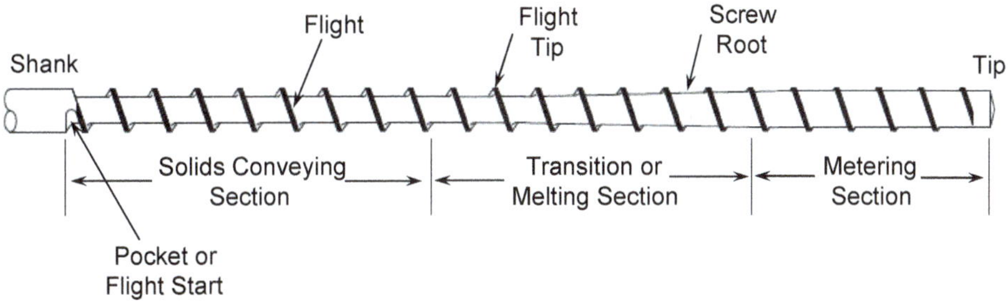

Figure 1.3 Schematic of a typical single-flighted screw (courtesy of Jeff A. Myers of Robert Barr, Inc.)

The flight is a helical structure that is machined into the screw and extends from the flight tip to the screw core or root. The flight has a width at the flight tip called the flight land. The small clearance between the flight land and the barrel wall minimizes the flow of polymer back toward the feed section. The polymer that does flow between the clearances supports the screw and centers it in the barrel. The radial distance between the flight tip and the screw root is referred to as the local flight height or channel depth. The feed section usually has a constant-diameter core that has the smallest diameter, the largest channel depth, and the largest cross-sectional volume in the screw. The deep channel conveys the relatively low bulk density feedstock pellets into the machine. The feedstock is conveyed forward into the transition section or melting section of the screw. The transition section increases in root diameter in the downstream direction, and thus the channel depth decreases. Here, the feedstock is subjected to higher pressures and temperatures, causing the feedstock to compact and melt. As the material compacts, its bulk density can increase by a factor of nearly two or more. As the feedstock compacts, the entrained air between the pellets is forced back and out through the hopper. For example, a pellet feedstock such as ABS resin can have a bulk density at ambient conditions of 0.65 g/cm^3 while the melt density at 250 °C is 0.93 g/cm^3. Thus for every unit volume of resin that enters the extruder, about 0.3 unit volumes of air must be expelled out through the voids in the solid bed and then out through the hopper. The transition section is where most of the polymer is converted from a solid to a fluid. The fluid is then conveyed to the metering section where the resin is pumped to the discharge opening of the extruder. In general, the metering section of a conventional screw has a constant root diameter, and it has a much smaller channel depth than the feed section. The ratio of the channel depth in the feed section to the channel depth in the metering section is often referred to as the compression ratio of the screw.

1.3.1 Screw Geometric Quantitative Characteristics

The book *Engineering Principles of Plasticating Extrusion* by Tadmor and Klein [4] has been used extensively in gaining an understanding of the fundamentals of extrusion processes. The following section endeavors to maintain the quality of the development of the screw geometry section of this classic text. Understanding the relationships between the screw geometry and the symbolic and mathematical representation of a screw is a critical beginning for understanding the rate, pressure, and temperature calculations. These functions related to the performance of single-screw extruders are developed later in this book and require an understanding of the screw geometry.

The geometry of a double-flighted screw and its nomenclature are presented in Figure 1.4 using the classical description from Tadmor and Klein [4]. The nomenclature has been maintained to provide consistency with the classical literature and to provide some generality in the development of the symbols and equations that are used in extruder analysis.

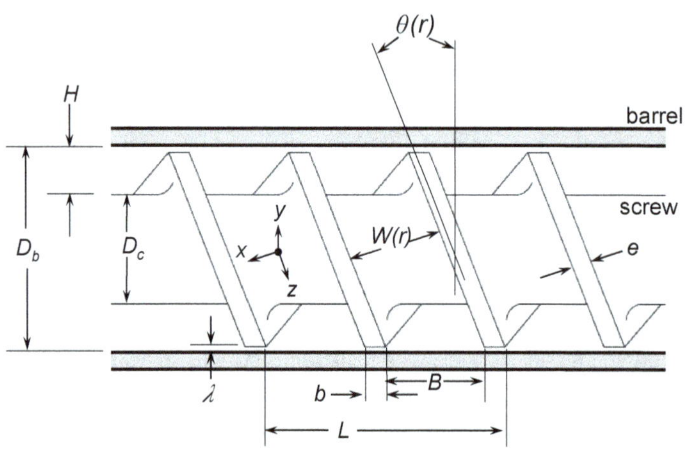

Figure 1.4 A schematic of a double-flighted screw geometry

Several of the screw geometric parameters are easily obtained by observation and measurement, including the number of flight starts, inside barrel diameter, channel depth, lead length, flight width, and flight clearance. The number of flight starts, p, for the geometry in Figure 1.4 is two. The inner diameter of the barrel is represented by D_b, and the local distance from the screw root to the barrel is H. The diameter of the screw core is represented by D_c. The mechanical clearance between the land of the screw flight and the barrel is λ. The mechanical clearance is typically very small compared to depth of the channel. The lead length, L, is the axial distance of one full turn of one of the screw flight starts. This is often constant in each section of the screw, but in some screws, such as rubber screws, it often continuously decreases along the length of the screw. A screw that has a lead length that is equal to the barrel diameter is referred to as square pitched. The flight width at the tip of the screw and perpendicular to the flight edge is e.

The remaining geometrical parameters are easily derived from the measured parameters presented above. Several of the screw parameters are functions of the screw radius. They include the perpendicular distance from flight to flight, $W(r)$; the width of the flights in the axial direction, $b(r)$; and the helix angle, $\theta(r)$, the angle produced by the flight and a plane normal to the screw axis. These parameters will be discussed later. At the barrel wall these parameters are subscripted with a b. The helix angle at the barrel wall is θ_b and is calculated using Equation 1.1. The helix angle at the barrel wall for a square-pitched screw is 17.7°.

$$\tan\theta_b = \frac{L}{\pi D_b} \quad \text{thus} \quad \theta_b = \arctan\frac{L}{\pi D_b} \tag{1.1}$$

The relationship between the width of the channel perpendicular to the flight at the barrel interface, W_b, and the axial distance between the flight edges at the barrel interface, B_b, is as follows:

$$W_b = B_b \cos\theta_b = \left(\frac{L}{p} - b_b\right)\cos\theta_b = \frac{L}{p}\cos\theta_b - e \tag{1.2}$$

$$e = b_b \cos\theta_b \tag{1.3}$$

As mentioned earlier, several of the geometric parameters are a function of the radial position (r) of the screw. These parameters include the helix angle and the channel widths. The length of an arc for one full turn at the barrel surface is πD_b. At the screw surface the length of the arc for one turn is $\pi(D_b - 2H)$; the lead length, however, remains the same. This leads to a larger helix angle at the screw root than at the barrel surface. This analysis is for a flight width that does not change with the depth of channel. As discussed in Chapter 10, for a properly designed screw, the flight width will increase as the root of the screw is approached due to the flight radii.

The helix angle and the channel widths at the screw core or root are designated with a subscript c, and they are calculated as follows:

$$\tan\theta_c = \frac{L}{\pi(D_b - 2H)} = \frac{L}{\pi D_c} \quad \text{thus} \quad \theta_c = \arctan\frac{L}{\pi D_c} \tag{1.4}$$

Thus the screw has a narrower normal distance between flights at the screw root because the helix angle is larger and because the lead remains the same.

$$W_c = B_c \cos\theta_c = \left(\frac{L}{p} - b_c\right)\cos\theta_c = \frac{L}{p}\cos\theta_c - e \tag{1.5}$$

$$e = b_c \cos\theta_c \tag{1.6}$$

For a generalized set of functions in terms of the radius, *r*, and the local diameter, *D*, the helix angle is calculated as follows:

$$\theta(r) = \arctan\frac{L}{\pi D} \tag{1.7}$$

In terms of the barrel dimensions and parameters:

$$\theta(r) = \arctan\left(\frac{D_b}{D}\tan\theta_b\right) \tag{1.8}$$

The channel width at any radius thus follows:

$$W(r) = B\cos\theta(r) = \left(\frac{L}{p} - b(r)\right)\cos\theta(r) = \frac{L}{p}\cos\theta(r) - e \tag{1.9}$$

The average channel width is used for many of the calculations in this book. This average channel width is represented as simply *W* here and is calculated using Equation 1.10. The use of the average channel width will be discussed in detail in Chapter 7.

$$W = \frac{W_b + W_c}{2} \tag{1.10}$$

Calculations in helical coordinates are very challenging. The procedure used in this text will be to "unwrap" the screw helix into Cartesian coordinates for the analysis. It is important to be able to calculate the helical length in the *z* direction at any radius *r* for the axial length *l*:

$$z(r) = \frac{l}{\sin\theta(r)} \tag{1.11}$$

1.4 Simple Flow Equations for the Metering Section

The efficient operation of a single-screw extruder requires that all three extruder sections, solids conveying, melting, and metering, must be designed to work efficiently and in coordination to have a trouble-free process. The specific analysis for each of these topics will be covered in later chapters. The simple flow calculations for the metering section should be performed at the start of any extruder troubleshooting process. It is presented at this time so that the reader may see how the equations developed in subsequent sections are used at the start of a typical troubleshooting problem.

For a properly operating smooth-bore single-screw extruder, the metering section of the screw must be the rate-limiting step of the process. Thus, calculation of the flows in the metering section of a process can be used to determine if the extruder is operating properly. A simple and fast method of estimating the flow components in a metering section was developed by Rowell and Finlayson [5] for screw pumps, and it was outlined in the Lagrangian frame by Tadmor and Klein [4]. These flows are based on a reference frame where the barrel is rotated in the opposite direction from the normal screw rotation, and they are historically called drag flow and pressure flow. Screw rotation analysis, as developed by Campbell and coworkers [6 - 15], is used here to arrive at the same flow equations while retaining the screw rotation physics of the extruder. For screw rotation, the flows are called rotational flow and pressure flow. This screw rotation analysis has been shown to provide a better understanding of the flow mechanism in the extruder, a better estimate of viscous dissipation and temperature increase in the extruder, and a better prediction of the melting characteristics. Several other methods are available for estimating the flow components in the metering section of a screw, but they can be more complicated and time consuming [16].

The method described here is the simplest of all methods. This simplicity is based on numerous assumptions (listed below), and thus it requires a minimum amount of data and computational effort. The calculation is only meant to give a quick and crude estimate of the flows, since in most cases assumptions 4 through 7 are violated. An improved yet more difficult method is introduced in Chapter 7. The assumptions for the simple calculation method are as follows:

1. Flow is fully developed.
2. Flow channels are completely filled.
3. No slip at the boundary surfaces.
4. No leakage flow over the flight tips.
5. All channel corners are square.

6. Flows are isothermal and Newtonian.
7. Channel dimensions are not changing in the metering section.

The fully developed flow components in a constant-depth metering section can be estimated using flow analysis for a long rectangular channel, as outlined in Chapter 7. For those calculations, a geometry transformation is first performed. That is, the channels of the screw are "unwound from the helix" and "straightened into a long trough." The barrel now becomes an infinitely large flat plate. Next, the screw or trough is moved at a fixed angle to the stationary barrel. These equations were developed using a Cartesian coordinate system; that is, the z direction is in the down-channel direction and parallel to the flight edge, the y direction is normal to the barrel surface, and x is the cross-channel direction and thus perpendicular to the flight edge.

The geometric parameters for a screw in the "wound" form are shown by Figure 1.5 for a single-flighted screw.

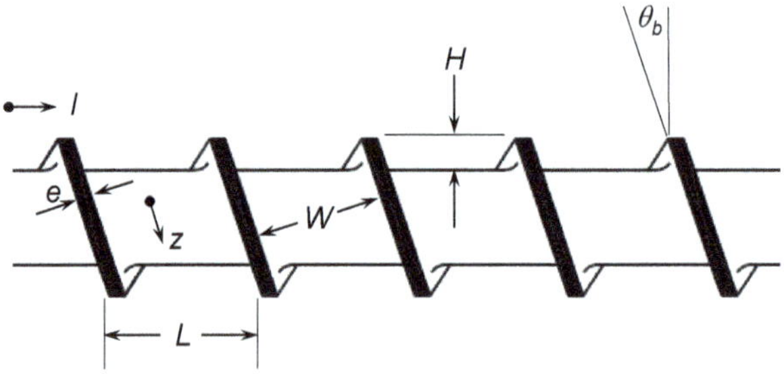

Figure 1.5 Geometric parameters for a single-flighted screw in the wound state

Two driving forces for flow exist in the metering section of the screw. The first flow is due just to the rotation of the screw and is referred to as the rotational flow component. The second component of flow is due to the pressure gradient that exist in the z direction, and it is referred to as pressure flow. The sum of the two flows must be equal to the overall flow rate. The overall flow rate, Q, the rotational flow, Q_d, and the pressure flow, Q_p, for a constant depth metering channel are related as shown in Equation 1.12. The subscript d is maintained in the nomenclature for historical consistency even though the term is for screw rotational flow rather than the historical drag flow concept.

$$Q = Q_d - Q_p \tag{1.12}$$

The volumetric rotational flow term (Q_d) depends on the several geometric parameters and rotation speed. Since most extruder rates are measured in mass per unit time, the term Q_{md} is defined as the mass rotational flow:

$$Q_d = \frac{pV_{bz}WHF_d}{2} \tag{1.13}$$

$$Q_{md} = \frac{p\rho_m V_{bz} WHF_d}{2} \tag{1.14}$$

where ρ_m is the melt density at the average fluid temperature of the resin, V_{bz} is the z component of the screw velocity at the barrel wall, H is the depth of the channel, and F_d is the shape factor for plane Couette flow. The analysis using plane Couette flow does not take into account the effect of the flights (channel helix) on the flow rate. The F_d term compensates for the reduction in flow rate due to the drag-induced resistance of the flights. For an infinitely wide channel with no flights, F_d would be equal to 1. As the channel width approaches the depth, F_d is about 0.5.

The analysis developed here is based on screw rotation physics [13], and thus several other definitions are developed here. The velocities at the screw core, indicated by the subscript c, in the x and z directions are as follows:

$$V_{cx} = \pi N D_c \sin\theta_c \tag{1.15}$$

$$V_{cz} = -\pi N D_c \cos\theta_c \tag{1.16}$$

where N is the screw rotation rate in revolutions per second. Cross-channel velocity for the screw in the laboratory frame (screw rotation frame) is

$$V_x = V_{cx}\left(1+\frac{H}{R_c}\right)\frac{y}{H}\left(2-3\frac{y}{H}\right)+V_{cx}\left(1+\frac{y}{R_c}\right) \tag{1.17}$$

where R_c is the screw core radius.

The down-channel velocity in the laboratory frame for very wide channels ($H/W < 0.1$) as a function of the height of the channel y is as follows:

$$V_{dz} = \frac{y}{H}\left|V_{cz}\right|\left(1+\frac{H}{R_c}\right)-\left|V_{cz}\right|\left(1+\frac{y}{R_c}\right) \tag{1.18}$$

Pressure flow velocity in the z direction for a very wide channel ($H/W < 0.1$) as a function of y is

$$V_{pz} = \frac{H^2}{8\eta}\frac{\partial P}{\partial z}\left[\left(\frac{y}{H}\right)^2-\frac{y}{H}\right] \tag{1.19}$$

The z component of the screw velocity at distance H from the screw root is computed as

$$V_{bz} = \pi N D_b \cos\theta_b \tag{1.20}$$

or

$$V_{bz} = V_{cz}\left(1+\frac{H}{R_c}\right) \tag{1.21}$$

The volumetric pressure flow term, Q_p, and the mass flow pressure flow term, Q_{mp}, are computed as follows:

$$Q_p = \frac{pWH^3F_p}{12\eta}\left[\frac{\partial P}{\partial z}\right] \tag{1.22}$$

$$Q_{mp} = \frac{p\rho_m WH^3F_p}{12\eta}\left[\frac{\partial P}{\partial z}\right] \tag{1.23}$$

where F_p is the shape factor for pressure flow, $\partial P / \partial z$ is the pressure gradient in the channel in the z direction, and η is the shear viscosity of the molten polymer at the average channel temperature and at an average shear rate, $\dot{\gamma}$:

$$\dot{\gamma} = \frac{\pi D_c N}{H} \tag{1.24}$$

The shear rate in the channel contains contributions from the rotational motion of the screw and the pressure-driven flow. The calculation of the shear rate, $\dot{\gamma}$, using Equation 1.24, is based on the rotational component only and ignores the smaller contribution due to pressure flow. For the calculations here, Equation 1.24 can be used.

The relationship between the pressure gradient in the z direction to the axial direction, l, is as follows:

$$\frac{\partial P}{\partial z} = \frac{\partial P}{\partial l}\sin\theta_b \tag{1.25}$$

The pressure gradient is generally unknown, but the maximum that it can be for a single-stage extruder screw is simply the discharge pressure, P_{dis}, divided by the helical length of the metering section. This maximum gradient assumes that the pressure at the start of the metering section is zero. For a properly designed process, the actual gradient will be less than this maximum, and the pressure at the start of the metering section will not be zero.

$$\frac{\partial P}{\partial z} = \frac{P_{dis}\sin\theta_b}{l_m} \tag{1.26}$$

where l_m is the axial length of the metering section. The shape factors F_d and F_p [4, 17] are computed to adjust for the end effect of the flights. The factors are from the summation of an infinite series as part of an exact solution for the constant temperature and constant viscosity solution in the z direction in the unwound screw channel. The factors are calculated as follows:

$$F_d = \frac{16W}{\pi^3 H} \sum_{i=1,3,5,...}^{\infty} \frac{1}{i^3} \tanh\left(\frac{i\pi H}{2W}\right) \quad (1.27)$$

$$F_p = 1 - \frac{192H}{\pi^5 W} \sum_{i=1,3,5,...}^{\infty} \frac{1}{i^5} \tanh\left(\frac{i\pi W}{2H}\right) \quad (1.28)$$

The shape factors range from 0 to 1 and approach 1 for shallow channels; that is, $H / W \approx 0$. It is important to include the shape factors when evaluating commercial screw channels. This becomes extremely important for deep channels where H/W does not approach 0. The total mass flow rate, Q_m, is calculated by combining the flow components as provided in Equation 1.29 for the total mass flow rate. As stated previously, the rate, rotational flow, and pressure flow calculations should be performed at the start of every troubleshooting project.

$$Q_m = \frac{p\rho V_{bz} WHF_d}{2} - \frac{p\rho WH^3 F_p}{12\eta}\left[\frac{\partial P}{\partial z}\right] \quad (1.29)$$

1.5 Example Calculations

Three examples are presented that introduce the use of the equations developed in this chapter. These calculations should be used at the start of the performance analysis of all troubleshooting problems. This analysis will be expanded in subsequent chapters through Chapter 7 using additional tools and understandings to complete the troubleshooting process.

1.5.1 Example 1: Calculation of Rotational and Pressure Flow Components

A manager decided to buy a new 88.9 mm (3.5 inch) diameter general purpose extruder to extrude products using several different low-density polyethylene (LDPE) resins. The die manufacturer has indicated that the die entry pressure will be 12.4 MPa. Both vendors used nominal viscosity data for a commercial LDPE resin at a temperature of 210 °C when doing their predictions because the manager had not yet settled on the resins or manufacturer for the new materials. The manager has specified that the extruder must be capable of running flood-fed at a maximum rate of 250 kg/h. Flood feeding refers to the operation of the extruder with resin covering the screw in the feed hopper. In this operation, increasing the screw speed will increase the rate of the process. The shear viscosity specified was

2000 Pa·s. The melt density for LDPE resin at 210 °C is 0.75 g/cm^3. The screw was specified with a 6 diameter long feed section with a constant channel depth of 16.51 mm, a 9 diameter long transition section, and a 9 diameter long metering section with a constant channel depth of 5.08 mm. The screw lead length is 1.2 times the screw diameter at 107 mm and the screw flight width perpendicular to the flight edge is 9.0 mm. The extruder manufacturer has stated that the extruder is capable of a maximum screw speed of 108 rpm. Will this extruder meet the desired rate expectations of the manager for the LDPE resin?

In order to address this key question, the information provided in this chapter will be used to calculate the geometrical and flow data for this analysis. When making the calculations for any engineering analysis it is absolutely imperative that the same units system be used for all the calculations. It is generally accepted today that the calculations are performed using the SI units system [18]: mass in kilograms, length in meters or some fraction thereof, time in seconds, energy in joules, pressure in Pascal, and viscosity in Pascal seconds.

To start the calculation, the geometrical parameters are calculated based on the known specifications for the metering channel of the screw. Only the metering channel is considered here since the metering channel is the rate-controlling operation for a properly designed screw and a smooth-bore extrusion process. The specifications and calculated values for the metering channel of the screw are provided in Table 1.1 along with the equations used for the calculations.

Table 1.1 Geometric Parameter Values for the Screw in Example 1

Parameter	Value	Equation
Barrel diameter, D_b	88.9 mm	
Core diameter, D_c	78.74 mm	
Lead length, L	107 mm	
Meter channel depth, H	5.08 mm	
Flight width, e	9.0 mm	
Flight starts, p	1	
Helix angle at the barrel, θ_b	21.0°	1.1
Helix angle at the screw core, θ_c	23.4°	1.4
Channel width at the barrel, W_b	90.9 mm	1.2
Channel width at the screw core, W_c	89.2 mm	1.5
Average channel width, W	90.0 mm	1.10
Channel aspect ratio, H/W	0.056	
Unwrapped channel length for one turn, z_b	299 mm	1.11
Total helical length of the metering section, Z_b	2.23 m	1.11
Shape factor for rotational flow, F_d	0.966	1.27
Shape factor for pressure flow, F_p	0.965	1.28

Since the manager is asking for an extruder and screw that will provide 250 kg/h, the expected maximum rate will be calculated at the maximum screw speed of 108 rpm. At a screw speed of 108 rpm (N = 1.80 rev/s), the z component of the flight tip velocity (V_{bz}) at H is calculated at 469 mm/s using Equation 1.20. Now all terms required to calculate the rotational mass flow rate are known and are calculated using Equation 1.14:

$$\begin{aligned} Q_{md} &= \frac{(750\ \text{kg/m}^3)(0.469\ \text{m/s})(0.090\ \text{m})(5.08\times10^{-3}\ \text{m})(0.966)}{2} \\ &= 0.077\ \text{kg/s} => 280\ \text{kg/h} \end{aligned}$$

For this calculation, the density of the molten LDPE resin is known to be 750 kg/m^3 at 210 °C. Next, the pressure flow term needs to be calculated based on the maximum pressure gradient possible in the metering channel. The maximum pressure gradient will occur when the pressure at the entry to the metering section is near zero. In practice, the pressure at the entry to the meter will be considerably higher. With zero pressure at the entry and 12.4 MPa at the discharge (die entry pressure), the maximum pressure gradient is estimated by dividing the pressure change by the helical length (Z_b) of the metering channel. This maximum pressure gradient $\partial P / \partial z$ is calculated at 5.56 MPa/m. Since the pressure is increasing in the helical direction, the pressure gradient is positive. Next, the mass flow rate due to the pressure gradient is calculated using Equation 1.23 and a shear viscosity of 2000 Pa·s as follows:

$$\begin{aligned} Q_{mp} &= \frac{(750\ \text{kg/m}^3)(0.090\ \text{m})(5.08\times10^{-3}\ \text{m})^3(0.965)}{12(2000\ \text{Pa}\cdot\text{s})}\left(\frac{5.56\times10^6\ \text{Pa}}{\text{m}}\right) \\ &= 0.0020\ \text{kg/s} => 7\ \text{kg/h} \end{aligned}$$

The overall mass rate expected from the extruder and screw specification was calculated using Equation 1.12 at 273 kg/h. The manager specified a rate of 250 kg/h for the process. Since the extruder is capable of 273 kg/h at the maximum screw speed, the extruder is specified properly from a rate viewpoint. Further calculations indicated that 250 kg/h of LDPE could be extruded at a screw speed of 99 rpm for a maximum pressure gradient of 5.56 MPa/m, discharging at 210 °C and a pressure of 12.4 MPa. The pressure mass flow Q_{mp} is very low, at about 2% of the rotational mass flow Q_{md}. If the screw geometry is kept constant, this ratio will increase if the discharge pressure increases or the viscosity decreases. As this book progresses, the solution for Example 1 will be expanded as new materials are introduced in subsequent chapters through Chapter 7.

1.5.2 Example 2: Flow Calculations for a Properly Operating Extruder

For this example, a plant is using a smooth-bore extruder to process an LDPE resin. At a screw speed of 100 rpm, the extruder is operating flood-fed with a rate of 800 kg/h. For these conditions, the plant personnel have measured the discharge pressure at 15.9 MPa and the discharge temperature at 210 °C. The plant manager wants to know if this extruder is operating properly and whether the metering section of the screw is controlling the rate. For this case, the extruder is 152.4 mm in diameter, D_b, the lead length, L, is 152.4 mm, the width of the flight, e, is 15.2 mm, and the channel depth, H, in the metering section is 6.86 mm. The metering section length is 76.2 cm or 5 turns. The melt density for LDPE resin at 210 °C is 0.750 g/cm^3.

The shear rate for this example is estimated using Equation 1.24 to be 106 1/s. The viscosity for this LDPE resin at 210 °C and a shear rate of 106 1/s is about 300 Pa·s. From the screw geometry, screw speed, and melt density, the rotational flow rate, Q_{md}, is computed at 888 kg/h. Since the rotational flow and pressure flow must equate to the total flow using Equation 1.29, the pressure flow rate, Q_{mp}, is 88 kg/h. The positive sign for the pressure flow rate term means that the pressure gradient is reducing the flow. Likewise, a negative pressure flow rate term would mean that the pressure gradient is causing the flow rate to be higher than the rotational flow rate. For this resin and a viscosity of 300 Pa·s, the pressure gradient in the channel is calculated using Equation 1.29 as 1.40 MPa/turn. Thus, the discharge end of the metering channel is at a higher pressure than the entry.

To answer the question as to whether the extruder is operating properly, several additional calculations are performed. For this screw, there are 5 screw turns in the metering section, which is calculated by dividing the axial length (76.2 cm) by the lead length (L = 152.4 mm). Multiplying the number of turns by the pressure gradient in the metering section reveals that the total pressure increase in the metering section is 7.0 MPa. To achieve the measured discharge pressure of 15.9 MPa, the pressure at the entry of the metering section must be 8.9 MPa. Because a positive pressure over the entire length of the metering section is occurring, the calculations indicate that the screw is full with resin and functioning properly. That is, the metering section is controlling the rate. The axial pressure profile for this case is shown in Figure 1.6.

1.5.3 Example 3: Flow Calculations for an Improperly Operating Extruder

An extruder with a metering section having the same geometry as the extruder in Example 2 is operating flood-fed at a rate of 550 kg/h. The screw speed is 100 rpm, the discharge temperature is 210 °C, and the discharge pressure is 10.3 MPa. Is this extruder operating properly so that its metering section is controlling the rate? Because the metering section geometry, screw speed, and discharge temperature have not changed, the method still calculates the rotational flow rate component as 888 kg/h. However, the pressure flow component is now calculated as 338 kg/h, and the pressure gradient corresponding to this flow is 5.4 MPa/turn. To determine if the extruder is operating properly, the pressure generating capacity needs to be calculated by multiplying the 5 screw turns in the metering section length by the pressure gradient. The result is a total pressure increase of 27 MPa in the metering section. This pressure increase is greater than the 10.3 MPa pressure measured at the discharge of the extruder. In fact, only 1.9 diameters of the metering section length are required to generate a discharge pressure of 10.3 MPa (calculated by dividing 10.3 MPa by 5.4 MPa/turn). This means that the remaining 3.1 diameters of the metering section are at zero pressure and are only partially filled. A partially filled system will always be at zero pressure. Because a positive pressure over the entire length of the metering section is not possible, the calculations indicate that the metering section is not full and is not functioning properly. A process upstream of the metering section is controlling the rate.

Figure 1.6 shows the axial pressures for the extruders described in Examples 2 and 3. In this figure, the solid lines were calculated while the dotted lines were estimated based on experience. For Example 2, where the extruder operates properly, the extruder pressure is positive at all axial positions. Thus, all the channel sections are operating full and under pressure as designed. For Example 3, where the extruder operates improperly, the extruder pressure is zero for portions of the melting and metering sections. In these portions of the extruder, the screw channel is not pressurized, and the extruder is operating partially filled. This means that the metering section is not controlling the rate as designed, and the screw is not operating properly. Extruders that are operated partially filled in the metering section can have low production rates, high scrap rates, and material degradation, and cause high labor costs.

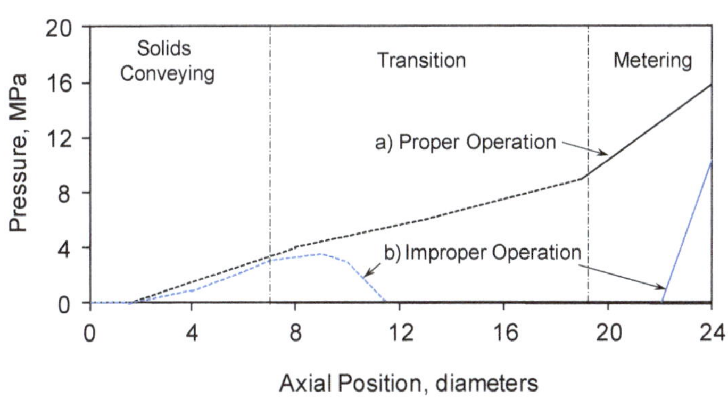

Figure 1.6 Axial pressure profiles for a) Example 2 where the extruder is operating properly (all channels are full and pressurized), and b) Example 3 where the extruder is operating improperly. For Example 3, the channel is not pressurized between diameters 12 and 22, indicating that the channels are partially filled at these locations

1.5.4 Metering Channel Calculation Summary

Calculation of the rotational flow rate and an estimate of the pressure profile in the metering channel should be performed for all design and troubleshooting projects. As shown in the example problems, the rate can be quickly and easily estimated for a new installation. The calculation method should always be performed for machines that are operating at low rates and have degradation products in the extrudate. As described in Section 1.5.3, the calculation is capable of predicting partially filled channels in the metering section. Partially filled metering channels will cause the resin to degrade, and the degradation products will eventually be discharged from the machine, contaminating the final product.

Nomenclature

b_b	axial flight width at the barrel wall
b_c	axial flight width at the screw root
B	axial channel width as a function of the radial position
B_b	axial channel width at the barrel wall
B_c	axial channel width at the screw core
D	local diameter
D_b	inner diameter of the barrel
D_c	diameter of the screw core
e	flight width of the screw and perpendicular to the flight edge
F_d	shape factor for rotational flow
F_p	shape factor for pressure flow
H	local distance from the screw root to the barrel
l	axial distance
l_m	axial distance for the metering section
L	lead length
N	screw rotation speed in revolutions/s
p	number of flight starts
P_{dis}	discharge pressure
P	pressure in the channel
Q	volumetric flow rate

Q_d volumetric rotational flow rate

Q_m mass flow rate

Q_{md} mass rotational flow rate

Q_{mp} pressure-induced mass flow rate

Q_p volumetric pressure flow rate

R_c radius of the screw core

V_{dz} down-channel velocity (z direction) as a function of y

V_{pz} z component of the velocity due to a pressure gradient

V_x cross-channel velocity (x direction) in the channel as a function of y

V_{cx} x component of velocity of the screw core

V_{cz} z component of velocity of the screw core

V_{bz} z component of velocity of the screw flight at the barrel wall

W average channel width

W_b channel width perpendicular to flight at the barrel wall

W_c channel width perpendicular to flight at the screw core

$W(r)$ channel width perpendicular to flight at radius r

x independent variable for the cross-channel direction perpendicular to the flight edge

y independent variable for the direction normal to the barrel surface (channel depth direction)

z independent variable in the down-channel direction (or helical direction)

z_b helical length of the channel at the barrel wall

$z(r)$ helical length of the channel at radial position r

Z_b helical length of the metering channel at the barrel wall

$\dot{\gamma}$ average shear rate in the channel

η shear viscosity of the polymer at the average channel temperature and average shear rate, $\dot{\gamma}$

θ_b helix angle at the barrel

θ_c helix angle at the screw core

$\theta(r)$ helix angle at radial position r

λ mechanical clearance between the top of the screw flight and the barrel wall

ρ_m melt density of the fluid

References

1. Kamal, M.R., Isayev, A.I., and Liu, S-J., "Injection Molding Technology and Fundamentals," Hanser Publications, Munich (2009)
2. Kirkland, C., "Damage Control Saturn Style," *Injection Molding Magazine*, September (1994)
3. Chung, C.I. and Barr, R.A., "Energy Efficient Extruder Screw," U.S. Patent 4, 405, 239 (1983)
4. Tadmor, Z. and Klein, I., "Engineering Principles of Plasticating Extrusion," Van Nostrand Reinhold Co., New York (1970)
5. Rowell, H.S. and Finlayson, D., "Screw Viscosity Pumps," *Engineering*, **114**, 606 (1922)
6. Campbell, G.A., Tang, Z., Wang, C., and Bullwinkel, M., "Some New Observations Regarding Melting in Single Screw Extruders," *SPE ANTEC Tech. Papers*, **49**, 213 (2003)
7. Campbell, G.A. and Tang, Z., "Solid Bed Melting in Single Screw Extruders - An Alternative First Order Mechanism," *SPE ANTEC Tech. Papers*, **50**, 162 (2004)
8. Campbell, G.A. Wang, C., Cheng, H., Bullwinkel, M., and te-Riele, M.A., "Investigation of Flow Rate and Viscous Dissipation in a Single Screw Pump-Extruder," *Int. Polym. Process.*, **16**, 323 (2001)
9. Campbell, G.A. and Dontula, N., "Solids Transport in Extruders," *Int. Polym. Process.*, **10**, 30 (1995)
10. Campbell, G.A., Spalding, M.A., and Carlson, F., "Prediction of Screw Temperature Rise in Single Screw-Pump Extruders," *SPE ANTEC Tech. Papers*, **54**, 267 (2008)
11. Campbell, G.A., Sweeney, P.A., and Felton, J.N., "Experimental Investigation of the Drag Flow Assumption in Extruder Analysis," *Polym. Eng. Sci.*, **32**, 1765 (1992)
12. Campbell, G.A., Sweeney, P.A., and Felton, J.N., "Analysis of an Alternative Extruder Screw Pump Design," *Int. Polym. Process.*, **7**, 320 (1992)
13. Campbell, G.A., Sweeney, P.A., Dontula, N., and Wang, Ch., "Frame Indifference: Fluid Flow in Single Screw Pumps and Extruders," *Int. Polym. Process.*, **11**, 199 (1996)
14. Campbell, G.A., Cheng, H., Wang, C., Bullwinkel, M., and te-Riele, M.A., "Temperature Rise in a Single Screw Pump-Extruder," *SPE ANTEC Tech. Papers*, **47**,152 (2001)
15. Campbell, G.A., Spalding, M.A., and Tang, Z., "An Alternative Analysis of Single-Screw Melting," *SPE ANTEC Tech. Papers*, **55**, 147 (2009)
16. Spalding, M.A., Dooley, J., Hyun, K.S, and Strand, S.R., "Three Dimensional Numerical Analysis of a Single-Screw Extruder, " *SPE ANTEC Tech. Papers*, **39**, 1533 (1993)
17. Squires, P.H., "Screw Extruder Pumping Efficiency," *SPE J.*, **14**, 24 (1958)
18. Conant, F.S., "Using the SI Units," *Polym. Eng. Sci.*, **17**, 222 (1977)

2 Polymer Materials

A brief discussion of polymer materials is presented here to introduce the extrusion practitioner to material properties and how they relate to extrusion and polymer processing. The discussion presented here is intended to introduce several characteristics of polymer materials that can be affected during processing using single-screw extruders. For example, different in-specification lots of the same resin may vary slightly in molecular weight and thus can require different power levels to extrude. Moreover, poorly designed extrusion systems can cause the polymer to degrade in the extruder, leading to black specks in a product. Styrenic resins can degrade to produce a styrene odor at the die, especially at high processing temperatures. Other materials such as polyester resins and polyurethane elastomers can have reduced mechanical properties if they are improperly extruded. Clearly, the design of a high-performance extrusion process is highly dependent on the characteristics of the materials processed.

This chapter will introduce a brief history of polymers, the structure of polymers, and several methods for classifying and quantifying their properties from a molecular and a macroscopic point of view. The general concept of what constitutes a polymer will be introduced, including the molecular characteristics of polymers that control viscous flow and the mechanical properties. The chapter also introduces the chemical reactions for producing polymers and the degradation reactions that commonly occur in an extruder. Polymer rheology and viscoelasticity will be presented in Chapter 3. Other physical properties related to processing are presented in Chapter 4. All of these properties are related to the polymer structure and strongly affect the extrusion and mixing of polymers.

2.1 Introduction and History

Polymers are a class of materials that have large molecules made up of many repeating units or mers. The relationship between a mer and polymers will become clearer as this chapter progresses. All synthetic polymers are macromolecules that are made from as many as 20,000 mers in each molecule. The reactant is typically only one mer and is thus referred to as the monomer. Some natural polymers have many more mers per molecule, such as natural rubber with a molecular weight of over 1 million kg/kg-mol when removed from the rubber tree. Inorganic polymers are an interesting subset of synthetic polymers. Polyphosphates, polysilicates, and polysilicones are classes of inorganic polymers that have substantial commercial interest as ion exchange resins, surfactants, and coatings. Inorganic polymers are generally not extruded, so they will not be discussed further in this book. Polymers of interest here are primarily composed of carbon, oxygen, nitrogen, and hydrogen, although other atoms can occur. These organic polymers are all macromolecules. Natural polymers are formed from peptide molecules, resulting in polypeptide polymers. The polypeptide repeating unit is shown in Figure 2.1. These polymers are found in animal hooves, skin, hair, tendons, and proteins. Although these polymers function well for the organism, their utility as commercial polymers is limited by their lack of diversity in properties, especially mechanical properties.

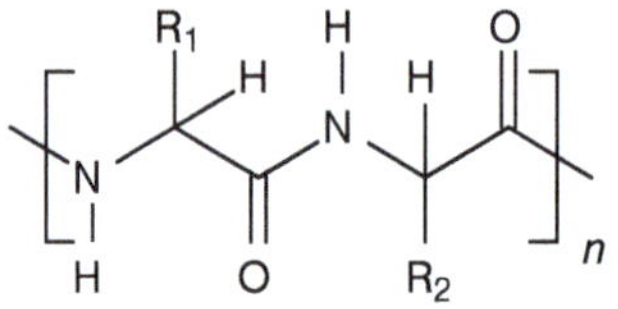

Figure 2.1 Chemical structure of polypeptides. The R_1 and R_2 group possibilities are shown in Figure 2.2, and n is the number of repeating units in the molecule

There has been a tremendous interest in polymers since World War II. In the US, consumption was 18 million metric tons in 1974, 25.7 million metric tons in 1984, and 41.3 million metric tons in 1994 [1]. Polymer production has increased from essentially zero at the end the World War II to about 101 million metric tons worldwide in 1993 [2] and 241 million metric tons in 2006 [3]. The reason for this increase is quite simple. Synthetic polymers are numerous in structure and are very diverse in their structure-property relationships. Polymers are used extensively in electrical applications, including insulators, capacitors, and conductors. They are also used in many optical applications, the biochemical industry, structural applications, packaging, and they are used extensively as thermal insulation [4].

2.1.1 History of Natural Polymers

Natural polymers have been used by humans since the dawn of history and civilization. Natural polymers have been used by early humans for making tools, weapons, clothing, shelter, and sporting objects. Many natural polymers came from the animals that were either hunted or domesticated. Our ancestors used the skin, bone, and horn in their daily existence. The Mayans used a rubber ball of coagulated latex in their national sport. As an incentive to play at peak performance, the Mayans would kill the losing team. Today, many natural polymers are used commercially, including wood, rubber, silk, cotton, leather, paper, oil-based paints, and casein (adhesives). Moreover, many active industrial programs are focused on making environmentally friendly polymers from starch [5, 6], cellulose, and lactic acid derived from plant sugar or corn. Lactic acid is then converted to polylactic acid (PLA) [7].

As expected, the mechanical properties of polymers are their most-utilized characteristic. Polymers play a very strong functional role in animals, and as indicated above these polymers are all polypeptides. The structure of these polymers is such that they have "backbone" repeating units that can be considered to be their mers. The term backbone is commonly used to describe the longest section of the polymer. All of these polypeptides have two side-chain groups on the mer, designated as R_1 and R_2. Several side groups are shown in Figure 2.2. These side groups can be neutral, negatively charged, or positively charged. The counter ions for the negatively charged group include Na^+ and K^+ while the counter ions for the positively charged group include Cl–. Other counter ions are known to exist.

$-H$ $-CH_3$ $-CH_2-OH$	$-CH_2-\overset{O}{\overset{\parallel}{C}}-O^-$	$-CH_2-CH_2-CH_2-CH_2-NH_3^+$
Neutral Groups	Negative Charged	Positive Charged

Figure 2.2 Several side groups for polypeptides

The side groups and the repeating structure of the side groups change the chemical and physical properties of the polymer, and this defines the chemical and physical characteristics of the different polypeptide molecules. Not all natural macromolecules, however, are polymers. For example, insulin is a natural macromolecule with a molecular weight of 5733 kg/kg-mol. Insulin has long linear chains that are connected by 21 sulfur crosslinks. When it is decomposed 51 residual molecules result. Insulin is not a polymer because it does not have repeating units of monomers.

Another important class of natural polymers is polysaccharides. The two most common are cellulose and starch. Cellulose has a molecular weight greater than 1 million kg/kg-mol while starch has a very similar structure but a molecular weight of only 10,000 to 40,000 kg/kg-mol. Cellulose comes from woody plants and represents about 50% of the plant mass. Cellulose is the material in pulp fibers that constitutes the paper in this book. The other half of the woody material is lignin. The lignin holds the cellulose together, and it is discarded in the pulping process. It is interesting to note that the structure of lignin has not been quantified. This is because it has to be chemically destroyed to remove it from the wood to obtain the cellulose. The chemical process of removing lignin alters the molecule in such a random way that the molecular structure of the original molecule has yet to be confirmed. The dominant commercial chemical that has been produced economically from lignin is synthetic vanilla for flavoring foods. Early manufacturing plants used the sulfite process to separate the cellulose from the lignin, and the lignin with large volumes of water was discharged into rivers. With the advent of the environmental movement the paper industry moved away from the sulfite process and has moved to a great extent to the Kraft pulping process [8]. This process is characterized by an organic sulfur odor in the vicinity of many current pulp mills. The reason that this process has become economically important is that the lignin spent liquor can be concentrated and burned to recover its energy. Cellulose is an important starting material for a number of commercial polymers, including cellulose acetate (rayon) fibers and composites [9, 10].

From an end-use point of view, natural polymers have consistent properties, but they lack diversity. That is, these polymers provide the consistent properties required by the organism. Many new commercial applications, however, were identified that could not be met by natural polymers, creating the need to develop a large number of synthetic polymers. By changing the chemistry, synthetic polymers can be made that are hard-glassy resins, soft-sticky adhesives, strong-tough textile fibers, highly extensive elastomers, and durable surface coatings.

2.1.2 The History of Synthetic Polymers

Some important everyday items that are made from polymers with widely different properties include billiard balls, plastic dishes, soda bottles, barrier and decorative films, egg cartons, polymeric drinking glasses, foam seats, and automotive tires. These applications for synthetic polymers have developed over about 150 years. As shown in Table 2.1, modern polymer material science and technology can be traced back to as early as 1770 [1]. Some important advances in the understanding of polymer production were developed before World War II.

Table 2.1 Important Dates for Development or Implementation of Polymeric Materials before World War II [1]

Year	Advancement
1770	Priestly is said to have given rubber its name because it erased pencil marks.
1838	Regnault polymerized vinylidene chloride (PVDC) using sunlight.
1839	Discovery of the vulcanization of rubber.
1868	Hyatt produced billiard balls from cellulose nitrate.
1907	Baekeland developed phenol-formaldehyde resin.
1910	First rayon plant built in the US.
1929	Carothers produced condensation polymers.
1930s	Development of polychloroprene, PMMA, PS, PA 66, PVDC.

In 1838 Macintosh and Hancock at Goodyear discovered how to take tacky natural rubber from rubber trees and react it with sulfur in the presence of heat to vulcanize the rubber to a nonstick compound that could be useful for items such as boots, rain coats, and tires. Synthetic rubber research started between World Wars I and II and progressed very quickly after World War II. The modern birth of solid synthetic polymers for commercial products may be traced to Hyatt in 1868. He discovered how to react cellulose nitrate and camphor to produce a hard plastic that was used to fabricate billiard balls because ivory had become scarce.

The growth in polymer science and technology was accelerated dramatically during and since World War II. For example, the understanding of science and engineering of synthetic polymers dramatically accelerated during World War II because of the GR-S (government rubber styrene) SBR (styrene-butadiene rubber) project. The GR-S project was a huge project initiated by the US government and industrial partners to develop SBR materials [11]. At the start of this project the US was not producing synthetic rubber, and all rubber articles were based on natural rubber. A maritime blockade by Japan during World War II caused the loss of access to natural latex and rubber from Asia. In 1940 the US produced no synthetic rubber. By the year 1945 the US was producing 700,000 tons per year of SBR. The military vehicles used during World War II needed to have rubber tires that would work in both hot and very cold environments. This war-driven project had a much more important influence on the polymer industry than just the production of synthetic rubber. More importantly the project trained people, provided insight into polymer characterization, and developed an understanding of the engineering necessary to purify and transport monomers and build production facilities. The development of polymer science and technology accelerated after World War II, as shown in Table 2.2.

Table 2.2 Important Dates for Development or Implementation of Polymeric Materials after World War II [1]

Year	Advancement
1945	Cellulose propionate
1946	Screw injection molding
1947	Epoxy resins
1948	ABS resins
1948–1950	Acrylic fibers
1949	Blow molding
1950	Ziegler and Natta developed vinyl polymers; Swarc developed living polymers
1954	Polyurethane foam in the US; SAN copolymers
1958	Rotational molding
1960	Spandex developed by Shivers
1962	Phenoxy resins; polyimide resins
1970s	Polybutene resins; reaction injection molding; interpenetrating networks; polyester beverage bottles; aromatic polyamides; UNIPOL™ gas phase polymerization process; LLDPE resins
1980s	Polysilanes; liquid crystalline polymers; poly(etheretherketone); conducting polymers; polyetherimide
1990s	Metallocene catalysts and resins; living cationic polymerization; syndiotactic polystyrene commercialized; styrene-ethylene copolymers; nanocomposites
2000s	Olefinic block copolymers (OBC) [12]

UNIPOL™ is a trademark of Univation Technologies, LLC

2.2 Characteristics of Synthetic Polymers

This introductory section develops some of the organic and physical chemistry of polymeric materials. The discussion is approached from the viewpoint of a process engineer's needs and thus does not use a fundamental chemistry mechanistic approach. With this approach, atoms can be considered to be spherical structures, with a very dense core of protons and neutrons, and a large fraction of unfilled space defined by the electron cloud. Around this dense core the atom's electrons rotate at several average distances called orbitals. The electrons in the outer orbital shell are called the valence electrons. Generally it is accepted that atoms bond to form molecules by the interaction of the electrons in the outer orbital. One important type of bond is the ionic bond, which involves electron transfer. This type of bond is present in common table salt or sodium chloride. Typical examples of this type of bond are shown in Figure 2.3, where sodium and chlorine react to form sodium chloride and magnesium and chlorine react to form magnesium chloride.

The subscripts by the atoms in this figure are the number of electrons in the first three orbital shells, starting with the shell nearest to the nucleus. These first two shells are full and contain two and eight electrons respectively. A sodium (Na) atom has one valence electron while magnesium (Mg) and chlorine (Cl) have two and seven valence electrons, respectively. In each of these reactions, electron(s) transfer from the metal to the nonmetal, allowing chlorine to fill its outer valance shell. As a result the chlorine has a net negative charge, having one more electron than it has protons, and the sodium has a net positive charge due to having one more proton than electrons.

$$Na_{2,8,1} + Cl_{2,8,7} \rightarrow Na^{+}Cl^{-}$$

$$Mg_{2,8,2} + 2(Cl_{2,8,7}) \rightarrow Mg^{+2}(Cl^{-})_2$$

Figure 2.3 Ionic bond reactions

These ionic reactions or electron transfer reactions are not what generally occur in the structure of both natural and synthetic polymers. In polymers it is the covalent bond that dominates, and in a covalently bonded structure there is no transfer of electrons from one atom to another. Instead the electrons are shared between the adjacent atoms in the molecule. The commercial polymeric materials that will be covered in this text will generally be based on seven atomic species: silicon, hydrogen, chlorine, carbon, oxygen, nitrogen, and sulfur. Figure 2.4 shows these atoms with the number of outer valance electrons.

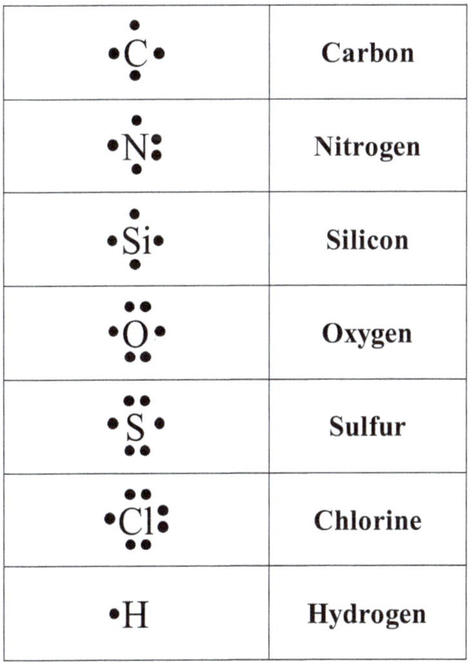

Figure 2.4 Atoms considered when discussing polymers in this book

As indicated in Figure 2.4, all of these atoms have at least one unpaired electron in the valence or outer electron shell. A covalent bond, as suggested by the word covalent, is a bond which shares at least one pair of valence electrons between two atoms. When examining the molecular structure of polymers, it is found that all commercial polymer molecules are formed from covalent bonds.

Examination of these atoms reveals there are valence electrons numbering from one to seven in these atoms. Except for hydrogen, which has a full shell of two electrons, all of the other atoms in Figure 2.4 require eight electrons to fill the valence shell. When these shells are filled in a covalent manner by polymerization, the covalent bonds can lead to the production of large polymer molecules.

For most commercial polymers, carbon and silicon are the two backbone building blocks, and they routinely form chains of like atoms. Generally, three types of covalent bonds are associated with carbon in the production of commercial vinyl polymers, as shown in Figure 2.5: the single, double, and triple carbon-to-carbon bonds.

C:C	Single bond
C::C	Double bond
C:::C	Triple bond

Figure 2.5 Covalently bonded carbon

As stated earlier, when polymer molecules form, valence electrons are shared by the associated atoms to form covalent bonds. Carbon always has four covalent bonds, each of which is made up of two shared electrons. In Figure 2.5 the hydrogen atoms that are associated with the carbon atom in these molecules have been omitted from the diagrams. In order to meet the four covalent bond requirement, the molecule ethane must have three other covalent bonds on each carbon, and these are covalent bonds with hydrogen. These three carbon examples represent a single carbon-carbon bond for ethane, a double bond for ethylene, and a triple bond for acetylene. The single bond has two electrons, the double bond has four electrons, and the triple bond has six electrons. This leads to each carbon in ethylene having two hydrogen atoms attached, and in the case of acetylene, one covalent bonded hydrogen is attached to each carbon.

All of these chemical species have importance in the production of polymeric materials. There are several shorthand techniques for writing down the structures of polymers. The carbon-based polymer molecules using the "stick" representation are made up of atoms connected by covalent bonds (represented here by the straight lines between the carbon and the hydrogen and the carbon-to-carbon molecules), as shown in Figure 2.6. To reiterate, carbon is always tetravalent, having four covalent bonds, and a schematic of the paired electrons for two of the incorporated carbon molecules can be seen in the bottom of Figure 2.6. Thus, each "stick" represents two electrons. For the two highlighted carbon atoms in the polyethylene molecule of Figure 2.6, the electron representation is shown, where there are four covalent bonds associated with each carbon and each bond is made up of two shared electrons represented by the black dots. This polymer molecule is made up of only carbon and hydrogen with no double bonds, and it represents a linear form of polyethylene typical of high-density polyethylene (HDPE) resin. The long chain representation of HDPE resin is shown in Figure 2.8(a). For the commercial polymers presented in this book, carbon can bond with oxygen, nitrogen, hydrogen, sulfur, chlorine, and silicon. Both carbon and silicon are chain formers while oxygen and nitrogen are typically incorporated atoms. Bonding with other atomic species, however, is possible but beyond the scope of this book. Moreover, some of these atoms have some restrictions when they react with carbon to form polymers.

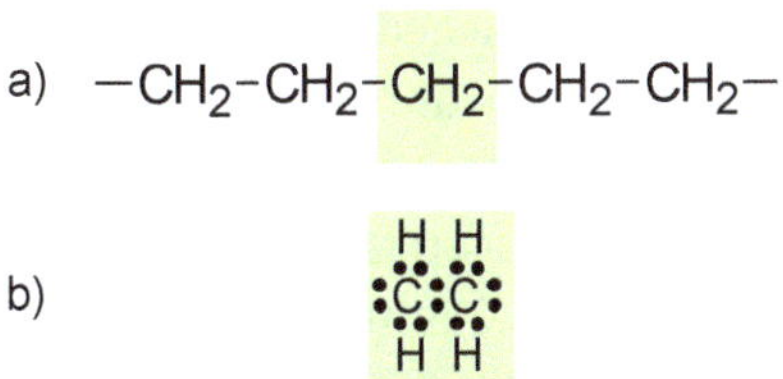

Figure 2.6 Representations for PE resin: a) stick method, and b) electron representation of the highlighted segment

Nitrogen and oxygen can be incorporated into the backbone such that they are surrounded by different atom types. For example, organic peroxides contain two covalently bonded oxygen atoms that form the peroxide linkage. These molecules are inherently unstable. Two covalently bonded nitrogen atoms are also similarly unstable. These unstable structures decompose to form smaller unstable molecules that are used to start the polymerization for some types of monomers. Thus, to be incorporated implies that the molecules are found only singularly in the backbone chain. Sulfur and silicon are considered to be chain formers. They can be found in the backbone in multiple units connected covalently to molecules of the same type or with carbon. Complete molecules with a silicon backbone are possible, and molecules with multiple sulfur links incorporated into the system are common, particularly in sulfur-crosslinked rubber.

2.3 Structure Effects on Properties

The polymer backbone can have functional groups that are attached to some of the carbon atoms. These groups change the chemical and physical properties of the polymer molecule. If these groups are relatively short then they are usually referred to as pendant groups. Several common pendant groups are shown in Figure 2.7.

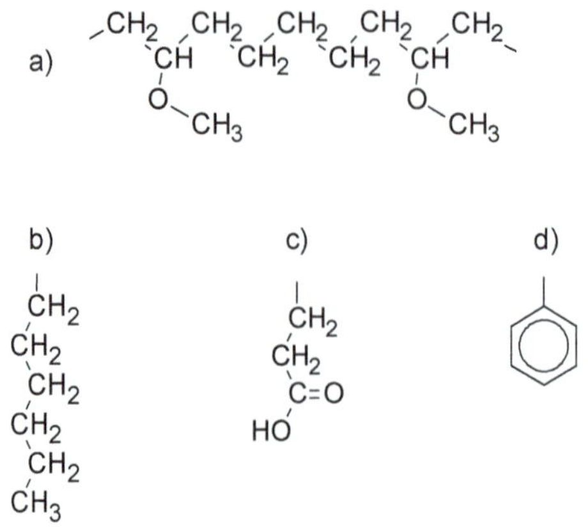

Figure 2.7 Pendant groups:
a) PE backbone with methoxy pendant groups,
b) hexyl pendant group,
c) propanoic acid pendant group, and
d) phenyl pendant group

The structure in Figure 2.7(a) shows two methoxy groups attached to the backbone of the polymer. The methoxy group consists of an oxygen atom attached to a methyl group. Another common pendant group shown is the hexyl group with six carbon atoms. The hexyl group is a prominent group in some commercial LLDPE resins, creating a level of short-chain branching as shown in Figure 2.8(b). The number of hexyl groups positioned on the backbone control the crystallinity (or solid density) and the modulus of the resin. An acid-containing group in Figure 2.7(c) is used to produce the potential for secondary reactions with inorganic salts or to form ester linkages. A phenyl group can also be a pendant group. If the phenyl group is on every other carbon then the polymer is polystyrene. Polystyrene is produced by polymerizing styrene monomer. Not shown in the figure is the CH_3 pendant group. If this group is on every other carbon atom the structure becomes polypropylene. Polypropylene is polymerized from propylene. Sometimes, pendant groups can be a large fraction of the length of the backbone. These pendant groups are then described as long-chain branches, as shown by Figure 2.8(c). Low-density polyethylene resin produced using a high-pressure reaction process has many long-chain branches. Recently, metallocene and single-site catalyst technology [13, 14] have allowed development of olefin polymers with controlled levels of long-chain branching along with short-chain branching, as depicted in Figure 2.8(d). These resins all flow when heated and are referred to as thermoplastics.

If a large number of branches exist that connect all of the backbone molecules into a three-dimensional network, the material will not flow when heated, and it is considered a thermoset resin. Vulcanized rubber is an example where the sulfur linkages create a three-dimensional network, converting the precursor rubber into a solid thermoset material. Crosslinked backbone chains are shown in Figure 2.8(e). When extruding many thermoplastics, the polymer can undergo chemical reactions to form small amounts of crosslinked material. Partial crosslinking is a problem with some PE resins that contain residual double bonds that are made using chromium catalysts. The crosslinks appear in the final product as a defect called a gel. The catalyst residue from chromium processes can also contribute to unwanted reactions in the extrusion process [15], leading to the formation of double bonds between carbons. Moreover, crosslinking of PE resins often occurs in the extruder, especially for poorly designed processes. In polyurethanes the crosslinks can come from partial hydrolysis of the urethane bonds and then from reaction with pendant hydrogens on either the remaining urethane bonds or on polyurea bonds. PVC [16] and PVDC [17] resins undergo dehydrochlorination at elevated process temperatures, creating HCl gas, a complex mixture of hydrocarbons, and black char. Also, from an extruder troubleshooting point of view localized crosslinks can lead to gels if the crosslinking results in unmeltable structures. These structures will cause defects, including gels in extruded films or black specks in the product. These two issues will be discussed in detail in Chapter 11.

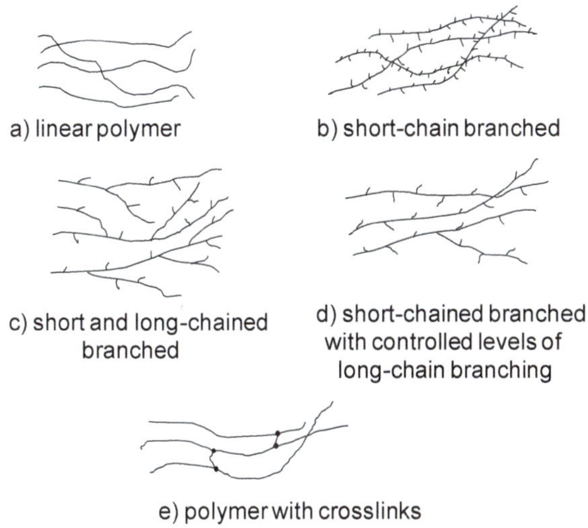

Figure 2.8 Schematic of common polymer chain structures

The straight-line representation of bond angles previously depicted is not very representative given the tetrahedral structure of carbon atoms. There is a generally accepted more stylized representation for the carbon atoms, represented in a zigzag fashion, as shown in Figure 2.9. In this representation, the hydrogen molecules are located out of the plane of the diagram. Although this comes closer to the structure of the carbon-based polymer molecule, this is still not completely representative of the structure since with single covalent bonds there is a thermal-induced rotation between the carbon bonds. These rotations at the carbon-carbon bonds control another important property of polymers, the glass transition temperature (T_g).

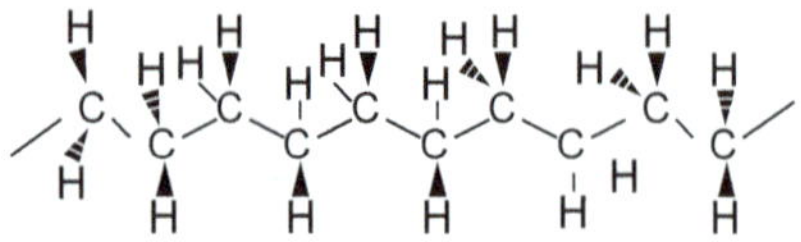

Figure 2.9 Polymer bond representations for a PE resin segment. All carbon atoms are fixed in the plane of the diagram while half of the hydrogen atoms are in front of the plane and half are behind the plane

The zigzag pattern does suggest that the carbon-carbon bond rotation must overcome a potential energy barrier to rotate around its closest carbon neighbors. This carbon-carbon backbone rotation can be strongly affected by the pendant group and by temperature, leading to different glass transition temperatures in polymer molecules.

2.3.1 Stereochemistry

Stereo structure of molecules has a strong effect on the properties and processing of materials. The concept of stereo structures is depicted in Figure 2.10 for the stereoisomers of PP where the pendant group is the methyl group. In this figure, the carbon and hydrogen atoms were omitted, and just the stick method that represents the carbon-to-carbon bonds is shown. In the case of syndiotactic polypropylene, every other methyl group is above or below the plane of symmetry; it has a melting temperature of 150 °C. In the case of isotactic polypropylene, every methyl group is on one side of the plane of symmetry. This is the form of most extrusion- and molding-grade polypropylene resins because it has the highest melting temperature of 187 °C. Finally, atactic polypropylene has the methyl groups randomly distributed above and below the plane of symmetry. This material has a glass transition temperature of -10 °C and no melting temperature. Atactic polypropylene is extensively used as a hot melt adhesive and has little or no crystalline structures. It should be emphasized that each of these different stereo structures provides very different physical properties even though the chemical structure of the monomer and polymer is the same. The different melting temperatures and properties are a result of the stereo structures of the polymers. The traditional process for making isotactic polypropylene also produces up to 20 percent atactic polypropylene, and this has to be removed before the high temperature product can be sold. Recently syndiotactic polystyrene has been evaluated as an injection-molding resin due to its high melting temperature and high modulus [18]. In this molecule the phenyl ring attached to every other carbon is alternately above and below the plane of symmetry. In general, the production of three different stereo structures from the same monomer is performed using different polymerization catalysts. Polyethylene does not have stereoisomers because the two hydrogen atoms attached to a single carbon atom in the backbone are indistinguishable.

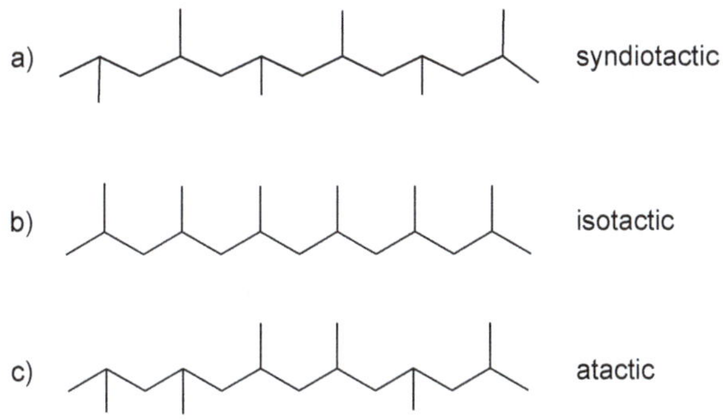

Figure 2.10 Structures of PP: a) syndiotactic, b) isotactic, and c) atactic

Stereoisomers are important in polymer production when carbon-carbon double bonds occur in the structure, as shown in Figure 2.11. This figure shows the classic chemical structure of the cis and trans isomers so well known in conventional chemistry and applied to polymers. They are sometimes known as the "boat" (cis) and "chair" (trans) configuration of molecular structure around the double bond. The location of these double bonds and the bulky units of the side chains can affect the average rotation time of the backbone bonds in these materials. As an example, butadiene can polymerize into a 1,2 or 1,4 configuration, and the structures will have very different chemical and physical properties. For the 1,4 polymerization product, the cis configuration has a melting temperature of 12 °C and the trans configuration at 142 °C. The cis and trans configurations of polyisoprene have different thermal properties. The cis configuration has a melting temperature of 28 °C while the trans configuration has a melting temperature of 80 °C. The trans configuration is used in many rubber compounds because its melting temperature is well above room temperature, and it is similar to natural rubber.

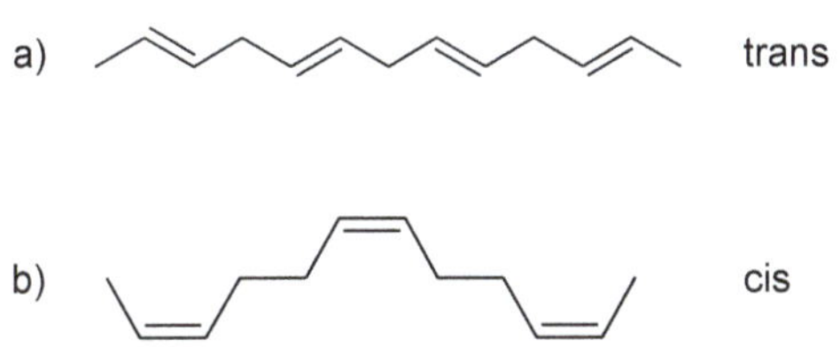

Figure 2.11 Double bonds in some rubbers: a) trans configuration, and b) cis configuration

2.3.2 Melting and Glass Transition Temperatures

The extruder temperature profile for a single-screw extruder is set such that the functions of the process convert the polymer from a "solid" to a "fluid." These two words are in quotation marks because for noncrystalline glassy (or amorphous) polymers there is no melting temperature. The melting temperature is defined as the peak temperature where most of the crystals are melted, and it will be discussed further in the next section. Since amorphous polymers do not contain crystal structures, they do not have a melting temperature. Amorphous polymers, however, undergo a change in properties at the glass transition temperature (T_g), and they are discharged from the extruder at a temperature substantially higher than the T_g. PS, PMMA, and PC resins appear as solids at room temperature because at room temperature the backbone carbon-carbon bonds rotate very slowly, and thus they only respond very slowly to an external stress. When they are below the T_g they act like a glass and not a fluid. As introduced earlier, the T_g is controlled by molecular motion as it relates to the backbone covalent bond frequency of rotation. In order for a carbon-carbon bond to rotate, the carbon that is rotating must overcome a potential energy barrier related to changing its relative position to the adjacent carbons. This energy barrier is due to the tetrahedral structure of the carbon covalent bonds. All of the structural changes discussed above in the molecular architecture (side chains and pendant groups), stereo structure, and double bond placement have a strong influence on the magnitude of this potential energy barrier and thus on the polymer's properties as a function of the environmental temperature and its glass transition temperature. The process for heating an amorphous resin from a temperature below the T_g to a temperature where the material is higher than the T_g is called devitrification.

At higher temperatures, the carbon-carbon backbone bonds will rotate at a higher frequency because there is sufficient molecular energy to counter the energy barrier that wants to keep the bond in a stable position. Thus, at higher relative temperatures the number of carbon-carbon backbone bond rotations per second increases. The increase in the rate of bond rotations is very small below the T_g due to the average bond potential energy being lower than the energy necessary for rotation. It is thus thought that the bonds only rotate essentially one at a time below T_g. When the bonds rotate many times per second, the polymer will be above its T_g, and it is usually thought that multiple groups of bonds have the energy to rotate simultaneously. The effect of side chains in general is to increase the potential energy barrier and thus require more thermal energy to increase bond rotation. It therefore follows that higher temperatures are required to get an equal rotation rate and multiple bonds rotating if molecules have large side chains. Natural rubber, poly 1,4 isoprene, has much smaller side chains than polystyrene. The trans stereo polymer of isoprene has a T_g of −50 °C. Natural rubber has many rotations at 25 °C but only a few rotations at −100 °C. At 25 °C it is well above its T_g, and it would flow under a slight force, or if slightly crosslinked it will stretch like a rubber band. At −100 °C it is below the T_g, and it would break like window glass if struck with a hammer. These differing responses to applied stresses relates to the ability of the molecule to dissipate energy and its ability to rotate the bonds in the backbone. In contrast, PS has many rotations at 150 °C and only a few rotations at 25 °C. The glass transition temperature for PS resin is about 100 °C. The difference in rotation rates at comparable temperatures controls many of the properties and characteristics of polymers. For extrusion troubleshooting, the T_g sets the lower temperature limit for extruder operation since we need fluid-like behavior in the metering end of the extruder. Most amorphous resins will not flow well, however, until the temperature is 50 °C or more above the T_g. An excellent discussion on T_g with much more detail can be found elsewhere [1].

Polymers can thus be defined in general terms as "glassy polymers" for those resins with a T_g that is above room or use temperature, and as semicrystalline polymers that have a T_g well below room temperature and get their solid-like properties at use temperature because they have a crystalline phase. Semicrystalline polymers have a portion of the material that is completely amorphous, and thus this portion has a T_g. The polymer industry has over the years worked to develop the "ultimate" engineering polymer, which usually is partially crystalline. Like all fields of engineering there are exceptions to this, and PC is one of several exceptions. PC has a relatively high T_g of about 155 °C and has good impact strength in the glassy state at or near room temperature, unlike most glassy polymers such as PS or PMMA that fracture easily when impacted. An engineering polymer must have a very broad temperature use range. This can loosely be defined as the difference in temperature between the melting temperature (T_m) and the T_g. Over this temperature range partially crystalline polymers can be thought of as a flexible matrix of material above the T_g, held together by the high-modulus crystallites of the polymer acting as high-modulus fillers and physical crosslinks. Extensive early investigation of homogeneous polymers led to a relationship between T_g and T_m. Using absolute temperature the relationship was found that T_m is about 1.4 to 2.7 times the T_g of all homogenous polymers. For most polymers the melting temperature is thus a bounded function of the T_g. So the use temperature range is more or less constant for homogenous polymers. Although not discussed here, random and block copolymers do not follow this relationship.

2.3.3 Crystallinity

As discussed above, many polymers contain some crystalline structures when they are solidified. These polymers are referred to as semicrystalline resins. These crystalline structures can be observed using microscopy as shown in Figure 2.12 for PP and sPS resins. As shown schematically in Figure 2.13 and discussed above, not all portions of the polymer chains are incorporated into the crystalline structure. Instead, the portions of the chains that are not crystallized make up the amorphous phase. Solid density is the most commonly used method for measuring the degree of crystallinity in a polymer, especially for olefinic resins. LLDPE resins are common semicrystalline polymers, and have solid densities ranging from about

0.915 to 0.935 g/cm^3. Resins with higher solid densities contain higher levels of crystallinity and lower levels of the amorphous phase. For LLDPE resins, the crystallinity level is in general controlled by the amount of comonomer, that is, the amount of alpha olefin reacted into the polymer chain to create pendant groups. All PE resins are classified by their density: HDPE with a solid density greater than 0.940 g/cm^3, MDPE with a density range of 0.925 to 0.940 g/cm^3, LDPE with a range of 0.915 to 0.925 g/cm^3, and ULDPE with a density less than 0.915 g/cm^3.

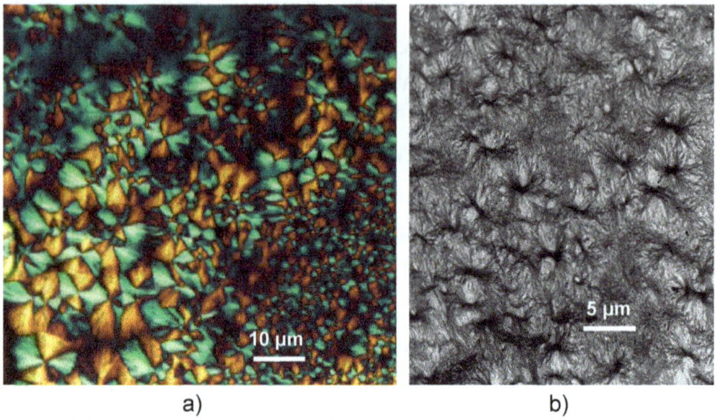

Figure 2.12 Photographs of polymer crystalline structures: a) optical photograph of PP resin using polarizing light and filters, and b) electron micrograph of sPS resin (courtesy of Robert C. Cieslinski of The Dow Chemical Company)

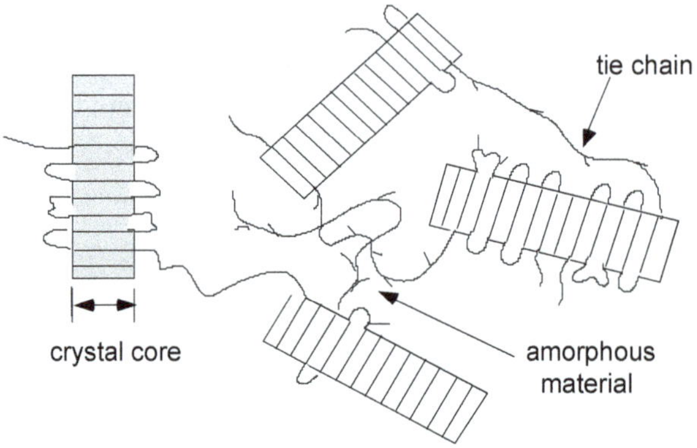

Figure 2.13 Schematic of the crystalline and amorphous regions of a semicrystalline polymer

Amorphous material often produces tie chains that connect two or more different crystals. These tie chains increase the properties of the solid resin by forming a temporary three-dimensional crosslinked system. As the resin is melted in an extruder, the crystals and the tie chains are destroyed, and the polymer acts like a normal thermoplastic resin. When the resin is cooled and solidified, the crystals and tie chains reform.

The level of short-chain (SCB) and long-chain (LCB) branches control the solid resin density of a PE resin. For example, the level of SCB is controlled by the amount of alpha olefin comonomer incorporated into LLDPE resin as a pendant group. The random positioning of the pendant groups disrupts the crystallization process when the polymer is cooled from the molten state, causing the level of crystallinity to decrease with increasing amounts of alpha olefin comonomer.

When semicrystalline polymers are heated in a differential scanning calorimeter (DSC), the temperature where the highest energy required for crystals to melt is referred to as the melting temperature (T_m). Figure 2.14 shows DSC heat flow curves for an LLDPE resin with a solid density of 0.935 g/cm^3 and a PP resin. The melting temperature of the PP resin (158 °C) is higher than that for the LLDPE resin (126 °C). As the polymer melts, a heat of fusion (area under the DSC curve during phase change) is determined, and the weight percent crystallinity can be calculated (100% crystallinity has a heat of fusion of 292 J/g for PE). Thus, a PE resin with heat of fusion of 100 J/g will contain about 34% crystalline material and 66% of the amorphous phase.

The modulus of PE resins increases with increasing solid density. Thus, a HDPE resin has a higher modulus than an LDPE resin, as shown by the data in Table 2.3. In general, resins with low solid densities feel "soft" to the touch while resins with high densities feel hard. The T_m and T_g for selected semicrystalline and amorphous materials are given in Table 2.3.

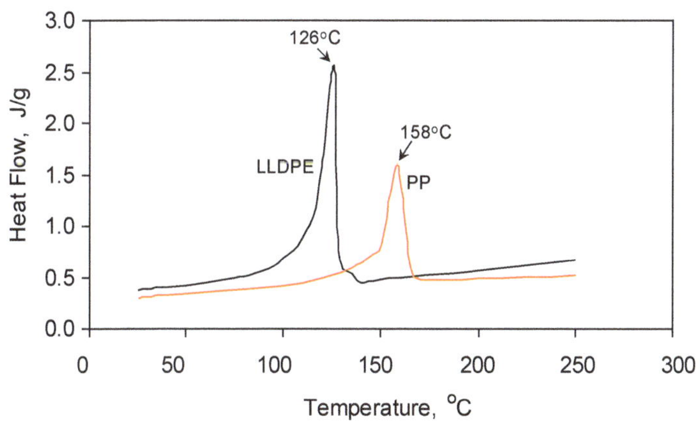

Figure 2.14 DSC heat flows for LLDPE and PP resins

Table 2.3 Physical Properties and Applications for Selected Resins

Resin	Tensile Modulus, MPa	Impact Strength, J/cm	Melting Temperature (T_m), °C (semicrystalline)	Glass Transition Temperature (T_g), °C (amorphous)	Applications
HDPE	1070-1090	0.2-2.1	135	(-110) to (-125)	Liners, sacks
HIPS	1100-2500	0.5-3.7		100	Packaging, specialty films
LDPE	170-280	No break	110	(-110) to (-125)	Films
LLDPE	260-520	No break	120	(-110) to (-125)	Films
PA 66	1600-3400	0.5-1.1	260	55 to 58	Barrier films
PET	2800-4100	0.13-0.37	240	78 to 80	Packaging films
PLA	3300-3800	0.03	160	55 to 60	Packaging, fibers
PP	1140-1550	0.2-0.75	170	-20 to -10	Packaging (sheet)
PS	3100-3300	0.20-0.24		100	Packaging, foam sheet
PVDC	340-550	0.1-0.5	165	-15	Oxygen and water barrier applications (food packaging)

The values given here are for resins with relatively low levels of additives. Resins with fillers or low molecular weight processing aids can have physical properties considerably different from these.

2.4 Polymer Production and Reaction Engineering

This section introduces simple polymer reaction chemistry used to produce many commodity polymers. Understanding this simplified approach to the chemistry of polymer production is important in troubleshooting many extrusion processes, especially those that are producing unwanted degradation products that contaminate the discharge resin. There are two general types of polymer production processes: 1) step or condensation reactions, and 2) addition or vinyl polymerization reactions. An overview of the reaction mechanisms will be presented in the next sections.

2.4.1 Condensation Reactions

The classification of a condensation polymer is historically based on the observation that during polymerization a small molecule, such as water, is condensed or removed as part of the reaction. There are a large number of polymers produced from condensation reactions and only a representative sample is presented in Figure 2.15. Reactors for producing these materials include batch, continuously vented, tray reactors, twin-screw extruders, and vented single-screw extruders. These production devices will not be covered in this text because they are of more interest to the manufacturing engineer than the extrusion process engineer.

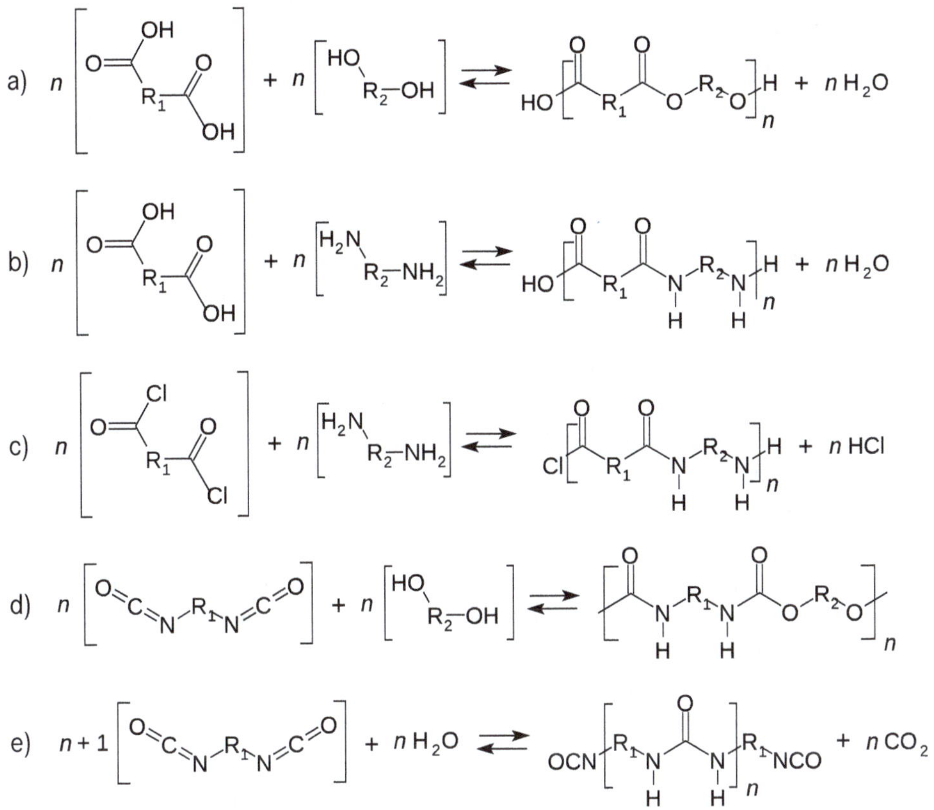

Figure 2.15 Reactions to produce typical condensation polymers: a) polyester, b) polyamide with water as a byproduct, c) polyamide with HCl as a byproduct, d) polyurethane, and e) polyurea

Not all of these polymers have a small molecule released during production, but they essentially all will react with water to reduce the polymer molecular weight during the extrusion process. Examining more closely the reactions in Figure 2.15, reaction (a) produces a polyester if the acid and base have two or more reactive acid and hydroxyl groups per molecule. If the number of reactive groups on each molecule is exactly two as shown then the product will be a linear polymer such as PET. If the average functionality is greater than two then an alkyd resin can be produced. In all of these reactions R_1 or R_2 represent the remainder of atoms that are attached to the reactive groups of the monomers. Reactions (b) and (c) produce polyamides. If R_1 and R_2 in reaction (b) are six-carbon linear chains and the monomer has reactive groups on both ends then the product will be PA 66. Reactions (d) and (e) are important in producing polyurethane thermoplastics and polyurethane foams. Reaction (d) produces a polyurethane by reaction of an isocyanate and an organic alcohol. If the reactants are multifunctional a polymer is produced. A functionality of two produces a linear polymer while more than two produces a crosslinked polymer. Reaction (e) produces a polyurea, which is the product of an isocyanate and water. The evolved CO_2 is the blowing agent used to produce polyurethane foams, both rigid and flexible.

All chemical linkages in a condensation polymer have the potential to react at all times. For example, an acid end group on a chain that is n units in length can react with an alcohol end group on a chain that is m units in length, creating a new polyester chain that is $n + m$ units in length. Moreover, since the ester reactions are highly reversible, the new chain can hydrolyze into two smaller chains with lengths different than n and m. Because all the bonds can react at any time, these reactions are characterized by Gaussian statistics, and thus the molecules grow or decompose as a result of random event polymerization or hydrolysis of the reactive groups, respectively. Thus, the final resin will have a distribution of chains with different molecular weights; the weight average will approach two times the number average molecular weight. The length and distribution will affect its end-use physical and rheological properties. High molecular weight polymers only come at very high conversion of the monomer to polymer, which is when the small condensation molecules have been removed from the reaction mixture. All of the process reactors for the polymers listed above are thus vented and usually attached to strong vacuum systems. Also, the production of high molecular weight polymers (high average n values) only occurs when the reactive groups are present at equimolar concentrations. That is, there is the same number of the two reactant species in the reactor at the beginning of the polymerization. Only when many of the molecules in the reacting mass are connected as a polymer molecule and when essentially all of the monomer has been reacted into the polymer is a high molecular weight polymer produced with good physical properties.

In addition to condensation reactions, ring-opening reactions are often used to produce polyethers and polyamides. The most well known polyether and polyamide are polyethylene glycol and PA 6, as shown in Figure 2.16. Both are considered condensation polymers and are produced from cyclic monomers. Polyethers are somewhat more stable to hydrolysis than most other condensation polymers, but they are, however, degraded by reaction with adsorbed water, and they thermodynamically rapidly decompose at temperatures above 500 °C. Reaction (a) in Figure 2.16 produces polyethylene glycol. Ring opening in Figure 2.16(b) is also used to produce PA 6 from caprolactam. Polytetrahydrofuran is produced from the ring opening of tetrahydrofuran, and it is a semicrystalline polyol used in many high-modulus polyurethanes. The mechanism for these reactions is quite complex and the reader should consult a text on polymer chemistry or polymer materials [1, 19] for a thorough discussion of ring-opening reactions. It is important to emphasize again that all of these polymers can be susceptible to degradation due to the attack of water molecules in the extruded polymer, leading to hydrolysis and the loss of physical properties in the finished product.

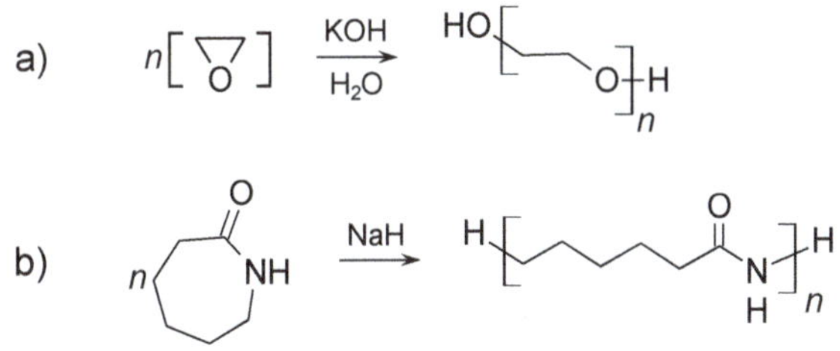

Figure 2.16 Ring-opening catalyzed condensation polymers: a) reaction for polyethylene oxide, and b) caprolactam ring opening to PA 6

2.4.2 Addition Reactions

Addition reactions are used to produce vinyl polymers, and several commercial monomers are presented in Figure 2.17 In this route to a high molecular weight polymer, the monomer has at least one double bond and often a larger number of nonreactive functional or pendant groups. This type of polymerization is often called addition or chain polymerization. During the reaction of a bulk polymerization of vinyl monomers, only certain activated species react, only a few chains grow at once, and a high molecular weight polymer is obtained quickly. Addition reactions are controlled by active centers. This mechanism is entirely opposite to the reaction mechanism for condensation polymers. As discussed previously, condensation reactions occur with many chains growing and decomposing at once, and high molecular weights depend on equimolar amounts of the two reactants.

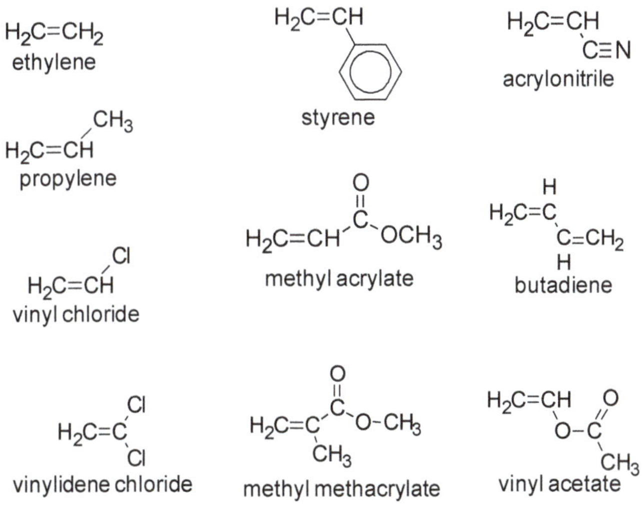

Figure 2.17 Typical vinyl monomers

The active centers can typically be free radicals, positive ions, negative ions, or a metallic catalyst. The metallic catalysts include Ziegler-Natta catalysts and chromium derivatives, and metallocene catalysts. Metallocene catalysts are typically cyclopentadienyl derivatives of a transition metal. Most of these polymers are traditionally made in bulk from vinyl monomer and free radicals. Many are polymerized in solvents with either homogenous or heterogeneous catalysts. Some polymers such as those produced by the low-pressure gas phase process use Ziegler-Natta catalysts that have the catalyst supported on a solid substrate. However, more and more of these vinyl polymers are now being produced using metallocene catalysts because this technology, which became commercially economical in the 1990s, leads to better control of the polymer molecular structure and thus the physical properties.

The use of free radicals, positive ions, or negative ions for the polymerization depends on the monomer and the pendant group. Monomers with strong electron-withdrawing groups can be polymerized with anions to form stable carbanions. Strong electron-donating groups on the vinyl monomer lead to polymerization with cations to form stable carboniums. Vinyl monomers with intermediate electron characteristics (not strongly withdrawing or strongly donating) polymerize with free radicals produced by molecules that decompose, leaving an unpaired electron, as shown in Figure 2.18. Some monomers such as ethylene, styrene, and butadiene will polymerize by essentially any of the above methods. Four typical free radical initiators are benzoyl peroxide, AIBN, and t-butyl-hydroperoxide, and potassium persulfate, as shown in Figure 2.18. Potassium persulfate is one of the few materials that will produce stable free radicals in water, and thus it is used extensively in emulsion polymerization. Due to the mechanism of growth of the polymer inside a soap micelle in emulsion polymerization, very high molecular weight polymers are produced because it is difficult for a second free radical to enter the micelle. The fragments of the free radical initiator molecule can thus terminate the resulting polymer molecule when the growing polymer chain encounters a free radical, as is the case almost exclusively in emulsion polymerization. In each of these cases, the free radical is indicated by (•), as shown in Figure 2.18.

The reaction sequence for a typical vinyl polymer has four steps. In the first step, a free radical must be produced from the initiator such as those shown in Figure 2.18 and Figure 2.19. These radical formation reactions are typically first order in rate and are promoted by the elevated temperature of the reaction. For some free radical initiators, light can also promote the reaction. Then a sequence of events in the reaction mixture occurs, including initiation of a chain, followed by propagation, and finally termination of the chain. Termination of the chain will be discussed later. The schematic steps to produce an addition polymer from bulk or solvent polymerization are detailed in Figure 2.19. The radical produced from the initiator reacts with the monomer in Step 2 to produce a new free radical by opening the double bond of a vinyl monomer. The active center of the new radical is at the opposite end of the activated mer. Propagation (Step 3) occurs when the active center reacts with more monomer and continues to add more monomer, one mer at a time. This propagation step continues until a termination reaction ends the process.

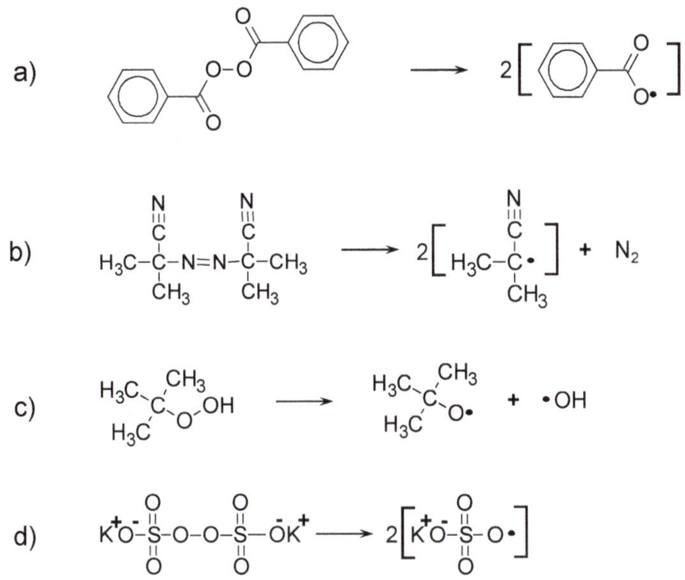

Figure 2.18 Four free radical initiators: a) benzoyl peroxide, b) azobisisobutylnitrile (AIBN), c) t-butylhydroperoxide, and d) potassium persulfate; (•) is the free radical

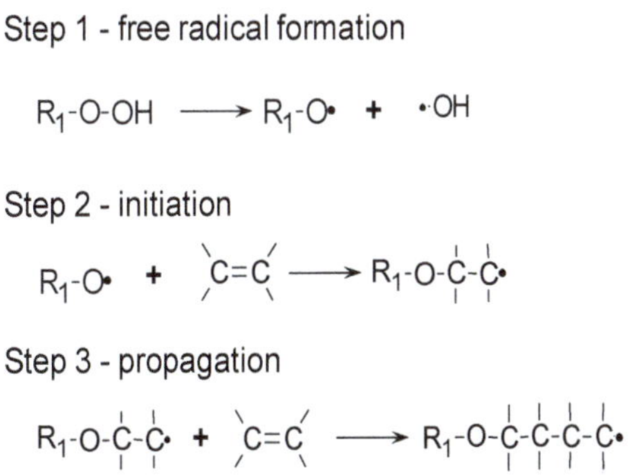

Figure 2.19 Addition polymer free radical formation, initiation of the monomer, and propagation

Termination (Step 4) can normally occur by either combination or disproportionation reactions, as shown in Figure 2.20. For a combination termination, two chains with active centers combine to form a new chain without an active center. The molecular weight of the chain produced is the sum of both radicals. If these are two relatively large growing chains, a very high molecular weight polymer is obtained.

If large and small chains are combined, then a much smaller molecule is produced. If one of these radicals is a growing polymer and the other a free radical derivative of the decomposition of the initiator, a molecule that is about the size of the propagating molecule terminated on both ends by initiator fragments is formed. For the disproportionation termination, a hydrogen atom with its electron is extracted from one of the radicals, creating a polymer chain that is saturated and a polymer chain with a double bond at the very end of the chain, as shown in Figure 2.20. These different termination mechanisms cause the bulk polymerized materials to have very broad molecular weight distributions. Polymerizations that produce terminal double bonds can cause problems during extrusion. The double bonds can react with oxygen to create a new free radical. The free radical can then continue to react, forming links, branching, and degradation products [12] in the extruder.

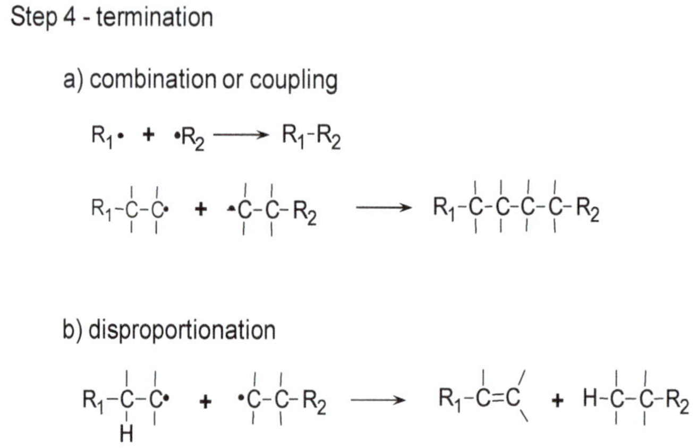

Figure 2.20 Termination of addition reactions

Many polymers are produced using addition reactions, including PE resins, PP, PS, to name just a few. Moreover, there are many different reactor styles and comonomers used to produce PE resins. These methods and the resin stuctures that they produce are beyond the scope of this work, but they can be obtained in reference [20].

2.5 Polymer Degradation

The polymer industry works very diligently to produce high-quality polymers. All of these polymeric materials, however, are subject to substantial degradation under the conditions found in most extruders unless precautions are taken to minimize the reactions. The degradation can be broadly classified into three areas. Degradation of addition polymers is associated with the loss of a hydrogen atom, usually due to high shear stress related to high polymer viscosity. The resulting radical reacts with oxygen from the atmosphere to form a peroxide, reacting with a double bond to form a branch and eventually a crosslinked gel. For condensation polymers the major form of degradation is attack by absorbed water that comes from the humidity in the air, causing a loss in molecular weight. Condensation resins can also undergo attack from free radicals, resulting in black colored char. Finally for chlorine- and fluorine-containing compounds such as PVC, PVDC, and PVDF, a major mechanism for degradation is the formation of HCl or HF and the subsequent crosslinking of the polymer into a black char. These reactions once initiated are very rapid and can be catalyzed by brass or other metals.

In order to minimize the degradation of a polymer when it is extruded, resin manufacturers add stabilization chemicals to the resin. Often the term resin is used for a polymer that contains additives. Moreover, ultraviolet stabilizers are often added to minimize the attack from light on the extruded and injection-molded parts. For example, resin manufacturers add oxidation stabilization packages to nearly all addition polymers to minimize degradation in normal extrusion operations. These compounds are free radical scavengers. These compounds, however, are "consumed" in the extruder over time because they are destroyed as they react with oxygen and free radicals in the polymer melt. Thus, these radicals have a minimum effect on the polymer as long as there is antioxidant still available in the polymer melt. Once these oxidation stabilizers are consumed by multiple passes of the polymer through the extruder, the polymer will potentially degrade in the extruder. Multiple passes typically occur with processes that utilize a very high regrind rate. Moreover, if the extrusion process is improperly designed such that stagnation regions exist, then the resin will be exposed to high temperatures for a very long time, creating the formation of degradation products. Another issue is "old" polymer that has been stored in a warehouse for extended periods of time. In this case, the antioxidants will slowly lose effectiveness. Commonly used antioxidants and their mechanism for protecting PE resins from degradation can be found in [21].

Improperly designed processes can be highly prone to discharging degradation products into the final product. If the process involves hydrolysis such as occurs with the inclusion of water for a condensation polymer, the resin in the final product will have a considerably lower molecular weight and thus deminished physical properties. In these cases, the final product may need to be discarded because the physical properties do not meet the product specification. If a region exists in the process where the flow is stagnant, then the resin in the region can have essentially an infinite residence time, allowing the resin to degrade via oxidation. The degradation products typically stay in the stagnant region until a minor process disruption occurs. The disruption will cause some of the degraded resin to move from the stagnant region into the main flow path. The degradation products will then flow downstream and contaminate the final product. The physical forms of the degradation products for several common polymers are shown in Table 2.4. Troubleshooting processes that have degradation products occurring in the final product are discussed in Chapter 11.

Table 2.4 Degradation Products for Common Polymers

Resin	Physical Products Formed
ABS, HIPS	Styrene monomer, black specks
Acetal	Loss of molecular weight – acid hydrolysis and oxidation
PA	Loss of molecular weight by hydrolysis with water contamination
PBT	Loss of molecular weight by hydrolysis with water contamination
PC	Loss of molecular weight by hydrolysis with water contamination, black specks
PE	Crosslinked resin (gels)
PEN	Loss of molecular weight by hydrolysis with water contamination
PEEK	Breaking of ether bonds, reducing molecular weight
PET	Loss of molecular weight by hydrolysis with water contamination
PMMA	Methyl methacrylate monomer, black specks
PP	Loss of molecular weight via chain scission
PPO/HIPS blend	Black specks from oxidation and styrene monomer
PS	Styrene monomer, brown-colored resin
PVC, PVDC	HCl generation by dehydrochlorination, black specks
PVDF	HF generation by dehydrofluorination, black specks
SAN	Styrene monomer, black specks
SBS	Crosslinked gels
SIS	Loss of molecular weight by chain scission
sPS	Styrene monomer
TPU	Loss of molecular weight with water contamination, resin can turn yellow

As shown in Table 2.4, PE resins will crosslink during the degradation process. Since the molecular architecture is very broad for this class of materials, the degradation process and rates are also broad. For example, LDPE resins produced from the high-pressure process are the most stable to oxidation. In Section 11.10.1 an improperly designed process operated without discharging degradation products for about 13 days, but after 13 days the process was discharging crosslinked gels to the product. Thus, the resin required about 13 days of residence time in this case to degrade the resin to a detectable defect. LLDPE resins are considerably less stable. The formation of gels at processing temperatures can occur in as little as 20 minutes in stagnated regions.

The next sections will present an overview of the relative susceptibility of different polymers to degradation reactions. For a more in-depth discussion the reader is encouraged to consult the open literature for the resin. Mitigating these types of degradation reactions is highly important in order to maintain the ultimate physical properties of the final product and eliminate degradation products in the extrudate. Several case studies relating to resin degradation and contamination will be presented in Chapter 11.

2.5.1 Ceiling Temperature

The heat of reaction for vinyl polymers affects the thermal stability of the polymer during extrusion, and the thermal stability is related to the ceiling temperature. The ceiling temperature is the temperature where the polymerization reaction equilibrium is shifted so that the monomer will not polymerize, or if kept at this temperature all the polymer will be converted back to monomer. From thermodynamics the equilibrium constant for any reaction is a function of the heat of reaction and the entropy of the reaction. For PS resin, the exothermic heat of reaction for polymerization is 70 kJ/gmol, and the ceiling temperature is 310 °C. Ceiling temperatures for select polymers are shown in Table 2.5.

Table 2.5 Ceiling Temperatures and Heats of Reaction (Exothermic) for Select Polymers

	Heat of Reaction, kJ/gmol	Ceiling Temperature, °C
Poly(α methyl styrene)	33	61
PMMA	54	220
PS	70	310
PE	94	400
PVC	109	-

Thus there is an equilibrium concentration between monomer and polymer at high temperatures such that both the monomer and polymer are present. In order to mitigate the formation of monomer, the extruder must be operated at temperatures considerably less than the ceiling temperature. The ceiling temperature is determined from the equilibrium thermodynamics of the reaction. For polymerization reactions that have negative enthalpy of polymerization, the temperature above which a monomer cannot be converted to a long-chain polymer is controlled by the Gibbs free energy ΔG:

$$\Delta G = 0 = \Delta H - T\Delta S \tag{2.1}$$

where ΔH is the heat of polymerization and ΔS is the entropy change. At equilibrium by a thermodynamic definition, the Gibbs free energy change ΔG is equal to 0. This leads to a simple calculation for the ceiling temperature where equilibrium is achieved:

$$T_c = \frac{\Delta H}{\Delta S} \tag{2.2}$$

where T_c is defined as the ceiling temperature for the polymerization in degrees Kelvin. Further details on the thermodynamic concept of free energy can be obtained in any thermodynamics text. The processing importance here is that the polymer will produce monomer in the extruder as the ceiling temperature is approached. In other words the polymer will rapidly depolymerize due to a thermodynamic driving force, and no stabilizer can prevent the reaction. From the data in Table 2.5 and Equation 2.2, the ceiling temperature is directly proportional to the heat of polymerization. All vinyl polymerizations have exothermic heats of reaction. If the heat of the polymerization reaction is known, then the ceiling temperature can be estimated for other polymers from the data in Table 2.5. The more heat that is evolved during the reaction to produce the polymer the higher the ceiling temperature and thus thermodynamic stability of the polymer. This equilibrium shift, however, does not suddenly occur at the ceiling temperature. Instead, the evolution of monomer will start to occur at temperatures considerably less than the ceiling temperature. For example, during the extrusion of PS resin at 200 °C, a small amount of styrene monomer can be detected via smell at the die, a temperature that is about 110 °C less than the ceiling temperature. For PMMA, methyl methacrylate monomer is detected at all reasonable extrusion temperatures. Both PS and PMMA resins degrade by the evolution of a monomer unit one at a time from the end of the chains. This is why the property loss is slower for these resins than that for the hydrolysis of condensation polymers where the chains are randomly broken. For many vinyl polymers the loss of monomer is more random in chain location and the properties will be lost more rapidly than in the case of PS resin. This is an important degradation mechanism for polymers and very often not understood by the extrusion engineer or operator as the cause of production problems. If recycled resin is passed through an extruder multiple times, the degradation reactions can quickly diminish the physical properties to an unacceptable level.

The exothermic heat of reaction for PVC is relatively high, and thus so is the ceiling temperature. PVC resins, however, will dehydrohalogenate at temperatures considerably lower than the ceiling temperature, forming HCl gas and charred material. In this case, thermal degradation reactions occur at temperatures less than the ceiling temperature.

The ceiling temperature for sPS resin is the same as that for atactic PS (or just PS). Due to the crystalline structure of sPS resin, the melting temperature is about 270 °C and the processing temperature is about 290 °C. At these temperatures and with a ceiling temperature of 310 °C, a considerable level of depolymerization and evolution of styrene monomer occurs during processing. Although the high melting temperature of sPS provides a material that has a high resistance to heat and chemicals, the processing window is very narrow since the melting temperature is only 40 °C lower than the ceiling temperature. Due to this small temperature processing window, it is important to minimize the time that the polymer experiences the high temperatures of the processing step.

2.5.2 Degradation of Vinyl Polymers

Oxygen radicals are well known for their attack on hydrogen atoms at the conditions existing in an extruder. These attacks are different for hydrogen atoms that are in different configurations in the polymer molecule. For example, the relative stability and thus ease of formation of a radical is higher for the abstraction of hydrogens attached to tertiary carbons of a PP resin as compared to hydrogens attached to secondary and primary carbons, as shown by Figure 2.21. Thus, the easiest mode of oxygen radical attack on a PP molecule will preferentially occur at the hydrogens attached to tertiary carbons. Preferential oxygen radical attacks on hydrogens of polypropylene glycol and trans polyisoprene are shown in Figure 2.21. Moreover, the relative rates of attack by oxygen radicals on the hydrogen atoms, shown by the arrows in Figure 2.21, are reported to be 6.5 for PP, 9 for polypropylene glycol, and 10 for trans polyisoprene [2]. The attack on the backbone hydrogen in PE resins by oxygen radicals is quite low unless there are residual double bonds (unsaturation), very long thermal residence times, or very high shear stresses such as found in the gap between the screw flight and the barrel. HDPE resin produced using a chromium-type catalyst in the Phillips process is reported to have a double bond in each molecule, so it is susceptible to attack by free radicals [15]. When metallocene-catalyzed polymers were introduced for the production of PE films, it was found that antioxidants were very important in producing stable products for extrusion processes.

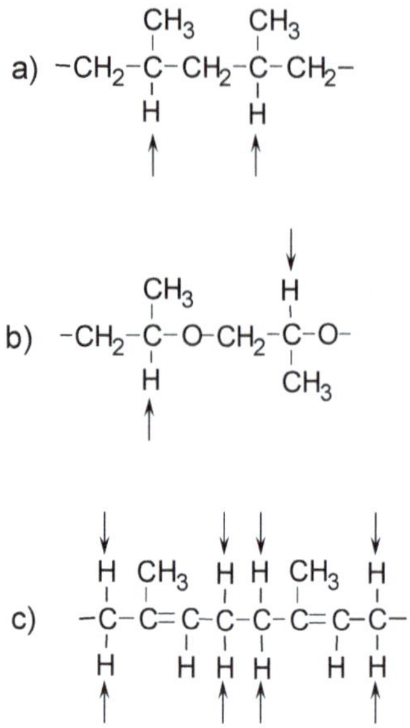

Figure 2.21 Most probable hydrogens attacked by oxidation: a) PP resin segment, b) polypropylene glycol segment, and c) trans polyisoprene segment. The arrows point to the hydrogens that are the most susceptible to attack and removal

In general the oxidation process for polymers in an extruder occurs in a sequential process similar to the process for producing the polymer. The oxidative degradation process for vinyl polymers is shown in Figure 2.22. For this process, a radical must be produced first. The hydrogen atoms shown in Figure 2.21 can be abstracted by high local shear stress such as found in the barrel-flight gap in the extruder, producing a process symbolized as a free radical in Step 1. Also, oxygen can attack to form a radical at high enough temperatures. The radicals can then propagate as shown in Step 2. High temperatures accelerate these attacks on the polymer structure. Moreover, high temperatures coupled with regions in the process with very long residence times will almost always lead to degradation products. These regions are often found in improperly designed flight radii, partially filled metering channels, and mixing sections. As discussed previously, oxygen is not a chain former, and thus two adjacent oxygen atoms in the polymer backbone will be unstable such as in Step 2. In Step 2 one of the propagation products is a peroxide, which can decompose on heating to form a new radical and further attack the polymer. The reaction in Step 3 can lead to crosslinks that appear as gels or to char-like materials that eventually appear as black specks in the finished product. The peroxide in Step 3 can also decompose on heating to produce more radicals. As stated previously, the resin manufacturers typically add antioxidants to the resins after they are produced so that oxidation is minimized in the extruder during processing. These stabilization packages can contain phenolic stabilizers, organophosphorus compounds, lactones, and hydroxylamine compounds as well as other materials.

Step 1– initiation and production of a free radical

$R\cdot$, $RO_2\cdot$

Step 2 - propagation

$$R\cdot + O_2 \longrightarrow RO_2\cdot$$

$$RO_2\cdot + R_1H \longrightarrow ROOH + R_1\cdot$$

Step 3 - termination

$$R_1\cdot + R\cdot \longrightarrow R_1\text{-}R$$

$$RO_2\cdot + R_1\cdot \longrightarrow ROOR_1$$

$$2RO_2\cdot \longrightarrow \text{non-radical products}$$

Figure 2.22 Oxidation mechanism for vinyl polymers

As introduced previously, the compounds in stabilizer packages are consumed with time and thus they will be eventually depleted. As expected, the consumption rate will depend on the temperature and residence times for the resin. Thus, it is important to minimize the number of recycles for a product stream and excessively high temperatures in the extruder. For example, LLDPE resin produced using a metallocene catalyst showed an increase in the molecular weight as measured by a decreasing flow index, yellowing of the sample, and consumption of stabilizers with multiple passes through an extruder [22, 23]. After the stabilizers are consumed, the rate of polymer oxidation to unwanted byproducts such as gels and black specks will accelerate. The type of degradation is dependent on the polymer structure. For example, PE and PP resins contain only carbon and hydrogen atoms, yet their degradation mechanism and products are drastically different. Free radical attack on PE resin almost always results in long-branched polymers and local crosslinks (gels in film production), and the continued degradation will produce a carbonaceous material usually described as black specks in the product. PP resins, however, will degrade in a very different manner via scission of the molecule by free radical attack, causing a substantial decrease in molecular weight and thus the physical properties. Chain scission using peroxides [24, 25] is a common manufacturing method of producing high melt flow PP resins from low melt flow reactor products.

2.5.3 Degradation of Condensation Polymers

As presented in Section 2.4.1, all condensation polymers will react with water, and these reactions are equilibrium reaction systems. Water is absorbed by all of these polymers due to the polar nature of the bonds that form the polymer. If the absorbed water is not removed prior to processing, the water can cause the polymer to hydrolyze at the high temperatures of processing, reducing the molecular weight. Equations (a) and (b) in Figure 2.15 are the reversible reactions used to form polyester and PA resins, respectively. These materials can absorb as much as 5% water when stored in high-humidity conditions in a warehouse. The hydrolysis reaction rate is very low at room temperature, so the highly absorbed water content does not in general hurt the polymer structure over normal storage times. But at the high temperatures in an extruder this reaction is quite fast, and the polymer will be destroyed unless it is dried before extrusion. It is reported that 5% water can be absorbed in PA 66 resin at nominal storage conditions [1]. This absorbed water would destroy about one-third of the amide bonds at equilibrium and reduce the polymer to a material having no usable physical properties. Polyesters, polyurethanes, polycarbonate, and polyacetals must also be dried prior to extrusion. Extruding any of these materials without proper drying will lead to a great loss in properties. Polycarbonate and polyacetal are somewhat more resistant to hydrolysis degradation than polyurethanes and polyesters. For PET it could be expected that the intrinsic viscosity (IV) of poorly dried polymer would easily decrease from about 0.9 dl/g to about 0.6 dl/g for one pass through the extruder. This decrease in IV indicates a substantial decrease in molecular weight and thus the physical properties of the final product.

References

1. Rodriguez, F., "Principles of Polymer Systems," 4th ed., Taylor & Francis, Washington, DC (1996)
2. Rodriguez, F., Cohen, C., Ober, C.K., and Arcer, L.A., "Principles of Polymer Systems," 5th ed., Taylor & Francis, Washington, DC (2003)
3. Walker, P., "Trends in Manufacturing Polymers: Achieving High Performance in a Multi-Polar World," Accenture (2009)
4. Paquet, A., Chum, S., and Rubens, L.C., "Extruded STYROFOAM™ and ETHAFOAM™ Products; Historical Perspective," *SPE ANTEC Tech. Papers*, **49**, 3801 (2003)
5. Avérous, L. and Halley, P.J., "Biocomposites Based on Plasticized Starch," *Biofuels, Bioprod. Biorefin.*, **3**, 329 (2009)
6. Xie, F., Yu, L., Liu, H., and Chen, L., "Starch Modification Using Reactive Extrusion," *Starch/Stärke*, **58**, 131 (2006)
7. Jacobsen, S., Degee, P.H., Fritz, H.G., Dubois, P.H., and Jerome, R., "Polylactide (PLA): A New Way of Production," *Polym. Eng. Sci.*, **39**, 1311 (1999)
8. Iliuta, M.C. and Larachi, F., "Solubility of Total Reduced Sulfurs (Hydrogen Sulfide, Methyl Mercaptan, Dimethyl Sulfide, and Dimethyl Disulfide) in Liquids," *J. Chem. Eng. Data*, **52**, 2 (2007)

9. Harris, P.J. and Smith, B.G., "Plant Cell Walls and Cell-Wall Polysaccharides: Structures, Properties and Uses in Food Products," *Int. J. Food Sci. Technol.*, **41**, 129 (2006)

10. Ito, H., Kumari, R., Takatani, M., Okamoto, T., Hattori, H., and Fujiyoshi, I., "Viscoelastic Evaluation of Effects of Fiber Size and Composition on Cellulose-Polypropylene Composite of High Filler Content," *Polym. Eng. Sci.*, **48**, 415 (2008)

11. Cieslinski, A., "An Introduction to Rubber Technology," Elsevier Science & Technology Books, St. Louis (2000)

12. Swogger, K.W., Carnahan, E.M., Hoenig, W.D., and Frencham, A.R., "The Development of a New Generation of Novel High Performance Olefin Elastomers: From Molecular Design to Market Development," *SPE ANTEC Tech. Papers*, **52**, 1008 (2006)

13. Swogger, K.W., "An Outlook for Metallocene and Single Site Catalyst Technology into the 21st Century," *SPE ANTEC Tech. Papers*, **44**, 1790 (1998)

14. Hiltner, A., Wang, H., Khariwala, D., Cheung, W., Chum, S., and Baer, E., "Solid State Structure and Properties of Novel High Performance Olefin Elastomers," *SPE ANTEC Tech. Papers*, **52**, 1000 (2006)

15. Epacher, E., Krohnke, C. and Pukhanszky, B., "Effect of Catalyst Residues on the Chain Structure and Properties of a Phillips Type Polyethylene," *Polym. Eng. Sci.*, **40**, 1458 (2000)

16. Wakeman, I.B. and Johnson, H.R., "Vinyl Chloride Formation from the Thermal Degradation of Poly(Vinyl Chloride)," *Polym. Eng. Sci.*, **18**, 404 (1978)

17. Hsieh, T-H. and Ho, K-S., "Thermal Dehydrochlorination of Poly(vinylidene chloride)," *J. Polym. Sci., Part A-1: Polym. Chem.*, **37**, 2035 (1999)

18. Schellenberg, J. and Leder, H-J., "Syndiotactic Polystyrene: Process and Applications," *Adv. Polym. Technol.*, **25**, 141 (2006)

19. Hiemenz, P.C. and Lodge, T.P., "Polymer Chemistry," 2nd ed., CRC Press, Taylor & Francis (2007)

20. Spalding, M.A. and Chatterjee, A.M., editors "Handbook of Industrial Polyethylene and Technology," Wiley - Scrivener Publishing, Hoboken, NJ, 2017.

21. Fay, J. and King, R.E. III, "Degradation and Stabilization of Polyethylene," Chapter 23 of "Handbook of Industrial Polyethylene and Technology," Spalding, M.A. and Chatterjee, A.M., editors, Wiley - Scrivener Publishing, Hoboken, NJ, 2017.

22. Hoang, E.M., Liauw, C.M., Allen, N.S., Fontan, E., and Lafuente, P., "Effect of Metal Stearate Antacid on the Melt Stabilization Performance of Phenolic/Phosphite Antioxidants in Metallocene LLDPE. Part 1: Melt Processing Stability," *J. Vinyl Addit. Technol.*, **10**, 3, 137 (2004)

23. Hoang, E.M., Liauw, C.M., Allen, N.S., Fontan, E., and Lafuente, P. "Effect of Metal Stearate Antacid on the Melt Stabilization Performance of Phenolic/Phosphite Antioxidants in Metallocene LLDPE. Part 2: Discoloration," *J. Vinyl Addit. Technol.*, **10**, 3, 144 (2004)

24. Tzoganakis, C., Vlachopoulos, J., and Hamielec, A.E., "Production of Controlled-Rheology Polypropylene Resins by Peroxide Promoted Degradation During Extrusion," *Polym. Eng. Sci.*, **28**, 170 (1988)

25. Scorah, M.J., Zhu, S., Psarreas, A., McManus, N.T., Dhib, R., Tzoganakis, C., and Penlidis, A., "Peroxide-Controlled Degradation of Polypropylene Using a Tetra-Functional Initiator," *Polym. Eng. Sci.*, **49**, 1760 (2009)

3 Introduction to Polymer Rheology for Extrusion

Polymers have numerous physical properties that are important to processing. The most important of these properties is their rheological response to flow. This chapter will introduce the basic concepts of rheology, the effect of molecular weight, molecular weight distribution, and structure on viscosity, and its measurement. As discussed in Chapter 1, the shear viscosity of a resin is highly important for calculating the pressure-driven flow in the channel in the metering section. Moreover, the shear viscosity also affects many other aspects of extrusion including the ability of the machine to generate pressure and dissipate energy, the discharge temperature, and the melting process. Process calculations, screw design, and troubleshooting all require viscosity data. Other physical properties related to processing will be presented in Chapter 4.

3.1 Introduction to the Deformation of Materials

When a material is extruded it is deformed, causing the material to change shape both macroscopically and microscopically. This deformation occurs in the extruder, in downstream processing equipment, and in the die. The study of the deformation of materials is called rheology. This chapter introduces several important aspects of polymer rheology that are required for extruder design and troubleshooting, that is, it provides an understanding of the relationship between the forces imposed on the polymer melt and the resulting deformation. The focus will be on engineering models for quantitative evaluation of the resistance to flow and deformation of polymer melts to applied forces. The chapter does not present the fundamental theories that relate polymer deformation to molecular structure. These details would be covered in a course or text on the fundamentals of rheology, and they are beyond the scope of this chapter.

The shear rate and shear stress associated with the resistance to flow of polymer melts in extruders are related to viscous energy dissipation, temperature changes, and the extrudate temperature. These interactions are important concepts used by the screw designer when seeking to evaluate the optimum operation of extruders and dies and in particular the pressure gradients and temperature changes due to the flow of the polymer in the screw channels and in the die cavity. In this chapter three aspects of polymer rheology will be discussed: Newtonian viscosity, shear-rate-dependent viscosity, and a very brief introduction to the concept of the dynamic viscoelastic properties of polymers. Viscoelasticity can have a strong effect on the mixing in some designs of single-screw processes, especially when the mixing section causes the polymer melt to be stretched or elongated.

Almost all aspects of polymer rheological properties are related to the polymer molecular weight, molecular weight distribution, and polymer molecular structure. A polymer's resistance to deformation is also a function of the chemical composition of the mer, the temperature, and the strain rate that is applied. In the previous chapter several important chemical-based properties of polymers were introduced and discussed. The concepts included molecular architecture, crystalline and amorphous structures, stereoisomers, pendant groups, branches, and crosslinks. In this chapter those basic polymer concepts are expanded to include polymer molecular weight and molecular weight distribution. Since the molecular weight of a polymer is a primary factor influencing many rheological characteristics of polymeric materials, a discussion of molecular weight is developed here as a background for relating polymer molecular structure to basic rheological responses of polymers.

3.2 Introduction to Basic Concepts of Molecular Size

The concepts of polymer molecular weight averages and molecular weight distribution will now be developed. Unlike small molecules like sodium chloride with a molecular weight of 58.45 kg/kg-mol and simple organic compounds like propane and ethanol with molecular weights of 44.09 and 46.07 kg/kg-mol, respectively, polymers essentially never have a single well-defined molecular weight. Instead the molecular weight of polymers must be discussed in terms of averages of distributions of molecular size. When working with polymers as they relate to extrusion there are several average molecular weights typically defined: number average molecular weight, M_n; weight average molecular weight, M_w; z average molecular weight, M_z; and the $z + 1$ average molecular weight, M_{z+1}. These different representations of polymer molecular weight, and thus average size, all have important bearings on how to characterize the properties of polymers. The relative size of these averages influences how the polymer responds to shear stress and shear rate. The use of some of these concepts is discussed more completely later in the chapter as they are the most important parameters related to the flow resistance of polymer melts.

3.2.1 Size Distribution Example

An example of size distribution is described here as an illustration of length, area, and volume distributions. A collection of particles, in this case large particles [1], will be analyzed regarding several distributions related to particle diameters. The following discussion regarding distributions is summarized in Table 3.1. How would you answer this question: would you be willing to catch 1000 steel balls with an average diameter of 6.1 cm [1]? Many would assume these balls are about the same size and say yes. It is important to understand that there are many types of averages that can be calculated. These particular particles are relatively large. They are a collection of 1000 steel balls. Nine hundred of the balls have a diameter of 2.54 cm. Fifty balls have a diameter of 12.7 cm and 50 have a diameter of 63.5 cm. A length dimension would be the number of balls of a particular size times the diameter. An area dimension would be the number of balls multiplied by the diameter squared, and a volume dimension would be the number of balls times the diameter cubed. These transformations are summarized in Table 3.1 along with the summations on the particular dimensional function. The length, area, and volume average diameters are calculated using Equation 3.1, Equation 3.2, and Equation 3.3, respectively.

Table 3.1 Summary of Distribution Explanation

Number of Balls N_i	Ball Diameter D_i, cm	Length N_iD_i, cm	Area $N_iD_i^2$, cm²	Volume $N_iD_i^3$, cm³
900	2.54	2286	5806	14,750
50	12.7	635	8065	102,420
50	63.5	3175	201,613	12,802,000
	Sum of column:	6096	215,484	12,919,170

$$D_L = \frac{\sum_{i=1}^{3} N_i D_i}{\sum_{i=1}^{3} N_i} = 6.1\,\text{cm} \quad \text{length average diameter} \tag{3.1}$$

$$D_A = \frac{\sum_{i=1}^{3} N_i D_i^2}{\sum_{i=1}^{3} N_i D_i} = 34.8\text{ cm} \quad \text{area average diameter} \tag{3.2}$$

$$D_V = \frac{\sum_{i=1}^{3} N_i D_i^3}{\sum_{i=1}^{3} N_i D_i^2} = 60.9\text{ cm} \quad \text{volume average diameter} \tag{3.3}$$

There then follows some appropriate averages. The length average diameter, D_L, is calculated to be 6.1 cm. The area average diameter, D_A, is 34.8 cm, and the volume average, D_V, is 60.9 cm. This last average suggests that the answer to the posed question might safely be no. A 60.9 cm steel ball has a mass of 930 kg. Note that the averages are related by particular summations. These same summation concepts are used to characterize polymer molecular weight averages.

3.2.2 Molecular Weight Distributions for Polymers

Commercial polymers are made up of molecules of differing molecular weight as discussed in Chapter 2. The shape of a typical commercial polymer molecule can be as a first approximation considered a "Gaussian spherical" shape in the undisturbed melt or solvent with a size proportional to the number of atoms in the molecule. Design engineers are interested in a number of properties: tensile strength, melt elasticity, and power to operate the production extruder; all are functions of different polymer average molecular weights. Thus it follows, when describing the molecular weight of a polymer, that the appropriate average must be used when dealing with the distribution of polymer molecular sizes in typical polymer melts, solutions, or solids in order to understand how the polymer responds to environmental forcing functions like shear, normal stress, strain or strain rate, and elongational strain.

The two most common average molecular weights of polymers are the number average, M_n, and the weight average, M_w, molecular weight. The number average molecular weight, M_n, is defined as the total weight divided by the total number molecules. The following summation is the mathematical representation of M_n:

$$M_n = \frac{\sum_{i=1}^{n} N_i M_i}{\sum_{i=1}^{n} N_i} \tag{3.4}$$

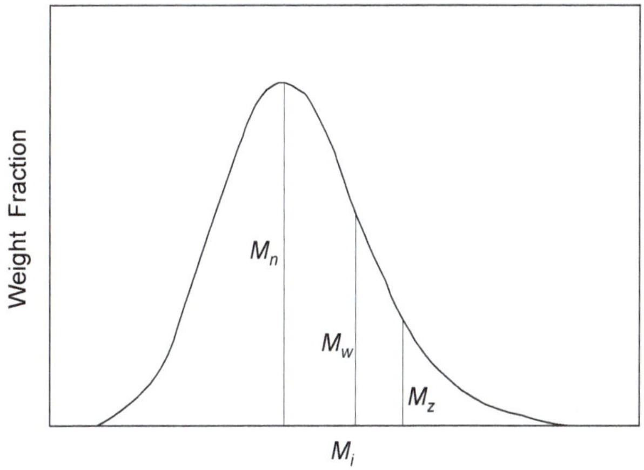

Figure 3.1 Schematic of a molecular weight distribution and the number average, weight average, and *z* average molecular weights

where N_i is the number of molecules with a molecular weight of M_i. As expected, the number average molecular weight is related to the number of molecules (the summation in the denominator), and it is therefore a colligative property of the material. It can thus be directly measured by techniques such as osmotic pressure, freezing point depression, and the boiling point increase of a good solvent. The number average molecular weight often correlates well with the tensile strength and impact properties of a polymer family. The number average molecular weight is also the first moment of the molecular weight distribution function. Discussion of moments of distributions can be found in texts on the introduction to statistics. The number average molecular weight is at the maximum of the differential representation of a molecular weight chromatogram, as shown in Figure 3.1. This figure is a representation that illustrates how the molecular weight averages are related to the molecular weight distribution; that is, the distribution of molecular sizes in a polymer sample. There are a few techniques for obtaining a distribution of the polymer molecular weights, including but not limited to gel permeation chromatography, high-pressure liquid chromatography, and temperature-rising elution fractionation for crystalline polymers. These techniques and others are presented for PE resins in reference [32]. The number average molecular weight is located at the maximum for the weight fraction, W_i, distribution, and the weight average and *z* average molecular weights are at higher molecular weights in distribution, as shown in Figure 3.1. The weight fraction is calculated using Equation 3.5, and the sum of the weight fractions must equal one by definition.

$$W_i = \frac{N_i M_i}{\sum_{j=1}^{n} N_j M_j} \tag{3.5}$$

$$1 = \sum_{i=1}^{n} W_i \tag{3.6}$$

The weight average molecular weight, M_w, is the second moment of the distribution and calculated from the distribution as follows:

$$M_w = \frac{\sum_{i=1}^{n} N_i M_i^2}{\sum_{i=1}^{n} N_i M_i} \tag{3.7}$$

The weight average molecular weight is always equal to or larger than the number average molecular weight. Since it is based on the size or the mass of the molecules, the weight average molecular weight can be directly measured with techniques that include light scattering and ultracentrifugation. The weight average molecular weight correlates with the Newtonian viscosity of many polymers. For example, the higher the weight average molecular weight of a polymer, the higher the Newtonian viscosity will be at limiting flow rates for the same number average molecular weight. This correlation will be discussed later in the chapter.

The third important average molecular weight is the third moment or the *z* average molecular weight, M_z. This average is calculated from the molecular weight distribution as follows:

$$M_z = \frac{\sum_{i=1}^{n} N_i M_i^3}{\sum_{i=1}^{n} N_i M_i^2} \tag{3.8}$$

The *z* average molecular weight has been found to correlate with the shear viscosity of polymer melts when the molecular weight distribution is very broad and where very large molecules appear to dominate the resistance to fluid flow.

Another important molecular weight concept is the polydispersity index (*PI*). This index is defined as the ratio of the weight average molecular weight divided by the number average molecular weight. It is a first-order attempt to characterize the broadness of the molecular weight distribution. This ratio is always greater than or equal to 1. Sometimes a polymer is labeled as monodispersed with a polydispersity index of exactly 1. These polymers, however, usually have a polydispersity index of 1.01 to 1.1. The polydispersity index is related to the effect of shear rate on the measured viscosity of polymer melts that have a molecular weight greater than the so-called critical molecular weight, M_c. As the polydispersity index increases the viscosity becomes shear-rate dependent at lower shear rates, or narrow distribution polymers have higher viscosity for a wider shear rate range.

Polymer molecular weight, branching, molecular weight distribution, and the rheological characteristics of polymers are therefore very interdependent.

$$PI = \frac{M_w}{M_n} \tag{3.9}$$

Finally, a solvent viscosity method is often used to measure the molecular weight of many polymers such as PET, and this viscosity average molecular weight is often close to the weight average molecular weight.

3.3 Basic Rheology Concepts

Materials are modeled classically as either viscous, such as water or molasses, or elastic in nature, such as steel beams or metal springs. In general, polymers are complex materials that behave in a combined response to strain with both viscous and elastic characteristics. Under conditions where the material exhibits both viscous and elastic rheological behavior, the polymers are described as viscoelastic. That is, polymers have substantial viscous and elastic characteristics when strained.

Although polymers exhibit both viscous and elastic responses at all temperatures, the elastic response is particularly strong at temperatures less than 50 °C above the glass transition temperature, particularly for polymers well above their critical molecular weight. Polymers are often considered to have dominant viscous rheological responses if they are stressed at temperatures over 100 °C above the glass transition temperature for amorphous polymers or 100 °C above the crystalline melting point for semicrystalline resins.

Elastic and viscous characteristics of materials can be visualized using a Cartesian material element, as shown in Figure 3.2. For this visualization the square shape in the *x*-*y* plane is deformed into a parallelogram. A force is applied to the material element parallel to one axis, in this case along the *x* axis at a distance *H* up the *y* axis. The material element is deformed away from the *y* axis by a distance *a* by the force *F*.

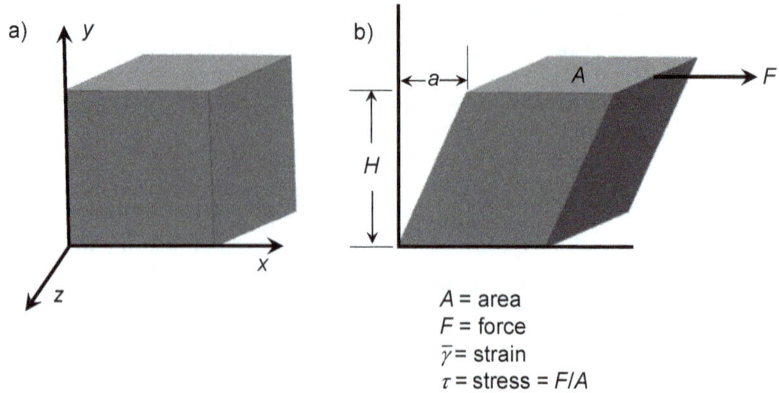

Figure 3.2 Response to elastic deformation of a solid: a) material element without an applied force, and b) solid deformation and the elastic response to an applied force

For this solid the velocity is zero. This is thus a static situation. The strain can be calculated along with the stress:

$$\bar{\gamma} = \frac{a}{H} \tag{3.10}$$

$$\tau = \frac{F}{A} \tag{3.11}$$

where A is the surface area for the exerted force F, $\bar{\gamma}$ is the strain, and τ is the stress. It is important to note that this analysis is only valid for infinitesimally small local deformations.

The evaluation of viscosity is similar to the evaluation of elastic deformation except the stress in the element changes due to the local velocity gradient. The time variable is defined as the strain rate. The element changes its shape as a function of time, while the strain-rate-induced stress is present. Viscosity is the local slope of the function relating stress in the element to strain rate. The usual functionality is found in Figure 3.3. The process can be visualized by a constant force on the top of the element that creates a strain rate throughout the element. This strain rate causes each molecular layer of the material to move relative to the adjacent layer continuously. Obviously the element is suspended in a continuum and material flows into and out of the geometric element.

For this simple geometry the shear rate, $\dot{\gamma}$, is equal to the difference between the velocity at the top of the element, U, and the velocity at the bottom of the element, zero, divided by the height of element H. The shear stress is again $\tau = F/A$, the element surface area divided by the force. The viscosity, η, is the ratio of shear stress, τ, divided by shear rate, $\dot{\gamma}$, at any shear rate, $\eta = \tau/\dot{\gamma}$. For Newtonian materials such as water, molasses, or gasoline at the nominal shear rates found in everyday life, the slope of the shear stress with shear rate curve is a constant and equal to the Newtonian viscosity.

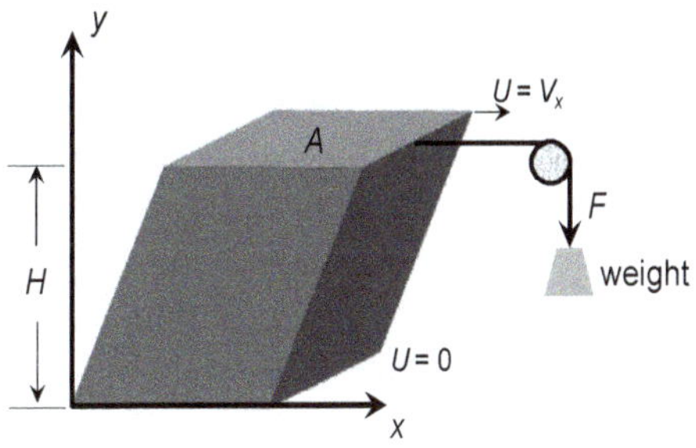

Figure 3.3 Viscosity in a simple fluid element
A = area
F = force
U = velocity
$\dot{\gamma}$ = shear rate = V_x/H
τ = stress = F/A

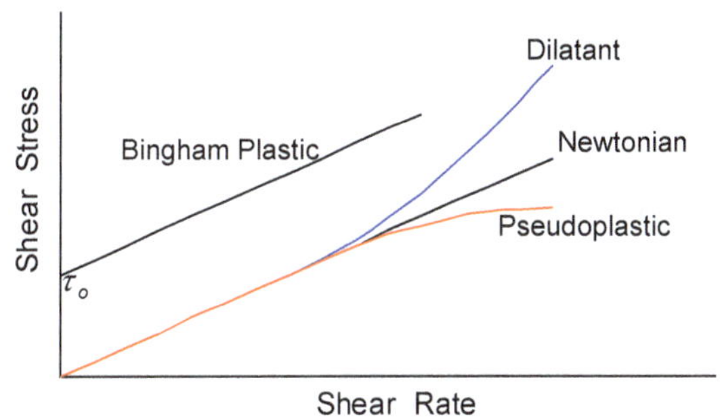

Figure 3.4 Shear stresses for several viscosity characteristics as a function of shear rate

Polymer rheology can respond nonlinearly to shear rates, as shown in Figure 3.4. As discussed above, a Newtonian material has a linear relationship between shear stress and shear rate, and the slope of the response is the shear viscosity. Many polymers at very low shear rates approach a Newtonian response. As the shear rate is increased most commercial polymers have a decrease in the rate of stress increase. That is, the extension of the shear stress function tends to have a lower local slope as the shear rate is increased. This is an example of a pseudoplastic material, also known as a shear-thinning material. Pseudoplastic materials show a decrease in shear viscosity as the shear rate increases. Dilatant materials increase in shear viscosity as the shear rate increases. Finally, a Bingham plastic requires an initial shear stress, τ_0, before it will flow, and then it reacts to shear rate in the same manner as a Newtonian polymer. It thus appears as an elastic material until it begins to flow and then responds like a viscous fluid. All of these viscous responses may be observed when dealing with commercial and experimental polymers.

Material rheology can also respond nonlinearly with the time of shearing, as shown in Figure 3.5. Two common time-dependent viscosity responses are shown in Figure 3.5, thixotropy and rheopexy. Thixotropy is characteristic of a material that decreases in viscosity as a function of time at constant shear rate. These time-dependent characteristics are often found in common household materials such as flowing ketchup from its container. Ketchup decreases in viscosity as the bottle is shaken; the longer it is shaken the lower the viscosity. Thixotropy is very often found to be the result of breaking hydrogen bonds in the material, which on standing will reform and increase the viscosity again. This attribute of shear-rate-dependent viscosity in thixotropic materials is used in epoxy body fillers for automobile repair, household paints, and many adhesives. These materials spread well while sheared and then maintain their shapes when the shear stops. Rheopexy is just the opposite effect and is often found in inorganic slurries such as those used in making synthetic sapphires. In this case the slurry particles are thought to form transient structures at moderate shear rates, causing an increase in the resistance to flow and thus an increase in the viscosity as a function of time.

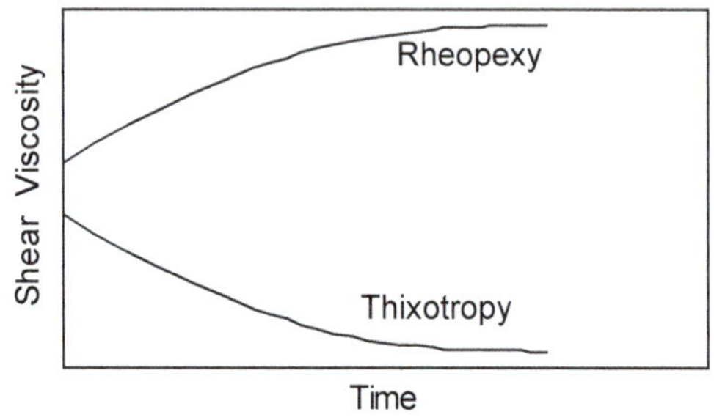

Figure 3.5 Schematic showing the rheopexy and thixotropic time responses for shear viscosity

Another important characteristic of all polymer systems relating to the viscosity is the temperature increase of the material due to flow. Temperature increase in the melt is important because as the temperature increases the polymer melt viscosity decreases. As mentioned previously, if the temperature is quite high relative to the glass transition or melting temperature the viscous effects of most polymers are dominant relative to the elastic effects. The temperature increase will also affect how well the screw pumps the polymer against the discharge pressure required to operate the die. Because of the high viscosity of polymers, energy dissipation is an extremely important characteristic in polymer systems when they are sheared. At a constant shear rate the higher the viscosity the more rapidly the polymer temperature will increase as a function of time. The shear rates induced by the polymer motion dissipate energy in the polymer due to the internal molecular friction. This dissipation can cause large temperature increases in the extruder channels and in the narrow channels of shaping dies.

The rate of energy dissipation, $\dot{Q}$ (W/m^3), is related to the shear stress, shear rate, and the viscosity as follows:

$$\dot{Q} = \tau\dot{\gamma} \tag{3.12}$$

$$\dot{Q} = \eta\dot{\gamma}^2 \tag{3.13}$$

$$\dot{Q} = \frac{\tau^2}{\eta} \tag{3.14}$$

The rate of energy dissipation, $\dot{Q}$, is related to the adiabatic polymer temperature increase as follows:

$$\frac{dT}{dt} = \frac{\dot{Q}}{\rho C_v} \tag{3.15}$$

where dT is the temperature increase for the time interval, dt, in the material element being deformed. C_v is the heat capacity of the material and ρ is the density.

The viscosity and the resulting dissipation are very important issues when troubleshooting the extrusion process, especially processes that are rate limited by the discharge temperature of the extrudate. Furthermore, viscous dissipation occurs in all melt processing operations when polymers are sheared. Table 3.2 indicates the nominal shear rate range for different types of polymer processing operations. In general, resins in compression-molding processes can experience shear rates as low as 0.001 1/s while resins used for high-speed injection molding can experience shear rates of 50,000 1/s or higher during the injection step. From Table 3.2, it is obvious that each of these processes would have very different shear heating levels for the same resin. Resin manufacturers, however, design the polymers for the application so as to control viscous dissipation via the rheological properties. For example, resins designed for an extrusion-calendering process where the shear rates are relatively low would have a higher shear viscosity than a similar resin designed for a high-speed injection-molding process where the shear rates are relatively high. Thus, it is important to understand the shear rheology of the resin for the shear rate range and temperature range for the application.

Table 3.2 Typical Shear Rates Encountered in Select Polymer Processing Operations

Process	Shear Rate Range, 1/s
Compression molding	0.001–10
Calendering	1–100
Extrusion	$10–10^3$
The injection step of injection molding	$10^3–10^5$

3.4 Polymer Solution Viscosity and Polymer Molecular Weight

Since polymer molecular weight and distribution control the viscosity response of polymers when they are sheared, this discussion of polymer viscosity begins with how solution viscosity can be used to determine polymer molecular weight. Changes in molecular weight will be related to many of the troubleshooting opportunities that will be developed in later chapters. Many of the problems found in extrusion troubleshooting are related to a decrease in intrinsic viscosity of polymers such as PET, since this relates to the properties of the final parts. Intrinsic viscosity is a polymer solution characteristic relating the molecular weight of the polymer to its polymer solution viscosity function.

Polymer solution viscosity is a troubleshooting tool for polymer processing. Solution viscosity may be utilized to determine the degradation of some polymers via a decrease in molecular weight, particularly for polymers that can be hydrolyzed by water. It can be a troubleshooting tool because the results can be obtained quite rapidly. Intrinsic viscosity is commonly known as IV when referring to polyesters. It is applicable, however, to any polymer that is soluble at moderate concentrations in a good solvent [2].

Many polymer systems are analyzed for the viscosity average molecular weight, M_v, using solution viscosity and an Ubbelohde-type rheometer like the one shown in Figure 3.6(a). This rheometer design has a large lower bulb so that the dilution can be accomplished in the rheometer, and the extra arm surrounding the capillary and the upper bulb is a siphon break. Thus the level change in the lower bulb does not affect the time reading used to calculate the viscosity. When substantial dilution is needed the working capillaries are attached to a large glass reservoir by glass blowers. For proper operation, the polymer solution is forced by using pressure on the lower bulb up the capillary into the upper two bulbs. The time needed to have the fluid flow from the calibration Line 1 to Line 2 by gravity is proportional to the solution viscosity. Because of the siphon brake, fluid level effects are minimized. The time, if there were no end effects, is proportional to η/ρ, the kinematic viscosity.

An Ostwald viscometer is similar to an Ubbelohde-type rheometer except that it is simpler in design and is less expensive. A schematic of an Ostwald viscometer is shown in Figure 3.6(b). It is characterized by a lower bulb that acts as a solution reservoir. A solution of known polymer concentration is placed in the lower bulb. A single capillary tube in which the measurement is taken is connected to the bottom of the bulb and to two small bulbs at the top of the capillary. Fluid is forced from the lower bulb through the capillary into the two small bulbs attached to the top of the capillary. There is a line between the two bulbs and at the exit of the lower bulb. The fluid is then allowed to drain back into the lower bulb through the capillary, and the time for the fluid to travel between the two lines is recorded. The time, if there were no end effects, is proportional to the kinematic viscosity (η/ρ).

Many operators use a vacuum to pull the fluid through the capillary and into the measurement bulbs at the start of the measurement. This can introduce a serious random error into the measurements because each time the measurement is taken some solvent will evaporate, thus changing the polymer concentration in the solution. Another important error in the measurement using the Ostwald viscometer is the fluid head, which is the driving force for fluid flow in the capillary. The fluid head is changing all the time due to both the change in the height of the fluid in the upper bulb and the height of the fluid in the lower bulb. This problem is increased if dilution is carried out in the bulb due to the increasing volume of the solvent system.

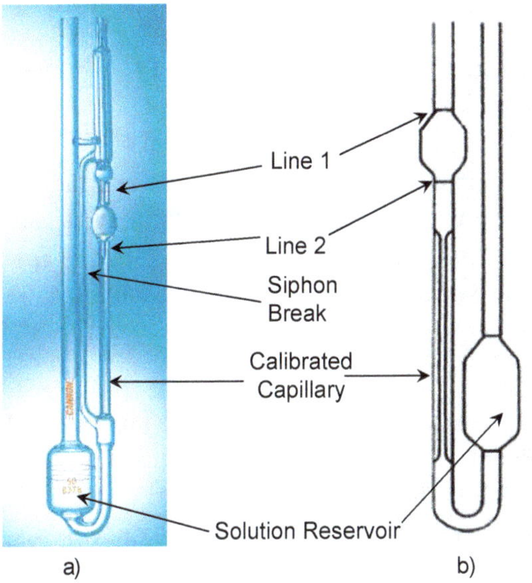

Figure 3.6 Solution capillary rheometers: a) Ubbclohde-type rheometer (courtesy of Cannon Instrument Company, USA), and b) a schematic of an Ostwald viscometer

The viscosity average molecular weight is determined through the use of the Mark-Houwink-Sakurada equation [3] using solution viscosity:

$$[\eta] = K'M_v^a \tag{3.16}$$

where K' and a are parameters dependent on the solvent-polymer pair and the temperature of the measurement. The theta condition is the temperature where the polymer segments just begin to associate more with each other than with the solvent. At this condition, a has a value of 0.5. A tabulated list of K' and a values may be found elsewhere [1, 4].

In many applications a capillary viscometer is used to measure the viscosity of polymer solutions of differing concentrations, and then the data are plotted to obtain the intrinsic viscosity. Several techniques have been used to determine intrinsic viscosity from solution. One is the Huggins equation for intrinsic viscosity [1]:

$$\frac{\eta_{sp}}{c} = [\eta] + K[\eta]^2 c \tag{3.17}$$

$$\text{where} \quad [\eta] = \lim_{c \to 0} \frac{\eta_{sp}}{c} \tag{3.18}$$

where $[\eta]$ is the intrinsic viscosity and has units of dl/g. The other equation relating solution viscosity to intrinsic viscosity is the Kraemer equation [1]:

$$\frac{\ln(\eta_r)}{c} = [\eta] + K''[\eta]^2 c \tag{3.19}$$

$$\text{where} \quad \eta_r = \frac{\eta}{\eta_s} \quad \text{and} \quad [\eta] = \lim_{c \to 0} \frac{\ln(\eta_r)}{c} \tag{3.20}$$

In both of these equations c is the polymer concentration in g/dl. The Kraemer function is based on the relative viscosity, which is a ratio of the solution viscosity (η) and the viscosity of the pure solvent (η_s). The Huggins function uses the specific viscosity, η_{sp}, which is defined in terms of the relative viscosity as follows:

$$\eta_{sp} = \eta_r - 1 \tag{3.21}$$

Intrinsic viscosity is related to the relative viscosity via a logarithmic function and to the specific viscosity by a simple algebraic relationship. Both of these functions can be plotted on the same graph, and when the data are extrapolated to zero concentration they both should predict the same intrinsic viscosity. The specific viscosity function has a positive slope and the relative viscosity function has a negative slope, as shown in Figure 3.7. The molecular weight of the polymer can be determined from the intrinsic viscosity, the intercept of either function, using the Mark-Houwink-Sakurada equation.

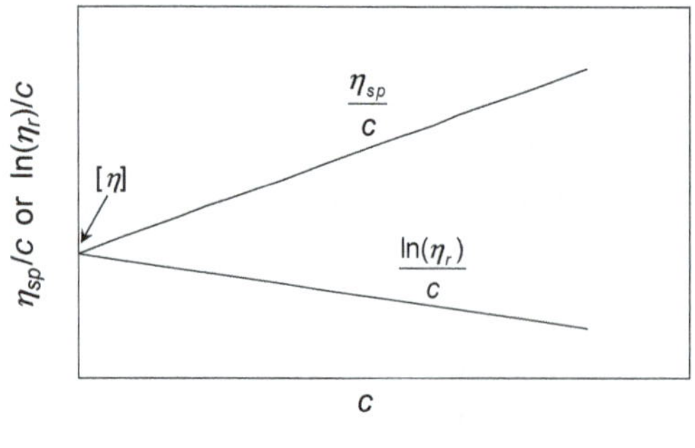

Figure 3.7 Functions leading to intrinsic viscosity

The molecular weight is calculated by rearranging the Mark-Houwink-Sakurada equation:

$$M_v = \left(\frac{[\eta]}{K'}\right)^{1/a} \tag{3.22}$$

This equation can be rearranged to incorporate the definition of the viscosity average molecular weight [2]:

$$M_v = \left(\sum_i w_i M_i^a\right)^{1/a} \tag{3.23}$$

where w_i is the weight fraction of the polymer with a molecular weight of M_i. Note as a approaches 1.0, the viscosity average molecular weight approaches the weight average molecular weight. For a values between 0.5 and 1 the viscosity average molecular weight will be between the number and weight average molecular weights.

3.4.1 Sample Calculation of Solution Viscosity

The Mark-Houwink-Sakurada constants for PMMA resin are a = 0.73 and K' = 7.1 × 10^{-5}. Table 3.3 contains solvent viscosity versus concentration data. Find the intrinsic viscosity using both the specific and relative viscosities and the viscosity average molecular weight.

Table 3.3 Concentration and Solvent Viscosity Data for the Sample Calculation

Concentration, g/dl	Solvent Viscosity, Pa·s
0.0	3.0 × 10^{-3}
0.275	3.51 × 10^{-3}
0.344	3.645 × 10^{-3}
0.896	4.887 × 10^{-3}
1.199	5.676 × 10^{-3}
1.604	6.99 × 10^{-3}
2.108	8.985 × 10^{-3}

The relevant equations are Equation 3.20 and Equation 3.21 for the relative viscosity and specific viscosity, respectively. The values for these terms and plotting terms are provided in Table 3.4.

Table 3.4 Calculation of the Relative Viscosity, Specific Viscosity, and Plotting Parameters for the Data in Table 3.3

Concentration, g/dl	Relative Viscosity, η_r from Equation 3.20	Specific Viscosity, η_{sp} from Equation 3.21	$\ln(\eta_r)/c$, dl/g	η_{sp}/c, dl/g
0.275	1.17	0.17	0.571	0.618
0.344	1.215	0.215	0.566	0.625
0.896	1.629	0.629	0.545	0.702
1.199	1.892	0.892	0.532	0.744
1.604	2.33	1.33	0.527	0.829
2.108	2.995	1.995	0.52	0.946

Next, a plot is made for the terms η_{sp}/c and $\ln(\eta_r)/c$ as a function of c as was shown in Figure 3.7. Extrapolation of the functions to a concentration of zero provides the intrinsic viscosity of 0.565 dl/g, as shown in Figure 3.8. Using Mark-Houwink-Sakurada Equation 3.22, the viscosity average molecular weight is calculated at 221,000 kg/kg-mole.

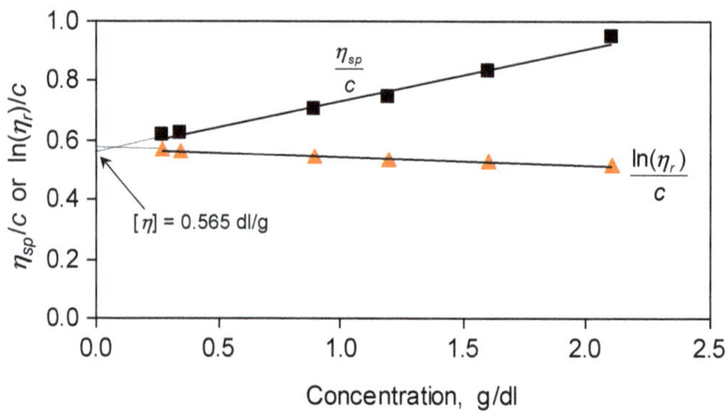

Figure 3.8 Intrinsic viscosity determination for the data in Table 3.3 and Table 3.4

3.5 Introduction to Viscoelasticity

When a polymer is extruded through an orifice such as a capillary die, a phenomenon called die swell is often observed. In this case, as the polymer exits the cylindrical die, the diameter of the extrudate increases to a diameter larger than the diameter of the capillary die, as shown in Figure 3.9. That is, it increases in diameter as a function of the time after the polymer exits the die. Newtonian materials or pure power law materials would not exhibit this strong of a time-dependent response. Instead they may exhibit an instantaneous small increase in diameter, but no substantial time-dependent effect will be observed. The time-dependent die swell is an example of the polymer's viscoelastic response. From a simplified viewpoint, the undisturbed polymer molecules are forced to change shape as they move from the large area of the upstream piston cylinder into the capillary. For short times in the capillary, the molecules remember their previous molecular shape and structure and try to return to that structure after they exit the die. If the time is substantially longer than the relaxation time of the polymer, then the molecules assume a new configuration in the capillary and there will be less die swell.

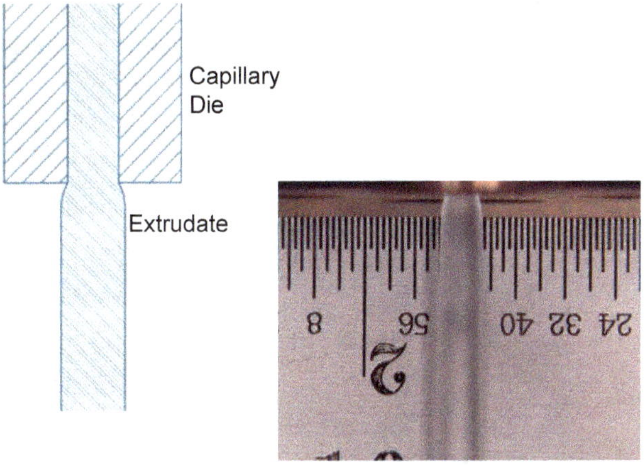

a) b)

Figure 3.9 Polymer die swell: a) schematic showing die swell, and b) a photograph of a PE resin discharging from a capillary die and swelling

As an introduction, the viscoelastic properties of polymers are represented as the summation of mechanical analog responses to applied stress. This introductory use of one dimensional mechanical analogs is intended to provide the reader with a comceptual understanding of time dependent properties. Any in-depth discussion of bulk polymer viscoelasticity involves the use of tensors, and this high-level mathematics topic is beyond the scope of what will be presented in this book. Earlier in the chapter the concept of elastic and viscous properties of polymers were briefly introduced. A purely viscous response can be represented by a mechanical dash pot, as shown in Figure 3.10(a). This purely viscous response is normally the response of interest in routine extruder calculations. For those familiar with the suspension of an automobile, this would represent the shock absorber in the front suspension. If a stress is applied to this element it will continue to elongate as long as the stress is applied. When the stress is removed there will be no recovery in the strain that has occurred. The next mechanical element is the spring (Figure 3.10[b]), and it represents a purely elastic response of the polymer. If a stress is applied to this element, the element will elongate until the strain and the force are in equilibrium with the stress, and then the element will remain at that strain until the stress is removed. The strain is inversely proportional to the spring modulus. The initial strain and the total strain recovery upon removal of the stress are considered to be instantaneous.

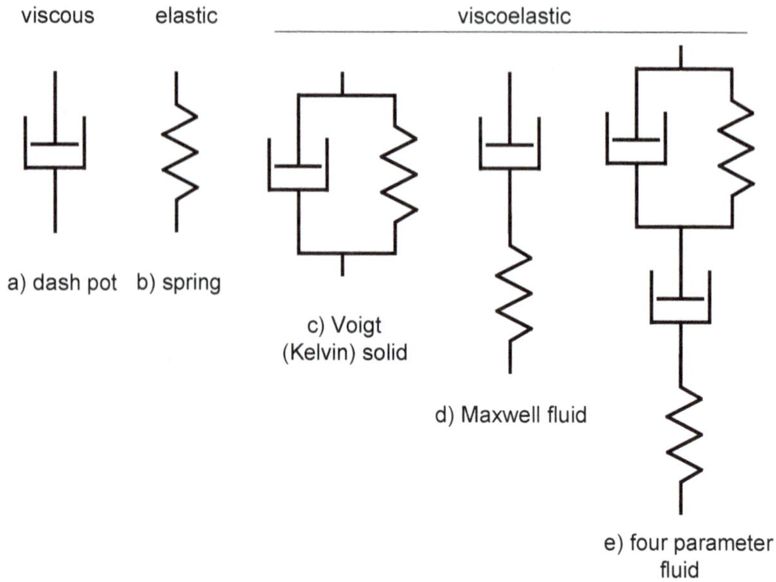

Figure 3.10 Basic mechanical elements for solids and fluids: a) dash pot for a viscous response, b) spring for an elastic response, c) Voigt or Kelvin solid, d) Maxwell fluid, and e) the four-parameter viscoelastic fluid

When dash pot and spring elements are connected in parallel they simulate the simplest mechanical representation of a viscoelastic solid. The element is referred to as a Voigt or Kelvin solid, and it is shown in Figure 3.10(c). The strain as a function of time for an applied force for this element is shown in Figure 3.11. After a force (or stress) elongates or compresses a Voigt solid, releasing the force causes a delay in the recovery due to the viscous drag represented by the dash pot. Due to this time-dependent response the Voigt model is often used to model recoverable creep in solid polymers. Creep is a constant stress phenomenon where the strain is monitored as a function of time. The function that is usually calculated is the creep compliance $J(t)$: $J(t)$ is the instantaneous time-dependent strain $\varepsilon(t)$ divided by the initial and constant stress σ_0.

$$J(t) = \frac{\varepsilon(t)}{\sigma_0} = \frac{1}{E_v}\left\{1 - \exp\left(\frac{-E_v t}{\eta_v}\right)\right\} \tag{3.24}$$

with the element relaxation time $\lambda_v = \left(\frac{\eta_v}{E_v}\right)$ (3.25)

Here η_v is the element viscosity, E_v is the element modulus, and λ_v is the relaxation time. If this model material is placed in a constant stress environment for a fixed time and then the stress is removed, then the time-dependent strain will have the recovery characteristics shown in Figure 3.11.

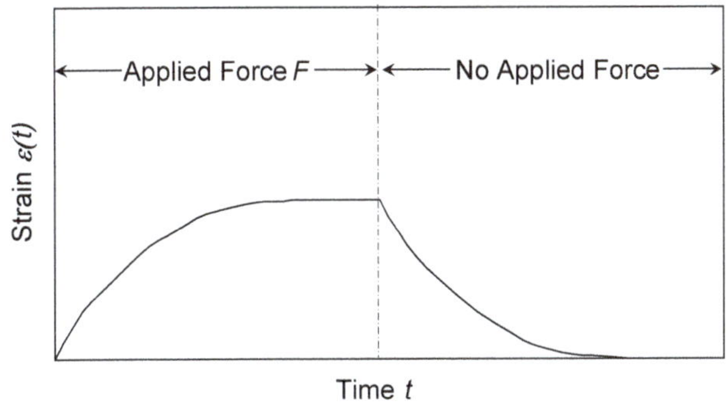

Figure 3.11 Qualitative creep function of a Voigt polymer solid

As shown by Figure 3.11 for an applied force, the creep strain is increasing at a decreasing rate with time because the elongation of the spring is approaching the force produced by the stress. The shape of the curve up to the maximum strain is due to the interaction of the viscosity and modulus. When the stress is removed at the maximum strain, the strain decreases exponentially until at an infinite time it will again be zero. The second half of this process is often modeled as creep recovery in extruded or injection-molded parts after they cool. The creep recovery usually results in undesirable dimensional changes observed in the cooled solid with time. Creep recovery response is due to freezing in local deformation of the polymer molecules when the polymer cools rapidly. The polymer molecules are frozen into shapes that distort their Gaussian spherical equilibrium shape. If the polymer is heated or allowed to relax over a very long time period there will be dimensional changes as the polymer molecules assume their thermodynamic equilibrium states (Gaussian spherical equilibrium shape).

When a spring and a dash pot are connected in series the resulting structure is the simplest mechanical representation of a viscoelastic fluid or Maxwell fluid, as shown in Figure 3.10(d). When this fluid is stressed due to a strain rate it will elongate as long as the stress is applied. Combining both the Maxwell fluid and Voigt solid models in series gives a better approximation for a polymeric fluid. This model is often referred to as the four-parameter viscoelastic model and is shown in Figure 3.10(e). A typical strain response as a function of time for an applied stress for the four-parameter model is found in Figure 3.12.

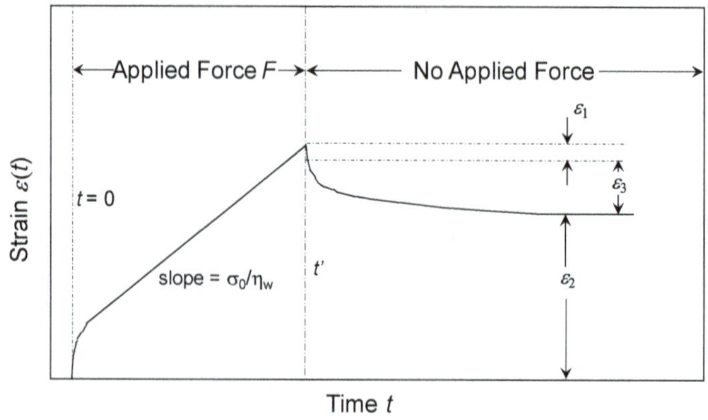

Figure 3.12 Strain response for a four-parameter viscoelastic fluid

The four-parameter model is very simple and often a reasonable first-order model for polymer crystalline solids and polymeric fluids near the transition temperature. The model requires two spring constants, a viscosity for the fluid component and a viscosity for the solid structured component. The time-dependent creep strain is the summation of the three time-dependent elements (the Voigt element acts as a single time-dependent element):

$$\varepsilon(t) = \varepsilon_1(t) + \varepsilon_2(t) + \varepsilon_3(t) \tag{3.26}$$

$$\varepsilon(t) = \frac{\sigma_0}{E_M} + \frac{\sigma_0}{\eta_M} + \frac{\sigma_0}{E_v}\left\{1 - \exp\left(\frac{-E_v t}{\eta_v}\right)\right\} = \sigma_0\left(1 + \frac{1}{\lambda_M}\right) - \frac{\sigma_0}{E_v}\left\{1 - \exp\left(\frac{-t}{\lambda_v}\right)\right\} \tag{3.27}$$

Referring to Figure 3.12, the instantaneous recoverable strain ε_1 is the result of the first term in the strain function. The time-dependent recoverable strain ε_3 is from the Voigt element, and the unrecoverable strain is the result of the Maxwell creep with the slope σ_0/η_M and is ε_2. The potential infinitely large unrecoverable strain of the Maxwell model is why it can be used to model a fluid in a shear environment. This is a one-dimensional model, and it can sometimes be utilized when dealing with simple deformation. Die swell can be explained from a simple point of view with the four-parameter model. When the polymer is extruded through a capillary in the rheometer, the fluid is forced to elongate as it is forced into the capillary. If the two time constants are very small compared to the time that the polymer is in the capillary, then the polymer will relax and not tend to retract much as it exits the capillary. If λ is large relative to the time in the capillary, then the strand will try to recover its original dimension. The diameter will increase as the polymer retracts after exiting the die due to a conservation of volume in the material elements, as the molecules tend to return to their original Gaussian spherical equilibrium shape. Viscoelastic analysis often needs to be analyzed in three dimensions to describe accurately die swell or other viscoelastic phenomena. As an example, the one-dimensional Maxwell fluid in simple shear flow can be expressed in a differential equation form:

$$\tau_{21} + \frac{\eta_0}{G}\frac{\partial \tau_{21}}{\partial t} = -\eta_0 \dot{\gamma}_{21} \quad \text{Maxwell's model for simple shear flow} \tag{3.28}$$

In the three-dimensional form:

$$\underline{\underline{\tau}} + \frac{\eta_0}{G}\hat{\underline{\underline{\tau}}} = -\eta_0 \dot{\underline{\underline{\gamma}}} \quad \text{Maxwell's model for general flow} \tag{3.29}$$

where $\underline{\underline{\tau}}$ and $\dot{\underline{\underline{\gamma}}}$ are the shear stress and deformation rate tensors and $\hat{\underline{\underline{\tau}}}$ is the co-deforming time derivative [5]. Notice that the parameters have changed from the tensile modulus E and the mechanical strain ε to G the bulk modulus and $\dot{\gamma}$ the shear rate. As pointed out previously tensor forms of rheological models are beyond the scope of this introductory treatment. For those interested in a more complete discussion of polymer rheology and viscoelastic models such as the Kaye-Bernstein, Kearsley, Zapas (K-BKZ), Oldroyd-B, Phan Thien-Tanner models, and others the reader should consult rheology texts [5–9].

Real polymers are more complex than these simple mechanical models. Qualitatively, when a real polymer is forced to flow through a contraction or expansion in an extrusion screw, it will exhibit viscoelastic behavior. The polymer molecules will be elongated if forced through a contraction, or they will retract when they flow into an expansion. The effect of viscoelastic behavior in a capillary rheometer is observed in the form of recirculation flow just before the polymer enters the capillary. The polymer is characterized by its relaxation time λ as provided by Equation 3.30. For $\lambda = 0$ the fluid is purely viscous with no elastic behavior, and for $\lambda = \infty$ the material is completely elastic and the stress does not decay with time. The viscoelastic stretching due to a contraction can aid in elongational mixing if the relaxation time is long and the stretching results in the polymer forming microscopic layers. If the relaxation time is short compared to the flow path of the fluid then the polymer will not achieve all of the potential mixing that has occurred as a result of the flow through the contraction. If the reader is interested in the viscoelastic effects of flow on mixing and structure than she/he should consult advanced rheology texts and the open literature.

$$\lambda = \frac{\eta_0}{G} \tag{3.30}$$

Moreover, real polymers are thought to have five regions that relate the stress relaxation modulus of fluid and solid models to temperature as shown in Figure 3.13. In a stress relaxation test the polymer is strained instantaneously to a strain

ε_0, and the resulting stress is measured as it relaxes with time. Below the T_g a solid model should be used. Above the T_g but near the T_g, a rubbery viscoelastic model should be used, and at high temperatures well above the rubbery plateau a fluid model may be used. These regions of stress relaxation modulus relate to the specific volume as a function of temperature and can be related to the Williams-Landel-Ferry (WLF) equation [10].

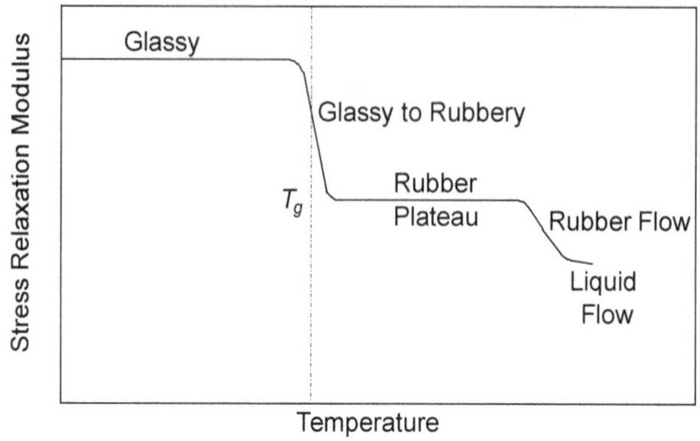

Figure 3.13 Polymer viscoelastic response as a function of temperature, showing the five regions: glassy, glass to rubbery, rubber plateau, rubber flow, and liquid flow

Because of equipment limitations in measuring stress and strain in polymers, the time-temperature superposition principle is used to develop the viscoelastic response curve for real polymers. For example, the time-dependent stress relaxation modulus as a function of time and temperature for a PMMA resin is shown in Figure 3.14. The data is for a very broad range of times and temperatures. The superposition principle is based on the observation that time (rate of change of strain, or strain rate) is inversely proportional to the temperature effect in most polymers. That is, an equivalent viscoelastic response occurs at a high temperature and normal measurement times and at a lower temperature and longer times. The individual responses can be shifted using the WLF equation to produce a modulus-time master curve at a specified temperature, as shown in Figure 3.15. The WLF equation is as shown by Equation 3.31 for shifting the viscosity. The method works for semicrystalline polymers. It works for amorphous polymers at temperatures (T) greater than T_g + 100 °C. Shifting the stress relaxation modulus using the shift factor a_t works in a similar manner.

$$\log \frac{\eta}{\eta_g} = \log a_T = \log \frac{t(T)}{t(T_g)} = \frac{-17.44(T - T_g)}{51.6 + (T - T_g)} \tag{3.31}$$

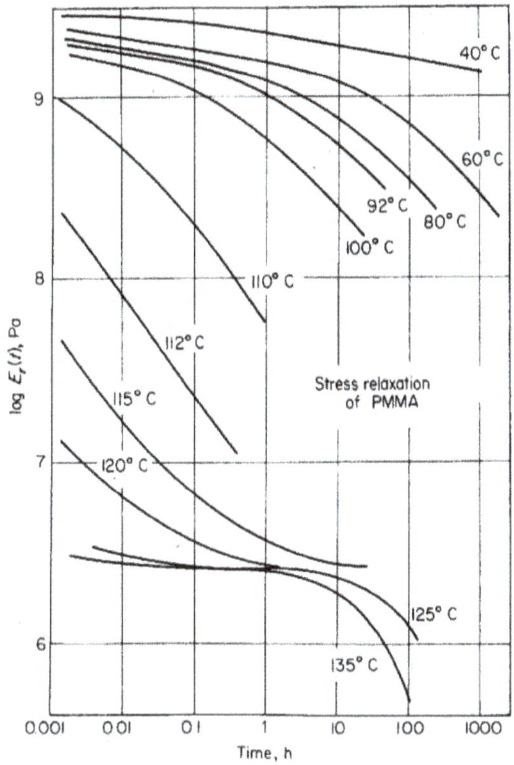

Figure 3.14 Time-dependent stress relaxation modulus as a function of time and temperature for a PMMA resin (Figure 8.11 in Rodriguez [1])

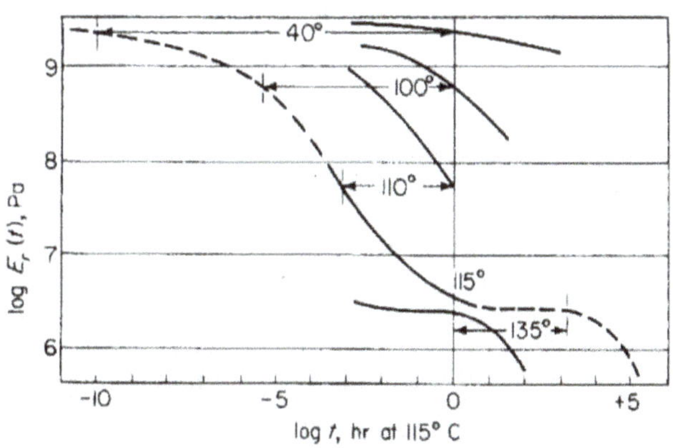

Figure 3.15 Modulus-time master curve at 115 °C based on time-temperature superposition of the PMMA resin data shown in Figure 3.14 (Figure 8.13 in Rodriguez [1]). The T_g for PMMA resin is 105 °C

Figure 3.15 demonstrates how the individual modulus curve segments from Figure 3.14 are shifted to a master curve. The effect of a very high strain rate at a temperature above the T_g will be to observe the polymer responding as if it were below its T_g at a low strain rate. For example, a fluid polymeric strand will break as if it were glassy if elongated at very high rates. This response is often observed in melt tension experiments. These concepts are presented to acquaint the reader with the time-temperature superposition polymer response in this very brief introduction to viscoelastic and time-dependent responses of polymers. If a more complete understanding is desired then the reader is encouraged to consult the polymer rheology texts referenced above.

Viscoelastic rheology is thought to be a dominant material characteristic in the shear mixing of highly viscoelastic polymers. Mixing of low viscosity fluids is usually dominated by microscopic eddies produced by turbulent flow found when the fluid has a Reynolds number in excess of 4000. The high viscosity of polymers in an extruder at normal rotation conditions produces a Reynolds number of less than 10. Mixing of polymers is thus dominated by slicing of adjacent fluid elements. In some mixing elements the polymer is stretched, and as a result the fluid elements are elongated in a manner such that the striations in the polymer matrix are diminished in dimension elastically. Any heterogeneous concentrations of the system are minimized in size thus making the material more homogeneous from a macroscopic point of view. If the relaxation time for the polymer is short then when the fluid element leaves the local mixing zone it will not have a tendency to recoil and recover the elastic strain, and it will remain in its elongated state thus reducing the dimension of the striations. Mixing is discussed in more detail in Chapter 8.

3.6 Measurement of Polymer Viscosity

The following section will now focus on experimental methods for determining viscosity and how the viscosity function relates to analyzing single-screw extrusion processes.

There are a number of techniques that are used to measure polymer viscosity. For extrusion processes, capillary rheometers and cone and plate rheometers are the most commonly used devices. Both devices allow the rheologist to simultaneously measure the shear rate and the shear stress so that the viscosity may be calculated. These instruments and the analysis of the data are presented in the next sections. Only the minimum necessary mathematical development will be presented. The mathematical derivations are provided in Appendix A3. A more complete development of all pertinent rheological measurement functions for these rheometers are found elsewhere [9].

3.6.1 Capillary Rheometers

Capillary rheometers are one of the most common instruments for measuring the viscosity of polymer melts. The instruments are designed as either batch laboratory units or more sophisticated continuous units. The device presented in Figure 3.16 is a batch laboratory device. There are many continuous capillary rheometers available with several capillary sizes and with multiple pressure transducers to measure pressure change through each capillary. Often the polymer is forced through the capillaries with a positive displacement pump. These continuous devices usually come with proprietary software and thus the measurement details and data corrections are not readily apparent to the user. In this chapter the focus will be on laboratory evaluation of rheological data using traditional devices. The batch unit shown in Figure 3.16 is constructed using a heated barrel and a tight-fitting plunger. At the base of the barrel is a removable capillary die with a precise inside diameter and length for the flow of molten polymer. The shear viscosity is measured by relating the volumetric flow to the force (or pressure) required to push the polymer through the capillary die.

Capillary rheometers are used extensively to measure viscosity in the intermediate to high shear rate range. The rheometer has for all practical purposes a lower limit in viscosity measurement because of the plunger seals. These seals are shown on the bottom of the plunger in Figure 3.16, and they induce a frictional resistance when they are pushed through the rheometer barrel. The piston force can be evaluated without polymer in the barrel, but it is always a source of error at low viscosities because of experimental variability. Moreover, barrel friction is one of the critical corrections that must be made when evaluating viscosity measurements using a capillary rheometer at low shear rates. Unfortunately, at low shear rates the frictional forces can often be of the same order as the polymer capillary forces. Thus for these shear rates the cone and plate rheometer is often indicated for use; it will be described in detail later.

To operate the rheometer, polymer pellets are first placed in the barrel and tamped as they melt to remove any entrained air and eliminate voids. The polymer is then heat soaked in the barrel with the plunger inserted, bringing the temperature of the resin to the barrel temperature. In a typical viscosity measurement experiment the plunger is moved downward at a constant velocity. A force transducer is attached to the driven end of the plunger. Using this force and the plunger diameter the pressure at the interface of the polymer and plunger can be calculated. The volumetric flow rate, Q, of the polymer through the capillary can be calculated by using the velocity of the plunger and the cross-sectional area of the plunger.

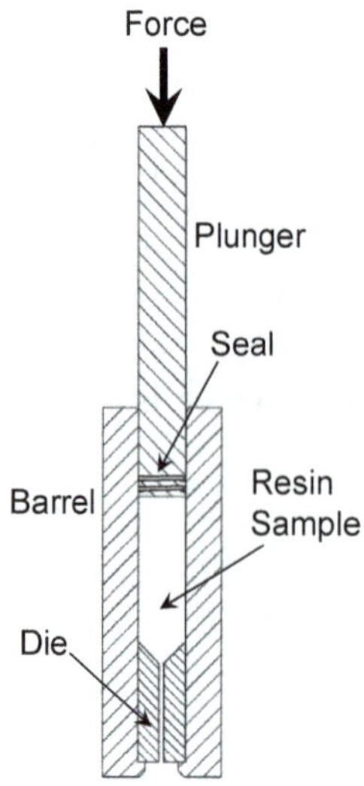

Figure 3.16 Schematic of a capillary rheometer

As discussed later, the calculation of the viscosity is based on a mathematical analysis for an infinitely long capillary, neglecting the entry and exit effects. In many cases, the entry and exit flow effects can be significant and affect the measurement process. For these cases, the Bagley correction is performed to account for the end effects at the entrance and exit of the capillary. It can be seen from Figure 3.16 that the typical capillary has an angular transition entrance and is flat at the end as the polymer exits to the air. In order to correct for the entrance and exit effects a series of experiments must be made to determine the magnitude of these corrections. For these experiments, the hot polymer is extruded through capillaries of constant diameter of various lengths and at several shear rates. Pressure changes are measured at a series of constant shear rates and plotted as a function of the length (L) to the radius (R) ratio, L/R, for the capillary diameter [8]. The pressure change with L/R is plotted and it leads to the Bagley correction when extrapolated to an $L/R = 0$. A Bagley plot for an HDPE resin at 190 °C is shown in Figure 3.17. First, the apparent shear rate at the capillary wall is calculated using a Newtonian viscosity model for flow in the capillary using the volumetric flow rate as follows:

$$\tau_w = \mu \dot{\gamma}_a \tag{3.32}$$

$$\dot{\gamma}_a = \frac{4Q}{\pi R^3} \tag{3.33}$$

where μ is the Newtonian viscosity, τ_w is the shear stress at the barrel wall, $\dot{\gamma}_a$ is the apparent shear rate at the capillary wall based on the assumption of a Newtonian polymer flowing in the capillary, and Q is the volumetric flow rate through the capillary. Before the viscosity can be calculated, the shear rate also will need to be corrected to be consistent with the shear thinning characteristics of the polymer. It is apparent in Figure 3.17 that as the shear rate increases from 100 to 1500 1/s the Bagley correction increases because the intercept of the line on the pressure axis increases with shear rate. It is also apparent that as the capillary length to radius ratio (L/R) increases the $\Delta P_{end\ effects}$ as a fraction of the measured pressure change decreases. It is generally thought that for a capillary L/D greater than about 50 the end effects become a small enough fraction of the measured pressure change that the end effects pressure change can be ignored.

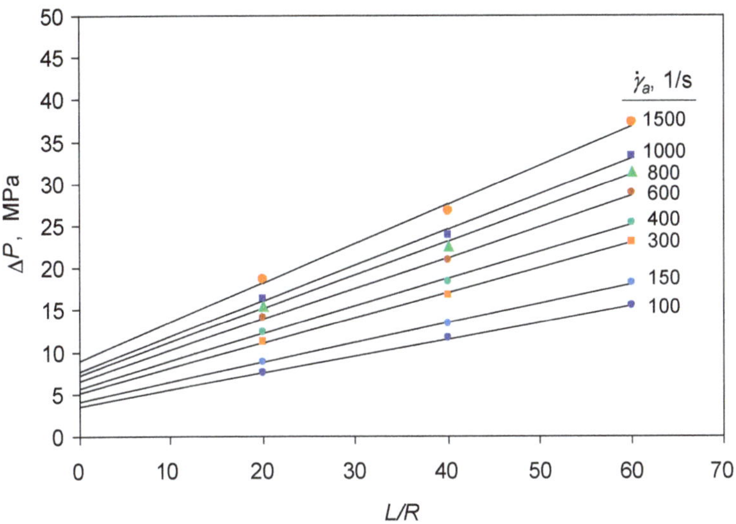

Figure 3.17 Bagley plot corrections for a capillary rheometer for an HDPE resin at 190 °C

The pressure change due to the entry effects for a set of experiments is the intercept ΔP_0 obtained through linear regression of the ΔP data as a function of L/R. This intercept ΔP_0 is then subtracted from the measured pressure to obtain the correct pressure change, ΔP_c. This pressure correction is known as the Bagley correction. The slope of the regression analysis from Figure 3.17 can be used to obtain the stress at the capillary wall, τ_w, as follows:

$$\tau_w = \frac{\Delta P_c R}{2L} = \text{(slope of Figure 3.17 divided by 2)} \tag{3.34}$$

Before the viscosity can be calculated from capillary data, as mentioned above, the apparent shear rate, $\dot{\gamma}_a$, must be corrected for the effect of the pseudoplastic nature of the polymer on the velocity profile. The calculation can be made only after a model has been adopted that relates shear stress and shear rate for this concept of a pseudoplastic shear-thinning material. The model choice is a philosophical question [11]; after rheologists tried numerous models, there are in general two simple models that have withstood substantial testing when the predictions are compared with experimental data [1]. The first is:

$$\dot{\gamma} = m\tau^{z} \tag{3.35}$$

where m is a constant and z is the slope of the $d\ln\dot{\gamma}$ with respect to $d\ln\tau_w$ curve. In this adaptation of the power law model z is 1 for a Newtonian fluid and tends toward infinity for a perfect pseudoplastic material, and the implicit assumption is that the stress is the independent variable. The Weissenberg-Rabinowitsch correction is used to calculate the value of z as function of shear stress [8]. This value is generally not always a constant over the shear rate range of interest. The function for z is initially about 1 at low shear rates and as the sheared material approaches pure pseudoplastic behavior, z tends toward infinity at high shear rates. Figure 3.18 shows the apparent shear rate as a function of τ_w for the HDPE resin at 190 °C.

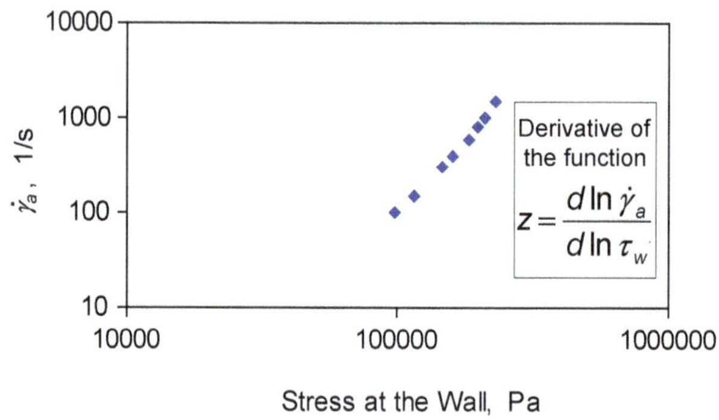

Figure 3.18 Apparent shear rate as a function of the wall stress (τ_w). The first derivative of the function is used to perform the Weissenberg-Rabinowitsch correction. The data are for the HDPE resin at 190 °C as shown in Figure 3.17

The calculation of the shear rate at the capillary wall, $\dot{\gamma}_w$, is computed from the function slope of Figure 3.18 and the apparent shear rate using Equation 3.36. The derivative of the function appears relatively constant over the shear stress range for Figure 3.18. Many resin systems will have derivatives that vary from point to point. The corrected viscosity can then be obtained by dividing the shear stress at the wall by the shear rate $\dot{\gamma}_w$. Equation 3.36 is known as the Weissenberg-Rabinowitsch equation [9].

$$\dot{\gamma}_w = \frac{4Q}{\pi R^3}\left[\frac{1}{4}\left(3 + \frac{d\ln\dot{\gamma}_a}{d\ln\tau_w}\right)\right] = \dot{\gamma}_a\left[\frac{1}{4}\left(3 + \frac{d\ln\dot{\gamma}_a}{d\ln\tau_w}\right)\right] \tag{3.36}$$

The second model used to correct the shear rate for pseudoplastic materials is shown by Equation 3.37:

$$\tau_w = m\dot{\gamma}_w^{n} \tag{3.37}$$

For this power law model n is the slope of the shear stress at the wall with respect to the apparent shear rate at the wall as shown in Figure 3.19. The value of n is also not always a constant, but can change point to point.

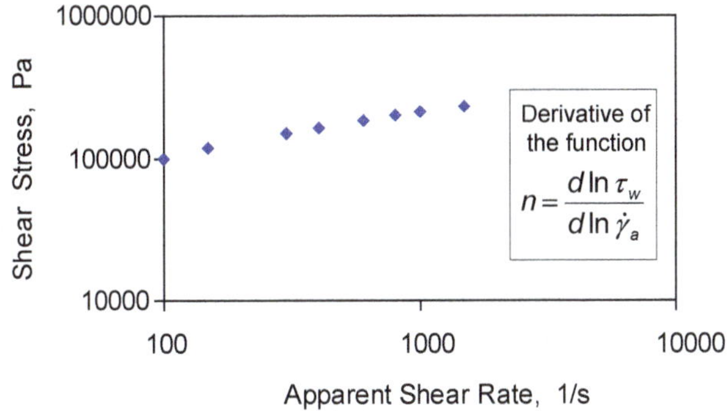

Figure 3.19 Calculation of power law index n from the shear stress at the wall (τ_w) as a function of the apparent shear rate ($\dot{\gamma}_a$)

The slope of this function is the Rabinowitsch [12] correction for the shear rate in terms of the power law parameter n:

$$n = \frac{d \ln \tau_w}{d \ln \dot{\gamma}_a} \tag{3.38}$$

The shear rate at the capillary wall is then calculated from the apparent shear rate at the wall as follows:

$$\dot{\gamma}_w = \dot{\gamma}_a \left(\frac{3n+1}{4n} \right) \tag{3.39}$$

where the function in the parentheses in Equation 3.39 is the Rabinowitsch correction for this viscosity model. The shear stress is a function of the shear rate to the power n. Both z and n are used to correct for the nonparabolic profile of the velocity for polymeric materials in capillary flow, depending on the model adopted. In fact n can be thought of as $1/z$. The slope is more or less constant for this data set for an HDPE resin, and it is about $n = 0.315$. The corrected shear rate at the capillary wall ($\dot{\gamma}_w$) is about 1.54 times larger than the apparent shear rate at the wall ($\dot{\gamma}_a$).

Since the viscosity is defined as:

$$\eta(\dot{\gamma}_w) = \left(\frac{\tau_w}{\dot{\gamma}_w} \right) \tag{3.40}$$

it follows that the power law viscosity calculated from capillary data is:

$$\eta(\dot{\gamma}_w) = m \left| \dot{\gamma}_w \right|^{n-1} \tag{3.41}$$

Note that if $n = 1$ then this is the Newtonian viscosity and therefore $m = \mu$, and the viscosity is not a function of the shear rate and does not need to be corrected. The viscosity as a function of the corrected shear rate for the HDPE resin at 190 °C is shown in Figure 3.20.

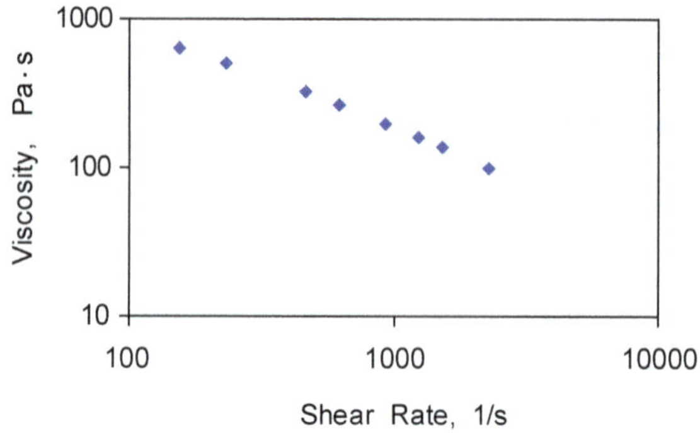

Figure 3.20 Viscosity as a function of the corrected shear rate for the HDPE resin at 190 °C

3.6.1.1 Capillary Viscosity Calculation Example

A new polymer was developed by the polymer scientists in a company. A sample has been provided to the process development laboratory to determine the viscosity of the polymer as a function of shear rate and temperature. The instrument available is an old capillary rheometer. The piston has a diameter of 9.525 mm, and a series of capillaries that fit the rheometer barrel have a diameter of 2.54 mm and lengths of 25.4, 50.8, 76.2, and 101.6 mm. The rheometer temperature was set at 270 °C. Shear viscosity data are needed to estimate process performance.

The flow rate from the motion of the piston in the 9.525 mm diameter cylinder is calculated first. One of the drive gears produces a piston velocity of 13.56 mm/min. The volume produced by the piston is as follows:

$$Q = 13.56 \frac{\text{mm}}{\text{min}} \frac{\text{min}}{60\ \text{s}} (\pi) \left(\frac{9.525\ \text{mm}}{2} \right)^2 = 0.0161 \frac{\text{cm}^3}{\text{s}} = 1.61 \times 10^{-8} \frac{\text{m}^3}{\text{s}} = 16.1 \frac{\text{mm}^3}{\text{s}}$$

The apparent shear rate at the capillary walls is calculated using Equation 3.33 as follows:

$$\dot{\gamma}_a = \frac{4Q}{\pi R_c^3} = \frac{4\left(16.1 \frac{\text{mm}^3}{\text{s}}\right)}{(3.1416)(1.27\ \text{mm})^3} = 10\ 1/\text{s}$$

Next the pressure change across the different capillaries must be calculated and corrected for the end-effect pressure loss using a Bagley correction.

Table 3.5 Summary of Capillary Rheometer Calculations

Capillary Length, mm	Pressure Change, MPa	Regressed Pressure Change, MPa	Corrected Pressure Change, MPa	Wall Shear Stress, Pa	Capillary Length Divided by Radius, *L/R*
25.4	9.023	9.023	7.523	1.881×10^5	20
50.8	16.546	16.546	15.046	1.881×10^5	40
76.2	24.069	24.069	22.569	1.881×10^5	60
101.6	31.592	31.592	30.092	1.881×10^5	80

The pressure change is measured for each capillary at the apparent shear rate $\dot{\gamma}_a$. Regression analysis is then used to obtain the slope and intercept for the function of pressure change (column two) in Table 3.5 with respect to the length divided by the radius (*L/R*, column six). The slope of the function is 0.376 MPa, and the intercept is 1.5 MPa. The regressed pressure change is obtained from the slope and the intercept, and the pressure change corrected for the end effects are as follows:

$$(\text{Pressure Change})_{\text{regressed}} = (0.376\ \text{MPa}) * (L\ /\ R) + (1.5\ \text{MPa})$$

$$(\text{Pressure Change})_{\text{corrected}} = (\text{Pressure Change})_{\text{regressed}} - (1.5\ \text{MPa})$$

The shear stresses at the wall of the capillaries are calculated using the corrected pressure change and Equation 3.34.

$$\tau_w = \frac{\Delta P_c}{2} \frac{1}{L\ /\ R} = 1.881 \text{ x } 10^5 \text{ Pa}$$

The shear rate now must be corrected for the non-Newtonian character of the polymer. In order to do this, data must be available at several shear stresses and shear rates. The shear stress as a function of the apparent shear rate at this temperature is graphed in Figure 3.21.

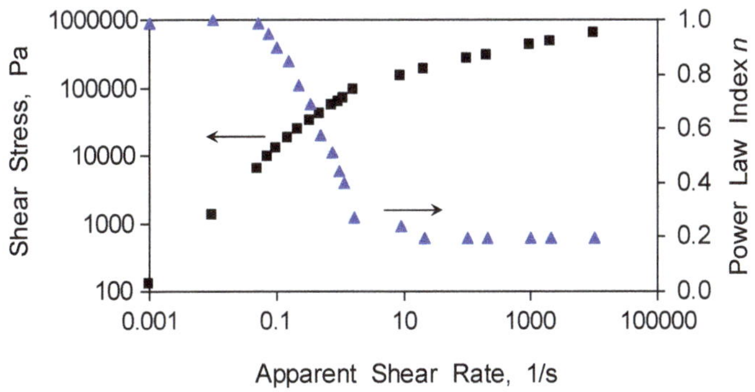

Figure 3.21 Shear stress and power law index as a function of the apparent shear rate for a developmental polymer at 270 °C

The local value for n at an apparent shear rate of 10 1/s is calculated using the data points at apparent shear rates of 50 and 5 1/s as follows:

$$n = \frac{\ln(2.595 \times 10^5) - \ln(1.458 \times 10^5)}{\ln(50) - \ln(5)} = 0.199$$

Next the corrected shear rate at the wall needs to be calculated using Equation 3.39:

$$\gamma_w = \dot{\gamma}_a\left(\frac{3n+1}{4n}\right) = \left(10\frac{1}{s}\right)\left(\frac{3(0.199)+1}{4(0.199)}\right) = 20.1 \text{ 1/s}$$

Notice that the corrected shear rate at the wall is about twice the apparent shear rate at the wall for this case. This set of calculations must be repeated with the other data sets until the total shear rate and temperature range is covered. A typical set of data is presented Table 3.6.

Table 3.6 Rheological Data at 270 °C

Shear Rate $\dot{\gamma}_w$, 1/s	Power Law n	Shear Stress, Pa	Viscosity, Pa·s
0.001	0.99	127.1	126,800
0.01	0.997	1269	126,800
0.05	0.986	6289	125,400
0.076	0.947	9380	123,300
0.103	0.899	12,420	120,800
0.157	0.851	18,010	115,000
0.216	0.762	23,480	108,900
0.334	0.688	32,580	97,520
0.474	0.575	41,590	87,730
0.743	0.513	53,920	72,610
0.922	0.441	61,570	66,790
1.11	0.396	68,310	61,830
1.67	0.272	89,860	53,850
8.952	0.24	145,800	16,290
20.08	0.199	188,100	9365
100.8	0.197	259,500	2574
201.8	0.197	297,700	1475
1,009	0.197	408,900	405
2,018	0.197	468,800	232
10,000	0.197	643,900	63.8

There would be a minimum of 80 data sets needed to generate this data for one temperature. Because of the time involved, usually about 10 to 15 shear rate data points are generated at each temperature. The plot of the viscosity as a function of shear rate at 270 °C is presented in Figure 3.22. The viscosity below a shear rate of 5 1/s would be best taken using a cone and plate rheometer. The wall friction for the capillary rheometer between the piston and the rheometer cylinder wall would likely cause a force on the piston of the same order as the force due to the flow stress.

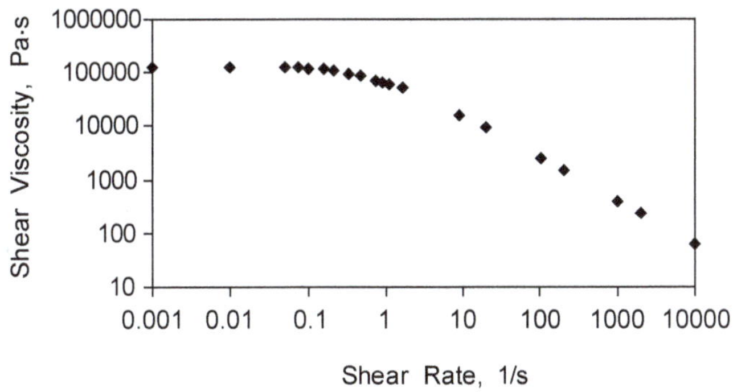

Figure 3.22 Shear viscosity data for the current example

Now the functions for doing simple power law-dependent simulations are developed. The zero-shear viscosity, η_0, is 1.268×10^5 Pa·s as shown by Figure 3.22 and the viscosity data in Table 3.6. This holds for all shear rates in the plateau range. For the power law fit, the last six entries in Table 3.6 are used to develop a regression fit, and then the line is extrapolated back to lower shear rates. The regression fit is as follows:

$$\ln\left(\eta_{Power\ Law}\right) = \ln(m) + (n-1)\ln\left(\left|\dot{\gamma}_R\right|\right)$$

From the linear regression fit:

$$\text{Slope}_{\text{Power Law}} = -0.802 = (n-1)$$

$$\text{Intercept}_{\text{Power Law}} = 11.55 = \ln(m)$$

$$\eta_{Power\ Law}(\dot{\gamma}) = e^{11.55}\left[\frac{\dot{\gamma}}{1/s}\right]^{-0.802}$$

These two functions have been superimposed on the data from Figure 3.22, and the composite is presented in Figure 3.23. The zero shear and power law functions as mentioned above are often used in flow simulations.

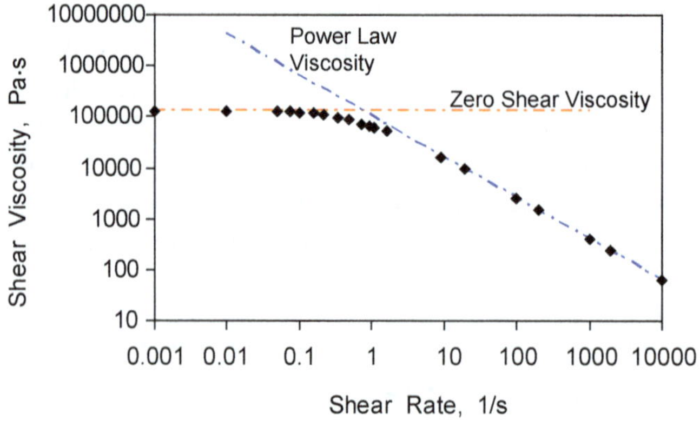

Figure 3.23 Power law and zero shear fit for the viscosity at 270 °C

The viscosity activation energy is often evaluated to determine the temperature dependence for these viscosity functions. Viscosities at two temperatures are provided in Table 3.7.

Table 3.7 Example Viscosity Data at Two Temperatures and Two Shear Rates. Data Shows the Effect of Temperature on Viscosity

Shear Rate, 1/s	Temperature		Viscosity, Pa·s
	°C	K	
0.01	230	503	1.74×10^6
0.01	270	543	1.26×10^5
20	230	503	2.40×10^4
20	270	543	9.36×10^3

In this example the temperature viscosity function used is based on the viscosity, η_R, at a reference temperature T_R which has units of Kelvin:

$$\eta(T) = \eta_R \, e^{\frac{\Delta E}{R_g}\left[\left(\frac{1}{T} - \frac{1}{T_R}\right)\right]}$$

$$\frac{\Delta E}{R_g}\left[\frac{1}{T_1} - \frac{1}{T_R}\right] = \ln \eta(T_1) - \ln(T_R)$$

First the temperature dependence of the limiting zero shear rate viscosity (Newtonian) is calculated at a shear rate of 0.01 1/s using the data in Table 3.7:

$$\frac{\Delta E_0}{R_g} = \frac{\ln(1.26\times10^5) - \ln(1.744\times10^6)}{\dfrac{1}{543\text{ K}} - \dfrac{1}{503\text{ K}}} = 1.794\times10^4 \text{ K}$$

$$\eta_0(T) = 1.268 \times 10^5 e^{\left[1.794 \times 10^4 \text{ K}\left(\frac{1}{T} - \frac{1}{543 \text{ K}}\right)\right]}$$

For the power law region at a shear rate of 20 1/s, the calculations lead to the temperature dependence of the power law viscosity function using the same method:

$$\frac{\Delta E_{Power\ Law}}{R_g} = \frac{\ln(9.365 \times 10^3) - \ln(2.405 \times 10^4)}{\dfrac{1}{543 \text{ K}} - \dfrac{1}{503 \text{ K}}} = 6.44 \times 10^3 \text{ K}$$

$$\eta_{Power\ Law}(T) = 9.365 \times 10^3 e^{\left[6.44 \times 10^3 \text{ K}\left(\frac{1}{T} - \frac{1}{543 \text{ K}}\right)\right]}$$

Notice that the activation energy in the power law regime is about one-third that for the zero shear rate viscosity regime. Combining the shear rate and temperature viscosity shifts:

$$\eta_{Power\ Law}(\dot{\gamma}, T) = 1.041 \times 10^5 \left[\dot{\gamma}\right]^{-0.802} \left[e^{\left[6.44 \times 10^3\left(\frac{1}{T} - \frac{1}{543}\right)\right]}\right]$$

This is the equation that will allow the prediction of the viscosity as a function of both shear rate and temperature in the power law region. Finally, the equations are checked for acceptability and errors by testing it with the experimental data. For the case above, the error was 2% or less. Most of the errors come from truncating the fitting constants. However, a maximum 2% error is usually quite acceptable.

3.6.2 Cone and Plate Rheometers

The rheometer most often used to measure viscosity at low shear rates is the cone and plate viscometer. A schematic of a cone and plate rheometer is found in Figure 3.24. The device is constructed with a moving cone on the top surface and a stationary plate for the lower surface. The polymer sample is positioned between the surfaces. Two types of experiments can be performed: the cone can be rotated at a constant angular velocity, or it can be "rotated" in a sinusoidal function. The motion of the cone creates a stress on the polymer between the cone and the plate. The stress transferred to the plate provides a torque that is measured using a sensor. The torque is used to determine the stress. The constant angle of the cone to the plate provides an experimental regime such that the shear rate is a constant at all radii in the device. That is, the shear rate is independent of the radial position on the cone, and thus the shear stress is also independent of the position on the cone.

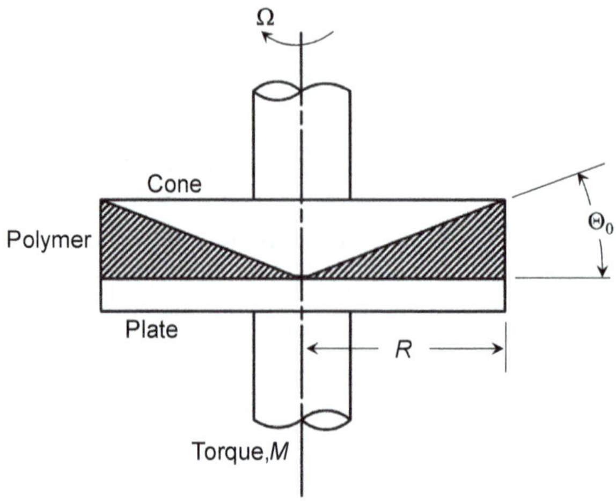

Figure 3.24 Schematic of a cone and plate rheometer

In this device a cone with constant angle Θ_0 is rotated at a constant rate of angular rotation Ω. At any radius the angular velocity is the same. This leads to the following working equations:

$$\sigma_\Omega = \left(\frac{3M}{2\pi R^3}\right) \tag{3.42}$$

$$\dot{\gamma} = \frac{\Omega}{\Theta_0} \tag{3.43}$$

$$\eta_{\dot{\gamma}} = \left(\frac{3M\Theta_0}{2\pi R^3 \Omega}\right) \tag{3.44}$$

where M is the torque to turn the cone, Ω is the angular rotation rate, Θ_0 is the cone angle, σ_Ω is the stress due to the angular rotation rate, and R is the cone radius. Because the shear rate and shear stress are independent of cone radius, no correction is necessary for pseudoplasticity with this device. Because of the elastic effects of the polymer melt, steady shear viscosity evaluations for polymer melts are usually measured below a few hundred 1/s. Elastic effects can cause a transverse or recirculation flow between the inner and outer radii of the sample. Experimentally these are quite easy to see because they manifest themselves as standing waves on the free surface of the polymer melt in the rheometer gap [9, 13]. The transverse flow increases the energy dissipated and thus causes an error in the measurement.

In order to extend the useful shear rate range of the cone and plate rheometer, the device is often operated in an oscillation mode. That is, the rotation rate Ω is driven with a function of $\sin(\omega t)$ at small rotation angles. The associated small angle of oscillation maintains the polymer in the linear viscoelastic regime, that is, with small local strains. There will thus be no transverse flow. Some typical data using this mode are presented in Figure 3.25 for GPPS resin at 225 °C. This technique is referred to as small angle oscillatory shear (SAOS). It relies on the assumption that under these conditions the polymer behaves as a linear viscoelastic fluid because the small strains in the polymer are recoverable.

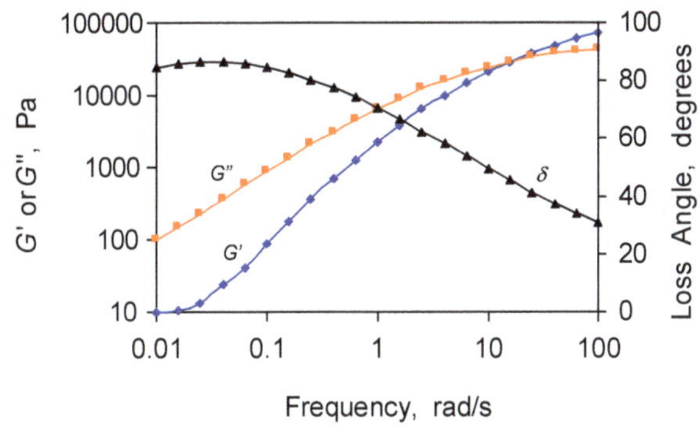

Figure 3.25 Storage modulus, elastic modulus, and loss angle for a cone and plate rheological measurement. The data are for a GPPS resin with an MFR of 1.5 dg/min (5 kg, 200 °C) measured at 225 °C

Using the cone and plate rheometer the angle Ω is forced in a sinusoidal manner, leading to "linear strain" being introduced in the polymer. The shear strain, γ, is a sinusoidal function of time t with a shear rate amplitude of $\dot{\gamma}_0$ as follows:

$$\gamma = \frac{\dot{\gamma}_0}{\omega}\sin(\omega t) = \gamma_0 \sin(\omega t) \tag{3.45}$$

The resulting instantaneous stress:

$$-\tau = \tau_0 \sin(\omega t + \delta) = \tau_0(\cos\delta)\sin(\omega t) + \tau_0(\sin\delta)\cos(\omega t) \tag{3.46}$$

Here τ is the resulting shear stress, δ is the phase shift often represented as $\tan(\delta)$, and ω is the frequency. The term δ is often referred to as the loss angle. The in-phase elastic portion of the stress is $\tau_0(\cos\delta)\sin(\omega t)$, and the out-of-phase viscous portion of the stress is $\tau_0(\sin\delta)\cos(\omega t)$. They are both measured and recorded. This leads to the definition of the complex modulus and viscosity, which can be used to extend the range of the data using the cone and plate rheometer [6]:

$$G' = \frac{\tau_0}{\gamma_0}\cos(\delta) \quad \text{elastic response} - \text{in phase with imposed strain} \tag{3.47}$$

$$G'' = \frac{\tau_0}{\gamma_0}\sin(\delta) \quad \text{viscous response} - \text{out of phase with imposed strain} \tag{3.48}$$

where G' is the storage modulus and G'' is the loss modulus. Using these data and the following relationship, a complex viscosity magnitude can be calculated from G', G'', and the frequency using Equation 3.49. The complex viscosity for the GPPS resin shown in Figure 3.25 is shown in Figure 3.26.

$$\left|\eta^*\right| = \frac{\sqrt{(G')^2 + (G'')^2}}{\omega} \tag{3.49}$$

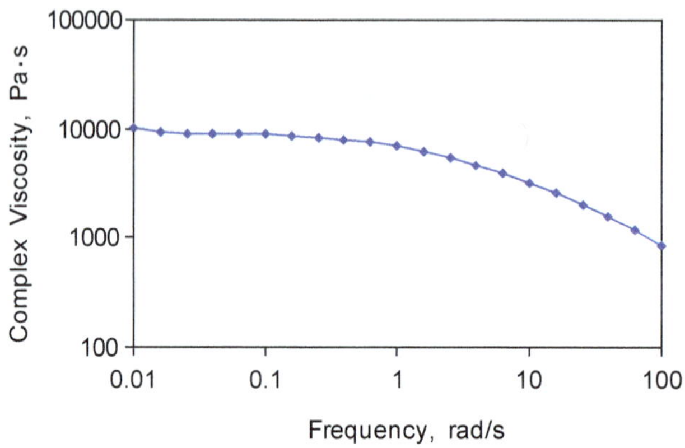

Figure 3.26 Complex viscosity measured using a cone and plate rheometer. The data are for a GPPS resin with an MFR of 1.5 dg/min (5 kg, 200 °C) measured at 225 °C. The data are from Figure 3.22

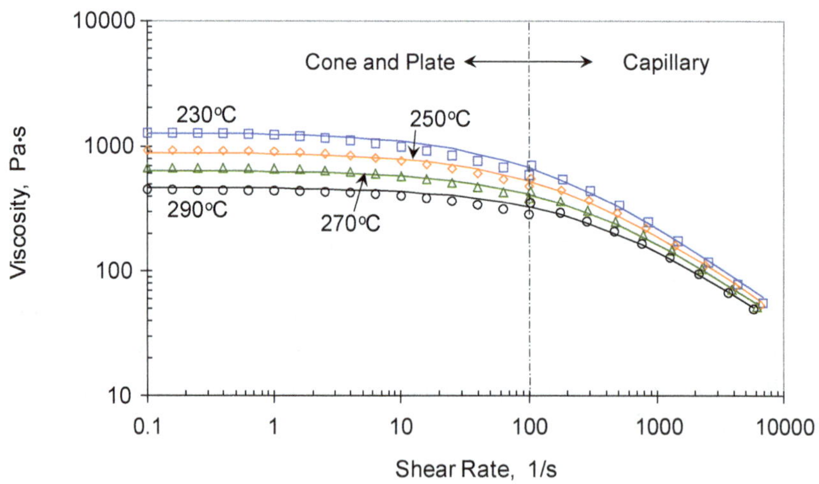

Figure 3.27 Cone and plate and capillary rheology data for a specialty PE resin. A slight shift in the rheology occurs between the two methods at a shear rate of 100 1/s

Rheologists sometimes use an empirical functionality called the Cox-Merz rule, which states that the complex viscosity value at a given frequency is equal to the steady shear viscosity at the same shear rate [14]:

$$\eta(\dot{\gamma}) = \left|\eta^*(\omega)\right| \quad \dot{\gamma} = \omega \tag{3.50}$$

Using two rheometers, the cone and plate and the capillary, the complete viscosity function can be developed relative to shear rate as shown by Figure 3.27. In general for highly viscous polymers the Newtonian viscosity plateau is obtained using a cone and plate rheometer while the power law region is usually obtained with a capillary rheometer. It is important that the data generated using the different rheometers overlap at the temperatures and shear rates used to take the viscosity data. Not all materials follow the Cox-Merz rule, so care must be taken to be sure that the data obtained is consistent in the overlap region with some steady-shear device. This can be very important when simulating the flow in an extruder channel. The data in Figure 3.27 has a very good agreement in the overlap region of the cone and plate and the capillary rheometer.

3.6.3 Melt Index and Melt Flow Rate

The preceding sections have discussed the shear rheology of polymers. As indicated by the discussion, the rheological behavior is extremely complex and depends on many flow conditions. Resin manufacturers and customers, however, often classify the resin rheology using a single index value referred to as the melt index (MI), melt flow index (MFI), or melt flow rate (MFR). The test is often used as a quality control indication of molecular weight for homogeneous families of polymers that have more or less the same shear thinning characteristics. The test is performed using a melt indexer, which is a specialized capillary rheometer operating with a constant shear stress at the capillary wall. A schematic of a melt indexer is shown in Figure 3.28. It is configured with a temperature-controlled barrel with an inside diameter of 9.5504 mm, a weighted plunger, and a die that has a diameter of 2.0955 mm and a length of 8.001 mm, as prescribed by ASTM Standard D1238-04c [15]. The index refers to the mass of the extrudate in grams that is pushed out of the die in a 10 minute period. The index is reported in units of g/10 min or dg/min and is accompanied by the conditions of the test, that is, weight and temperature. For most PE resins, the test is performed using a load of 2.16 kg and a temperature of 190 °C. A list of the ASTM conditions is shown in Table 3.8. For PP resins, the test is performed using the same load but the measurement is carried out at 230 °C. A resin with a low melt index corresponds to a highly viscous resin and thus a high molecular weight. A resin with a high melt index has a relatively low viscosity and thus a low molecular weight. When the melt index is less than one, the material is said to have a fractional melt index.

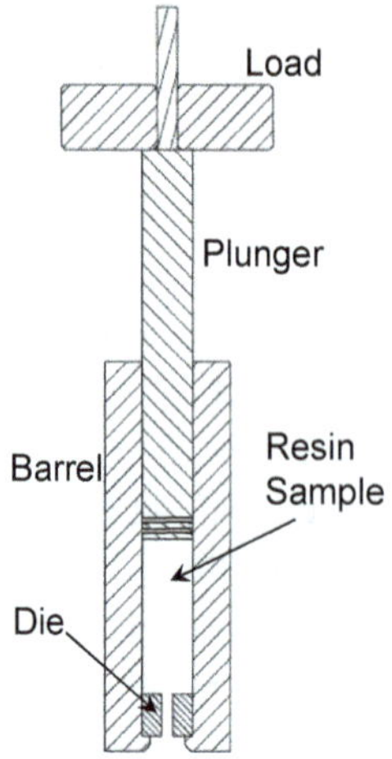

Figure 3.28 Schematic of a melt indexer

The melt index test measures the flow property at a fixed wall shear stress in the capillary. The shear stress depends on the load specified for the condition and it is provided in Table 3.8. The apparent shear rate at the capillary wall that the resin experiences depends on the MI value measured, and it can be calculated using a modification of Equation 3.33 as follows:

$$\dot{\gamma}_a = \left[\frac{1}{6 \times 10^5}\right]\frac{4MI}{\pi \rho R^3} \tag{3.51}$$

where ρ is the melt density at the test condition in kg/m^3 and R is the radius of the capillary. For the test, R is 1.048×10^{-3} m. The reduced equation is as follows:

$$\dot{\gamma}_a = 1845\frac{MI}{\rho} \tag{3.52}$$

For a PE resin with an MI of 10 dg/min and a melt density of 766 kg/m^3 at 190 °C, the apparent shear rate that the resin experienced was 24 1/s.

As previously discussed, a melt indexer measures the flow property at a constant stress. The stress at the die wall can be calculated using a modification of Equation 3.34. The pressure change for this calculation is simply the load weight divided by the area of the plunger. The modified equation is as follows:

$$\tau_w = \frac{[Mg/(\pi r_p^2)]R}{2L} \tag{3.53}$$

where M is the total mass or load on the plunger in kg, g is the gravitational acceleration constant (9.807 m/s^2), r_p is the radius of the plunger, R is the radius of the capillary die, and L is the length of the capillary die. For the melt indexer geometry as specified by ASTM D1238-04c, Equation 3.53 simplifies as follows:

$$\tau_w = \left[\frac{8963 \cdot \text{Pa}}{\text{kg}}\right] M \tag{3.54}$$

The shear stresses at the wall of the capillary die for the ASTM conditions are provided in Table 3.8. The apparent shear rate and stress at the capillary wall are provided here to show the magnitudes of the conditions and test results. They should not be used to derive viscosity measurements for anything other than a comparative analysis since the MI measure has some variability between laboratories and is highly unreliable [16, 17]. The indexer has a very short L/R ratio of about eight and for many resins the pressure end effects are significant. The end effects are not removed from the index. Although the MI is used to identify a resin, it has little value when troubleshooting an extruder. It is strongly suggested that the viscosity of the resin be measured as a function of shear rate and temperature for the operating range.

Table 3.8 Test Conditions and Shear Stress at the Capillary Wall for Melt Index Testing as Prescribed by ASTM D1238-04c [15]

Condition	Temperature, °C	Total Load with Piston, kg	Shear Stress at the Capillary Wall, kPa
(125 °C, 0.325 kg)	125	0.325	2.91
(125 °C, 2.16 kg)	125	2.16	19.4
(150 °C, 2.16 kg)	150	2.16	19.4
(190 °C, 0.325 kg)	190	0.325	2.91
(190 °C, 2.16 kg)	190	2.16	19.4
(190 °C, 21.6 kg)	190	21.6	194.0
(200 °C, 5.0 kg)	200	5.0	44.8
(230 °C, 1.20 kg)	230	1.20	10.8
(230 °C, 3.80 kg)	230	3.80	34.1
(265 °C, 12.5 kg)	265	12.5	112.0
(275 °C, 0.325 kg)	275	0.325	2.91
(230 °C, 2.16 kg)	230	2.16	19.4
(190 °C, 1.05 kg)	190	1.05	9.41
(190 °C, 10.0 kg)	190	10.0	89.6
(300 °C, 1.20 kg)	300	1.20	10.8
(190 °C, 5.0 kg)	190	5.0	44.8

3.7 Viscosity of Polymers as Functions of Molecular Character, Temperature, and Pressure

The previous sections showed how polymers responded to deformation and in particular shear rate. Polymer deformation in shear depends on many factors, including molecular weight, molecular weight distribution, molecular branching, temperature, and pressure. Additives such as fillers, plasticizers, and blowing agents also have strong effects on the rheological response to deformation. This section will discuss how the shear viscosity of a resin is affected by molecular weight, molecular weight distribution, temperature, and pressure. The effects of some additives and molecular branching are beyond the scope of this chapter, but they can be studied elsewhere [1, 9].

The viscosity of some polymers at constant temperature is essentially Newtonian over a wide shear rate range. At low enough shear rates all polymers approach a Newtonian response; that is, the shear stress is essentially proportional to the shear rate, and the linear slope is the viscosity. Generally, the deviation of the viscosity response to a pseudoplastic is a function of molecular weight, molecular weight distribution, polymer structure, and temperature. A model was developed by Adams and Campbell [18] that predicts the non-Newtonian shear viscosity behavior for linear polymers using four parameters. The Adams-Campbell model is as follows:

$$\eta(T,\dot{\gamma}) = \left[\exp\left[\ln \eta_0(T) + \frac{s\sigma}{\sqrt{2}}\left[\Theta\left(1 - erf\left(\Theta\right)\right) - \frac{1}{\sqrt{\pi}}\exp\left(\Theta^2\right)\right]\right]\right] \tag{3.55}$$

$$\Theta = \left(\frac{\ln \mu(T) - \ln(\dot{\gamma})}{\sqrt{2}\,\sigma}\right) \tag{3.56}$$

where s, σ, $\ln\mu(T)$, and $\ln\eta_0(T)$ are the model parameters. These parameters are based on molecular weights, T_g, and temperature, and they will be discussed next. The model is proven to predict the difference in temperature sensitivity in the Newtonian and power law regions, the effect of molecular weight, molecular weight distribution through polydispersity, and the effect of solvent-induced plasticization on the viscosity-shear rate relationship. In order to predict these effects, substantial viscosity shear rate data over a wide range of molecular weight, molecular weight distribution, and temperature were analyzed [18] to develop functional relationships for the four model parameters. The basis of this engineering analysis equation was the recognition that the $\ln(\eta)$ versus $\ln(\dot{\gamma})$ function has a second derivative that is the same as the log normal distribution function. The second derivative was integrated twice to obtain the above equation. Then the model parameters were statistically fitted to the molecular weight, molecular weight distribution, and temperature data. The four parameters have been found to fit the following forms [18]:

$$s = \left(\frac{M_w - M_c}{M_w} \right) \tag{3.57}$$

$$\sigma = 1.393 \sqrt{1.06 - \frac{M_n}{M_w}} \tag{3.58}$$

$$\ln \mu(T) = \exp\left(-2.7 + \left[39.52 - 15.83\sigma^2 \right] \left(\frac{T - T_g}{T_g} \right) \right) \left(\frac{M_w}{M_c} \right)^{-3.4} \tag{3.59}$$

$$\ln \eta_0(T) = \exp(-1.33) \left(\frac{M_w}{M_c} \right)^{3.4} \left(\frac{T - T_g}{T_g} \right)^{-10.09} \tag{3.60}$$

where M_c is the critical molecular weight for entanglement and T and T_g are the temperature and glass transition temperature in degrees Kelvin. The solvent sensitivity appears because several of the parameters have T_g as part of the temperature shift. The temperature dependence is in the form found in renormalization group theory, differing from the conventional Arrhenius temperature function, and it has been shown to predict the effect of solvents on polymer melt viscosity through the solvent effect on T_g. This model was used to produce the viscosity curves in Figure 3.29 and Figure 3.30 by only changing the temperature for two polymers with M_w = 300,000 kg/kg·mole and with two polydispersities. The critical molecular weight for entanglement of PS resin is 32,000 kg/kg·mole [19].

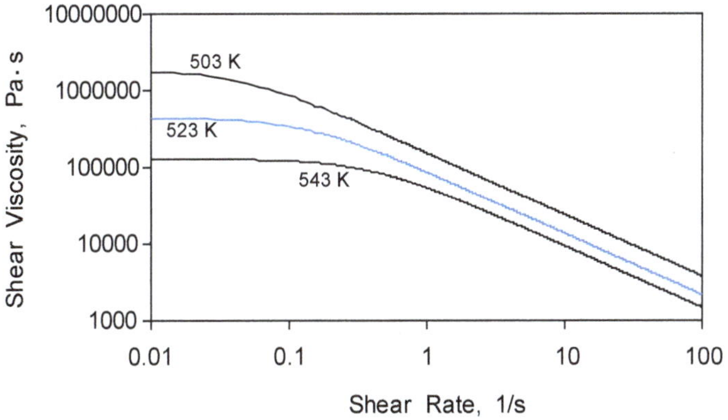

Figure 3.29 Effect of temperature and shear rate for a GPPS resin with a polydispersity (M_w/M_n) of 2.4 and a weight average (M_w) of 300,000 kg/kg·mole using the Adams-Campbell model [18]

Observing the viscosity in Figure 3.29, the transition point moves to higher shear rates from Newtonian to power law viscosity as the temperature is increased. At a temperature of 543 K the transition is about 0.1 1/s for this polymer with M_w/M_n = 2.4. If the data were terminated below 0.1 1/s shear rate it might be suggested that the polymer at 503 K is only power law in nature. In general all homopolymers have a transition from Newtonian to power law at some low shear rate.

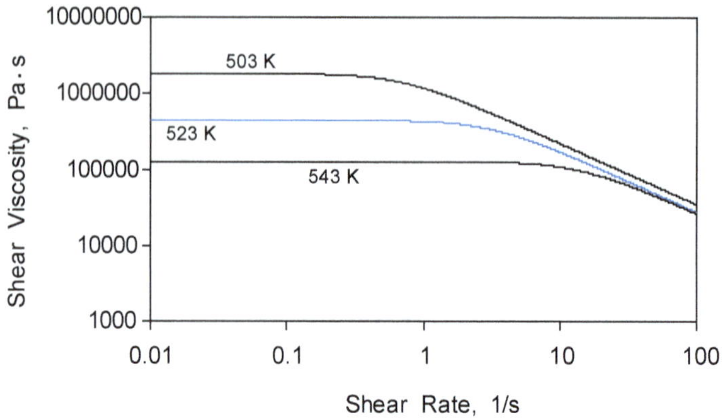

Figure 3.30 Effect of temperature and shear rate for a GPPS resin with a polydispersity (M_w/M_n) of 1.5 and a weight average (M_w) of 300,000 kg/kg·mole using the Adams-Campbell model [18]

The shear viscosity shown in Figure 3.30 is for a polymer with M_w/M_n = 1.5, and it is for the same weight average molecular weight and temperature as in Figure 3.29. At a temperature of 543 K, the resin shown in Figure 3.30 has the Newtonian to power law transition beginning at about 10 1/s. By decreasing the polydispersity the transition has moved almost two orders of magnitude higher. The narrow distribution polymer would require more power to process at low shear rates compared to the polymer in Figure 3.29. This should be kept in mind when changing resins on an extrusion line where the extruder may be power or torque limited. It should be noted that since the weight average molecular weight is the same, the Newtonian viscosity is essentially identical at the three temperatures plotted on Figure 3.29 and Figure 3.30.

The temperature dependence at a given shear rate is very different in the Newtonian and power law regions of Figure 3.29 and Figure 3.30. In general the activation energy is higher in the Newtonian region than in the power law region. And the power law activation energy usually decreases as the polydispersity decreases. The Adams-Campbell model compensates for this phenomenon. One of the parameters, σ, in the model correlates with the molecular weight distribution. Recently Wood-Adams and Dealy [20, 21] published several papers that show a more fundamental rheological analysis that correlates with polymer molecular weight and molecular weight distribution.

The Newtonian viscosity of some polymers increases essentially linearly with the weight average molecular weight, and for other polymers the Newtonian viscosity increases with an exponential power of the molecular weight. The exponential power is found to be about 3.4, but this power does deviate for some polymers. These two transitions, Newtonian to pseudoplastic and linear to 3.4 power in the Newtonian range are often related to molecular structure as demonstrated in Figure 3.31 [22]. The polystyrene data used to develop the Adams-Campbell viscosity function showed almost no shear thinning at M_c [18]. That is why the power law slope, s, is a function of M_w and M_c. At M_c the slope is zero and the material would be essentially Newtonian.

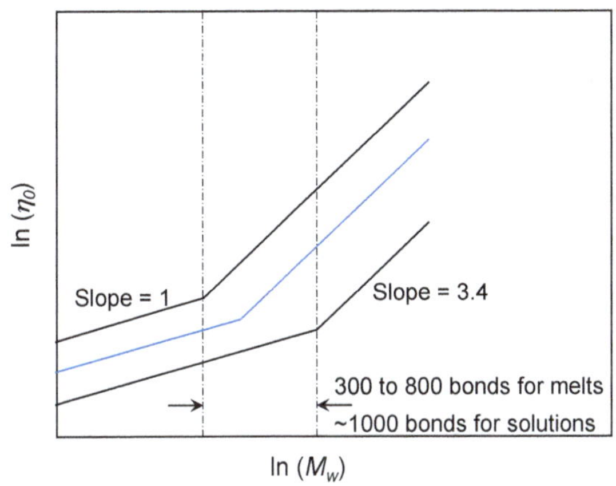

Figure 3.31 Effect of the number of polymer backbone bonds on the Newtonian viscosity

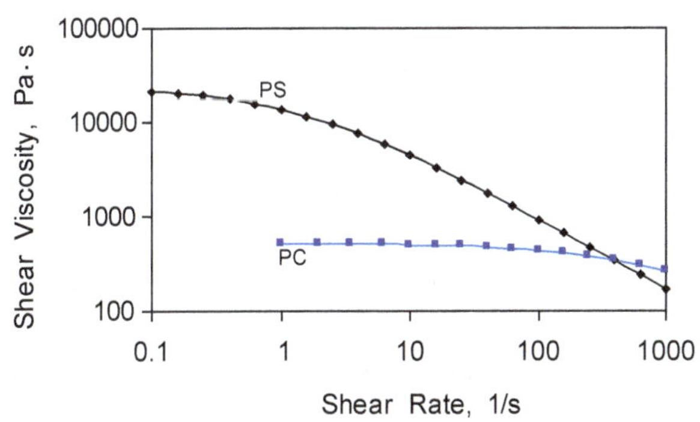

Figure 3.32 Shear viscosities for PC and PS resins at processing conditions of 280 and 210 °C, respectively. The PC resin is nearly Newtonian in the measurement range

The Newtonian melt viscosity of a polymer increases in an essentially linear manner as the molecular weight increases up to a molecular structure that has between 300 and 800 rotating bonds in the backbone of the polymer, as shown by Figure 3.31. This molecular weight is on the same order as the critical molecular weight for entanglement M_c, and for polystyrene it is about twice the reported M_c [23].

Polymers like PET and PC have many nonrotating bonds in their backbone at the modestly high molecular weights found in some commercial polymers. In general their number of rotating bonds is only on the order of the transition zone between the dotted vertical lines in Figure 3.31, and these polymers are thus more Newtonian-like than most commodity polymers. They show only minor shear thinning. For example, shear viscosities for PC and PS resins at processing temperatures are shown in Figure 3.32. In the shear rate measurement range, the PC resin is essentially Newtonian. The Newtonian, limiting shear, viscosity is reported to increase with the 3.5±0.3 [2, 20, 24, 25] power of the weight average molecular weight as the polymer is polymerized above the critical molecular weight. This break between linear and power functions of molecular weight does not occur generally in solutions until there are about 1,000 rotating bonds in the backbone of the polymer. The analytical equations for calculating the Newtonian melt viscosity function can thus be expressed as follows:

$$\eta_0 = BM_w^A \tag{3.61}$$

where B is a prefactor determined experimentally, M_w is the weight average molecular weight, and A is a constant with a value of 3.5±0.3 for polymers above the critical molecular weight.

Polymer viscosity is often classically modeled as an Arrhenius exponential function of temperature as follows:

$$\eta = \eta_R e^{\frac{\Delta E}{Rg}\left(\frac{1}{T}-\frac{1}{T_R}\right)} \tag{3.62}$$

This relationship for Newtonian viscosity is valid normally for temperatures higher than 50 °C or more above the T_g. The utility of the Arrhenius correlation can be limited to a relatively small temperature range for accurate predictions. The viscosity is usually described in this exponential function form in terms of an activation energy, ΔE, absolute temperature T in Kelvin, the reference temperature T_R in Kelvin, the viscosity at the reference T_R, η_R, and the gas law constant R_g. As the temperature approaches T_g for PS (T_g = 100 °C), which could be as high as 150 °C, the viscosity becomes more temperature sensitive and is often described by the WLF equation [10]:

$$\text{Log}\frac{\eta}{\eta_g} = \frac{-17.44(T-T_g)}{51.6+(T-T_g)} \tag{3.63}$$

This is a theoretical equation that was derived from free volume theory. If extruding materials at lower than normal temperatures, the higher sensitivity of the viscosity to temperature is an issue that needs to be considered. The engineering-based viscosity equation developed by Adams and Campbell [18] has been shown to hold for all nominal processing temperatures, from within a few degrees of T_g [26, 27] to conventional extruder melt temperatures. The Adams-Campbell model limiting shear temperature dependence is:

$$\eta_0 = A\left(\frac{T_g}{T - T_g}\right)^B \tag{3.64}$$

where A and B are experimentally determined constants.

Polymer viscosity is also affected by the local pressure:

$$\eta(\dot{\gamma}, P) = \eta_R \exp\left(\beta(P - P_R)\right) \tag{3.65}$$

where η_R is the reference viscosity at the reference pressure P_R, and β is the log linear fitting function for the log viscosity versus pressure experimentally determined. Usually this factor relates to compressibility [4]. For most cases the pressures in single-screw extruders are less than 35 MPa and seldom are higher than 70 MPa. For pressures in this range, the pressure effect for the shear viscosity can be ignored.

3.8 Historical Models for Non-Newtonian Flow

Performing numerical simulations of the extrusion process requires that the shear viscosity be available as a function of shear rate and temperature over the operating conditions of the process. Many models have been developed, and the best model for a particular application will depend on the rheological response of the resin and the operating conditions of the process. In other words, the model must provide an acceptable viscosity for the shear rates and temperatures of the process. The simple models presented here include the power law, Cross, and Carreau models. An excellent description of a broad range of models was presented previously by Tadmor and Gogos [4].

The power law viscosity model was developed by Ostwald [28] and de Waele [29]. The model has been used in the previous sections of this chapter and it has the form shown in Equation 3.66. The model works well for resins and processes where the shear rate range of interest is in the shear-thinning domain and the $\log(\eta)$ is linear with the $\log(\dot{\gamma})$. Standard linear regression analysis is often used to relate the log of the shear rate to the log of the viscosity for these power law fluids. The model can be expanded to include temperature dependency as shown by Equation 3.67.

$$\eta(\dot{\gamma}) = K\left|\dot{\gamma}\right|^{n-1} \tag{3.66}$$

$$\eta(\dot{\gamma},T) = \eta_R \left[e^{\frac{\Delta E}{R_g}\left(\frac{1}{T}-\frac{1}{T_R}\right)} \right] \left|\dot{\gamma}\right|^{n-1} \tag{3.67}$$

The disadvantage of the power law model is that it cannot predict the viscosity in the zero-shear viscosity plateau. When the zero-shear viscosity plateau is included, a nonlinear model must be specified with additional fitting parameters. A convenient model that includes the zero-shear viscosity and utilizes an additional parameter is the Cross model [30]:

$$\eta(\dot{\gamma},T) = \left[\frac{\eta_0(T)}{1+\left[\frac{\eta_0(T)\dot{\gamma}}{\tau^*}\right]^{n-1}} \right] \tag{3.68}$$

$$\eta_0(T) = \eta_0(T_R)\left[e^{\frac{\Delta E}{R_g}\left[\frac{1}{T}-\frac{1}{T_R}\right]} \right] \tag{3.69}$$

where n is the power law index, η_0 is the Newtonian viscosity at temperature T in Kelvin, T_R is the reference temperature in Kelvin, τ^* is the critical stress where η transitions from η_0 to the power law regime, ΔE is the activation energy for the temperature response, and R_g is the ideal gas constant.

The Carreau [31] model extends the shear rate range beyond the power law regime and into the infinite shear rate range. For polymer melts, operating at shear rates into the upper Newtonian (η_∞) region is difficult due to high levels of energy dissipation, high temperatures, and resin degradation. The model, however, allows the viscosity response at very high shear rates to deviate from a power law model. The Carreau model is expressed as follows:

$$\frac{\eta(\dot{\gamma})-\eta_\infty}{\eta_0-\eta_\infty} = \frac{1}{\left[1+(\lambda\dot{\gamma})^2\right]^{\frac{(n-1)}{2}}} \tag{3.70}$$

where n is a fitting parameter and λ is a characteristic relaxation time.

In many extrusion simulation calculations it is often adequate to approximate the polymer viscosity data using two straight-line functions. One line describes the power law region, and the other line is for the zero-shear viscosity. In the simulation, check both functions during the viscosity calculation of the simulation, where the viscosity is introduced in the program. The proper value is the smaller of the two viscosity values, Newtonian or power law, at any particular shear rate.

3.9 Power Law and Viscosity Shear Rate Dependence

3.9.1 Shear Stress from Newtonian to Infinite Shear

At this point in the discussion of the effect of rheology, the concept of viscosity and resistance to flow of polymers as a function of shear rate will be discussed because there is often a misconception regarding the dissipation in an extruder at high shear rates when the viscosity is in the power law region. As previously discussed, the viscosity decreases as a function of increasing shear rate as shown in Figure 3.23. Often this reduction in viscosity is misinterpreted as a reduction in the amount of power needed when the polymer is sheared at high rates. For an extruder, the misinterpretation would be that less motor power would be required to operate the machine at higher screw speeds. It is shear stress, however, that is the important parameter in the analysis of fluid flow resistance and energy dissipation, and it is apparent from Figure 3.33 that it does not decrease as the shear rate increases. Instead, the shear stress is always increasing in all regions. Thus, in the shear-thinning pseudoplastic region, dissipation will continue to increase. Recall that energy dissipation is modeled as $\dot{Q} = \tau\dot{\gamma}$, and since the shear stress increases with shear rate, energy dissipation will also increase. Even in the shear-thinning range the faster the extruder screw is rotated, the greater the energy required by the motor. The region where the shear stress increases at a non-linear rate with increasing shear rate is defined as the power law region Figure 3.33. With shear rates both below and above this region the polymer reacts in a Newtonian manner as a function of shear rate.

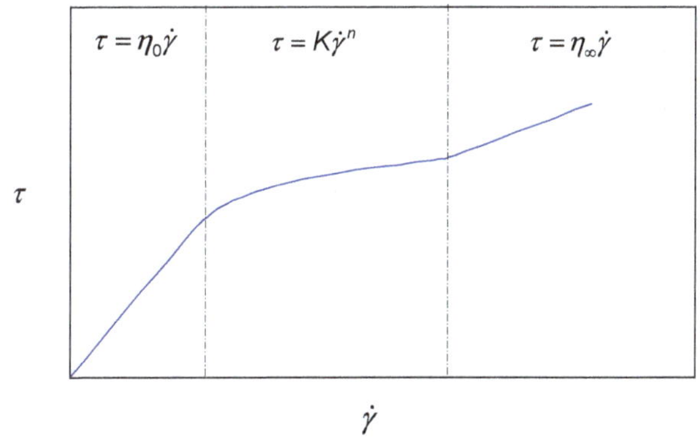

Figure 3.33 Shear stress as a function of shear rate for a typical polymer. The polymer has a zero-shear Newtonian plateau, a power law intermediate region, and an infinite-shear Newtonian plateau

3.9.2 Viscosity as a Function of Shear Rate

The general relationship for the viscosity as a function of shear rate in shear thinning slurries is illustrated in Equation 3.36. At very low shear rates, the polymer viscosity acts in a Newtonian manner being independent on the shear rate. The viscosity often decreases as the shear rate is increased with the viscosity transitioning into the fully developed power law region. The polymer dispersion at high shear rates can become Newtonian again but at a much lower value: the "infinite" shear viscosity. The power-law region found in Figure 3.34 is indicative of this rheological response where the viscosity decreases as the shear rate increases. For any given high molecular weight polymer of a solid particle dispersion system, the power-law constant n is usually found to be consistent regardless of the rheometer used: capillary, cone and plate, bob and cup, or parallel plate. For simple, one-velocity component flow, the viscosity-power law relationship is written as follows:

$$\eta(\gamma) = k\left|\dot{\gamma}\right|^{(n-1)} \tag{3.71}$$

where η is the viscosity, $\dot{\gamma}$ is the one component shear rate, and n is the power-law constant. The $|\dot{\gamma}|$ bracket is used because viscosity is positive definite. If $n = 1$, then the consistency index k is the Newtonian viscosity.

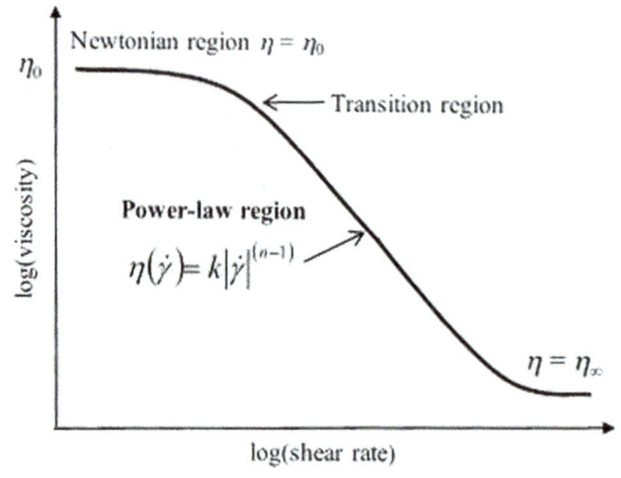

Figure 3.34 Typical shear thinning viscosity versus shear rate response with the power law region

3.9.3 The Power Law and Process Dissipation

The classical power law model was empirically based [28, 29], unlike the model developed here that is based on thermodynamic principles.

The viscosity and power law of filled or unfilled polymers are important issues relating to the extrusion of polymers. The power law has historically been empirically defined to depend on polymer molecular weight and the addition of fillers to the base polymer. There are many industrially important applications that incorporate particle filled polymeric dispersions. A few examples are sound mitigation systems in vehicles and structures, thermal interface materials in electronics, composites used in 3D printing that are often based on carbon nano-tubes, and rubber fillers used in HIPS and ABS resins. Recently a percolation-theory-based mechanism for power law shear thinning was developed for mono-dispersed Newtonian carrier-based slurries by Campbell and Radhakrishnan [33], and Campbell, Zakand, and Wetzel [34] which can be applicable to filled thermoset polyurethanes, polyester-styrene sheet molding resins, or filled epoxies. The analysis was extended to polymeric systems [35–38].

3.9.4 Viscosity, Shear Rate, and Dissipation

Hypothetical model power law viscosities and their viscous dissipation data, typical of what is seen as polymer molecular weight increases or with the addition of fillers above the percolation concentration, are plotted in Figure 3.35 and Figure 3.36. These relationships ultimately led to the definition of the power law constant n as the ratio of energy dissipation of a power law fluid relative to the dissipation of a Newtonian fluid being sheared in an equivalent manner.

It was desired to develop a theory for practitioners working with pure polymers or multiphase systems that could lead to an understanding of what the power law constant physically represents. An idealized set of viscosity relationships as a function of both shear rate and power law constant is shown in Figure 3.35. In each case the power law relationship is initiated at a shear rate of 100 1/s. The power law constant ranges from 1, which yields a Newtonian shear rate independent viscosity, to 0.01, which is very close to a perfect pseudoplastic fluid with $n = 0$. From Figure 3.35, it is observed that as n decreases toward 0, the rate of decrease of the viscosity with shear rate increases.

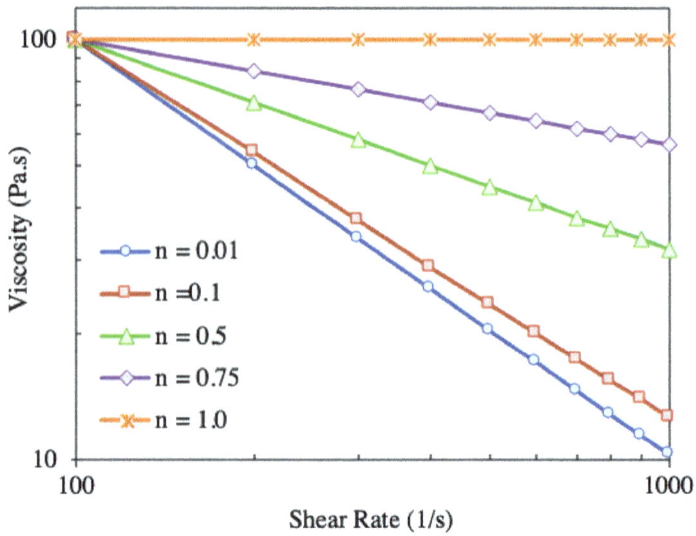

Figure 3.35 Idealized model viscosity as a function of power law *n*. Symbols are simulated data

In many industrial processes, such as extrusion, blow molding, and injection molding, it is the shear rate due to the fluid velocity and shear stress which combine to dictate the power required to make the process operable.

The shear stress for a power law fluid is defined as:

$$\tau(\gamma) = k|\dot{\gamma}|^{(n-1)} \quad \gamma = k|\dot{\gamma}|^{(n)} \tag{3.72}$$

In shear rate dominated processes it is the change in viscous dissipation with shear rate that determines the power needed to increase the production rate, i.e., material rate as a function of process time. The energy dissipation $\dot{\omega}$ is defined as the product of the shear stress and shear rate:

$$\dot{\omega} = \tau(\dot{\gamma})\dot{\gamma} \tag{3.73}$$

Analysis of the data in Figure 3.35 with an understanding of the importance of viscous dissipation in industrial processes led to comparing the dissipation rate at differing power law constant fluids to the dissipation rate of a Newtonian fluid at the same shear rate. This can be expressed as:

$$\frac{\dot{\omega}(n)}{\dot{\omega}(n=1)} = \frac{\tau(\dot{\gamma})_n \dot{\gamma}}{\tau(\dot{\gamma})_{n=1} \dot{\gamma}} \tag{3.74}$$

Figure 3.36 represents this concept for the hypothetical materials shown in Figure 3.35. The model data plotted in Figure 3.36 were calculated at 1000 1/s in the steady state power law region. The stress function has little change with shear rate for $n \leq 0.5$. A linear fit of the calculated dissipation ratio has a good correlation with the power law n. As the power law decreases, the dissipation ratio decreases at essentially the same rate.

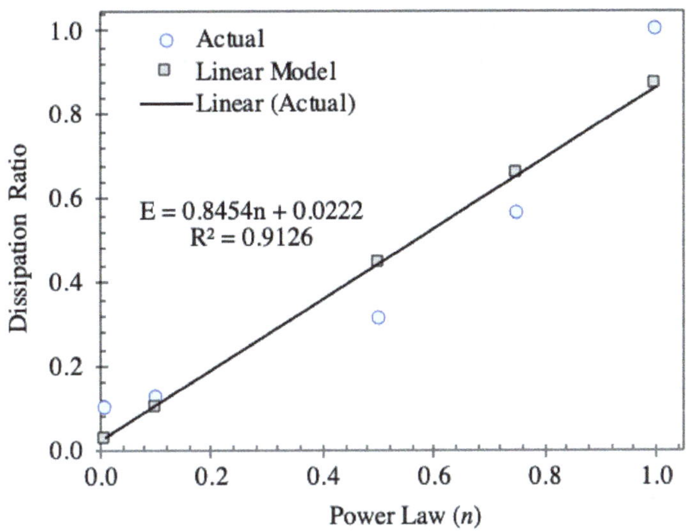

Figure 3.36 The dissipation ratio - power law divided by Newtonian viscosity at 1000 1/s

To relate the hypothetical materials described in Figure 3.35 and Figure 3.36 to tube flow experiments, the theoretical velocity characteristics of these materials were calculated based on tube flow. For all of these power law fluids the velocity approaches a constant value as a function of radius as the radius approaches the tube center.

In order for the fluid velocity to be "constant" at the center of the tube, it is thought that for high polymers and dispersions a structure is formed in this annulus of flow consistent with the predictions of percolation theory. For a power law fluid the velocity is of course not really a constant; however, this is a reasonable assumption so that the appropriate dissipation rates can be approximated. This observation was instrumental when selecting the Bingham model relating n for the flowing fluids to Newtonian to power law energy ratio, as developed below. The steady shear viscosity for both unfilled and filled polymers is also related to percolation theory in the analysis below.

3.9.5 Percolation in Structured Systems

Percolation theory is usually based on a site-bond analysis. For site-percolation, the critical percolation concentration is a function of the lattice structure [39]. As presented in [39], a simple cubic packing has a percolation probability of 0.31 and body centered cubic has a percolation probability of 0.24. For a random 3D filling distribution of spherical particles, the percolation concentration is in the range 0.154 to 0.186 volume fraction for randomly packed spheres [39, 40]. Percolation characteristics as a function of lattice coordination number are provided in Table 3.9 [39].

Table 3.9 Percolation Characteristics as a Function of Lattice Coordination Number [39]

Packing Structure	Coord. Number	Filling Factor	Critical Percolation Probability	Critical Percolation Density
fcc, hcp	12	0.7405	0.195	0.144
bcc	8	0.6802	0.24	0.163
sc	6	0.5236	0.31	0.162
diamond	4	0.3401	0.46	0.146

3.9.6 Tube Flow Data and Data Analysis

When filler is added to a Newtonian fluid the viscosity becomes a power law function when the filler exceeds the critical percolation concentration, and the rheological response seen in Figure 3.37 is typical.

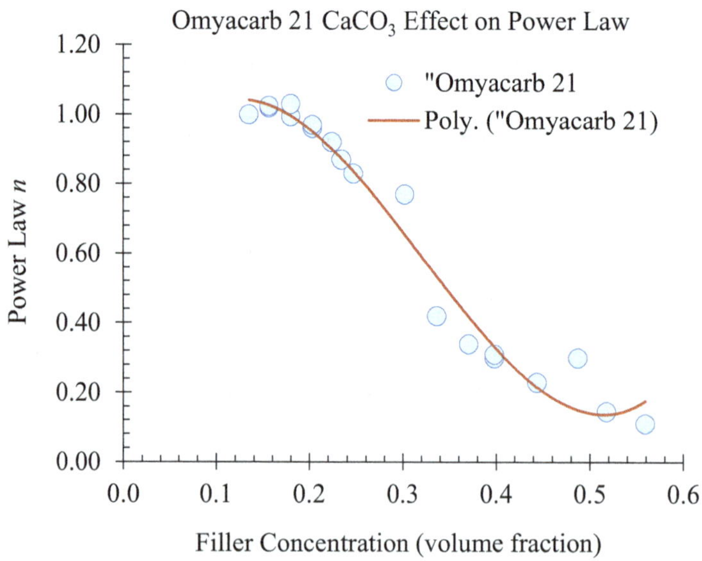

Figure 3.37 Effect of filler concentration on power law

This type of power law response has a different shape when analyzed using percolation concepts. Percolating material functions are often represented as power functions of the independent variable. The relationship between the power law and dissipation for PMMA slurries flowing in a tube was first reported by [33]. Prior to that, the percolation function derived by [41] fits the data in the power law region for calcium carbonate slurries (Equation 3.75). This same function also fits the PMMA slurries as a function of filler concentrations:

$$n = \frac{\left(\dfrac{\phi}{\phi_m}\right)^{-2/3} - 1}{\left(\dfrac{\phi_c}{\phi_m}\right)^{-2/3} - 1} \tag{3.75}$$

For this percolating system, the exponent relating the filler volume fraction divided by the maximum packing fraction was found to be −2/3; ϕ is the filler concentration, ϕ_m is the maximum concentration, ϕ_c is the percolation onset concentration, and n is the power law. The utility of this equation is demonstrated for two sets of PMMA dispersion data sets in Figure 3.38. These data will be utilized when determining the utility of the model for the power law based on the dissipation ratio, utilizing the concept demonstrated in Figure 3.36.

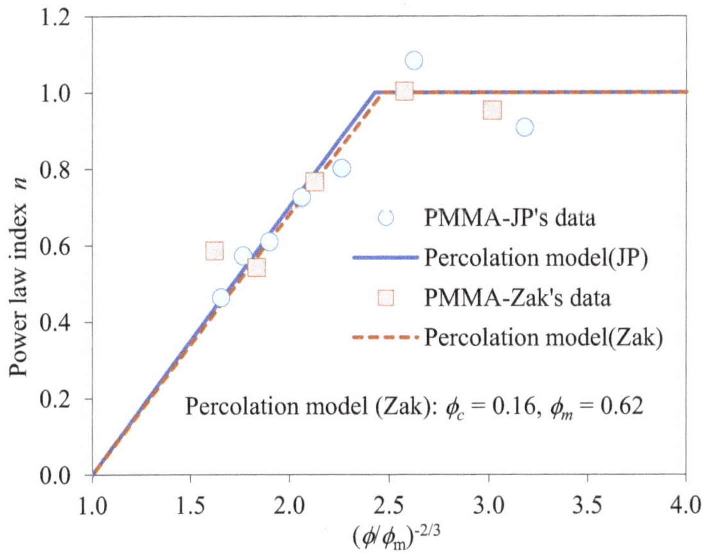

Figure 3.38 Power law data from pressure drop for PMMA slurries [41, 42]

3.9.7 Dispersion Based Power Law Constant *n*

The Clarkson research group developed an experimental tube flow system that provided the power law using pressure drop and the velocity profile of the flowing dispersion simultaneously. In tube flow, it was proposed to use a Bingham model of the stress tensor to represent the flow field for modeling purposes [33, 41, 42]. This modeling approach was undertaken to determine if the observation that the dissipation ratio of the slurries was well correlated to the power law as depicted in Figure 3.36. By using this Bingham model assumption, the central core is assumed to have no velocity gradient and therefore no dissipation.

Starting with the concept proposed by Brenner [43] and using the ratio described above, the power law constant n is defined using Equation 3.76:

$$n=\frac{\int_0^{V-V_c}\tau_{ij}d_{ij}\,dV}{\int_0^{V}\tau_{ij}d_{ij}\,dV} \quad \text{or for tube flow} \quad n=\frac{\int_{R_c}^{R}\tau_{ij}d_{ij}\,rdr}{\int_0^{R}\tau_{ij}d_{ij}\,rdr} \tag{3.76}$$

This hypothesized, generalized model predicts $n = 1$ when the volume of the non-deformable cluster $V_c = 0$, which is Newtonian flow. It also predicts that n will approach zero in the limit when V_c reaches V, which is perfect pseudoplastic flow. This definition would be applicable to all shear fields, leading to the assumption that the power law should be the same regardless of the rheometer type used [33].

For tube flow, using the Bingham viscosity model to evaluate the integrals between R_c, the core radius, and the tube radius, R, and cancelling like terms, the expression for the power law index can be determined [33]. In the case of tube flow the power law n is only a function of the size of the non-dissipating plug which is considered to be the percolating infinite cluster (Equation 3.77):

$$n=1-\frac{4}{3}\varepsilon+\frac{1}{3}\varepsilon^4 \tag{3.77}$$

Defining the volume fraction of the fluid in the infinite cluster(s) as Equation 3.78,

$$v=\varepsilon^2 \tag{3.78}$$

and substituting in Equation 3.77 results in Equation 3.79:

$$n=1-\frac{4}{3}\sqrt{v}+\frac{1}{3}v^2 \tag{3.79}$$

This volume-based relationship, where v is the non-dissipating volume, should be valid for any flow field. This leads to a theoretical expression for the power law n based on the ratio of the dissipation in the composite to that in the equivalent Newtonian fluid.

3.9.7.1 Testing the Theory Using Newtonian Carrier Fluid Slurries

As with all new theories, this one needed to be tested against real data. In this case the predicated power law from the flow velocity fields using Equation 3.77 is compared to the conventionally determined power law constant from pressure drop data. Figure 3.39 is a parity plot of the PMMA mono-disperse slurry power law from the Bingham rheology-based theory and the conventional pressure drop measurements [41, 42]. For both data sets, the slope is essentially 1 and the intercept is essentially 0. There is an excellent correlation between the new Bingham model-based power law constant theory, which is based on observed velocity profiles in the tube, and the experimental measurement of the tube flow power law using pressure drop. It is concluded that the power law constant n effectively relates with the dissipation ratio from Equation 3.77 through Equation 3.79 using a Bingham rheological model to represent the central non-dissipating cluster and deformation in the slurry fluid in the annulus of the tube. It has been recently demonstrated that the constants for the Bingham model can be used to develop the velocity function for the power law model [58].

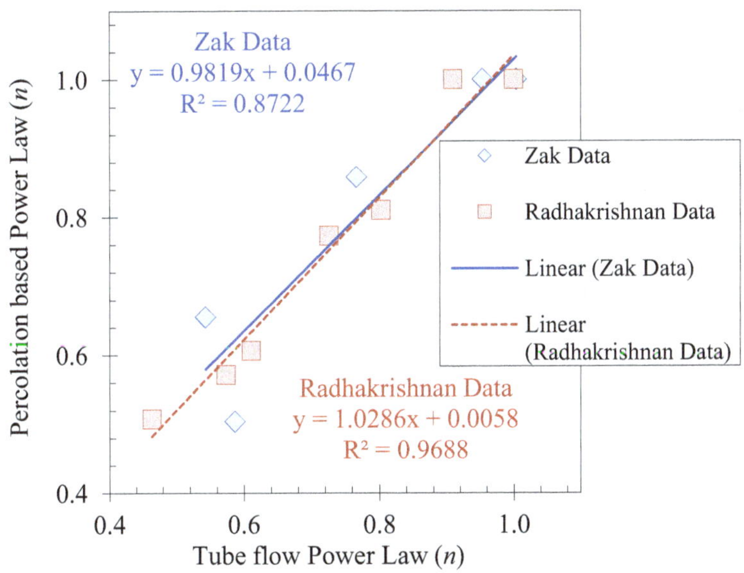

Figure 3.39 Parity plot of predicted power law versus experimental power law

The power law from the cluster volume based power law theory using Equation 3.79 was then tested as a function of filler cluster concentration [33] for the PMMA slurry data discussed previously. Also, data from Karnis et al. [44, 45] and Altobelli et al. [46] was converted into the form needed to evaluate the utility of Equation 3.79 [41]. It is apparent from the fit in Figure 3.40 that the power law theory n calculated from the plug size and velocity profiles reported in the literature data agree over a wide range of plug sizes, corresponding to the range of filler concentrations.

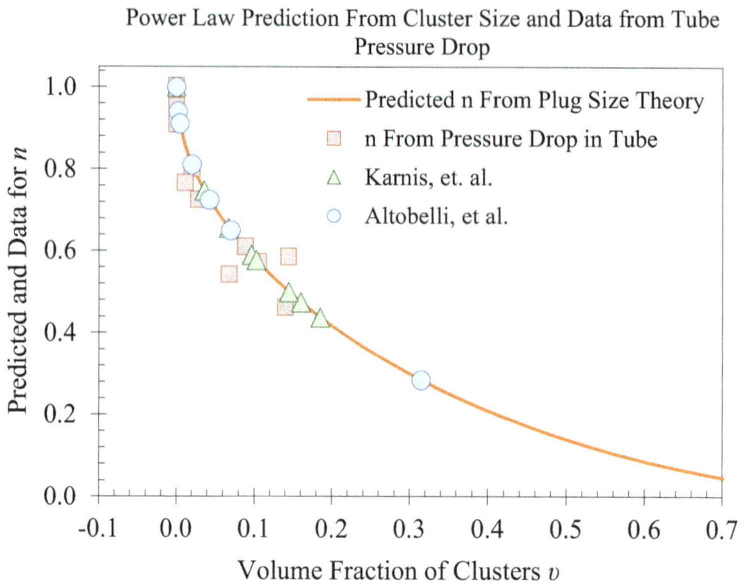

Figure 3.40 Comparison of the three sources of experimental data fit to the power law theory line with *n* from pressure drop in tube [33]

It has been demonstrated that the power law *n* for rigid particle-filled Newtonian carrier slurries, represents a measure of energy needed to overcome the viscous flow as a function of increased process shear rate. The increased process shear rate would of course lead to more production in a fixed amount of time. It has also been demonstrated that as the power law *n* decreases the dispersion has a non-linear increase in the fraction of the dispersion that is in non-dissipation clusters. In the next section data on unfilled and filled polymers will be presented and analyzed to determine if these systems also have responses that are similar to those for the power law theory discussed throughout Section 3.9.

3.9.7.2 Application of the Theory to Polymer Melts

The experimental and model prediction results for filled power law Newtonian fluid-based composites led to an investigation to determine if the proposed dissipation mechanism could be used to interpret the development of the power law for PE particulate composites. Four PE resins, one LDPE and three LLDPE resins, and three fillers, one pigmentary TiO_2 and two $CaCO_3$ particulate solids, were used in the polymer melt processing experiments. The initial polymer-based evaluation of a filled polymer melt system was previously reported and it focused on a TiO_2 filled LDPE [35, 38]. The investigation was then expanded to other PE materials and $CaCO_3$ fillers [36, 37]. Some typical characteristics of the polymeric materials and particulate fillers used in these investigations are listed in Table 3.10, Table 3.11, and Table 3.12.

Table 3.10 Polyethylene Polymers

Polymer	ρ (g/cc)	MI (g/10 min) (190 °C, 2.16 kg)	T_{MP} (°C)	Power Law n (>100 s^{-1})
Equistar LDPE Petrothene NA-206	0.018	13.5	~110	0.341
Dow LLDPE DNDA-1082	0.933	160.0	125	0.700
Dow GRSN-9820 NT-7	0.924	20.0	123	0.598
Dow GRSN-2070 NT	0.920	0.7	121	0.393

Table 3.11 Fillers Used with the Polyethylene Polymers

Filler	ρ (g/cc)	d_{50} (μm)	Surf. Mod.	Dc/*a*
Chemours Ti-Pure R-104 TiO_2	3.900	0.20 to 0.25	yes	4000.0
OMYACARB 40-UL $CaCO_3$	2.700	45.00	no	22.2
OMYACARB 5-FL $CaCO_3$	2.700	5.00	yes	200.0

These PE materials had a wide molecular weight range as indicated by the melt index, Table 3.10, and confirmed for three of the resins using GPC as listed in Table 3.12.

Table 3.12 Molecular Characteristics of Three of the LLDPE Polymers: Apparent Molecular Weight Averages from GPC

PS Equivalent MW Averages (kg/mol)					
Polymer	M_n	M_w	M_z	M_w/M_n	**Peak Area/Conc.**
DNDA-1082	14.40	84.80	335.0	5.89	7298
GRSN-9820 NT-7	19.60	346.00	2348.0	17.65	6609
GRSN-2070 NT	28.10	152.00	487.0	5.41	8112
PS Equivalent MW Averages (kg/mol), Converted from PS Equivalent MW, M(PE) = A*M(PS), and A ~ 0.43					
Polymer	M_n	M_w	M_z	M_w/M_n	MI
DNDA-1082	6.19	36.46	144.1	5.89	160.0
GRSN-9820 NT-7	8.43	148.78	1009.6	17.65	20.0
GRSN-2070 NT	12.08	65.36	209.4	5.41	0.7

Caution with reported data: GPC condition may not be optimal for these samples. However, relative comparison should still be valid.

Equistar Petrothene LDPE NA-206 resin was characterized regarding its viscosity as a function of shear rate at three temperatures. The Cross-WLF constitutive equations were used to fit the data and are plotted in Figure 3.41. Using the last eight data points, with shear rates from 223 to 4999 1/s, the power law index n was estimated to be 0.361 for this resin.

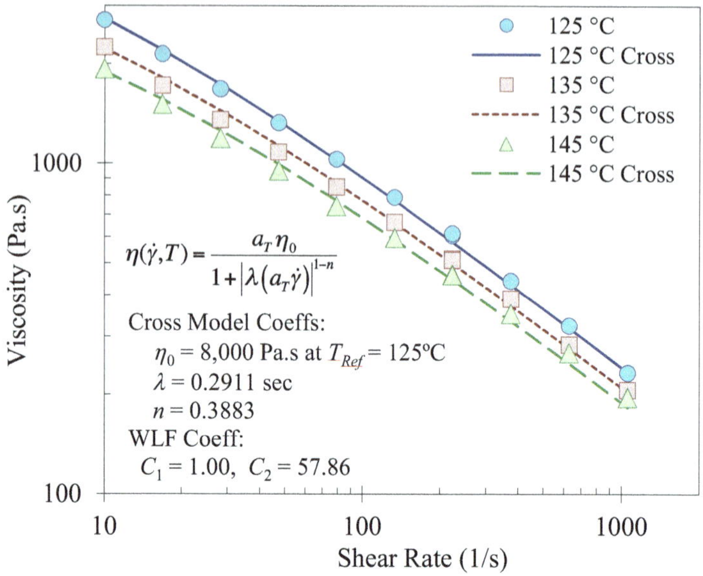

Figure 3.41 Melt viscosity data and Cross and WLF models versus shear rate for LDPE from 125 °C (389 °K) to 145 °C

The Petrothene NA-206 base resin was then filled with TiO_2 from 0.0265 to 0.495 volume fraction. Examining Figure 3.42, the shear thinning response and power law n decrease with increased filler concentration was dramatically different from that observed in Figure 3.37. For the LDPE case there was almost no change in the power law as the TiO_2 was added. The power law rapidly decreased as filler was added to the Newtonian carrier above the percolation concentration. The relatively constant power law for this TiO_2 filled PE from the lowest concentration of 0.0265 volume fraction to the highest concentration of 0.495 suggests that the increase in the resistance to cluster deformation seen in the slurries discussed in the previous sections resulting from the increase in solid particle concentration is in some manner offset by what occurs in the polymeric resin. A similar response was observed when calcium carbonate was added to the three LLDPE resins as shown in Figure 3.43.

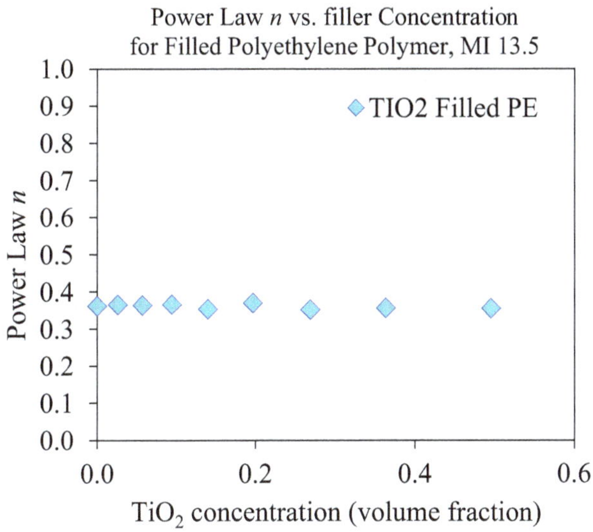

Figure 3.42 Power law as function of filler concentration for a Newtonian carrier LDPE

When the filler was added to a Newtonian carrier, the power law decreased rapidly in a sigmoid shaped function. The power law was found to be almost constant when the raw data was examined for the TiO_2 in LDPE. A speculative proposal was previously published for TiO_2 filled PE [35] that may help to explain the unexpected observation of the effect of filler on the shear thinning characteristics of high molecular weight polymers. It is well known that at or below the critical molecular weight M_c, PE has essentially no shear thinning behavior.

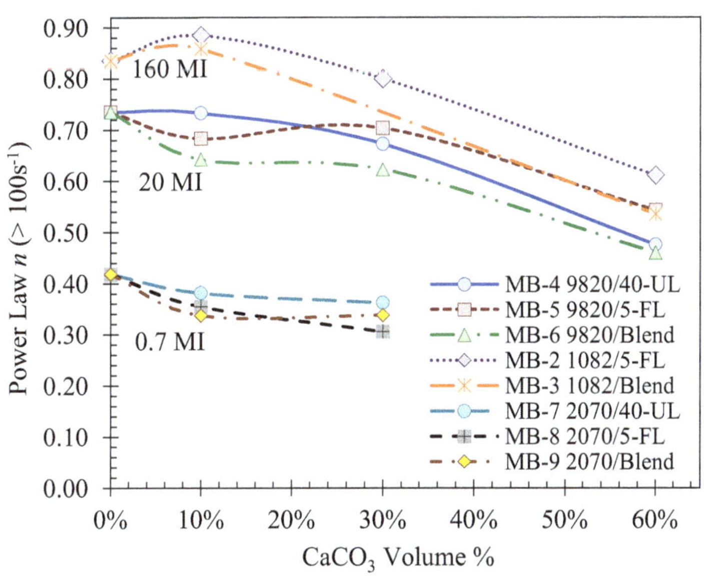

Figure 3.43 Effect of $CaCO_3$ filler on power law n for three LLDPE resins

It is speculated that under shear flow, polymer chains are thought to disentangle from one another [35]. It was further speculated that the low molecular fraction of the polymer, that fraction with molecular weight below M_c, might be considered to be a theta solvent at all temperatures for the higher molecular weight polymer molecules. As the molecular weight increases, the fraction of this solvent decreases such that the large molecules, ~10 nm, become insoluble or swollen since they have collapsed into a Gaussian sphere per the literature. The large molecules are proposed here to act in the molten polymer as deformable "fillers" with the critical percolation concentration reached at or near M_c. This hypothesis could be considered to be an extension of the rheological model developed by Williams and co-investigators [47–49].

When the molten polymer is sheared, the large molecules were proposed to decrease the volumetric dissipation in a similar manner that was observed for the Newtonian fluid slurries resulting in the power law response [36]. That is, from a percolation mechanism the large molecules interact as infinite clusters.

If these concepts are to be potentially applicable to other polymer systems, it is necessary to determine if the new proposal effectively relates the power law n with a filler-like response of the high molecular weight polymer molecules applies to other systems. That is, that polymers above M_c act as an effective filled system in a way that correlates with the Newtonian fluid-based slurry systems previously discussed. Several investigators have developed relationships between molecular properties of polymers and the power law [18, 20, 21, 50, 51]. A set of PS resin power law and molecular weight data was available [18, 51] in addition to the power law n data for the PE resins that were obtained from the as-received resins at Dupont [37, 38]. Table 3.13 lists the power law data for four PS materials with different molecular weights.

Table 3.13 Power Law Estimates for a Series of PS Resins [18], and Estimated Concentration of Polymer Filler

n	Concentration Volume Fraction	Molecular Weight
0.59	0.27	48000
0.39	0.35	117000
0.23	0.42	179000
0.19	0.43	217000

These data and the PE data for the four unfilled polymers evaluated in this work are plotted in Figure 3.44. The polymeric filler concentration for all eight polymers was determined using a filler versus power law function [36]. For this analysis using the concept of "polymeric filler concentration" it was assumed that these materials were structured in a similar manner to that developed by Williams and co-workers [47–49]. The percolation concentration function, Equation 3.75, was used in this case. All data for both the PE and PS resins fall along the same curve. This is an unexpectedly good correlation for the combined results for all eight polymers with two completely different molecular structures.

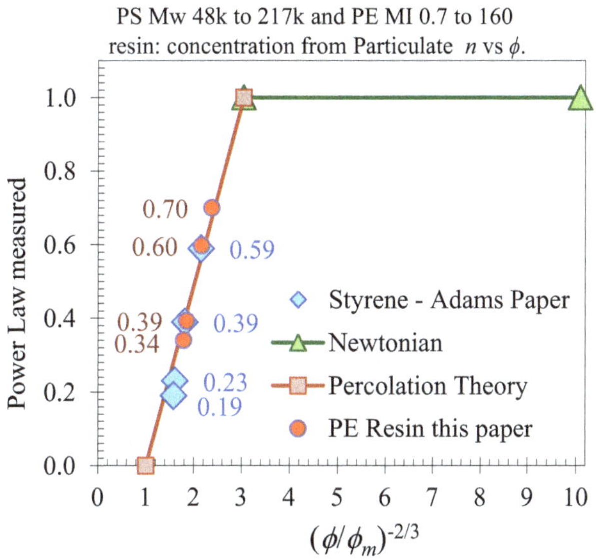

Figure 3.44 Percolation correlation for the power law for both PS and PE polymer melts

The remainder of this section is focused on the effect of fillers on viscosity, the power law, and the onset of the second Newtonian plateau for filled polymer systems, which have ramifications for the processing of filled systems.

The data from both polymer sets fall on the percolation prediction developed for Newtonian carrier fluid slurries as a function of polymeric melt filler concentration. It is interesting that these experimentally obtained polymer melt power laws, n, can be related to the effects observed for shear thinning slurries based on Newtonian liquids. This unexpected observation suggests that the underlying processes that relate to power law behavior may be more general than initially thought and should be further investigated.

It was further suggested that as a particulate solid, nano-size TiO_2, is added to the melt, it displaces some of the deformable filler, and because it has a nanomolecular size, the polymeric "filler" can fill the gaps between the solid filler or filler clusters. This could cause a different response from the Newtonian carrier composites as shown in Figure 3.38. Thus, the power law does not decrease as rapidly as it does when solid filler is added to a Newtonian fluid.

For the LDPE polymer system shown in Figure 3.42, the dissipation of the base polymer and of the TiO_2 slurry at low filler concentrations is thought to be dominated by the proposed cluster formed by the large polymer molecules in the base resin. The LDPE polymer/TiO_2 system was found to have an effective volume fraction of 0.375 for the "polymeric filler" leading to percolating clusters. Consistent with this assumption, as the TiO_2 filler was added, the components of the slurry, polymer fluid, polymer "filler", and TiO_2 solid did not appreciably increase the volume of the percolating clusters. The detailed discussion of this proposal can be found elsewhere [36].

Using the estimated "total solids" concentration and the measured neat LDPE power-law exponent n, the TiO_2 filled polymer values for n show a remarkable fit to the percolation power law model as a function of concentration, as shown by Figure 3.45 [36]. The unfilled polymer has a "filler concentration" consistent with the 3D percolation correlation. As filler is added, the data indicate that the percolation concentration may be higher than the 3D percolation. The power law n does not decrease substantially as the filler loading is increased to 49.5 vol. %. It has been reported that for 2D percolation, the percolation concentration can lie between 42 and 50 vol. % for "spherical" fillers. Continuing to address these data as percolation phenomena, as TiO_2 is added to the system, the percolating structure appears to transition from 3D to 2D percolation. For this nano-size filler two important observations stand out: 1) unlike the Newtonian system, the power law is only a weak function of the filler content, and 2) the flow field may transition from 3D to 2D as the filler is added. These intriguing results for TiO_2 in a PE resin led to further research being conducted on three other PE polymeric fluids and two $CaCO_3$ grades listed in Table 3.10, Table 3.11, and Table 3.12.

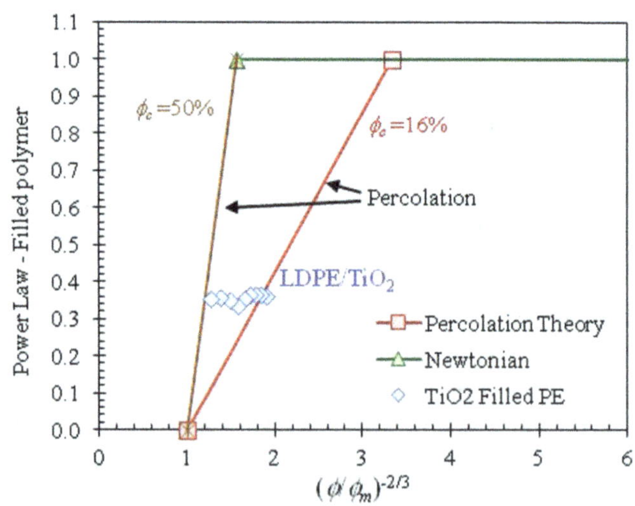

Figure 3.45 LDPE/TiO_2 blends power law n plotted with percolation theory and Newtonian fluids

Given the background information and concepts detailed above, three new LLDPE resins with varying molecular weights were compounded with two grades of $CaCO_3$ which were not nano-size and thus much larger than the base polymer molecules. The analysis is below.

For the compounded PE/filler dispersions, it is important to evaluate the effect of the compounding, residence time, and shear in the twin-screw extruder on the estimated power law constant. It was found that in all cases, *n* increased as a result of the compounding experience. This is consistent with an expectation of reduced molecular weight of the PE as a result of the time, melt temperature, and stress histories. The $CaCO_3$ dispersion data were analyzed in the same manner as was developed and utilized in Figure 3.45. As with TiO_2 in LDPE, the effect of $CaCO_3$ concentration on the change in *n* from 0 to 60 vol. % was not substantial as shown in Figure 3.43. Overall, *n* decreased for all systems with increasing $CaCO_3$ loading, and the maximum change in *n* was only about 0.2 units over the entire filler concentration range. This is much less than that observed for Newtonian fluid-based dispersions per Figure 3.37.

When evaluating polyethylene viscosity characters it is important to be aware of the polymer tendency to stick-slip on the rheometer or die walls as the shear rate increases. The melt viscosity of the LLDPE resins, the $CaCO_3$ masterbatches, and batch mixer letdown samples were measured in accordance with ASTM D 3835 using a Dynisco LCR 7001 capillary rheometer [52]. The viscosity of the three LLDPE resins were tested at 170, 190, and 210 °C and at shear rates from at least 250 down to 1 1/s using a 1.0 mm diameter die with a length to diameter ratio *L*/*D* = 30 and a 180° inlet angle. From end effect analysis, this would potentially produce a 10% maximum error in the viscosity. However, there would be essentially a consistent error for all of the shear rates. Thus, the error should not substantially affect the power law slope calculations. Duplicate tests with fresh samples were run at all temperatures. The test-to-test repeatability, expressed in the terms of the repeatability coefficient of variation, was less than 3% at all shear rates. Since only one capillary diameter was used, the wall slip could not be evaluated in a traditional manner per the analysis of Mooney [53] and Jastrzebski [54]. It has been proposed that the onset of wall slip may occur at wall shear stress values of 0.2–0.3 MPa [55]. The onset of wall slip is also indicated by a discontinuity or substantial change in the shear rate versus shear stress slope [35, 37, 57]. In general, the wall shear stress was less than 0.3 MPa, for many of the samples evaluated in this chapter. No shear stress-shear rate deviation from a smooth curve was noted in any of these samples. Also, there were no apparent surface defects in the rheometer extrudates. This does not preclude the existence of some minor slip at the higher shear rates.

Another issue is the migration of fillers in shear flow. For these systems this issue was extensively reviewed by Campbell and Wetzel [35, 37} and it was determined that this was not a significant issue for this analysis.

At very low shear rates, before the onset of the power law response, the neat polymer and the composites act in a Newtonian manner, as in Figure 3.30. It was previously proposed that at low shear rates, particles in percolating clusters have time to rearrange, allowing the cluster to deform as a result of the imposed flow field [57]. The particles rearrange themselves by moving into existing holes within a cluster or past neighbors. This allows for stress to be transferred throughout the suspension and for the fluid to perform work on all of the particle surfaces as well as the container surface, thus yielding a Newtonian viscosity.

Percolation theory was used to develop the following equation [57] to predict the effect of filler concentration on the slurry Newtonian viscosity:

$$\eta_r = e^{\frac{\phi_m - \phi_c}{\phi_m - \phi}} - 1 \tag{3.80}$$

where η_r is the Newtonian viscosity ratio, filled vs. fluid, φ_m is the maximum packing concentration, φ_c is the percolation concentration, and φ is the particulate solid concentration. Equation 3.80 has been previously shown to fit data for slurries used to determine flow dynamics of slurries in tube flow [33].

The shear stress, viscosity, and estimates of the yield stress for the highest filler concentrations in two of the polymers used in this current investigation may be found in Table 3.14 at two shear rates.

The effect of the filler on the viscosity is very important when investigating shear dependent processes such as die extrusion and injection molding. A typical rheological response, as represented by the viscosity ratio, $\eta_{compound}/\eta_f$, for the three PE resins and two fillers is plotted in Figure 3.46. The percolation model of Equation 3.80 was used to fit for the two $CaCO_3$ grades and a 70% ~45 µm/30% ~5 µm blend pre-compounded in LLDPE as a masterbatch and then let down with the LLDPE in a batch mixer [52]. MB-7 was the response of the 0.7 MI resin and it showed a relatively small increase in viscosity as the filler was added and a marginal fit as indicated by the sum of squares error. A preliminary analysis of the viscosity data used in this power law and the zero to infinite shear viscosity analysis can be found elsewhere [33, 52]. The percolation theory model in general fit the low shear rate data of these filled polymers quite well.

Table 3.14 Rheological Property Estimates of Several cf the Base Polymers and Particle Masterbatches

MB Sample	MB1 LD1	MB2 LD7	MB3 LD7	MB2 LD7	MB3 LD7	MB4 LD1	MB5 LD7	MB4 LD7	MB6 LD7	MB5 LD7	MB4 LD7	MB6 LD7
LLDPE	1082 (160 MI)					9820 (20 MI)						
$CaCO_3$	–	5-FL	Blend	5-FL	Blend	–	5-FL	40-UL	Blend	5-FL	40-UL	Blend
Vol. %	0	60	60	60	60	0	60	60	60	60	60	60
$\dot{\gamma}_A$ (s^{-1})	75	75	75	20	20	75	75	75	75	20	20	20
η_A (Pa·s)	66.2	1559.8	1597.3	3100.6	3266.7	389.1	1376.2	3150.2	3606.7	2476.8	5599.9	7210.2
τ (Pa)	4965	116985	119798	62043	65367	29183	103215	236265	270503	49561	112054	144276
n'	0.776	0.540	0.503	0.408	0.379	0.656	0.560	0.561	0.482	0.555	0.553	0.452
$\dot{\gamma}_W$ (s^{-1})	80.4	90.9	93.5	27.3	28.2	84.8	89.7	89.7	95.2	24.0	24.1	26.1
τ_0 extrap. (Pa)	0.140	77.8	133.7	430.1	651.1	4.10	53.0	119.3	398.5	56.2	130.9	582.6
τ_0 model (Pa)	0.005	0.110	0.110	0.110	0.110	0.036	0.214	0.214	0.214	0.214	0.214	0.214

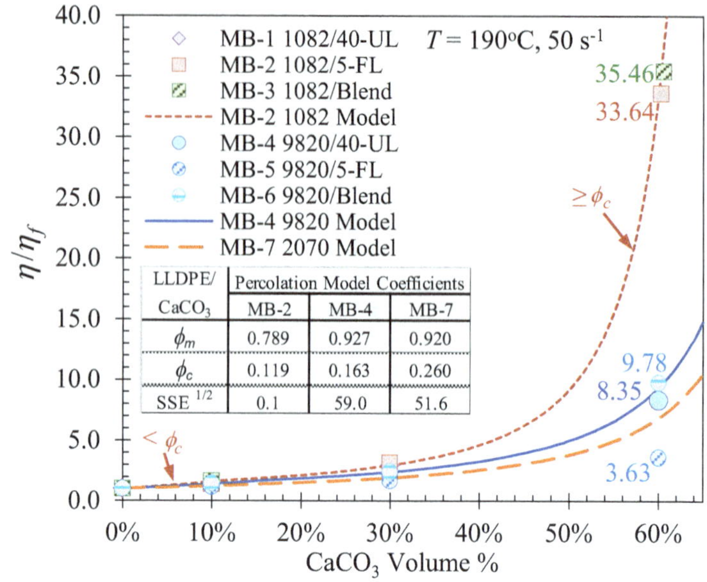

Figure 3.46 A relative viscosity plot of data and the percolation model of Equation 3.80 with φ_c and φ_m as fitting parameters

The effects of particle size and concentration were analyzed using a LLDPE resin with a MI of 20 dg/min, as shown in Figure 3.47 and Figure 3.48. The percolation concentration response of $CaCO_3$ of two different particle sizes in 20 dg/min MI LLDPE on *n* is shown in Figure 3.47. The percolation point is considered to be the onset of 3D percolation at 16 vol. % filler to the maximum concentration. The power law exponent for these pure resin systems initially fit a 3D percolation model. This is the same value that correlates the melt data for both PS and PE per Figure 3.44. The percolation point at 50% found in Figure 3.47 can be considered as the onset of two-dimensional percolation and is consistent with the upper concentration limit proposed in the literature for 2D percolation. The remaining data listed in Table 3.14 acted in a similar manner to Figure 3.47 with the data also transitions from 3D to 2D percolation [34, 36]. Previously in this chapter, it was observed that all of the melt data for the four PE resins, MI 0.7, 13.5, 20, and 160 dg/min, fall on the 3D percolation line per Figure 3.44. Given that the resin seams to exhibit an "apparent filler" concentration, the results suggest that one may wish to start looking at a high molecular weight polymer as acting like filled system with the polymer above the critical molecular weight acting as deformable filler [35].

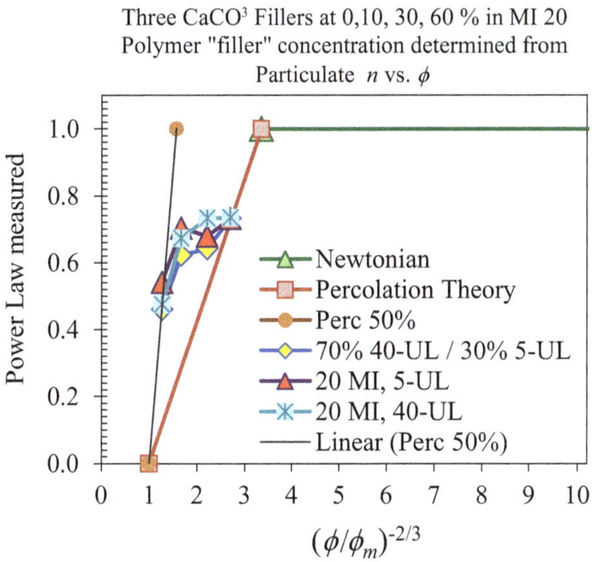

Figure 3.47 Effect of $CaCO_3$ filler in LLDPE on the composite power law

All of the filler systems, single size 5 and 45 µm and a 70/30% (45/5 µm) mixture, show a similar reduction in power law as a function of total concentration and show a strong tendency to migrate toward 2D percolation as the $CaCO_3$ concentration is increased. This was unexpected, particularly for the 70/30 mixed filler, as the 5 µm filler would normally have been expected to fill the voids between the larger 45 µm particles and thus the total concentration would have a smaller effect than would the single size particles alone as the total concentration was increased. This may suggest that the particles organize in macroscopic clusters in the rheometer and slide by one another with a resin lubrication layer between the cluster layers. This effect was not observed in the PMMA or $CaCO_3$ slurries discussed earlier where the suspending fluid was Newtonian. This unexpected observation for filled polymers may help explain some of the anomalies found when extruding highly filled resins, such as jetting in certain regions of an extrusion die or through the gate in an injection mold at lower than expected filler concentrations.

The change in power law as a function of cluster volume is plotted in Figure 3.48. The power law data for all three resins and the two fillers follows the power law function related to cluster volume per Equation 3.79, just as the other Newtonian fluid-based composites did in Figure 3.40. Even though the power law relationship appears to transition from 3D to 2D percolation as a function of total filler content, as shown in Figure 3.47, the ratio of the dissipations for the slurry as represented by n follows the same trend as the low viscosity Newtonian fluid based slurries and follows the power law theory based on the cluster volume, which is of course less than the total filler volume [33].

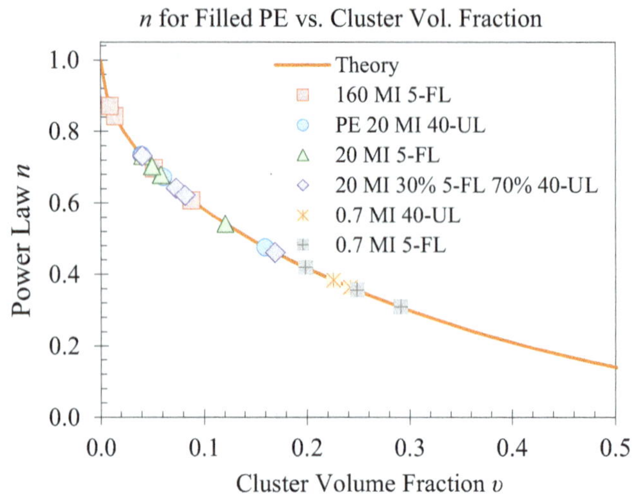

Figure 3.48 Fit of $CaCO_3$ filler in PE on the composite power law and fit to theory

3.9.8 Rheological Implictions for Extrusion and Molding Processes

3.9.8.1 Infinite Shear Newtonian Plateau

The change in power law as a function of filler content or polymer molecular weight has a strong implication regarding the ability to increase productivity by increasing the rate in extrusion dies or injection molding flow passages. The smaller the power law constant of the polymer system the less additional energy it will take to increase incrementally the production rate. The end of the power law region as shown in Figure 3.33 and Figure 3.34 is where the Newtonian infinite viscosity plateau begins. To understand the transition from power law to the second Newtonian plateau, data from the 11 literature sources were used as shown in Figure 3.49 [33]. The ratio of limiting shear viscosity η_0 to the infinite shear viscosity η_∞ is essentially a cubic power function of the power law n for these Newtonian fluid based slurries. As discussed in the subsequent section, the transition from power law to infinite shear viscosity has important implications regarding injection molding.

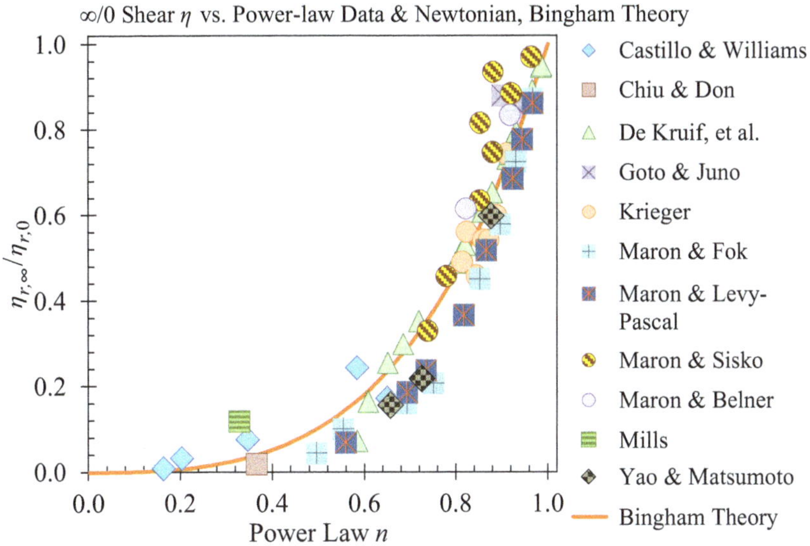

Figure 3.49 Effect of power law value on transition to second Newtonian plateau [33, 34]

3.9.8.2 Injection Molding and Power Law Rheology

As illustrated in Figure 3.33 and Figure 3.34, processing through the power law region requires little increase in equipment power because the shear stress increases at a much lower rate than the shear rate. When the second Newtonian plateau is reached, the power needed to increase the velocity through a die or mold runners increases linearly with shear rate and flow rate, since shear stress and shear rate are again linearly related. One of the authors (GAC) was qualitatively made aware of this transition when he worked at General Motors Research. One of the research injection molding machines was converted into a high shear rate rheometer by attaching an instrumented capillary die to the barrel. The onset of η_∞ to the GMR researchers was indicated by the observation that setting the injection speed control to a higher value resulted in essentially no increase in injection rate. Although at that time the correlation to power law was not known, it was observed that a polystyrene with n = 0.65 had a maximum shear rate of about 50,000 1/s while a polypropylene resin with an n = 0.08 had a maximum shear rate of about 200,000 1/s.

From a process engineering perspective, several industrial systems are now examined that can be affected by the increase in viscosity and decrease in power law. It is recognized that many polymer processes are dominated by infinite cluster formation above the percolation concentration as discussed in previous sections. As the power law decreases toward zero, the transition to the infinite shear Newtonian plateau occurs at a lower apparent viscosity and at a higher shear rate, as shown as an illustrative example in Figure 3.50.

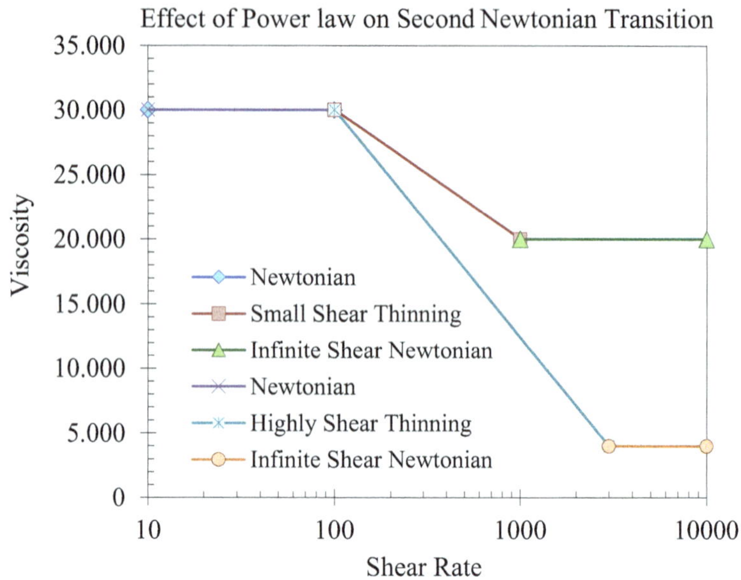

Figure 3.50 Effect of power law on transition to second Newtonian plateau for injection molding

With both materials having the same η_0, the "higher power law" resin reached the second Newtonian plateau at about 20,000 Pa·s and a shear rate of about 1000 1/s. Beyond this shear rate, it would be very difficult to increase the production rate by increasing the flow rate in a die or mold because one would rapidly exceed the equipment's available power because in the high shear Newtonian plateau the shear stress increased linearly with the shear rate. The higher shear thinning resin, smaller n, did not reach the second Newtonian plateau until about 3,000 1/s with an apparent viscosity of 3000 Pa·s. Since the shear stress increases much more slowly in the pseudoplastic viscosity region with shear rate, the resin with the greater power law value cannot be processed at a high material velocity or shear rate, and thus would have a potentially lower productivity than the one with the smaller n.

3.9.8.3 Extrusion Affected by the Rheology of Filled Polymers

Highly filled polymers are important in a number of applications today including heat conduction in electronic devices, sound dampening in automotive and other industries, and many more industrial applications as mentioned in Section 3.9.3. In all of these applications, the fillers of interest are compounded into a polymer matrix, and the dispersion is cast, extruded, or molded into the required shape. One real industrial problem that the authors are aware of is reviewed here: the production of sound dampening material to be used in auto manufacturing. Since sound dampening is a function of the mass of the material between the origination of the sound waves and the area to be dampened, it is natural to start with a high-density filler such as $BaSO_4$ (ρ = 4.5 g/cm³). However, $BaSO_4$ is costly, so there is an economic incentive to use $CaCO_3$ (ρ = 2.7 g/cm³). At a major polymer materials supplier, the manager of the group charged with the development of this composite material instructed the group members to substitute $CaCO_3$ for the $BaSO_4$ in the formulation. Sound deadening is a linear function of the mass of the filler in the composite, so it takes more $CaCO_3$ by volume to achieve the same mass in the composite. Most industrial product development professionals tend to think in weight or mass since that is how they measure when adding solid to solids. As pointed out above, the low shear rate melt viscosity is an exponential function of volume of the filler in the composite per Equation 3.80. Extrusion dies for this process are operated at a relatively low shear rate because of the high composite viscosity. Figure 3.51 plots low shear rate melt viscosity and the associated high shear power law n as a function of filler weight fraction. As the weight fraction approaches that required for effective sound dampening, 80 weight %, the viscosity of the $CaCO_3$ system is about five times higher than the $BaSO_4$ system. Also, the power law approaches the value of 1.0 at a lower concentration, indicating that the structure in the composite is almost all deformable cluster(s) at low shear rates. In all likelihood, the compounding extruder could over-torque or over-pressurize with these highly filled $CaCO_3$ compounds. At higher shear rates, these highly filled $CaCO_3$ compounds have a power law constant of a little less than 0.1, and from the discussion of the data in Figure 3.47 these systems may transition to 2D percolations and have a probability to jet in the extrusion die.

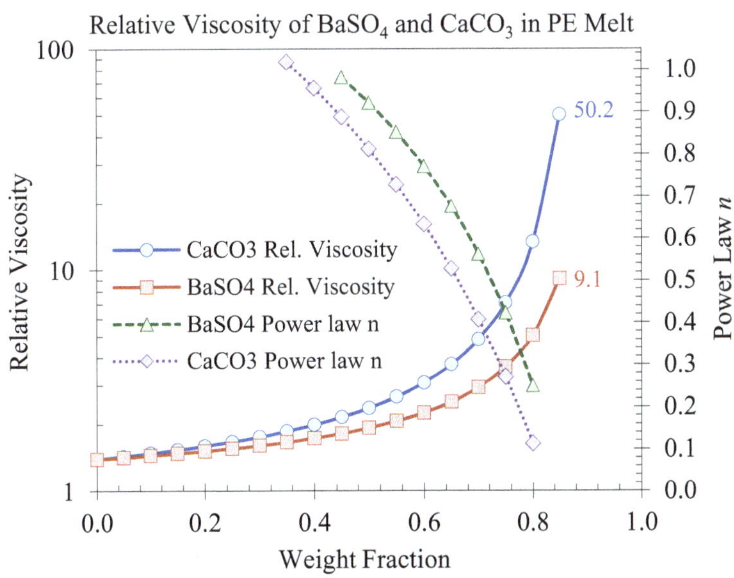

Figure 3.51 Effect of filler content on relative viscosity and the power law n

Furthermore, the critical wall stress can be easily exceeded. The wall stress for TiO_2 filled LDPE as a function of TiO_2 concentration is plotted in Figure 3.52 using the percolation convention, $(\varphi/\varphi_m)^{-2/3}$; a lower number on the *x*-axis implies higher filler concentration. As the filler concentration is increased, both the wall stress and the plug stress at the center of the channel increase. The power law rapidly approaches about 0.05, indicating again that the composite has a very high non-deformable, non-dissipating cluster content, from Equation 3.79. Thus, the shear rate would be very high at the die boundary. This can lead to jetting in flat dies and in gates in injection molds. For the case of the sound dampening project, the critical wall stress for the PE was exceeded and the $CaCO_3$ filled composite jetted from the die so much so that no useful product could be produced.

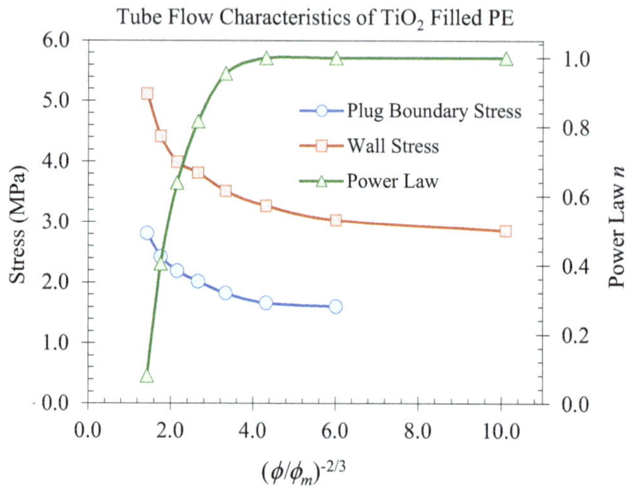

Figure 3.52 Power law, wall stress, and plug stress relating to filler volume fraction

Nomenclature

a	parameter for Equation 3.16
a_t	time-temperature shift factor for the WLF equation
c	polymer concentration in a solvent in g/dl
C_v	heat capacity
D_A	area average diameter for example in Section 3.2.1
D_i	diameter of the ith ball set for the example in Section 3.2.1
D_L	length average diameter for example in Section 3.2.1
D_V	volume average diameter for example in Section 3.2.1

E_v	element modulus for a basic mechanical element of Figure 3.10
g	gravitational acceleration constant
G	bulk modulus
G'	storage modulus as measured using a cone and plate rheometer
G''	loss modulus as measured using a cone and plate rheometer
$J(t)$	creep compliance
K	prefactor for the power law model
K'	parameter for Equation 3.16
L	length of the capillary die for a capillary rheometer
M	torque
M_c	critical molecular weight for entanglement
M_i	molecular weight of the ith molecule group in the polymer
M_n	number average molecular weight
M_v	viscosity average molecular weight
M_w	weight average molecular weight
M_z	z average molecular weight
M_{z+1}	$z + 1$ average molecular weight
n	power law index
N_i	number of molecules with a molecular weight of M_i
PI	polydispersity index
Q	volumetric flow rate
$\dot{Q}$	rate of energy dissipation
R	radius of the capillary die flow path for a capillary rheometer or the radius of a cone and plate rheometer
R_g	ideal gas law constant
t	independent variable for time
T	temperature
T_g	glass transition temperature
T_R	reference temperature
W_i	weight fraction of all molecules with a molecular weight of M_i
$\overline{\gamma}$	strain
$\dot{\gamma}$	shear rate
$\dot{\gamma}_a$	apparent shear rate at the capillary wall

$\dot{\gamma}_w$	shear rate at the capillary wall
$\dot{\gamma}_0$	shear rate amplitude for a cone and plate rheometer
δ	phase shift or loss angle
ΔE	activation energy
ΔP_c	corrected pressure change for capillary flow calculations
ε	time-dependent strain
η	solution viscosity or shear viscosity
η_g	shear viscosity at the glass transition temperature
η_R	shear viscosity at the reference temperature T_R
η_s	solvent viscosity
η_{sp}	specific viscosity
η_r	relative viscosity
η^*	complex viscosity
η_0	zero shear rate viscosity
η_∞	viscosity at infinite shear rate
$[\eta]$	intrinsic viscosity
λ	characteristic relaxation time for the Carreau viscosity model or relaxation time for Equation 3.30
μ	Newtonian viscosity
ρ	density
σ_0	initial stress applied to a basic mechanical element of Figure 3.10
σ_Ω	stress due to the angular rotation rate for a cone and plate rheometer
τ	shear stress
τ_w	stress at the wall
τ^*	critical stress where η transitions from η_0 to the power law regime for the Cross viscosity model
Φ	parameter in the Adams-Campbell rheology model
Φ_0	cone angle for a cone and plate rheometer
ω	oscillation frequency for a cone and plate rheometer
Ω	angular rotation rate

References

1. Rodriguez, F., "Principles of Polymer Systems," 4th ed., Taylor & Francis, Washington, DC (1996)
2. Rodriguez, F., Cohen, C., Ober, C.K., and Arcer, L.A., "Principles of Polymer Systems," 5th ed., Taylor & Francis, Washington, DC (2003)
3. Kurata, M. and Tsumashima, Y., "Viscosity-Molecular Weight Relationships," in "Polymer Handbook," 3rd ed., Brandrup, J. and Immergut, E.H., (Eds.), John Wiley & Sons, New York (1989)
4. Tadmor, Z. and Gogos, C.G., "Principles of Polymer Processing," 2nd ed., John Wiley & Sons, New York (2006)
5. Bird, R.B., Hassager, O., Curtiss, C.F., and Armstrong, R.C., "Dynamics of Polymeric Liquids," vol. 2, John Wiley & Sons, New York (1987)
6. Dealy, J.M. and Saucier, P.C., "Rheology in Plastics Quality Control," Hanser Publishers, Munich (2000)
7. Dealy, J.M. and Larson, R.G., "Structure and Rheology of Molten Polymers," Hanser Publishers, Munich (2006)
8. Morrison, F.A., "Understanding Rheology," Oxford University Press, New York (2001)
9. Macosko, C.W., "Rheology, Principles, Measurements, and Applications," VCH Publishers, New York (1994)
10. Williams, M.L., Landel, R.F., and Ferry, J.D., "The Temperature Dependence of Relaxation Mechanisms in Amorphous Polymers and Other Glass-Forming Liquids," *J. Am. Chem. Soc.*, **77**, 3701 (1955)
11. Middleman, S., "Fundamentals of Polymer Processing," McGraw-Hill, New York (1977)
12. Rabinowitsch, B., "The Viscosity and Elasticity of Sols," *Z. Phys. Chem.*, **A145**, 1 (1929)
13. Broyer, E. and Macosko, C.W., "Comparison of Cone and Plate. Bicone and Parallel Plates Geometries for Melt Rheological Measurements," *SPE ANTEC Tech. Papers*, **21**, 343 (1975)
14. Cox, W.P. and Merz, E.H., "Correlation of Dynamic and Steady Flow Viscosities," *J. Polym. Sci.*, **28**, 619 (1958)
15. ASTM D1238-04c, "Standard Test Method for Flow Rates of Thermoplastics by Extrusion Plastometer," (2004)
16. Gleissle, W., "Influence of the Viscoelastic Functions on the Melt Index," *J. Rheol.*, **4**, 138 (1994)
17. Gleissle, W., "Influence of the Measuring Apparatus and Method on the Value of the Melt Index," *J. Rheol.*, **4**, 13 (1994)
18. Campbell, G.A. and Adams, M.E., "A Modified Power Law Model for the Steady Shear Viscosity of Polystyrene Melts," *Polym. Eng. Sci.*, **30**, 587 (1990)

19. Meister, B. J., "Rheology," in "Styrene Polymers, Encyclopedia of Polymer Science and Engineering," Moore, E. R., (Ed.), **16**, 117 (1989)

20. Wood-Adams, P. M. and Dealy, J. M., "Use of Rheological Measurements to Estimate the Molecular Weight Distribution of Linear Polyethylene," *J. Rheol.*, **40**, 761 (1996)

21. Wood-Adams, P. M., Dealy, J. M., deGroot, A. W., and Redwine, O. D., "Effect of Molecular Structure on the Linear Viscoelastic Behavior of Polyethylene," *Macromolecules*, **33**, 7489 (2000)

22. Fox, T. G., Gratch, S., and Loshaek, S., Chapter 12 in "Rheology," Eirich, F. R., (Ed.), vol. 1, Academic Press, New York (1956)

23. Ito, Y. and Shishido, S., "Capillary End-Correction in the Upper Newtonian Region," *J. Polym. Sci., Polym. Phy. Ed.*, **10** (11), 2239 (1972)

24. Berry, G. C. and Fox, T. G., "The Viscosity of Polymers and Their Concentrated Solutions," *Adv. Polym. Sci.*, **5**, 261 (1968)

25. Ferry, J. D., "Viscoelastic Properties of Polymers," Wiley, New York, Chapter 10 (1980)

26. Ramesh, N. S., Rasmussen, D. H., and Campbell, G. A., "The Heterogeneous Nucleation of Microcellular Foams Assisted by the Survival of Microvoids in Polymers Containing Low Glass Transition Particles. Part I: Mathematical Modeling and Numerical Simulation," *Polym. Eng. Sci.*, **34**, 1685 (1994)

27. Ramesh, N. S., Rasmussen, D. H., and Campbell, G. A., "The Heterogeneous Nucleation of Microcellular Foams Assisted by the Survival of Microvoids in Polymers Containing Low Glass Transition Particles. Part II: Experimental Results and Discussion," *Polym. Eng. Sci.*, **34**, 1698 (1994)

28. Ostwald, W., *Kolloid-Z.*, **36**, 99 (1925)

29. de Waele, A., "Viscometry and Plastometry," *J. Oil Colour Chem. Assoc.*, **6**, 23 (1923)

30. Cross, M. M., "Rheology of Non-Newtonian Fluids: A New Flow Equation for Pseudoplastic Systems," *J. Colloid Sci.*, **20**, 417 (1965)

31. Carreau, P. J., Ph. D. Thesis, University of Wisconsin, Madison (1968)

32. deGroot, A. W., Gillepsie, D., Cong, R., Zhou, Z., and Paradkar, R., "Molecular Structural Characterization of Polyethylene," Chapter 5 of "Handbook of Industrial Polyethylene and Technology", edited by Spalding, M. A. and Chatterjee, A. M., Wiley - Scrivener Publishing, Hoboken, NJ, 2017

33. Campbell, G. A., Zak, M. E., and Radhakrishnan, J. S.,"Development of a Predictive Power Law Relationship for Polymer Composites Based on Newtonian Carrier Concentrated Slurries," *Polym. Compos.*, **39**, 1172 (2016)

34. Campbell, G. A., Zak, M. E., and Wetzel, M. D., "Newtonian, Power Law, and Infinite Shear Flow Characteristics of Concentrated Slurries Using Percolation Theory Concepts," *Rheol. Acta*, **57**, 197 (2018)

35. Campbell, G. A. and Wetzel, M. D., "Investigation of the Effect of Filler Concentration on the Flow Characteristics of Filled Polyethylene Melts," *Int. Polym. Process.*, **33**, 619 (2018)

36. Wetzel, M.D. and Campbell, G.A.,"Characterizing the Flow of Slurries using Percolation Theory Based Functions," *Polym. Eng. Sci.*, **57**, 403 (2017)
37. Wetzel, M.D. and Campbell, G.A., "A Study of Concentrated Suspensions in Polyethylene Melts and the Impact on Viscosity and Polymer Processing Operations," *Int. Polym. Process.*, **33**, 574 (2017).
38. Wetzel, M.D., Pettitt, Jr., D.R., and Campbell, G.A., "Development of a Predictive Power Law Relationship for Concentrated Slurries, Part 2: Experiment and Processing Implications," *SPE ANTEC Tech. Papers*, 735 (2016)
39. Scher, H.H. and Zallen, R., "Critical Density in Percolation Processes," *J. Chem. Phys.*, **53**, 3759 (1970)
40. Powell, M.J., "Site Percolation in Random Networks," *Phys. Rev. B*, **21**, 3725 (1980)
41. Zak, M.E., "An Investigation of the Shear-Thinning Behavior an Power-Law Modeling of Concentrated, Non-Colloidal Suspensions," Ph.D. Dissertation, Clarkson University (1994)
42. Radhakrishnan, J.S., "Investigation of Poiseuille Flow of Concentrated Monodisperse and Bidisperse Suspensions," Ph.D. Dissertation, Clarkson University (2002)
43. Brenner, H., "Dissipation of Energy due to Solid Particles Suspended in a Viscous Liquid," *Phys. Fluids*, **1**, 338 (1958)
44. Karnis, A., Goldsmith, H.L., and Mason, S.G., "The Kinetics of Flowing Dispersions, 1. Concentrated Suspensions of Rigid Particles," *J. Coll. Int. Sci.*, **22**, 531 (1966)
45. Karnis, A., Goldsmith, H.L., and Mason, S.G, "The Flow of Suspensions Through Tubes: V. Inertial Effects," *Can. J. Chem. Eng.*, **44**, 181 (1966)
46. Altobelli, S.A., Givler, R.C., and Fukushima, E., "Velocity and Concentration Measurements of Suspensions by Nuclear Magnetic Resonance Imaging," *J. Rheol.*, **35**, 721 (1991)
47. Bazua, E.R. and Williams, M.C., "Rheological Properties of Internal Viscosity Models with Stress Symmetry," *J. Polym. Sci. (Phys.)*, **12**, 825, (1974)
48. Manke, C.W. and Williams, M.C., "Internal Viscosity of Polymers and the Role of the Solvent Resistance," *Macromolecules*, 18, 2045, (1985)
49. Manke, C.W. and Williams, M.C., "Transient Stress and Strain Responses Predicted by the Internal Viscosity Model in Shear Flow," *J. Rheol.*, **33**, 949 (1989)
50. Nichetti, D. and Manas-Zloczower, I., "Viscosity Model for Polydisperse Polymer Melts," *J. Rheol.*, **42**, 951, (1998)
51. Adams, M.E., "A Predictive Model for the Steady Shear Viscosity of Polymer Melts," M.S. Thesis, Clarkson University, (1987)
52. Wetzel, M.D., Howe, J.C., Sterling, M.T., and Campbell, G.A., "The Rheology of Concentrated Slurries: Experimental Evaluation and the Effects on Polymer Processing," *SPE ANTEC Tech, Papers*, 1144 (2017)
53. Mooney, M., "Explicit Formulas for Slip and Fluidity," *J. Rheol.*, **2**, 210 (1931)
54. Jastrzebski, Z.D., "Entrance Effects and Wall Effects in an Extrusion Rheometer During Flow of Concentrated Suspensions," *Ind. Eng. Chem. Fund.*, **6**, 445, (1967)

55. Hatzikiriakos, S. G. and Dealy, J. M.,"Wall Slip of Molten High Density Polyethylenes. II. Capillary Rheometer Studies," *J.Rheol.*, **36**, 703 (1992)

56. Seto, R., Mari, R., Morris, J. F., and Denn, M. M., "Discontinuous Shear Thickening of Frictional Hard-Sphere Suspension," *Phys. Rev. Lett.*, **111,** :218301 (2013)

57. Campbell, G. A. and Forgacs, G., "Viscosity of Concentrated Suspensions: An Approach Based on Percolation Theory," *Phys. Rev. A*, **41**, 4570 (1990)

58. Campbell, G. A., Radhakrishnan, J. S., and Wetzel, M. D. "Comparing the Power Law Constant (n) for Mono- and Bi-Dispersed Filled Slurries; Using Percolation Theory Concepts," *Rheol. Acta*, **59** (2020)

4 Resin Physical Properties Related to Processing

In Chapter 3, the rheological properties of resins and how they relate to processing were presented. These rheological properties are the most important resin properties for many processes. Other physical properties related to processing, however, are keys to the functions of single-screw extruders. For example, proper solids conveying depends on many physical properties, including the bulk density and compaction of the resin, the coefficient of dynamic friction (or stress at a sliding interface), and the lateral stress ratio. As the solids are conveyed downstream they increase in temperature due to conduction and the dissipation of energy at the polymer-metal interfaces. The temperature increase is controlled by these same physical properties and also the heat capacity and thermal conductivity of the resin. Moreover, the metering section of the screw is a volumetric pump. Since processors require rates in mass units per time, melt density is an important physical property to know.

The screw geometry and the process conditions chosen are highly dependent upon these physical properties. That is, to design an economically attractive and stable process the screw and process conditions must be adapted to the resin feedstock type and form. For example, a screw designed for use with just a pellet feedstock will have a feed channel that is shallower than a screw designed for the same resin but containing a level of an in-plant, low bulk density recycle resin stream such as ground sheet or film. Alternatively, if the screw was originally designed for a pellet feedstock and plant personnel later decide to add a lower bulk density recycle stream at a feed composition of 50%, the extruder will likely not operate properly because the solids-conveying channel will not be able to feed the low-density feedstock blend at a high enough rate to maintain the metering channel full of resin. In this case, the plant could experience resin degradation in the product and flow surging.

In this chapter the physical properties of resins related to processing will be described. The chapter and Appendix A4 include physical properties for many resins. These properties include bulk density and compaction, lateral stress ratio, stress at a sliding interface, melting flux, heat capacity, thermal conductivity, and melt density. Some of these properties are easy to measure by many laboratories while others such as the melting flux and stress at a sliding interface can be measured in only a few places using highly specialized equipment.

4.1 Bulk Density and Compaction

The bulk density of a resin is a key property for the design of a screw. As presented earlier in this chapter, proper solids conveying is highly dependent on a match between the feed channel geometry and the bulk density of the feedstock resin. The bulk density is defined as the mass of resin that occupies a specific volume. The bulk density is a function of temperature, pressure, and form of the feed. For example, the bulk density of a PC resin pellet feedstock at ambient temperature and pressure is about 0.72 g/cm^3. Since the solid density of PC resin is 1.20 g/cm^3, about 40% of the test volume is occupied by air. For this resin feedstock, the pellets enter the feed channel at a bulk density of 0.72 g/cm^3. As the resin is conveyed downstream in the extruder, energy dissipated from the solids-conveying and melting processes causes the temperature of the resin and the pressure in the channel to increase. The increasing temperature and pressure causes the bulk density of the resin to increase, forcing the entrained air out of the extruder through the hopper. The bulk density of the solid bed approaches the solid density of the resin just before melting (or devitrification) is completed.

To illustrate the compaction process that occurs in an extruder, a Maddock solidification [1] experiment (described in detail in Section 10.3.1) was performed using a 63.5 mm diameter machine [2]. The extruder was operated at a screw speed of 60 rpm with a poly(vinylidene chloride) copolymer (PVDC) powder. After the extruder reached a steady-state operation, screw rotation was stopped and full cooling was applied to the extruder. After several hours of cooling, the screw and PVDC resin were removed from the extruder and the density of the bed was measured using Archimedes's principle. The compaction phenomenon in the extruder is shown by the density measurements of the solid bed in Figure 4.1. As shown in this figure, the density of the solid bed increased from the feedstock bulk density of 0.73 g/cm^3 to nearly the solid density of 1.7 g/cm^3.

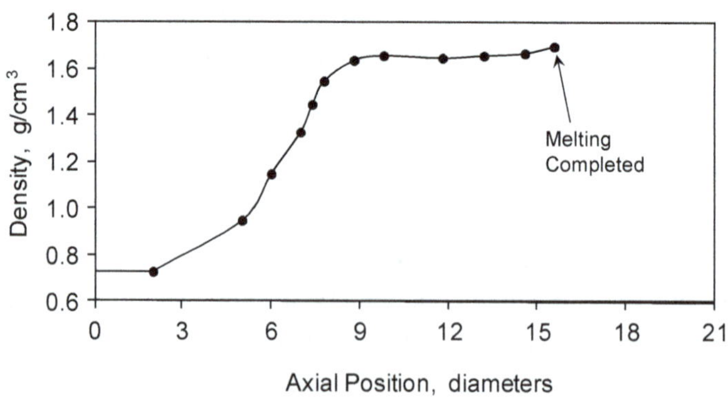

Figure 4.1 Density of the solid bed from a Maddock solidification experiment running PVDC resin [2]. The extruder was 63.5 mm in diameter and had an *L*/*D* of 21

4.1.1 Measurement of Bulk Density

The bulk density of the feedstock at ambient temperature and pressure should be measured prior to the design of a new screw, especially if it contains in-plant recycle resin. The measurement method is extremely simple and requires only a calibrated cell and a scale. A calibrated measuring cell with a volume of 500 cm^3 can easily be constructed by welding a thin-walled metal pipe to a flat sheet of metal, as shown in Figure 4.2. The bulk density is measured by filling the cell with feedstock, leveling the top with a steel ruler, and then weighing the cell contents. A more formal measurement technique was developed by ASTM as standard method D1895.

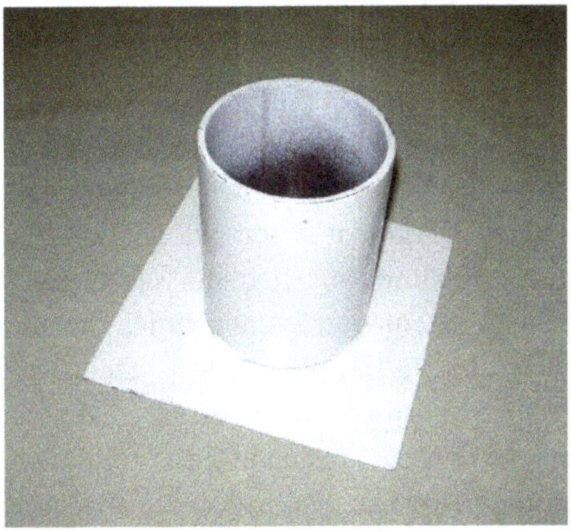

Figure 4.2 A calibrated cell with a volume of 500 cm^3 for measuring the bulk density of a resin feedstock at ambient temperature and pressure

Many processes, including most types of extrusion, blow molding, injection molding, and thermoforming, create a scrap stream. For thermoforming, the scrap level from off-specification parts and edge trim can be over 50%. In order for most processes to be profitable, the scrap material must be ground and recycled back into the feedstock. The bulk density of this in-plant recycle can be as much as 80% of the bulk density of the virgin pellets for the grinding of thick parts. For films or foamed products, the ground recycle can have an extremely low bulk density with values of less than 5% of the virgin pellets. Blending high levels of a low-density, in-plant recycle with virgin pellets can cause the bulk density of the resulting feedstock to be considerably less than that of the pellets. Typically, screws with deeper feed channels are required to properly extrude these blends. Adjusting channel depths to accommodate the addition of in-plant recycle is presented in Sections 6.1 and 12.7.7. In extreme cases where the density of the recycle is very low, such as ground foam sheet, the recycle stream may need to be densified using a specially designed repelletization extrusion process. The repelletized material is then blended with virgin resin in the main extruder.

4.1.2 Measuring the Compaction Characteristics of a Resin

As shown in Figure 4.1, resin feedstocks have a considerable level of interparticle space that is occupied by air. This level of space and thus the bulk density of the feedstock depend on the temperature, pressure, pellet (or powder) shape, resin type, and the level and shape of the recycle material. For a specific resin feedstock, the bulk density increases with both temperature and the applied pressure. Understanding the compaction behavior of a resin feedstock is essential for both screw design and numerical simulation of the solids-conveying and melting processes. Screw channels must be able to accommodate the change in the bulk density to mitigate the entrainment of air and the decomposition of resin at the root of the screw. Typically, screw channels are set by using an acceptable compression ratio and compression rate for the resin. These parameters will be discussed in Section 6.1.

The compaction of the resin can be easily measured using a specially designed cell [3–6]. The cell developed at The Dow Chemical Company, shown in Figure 4.3, consisted of a cylinder with a 2.540 cm diameter bore and a piston with a slip fit. A load cell positioned between the piston and the loading mechanism was used to measure the applied force. Loads were applied to the piston by mechanical means using a pneumatic system. The pressure delivered to the pneumatic system was varied by an adjustable pressure regulator. The cell was always operated at constant force. Polymer samples were contained between the piston and a bottom plug inserted into the cylinder. The force transmitted from the piston through the sample and plug was measured by a load cell. The piston position and hence sample volume was measured using a linear variable-differential transformer (LVDT). The cylinder was heated by electrical band heaters and temperature controlled using a thermocouple embedded in the cylinder wall.

At the start of the experiment, 11.5 g of feedstock resin is placed into the cylinder and the piston is inserted. The resin is then allowed to warm up to the preset cylinder temperature. In general, 10 minutes is adequate for thermal equilibration as measured by density fluctuations. The position of the piston is then recorded for each pressure setting, starting at the lowest and increasing to the maximum pressure. The piston position is recorded only after movement ceases.

The force applied to the top of the solid bed is transmitted through the bed to the bottom plug. The resulting force on the plug is detected by the bottom load cell. If there is no friction at the wall and no internal losses, then the transmitted force will equal the applied force. The solid bed, however, touches the wall and exerts a force, causing drag and a pressure gradient in the solid bed. When a constant force is applied, the bed compacts to the point where the frictional force at the wall balances the net force on the differential plug due to the pressure gradient.

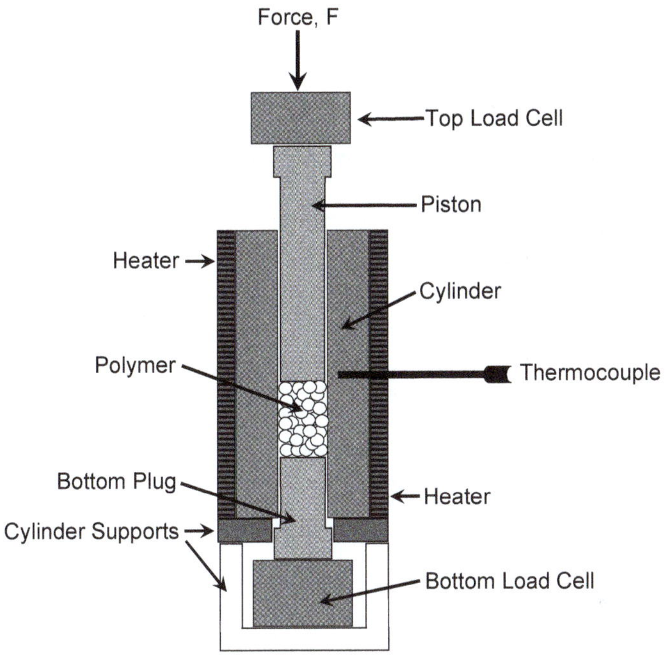

Figure 4.3 Schematic of a compaction testing cell [3]

Bulk densities for LDPE and HIPS resins as a function of temperature and pressure were measured using the cell shown in Figure 4.3. As shown in Figure 4.4, semi-crystalline LDPE resin pellets compact very evenly over the temperature range of 25 °C to near the melting temperature. The bulk density at 25 °C and zero pressure was measured using the cell shown in Figure 4.2. As the temperature and pressure increase, the air entrained between the pellets was expelled out of the cell as the pellets deformed. At high pressure and temperature, most of the air is expelled and the bulk density is essentially the solid density at the measured temperature. The solid density of the LDPE resin at 100 °C is about 0.87 g/cm^3. The solid density at ambient temperature is 0.92 g/cm^3. The solid density of the resin decreased at 100 °C due to the reduction in the crystallinity and the thermal expansion of the amorphous phase. This LDPE resin was used for the solids-conveying studies using the solids-conveying device at Dow. The solids-conveying data and device are presented in Section 5.3.2 and Appendix A5.

Amorphous HIPS resin pellets compact very little in the temperature range of 25 °C to about 75 °C, a temperature that is about 25 °C below T_g, as shown in Figure 4.5. The HIPS resin compacts to a much higher degree for temperatures of 25 °C below T_g up to T_g. Like the LDPE resin, the bulk density at 25 °C and zero pressure was measured using the cell shown in Figure 4.2 at 0.62 g/cm^3. At temperatures 25 °C below the T_g (100 °C), the bulk modulus of the resin is relatively high, and thus the pellets do not deform easily under pressure. At higher temperatures, however, the bulk modulus decreases rapidly, allowing the pellets to deform and compact. This type of behavior is typical for amorphous resins.

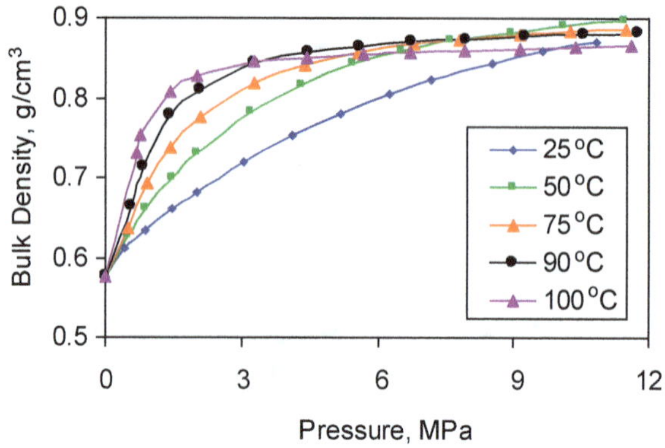

Figure 4.4 Bulk density as a function of temperature and pressure for an LDPE resin in the pellet form. This is the same LDPE resin used in the solids-conveying studies of Section 5.3.2

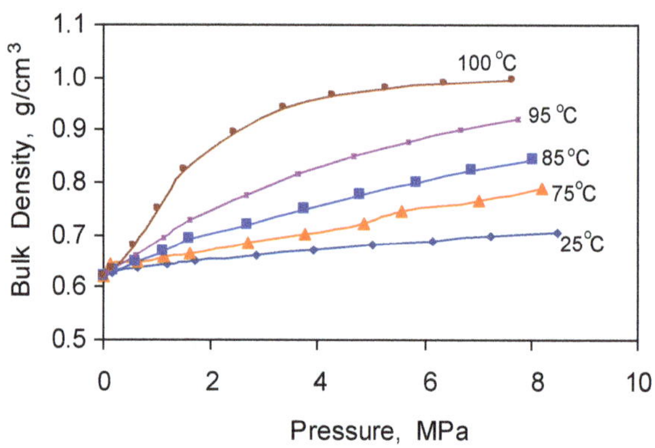

Figure 4.5 Bulk density as a function of temperature and pressure for a HIPS resin in the pellet form

As previously discussed, in-plant recycle can have a strong effect on the bulk density and compaction of the feedstock resin. For example, a thermoforming process running a HIPS resin was creating an in-plant recycle stream. The sheet was fairly thin and was ground to a manageable size. The bulk density of this recycle material was 0.43 g/cm^3, a bulk density that was about 30% less than that of the pellets. In order to be economically attractive, the plant must be able to recycle the in-plant material, causing the recycle composition of the feedstock to the extruders to be 50% or more. As an example, a feedstock containing 10% pellets and 90% recycle material had a bulk density at ambient conditions of about 0.45 g/cm^3. The compaction characteristic for this resin is shown in Figure 4.6. As shown in this figure, the compaction of the feedstock is relatively easy at low pressures due to the relatively low bulk modulus of the recycle material, and then it compacts to nearly the same level as the sample containing 100% pellets (Figure 4.5). The geometry of the feed section of the screw must be adjusted to accept high levels of a low-density recycle stream.

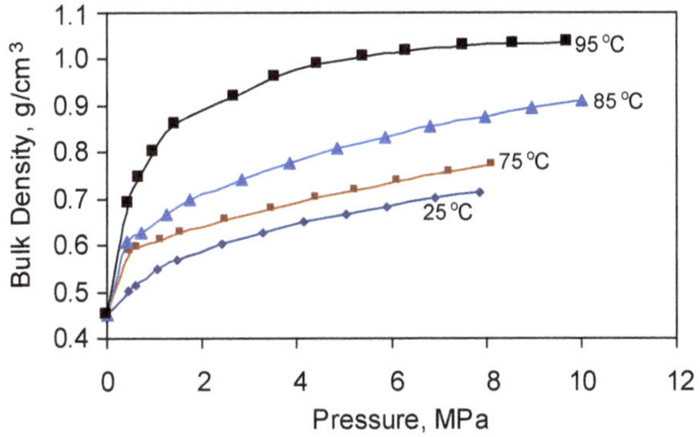

Figure 4.6 Bulk density as a function of temperature and pressure for a feedstock containing 10% HIPS pellets and 90% of an in-plant HIPS recycle stream

4.2 Lateral Stress Ratio

The original solids-conveying model developed by Darnel and Mol [7] assumed that the pressure (or stress) in the solid bed is isotropic. This assumption was made to simplify the mathematics and because of the lack of stress data for solid bed compacts. Previous research, however, showed that stresses in solid compacts are not isotropic [8]. Anisotropic stresses can be represented by the lateral stress ratio. It is defined as the ratio of the compressive stress in the secondary direction to the compressive stress in the primary direction, as shown in Figure 4.7 and Equation 4.1.

$$K_{xy} = \frac{\sigma_x}{\sigma_y} \quad (4.1)$$

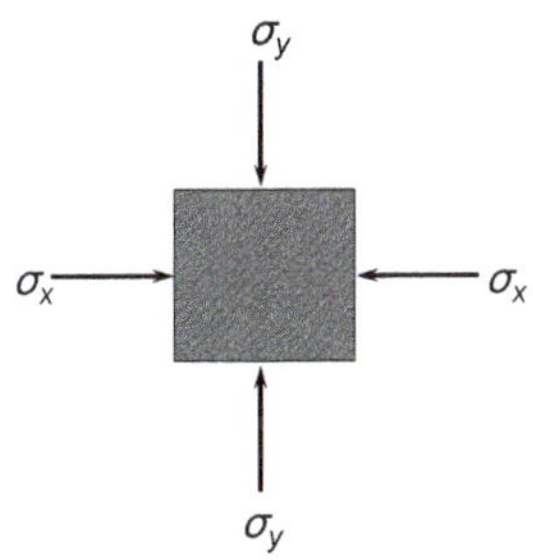

Figure 4.7 Schematic of the stresses on a volume of the solid bed

where K_{xy} is the lateral stress ratio, σ_x is the stress in the x direction, and σ_y is the stress in the y direction. The lateral stress ratios have been measured for select polymers [2, 8–11], but generally the ratio is rarely available. The stresses are not isotropic due to the complicated nature of pellet-to-pellet friction and the geometry of the feedstock.

4.2.1 Measuring the Lateral Stress Ratio

The lateral stress ratio depends on the resin type and shape, surface treatments such as additives, temperature, and pressure. The ratio is measured using a compaction cell [2], as shown in Figure 4.8. This cell is very similar to one shown in Figure 4.3 except the piston for the lateral stress ratio cell is octagonal in cross section and a pressure sensor is mounted in the cylinder wall. The stress ratio is calculated by dividing the pressure measured at the side of the cylinder by the calculated pressure in the axial direction at the height of the sensor. The calculation method can be found elsewhere [2]. The lateral stress ratio for select resins at 25 °C and 2.5 MPa are provided in Table 4.1.

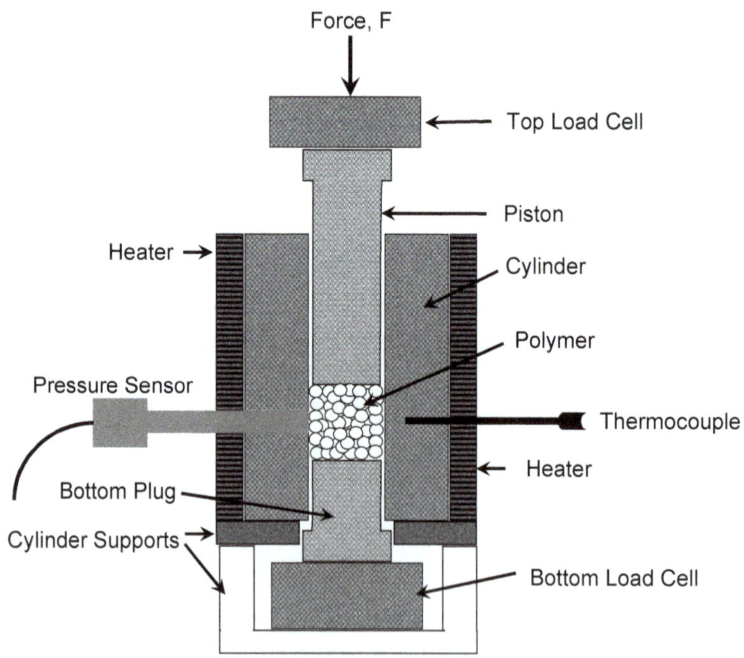

Figure 4.8 Schematic of a lateral stress compaction testing cell [2]

Table 4.1 Lateral Stress Ratio at 25 °C and 2.5 MPa for Select Resins [2]

Resin	Form	Lateral Stress Ratio at 25 °C and 2.5 MPa
PVDC	Powder	0.37
ULDPE	Pellets	0.33
LDPE	Pellets	0.67
LLDPE	Pellets	0.34
HDPE	Pellets	0.31

The lateral stress ratio and two other stress ratios can also be measured using a triaxial testing cell that was designed for geomaterials [12]. For this test, the resin sample is placed inside a cylindrical latex membrane as shown in Figure 4.9. An axial load is applied to the top of the sample using a piston, and the strain is measured using an LVDT device. The stress in the hoop direction is applied to the outside of the latex membrane by setting a pressure inside the cell. The applied strain rate is typically maintained at a rate of 0.5 percent per minute. The resulting load is measured on the piston. The lateral stress ratio is simply the set stress in the hoop direction divided by the measured stress in the axial direction. The lateral stress ratio was measured for an LLDPE resin at 0.343 [12]. This resin was considerably different than the LLDPE resin used in the testing reported in Table 4.1.

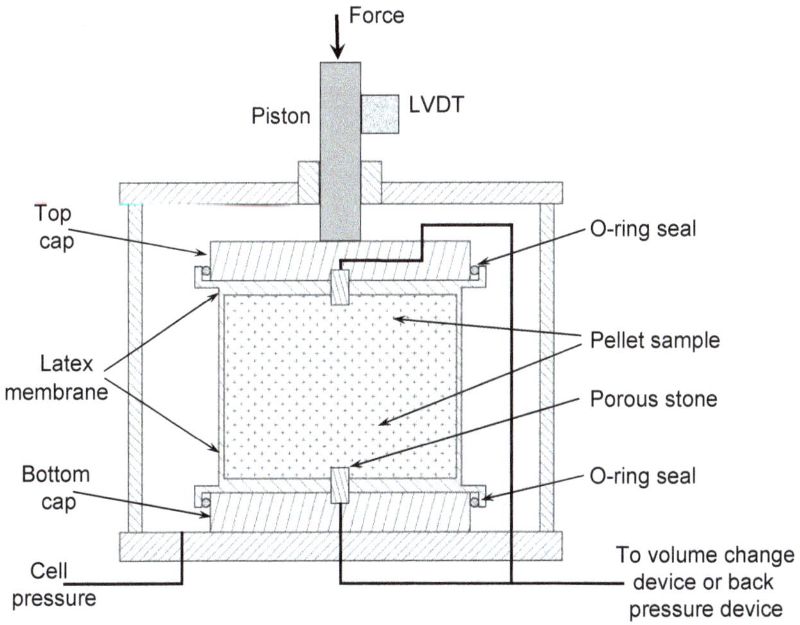

Figure 4.9 Schematic of a triaxial stress testing cell [12]

The lateral stress ratio is defined for particle systems that are consolidated under pressure, and thus the particles are essentially locked in place and unable to move relative to the neighboring particles. This type of consolidation occurs in the later section of the solids-conveying zone. In the early sections of the conveying zone, where the pressure is relatively low, the particles are known to move [13, 14], and thus the lateral stress ratio is not pertinent in this section. The mechanism for solids conveying will be presented in detail in Chapter 5.

4.3 Stress at a Sliding Interface

The dynamic coefficient of friction is an important physical property for conveying solids in the feed section of the extruder. That is, the forces responsible for solids conveying originate from the coefficient of dynamic friction between the solids bed and the metal surfaces of the machine. The coefficient is defined as the ratio of the applied force from the pellet bed normal to the metal surface of the machine to the force required to keep the solid bed moving. The coefficient of dynamic friction is a solid-state physical property and thus is defined for temperatures up to the melting temperature for semicrystalline resins and up to the devitrification temperature (T_g) for amorphous resins. Solids conveying also depends on the forces generated by metal surfaces that are at temperatures higher than the melting or devitrification temperature. These forces originate from a viscous mechanism. To view both forces, the coefficient of dynamic friction can be converted to the stress at the interface using Equation 4.2.

$$\tau = fP \tag{4.2}$$

where τ is the shear stress at the polymer-metal interface, f is the coefficient of dynamic friction, and P is the pressure.

As will be discussed in Chapter 5, proper solids conveying depends on a favorable balance of the forwarding forces at the barrel wall and the pushing flight of the screw and the retarding forces at the root of the screw and trailing flight. That is, the temperature of the screw and barrel in the solids-conveying zone must be controlled such that the forwarding forces are maximized and the retarding forces are minimized. Most screws are designed such that the solids-conveying section can deliver resin at a higher rate than the metering section, maintaining the metering section of the screw as the rate-determining step of the process. If the temperature of the metal surfaces in the solids-conveying section is outside of the acceptable processing window, solids-conveying rates can be reduced to a level where it controls the rate of the process and not the metering section. When the solids-conveying section controls the rate, many problems can occur, including reduced rates, high discharge temperatures, resin degradation, and flow surging. All of these problems will reduce the profitability of a plant and also have the potential to shut the line down in severe cases. Although this physical property is highly important, the shear stress at a sliding interface is rarely known due to the limited number of measuring devices [15, 16] and the high cost of the measuring device.

For a specific resin, the shear stress at the interface depends on the temperature of the interface, pressure, and the sliding velocity. It also depends on resin type, additives and additive levels, and the rheological properties of the resin. Stresses at the interface and the coefficients of friction for numerous resins have been published previously from two sources, and the data can be found in the references [15–31]. Additional stress data are provided in Appendix A4 and in several of the case studies in Chapter 12.

4.3.1 The Screw Simulator and the Measurement of the Stress at the Interface

The stress at the interface, coefficients of dynamic friction, and melting fluxes (to be discussed next) are measured using a device known as the Screw Simulator. The device was developed and designed by Chung in 1977 [15]. The device at Dow [16] was based on Chung's unit, and it consisted of a 30.5 cm diameter roll, a sample chamber, and a plunger, as shown in Figure 4.10 and Figure 4.11. Pellets are placed in the sample chamber and are then forced onto the rotating roll with a fixed force using the plunger. The torque required to rotate the roll at a constant speed is measured and used to compute the shear stress at the polymer-metal interface. The roll can be heated up to 300 °C. The temperature was measured by thermocouples embedded in the roll and located three mm away from the sliding interface. Frictionally dissipated energy causes the interface temperature to be higher than the temperature measured by the embedded thermocouples. For this case, an implicit, finite-difference technique was used to predict the polymer-metal interface temperature [16].

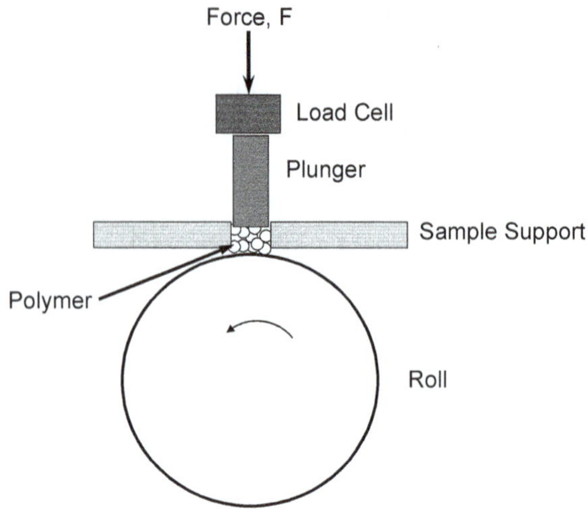

Figure 4.10 Schematic of the Screw Simulator [16]

Figure 4.11 Photograph of a Screw Simulator [16]

The shear stress at the interface for a PET resin as a function of temperature and sliding velocity at a pressure of 0.7 MPa is shown in Figure 4.12. As shown in this figure, the stress is highly dependent on the temperature of the roll surface. The stress level is relatively low at temperatures between 25 and 150 °C, and it increases to a maximum value at a temperature of 230 °C, near the melting temperature. At higher temperatures, the stress is due to rheological forces, and it decreases with increasing temperature. For this resin, the stress is essentially independent of sliding velocity. In order to maximize the forwarding forces at the barrel wall, the temperature of the inside barrel wall in the solids-conveying section should be set at about 230 °C. Maintaining the inside barrel wall at 230 °C throughout the entire conveying zone, however, is not practical and typically not measured. For example, the control thermocouple for the first barrel zone is typically positioned midway into the barrel wall and thus relatively far away from the inside surface. Moreover, the feed casing is normally cooled with water, and the second barrel temperature is likely higher than the first zone. The retarding forces at the root of the screw and the trailing side of the flight must be minimized while maintaining the resin in a solid state. As shown in Figure 4.12, the screw should be operated at temperatures of 130 °C or less for this application. Maximizing the forwarding force at the pushing flight of the screw is not possible since this region of the screw is also at essentially the same temperature as the root of the screw and the trailing flight.

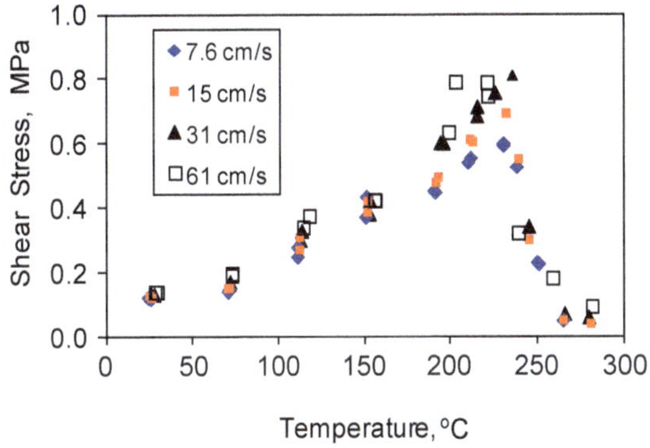

Figure 4.12 Shear stress at the interface for a PET resin at 0.7 MPa

4.4 Melting Flux

The melting flux is defined as the melting (or devitrification) rate for a specific unit of area. This is an important physical property for a resin since the maximum rate that an extruder can operate at is often limited by the melting capacity of the machine. The melting flux is measured using a Screw Simulator such as that shown in Figure 4.10 and Figure 4.11. The flux is measured by the rate of downward movement of the plunger with time. The details were presented earlier [22]. Like stress for the interface, very little data are available for the melting flux due to the unavailability and high cost of a Screw Simulator. Melting flux data for selected resins are available in Appendix A4.

The melting flux of a resin depends on the temperature of the polymer-metal interface, the sliding velocity, and pressure. The melting flux for a PC/ABS resin blend with an MFR of 1.7 dg/min (260 °C, 3.8 kg) is shown in Figure 4.13 as a function of temperature and velocity for a pressure of 0.7 MPa. As shown in this figure, mechanical melting started at a temperature of about 180 °C, and the melting flux increased with increasing temperature and sliding velocity. Melting flux also increases with increasing pressure. The melting flux is controlled by a viscous energy dissipation mechanism [17, 22], and the shear stress depends on the viscosity and the velocity gradients in the thin melt film between the solids and the barrel wall. For low viscosity materials, low shear stresses are produced and less energy is dissipated in the thin melt film. Therefore, the melting flux might be expected to increase with increasing viscosity for the same resin type. Previous research, however, clearly showed that the melting rate increases with decreasing shear viscosity, that is, increasing flow index [25, 28, 30]. The relationship between the flow index and melting flux is not straightforward and not well understood.

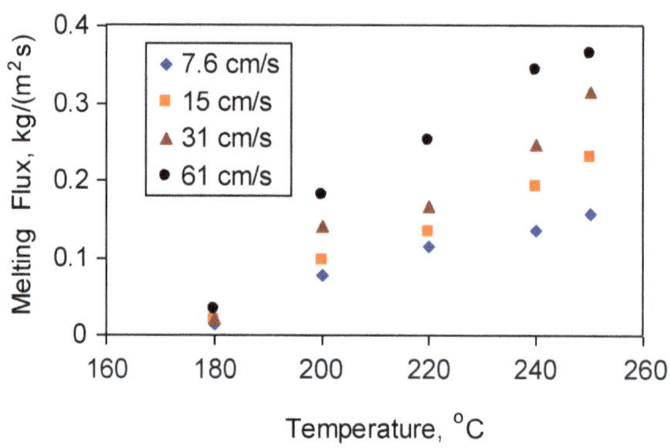

Figure 4.13 Melting flux for a PC/ABS blend with an MFR of 1.7 dg/min (260 °C, 3.8 kg). The data were measured using a pressure of 0.7 MPa

If an extruder is not limited by screw speed or the availability of power, the melting capacity of the machine will likely be the rate-limiting step. That is, as the screw is rotated faster, a speed will occur where solid polymer fragments are discharged with the extrudate. This effect is evident in the melting flux data for the PC/ABS resin in Figure 4.13. Previous research [32] showed that the solids-conveying rate for a screw doubles when the screw speed is doubled. Moreover, the rotational flow in the metering section will double if the screw speed is doubled. If the screw speed is doubled, the velocity of the solids rubbing on the barrel wall will also double. If the barrel wall in the melting section of the barrel for operation with PC/ABS resin is at 240 °C, the melting flux increases from 0.25 kg/(m^2·s) at a velocity of 31 cm/s to 0.34 kg/(m^2·s) at a velocity of 61 cm/s. Thus, the melting flux increased by about 1.4 times. If the rate doubles with the increase in screw speed and the melting flux only increases by about 1.4 times, then more barrel surface area will be required to melt the resin, pushing solids closer to the tip of the screw. If the screw speed is high enough, then solids will be discharged with the extrudate.

The melting flux can vary significantly between different polymer types. For example, Figure 4.14 shows the melting fluxes for three resins with comparable shear viscosities. The melting flux at 225 °C for the LDPE resin is about four times higher than that for the PP resin and 25% higher than the LLDPE resin. The design of the screw needs to take the melting flux into account. Typically an LDPE screw will have a relatively short transition section while screws designed for LLDPE and PP resins will have longer transition sections.

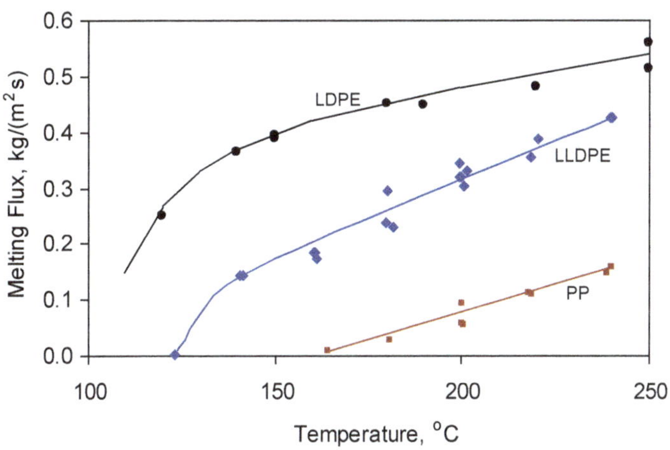

Figure 4.14 Melting flux for PP, LLDPE, and LDPE resins at a velocity of 31 cm/s and a pressure of 0.7 MPa

■ 4.5 Heat Capacity

The heat capacity is the amount of energy required to increase the temperature of a unit mass of material. It is commonly measured using a differential scanning calorimeter (DSC). The heat capacity depends on the resin type, additives such as fillers and blowing agents, degree of crystallinity, and temperature. A temperature scan for the resin will reveal the T_g for amorphous resins and the peak melting temperature and heat of fusion for semicrystalline resins. The heat capacities for LDPE and PS resins are shown in Figure 4.15.

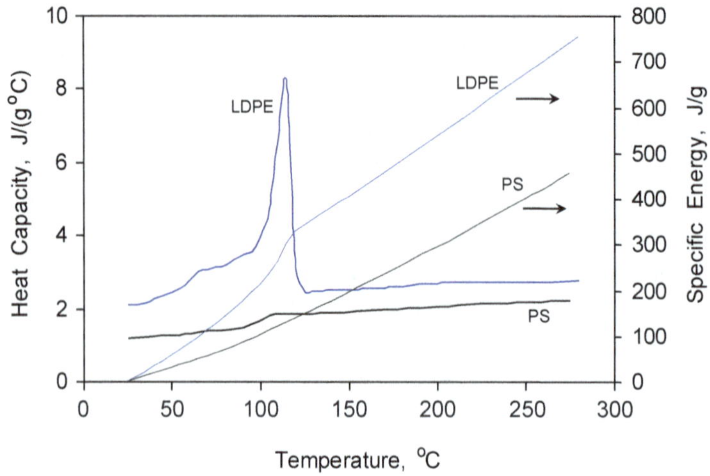

Figure 4.15 Heat capacity and specific energy for LDPE and PS resins as a function of temperature. The LDPE resin had a solid density of 0.922 g/cm^3

The specific energy is the amount of energy required to increase a unit mass of material from ambient conditions to the process temperature. It is easily calculated from the heat capacity by integrating as follows:

$$E_R(T) = \int_{25}^{T} C_v \, dT' \tag{4.3}$$

where C_v is the heat capacity, T is the specified temperature in degrees Celsius, and $E_R(T)$ is the specific energy for the resin. The specific energies for the LDPE and PS resins were calculated from the heat capacity and are shown in Figure 4.15. For example, if an extruder is plasticating LDPE resin and discharging it at a temperature of 220 °C, the specific energy required to increase the resin to this temperature is about 590 J/g. If the extruder motor inputted 700 J/g via the screw, then 110 J/g must have been transferred out through the barrel wall as an energy flux. Energy balances like this are very important for troubleshooting and scale-up since the ability to transfer energy through the barrel wall decreases as the size of the machine increases. Heat capacities for several resins are given in the references [27, 33, 34].

4.6 Thermal Conductivity and Heat Transfer

Thermal conductivity is the ability of a material to transfer energy via an imposed thermal gradient. For polymers, the thermal conductivity is relatively low and thus provides very desirable high thermal insulating properties. The thermal conductivity for polymers is difficult to measure, and thus the measurements tend to vary between methods and researchers. Table 4.2 shows typical thermal conductivities for carbon steel, stainless steel, glass, and several polymers in the solid state. In general, the thermal conductivities in the molten state are not much different, and these reported values can be used for most simulation and analysis work. As indicated by the data in the table, the thermal conductivity of carbon steel is nearly 300 times higher than that for LDPE resin. For processing, the low thermal conductivities of polymer melts limit the rate of heat transfer in and out of the equipment [35–38], and it strongly affects temperature measurements of molten polymer streams [39]. Thermal conductivity depends on the amount and type of fillers, additives, and blowing agents.

For most processing equipment, the low thermal conductivity of polymers strongly influences the overall heat transfer coefficient between the bulk of the polymer and the contacting metal surfaces, creating limitations in heat transfer rates. Heat transfer rates between processing equipment and the polymer depend on many factors, including thermal conductivity, machine clearances, and screw speeds, and they range between 160 $W/(m^2 \cdot °C)$ for flow through transfer pipes [42] to 600 $W/(m^2 \cdot °C)$ for twin screw extruders [35]. Heat transfer coefficients will be presented in more detail in Section 7.7.4.

Table 4.2 Thermal Conductivities for Selected Resins, Steel, and Glass

Material	Thermal Conductivity, W/(m·°C)	Reference
ABS	0.25	[40]
HDPE	0.24	[17]
LDPE	0.18	[17]
PC	0.26	[17]
PET	0.20	[40]
PMMA	0.19	[17]
PP	0.19	[40]
PS	0.15	[34]
PVC (rigid)	0.17	[40]
Glass	1.3	[41]
Stainless steel	17	[41]
Carbon steel	52	[41]

The temperature of molten polymer process streams is commonly measured using a thermocouple positioned through a transfer line wall and partially immersed in the polymer stream. Process stream temperature measurements that use an exposed-tip thermocouple, however, can be misleading since the temperature of the thermocouple junction is a balance between the heat transferred from the polymer stream and from the thermocouple assembly [39]. Due to the low heat transfer rate between the polymer and the exposed tip and the high thermal conductivity of the thermocouple sheath, the temperatures measured can be different by up to 35 °C depending on conditions. Extrudate temperatures, however, can be accurately measured using a preheated, handheld thermocouple probe. This method minimizes thermal conduction through the probe sheath.

4.7 Melt Density

Single-screw extruders and most other processing machines are volumetric metering devices. That is, the extruder will discharge a volume of resin for each revolution of the screw. Since the processor requires rate data in mass units (kg/h), the melt density is a needed physical property. The melt density for polymers is always less than the solid density of the material, and the melt density decreases further with increasing temperature. The melt density is mildly dependent on pressure, and at the relatively low operating pressures of an extruder the dependency is usually ignored. For example, the melt density of HDPE resin at 250 °C is 0.712 and 0.735 g/cm^3 at 0 and 20 MPa [43], respectively, a change of about 3% over the pressure range. It is also dependent on the amount and type of fillers, additives, and blowing agents. Melt densities (without blowing agents) can be easily measured using a melt indexer by weighing a known volume of extrudate. The melt densities for several common polymers as a function of temperature are shown in Figure 4.16. Additional melt density data can be found elsewhere [43].

Melt density can be measured for complex systems containing dissolved blowing agents and fillers using an extrusion system with a downstream gear pump [44]. For this method, the resin feedstock is plasticated and then flowed to a calibrated gear pump (volume/revolution). Pressure valves located before and after the pump are used to eliminate the leakage flow over the gear teeth due to a pressure gradient. The mass rate extruded is measured at the die and then divided by the volume rate based on the rotation level of the gear pump.

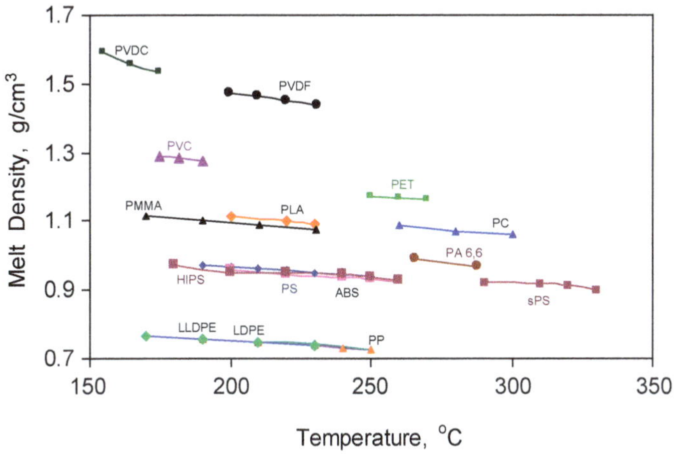

Figure 4.16 Melt densities for select resins as a function of temperature. PVC, PMMA, and PVDF resin data are from Zoller and Walsh [43]. All resins contain only extremely low level of additives

Changing resin types on processing equipment can change the rate and takeoff equipment velocities. For example, if a HIPS resin was replaced with a PLA resin, the line rate would change due to a change in the melt density. For this case and at a discharge temperature of 230 °C, the melt densities for HIPS and PLA are 0.95 and 1.09 g/cm^3, respectively. Thus based on melt density alone and ignoring all other factors, the specific mass rate for PLA would be expected to be about 15% higher than that for HIPS. Moreover, for the same mass flow rate the linear velocity at the discharge of a die would be about 15% less for PLA as compared to HIPS.

Filled resins and modified resins via blending are very common in the industry. If the components are different in density and immiscible, then the melt density can be easily estimated using a volumetric blending rule using Equation 4.4.

$$\rho = \frac{1}{\left[\dfrac{x_f}{\rho_f} + \dfrac{(1 - x_f)}{\rho_r}\right]} \tag{4.4}$$

where x_f is the mass fraction of filler in the resin blend, ρ_f is the solid density of the filler, ρ_r is the melt density of the resin, and ρ is the melt density of the blend.

As an example, a PE resin containing 8 weight percent talc was used in an extrusion application. The solid density of talc is 2.75 g/cm^3 [45], and at 190 °C, the melt density of PE resin is 0.760 g/cm^3. Using Equation 4.4, the melt density of the 8% blend is 0.807 g/cm^3. The melt density of the blend was measured using a melt indexer at 0.817 g/cm^3. Here the numerical estimation is likely good enough for rate estimates and numerical simulations.

Nomenclature

C_v	heat capacity
$E_R(T)$	specific energy for the resin as a function of temperature
f	coefficient of dynamic friction
K_{xy}	lateral stress ratio
P	pressure
T	temperature
x	independent variable in the horizontal direction (Equation 4.1)
x_f	weight fraction of the filler (Equation 4.4)
y	independent variable in the vertical direction (Equation 4.1)
ρ	melt density of the filled resin blend
ρ_f	solid density of the filler component
ρ_r	melt density of the resin
σ_x	stress in the x direction
σ_y	stress in the y direction
τ	shear stress at the polymer-metal interface

References

1. Maddock, B.H., "A Visual Analysis of Flow and Mixing in Extruder Screws," *SPE J.*, **15**, 383 (1959)
2. Spalding, M.A., Hyun, K.S., and Hughes, K.R., "Stress Distribution in Solid Polymer Compacts," *SPE ANTEC Tech. Papers*, **42**, 191 (1996)
3. Hyun, K.S. and Spalding, M.A., "Bulk Density of Solid Polymer Resins as a Function of Temperature and Pressure," *Polym. Eng. Sci.*, **30**, 571 (1990)
4. Derezinski, S., "The Compression of the Resin Solid Feed Bed in Extrusion," *J. Plast. Film Sheeting*, **5**, 271 (1989)
5. Smith, N.M. and Parnaby, J., "Bulk Density Versus Hydrostatic Pressure Characteristics of Plastics in Powder and Pellet Form," *Polym. Eng. Sci.*, **20**, 830 (1980)
6. Qiu, D.Q. and Prentice, P., "Influence of Pressure on Bulk Density of Polymer Solid Granules at Different Temperatures," *Adv. Polym. Technol.*, **17**, 23 (1998)
7. Darnell, W.H. and Mol, E.A.J., "Solids Conveying in Extruders," *SPE J.*, **12**, 20 (1956)
8. Schneider, K., *Chem. Ing. Tech.*, **41**, 142 (1969)

9. Umeya, K. and Hara, R., "Rheological Studies of Compaction for Polystyrene Powder," *Polym. Eng. Sci.*, **18**, 366 (1978)

10. Crawford, R.J. and Paul, D.W., "Radial and Axial Die Pressures During Solid Phase Compaction of Polymeric Powders," *Eur. Polym. J.*, **17**, 1023 (1981)

11. Kwade, A., Schulze, D., and Schwedes, J., "Determination of the Stress Ratio in Uniaxial Compression Tests – Part 2," *Powder Handl. Process.*, **6**, 2, 199 (1994)

12. Penumadu, D., Yamamuro, J., Abrantes, A.E., and Campbell, G.A., "Stress-Strain Behavior of Polymer Pellets," *SPE ANTEC Tech. Papers*, **43**, 224 (1997)

13. Thompson, M.R., Rahim, S., and Alotaibi, A.S., "Cohesion-Driven Recirculation of a Polymer Solids Bed in a Screw Simulator," *SPE ANTEC Tech. Papers*, **55**, 142 (2009)

14. Campbell, G.A., Nagarajan, G., Small, J.D., and Staples, J., "Solids Conveying in Low Friction and Starved Fed Extruders," *SPE ANTEC Tech. Papers*, **41**, 268 (1995)

15. Chung, C.I., Hennessey, W.J., and Tusim, M.H., "Frictional Behavior of Solid Polymers on a Metal Surface at Processing Conditions," *Polym. Eng. Sci.*, **17**, 9 (1977)

16. Spalding, M.A., Kirkpatrick, D.E., and Hyun, K.S., "Coefficients of Dynamic Friction for Low Density Polyethylene," *Polym. Eng. Sci.*, **33**, 423 (1993)

17. Mount, E.M. and Chung, C.I., "Melting Behavior of Solid Polymers on a Metal Surface at Processing Conditions," *Polym. Eng. Sci.*, **18**, 711 (1978)

18. McClelland, D.E. and Chung, C.I., "Shear Stress at Polymer/Metal Interface During Melting in Extrusion," *Polym. Eng. Sci.*, **23**, 100 (1983)

19. Kirkpatrick, D.E., Hyun, K.S., and Chung, C.I., "Friction and Its Effect on the Mechanical-to-Thermal Energy Conversion During Extrusion of Poly(Vinylidene Chloride)," *Polym. Eng. Sci.*, **33**, 1261 (1993)

20. Spalding, M.A. and Hyun, K.S., "Coefficients of Dynamic Friction as a Function of Temperature, Pressure, and Velocity for Several Polyethylene Resins," *Polym. Eng. Sci.*, **35**, 557 (1995)

21. Spalding, M.A., Hyun, K.S., Jenkins, S.R., and Kirkpatrick, D.E., "Coefficients of Dynamic Friction and the Mechanical Melting Mechanism for Vinylidene Copolymers," *Polym. Eng. Sci.*, **35**, 1907 (1995)

22. Spalding, M.A., Hyun, K.S., and Cohen, B.R., "An Experimental Study of the Frictional and Viscous Forces Responsible for Solids Conveying and Melting in Single-Screw Plasticating Extruders," *SPE ANTEC Tech. Papers*, **43**, 202 (1997)

23. Hogan, T.A., Spalding, M.A., Hyun, K.S., and Hall, M.J, "Extrusion of ISOPLAST Engineering Thermoplastic Polyurethanes," *SPE ANTEC Tech. Papers*, **45**, 124 (1999)

24. Spalding, M.A., Powers, J.R., Wagner, P.A., and Hyun, K.S., "Flow Surging in Single-Screw, Plasticating Extruders," *SPE ANTEC Tech. Papers*, **46**, 254 (2000)

25. Hong, B.K., Kim, H.S., and Chung, C.I., "Dependence of Melting Behavior on Melt Index," *SPE ANTEC Tech. Papers*, **47**, 272 (2001)

26. Spalding, M.A., Nelb, R.G., Patterson, J., Youngson, C., and Zawisza, M.J., "Processing Ignition-Resistant Polystyrene Resins In Injection Molding Machines," *SPE ANTEC Tech. Papers*, **47**, 476 (2001)

27. Stangland, E.E., Dooley, J., Spalding, M.A., Kim, E.K., and Hyun, K.S., "Fundamental Characterization of Polypropylene Extrusion," *SPE ANTEC Tech. Papers*, **48**, 302 (2002)

28. Hogan, T.A., Spalding, M.A., Cho, K.S., and Chung, C.I., "The Melting Characteristics of Polycarbonate Resins," *SPE ANTEC Tech. Papers*, **48**, 384 (2002)

29. Altinkaynak, A., Gupta, M., Spalding, M.A., and Crabtree, S.L., "An Experimental Study on Shear Stress Characteristics of Polymers in Single-Screw Extruders," *SPE ANTEC Tech. Papers*, **53**, 395 (2007)

30. Altinkaynak, A., Gupta, M., Spalding, M.A., and Crabtree, S.L., "Shear Stress and Melting Flux for Polycarbonate Resins," *SPE ANTEC Tech. Papers*, **54**, 1405 (2008)

31. Kim, H.S., Chung, C.I., and Butler, T.I., "Rubbing Mechanisms of Polymers on Metal Surfaces Relevant to Extrusion," *SPE ANTEC Tech. Papers*, **46**, 124 (2000)

32. Hyun, K.S. and Spalding, M.A., "A New Model For Solids Conveying In Single-Screw Plasticating Extruders," *SPE ANTEC Tech. Papers*, **43**, 211 (1997)

33. Tadmor Z. and Gogos, C.G., "Principles of Polymer Processing," John Wiley & Sons, New York (1979)

34. Moore, E.R., "Styrene Polymers," in "Encyclopedia of Polymer Science and Engineering," **16**, 1 (1989)

35. Todd, D.B., "Heat Transfer in Twin Screw Extruders," *SPE ANTEC Tech. Papers*, **34**, 54 (1988)

36. Jepson, C.H., "Future Extrusion Studies," *Ind. Eng. Chem.*, **45**, 992 (1953)

37. Kramer, W.A., "Extruder Barrel Cooling," *SPE ANTEC Tech. Papers*, **33**, 34 (1987)

38. Lobo, H. and Cohen, C., "Measurement of Thermal Conductivity of Polymer Melts by the Line-Source Method," *Polym. Eng. Sci.*, **30**, 65 (1990)

39. McCullough, T.W. and Spalding, M.A, "Predicting Actual Temperature Distributions in a Polymer Stream Using an Adjustable-Depth, Exposed-Tip Thermocouple Assembly," *J. Reinf. Plast. Compos.*, **16**, 1622 (1997)

40. Encyclopedia, 98, *Modern Plastics*, **74**, 13, (1998)

41. "Handbook of Chemistry and Physics," 55th ed., R.C. Weast (Ed.), CRC Press, Cleveland, 1974

42. Sato, S., Oka, K., and Murakami, A., "Heat Transfer Behavior of Melting Polymers in Laminar Flow Field," *Polym. Eng. Sci.*, **44**, 423 (2004)

43. Zoller, P. and Walsh, D.J., "Standard Pressure-Volume-Temperature Data for Polymers," Technomic Publishing Co., Inc., Lancaster, PA, 1995

44. Park, S., Park, C.B., Ladin, D., Naguib, H.E., and Tzoganakis, C., "On-Line Measurement of Polymer Melt Specific Volume Using a Gear Pump," *SPE ANTEC Tech. Papers*, **45**, 2471 (1999)

45. Dakota Matrix Minerals, "Talc Mineral Data," *http://webmineral.com/data/Talc.shtml#.XHakDK_f071*

5 Solids Conveying

Solid resin feedstock in the form of pellets, powders, and ground recycle must be conveyed at a high rate from the base of the hopper to the start of the melting section of the screw. This section is commonly referred to as the solids conveying or feed zone. A properly designed solids conveying process will provide a stable flow of material to the downstream sections of the process. As discussed in Chapter 1, the metering section of the screw must be the rate controlling step of the extrusion process for smooth bore machines, and all sections upstream, including the solids conveying section, must be able to supply the required rate. If the solids conveying process is not designed properly, many problems can occur including reduced rate, flow surging, and degradation of the resin. These problems will be discussed in detail in subsequent chapters. Historically, the solids conveying zone has been studied much less extensively when compared to the metering zone [1]. This follows in part from the notion that the solids conveying process can always operate at high enough rates to maintain the metering zone as the rate-limiting step of the process. However, many of the common problems that occur in production extruders have a root cause linked to the solids conveying zone design [2–6]. These difficulties are often due to a lack of understanding of the contribution of the solids conveying zone when considering screw design and process development.

This chapter provides a description of the solids conveying process and the theoretical models in the literature. The literature models will be presented before the experimental solids conveying data because only recently has experimental data become available for this process section. That is, the early theoretical models were developed without actual solids conveying data. Data will be presented regarding the temperature and forces that are associated with solids conveying of different polymers. Next, a comparison of the models with the experimental data will be provided. Both smooth bore and grooved barrel feed sections will be presented. The field experiences of the authors, however, are dominated by smooth bore extruders.

5.1 Description of the Solids Conveying Process

To review what has been presented in previous chapters, a typical single-screw extruder consists of a single screw within a barrel [7], schematically represented in Figure 5.1, and it is fed through a feed hopper. Although the feed hopper is not the focus of the chapter, a mathematical description can be obtained elsewhere [8, 9]. The slope of the hopper wall should be about 45° for polymers such as polyolefins and many amorphous thermoplastics. For materials that have a high pellet-to-pellet friction coefficient or are powders, the necessary angle is sometimes as high as 80 to 85°. The screw has four major sections that serve different functions in the extrusion process. Sections 1 and 2 compose the solids conveying zone of the extruder. These zones are responsible for the conveying of the small, solid polymer pellets, powder, or ground recycle from the hopper to the end of the solids conveying section of the screw where the solid bed is consolidated and the surface of some of the pellets are heated to near the glass transition or melting temperature. Section 1 has a water cooled casing and the resin particles are very loosely packed and under very low pressure (smooth bore). The barrel segment in Section 2 is heated and solids conveying occurs with a more tightly packed bed with moderate pressure [10]. Section 3 is referred to as the melting zone where the pellets are melted via dissipation from the motion of the screw and often assisted by external barrel heaters. As the pellets or particles are being compressed in the solids conveying and melting zones, the gas (for most processes this is air) in the void spaces between the particles is expelled out through the hopper as the pressure increases with the downstream direction. The molten resin is then fed to Section 4, the metering zone. By the time the molten resin reaches the discharge of the extruder, the resin should be a homogeneous fluid that can be at great pressure due to the flow constriction of the die. The screw geometry in Sections 1 and 2 is usually a relatively deep channel for smooth barrel machines where the polymer pellets are conveyed forward by the motion of the screw helix.

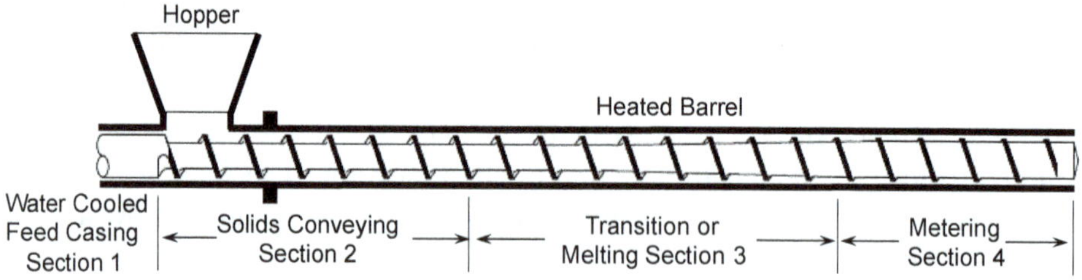

Figure 5.1 A typical plasticating extruder with a smooth barrel solids conveying section

Two basic styles of solids conveying zones were developed. The first style is referred to as a smooth-bore or smooth barrel extruder. This type of machine is the dominant style in North America, but they are also common in Europe and Asia. As the name implies, the entire feed casing and barrel length are smooth. The solids conveying occurs due to the motion of the flights and a higher level of dynamic frictional forces at the barrel surface as compared to the forces on the screw. The feed opening can be constructed using a water cooled feed casing. The casing is typically 2.5 diameters in length, often constructed with a liner, and is typically designed at the same diameter and tolerance as the barrel. The screw diameter in the casing is typically undercut such that the clearance is 3 to 4 times the normal clearance of the downstream screw sections in the barrel, although it is not standard practice by all extruder manufacturers. The larger clearance is to guarantee that the screw does not rub against the casing wall. The nominal length-to-diameter ratio of the extruder can be for just the barrel excluding the feed casing length or it can include the feed casing, depending on the specification practice of the extruder manufacturer. Some manufactures do not use a water-cooled feed casing as shown in Figure 5.1. Instead, an integral feed section is designed where the barrel extends back to the gearbox, an elliptical or rectangular hole is cut into the barrel for feed entry, and a hopper and water cooled heat exchanger are strapped onto the barrel segment. The heat exchanged maintains relatively cool temperatures near the feed throat such that bridging of the pellets does not occur. Integral feed sections often operate at higher temperatures than water cooled feed casing designs because of energy migration from the heated barrel zones. The high temperature in the integral feed section can provide improved solids conveying rates and higher extrusion rates, especially for PP resins [11]. Because of the potential higher rates, slightly larger motors are often specified for extruders with integral feed casings.

The second machine style is referred to as a grooved barrel extruder because it has relatively deep grooves cut into the water cooled feed casing. These grooves are typically deep near the feed opening and become shallower in the downstream direction until the barrel transitions into a smooth bore. The groove length is between 3 and 5 diameters in length with 3.5 diameters being typical. The grooves can be positioned in the axial direction or they can be designed in a helical pattern [12, 13]. Grooved barrel machines are common in Europe with German manufacturers leading the technology development. Solids conveying occurs due to the motion of the flights. The grooves in the feed casing keep the solids from rotating with the screw and force the solids downstream. This solids conveying mechanism has the capability of generating very high pressures and dissipating a high level of energy in the grooved section. Cooling on the feed casing is highly important so that melting does not occur in the grooves. If melting does occur in the grooves, the grooves can become ineffective and the solids conveying can be compromised. The screw design for a grooved barrel extruder is considerably different due to the development of the high pressures created by the grooves.

The next sections will describe the smooth bore and grooved barrel solids conveying mechanisms and designs. Due to the North American preference of smooth bore machines, the authors have extensive experience in these machines with limited experience in grooved barrel machines. Moreover, the melting and metering section chapters to follow provide experimental evidence first and then the models. But since solids conveying data did not exist in the early years, model development will be presented first with modern experimental evidence following.

5.2 Literature Review of Smooth-Bore Solids Conveying Models

The traditional analysis for modeling the solids conveying section of an extruder is based on the derivation by Darnell and Mol [14]. Their approach treated the polymer as a solid incompressible plug that traveled down the screw due to the lack of balance between pressure, contact forces, and torques that act on the polymer bed. The axial velocity of the solid plug depends on the screw geometry and the solids forwarding angle ϕ. Once the plug velocity and solids forwarding angle are determined, then force balance equations can be used to estimate the pressure generation. The derivation of the plug velocity equation is widely accepted in these types of conveying models. Model development will start here with the derivation of the solid plug velocity equation.

In the following development for the plug velocity, all velocities are measured relative to the screw. The plug velocity in the tangential direction, $V_{p\theta}$, and the plug velocity in the z direction, V_{pz}, are measured at the barrel surface. The plug velocity in the axial direction, V_{pl}, is independent of channel depth [1]. These velocity components are shown in Figure 5.2. The z direction is consistent with the unwrapped configuration found in Chapter 1. The volumetric flowrate, Q_s, for the solids in an extruder can be expressed as a function of the velocity of the plug in the axial direction (V_{pl}), and the apparent cross-sectional area of the channel (an integral on the radius from the screw surface R_c to the flight tip R_b), as provided by Equation 5.1.

$$Q_s = V_{pl} \int_{R_c}^{R_b} \left(2\pi r - \frac{pe}{\sin\theta} \right) dr \tag{5.1}$$

where R_c and R_b are the screw core radius and the barrel radius, respectively, e is the flight width, θ the helix angle, and p is the number of flight starts. Recall from Chapter 1, the helix angle is a function of the radius as the depth of the channel goes from 0 at the screw surface to the maximum height at the barrel wall of H.

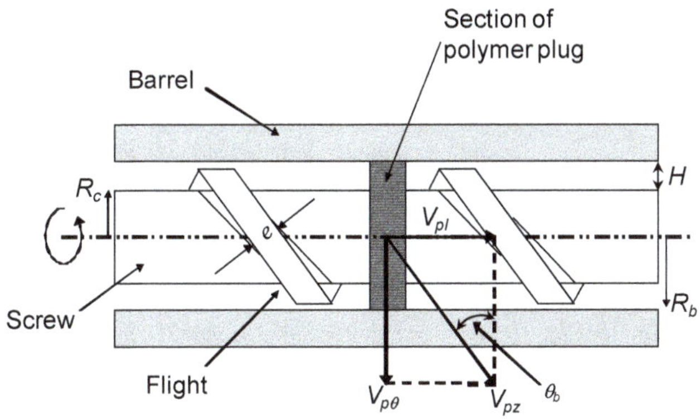

Figure 5.2 Schematic of a typical screw-barrel profile [1]

The r dependence of θ is then usually approximated as an average. Here, the average helix angle is designated as θ while θ_b and θ_c are the helix angles at the barrel and screw core, respectively. Equation 5.1 can be integrated and rewritten as follows:

$$Q_s = V_{pl}\left[\frac{\pi}{4}\left(D_b^{\,2} - D_c^{\,2}\right) - \frac{peH}{\sin\theta}\right] \tag{5.2}$$

where D_b is the barrel diameter and D_c is the screw core diameter.

The velocity of the plug in the down channel (z) and tangential (θ) directions can be vectorially visualized in Figure 5.3. Notice that the volumetric flow rate, Q_s, as well as all of the plug velocities are dependent upon the unknown solids forwarding angle ϕ.

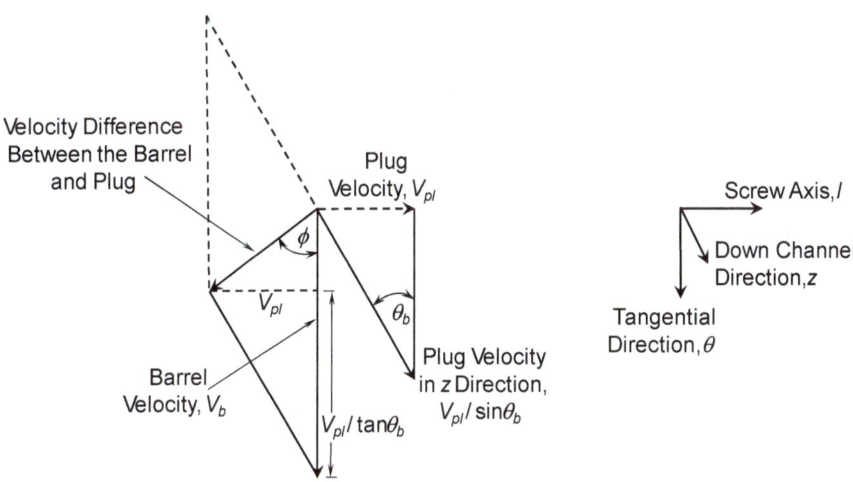

Figure 5.3 Vector diagram for the various velocities

From Figure 5.3, it is a straight forward process to determine that the following relationship exists for the solids forwarding angle φ:

$$\tan\varphi = \frac{V_{pl}}{V_b - \dfrac{V_{pl}}{\tan\theta_b}} \tag{5.3}$$

This can be rearranged to give:

$$V_{pl} = V_b \frac{\tan\varphi \tan\theta_b}{\tan\varphi + \tan\theta_b} \tag{5.4}$$

where V_b is the linear velocity of the screw tip at the barrel surface and is calculated as follows:

$$V_b = \pi N D_b \tag{5.5}$$

where N is the screw rotation rate in revolutions/s. Next, $D_b^2 - D_c^2 = 4H(D_b - H)$, is substituted into the volumetic solid flow rate equation Equation 5.2 as follows:

$$Q_s = \pi^2 NHD_b\left(D_b - H\right)\frac{\tan\varphi \tan\theta_b}{\tan\varphi + \tan\theta_b}\left[1 - \frac{pe}{\pi\left(D_b - H\right)\sin\theta}\right] \tag{5.6}$$

With another substitution, the above formula for Q_s can be simplified to:

$$Q_s = \pi^2 NHD_b\left(D_b - H\right)\frac{\tan\varphi \tan\theta_b}{\tan\varphi + \tan\theta_b}\left[\frac{W}{W + e}\right] \tag{5.7}$$

where the average channel width W is given by Equation 5.8. To be consistent with all chapters in this book, the channel width at the barrel and at the screw core are labeled W_b and W_c, respectively.

$$W = \frac{\pi}{p}\left(D_b - H\right)\sin\theta - e \tag{5.8}$$

The formulation for Q_s given in Equation 5.7 is widely accepted for all force balance literature models. The differentiation from model to model occurs in the assumptions made to resolve the unknown solids forwarding angle φ. The value of φ is typically between 2 and 7 degrees for commercial extrusions.

5.2.1 Darnell and Mol Model

The earliest solids conveying model for smooth-bore extruders that provided qualitative content was developed by Darnell and Mol [14] in 1956. They neglected the width of the flight and assumed that the resin moved as a plug and was in contact with all surfaces of the screw and barrel. The force balance on the plug provided the relationship between pressure and the solids forwarding angle φ. The forces were generated by the differential pressure gradient in the z direction and friction between the plug and the barrel and screw surfaces. Their derivation set the coefficient of friction at the barrel equal to that of the screw in all but one term. The solids forwarding angle for Darnell and Mol's model is as follows:

$$\cos\varphi = K\sin\varphi + \frac{D_c}{D_b}\sin\theta_c\left(K + \frac{D_c}{D_b}\cot\theta_c\right) + \frac{2H}{L}\left(K\frac{D_c}{D_b}\tan\theta_c + \left(\frac{D_b - H}{D_b}\right)^2\right)$$
$$+\frac{H(D_b - H)}{Z_b f_b D_b}\left(\frac{(D_b - H)}{D_b}\cos\theta + K\sin\theta\right)\ln\left(\frac{P_z}{P_0}\right) \quad (5.9)$$

where

$$K = \frac{(D_b - H)}{D_b}\frac{\sin\theta + f_s\cos\theta}{\cos\theta - f_s\sin\theta} \quad (5.10)$$

and where f_s and f_b are the dynamic coefficients of friction at the screw and barrel surfaces, respectively, Z_b is the helical length of the solids conveying section at the barrel wall, P_0 is the pressure at the start of the solids conveying channel due to the height of the pellets in the hopper, and P_z is the pressure at the end of the solids conveying section.

Darnell and Mol neglected the width of the flight ($e = 0$) so the formulation of Q_s was reduced in their model as follows:

$$Q_s = \pi^2 NHD_b(D_b - H)\frac{\tan\varphi\tan\theta_b}{\tan\varphi + \tan\theta_b} \quad (5.11)$$

Because Darnell and Mol neglected the width of the flight and allowed the dynamic coefficient of frictions to be set equal for part of the derivation, the model is rarely used. The model adaptation of Tadmor and Klein is used instead. This model will be described next.

5.2.2 Tadmor and Klein Model

As previously stated, the traditional analysis for modeling the solids conveying section of an extruder is based on the derivation by Darnell and Mol [14]. Their approach treated the polymer as a solid incompressible plug which traveled down the screw due to the lack of balance between the pressure gradient, contact forces, and torque forces which act on the polymer bed [14]. After Schneider's published work in 1969 on measuring the friction coefficient for beds of solid polymer pellets [15], Tadmor and Klein [1] reevaluated the original work by Darnell and Mol. In order to better conceptualize the movement of the solid pellets, the frame of reference was changed so that the barrel is the component of the extruder that rotates and not the screw. They then rationalized that the driving force for the movement of the plug was the friction force between the barrel surface and the solid plug [1]. The frictional forces at the screw core and flights on the plug retarded the motion and thus resisted the movement of the plug toward the end of the screw [1].

Tadmor and Klein [1] included an undetermined force, F^*, in their analysis, which was evaluated using a torque balance on the polymer bed. This force acted perpendicular to the polymer plug, at the pushing flight interface. The resulting model did not predict the modern experimental data well as will be shown later. At the time that this model was developed, experimental data for verification was not available. The derivation of the model was provided in the original writing [1] and it will not be presented here. The solids forwarding angle was related to the pressure gradient as follows:

$$\cos\varphi = K\sin\varphi + 2\frac{H}{W_b}\frac{f_s}{f_b}\sin\theta_b\left(K + \frac{D}{D_b}\cot\theta\right) + \frac{W_c}{W_b}\frac{f_s}{f_b}\sin\theta_b\left(K + \frac{D_c}{D_b}\cot\theta_c\right) + \frac{W}{W_b}\frac{H}{z_b}\frac{1}{f_b}\sin\theta\left(K + \frac{D}{D_b}\cot\theta\right)\ln\left(\frac{P_z}{P_0}\right) \tag{5.12}$$

where

$$K = \frac{D}{D_b}\frac{\sin\theta + f_s\cos\theta}{\cos\theta - f_s\sin\theta} \tag{5.13}$$

The model could be used to either calculate the pressure gradient or pressure at the end of the conveying section for a specified rate, or it can be used to calculate the rate for a specified pressure gradient or section exit pressure.

Two issues have been found with this model that limits its utility: 1) the model is very sensitive to the ratio of the barrel and screw friction factors and does not work well if at all when the screw friction factor is higher than that for the barrel, and 2) there must be a pressure at the inlet for the model to work effectively, this is of course not the case in lightly starve-fed extruders. This entry pressure limitation is a common problem for all the force balance models.

Tadmor and Broyer [16, 17] produced two more solids conveying models: a modified version of Tadmor's earlier model [1], which assumed isothermality [16], and a second model that allowed for thermal effects [17]. Later, a model by Strand et al. [18] was developed based on the Tadmor-Klein model for starve-fed solids conveying.

Like the original Darnell-Mol model, the Tadmor-Klein model was developed with only very limited rate data because solids conveying measuring devices were not available at that time. Moreover, dynamic coefficient of friction data were also unavailable during their model developments. Complete plasticating extrusion simulations, however, were developed with the Tadmor-Klein model using engineering friction coefficients to estimate the characteristics in the process.

5.2.3 Clarkson University Models

Several solids conveying models were developed by Campbell and his students at Clarkson University [19, 20]. These models will be referred to as either the Clarkson University models or the Campbell models. They proposed that the movement of the screw flight was pushing the polymer bed as the screw turns rather than the frictional force at the barrel moving the polymer pellets down the screw. For these models, they assumed that the solid bed behaved more like an elastic fluid rather than a solid and removed the torque balance constraint. Campbell and Dontula [20] reasoned that because the solid polymer pellets more closely resemble an elastic particulate fluid, no torque balance in the bed would be necessary. They further assumed that the force normal to the pushing flight was due to a combination of the force due to the pressure in the channel and a force proportional to the frictional force exerted at the barrel by the solid bed. The Campbell-Dontula model was first published as:

$$\sin\varphi = \cos\theta_b - \left(2\frac{H}{W_b}\frac{f_H}{f_b} + f_H + \frac{W_c}{W_b}\frac{f_c}{f_b}\right)\sin\theta_b - \frac{HW\sin\theta}{W_b f_b Z_b}\ln\left(\frac{P_z}{P_0}\right) \tag{5.14}$$

where f_H and f_c are the dynamic coefficients of friction at the flights and screw core, respectively. Equation 5.14 was written using constant physical properties over the entire solids conveying zone with a helical length of Z_b. It could be rewritten in a differential form to allow variable physical properties. The derivation of this and two other more recent models are found in Appendix A5.

While this model rendered a more accurate prediction of the solids rate as a function of exit pressure, screw and barrel parameters, and pellet physical properties, it still contained significant deviations from experimental conveying data at high discharge pressures [21]. This deviation was a result of the assumption in this adaptation of the model that the lateral stress ratio was 1.0, and was not included as a parameter due to lack of data in the literature. The lateral stress ratios in Cartesian coordinates are defined as:

$$K_{yz} = \frac{\sigma_y}{\sigma_z} \tag{5.15}$$

$$K_{xz} = \frac{\sigma_x}{\sigma_z} \tag{5.16}$$

Here, K_{ij} is the lateral stress ratio, σ_x and σ_y are the stresses in the x and y directions, and σ_z, in the case of the single-screw extruder, is assumed to be the local downstream pressure P. Lateral stress ratios were discussed in Section 4.2.

Unlike the previous models by Darnell and Mol [14] and Tadmor and Klein [1], which are based upon the assumption of isotropic stress conditions, Campbell's model [20] considered anisotropic stress conditions, as suggested by Schneider [15], but it was assumed to be 1.0 due to the lack of published experimental data on the subject. Variations on the model set forth by Campbell and Dontula [20] include a modification to incorporate the lateral stress ratio [19, 22], and other modifications discussed by Hyun et al. [21, 23]. A modified Campbell-Dontula model with a homogeneous lateral stress K_{xz} is as follows:

$$\sin\varphi = \cos\theta_b - \left(2\frac{H}{W_b}\frac{f_H}{f_b} + f_H + \frac{W_c}{W_b}\frac{f_c}{f_b}\right)\sin\theta_b - \frac{HW\sin\theta}{K_{xz}W_b f_b Z}\ln\left(\frac{P_z}{P_o}\right) \tag{5.17}$$

As stated in Appendix A5, if F_N is redefined in terms of an internal angle, then the following model results and is referred to as the Campbell-Spalding model in the appendix:

$$\begin{aligned}\sin\varphi = {} & \cos\theta_b \sin(\frac{\pi}{2}+\varphi+\theta) - \left(2\frac{H}{W_b}\frac{f_H}{f_b} + f_H \sin(\frac{\pi}{2}+\varphi+\theta) + \frac{W_c}{W_b}\frac{f_c}{f_b}\right) \\ & \sin\theta_b - \frac{HW\sin\theta}{K_{xz}W_b f_b Z}\ln\left(\frac{P_z}{P_o}\right)\end{aligned} \tag{5.18}$$

Hyun et al. [21] evaluated both the original model by Darnell and Mol [14] and the model by Campbell and Dontula [19] for accuracy against experimental data, and determined that the Darnell-Mol model was less accurate than that of the Campbell-Dontula model. The incorporation of the lateral stress ratio in the calculations supported their conclusions even more. At the time of the work by Hyun et al., however, the physics for screw rotation was not well appreciated, and the evaluations for the Campbell models [23] were performed with coefficient of friction factors based on sliding velocities using barrel rotation. These different sliding velocities will be discussed in detail in Section 5.5.

Yamamuro, Penumadu, and Campbell [22] further revised the work of Campbell and Dontula [20] by including the normal and tangential stresses on the polymer plug using K_{xz} within the model [24]. Although the parameter K_{xz} could be experimentally determined [24], it was modified to an optimal value for each system; a value that yielded the best fit for the experimental data [24]. They also included a skew angle α in degrees which was reported to be primarily a function of barrel temperature as follows:

$$\alpha = 33 + 0.0004\left(T_b - 50\right)^{2.5} \tag{5.19}$$

where α denotes the angle that the force on the solid bed interacts with the advancing flight, and T_b is the temperature of the barrel in Celsius. The parameters in Equation 5.19 have units. These units have been omitted for clarity of the equation. The Yamamuro-Penumadu-Campbell model is as follows:

$$\sin\varphi = \sin(\alpha+\varphi) - 2\frac{H}{W_b}\frac{f_H}{f_b}\sin\theta_b - f_H\sin(\alpha+\varphi+\theta_b)\sin\theta_b - \frac{W_c}{W_b}\frac{f_c}{f_b}\sin\theta_b - \frac{HW\sin\theta}{K_{xz}W_b f_b Z_b}\ln\left(\frac{P_z}{P_o}\right) \tag{5.20}$$

As mentioned earlier in this section, each of these models must be solved for either $\sin\varphi$ or $\cos\varphi$ and $\tan\varphi$ as needed for Equation 5.7 to predict the solids flow rate Q_s. The following two identities are used for the transformation:

$$\tan\varphi = \frac{\sin\varphi}{\sqrt{1-\sin^2\varphi}} \tag{5.21}$$

$$\tan\varphi = \frac{\sqrt{1-\cos^2\varphi}}{\cos\varphi} \tag{5.22}$$

These continuum statics based models have very limited utility because solids conveying is a dynamic process. Of these three models developed in the past 15 or so years, the model by Yamamuro, Penumadu, and Campbell has the most generality because of the parameters K_{xz} and α in Equation 5.20. This parameter α may aid the model to account in part for the recent determination that the bed does not contact the trailing flight with the same force as expected from the lateral stress coefficient. The recent publications by Moysey and Thompson [25–27] using discrete element methods demonstrate some of the short comings of these models. These publications present theoretical results that indicate that the bed does not transfer the expected stress to the trailing flight.

5.2.4 Hyun and Spalding Model

The Hyun and Spalding [23] model presented here is similar to the ones presented previously except that the force at the pushing flight has a magnitude and acts at an angle that satisfies two torque balances, the model includes the lateral stress ratio and energy balances, and it utilizes realistic physical property data. The energy balances are very similar to those developed by Tadmor and Broyer [16] and they will not be presented here. The equations were developed using the following basic assumptions: the compacted solids act like a rigid body and flow as a plug, the solids contact all sides of the channel and barrel wall, and the pressure is anisotropic in the plane perpendicular to the flight tip. A single lateral stress ratio was assumed to describe adequately the anisotropic pressure distribution, but stress ratios in multiple directions could have been included; that is, like the model developed by Schneider [15]. The standard assumptions of constant bed velocity, constant bulk density, and coefficients of dynamic friction that are independent of pressure (i. e., constant coefficients at both the barrel and screw surfaces) were not used here. Instead, the bulk density was allowed to be a function of temperature and pressure, and the coefficient of dynamic friction was a function of temperature, pressure, and velocity as discussed in Chapter 4. The solid bed velocity was permitted to change to satisfy mass conservation.

The solids forwarding angle φ is related to rate by Equation 5.7, and the force and torque balances provide the relationship between and the pressure gradient and φ using Equation 5.23. Here, the equation is written in differential form so that integration can be performed numerically in the z direction using variable physical property data.

$$\cos\varphi = K\sin\varphi + 2\frac{Hf_s}{W_b f_b}\sin\theta_b\left[K + \frac{D}{D_b}\cot\theta\right] + \frac{W_c f_s}{W_b f_b}\sin\theta_b\left[K + \frac{D_c}{D_b}\cot\theta_c\right] + \frac{WH}{K_{xz} f_b P W_b}\sin\theta\left[K + \frac{D}{D_b}\cot\theta\right]\frac{\partial P}{\partial z} \tag{5.23}$$

where

$$K = \frac{D}{D_b}\tan\theta \tag{5.24}$$

Derivation of Equation 5.23 is provided in Appendix A5.

The Hyun-Spalding model was developed using barrel rotation physics. For this process, the relative sliding velocity between the solid bed and the barrel wall is considerably higher than that between the solid bed and the screw surfaces. These sliding velocities will be discussed in Section 5.5. For the higher sliding velocity at the barrel wall, f_b was always higher than f_c for the original publication [23], and the model appeared to predict the solids conveying performance of the experimental data. With the discovery that screw rotation physics provides different sliding velocities for solids conveying, the value of f_b was at most equal to f_c and likely lower. Screw and barrel rotation physics will be covered in detail in Chapters 6 and 7. With screw rotation physics, the model did not generate pressure very well for this LDPE resin system. The modified Campbell-Dontula model is considered the best smooth bore solids conveying model developed.

5.2.5 Moysey and Thompson Model

A three-dimensional, non-isothermal discrete particle model was developed by Mosey and Thompson [25–27]. The model considers the interactions and collisions of all pellets (or particles) in the feed hopper and the solids conveying section of the screw using a discrete element method. The method did not show an exponential increase in pressure as predicted by the models discussed previously. This result was believed to be caused by the compressibility of the solids in the bed. An anisotropic stress field predicted the highest pressure in the screw channel to occur at the screw root, and the lowest pressure at the trailing flight. The model also predicted recirculation flow of pellets in the feed throat. Although the model provided many insights into the process, the high computational effort has limited its use.

5.3 Modern Experimental Solids Conveying Devices

The Darnell-Mol [14], Schneider [15], and Tadmor-Klein [1] models were developed with only solids conveying data from fully operating extruders. That is, their data for verification of their models were confounded with melting and metering operations. Because of this lack of data and experimental rigs, the understanding of the solids conveying process has developed at considerably slower rates than other sections of the screw. Even today, solids conveying is the least understood section of the machine. Several new devices have been utilized recently to gain first order trends that are related to the solids conveying process using a conventional solids conveying screw section. One of these devices allows the particle motion to be video tracked; two other devices evolved to evaluate the rate, discharge pressure, torque and thrust on the barrel, and the heat generation at the polymer barrel interface as a function of screw rotation rate. The devices and research were performed independently at Clarkson University and at Dow by the authors.

Prior to these devices, Zhu and Chen [28] built a 45 mm diameter extruder with glass windows positioned in the barrel that allowed observation of the flows of a fully functioning machine. The extruder had a length-to-diameter ratio of 25 and thus it allowed the observation of all sections of the process. In the solids conveying section, they observed that the pellets had relative movements between them and that the pellets did not behave as a ridged solid plug. This result is consistent with Moysey and Thompson [25–27].

5.3.1 Solids Conveying Devices at Clarkson University

Two solids conveying devices were built at the Clarkson Polymer Extrusion Consortium. The first device was used to visualize the flow in the section without a discharge pressure. The second device was built such that a relatively high discharge pressure could be applied to the process.

As mentioned in the previous section, Moysey and Thompson [25–27] found that by using the discrete element method, the pellets were not fully compressed against the trailing flight of the extruder. As part of the effort of the Clarkson Polymer Extrusion Consortium, data were taken to evaluate the interaction of the solid polymer pellets with the screw and barrel interface. The photographs presented in Figure 5.4 were taken in 1992 using a deep channel screw enclosed in a glass barrel. These data are consistent with this theoretical prediction by Moysey and Thompson. The arrows in Figure 5.4 highlight the motion of a tracer particle (black pellet) in the solid bed as the screw is rotated. The tracer particle of interest is the one approaching the trailing flight, indicated by the arrow in Figure 5.4(a).

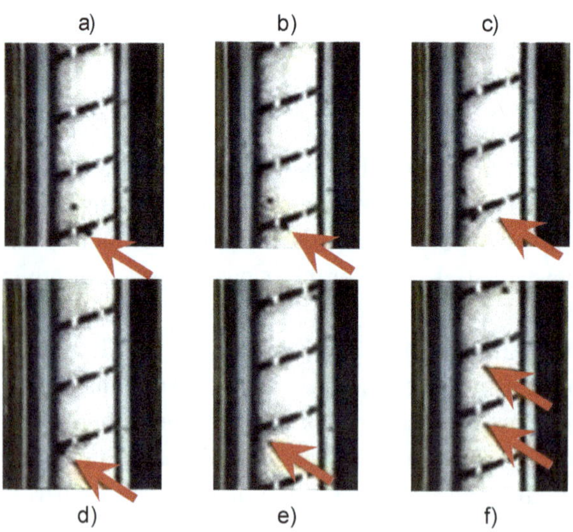

Figure 5.4 Polymer pellet motion in a single screw. The zero pressure solids conveying device at Clarkson was used for these experiments

The tracer particle in Figure 5.4(a) is transported closer to the trailing flight in Figure 5.4(b). In Figure 5.4(c), the particle begins to interact with the flight and in Figure 5.4(d), it is seen to rotate as the bed pushes it onto the flight. In Figure 5.5(e) the particle is no longer visible because it has been covered by white pellets. The two arrows in Figure 5.4(f) point out a small depression in the bed next to the trailing flight; this is consistent with the discrete element modeling [25–27] and strongly suggests that the force on the trailing flight is less than that on the advancing flight which is pushing the pellets helically up the extruder. Examining the path of the tracer particle just discussed shows that the particle is on a helical path very closely resembling the vector normal to the advancing flight. This indicates that the solid bed is being transported up the screw primarily by the normal force of the flight.

The zero pressure solids conveying device was used to study the dynamics of the pellets. The photographs presented in Figure 5.5 show the dynamics of the polymer pellets as they interact with the screw solid surfaces. In the experiment the extruder was empty initially and the solid pellets with tracers were fed into the hopper using a weigh feeder. The rate of feed was constantly increased until the screw was full. Figure 5.5 represents the time in the experiment where the screw was just full enough so that the screw core surface could act adhesively as a result of the polymer metal friction on the pellets to cause them to be moved. As the screw filled, it acted as a plug flow device with no pellet back flow, until the situation found in Figure 5.5 where the channel fill was slightly higher, about one pellet diameter, than the top of the screw core. The arrow in frame Figure 5.5(a) is pointing at the solid bed flight interface. Note that there is a round reflection on top of the screw core from the lights used during the videotaping in Figure 5.5(a).

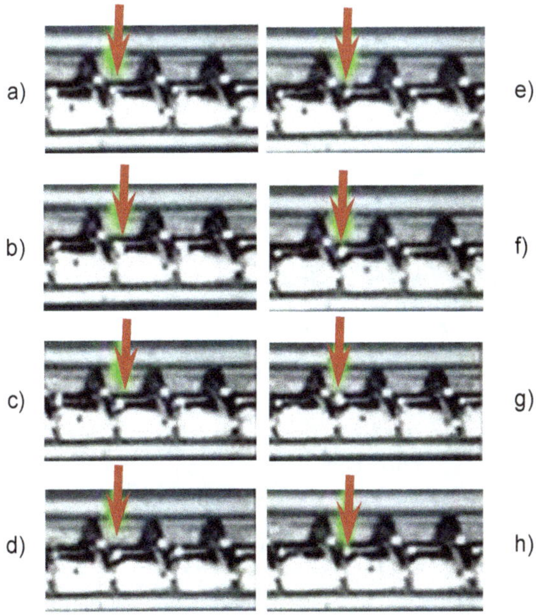

Figure 5.5 Effect of screw-particle adhesion on the motion of polymer pellets. The zero pressure solids conveying device at Clarkson was used for these experiments

In Figure 5.5(b) a pellet has been dragged back out of the bed and has begun to cover the reflection. In Figure 5.5(c) the pellet completely covers the reflection and in Figure 5.5(d) the pellet is being moved by the flight and the core into the empty side of the screw. This effectively moves the pellet a short distance back toward the inlet and it possibly accounts for the recirculation flow that caused the tracer to move as demonstrated in Figure 5.5(f), (g), and (h) that document another pellet being dragged out of the bed. The photographs presented in Figure 5.4 and Figure 5.5 are in agreement with the discrete particle simulations of Moysey and Thompson [25–27] and suggest that the assumption in all the models presented above that the lateral stress coefficient is homogeneous in the solids conveying channel is suspect. Therefore, when using any of these models the results should be used only with great care.

The Clarkson group also developed a device that allowed for a relatively high pressure at the discharge of the device [10, 29, 30]. As with the device discussed previously, this solids conveying device was very similar to the extruder in Figure 5.1, where Sections 3 and 4 have been removed, and a flow-restricting device is put in their place to limit conveying rates and vary the exit pressure of the polymer bed. A schematic of the device is shown in Figure 5.6 and a photograph shown in Figure 5.7. The device was fabricated from a 58 mm diameter extruder, with a shorter, floating barrel made so that the effective *L/D* of the extruder was 3.5. The barrel temperature was controlled by a four-zone, forced-air system, and the screw was core drilled and temperature controlled by a single-zone forced air stream introduced just under the beginning of the hopper. The barrel thickness was kept to a minimal 3 mm for optimal heat transfer characteristics. The screw was designed with a lead length of 58 mm (square pitched) and a channel depth of 8 mm. A 2.2 kW variable speed direct current (DC) main drive motor used a chain drive to turn the screw that conveys the solid polymer pellets from the hopper, through the barrel, to a pressure discharge control unit. The pellets are then sent to a material collector for weighing.

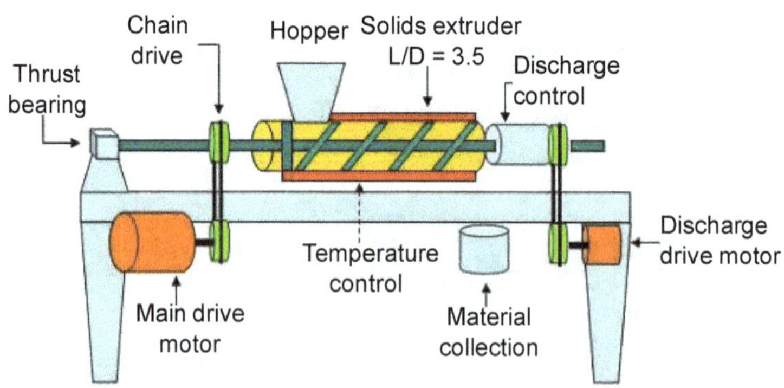

Figure 5.6 Schematic of the solids conveying device at Clarkson with the capability of applying a pressure at the discharge end of the device

Figure 5.7 Photograph of the Clarkson solids conveying device. The device is shown schematically in Figure 5.6

Figure 5.8 Solid pellet discharge control device. The device is shown schematically in Figure 5.6

The mass flow control device positioned at the discharge end of the extruder provided a controlled force or pressure to the discharge. The flow control device was just downstream of the red semicircle shown in Figure 5.7. It was designed with the goal of overcoming the flow restrictor limitations in the literature devices [10, 29]. A photograph of the flow control is presented in Figure 5.8. The flow control was machined similar to a continuous strand cutter with about half of the working surface machined away into scalloped cavities. The cutter was supported inside of a smooth tube such that it could move in the down screw direction to facilitate the exit pressure measurement. The drive on the cutter was reversible and was operated over the full range in order to take the pellet flow data at different pressures.

The barrel "floats" on the screw, rather than being permanently affixed to the frame of the device itself. This facilitates the measurements of barrel torques and thrusts as a function of operating conditions and the type of polymer being conveyed. Figure 5.9 schematically details the extruder section of the solids conveying device and the instrumentation location [10]. Dark red arrows indicate force measurements, and light yellow rectangles represent temperature and heat flux measurements. A total of five force devices allowed the measurement of the screw thrust in the backwards direction, barrel torque, barrel thrust, exit pressure, and the conveyed pellet weight at the discharge. These resultants were measured continuously using a high-speed data acquisition system. In addition, five temperature and heat flux measurements were taken by a computer data acquisition system at one second intervals and stored for later analysis. The temperature and heat flux devices were spaced regularly down the outside of the barrel.

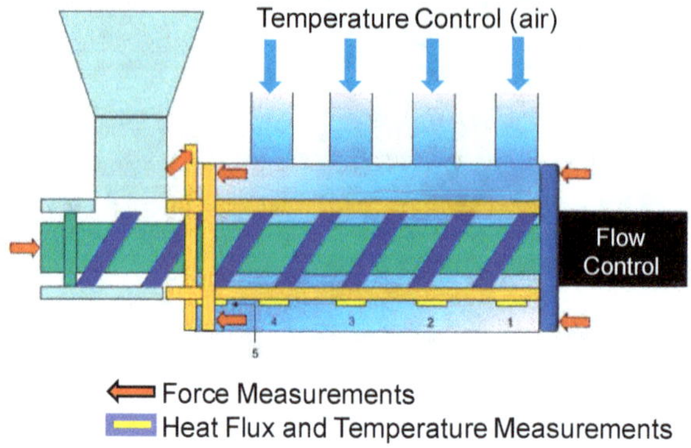

Figure 5.9 Details of extruder and data collection locations

Commercially available heat flux sensors with thermopiles sandwiched at the interface were used to measure the local temperatures and heat fluxes; that is, Omega Corporation, Model HFS-4 devices. The total thickness of the sensors was nominally less than 0.18 mm, and a schematic of the device is shown in Figure 5.10. By measuring the temperature difference across the center film (ΔT) and assuming one-dimensional heat transfer, then the heat flux can be measured using the temperature difference and the thermal conductivity of the film. The local temperature is recorded using the thermocouple nearest the barrel. The sensors were calibrated at ambient condition with zero heat flux.

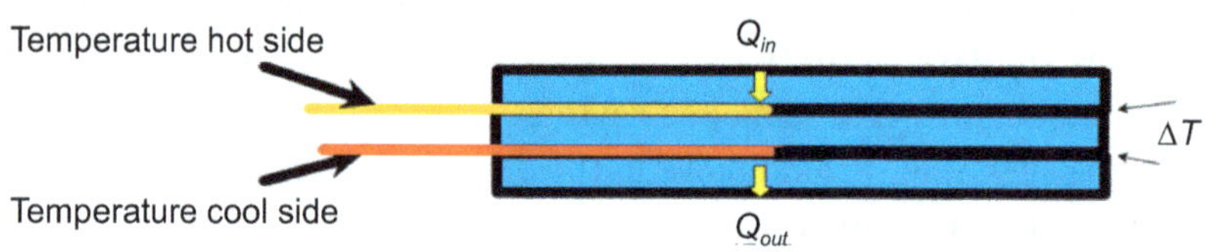

Figure 5.10 Heat flux and temperature sensor. Here the heat flux in and out of the film are designated as Q_{in} and Q_{out}, respectively

In order to see if the system could provide the flexibility to evaluate high modulus polymer pellets, an initial comparison of an HDPE and PS resins were evaluated. The experiments were conducted at a screw speed of 50 rpm and barrel temperatures of 24 and 35 °C for the HDPE and PS resins, respectively. The temperature of the screw was maintained at the barrel temperature. These conditions are similar to what might be observed in the feed casing and very early into Section 2. As shown in Figure 5.11, the HDPE resin exhibits lower conveying rates for discharge pressures greater than about 0.3 MPa as compared to the PS resin, even though the two have been reported to have roughly the same friction factor [29] at these low processing temperatures. Moreover, the solids conveying rates decreased as the discharge pressure increased. The decreasing rate with increasing discharge pressure was observed on almost all experiments. For a complete extruder, a 58 mm diameter machine would typically have a meter channel depth of about 3 mm. Based on the calculations presented in Chapter 1, this metering channel depth would have a rotational flow rate of 28 and 36 kg/h for the HDPE and PS resins, respectively. From Figure 5.10, this process would be able to maintain discharge pressures of 0.5 and 0.7 MPa for the HDPE and PS resins, respectively. These pressures are in the acceptable range for the just the feed casing. Additional pressure would be developed in the solids conveying in Section 2 where the barrel temperatures would be considerably higher and different for these resins.

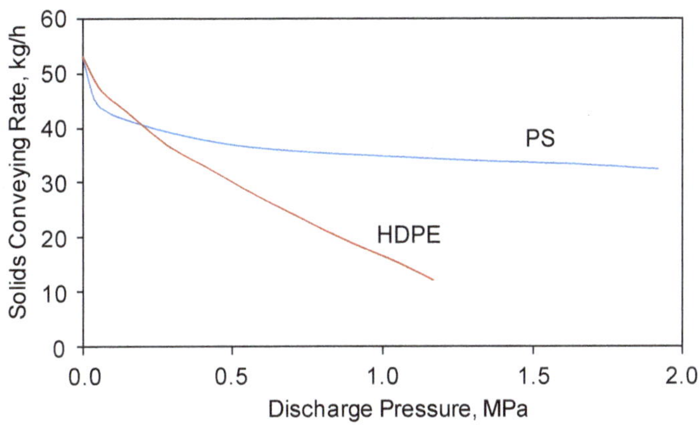

Figure 5.11 Solids conveying as a function of discharge pressure for HDPE and PS resins using the Clarkson solids conveying device. The data were collected at a screw speed of 50 rpm and barrel temperatures of 24 and 35 °C for the HDPE and PS resins, respectively. The temperature of the screw was maintained at the temperature of the barrel

Many hypotheses have been presented regarding this observation of differing feed rates for differing materials; however, none to date have proved conclusive. In order to develop a better semiquantitative understanding of this material-dependent pellet conveying phenomena, a series of experiments were conducted in Clarkson using the device shown in Figure 5.7 [29]. Several different polymers were evaluated in the solids conveying device, including LDPE, LLDPE, HDPE, PP, HIPS, and PS. A sampling of the data obtained is presented in the following paragraphs.

The goal of this preliminary investigation was to gain more knowledge of the interaction between solid polymer physical properties and their relation to extruder rate and energy dissipation characteristics in the solids conveying zone. The data presented here will focus on the effects of temperature and polymer type on solids conveying characteristics for a constant screw rotation rate of 50 rpm. The experiments were conducted maintaining screw rotational velocities and barrel and screw temperatures for any given run. To achieve this, variables such as air temperature and flow rate, and main and discharge motor drive power had to be adjusted for each run. After a thorough cleaning of the system (if the material is changed), the hopper, which had a slide valve just above the barrel, was loaded with the material to be tested (but the valve was not opened), the screw main drive and discharge drive motors were turned on, and power was varied to set the proper screw rotational speed (50 rpm). Then, the heat and/or cooling air is turned on, along with the data acquisition monitoring software, so that the operator could determine when a steady state has been achieved. Once the temperature has remained steady for a given period of time (normally one to two minutes), the hopper was opened and the material was conveyed through the extruder to the collection bucket. Once a steady state was achieved as determined by constant rate, barrel temperature, and the heat flux, the extruder was allowed to run for ten to fifteen minutes while the run data were acquired to allow for good time-averaging of the results. During this process, the barrel temperature profile, the screw rotational speed, and the exit pressure must be monitored and adjusted, if necessary, to maintain relatively constant steady-state values. Some variation was always observed due to the granular nature of the feed and its compaction characteristics and passage of the resin through the discharge device.

If the material has been changed, a thorough cleaning will not effectively remove all polymer residues from the barrel and screw surfaces as suggested by Schneider [15]. Schneider observed that the apparent friction factor for a given material (on carbon steel) would change gradually after materials were changed, and it would level off to a constant value after roughly 10,000 cycles. He termed this the "rubbing in" period of the screw and barrel, where a very thin layer of the polymer is deposited on the metal surface. Because of this effect, upon changing materials, the machine was operated for roughly 3,000 rotation cycles (one hour) at a modest exit pressure to "rub-in" the entire surface area of the screw and barrel. This same type of conditioning was required for measuring the coefficient of dynamic friction using a Screw Simulator [3].

The heat fluxes and temperatures obtained were used to calculate the interfacial temperature between the polymer and the barrel. Equation 5.25 and Equation 5.26 from Çengel [31] give the relationship needed to calculate the interface temperatures:

$$Q_{cyl} = \frac{T_o - T_i}{R_{cyl}} \tag{5.25}$$

$$R_{cyl} = \frac{\ln(r_o / r_i)}{2\pi L_b k} \tag{5.26}$$

where Q_{cyl} is the appropriate rate of heat transfer, T_o is the measured temperature on the outside barrel surface, T_i is the solid plug-barrel interface temperature, R_{cyl} is the thermal resistance of the barrel, r_o and r_i are the radii of the outer and inner surfaces of the barrel, L_b is the axial length of the barrel, and k is the thermal conductivity of the barrel. Appropriate values for k (mild carbon steel) were obtained from the following sources: Kreith and Bohn [32] and Cubberly [33]. A value of 52 W/(m·°C) is reported in Table 4.2.

Equation 5.25 is the appropriate relationship to determine the temperature gradient in the barrel assuming one-dimensional conductive heat transfer. However, the data collected was not simply a rate of heat transfer, it is a heat flux. Therefore, the following will hold:

$$Q_{cyl} = 2\pi r_o L_b q \tag{5.27}$$

where q is the locally measured value of the heat flux and it is positive for a flux into the barrel. This may be substituted into Equation 5.25, and simplified to yield an equation explicit in T_i,

$$T_i = T_o - \left[\frac{r_o q}{k}\right] \ln\left[\frac{r_o}{r_i}\right] \tag{5.28}$$

While a one-dimensional model for steady-state heat conduction might seem an oversimplifying assumption, values for a typical heat flux range from 0 up to 10 kW/m^2. With a one-dimensional model, this translates to about a maximum 0.5 °C temperature gradient within the barrel metal in the radial direction. If a two-dimensional model is used, the temperature gradient will decrease even further, and thus have virtually no impact on the interfacial temperature calculations.

A couple points are emphasized: 1) the authors have found no published effort to determine the forces on the barrel as a function of material type and solids conveying discharge pressure, and 2) unlike conventional barrels that are thick to induce a maximum thermal capacitance and strength, the barrel in this device was machined to a minimum thickness in order to enhance heat transfer and thus allow a good estimate of the inner wall temperature from the outside temperature measurements and the local heat flux.

5.3.1.1 Flow Rate and Exit Pressure

As previously discussed, the solids conveying rate decreases as the discharge pressure increases for all materials studied. For HDPE resin, the solids conveying rate decreased almost logarithmically with discharge pressure, as shown in Figure 5.12. This is a characteristic of most crystalline polymers, and has been well documented in prior literature [21] as well as predicted by all of the major models [1, 14, 19, 20, 22, 23]. The HDPE resin demonstrated the greatest solids conveying pellet flow differences between isotherms (data at different temperatures) when compared with the rest of the polymers examined. The data were collected at different barrel and screw temperatures. For the data in Figure 5.12, the temperature of the screw was maintained at the same temperature as the barrel.

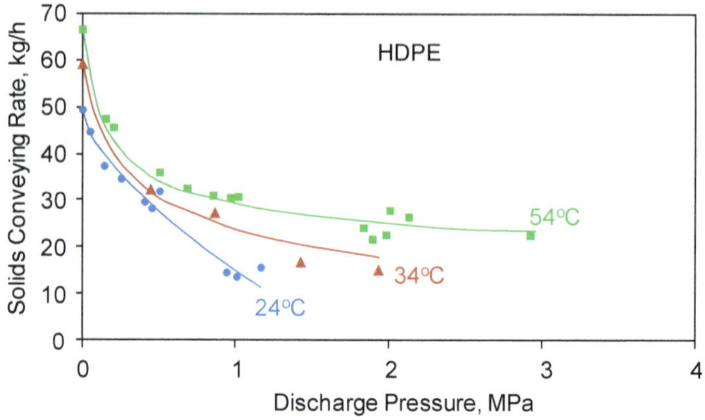

Figure 5.12 Effect of barrel and screw temperature on HDPE resin solids conveying. The temperature of the screw and barrel were maintained at the same temperature

The decreasing solids conveying rate with increasing discharge pressure is attributed to two factors: 1) higher restrictions or discharge pressures always reduce the flow rate through an orifice, and 2) the dynamic coefficient of friction depends on temperature. As the discharge pressure increases, the frictional heating at the inside barrel wall causes the surface temperature to increase. Many polyolefins, but not all, have coefficients of friction that decrease with increasing temperature in this temperature range [3, 34].

The dependence of rate on the discharge pressure was considerably less for HIPS resin as compared to HDPE resin. As shown in Figure 5.13, the solids conveying rate for HIPS resin decreased to a lesser extent with increasing discharge pressure, and the rate did not seem to be dependent on temperature in this temperature range. This result is consistent with the dynamic coefficient of friction for HIPS resin, as shown by Figure 12.17. As shown in Figure 12.17, the coefficient did not depend to a high level on temperature or velocity in this low temperature range.

Several of the most commonly used resins were studied at a screw and barrel temperature of 35 °C. As previously discussed, this temperature condition is comparable to the conditions in the feed casing or Section 1. This is just the start of solids conveying as conveying continues into Section 2 where the inside barrel wall temperatures are considerably higher. In order to visualize the contrast between the six different polymers tested, the solids conveying rates as a function of discharge pressure for these select resins are presented in Figure 5.14.

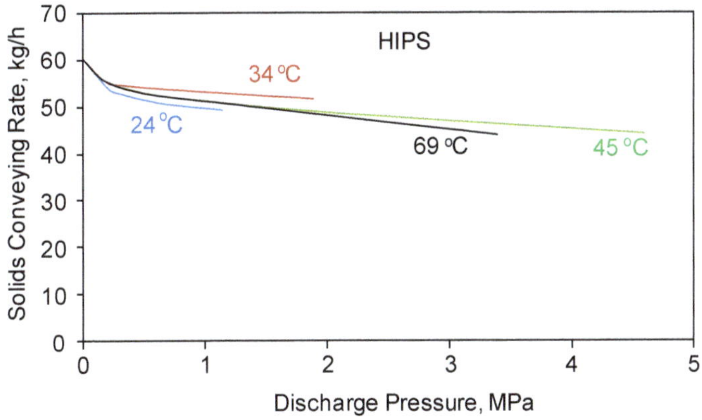

Figure 5.13 Effect of barrel and screw temperature on HIPS solids conveying. The temperature of the screw and barrel were maintained at the same temperature

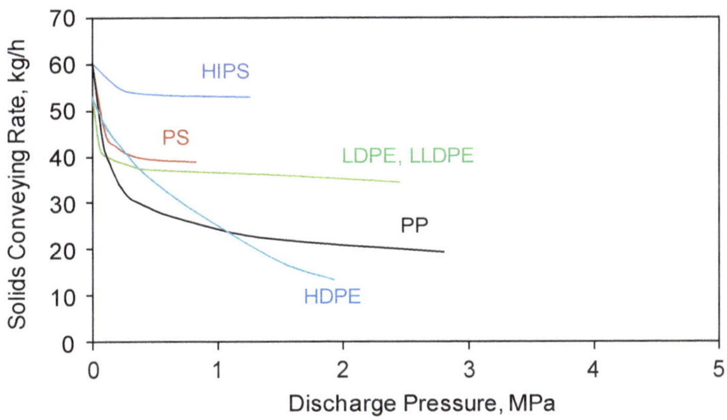

Figure 5.14 Comparison of the effect of exit pressure on solids polymer conveying for select resins. The screw and barrel were maintained at 35 °C

The polyolefin resins were more pressure sensitive than the two amorphous resins, both of which had a T_g of about 90 to 100 °C. HDPE was the most sensitive to pressure and HIPS the least with regard to solids conveying at this temperature. These data are another graphic demonstration that the concept of a general purpose screw is fraught with potential difficulty when changing polymers. Different polymers solids convey at very different rates under the same process conditions. Consider a case if the extruder is being used to convey PP resin with an expected pressure at the end of Section 1 of 1.2 MPa, and the decision is made to now use the extruder for PS; the solids conveying rate will be about 100% higher and there is a good chance that the product will have unmelted solids in the extrudate because the solids could be forced through the melting zone at a rate such that they will not have time to melt completely. If a properly designed screw for ABS resin is switched to run HIPS resins, the screw will likely be too shallow in the feed section to run HIPS effectively [35]. Although the equipment required to extrude these two resins are nearly identical, a screw designed for ABS will run HIPS at reduced rates, especially at high screw speeds. The difference in the rates is caused by the solids conveying behavior of the resins. These solids conveying properties will be discussed in Chapter 12. The LDPE and LLDPE resins have similar solids conveying rates because the additive levels and the solid densities were nearly identical at 0.92 g/cm^3. If these resins were compared to an LLDPE resin with a solid density of 0.89 g/cm^3, the solids conveying rates would be considerable higher for the resin with the lower solid density.

5.3.1.2 Thermal Effects

A series of experiments were performed where the barrel was at ambient temperature at the start of the run. The discharge control device was set so as to produce a high exit pressure. The motion of the screw and high discharge pressure caused a high level of internal particle-particle friction as well as interfacial friction at the pellet-metal surfaces. The frictional dissipation will cause the polymer to increase in temperature until a steady-state heat transfer has been achieved, or the resin melts and the device seizes. The temperature and heat flux responses in the barrel zones for the HIPS resin is shown in Figure 5.15 and Figure 5.16, respectively. Figure 5.15 shows the outside barrel temperature response at the five measurement positions of the solids conveying device while conveying and discharging at a pressure of about 1 MPa and at a screw speed of 50 rpm. As described earlier, temperature and Flux Position 1 was at the exit of the device while Position 5 was at the hopper. As the screw rotates and the axial pressure increases, frictional energy is dissipated, increasing the temperature of the resin and the barrel, as shown by Figure 5.15. As expected, the Position 1 temperature increases at a higher rate and to a higher level. This is because the pressure is the highest in the device at this position. After about 200 s, a new steady state for temperature was achieved where no appreciable change in barrel temperature occurs over time although some change in heat flux was observed, as indicated in Figure 5.16. This experiment was performed to investigate the response of the barrel steel to a temperature gradient and to ensure that the interfacial temperature calculations were performed correctly. Given the physical properties of the barrel steel, a local heat flux of 10 kW/m^2 was measured and it provided a modest 0.5 °C temperature gradient across the barrel (in the radial direction). However, considering that the barrel is only 3 mm thick, and is constructed of mild-carbon steel, the low temperature difference is understandable. This is demonstrated in Figure 5.15 and Figure 5.16. If the barrel had a significant thermal capacitance, the response of the temperature profile to a change in heat flux would be substantially delayed.

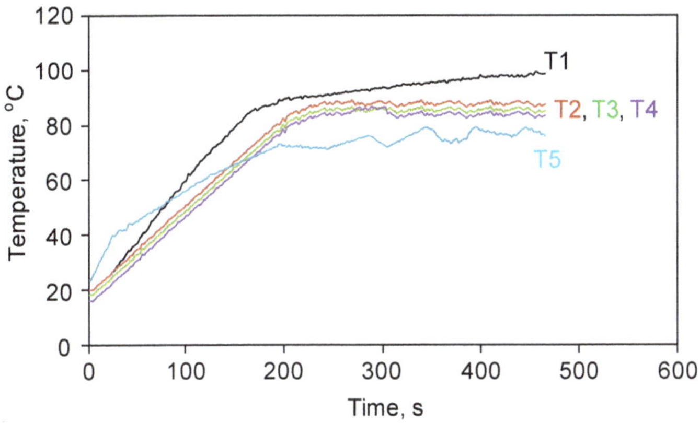

Figure 5.15 Outside barrel zone temperature measurements for a solids conveying experiment using HIPS resin and a screw speed of 50 rpm and a discharge pressure of 1 MPa

The energy dissipation is the highest at the discharge of the device where the pressure is the highest as measured by the increase in zone temperature. The energy dissipation flux is described by Equation 5.29 below:

$$F_e = fPV \tag{5.29}$$

where F_e is the energy dissipation flux and with units of J/(m^2·s), P is the local pressure, V is the velocity of the pellets sliding across the barrel wall, and f is the local coefficient of dynamic friction. At the midposition of the barrel, the pressure is likely about 0.3 MPa. For a screw speed of 50 rpm, the tangential barrel velocity is about 0.15 m/s, and at a temperature near 80 °C the coefficient of dynamic friction for HIPS is about 0.25 (Figure 12.17). These data and Equation 5.29 indicated that a flux of about 11 kW/m^2 was dissipated at the pellet-barrel interface midway downstream in the device. Since the major energy dissipation is produced at the pellet-barrel wall interface and because the thermal conductivity of the barrel is about 400 times higher than that for the pellets, a large portion of the dissipated energy transfers through the barrel as an energy flux. The energy flux through the barrel wall was measured at 10 kW/m^2. Thus, about 90% of the energy was transferred through the barrel wall and 10% was inputted to the resin. This ratio of energy transferred through the barrel wall as compared to that transferred to the resin is typical for many plasticating extrusion processes.

At high discharge pressures, the pellets exit the device hot enough to form a sintered mass, as shown in Figure 5.17(a) for PS resin. In this case, the temperature of the outside of the pellets exceeded the T_g of the resin and as the resin exited the device the pellets cohesively attached to one another. In most cases, the device was stopped and the melted masses were removed.

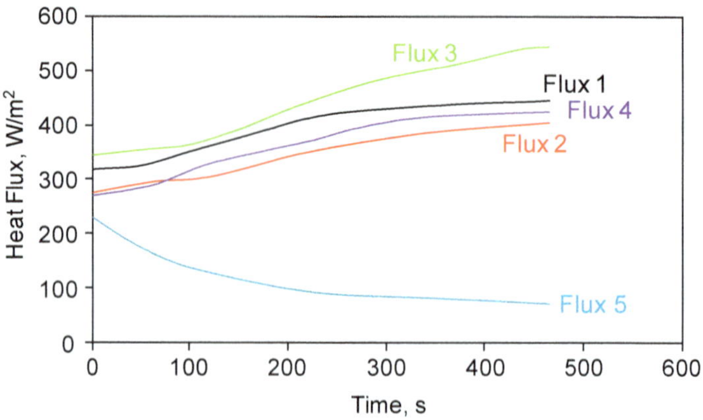

Figure 5.16 Heat flux response for HIPS. The positive heat flux shown here is for energy transferred from inside of the barrel to the outside. The discharge pressure and screw speed were maintained at 1 MPa and 50 rpm, respectively

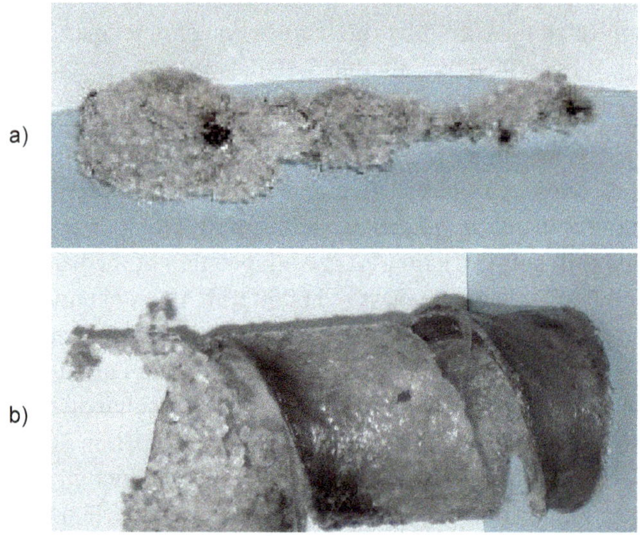

Figure 5.17 Photographs of sintered PS resin masses: a) from the discharge end of the solids conveying device, and b) mass removed from the screw

A similar experiment was carried out with PS resin and a discharge pressure of about 1.5 MPa. Due to frictional dissipation, the polymer heated to such a degree that it melted in the device. When the polymer stopped coming out at the discharge, the power to the drive and pressure devices were turned off. It was these temperature effects in part that dictated the maximum discharge pressure that could be used on the specific polymer during the conveying experiments. After the pressure control device was removed, the solids conveying device was heated up with the hot air control system and the material was extruded out of the device. A photograph of the melted mass from the channels is shown in Figure 5.17(b). Note that the resin partially melted all the way back to the hopper, however there was a section of unmelted and loose individual pellets.

5.3.1.3 Torque and Thrust

As the screw rotates, the pellets are conveyed and pressurized. A torque must be applied to the barrel to keep it from rotating in the same direction as the screw. Moreover, the motion of the pushing flight and the discharge pressure at the end of the device creates a force that pushes the screw backwards. In a commercial extruder, this force is balanced with another force that acts on the thrust bearing at the shank of the screw. For the solids conveying device here, the screw is free floating and the counter thrust is provided by the barrel wall, as shown in Figure 5.9. Figure 5.18 and Figure 5.19 are plots of the barrel torque and thrust as a function of discharge pressure for each resin at 35 °C, respectively. These data provide a qualitatively understanding of the conveying data previously presented, and they follow essentially the same trends as the conveying rate as a function of discharge pressure.

The effectiveness of a material to transfer stress to the walls of the barrel can thus be seen by studying Figure 5.18 and Figure 5.19. The steeper the slope for a given material (for both figures) the more effective that material is at transferring stress to the barrel wall. Therefore, in relative terms, the following ranking would be valid: PS (most effective), HIPS, LDPE (LLDPE), PP, and HDPE (least effective). As previously stated, the analysis is based on low barrel temperatures that would exist in the feed casing of the extruder (Section 1). Solids conveying rates are known to be higher for LDPE resin as the Section 2 temperature approaches 125 °C.

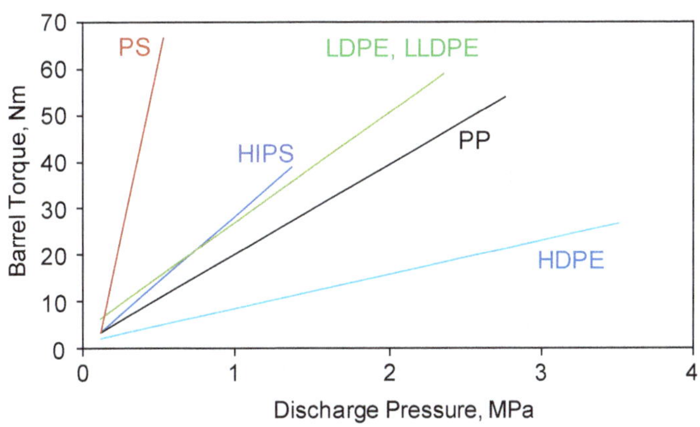

Figure 5.18 Torque as a function of discharge pressure for select resins. The temperature of the barrel and screw were maintained at 35 °C. The screw speed was maintained at 50 rpm

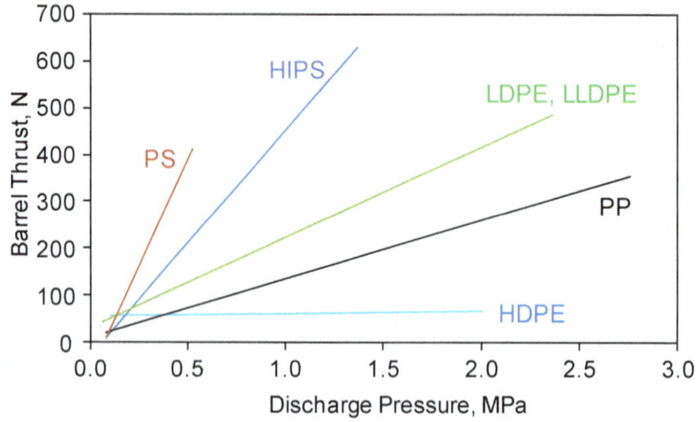

Figure 5.19 Barrel thrust as a function of discharge pressure for select resins. The temperature of the barrel and screw were maintained at 35 °C and at a screw speed of 50 rpm

5.3.2 The Solids Conveying Device at Dow

The first solids conveying device capable of producing reproducible discharge pressures was built by Gogos et al. [35] in 1993 at Polymer Processing Institute (PPI). The device was constructed by modifying an existing 50.8 mm diameter extruder with short barrel segments. A restriction was positioned at the discharge to create pressure. Strain gauges and load cells were used to measure the barrel stresses and discharge pressure, respectively. Based on the PPI device, a solids conveying device with an improved discharge restriction device was designed and built at Dow by Hyun and Spalding [21]. Photographs of the Dow solids conveying device are shown in Figure 5.20 and a schematic is shown by Figure 5.21. This work provided the first reproducible and useful data set for solids conveying. Collecting the data was very time consuming and required more than 10,000 kg of LDPE resin. These data are probably the best data sets to date in the literature. Evaluation of the results from these data concluded that most solids conveying models of the time generally predicted the trends seen in the data. However, the traditional models did not accurately predict rates at exit pressures greater than 0.1 MPa and at the various screw and barrel temperatures. However, the device, like all breakthrough concepts, had some limitations. The matched plate and torpedo design of the discharge control device often limited the scope of data that could be taken, as the polymer would often begin to melt in the barrel; an extremely undesirable effect when studying only solids conveying.

The solids conveying device at Dow was built from a standard 63.5 mm diameter extruder. It was built by replacing the standard barrel with a short barrel, providing an *L/D* ratio of 4.5. The feed hopper casing had an effective barrel length of 1.9 diameters and was water cooled. The feed opening was 1.5 diameters in length;

the enclosed solids conveying length by the barrel and casing was 3 diameters. The feed casing and barrel were smooth bore in design. The barrel temperature was controlled using a single-zone, 2000 Watt heater. The actual temperature of the barrel was measured using four strategically placed thermocouples at different axial positions.

Three screw geometries were studied using this device. Two of the screws will be described here. Both screws were square-pitched, had flight widths perpendicular to the flights of 6.4 mm, and flight clearances of 0.07 mm. The screws had constant channel depths of either 8.89 or 11.1 mm. These screws were labeled as the shallow screw and deep screw, respectively. The flight radii on the pushing and trailing flights were 6 and 16 mm, respectively. The shanks were bored for internal screw temperature control and were the entire length of the screw. The temperature of the screw was controlled by flowing pressurized and heated water into the shank end of the screw using a rotary union assembly.

a)

b)

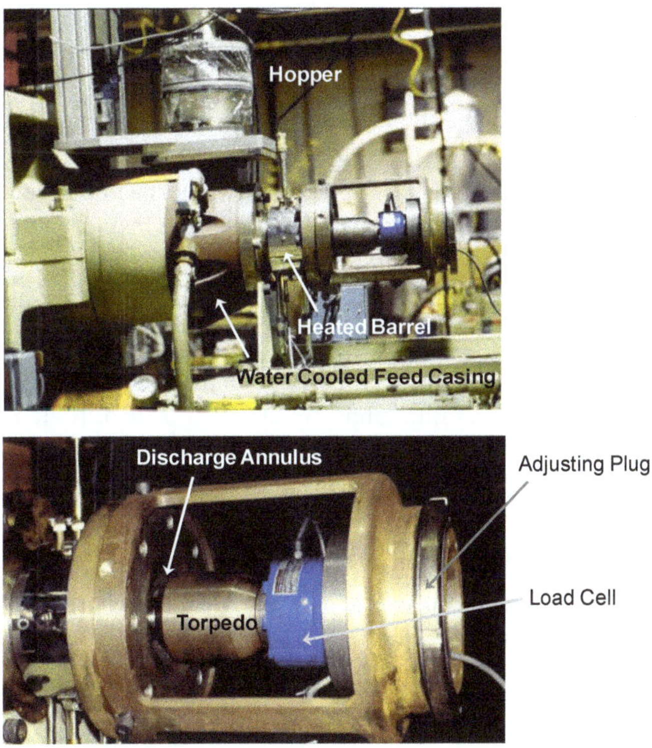

Figure 5.20 Photographs of the solids conveying device at Dow: a) full view of the device, and b) the pressure discharge unit. Pellets were discharged through an annular opening between the discharge (matched) plate and the torpedo

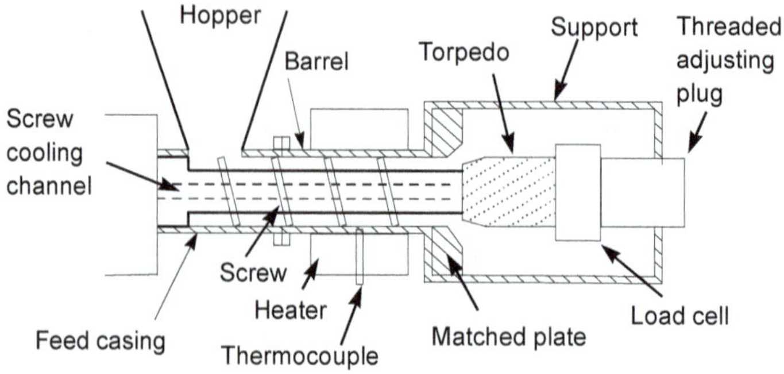

Figure 5.21 Schematic of the solids conveying device at Dow

Pressure was applied to the discharge end of the device using a rotating torpedo positioned on a spindle and a thrust bearing, as shown by Figure 5.21. The gap between the torpedo and a matching plate attached to the barrel controlled the amount of pressure at the discharge end. Typically, a gap that was slightly larger than the diameter of the pellets permitted the material to discharge at zero pressure. The gap between the torpedo and the matched plate was set by adjusting the position of a threaded plug. The force generated was measured using a load cell positioned between the torpedo and the plug. Pressure at the discharge end of the device was calculated by dividing the force at the torpedo by the cross-sectional area of the annulus created by the screw root and the inside barrel diameter. The feed casing temperature was maintained at about 30 °C using cooling water.

An LDPE resin was used for this study. The resin had a melt index of 2.0 dg/min (2.16 kg, 190 °C) and a solid density of 0.922 g/cm^3. The shear viscosity was reported previously [37]; thermal properties are provided in Chapter 4; bulk density as a function of temperature and pressure is provided in Figure 4.4; and the coefficients of dynamic friction are provided in Appendix A5. The lateral stress ratio was measured at 0.7 [38] using the device shown in Figure 4.8.

5.3.2.1 Flow Rate and Exit Pressure

Solids conveying rates were measured as a function of discharge pressure at screw speeds of 50 and 80 rpm, screw temperatures of 50, 75, and 100 °C, nominal barrel temperatures between 75 and 150 °C, and for the shallow and deep screws discussed previously. At most conditions, the pellets discharged were essentially identical to those of the feed. At high discharge pressures and high barrel temperatures, however, some of the pellets were melted by the time they were discharged. As indicated by Figure 5.22 through Figure 5.26, the solids conveying rates were complicated functions of the screw and barrel temperatures, screw speed, channel depth, and discharge pressure. The effect of these variables will be discussed next. The complete set of solids conveying data for this LDPE resin are provided in Appendix A5.

Solids conveying rates for a discharge pressure of zero and as a function of screw and barrel temperature are shown in Figure 5.22 for the shallow screw at a speed of 50 rpm. In all cases, the solids conveying rate increased with increasing barrel temperature in the range studied. For example, the rates for a screw temperature of 50 °C were 75, 94, and 98 kg/h at barrel temperatures of 50, 100, and 125 °C, respectively. This result is consistent with commercial extrusions of this resin where the optimal feed zone barrel temperature is in the range of 150 to 220 °C. The optimal feed zone barrel temperature must be determined experimentally using the technique provided in Section 10.9, and the temperature is dependent on the configuration of the machine. For this data set, the solids conveying rate increased with decreasing screw temperature. That is, as the temperature of the screw increases, the ability to convey solids is reduced. If the screw temperature is high enough, solids conveying rates can become unstable, causing flow surging at the discharge. Flow surging due to a high screw temperature is discussed in Section 12.7.3. Excessive screw cooling in the solids conveying zone can lead to problems [39] by solidifying a thin coating of resin in the channel near the end of the feed zone and early in the transition section. This coating can reduce the process of solids conveying, and in extreme cases, cause flow surging and material degradation. Moreover, reducing the screw temperature to a very low temperature can prevent the resin from increasing in temperature at the proper rate. This is problematic for most barrier screw designs where the solid bed must be at a high enough temperature to deform and flow past the entry of the section. This type of problem is presented in Section 11.10. For LDPE resins, a screw temperature near 75 °C is acceptable. For most commercial processes, screw cooling will not be used. Instead, the screw will be operated in a neutral condition where the screw is heated from the upstream sections of the process. For the solids conveying device at Dow, the neutral screw process behaves like a temperature-controlled screw near 75 °C, as shown in Figure 5.22.

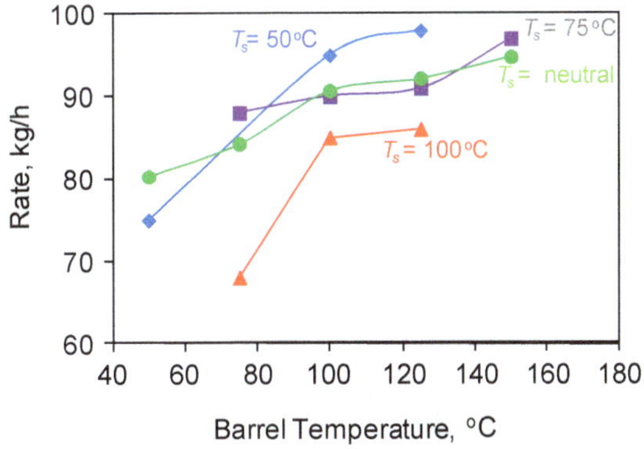

Figure 5.22 Solids conveying rate as a function of barrel and screw temperature for the shallow screw (8.89 mm) at a screw speed of 50 rpm and at zero discharge pressure

After the solids conveying device came to a steady state with zero discharge pressure, the torpedo mechanism was positioned such that the annular flow opening was restricting flow and thus causing a pressure at the discharge. The rates were measured as a function of discharge pressure until melting occurred or the rate become very low. Solids conveying rates for the shallow screw at a screw temperature of 50 °C as a function of discharge pressure and barrel temperature are shown in Figure 5.23 at a screw speed of 50 rpm. As expected, the rate was the highest at zero discharge pressure and it decreased as the discharge pressure increased. This is identical to the behavior observed by the Clarkson group, as shown in Figure 5.12 to Figure 5.14. Conveying rates were the highest for a barrel temperature of 125 °C, and the rates decreased with decreasing barrel temperature. The decrease in rate with increasing discharge pressure was observed for both screws and at all conditions. Similar rates and performance were observed for a screw temperature of 75 °C as shown in Figure 5.24. The solids conveying rates were slightly higher for a screw temperature of 75 °C as compared to those at 50 °C.

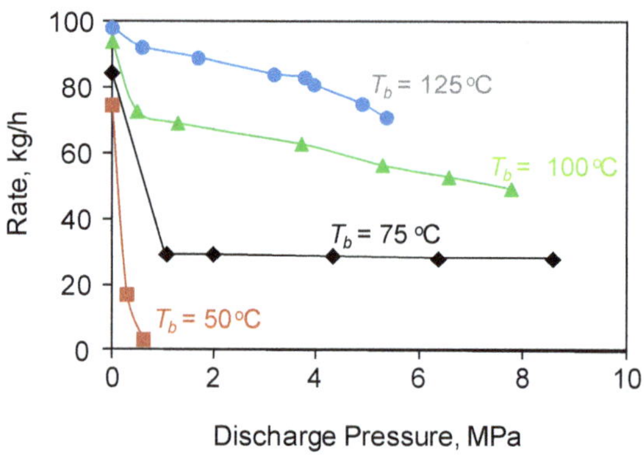

Figure 5.23 Solids conveying rate as a function of discharge pressure and barrel temperature for the shallow (8.89 mm) screw. The screw was maintained at a temperature of 50 °C and a speed of 50 rpm

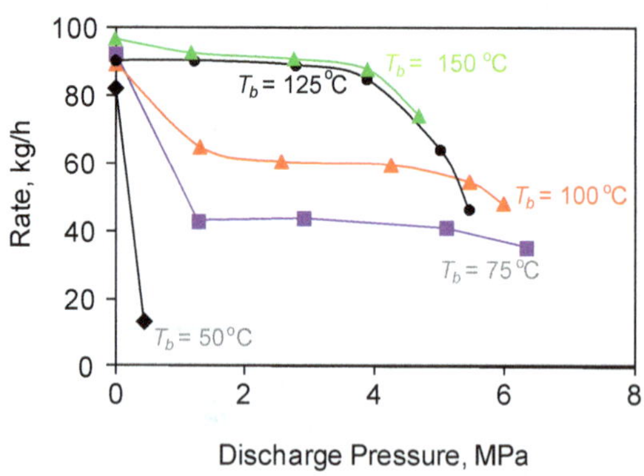

Figure 5.24 Solids conveying rate as a function of discharge pressure and barrel temperature for the shallow (8.89 mm) screw. The screw was maintained at a temperature of 75 °C and a screw speed of 50 rpm

Some melting occurred at high discharge pressures and at high barrel temperatures. At these conditions the melting was accompanied by a rate decrease with increasing pressure that was larger than expected. This observation is shown by the rapid rate decrease for barrel temperatures of 125 and 150 °C at discharge pressures between 4 and 5.5 MPa, as indicated in Figure 5.24. Whenever these fast rate losses occur at high pressures, it is always a result of the resin melting towards the end of the process.

The solids conveying rate is directly proportional to the screw speed as expected and as shown in Figure 5.25. Here the specific rate is nearly identical for the two screw speeds for a given discharge pressure for the shallow screw. For example, at a discharge pressure of 3.7 MPa the specific rates were 1.34 and 1.31 kg/(h·rpm) at screw speeds of 50 and 80 rpm, respectively. Moreover, a specific rate of 1.3 kg/(h·rpm) is considerably higher than the specific rate expected from the rate controlling downstream metering section. For example, a typical depth of a metering channel of a 63.5 mm diameter screw would be about 3.2 mm, providing a specific rotational flow rate of 0.71 kg/(h·rpm). Since the solids conveying section of this screw has the ability to operate at higher specific rates than the metering channel, the metering channel will always be full of resin and pressurized. Since both sections have to operate at the same specific rate, pressures in the machine will develop to satisfy this constraint.

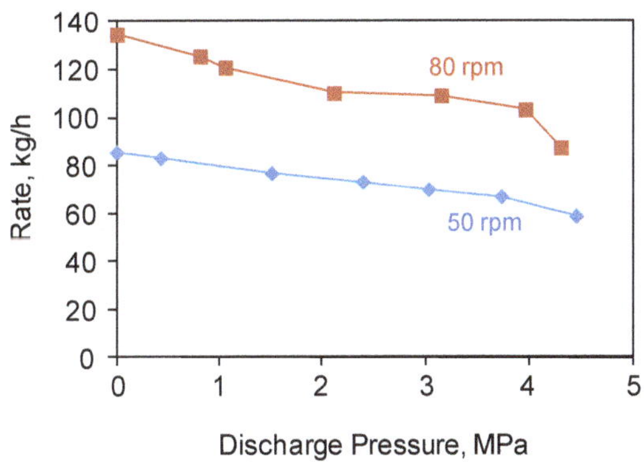

Figure 5.25 Solids conveying rate as a function of discharge pressure and screw speed for the shallow (8.89 mm) screw. The barrel and screw were maintained at temperatures of 125 and 100 °C, respectively

The effect of channel depth on solids conveying rate is shown in Figure 5.26 for screw and barrel temperatures of 75 and 125 °C, respectively, and at a screw speed of 50 rpm. At zero discharge pressure, the solids conveying rates were nearly proportional to the depth of the screw channel (or cross-sectional area perpendicular to the flight). For example, the conveying rates were 91 and 125 kg/h for the 8.89 and 11.1 mm deep screws, respectively. For these screws, the cross-sectional areas perpendicular to the flights were calculated at 420 and 530 mm^2, an area increase of 26% for the deep screw. The rate at zero discharge pressure was increased by 37% with the increase in channel depth. Similar results were observed at other screw and barrel temperatures and at a screw speed of 80 rpm. At high discharge pressures, however, the rates decreased more with increasing pressure for the deeper screw. For discharge pressure of 4 MPa, the rates were 83 and 93 kg/h for the 8.89 and 11.1 mm deep screws, respectively. In general, this result was observed at all conditions. Thus, the rate was considerably more sensitive to discharge pressure for the deeper screw. This result was expected based on extrusion experience.

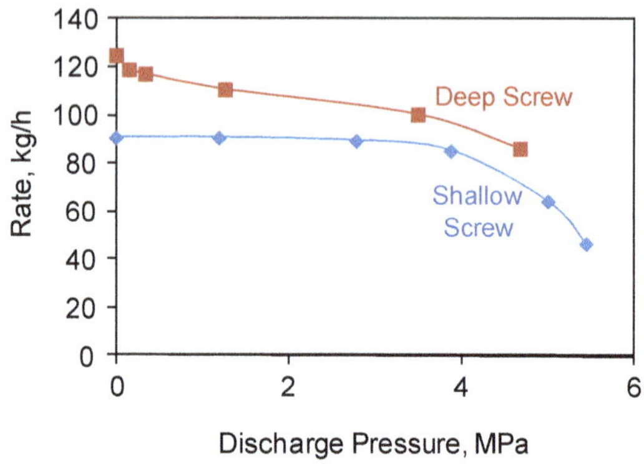

Figure 5.26 Solids conveying rate as a function of discharge pressure for the shallow (8.89 mm) and deep (11.1 mm) screws at a screw speed of 50 rpm. The screw and barrel temperatures were maintained at 75 and 125 °C, respectively

5.3.2.2 The Effect of Flight Radii Size

The effect of flight radii size on solids conveying was determined using two screws of different flight radii size [40]. Both screws were square-pitched, had flight widths perpendicular to the flights of 6.4 mm, flight clearances of 0.07 mm, and a constant channel depth of 11.1 mm. The small-radii screw had pushing and trailing flight radii of 6 and 16 mm, respectively. The small-radii screw was the deep screw used for the studies in the previous section. The large-radii screw had radii of 19 mm for both the pushing and trailing flights. The same LDPE resin as used in the studies above was used here.

Solids conveying rates were measured as a function of discharge pressure at a screw speed of 50 rpm, a screw temperature of 50 °C, and nominal barrel temperatures of 75, 100, and 125 °C. As indicated in Figure 5.27, Figure 5.28, and Figure 5.29, the solids conveying rates were complicated functions of the barrel temperature, flight radii size, and discharge pressure. The screw with the large flight radii had an unexpected behavior with increasing discharge pressure. As will be discussed next, the conveying rate for the large-radii screw showed a decrease in rate with increasing discharge pressure of up to 1 to 2 MPa, and then an increase in rate with increasing pressure for all but the low screw temperature experiment.

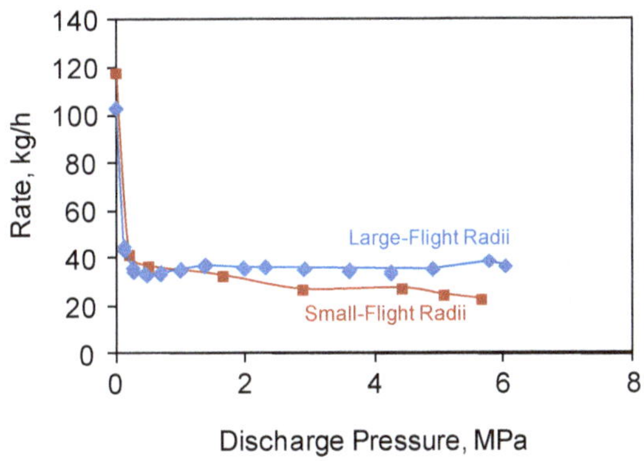

Figure 5.27 The effect of flight radii size on solids conveying rates for a barrel temperature of 75 °C, a screw temperature of 50 °C, and a screw speed of 50 rpm

Figure 5.27 shows the conveying rates for both screws at a barrel temperature of 75 °C. At a discharge pressure of zero, the rates were 118 and 103 kg/h for the small-radii and large-radii screw, respectively; this is a decrease in rate of about 13% for the large-radii screw. This rate decrease is most likely due to the lower cross-sectional area available for conveying for the screw with the large radii. The cross-sectional areas of the channels perpendicular to the flights were calculated at 550 and 490 mm^2 for the small-radii and large-radii screws, respectively; an area decrease of 11% for the large-radii screw. As the discharge pressure was increased to about 0.5 MPa, the rates decreased to about 37 and 34 kg/h for the small and large-radii screws, respectively. Thus, the large-radii screw had a rate that was 8% less than that for the small-radii screw. As the discharge pressure was increased further, the rate for the small-radii screw decreased nearly linearly to about 23 kg/h at 5.7 MPa. The rate for the large-radii screw, however, increased slightly to 37 kg/h as the pressure was increased to 1.4 MPa. For pressures between 1.4 and 6.0 MPa, the rates were essentially constant at about 37 kg/h for the large-radii screw. This was the first time that a rate increase with increasing pressure was observed for this solids conveying device.

Similar solids conveying rate data were observed for a barrel temperature of 100 °C, as shown in Figure 5.28. As indicated by this figure, the rates were 110 and 103 kg/h for the small- and large-radii screws, respectively; a decrease of about 6% for the large-radii screws. Like the data at a barrel temperature of 75 °C, the rate decrease was believed to be caused by the 11% lower cross-sectional area of the large-radii screw. As the pressure was increased to about 1.3 MPa, the rates for both screws decreased. But as the pressure was increased further, the rates continued to decrease for the small-radii screw but increased for the large-radii screw. At a discharge pressure of 5 MPa, the rate for the large-radii screw was about the same as that for the small-radii screw.

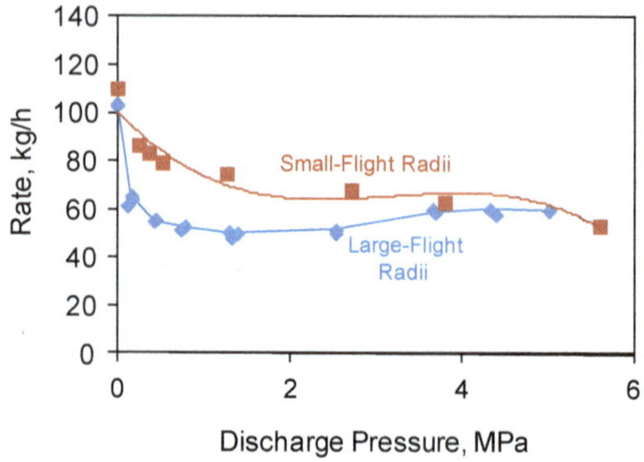

Figure 5.28 The effect of flight radii size on solids conveying rates for a barrel temperature of 100 °C, a screw temperature of 50 °C, and a screw speed of 50 rpm

For a barrel temperature of 125 °C, the conveying rates at zero discharge pressure were 124 and 107 kg/h for the small- and large-radii screws, respectively. As the discharge pressure was increased, the rates for both screws decreased in a very similar manner, as shown in Figure 5.29. For discharge pressures less than 4 MPa, the large-radii screw had rates that were about 15% less than those for the small-radii screw. For both screws, the rates decreased as the discharge pressure was increased. Partial melting of some of the pellets was observed for both screws at this barrel temperature. For the small-radii screw, the onset of melting was observed at a discharge pressure of about 5 MPa. The onset of melting for the large-radii screw, however, occurred at a lower pressure of 4 MPa. The data suggests that a thermal effect is occurring at the barrel wall and that it is sensitive to pressure.

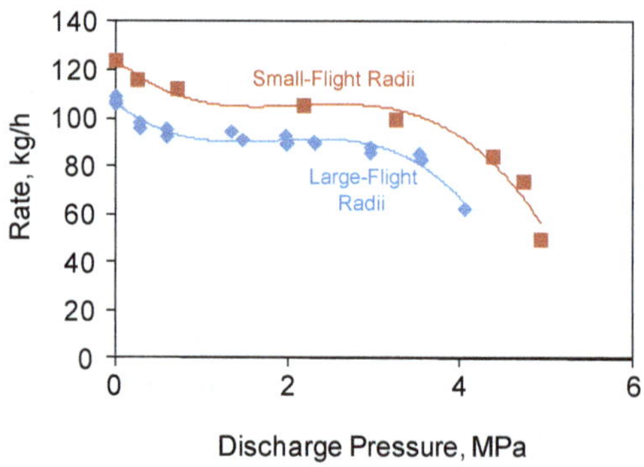

Figure 5.29 The effect of flight radii size on solids conveying rates for a barrel temperature of 125 °C, a screw temperature of 50 °C, and a screw speed of 50 rpm

The solids conveying data indicate that two competing factors are controlling the rates. The rate reduction caused by a decrease in the cross-sectional area perpendicular to the flight for the large-radii screw is the most obvious factor. As the radii of the flights are increased, the area perpendicular to the flight tip decreases and thus the volume available to convey material decreases. This factor is most apparent when the forwarding and retarding forces are minimal; that is, at zero discharge pressure and at low screw temperatures (T_s = 50 °C). At high discharge pressures, however, it is postulated that a channel with a large radius at the pushing flight will change the forwarding forces, as indicated in Figure 5.30. For a channel with a small pushing radius, the F^* force (Hyun-Spalding model) is directed through a surface that is parallel to the surface of the barrel, as shown in Figure 5.30(a). For this case, the normal force at the barrel is due to the pressure in the channel, F_p; F_p produces the F_1 forwarding force. For a channel with a large pushing radius, the F^* force is partially directed towards the barrel, like that shown in Figure 5.30(b). To counteract the component of the force directed at the barrel, the normal force at the barrel wall will increase by $F^* \sin\alpha$, causing the forwarding force F_1 (labeled $F_{1,\alpha}$ in Figure 5.30[b]) to increase. The angle α is the angle between the acting force F^* and a surface parallel with the barrel wall as shown in Figure 5.30(b). The higher forwarding force for the large-radii screw for high discharge pressures, $F_{1,\alpha}$, causes the rates to be higher than those for the small-radii screw. The larger normal force at the barrel wall for the large-radii screw causes larger amounts of energy to be dissipated at the interface, which causes the onset of melting to occur at lower pressures. The F_1 force is dependent on the coefficient of dynamic friction between the polymer and the barrel wall. As previously discussed, this coefficient has been shown to be a function of temperature, pressure, and sliding velocity. It is interesting to note that Derezinski [41] observed this behavior a few years earlier.

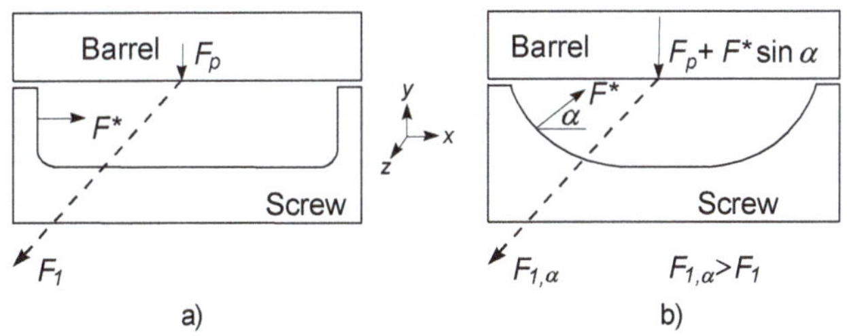

Figure 5.30 a) Schematic of the forwarding forces for a channel with small radii, and b) schematic of the forwarding forces for a channel with large radii

Although not investigated here, the effect of the radius on the trailing side of the channel is thought to only reduce the area available for conveying and not the forwarding or retarding forces. Thus, reducing the trailing flight radius will increase the area and should improve solids conveying.

5.4 Comparison of the Modified Campbell-Dontula Model with Experimental Data

This section will compare the modified Campbell-Dontula model to the experimental data for the LDPE resin. The physical properties for this resin were provided in Section 5.3.2. The dynamic coefficients of friction are available in Appendix A5 and are reported as stress at a sliding interface. The coefficient of friction is related to the stress at a sliding interface as follows:

$$\tau = fP \tag{5.30}$$

The modified Campbell-Dontula model was developed using the LDPE resin friction data as applied to an empirical model. The empirical model is shown by Equation 5.31 for the temperature range of 25 to 110 °C. The coefficients of dynamic friction using Equation 5.31 is shown in Figure 5.31.

$$f = -0.15 + \frac{V^{0.133}}{2.5} - \left(\frac{V^{0.133}}{2.5} - 0.05\right)\left(1 + erf\left(\frac{P^{0.3}(T^{1.08} - 180)}{90}\right)\right)0.5 \tag{5.31}$$

$$\text{for } 25\ °\text{C} < T < 110\ °\text{C}$$

Here, V is the sliding velocity in cm/s, P is the pressure in MPa, and T is the interface temperature in degrees Celsius. The constants in the equation have units and these units have been omitted for clarity. At temperatures less than about 110 °C, the stress at the interface is due to a frictional force mechanism. At temperatures higher than 110 °C, the forces are from a viscous mechanism. The friction coefficients provided by Equation 5.31 have the proper trends. Coefficients derived from the equation, however, need to be used with care because of its empirical nature and the difficulties in measuring frictional data.

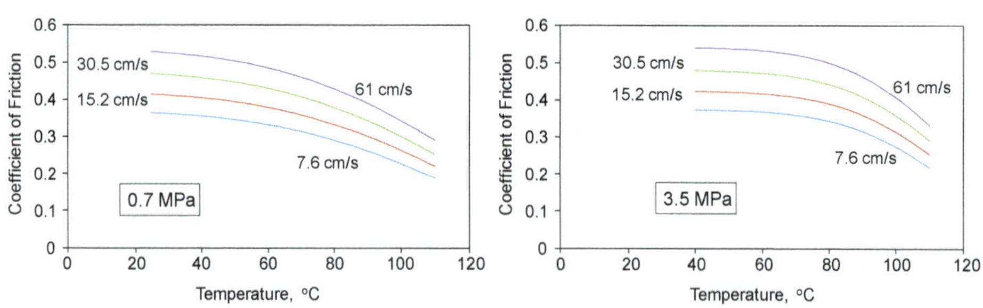

Figure 5.31 Coefficients of dynamic friction for the LDPE resin used in the solids conveying experiments using Equation 5.31

It is apparent that the coefficient of friction for this resin is very much dependent on the velocity of the moving surface in contact with the polymer, the local pressure, and the temperature. Thus, the dynamic friction factors change as the bed is moved through the feed section from the hopper to the point where the polymer starts to melt. Moreover, the bulk density of the pellets increases with increasing temperature and pressure. Thus, these physical properties are changing as the pellets are conveyed downstream. All force balance models except the Hyun-Spalding model were developed and integrated assuming constant physical properties. These integrated models could have been reformulated in differential form such that variable physical properties could be used. Here, the integrated form of the modified Campbell-Dontula model as provided by Equation 5.17 will be presented using coefficients of friction data at the average pressure and velocity.

The solids conveying rate data as a function of discharge pressure at a screw speed of 50 rpm and barrel and screw temperatures of 75 °C are provided in Table 5.1.

Table 5.1 Experimental Solids Conveying Rates (Figure 5.24) as a Function of Discharge Pressure at a Screw Speed of 50 rpm and Barrel and Screw Temperatures of 75 °C for the LDPE Resin. Data were Collected Using the Shallow Screw for the Dow Solids Conveying Device

Discharge Pressure, MPa	Rate, kg/h
0	93
1.3	43
2.9	43
5.1	41
6.3	35

The coefficients of dynamic friction need to be determined first for this analysis. The average pressure and temperatures are specified, but the velocities at the sliding interfaces need to be determined. The sliding velocities need to be calculated based on screw rotation physics. The sliding velocity at the barrel interface is as follows:

$$V_{bs} = \sqrt{V_{bx}^2 + V_{sz}^2} \quad \text{for screw rotation} \tag{5.32}$$

$$V_{bx} = \pi N D_b \sin\theta_b \tag{5.33}$$

$$V_{bz} = \pi N D_b \cos\theta_b \tag{5.34}$$

$$V_{sz} = \frac{Q_m}{\rho_b H W} \tag{5.35}$$

where V_{bs} is the sliding velocity for the solid plug relative to the barrel wall, V_{bx} and V_{bz} are the x and z components of the velocity at the screw tip, V_{sz} is the velocity of the solid bed, Q_m is the mass rate of the process, and ρ_b is the bulk density of the resin. The sliding velocity at the screw surface is calculated as follows:

$$V_{ss} = |V_{cz}| + V_{sz} \quad \text{for screw rotation} \tag{5.36}$$

$$V_{cz} = -\pi N D_c \cos\theta_c \tag{5.37}$$

where V_{ss} is the sliding velocity for the solid plug relative to the screw surfaces and V_{cz} is the z component of the velocity at the screw core for screw rotation. The V_{cz} velocity is in the negative z direction. This velocity will be discussed in detail in Chapters 6 and 7. At a mass rate of 43 kg/h and a bulk density of 577 kg/m^3, the solid bed velocity (V_{sz}) is calculated at 4.4 cm/s for the shallow screw at a screw speed of 50 rpm. The sliding velocities V_{bs} and V_{ss} were calculated at 6.7 and 15.3 cm/s, respectively. For this case and at a rate of 43 kg/h, the sliding velocity at the screw surface is higher than that at the barrel wall. At a pressure of 0.7 MPa, the coefficients of dynamic friction using Equation 5.31 are 0.30 for the barrel surface and 0.35 for the screw surface. To calculate the rates as a function of discharge pressure, the average pressure should be used for integrated equations and the local pressure for equations in their differential form.

As discussed previously in Section 5.2.4, screw rotation physics need to be used in order to calculate the sliding velocity of the solid bed relative to the barrel and screw surfaces. For barrel rotation physics, the sliding velocities at the barrel and screw surfaces are considerably different than that for screw rotation. At the barrel wall, the z component of motion must be corrected for the moving velocity of the barrel wall, as provided in Equation 5.38. For the example above, V_{bs} = 12.5 cm/s. Because the screw is stationary for barrel rotation physics, V_{cz} = 0, and the sliding velocity at screw surface using Equation 5.39 sets V_{ss} = V_{sz} = 4.4 cm/s. At a pressure of 0.7 MPa and screw and barrel temperatures of 75 °C, the coefficient of dynamic friction at the barrel surface is f_b = 0.33 and at the screw surface f_c = 0.27 for barrel rotation physics. Recall that f_b = 0.30 and f_c = 0.35 for screw rotation. The solids conveying models will respond quite differently if barrel rotation velocities are incorrectly used rather than screw rotation velocities.

$$V_{bs} = \sqrt{V_{bx}^2 + \left(V_{bz} - V_{sz}\right)^2} \quad \text{sliding velocity at the barrel for barrel rotation} \tag{5.38}$$

$$V_{ss} = V_{sz} \quad \text{sliding velocity at the screw for barrel rotation} \tag{5.39}$$

To visualize the difference between screw rotation and barrel rotation, a simple cardboard paper towel roller can be used to model the screw core and a wood block to model the solid plug. For barrel rotation, the roller is held constant and the block is moved downstream at a velocity of V_{sz}. Here the sliding velocity between the block and roller is simply V_{sz}. For screw rotation, the roller is rotated counterclockwise while the block is moved downstream with a velocity of V_{sz}. The observer will see that the core of the screw is moving in the negative z direction at a velocity of V_{cz}. The sliding velocity for this interface is provided by Equation 5.36.

The modified Campbell-Dontula solids conveying model was compared to the Dow experimental data at conveying conditions provided in Table 5.1 for the shallow screw (8.89 mm deep channel) and at a screw speed of 50 rpm. For the conditions in the table, the coefficients of friction at the barrel and screw surfaces were 0.30 and 0.35, respectively, using Equation 5.30; the conditions were screw and barrel temperatures of 75 °C, mid-rate of 43 kg/h, 50 rpm, and average pressure of 0.65 MPa. The calculated solids conveying rates as a function of discharge pressure using Equation 5.14 are graphed in Figure 5.32 along with the data set found in Table 5.2. As shown by Figure 5.32, the model fits the experimental data fairly well for a K_{xz} between 0.3 and 0.25. The value of K_{xz} was measured by Yamamuro et al. [22] as 0.5 and by Spalding et al. [38] as 0.7. For these conditions, the model and the measured physical properties predict the rates as a function of discharge pressure fairly well.

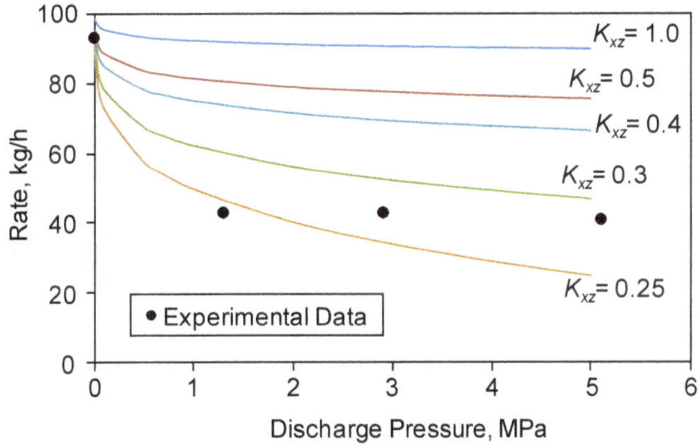

Figure 5.32 Solids conveying rate data calculate using the modified Campbell-Dontula model for the Dow solids conveying process using the shallow screw, 75 °C barrel and screw temperatures, and a screw speed of 50 rpm. The solids conveying rates measured from the experimental device are provided

A number of issues relative to the prediction of solids conveying in smooth bore single-screw extruders are exposed from the theoretical fits to the data in Figure 5.32. First, the data needed to carry out an effective simulation is difficult to take and is very time consuming, and only a few labs have the proper equipment; that is, bulk density measurement, dynamic friction data, lateral stress, and solids conveying data. Moreover, care must be taken to develop an accurate representation of the surface temperature for the barrel and screw as a function of the axial position. This would be quite difficult in a traditional extruder with only a control thermocouple to measure the temperature at the midpoint of the barrel thickness. Second the analytical models that are available only give an approximate solution because they are based on static analysis and the solids conveying is a dynamic problem. The model by Hyun and Spalding has to be integrated down the channel with local compaction and temperature increases for the bed calculated at each point, which is a difficult problem for many not familiar with advanced computational methods. The approach is complicated by our lack of understanding of where and to what degree the bed compacts. The model by Yamamuro, Penumadu, and Campbell [21] relies on a fitting factor which for the Dow LDPE conveying data, was a unique function of temperature. The numerical method of Thompson would likely be the most accurate, but it is extremely time consuming and requires advanced computational and mathematical skills. In addition this model also needs very accurate wall and internal friction data. The current models can be used to show trends but should probably only be used as a qualitative tool. The screw design engineer should rely on design experience when determining the characteristics of the solid conveying section of an extruder until better models are developed.

5.4.1 Solids Conveying Example Calculation

Referring back to the example in Section 1.5.1, it is desired to determine the solids conveying performance of the design at a screw speed of 108 rpm for an LDPE resin. As provided in this example, the screw diameter is 88.9 mm, the feed channel depth is 16.5 mm, and the flight width at the barrel is 9 mm. The feed channel is 5.5 diameters in length with the feed opening of 1.5 diameters. Thus, the axial length for solids conveying is 4 diameters. Further discussion at the plant indicates that the feed barrel section will be heated to 90 °C and the cored screw will be maintained at 75 °C. The LDPE resin has a bulk density of 577 kg/m^3 at ambient conditions. Using the modified Campbell-Dontula model developed in this chapter, determine the solids conveying rate if the discharge pressure is 3.4 MPa.

Table 5.2 Geometric Parameter Values for the Screw Used in the Example in Section 1.5.1

Parameter	Value	Equation
Barrel diameter, D_b	88.9 mm	
Core diameter, D_c	55.9 mm	
Average diameter, D	72.4 mm	
Lead length, L	107 mm	
Meter channel depth, H	16.5 mm	
Flight Width, e	9.0 mm	
Flight starts, p	1	

Parameter	Value	Equation
Helix angle at the barrel, θ_b	20.96°	1.1
Helix angle at the screw core, θ_c	31.35°	1.4
Average helix angle, θ	25.19	1.7
Channel width at the barrel, W_b	90.2 mm	1.2
Channel width at the screw core, W_c	82.4 mm	1.5
Average channel width, W	86.6 mm	1.10
Total helical length of the solids conveying section (4 diameters) at the barrel, Z_b	0.99 m	1.11

For this problem, the pressure at the end of the solids conveying section is defined. Thus, the solids forwarding angle can be computed using Equation 5.14. Once the solids forwarding angle is known, the rate is then calculated using Equation 5.11. Three values in Equation 5.14 are not known at this point in the analysis: the coefficient of dynamic friction at the barrel and screw surfaces, and the pressure P_o at the start of the solids conveying section. Although the hopper design is not specified, P_o is on the order of 1000 Pa using the method outlined by Agur and Vlachopoulos [8]. The coefficient of dynamic friction needs to be determined at the resin-metal interfaces, at the interface surface temperatures, at the sliding velocity, and at the average pressure of 1.7 MPa. The interface temperatures were specified in the problem statement as 75 and 90 °C for the screw and barrel surfaces, respectively. The sliding velocity at the barrel surface and screw surface are estimated by calculating the velocity components of Equation 5.32 and Equation 5.36, respectively. The velocity of the solids bed is estimated knowing that the extrusion rate should be near 273 kg/h.

$$V_{bx} = \pi N D_b \sin\theta_b = \pi\left(1.8 \text{ rev/s}\right)\left(0.0889 \text{ m}\right)\sin\left(20.96°\right) = 0.180 \text{ m/s}$$

$$V_{sz} = \frac{Q_m}{\rho_b H W} = \frac{0.075 \text{ kg/s}}{(577 \text{ kg/m}^3)(0.0165 \text{ m})(0.0866 \text{ m})} = 0.091 \text{ m/s}$$

$$V_{cz} = -\pi N D_c \cos\theta_c = -\pi\left(1.8 \text{ rev/s}\right)\left(0.0559 \text{ m}\right)\cos\left(31.35°\right) = -0.270 \text{ m/s}$$

$$V_{bs} = \sqrt{V_{bx}^2 + V_{sz}^2} = \sqrt{\left(0.180 \text{ m/s}\right)^2 + (0.091 \text{ m/s})^2} = 0.202 \text{ m/s} => 20.2 \text{ cm/s}$$

$$V_{ss} = \left|V_{cz}\right| + V_{sz} = 0.270 \text{ m/s} + 0.091 \text{ m/s} = 0.361 \text{ m/s} => 36.1 \text{ cm/s}$$

At 1.7 MPa pressure, a sliding velocity of 20.2 cm/s, and a barrel surface temperature of 90 °C, the coefficient of dynamic friction is estimated at f_b = 0.35 using Equation 5.31. The equation is then evaluated at the conditions at the screw root of

1.7 MPa pressure, 75 °C, and a sliding velocity of 36.1 cm/s for a coefficient of friction at the screw surface of f_s = 0.44. Here, $f_s = f_h$.

For the modified Campbell-Dontula model, Equation 5.13, φ is calculated as follows for K_{xz} = 1:

$$\sin\varphi = \cos\theta_b - \left(2\frac{Hf_H}{W_b f_b} + f_H + \frac{W_s f_s}{W_b f_b}\right)\sin\theta_b - \frac{HW\sin\theta}{K_{xz}W_b f_b Z_b}\ln\left(\frac{P_z}{P_o}\right) = 0.047$$

$\varphi = 2.7°$

The solids conveying rate can be now be estimated using Equation 5.7 adapted for mass flow:

$$Q_m = \pi^2 NHD_b\left(D_b - H\right)\frac{\tan\varphi\tan\theta_b}{\tan\varphi + \tan\theta_b}\left[\frac{W}{W+e}\right]\rho_b = 0.0413 \text{ kg/s} => 149 \text{ kg/h}$$

The estimated solids conveying rate of 149 kg/h is less than the calculated rotation rate for the metering section of 280 kg/h in Section 1.51. The model is indicating that the channel dimensions and conditions are incapable of conveying at 273 kg/h and at a pressure at the end of the conveying section of 3.4 MPa. In Example 1, the rate was determined at 273 kg/h for the maximum pressure gradient in the metering section. Next, the pressure at the end of solids conveying should be estimated at the conditions of the process. For a rate of 273 kg/h, the pressure at the end of the solids conveying section is calculated by first determining φ and then using Equation 5.13 to determine P_z. The pressure at the end of the solids conveying section and φ were determined to be 0.27 MPa and 5.50°, respectively. This pressure at the end of the solids conveying section may not be high enough to maintain a stable process. A pressure near 4 MPa would have been better.

The modified Campbell-Dontula model provides an acceptable prediction of the rate at the section exit pressure. For many cases, the coefficients of friction are adjusted until an acceptable performance is obtained. This model and the other models should always be used with caution. As previously discussed, these models use a static force balance to approximate a dynamic process.

■ 5.5 Grooved Bore Solids Conveying

Grooved bore extruders offer several advantages over smooth bore machines [42]. These advantages include higher rates and specific rates, and lower variations in the discharge pressure; rate is essentially independent of the discharge pressure, and lower discharge temperatures. The increased specific rate allows lower screw speeds at the same rate, which allows the lower discharge temperatures. The grooved bore process is designed such that relatively high pressures exist at the end of the grooved barrel section. The high pressure at the end of the solids conveying zone for some designs can be as high as 100 to 150 MPa [43]. This high pressure is typically reduced in a downstream decompression section of the screw. The specific rates can be as much as twice that of the same diameter smooth bore extruder. The melting capacity of the screw, however, can limit the maximum rate of the process as outlined by Ingen Housz and Meijer [44].

Grooved bore extruders have several disadvantages [42]. These include high levels of wear in the feed casing and especially on the edges of the grooves. The wear is caused by the very high pressures in the system. The high pressure also causes a high level of energy dissipation at the contact surface as provided by Equation 5.29. The high levels of energy must be removed via the cooling system to prevent the resin from melting in the grooves. If molten polymer is in the grooves, then the solids conveying rate of the process decreases substantially. Low bulk density and fluffy recycle streams cannot be added back to a grooved barrel extruder. Instead the recycle material must be pelletized using a secondary operation.

As stated earlier, grooved barrel extruders are designed to increase the solids conveying rate of single-screw extruders by placing axial or spiral grooves in the first 3 to 5 diameters of the feed section; 3.5 diameters of grooving appear to be optimal for many applications. These grooves reduce the tendency of the pellets to rotate with the screw, forcing them to flow downstream due to the motion of the helix. The grooves are typically built into a water-cooled feed casing although advanced casings have been designed with air cooling and electrical heating [45]. A schematic of a water-cooled grooved feed casing is shown in Figure 5.33. The grooves are typically tapered such that they are the deepest at the feed opening and the barrel becomes perfectly smooth 3 to 5 diameters downstream. Sikora [46] has shown that rate of taper of the grooves is highly important to the solids conveying performance in the section. The optimal temperature of the casing and the grooves depends on the resin. The temperature can range from 15 °C for LDPE resins [47] and other polyolefins, and up to 250 °C for engineering resins such as PA and PC [45] in order to mitigate high motor torque. Grünschloß [48] points out, however, that increasing the temperature of the grooved barrel segment can lead to flow surging. Instead, he states a redesign of the screw will result in a stable, high-speed process at a lower feed casing temperature. The redesign typically reduces the solids conveying rate or increases the melting capacity of the screw. The heated barrel should be thermally isolated from the water-cooled feed casing.

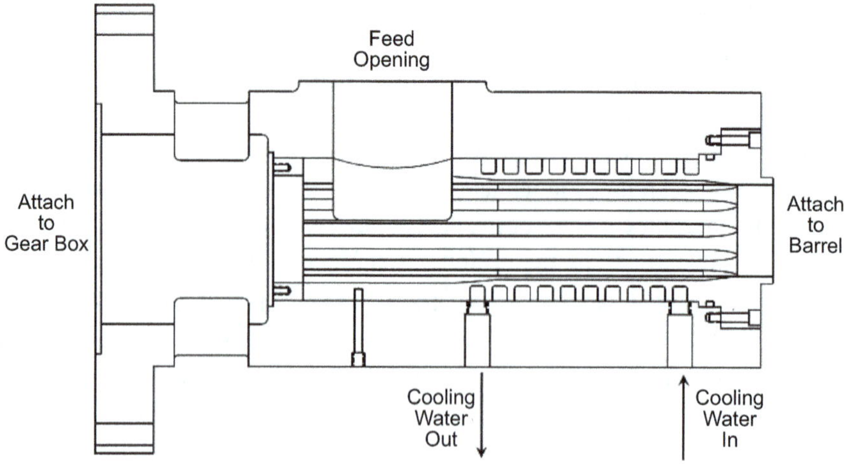

Figure 5.33 Schematic of a water-cooled grooved barrel feed casing. The grooves are axial (courtesy of Jeff A. Myers of Robert Barr, Inc.)

The design of the grooves and the feed channel depth are key features to a successful design. The range of the channel depths and groove design were summarized by Franzkoch and Menges [49] as a function of barrel diameter in Table 5.2. The feed channel depth is typically about 4 to 8% of the diameter for machines with diameters less than 100 mm. As a comparison, the feed channel depths for smooth bore extruders typically are in the range between 9 to 25% for pellet feedstocks, and even deeper for processes running low bulk density recycle. Practical experience has indicated that the maximum number of grooves is typically the barrel diameter in mm divided by 10. Thus, a 100 mm diameter machine would have at the maximum 10 grooves. Franzkoch and Menges caution not to design more than this number of grooves. The width of the grooves should be larger than the largest particle size fed to the machine. The width of the grooves also increases with increasing barrel diameter as shown in Table 5.2. The distance from the flight land to the root of the groove at the start of the feed casing should be no larger than the diameter of a pellet. The groove depth decreases in the downstream direction until the bore is smooth. The lead length is typically between 0.8 and 1.0 times the barrel diameter. For many applications, a double-flighted helix is used.

Table 5.3 Approximate Feed Channel Depths and Barrel Groove Dimensions as a Function of Barrel Diameters [49]

Barrel Diameter, mm	Feed Channel Depth, mm	Groove Width, mm	Maximum Number of Grooves
45	2.4 to 4.0	8	4
70	3.5 to 5.4	9	7
100	5.6 to 8.0	10	10
120	5.6 to 8.0	12	12

Ronaghan and Christiano [50] performed extrusion studies using an 88.9 mm diameter extruder that was fitted with a removable grooved bore liner. Three different styles of axial grooves were studied: a rectangular style, a ratchet style groove, and a semicircular design. The standard rectangular design and the semicircular design provided the highest solids conveying performance for an HDPE resin as measured by rate and pressure at the end of the grooved section. The performances of the three designs, however, were essentially the same when an LLDPE resin was extruded. Here the less aggressive ratchet design provided identical performance and may reduce the long-term wear of the machine.

Grooved bore extruders are common on European extruders up to barrel diameters of about 120 mm and on some applications up to 150 mm diameter [44]. For a 60 mm diameter machine, the maximum feed channel depth for a PP resin application is about 5 mm and 6 mm for HDPE resin applications; channel depths are not much larger than this for larger diameter machines. Ingen Housz and Meijer [44] state that this is due to the belief that the feed channel depth of grooved bore machines must be kept shallow to function properly. Based on their work, they state that feed channel depth needs to be optimized to match the melting capacity of the screw. If the melting capacity is not high enough, then the solids conveying section channel depth needs to be reduced to compensate. A troubleshooting case at Dow examined the low rates of a 250 mm diameter grooved bore extruder for LLDPE resin pellets. This extruder was designed with a square pitch, double flights, and a feed section depth of 5 mm. The axial barrel grooves were 6 diameters in length. This extruder design was incapable of conveying solids at a high enough rate to keep the downstream sections of the screw full of resin. The screw was removed and the feed channel depth was increased to 7.5 mm, but the rate did not increase appreciably. It was concluded that an extruder of this diameter was too large for effective groove feeding. This design problem will be covered in detail later in Section 5.5.1.2.

The *L/D* of grooved bore extruders is typically longer than typical smooth bore machines. The longer barrel lengths are used to increase the melting capacity of the machine. If shorter barrel lengths are used, then the ability to operate at high specific rates and low discharge temperatures may be compromised with solid polymer fragments in the extrudate. The extra barrel lengths along with barrier melting sections and dispersive mixers have been very effective at maintaining high rates and high quality extrudates. Larger motors are attached to these extruders to accommodate the extra barrel lengths. As discussed by Ingen Housz and Meijer [44], the feed section channel depth is in the range of 5 to 6 mm and possibly slightly deeper for a 60 mm diameter machine. At the end of this section, the pressure can be very high. The pressure, however, is typically relieved by increasing the volume of the channel [43, 49] by a combination of deepening the channel or increasing the lead length. These screws are often referred to as decompression or two-stage screws; not to be confused with a two-stage, vented extruder. Without a decompression section, the pressure at the end of the solids conveying section can be as high as 100 MPa, and the torque required to turn the screw can increase to a level high enough to cause a catastrophic failure of the screw; that is, the screw breaks in the feed section due to torsion. The decompression section typically decreases the pressure to a level similar to the discharge pressure. Downstream of the decompression section is the melting section. Design of the feed section is often a balance between pressure generation, machine wear, and matching the melting capacity. Barrier melting sections have been used extensively to maximize the melting capacity although conventional flighted melting sections are still used. Often a double-flighted barrier section (two solids channels and two melt flow channels) is used to balance the side forces on the screw. Balanced side forces will help reduce machine wear. Because pumping and pressurization is performed mainly by the grooved, solids conveying section, the requirements for a metering section are reduced. As such, the metering sections can be relatively short. Dispersive and distributive mixing sections are often used downstream of the barrier section [43, 48].

Franzkoch and Menges [49] stated that the tip velocity of the flight should be in the range of 0.2 to 1.0 m/s for optimal grooved barrel performance. Designs should not be made with tip speeds higher than 1.0 m/s. For a 100 mm diameter extruder, a tip speed of 1.0 m/s corresponds to a screw speed of 190 rpm.

Unlike smooth bore extruders where the metering section of the process controls the rate, the rate for grooved barrel extruders is controlled by the grooved section of the machine. Thus, the rate calculations for a process should be based on design of the feed channel, groove design, and the physical properties for the resin. The design of the downstream melting section, metering section, and mixers must be suitable for the rate provided by the solids conveying design.

5.5.1 Grooved Barrel Solids Conveying Models

Because the grooved portion of the extruder controls the rate of the entire process, establishing acceptable models is highly important for their successful design. Unlike smooth bore models, the accuracy of the grooved bore models can be obtained by comparing the model to actual extrusion rate data. Thus, solids conveying devices like those in Figure 5.6 and Figure 5.21 are not needed. Two models are presented here that provide an acceptable prediction of rate. Other models, however, are available [51].

5.5.1.1 The Peiffer Equation

Peiffer [52, 53] developed an empirical relationship between the geometry of the grooved feed section and rate as follows:

$$\frac{Q_m}{\rho_b N D_b^3} = 1.82\left(\frac{\bar{f}_b}{f_b^{1.2}} - 1.3\right)^{\frac{1}{6}}\left(\frac{H}{D_b}\right)\left(\frac{L}{D_b}\right)\left(1 + 0.9\frac{H_{gr}}{D_b}\right)\left(1 - 2.3\frac{e}{D_b}\right) \tag{5.40}$$

where Q_m is the mass rate of the extrusion, ρ_b is the bulk density of the feedstock resin, N is the screw speed (rev/s), f_b is the coefficient of dynamic friction at the conditions of the barrel, $\bar{f}_b$ is the average friction coefficient at the barrel wall, and H_{gr} is the maximum depth of the grooves in the barrel. The average friction coefficient at the barrel wall is considered a combination of the polymer-polymer internal friction coefficient f_i, the coefficient of dynamic friction at the conditions of the process, and the geometry of the grooves. The empirical relationship is as follows [54]:

$$\bar{f}_b = f_b + (f_i - f_b)\left(\frac{N_{gr} W_{gr}}{\pi D_b}\right)\left(1 - \exp\left[-5\left(\frac{H_{gr}}{W_{gr}}\right)\right]\right)^{0.9} \tag{5.41}$$

where N_{gr} is the number of grooves and W_{gr} is the width of the grooves. The quality of Equation 5.40 was determined by comparing it to extrusion rates using a parity plot, as shown in Figure 5.34. The calculated rates are within about ±10% of the experimentally determined values.

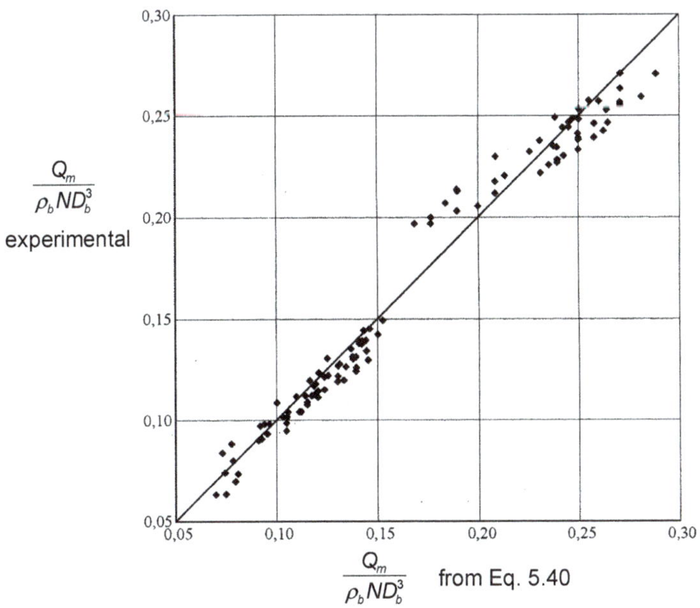

Figure 5.34 Parity plot showing the quality of the rate prediction using Equation 5.40 and experimental rate data [53]

5.5.1.2 Potente-Schöppner Model

Potente and Schöppner [55] developed a model based on different criteria for the channel and groove depths in relation to the pellet diameter for flow in the grooves and main channel. For the case where the depth of the groove is larger than the diameter of the pellet, a frictionally engaged solids flow occurs similar to that of a smooth bore machine. When the pellets are larger than the depth of the groove, the motion of the passing flight pushes the pellets in the groove downstream and the pellet segment protruding into the channel pushes material in the channel downstream at high rates. In order for the rate in the channel to be high, the depth of the channel needs to be less than about twice the characteristic pellet diameter. If the channel depth is larger than twice the pellet diameter, then solids conveying rates are diminished in the channel. The model is rather complex and omitted here. The model is reported to be within ±10% of the experimentally determined values.

The Potente-Schöppner model provides evidence for why extruders over 150 mm in diameter are not designed with grooved barrel feed sections. In general, resin suppliers produce pellets with a characteristic diameter between 3 and 4 mm; pellets with larger diameters, however, have been produced commercially. Resin pellets with diameter of 4 mm, limit the maximum feed channel depth to about 8 mm. For extruders of up to about 120 mm, the 8 mm feed channel depth is acceptable in design. For extruders over 150 mm in diameter, the feed channel would need to be deeper to maintain the downstream sections functioning properly. This higher depth would violate, however, the model rule that the screw channel should be less than twice the diameter of the pellets. This explains why the 250 mm diameter extruder described earlier was unsuccessful using a grooved bore feed section.

5.6 Solids Conveying Notes

As previously discussed, optimal solids conveying for a smooth-bore extruder will occur when forwarding forces are maximized and the retarding forces are minimized. For a specific screw design and resin, the conditions for the forces will be controlled by the screw and barrel temperatures. If these forces are not optimized, the solids conveying rate can be the rate-limiting step of the process instead of the metering section. This condition can lead to reduced rates, flow surging, and resin degradation. Numerous examples are provided in the latter chapters of this book. Optimizing the barrel temperature for a specific extruder is best performed experimentally, and a procedure is provided in Section 10.9. For screw diameters less than about 140 mm, the temperature of the screw typically does not need to be controlled for optimal solids conveying. For screws 150 mm in diameter and larger, the screw can become too hot for optimal solids conveying and cooling is often required. Screw cooling is provided using cooling water and a rotary union as described in Section 12.7.3 and Figure 10.8. For new installations and processes, adding a cooling bore to the shank of the screw is an excellent practice to mitigate risk at startup. Ideally the cooling hole will never be used, but if solids conveying rates become limited by a hot screw, cooling can be quickly added to the process.

Energy balances in Section 5.3.1.2 indicate that a rather large amount of the frictional energy dissipated at the barrel wall is transferred out through the barrel and lost to the environment. In most plasticating extrusion processes, the motor inputs 20 to 30% more energy needed to bring the extrudate up to discharge temperature. In some processes where the motor is operating at full motor current, the energy consumed in the solids conveying zone can be reduced and often a rate increase can be obtained. For bimodal HDPE resins [56], the amount of energy required to heat the resin from room temperature to 230 °C is about 650 to 700 J/g. For this example, the motor was inputting about 900 J/g, and thus, there was excess energy that goes in from the motor and leaves through the barrel wall. Much of this energy leaves through the first sections of the extruder; that is, Barrel Zones 1 and 2. One way to mitigate the consumption of this wasted energy is to reduce the dynamic coefficient of friction at the barrel wall for smooth bore extruders by increasing the temperature of these zones. For example, the Zone 1 temperature for one case was increased from 190 to 230 °C in 10 °C increments. The motor load was reduced by about 8% while very little loss in rate was observed. This additional motor capacity can be used for a rate increase or for other functions downstream such a barrel cooling over the metering zone [56].

Extruding resins that have high rubber contents or are very soft and rubber-like can cause an improperly designed feed section to over convey material to the melting section of the screw. In these cases, the melting section is not capable of melting the resin at a rate fast enough to keep up with the solids conveying section, causing process instabilities and solid polymer fragments in the extrudate. Reducing the feed channel depth is one remedy to mitigate this problem. Two case studies are provided where overfeeding occurred for an ABS resin (Section 12.7.6) and a TPU resin (Section 13.4.2).

Materials compounded with low to medium levels of low molecular weight liquid additives or waxy materials may experience solids conveying problems in smooth bore machines. At startup, the extruder will typically operate at the designed specific rate with the metering channel controlling the rate. But after a short period of time (minutes to hours), the specific rate of the extruder will decrease and the solids conveying section of the extruder will become the rate-limiting step. Here the liquid additives migrate from the feedstock and coat the feed zone barrel and feed casing such that the feedstock rotates with the screw and does not forward. Flow surging is often observed. Adding the liquid feedstock via injection through the barrel downstream of the solid conveying zone will eliminate the problem.

Getting the additive mixed into the resin matrix, however, is now the challenge for the process designer. Distributive and dispersive mixing sections on the screw are required downstream of the injection point.

Extensive wear to the feed casing and the feed zone of the barrel can occur for processes with extended service time. When the problem is discovered, the processor will typically replace the barrel and screw and assume that the feed casing is acceptable. When the new barrel and screw are installed, the barrel will have a smaller diameter than the feed casing, creating a ledge at the connection flange between the two segments. If this ledge is large enough it can impede solids conveying and cause the extruder to flow surge. A case study with this type of ledge is presented in Section 12.7.7. The process can be restored to the original performance by tapering the barrel flange such that an abrupt ledge does not exist.

Several manufacturers offer a low level of grooving to smooth bore extruders. These extruders are operated with screws designed for smooth bore extruders. They are not considered grooved bore extruders. The low level of grooving provides an improved level of solids conveying, allowing some processes to operate at higher specific rates. The grooves are made by scoring the smooth bore barrel and feed casing.

Two-stage, vented extruders must not be equipped with a grooved barrel. For a two-stage machine, the axial pressure gradient in the first-stage metering section is negative while the pressure gradient in the second stage metering section is positive. The negative pressure gradient in the first stage will cause the first metering stage to operate at specific rates higher than the calculated specific rotational rate. In order for the second stage to pump and pressurize the resin, its channel depth is 1.2 to 1.5 times deeper than the first-stage meter; that is, the pump ratio. If a grooved feed section is used, the negative pressure gradient in the first-stage meter can be very high and can cause the specific rate to be higher than what can be pressurized in the second stage. If the second stage cannot convey and pressurize the resin at a high enough rate, then vent flow will occur.

The upstream portions of a process and the hopper can cause compositional variations for compounding operations. For example, a simple compounding operation was being performed using a 114.3 mm diameter single-screw extruder and downstream pelletizing equipment. Upstream of the extruder was a high-intensity batch mixer where the component pellets were mixed. The discharge from the mixer went to the hopper of the extruder. During a trial, all feedstock materials were carefully weighed and added to the mixer and mixed at high intensity, yet the composition of the additives in the pelletized product was not consistent between samples, as shown in Figure 5.35. The cause of the problem was that the high intensity of the mixer fractured one of the brittle masterbatch concentrate pellets into fragments and a fine powder, and these materials segregated in the hopper between the mixer and the extruder. When a slight process upset occurred, the small diameter fragments (or fines) attached and collected in the hopper dislodged and flowed to the extruder, creating the high additive compositions shown in Figure 5.35. Many other samples were collected from this process and the trend was the same as that shown in Figure 5.35. For this case, the impulse addition of the powders to the extruder did not cause instabilities in the solids conveying process. However, when extruding fine powder feedstocks, segregation of the fine particles (less than 30 μm) from the majority particles (300 μm) can cause solids conveying instabilities if liquid additives were mixed into the powder prior to extrusion.

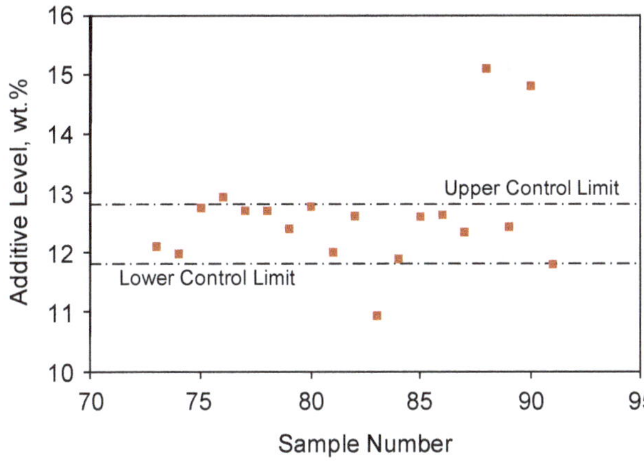

Figure 5.35 Measured additive level as a function of sample number for a simple compounding process. The samples were taken every 15 minutes. As indicated by the figure, most samples were in control, but occasional samples were far out of specification

Nomenclature

D	average diameter of the barrel (D_b) and screw core (D_c)
D_b	diameter of the barrel
D_c	diameter of the screw core
e	flight width at the tip of the screw and perpendicular to the flight edge
f	coefficient of dynamic friction
f_b	coefficient of dynamic friction at the barrel wall
$\overline{f}_b$	average friction coefficient at the barrel wall
f_h	coefficient of dynamic friction at the flight surfaces
f_s	coefficient of dynamic friction at the screw surfaces
f_i	coefficient of internal friction (pellet to pellet)

H	distance from the screw root to the barrel
H_{gr}	maximum depth of the barrel grooves in a grooved barrel extruder
K	thermal conductivity of the barrel
K	equation constant for several solids conveying models
K_{xz}	lateral stress ratio for stress in the x direction due to stress in the z direction
K_{yz}	lateral stress ratio for stress in the y direction due to stress in the z direction
L_b	axial length of the barrel in the solids conveying section
L	lead length
N	screw rotation speed
N_{gr}	number of grooves in a grooved barrel
p	number of flight starts
P	pressure
P_o	initial pressure at the base of the hopper and at the start of the solids conveying section
P_z	pressure at the end of the solids conveying section
Q_{cyl}	heat transfer rate
Q_s	volumetric flow rate
Q_m	mass flow rate
r	independent variable in the radial direction
r_i	radius of the inner surface of the barrel
r_o	radius of the outer surfaces of the barrel
R_c	radius of the screw core
R_b	radius of the barrel wall
R_{cyl}	thermal resistance of the barrel
T	temperature
T_b	temperature of the barrel surface
T_i	solid plug-barrel interface temperature
T_o	measured temperature on the outside barrel surface
V	sliding velocity for a solid plug on an interface
V_{cz}	z component of velocity of the screw core
V_b	velocity at the barrel wall
V_{bx}	x component of velocity of the screw flight at the barrel wall
V_{bz}	z component of velocity of the screw flight at the barrel wall

V_{bs}	sliding velocity of the solid bed against the barrel wall
V_{pl}	solid plug velocity in the axial direction
V_{pz}	solid plug velocity in the z direction
$V_{p\theta}$	solid plug velocity in the tangential direction
V_{ss}	sliding velocity of the solid bed against the screw surfaces
V_{sz}	velocity of the solid bed in the z direction
W	average channel width perpendicular to flight
W_b	channel width perpendicular to flight at the barrel wall
W_c	channel width perpendicular to flight at the screw core
W_{gr}	width of the grooves in a grooved barrel
z	independent variable in the down-channel direction (or helical direction)
Z_b	helical length of the solids conveying channel at the barrel wall
α	skew angle
θ	average helix angle; i.e., evaluated at D
θ_b	helix angle at the barrel wall; i.e., evaluated at D_b
θ_c	helix angle at the screw core; i.e., evaluated at D_c
ρ_b	solid bulk density of the resin
σ_x	normal stress on a solid plug in the x direction
σ_y	normal stress on a solid plug in the y direction
σ_z	normal stress on a solid plug in the z direction
φ	solids forwarding angle for the solid plug

References

1. Tadmor, Z. and Klein, I., "Engineering Principles of Plasticating Extrusion," Van Nostrand Reinhold Co., New York, 1970
2. Hyun, K.S. and Spalding, M.A., "Debugging and Optimizing Extruder Performance Using Process Data Collected With a Data Acquisition System," *SPE ANTEC Tech. Papers*, **40**, 162 (1994)
3. Spalding, M.A., Kirkpatrick, D.E., and Hyun, K.S., "Coefficients of Dynamic Friction for Low Density Polyethylene," *Polym. Eng. Sci.*, **33**, 423 (1993)
4. Hyun, K.S. and Spalding, M.A., "Use of Process Data Obtained from a Data Acquisition System for Optimizing and Debugging Extrusion Processes," *Adv. Polym. Tech.*, **15**, 29 (1996)

5. Spalding, M. A., Powers, J. R., Wagner, P. A., and Hyun, K. S., "Flow Surging in Single-Screw, Plasticating Extruders," *SPE ANTEC Tech. Papers*, **46**, 254 (2000)
6. Hyun, K. S. and Spalding, M. A., "Bulk Density of Solid Polymer Resins as a Function of Temperature and Pressure," *Polym. Eng. Sci.*, **30**, 571 (1990)
7. Baird, D. G. and Collias, D. I., "Polymer Processing: Principles and Design," John Wiley & Sons, New York, (1998)
8. Agur, E. E. and Vlachopoulos, J., "Numerical Simulation of a Single-Screw Plasticating Extruder," *Polym. Eng. Sci.*, **22**, 1084, (1982)
9. Walker, D. M., "An Approximate Theory for Pressures and Arching in Hoppers," *Chem. Eng. Sci.*, **21**, 975 (1966)
10. Campbell, G. A., Baird, J. C., Gilette, L., Bullwinkel, M., and Stowe, J., "An Experimental Investigation of Solids Conveying - Flow Rate and Temperature Rise in Single Screw Extruders," *SPE ANTEC Tech. Papers*, **48**, 328 (2002)
11. Smith, W. S., Sickles, R. A., Miller, L. A., and Womer, T. W., "An Experimental Investigation into Solids Feeding Characteristics of a Single Piece Barrel with Integral Feed-port Design vs a Standard Two Piece Water Cooled Feedblock and Barrel Configuration," *SPE ANTEC Tech. Papers*, **53**, 390 (2007)
12. Steward, E., "Understanding Grooved Feed," *Plast. Technol.*, **46** (4), 55 (2000)
13. Schlafi, D. and Zweifel, Y., "Barrier Screws in Helically Grooved Barrels: Operating Characteristics and Implications for Simulation Models," *SPE ANTEC Tech. Papers*, **47**, 195 (2001)
14. Darnell, W. H. and Mol, E. A. J., "Solids Conveying in Extruders," *SPE J.*, **12**, 20 (1956)
15. Schenider, K., *Kunstoffe*, **59**, 97 (1969)
16. Broyer, E. and Tadmor, Z., "Solids Conveying in Screw Extruders Part I: A Modified Isothermal Model," *Polym. Eng. Sci.*, **12**, 12 (1972)
17. Tadmor, Z. and Broyer, E., "Solids Conveying in Screw Extruders Part II: Non Isothermal Model," *Polym. Eng. Sci.*, **12**, 378 (1972)
18. Strand, S. R., Spalding, M. A., and Hyun, K. S., "Modeling of the Solids-Conveying Section of a Starve-Fed Single-Screw Plasticating Extruder," *Plast. Eng.*, **48**, 7, 17 (1992)
19. Campbell, G. A. and Wang, C., "Recent Advances in Understanding of Extruder Dynamics - Fluid and Solid Transport," Proceedings of the 1995 JSPP Annual Conference, Tokyo, Japan, p 19, June (1996)
20. Campbell, G. A. and Dontula, N., "Solids Transport in Extruders," *Int. Polym. Process.*, **10**, 30 (1995).
21. Hyun, K. S., Spalding, M. A., and Hinton, C. E., "Theoretical and Experimental Analysis of Solids Conveying in Single-Screw Extruders," *SPE ANTEC Tech. Papers*, **42**, 199 (1996)
22. Yamamuro, J. A., Penumadu, D., and Campbell, G. A., "Modeling Solids Conveying in Polymer Extruders," *Int. Polym. Process.*, **13**, 3 (1998)
23. Hyun, K. S. and Spalding, M. A., "A New Model For Solids Conveying In Single-Screw Plasticating Extruders," *SPE ANTEC Tech. Papers*, **43**, 211 (1997)

24. Penumadu, D., Yamamuro, J., Abrantes, A.E., and Campbell, G.A., "Stress-Strain behavior of Polymer Pellets," *SPE ANTEC Tech. Papers*, **43**, 224 (1997)

25. Moysey, P.A. and Thompson, M.R., "Plastic Contact Mechanics and Its Impact on DEM Simulations of Solids Transport in Extruders," *SPE ANTEC Tech. Papers*, **52**, 882 (2006)

26. Moysey, P.A. and Thompson, M.R., "Investigation of Solids Transport in a Single-Screw Extruder Using a 3-D Discrete Particle Simulation," *Polym. Eng. Sci.*, **44**, 2203 (2004)

27. Moysey, P.A. and Thompson, M.R., "Discrete Particle Simulation of Solids Compaction and Conveying in a Single-Screw Extruder," *Polym. Eng. Sci.*, **48**, 62 (2008)

28. Zhu, F. and Chen, L., "Studies on the Theory of Single Screw Plasticating Extrusion. Part I: A New Experimental Method for Extrusion," *Polym. Eng. Sci.*, **31**, 1113 (1991)

29. Campbell, G.A. and Baird, J.C., "Single Screw Solids Conveying-Different Polymers and a General Purpose Screw," *SPE ANTEC Tech. Papers*, **49**, 71 (2003)

30. Baird, J.C., Chemical Engineering M.S. Thesis, Clarkson University, (2002)

31. Çengel, Y.A., "Heat Transfer: A Practical Approach," McGraw-Hill, Boston, 149, (1998)

32. Kreith, F. and Bohn, M.S., "Principles of Heat Transfer," 5th ed., West Publishing Co., p. 5, (1993)

33. Cubberly, W.H., "Metals Handbook: Metallography and Microstructures, Volume 9," 9th ed., American Society for Metals, Metals Park, 148, (1985)

34. Altinkaynak, A., Gupta, M., Spalding, M.A., and Crabtree, S.L., "An Experimental Study on Shear Stress Characteristics of Polymers in Plasticating Single-Screw Extruders," *Polym. Eng. Sci.*, **49**, 471 (2009)

35. Powers, J.D., Spalding, M.A., and Hyun, K.S., "Extruding HIPS on ABS Screws," *SPE ANTEC Tech. Papers*, **40**, 151 (1994)

36. Gogos, C.G., Zafar, M., Sebastian, D., Todd, D.B., Hyun, K.S., and Spalding, M.A., "A Device to Study Solids Compaction and Conveying in Single Screw Extrusion," *SPE ANTEC Tech. Papers*, **40**, 288 (1994)

37. Dooley, J., "Simulating the Flow In a Film Die Using Finite Element Analysis," *SPE ANTEC Tech. Papers*, **36**, 168 (1990)

38. Spalding, M.A., Hyun, K.S., and Hughes, K.R., "Stress Distribution in Solid Polymer Compacts," *SPE ANTEC Tech. Papers*, **42**, 191 (1996)

39. Frankland, J., "Salt and Screw Cooling," *Plastics Technol.*, **58**, 5, 31 (2012)

40. Spalding, M.A., Dooley, J., and Hyun, K.S., "The Effect of Flight Radii Size on the Performance of Single-Screw Extruders," *SPE ANTEC Tech. Papers*, **45**, 190 (1999)

41. Derezinski, S.J., "Increased Barrel Stress Extruder," US Patent 5,314,246 (1994)

42. Berghaus, U., "Extrusion of Thermoplastics," Chapter 3.2 in "Screw Extrusion," White, J.L and Potente, H. (Eds.), Hanser Publishers, Munich, 2002

43. Davis, B.A., Gramann, P.J., Del P Noriega E., M., and Osswald, T.A., "Grooved Feed Single Screw Extruders-Improving Productivity and Reducing Viscous Heating Effects," *Polym. Eng. Sci.*, **38**, 1199 (1998)

44. Ingen Housz, J.F. and Meijer, H.E.H., "The Melting Performance of Single-Screw Extruders. II," *Polym. Eng. Sci.*, **21**, 1156 (1981)

45. Wortberg, J., "Screw and Barrel Design for Grooved Feed vs Smooth Bore Extruders," *SPE ANTEC Tech. Papers*, **48**, 80 (2002)

46. Sikora, J.W., "The Effect of the Feed Section Groove Taper Angle on the Performance of a Single-Screw Extruder," *Polym. Eng. Sci.*, **41**, 1636 (2001)

47. Womer, T.W., Wagner, J.R., Harrah, G., and Reber, D., "An Experimental Investigation on the Influence of Barrel Temperatures on the Output of a Constant Depth Screw with Grooved Barrel Feeding," *SPE ANTEC Tech. Papers*, **46**, 259 (2000)

48. Grünschloß, E., "A New Style Single Screw Extruder with Improved Plastification and Output Power," *Int. Polym. Process.*, **17**, 291 (2002)

49. Franzkoch, B. and Menges, G., "Grooved Forced-Feeding Zones can Improve Extruder Performance," *Plast. Eng.*, **34**, 7, 51 (1978)

50. Ronaghan, C. and Christiano, J.P, "Effect of Groove Geometry on the Performance of Grooved Feed Extruders," *SPE ANTEC Tech. Papers*, **48**, 323 (2002)

51. Pan, L., Jia, M.Y., Xue, P., Wang, K.J., and Jin, Z.M., "Studies on Positive Conveying in Helically Channeled Single-Screw Extruders," *eXPRESS Polym. Letters*, **6**, 543 (2012)

52. Peiffer, H., Dr.-Ing. Dissertation, RWTH Aachen (1981)

53. Potente, H., "Single Screw Extruder Analysis and Design," Chapter 5 in "Screw Extrusion," White, J.L. and Potente, H. (Eds.), Hanser Publishers, Munich (2002)

54. Potente, H., *Kunststoffe*, **75**, 439 (1985)

55. Potente, H. and Schöppner, V., "A Throughput Model for Grooved Bush Extruders," *Int. Polym. Process.*, **10**, 289 (1995)

56. Patterson, S.E. and Spalding, M.A., "Design Innovations of High Density Polyethylene Pipe Grade Resins," *SPE ANTEC Tech. Papers*, **58** (2012)

6 The Melting Process

The melting process is a highly important operation in plasticating extrusion because incomplete melting will allow solid resin particles to discharge with the extrudate, often causing defects in the final product. For example, small solid particles as gels often contaminate film products. These particles can be caused by the incomplete melting of the resin in the transition section of the screw and the lack of an adequate secondary dispersive mixing device to trap and melt the particles. The solid particles in the discharge are most likely to occur at high screw speeds and high rates. Solid resin particles in final products are discussed in Section 11.10.2.

As previously discussed and as shown in Figure 1.3, a single-screw extruder is basically divided into three sections: solids conveying, melting, and melt conveying. In the solids-conveying section, as discussed in Chapter 5, plastic materials, usually in the form of pellets or powders, are fed continuously through a hopper into the extruder. Then the polymer moves down the channel pushed by the rotating flights of the screw. Because of heat conduction through the barrel wall and mechanical friction, the solid polymers are heated and softened. As soon as the temperature of solid polymer reaches its melting point (for crystalline polymers) or softening point (for amorphous polymer), a polymer melt is formed. In the melting section, polymers are plasticated into melts by heat conduction through the barrel wall and by viscous dissipation in melt films. The dissipation is caused by the shearing action in the films due to the motion of the screw. This energy is then conducted into the solids, melting the resin. In the melt-conveying (metering) section, polymer melt is pressurized and readied to be pumped through the die. In a properly designed screw and process (smooth bore extruder), the metering section is the rate-controlling part of the screw, and it will be discussed in Chapter 7.

As discussed in Chapter 5, the solids-conveying rate for a specific screw is directly proportional to screw speed. That is, if the screw speed is increased by a factor of two the solids-conveying rate will nearly double. The melting flux of the screw as measured in $kg/(h \cdot m^2)$ at the barrel wall, however, will not increase at the same rate. Typically, the melting flux will increase by 40% for a doubling of the screw speed [1]. In order to complete the melting process, additional area at the barrel wall and screw root will need to be used for the melting process, pushing the solids downstream into the metering section. If the screw speed is high enough, solid polymer particles will be discharged with the extrudate. Although not popular with a production plant, often the only short-term solution to mitigating solid polymer particles from the extrudate is to decrease the rate of the line. As mentioned above and described in Chapter 8, these solid polymer particles can be trapped and melted using dispersive mixing devices or in screen packs.

Barrier melting sections are one of many ways to increase the melting rate of a process while mitigating the occurrence of solid polymer particles in the extrudate. The design of barrier melting sections is complex and many different variations exist. For these designs, a secondary flight is added to the transition section, creating two channels. The channel located adjacent to the trailing flight is for the transport of solids. The melting process occurs primarily between the compacted solids and the barrel wall. The secondary flight is undercut from the primary flight to allow newly melted resin to flow into the melt-conveying channel adjacent to the pushing flight. A schematic of a barrier melting section is shown in Figure 6.1, and the design of commercially available devices is covered in detail in Chapter 14. Barrier melting sections typically allow rates that are about 20% higher than those for conventional flighted screws.

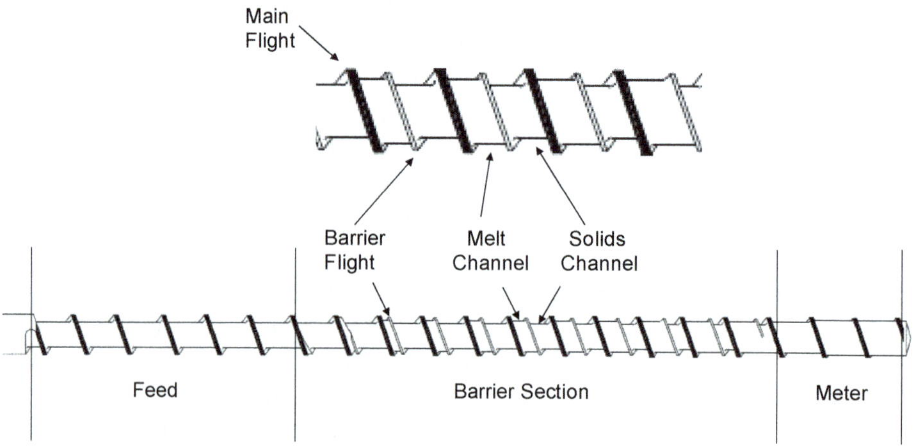

Figure 6.1 Schematic for a barrier melting section (courtesy of Jeff A. Myers of Robert Barr, Inc.). The barrier flight is undercut from the main flight to allow molten resin to transfer from the solids channel to the melt channel

The melting process is the primary mode for mixing in single-screw machines. This concept can be observed from Maddock solidification experiments. Melting as a method for mixing is presented in Section 8.4.

This chapter describes the processes that occur in the transition section or melting section of the screw. First, the historical melting process as initially outlined by

Maddock will be presented. Based on his experimental observations, melting models were developed by numerous researchers. These early models were all based on barrel rotation physics. Next, new melting models based on the screw rotation physics will be presented for conventional-flighted melting sections, barrier-flighted melting sections, and a special case of melting known as one-dimensional melting.

6.1 Compression Ratio and Compression Rate

Conventional transition sections are constructed by simply decreasing the depth of the channel in the down-channel direction. The amount and rate of the depth change sets the performance of the melting process and the removal of entrained air that resides between the feedstock pellets or powders. The compression ratio sets the amount of compression while the compression rate sets the rate of the compression. The compression ratio and compression rate are calculated as follows for conventional-flighted transition sections:

$$C = \frac{H_f}{H} \tag{6.1}$$

$$R = \frac{(H_f - H)\sin\theta_b}{ML} \tag{6.2}$$

where C is the compression ratio, H_f is the channel depth of the feed section, H is the depth of the metering channel, R is the compression rate in the transition section, M is the number of turns in the transition section, θ_b is the helix angle at the barrel wall, and L is the lead length; ML is the axial length of the transition section. As the compression occurs, the bulk density of the feedstock resin increases until the density of the solid bed is equal to the solid density of the resin. This process is responsible for eliminating the entrained gas between the pellets or powders. For most processes this gas is air, but it could be an inerting gas used to exclude oxygen from the extruder. Trace amounts of surface moisture could also be expelled. An example of the bulk density of the solid bed in the channels of the screw was shown previously in Figure 4.1 for a PVDC resin. For this case, the bulk density of the resin increased from 0.73 g/cm^3 at the start of the screw to 1.7 g/cm^3 for the fully compacted bed, expelling 0.6 volumes of gas for every volume of feedstock entering the extruder. In most cases, the entrained gas must egress through the hopper.

The melting performance is also highly influenced by the compression ratio and rate. If the compression ratio is too low, the ability of the section to develop pressure in the downstream direction can be limited. Lower than desired pressures can limit the melting rate of the resin and potentially limit the rate of the extruder. Moreover, if the compression ratio is too low the feed section may not be deep enough to maintain the metering section completely full of resin. Alternatively, if the compression ratio is too high, the pressure at the end of the transition section can be very high, causing local degradation of the resin for thermally sensitive resins, severe mechanical wear to the flights, and process instabilities such as flow surging. A specialized form of this type of flow surging is presented in Section 12.7.6 for an ABS resin. Resin degradation can occur at the root of the screw at the end of the transition section.

The compression ratio for pellet feedstocks typically ranges from about 1.8 for PMMA resins [2] to 4.5 for LDPE resins [3], although many exceptions are practiced. Compression ratios for several common resins are reported by Giles, Wagner, and Mount [2]. If high levels of a low-density recycle stream are added back into the feed hopper, then the compression ratio and compression rate will need to be increased as follows:

$$C_f = \frac{\rho_{pellets}}{\rho_f} C_{pellets} \tag{6.3}$$

$$R_f = \frac{\rho_{pellets}}{\rho_f} R_{pellets} \tag{6.4}$$

where $C_{pellets}$ and $R_{pellets}$ are the compression ratio and compression rate used for a pellet feedstock, $\rho_{pellets}$ is the bulk density of the pellets at ambient conditions, ρ_f is the bulk density of the feedstock mixture at ambient conditions, and C_f and R_f are the compression ratio and compression rate that should be used for the lower density feedstock, respectively. If the adjustment is not made to accommodate the low bulk density feedstock, the lower compression ratio may cause the feed section to be the rate-limiting step of the process instead of the metering section. An example for processing a low-density feedstock is provided in Section 12.7.7.

Compression rates typically vary between 0.0025 and 0.0055 for pellet feedstocks. If the compression rate is too low, then the compaction rate of the solid feedstock may not be high enough to force the entrained gas out through the hopper. If the compression rate is too high, poor melting performance and solid bed breakup can occur. For example, LDPE resins can be melted very easily using a screw with a compression rate of 0.0055, while LLDPE resins perform best with a compression rate near 0.0030 [3, 4]. If an LLDPE is extruded using a screw with a compression rate of 0.0055, then solid bed breakup and solid polymer particles in the extrudate are likely to occur.

6.2 The Melting Process

Experiments for studying the melting mechanism were first carried out by Maddock [5] and then by Street [6]. The melting mechanism was observed by performing Maddock solidification experiments as detailed in Section 10.3.1. Maddock first proposed the physical description of this melting process in 1959. In his description, the energy for the melting process came from two sources: energy conduction through the barrel wall, and viscous dissipation from the film at the inner barrel surface. For these experiments, the extruder was fed with a mixture of natural and colored (1 to 3%) polymer. When steady state was achieved, screw rotation was stopped, and full cooling was applied to the barrel, solidifying the polymer in the channels. Usually, the original conditions existing in the extruder were retained and observable in the solidified material. By examining the cross sections of the solidified polymer after taking it out of the screw channel, a clear picture of the melting behavior was obtained, as shown in Figure 6.2. For this figure, the region that is essentially white (base color of the ABS) was solid just prior to stopping the screw. The regions in the photograph that were mostly black or were black with white flow streaks were molten resin just prior to stopping the screw. The mostly black regions reveal the mixing abilities of the melting process. Maddock reported that energy from the shearing action occurring in the melt film and energy conduction through the barrel wall was used to melt the resin at the solid bed-melt film interface. The motion of the screw then caused the flow of material to be dragged out of the melt film to the pushing flight. The pushing flight then scraped the molten material from the barrel and forced a recirculation pattern in the melt pool. These circulation flows are clearly visible in the melt pool of Figure 6.2.

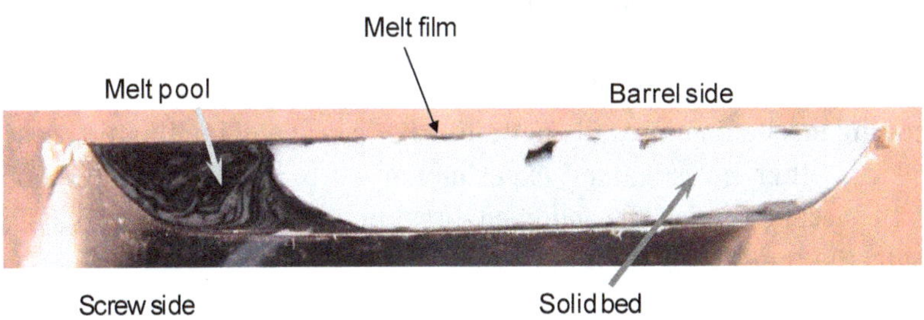

Figure 6.2 Photograph of resin solidified in the transition section after a Maddock solidification experiment for an ABS resin. The pushing flight is on the left side of the photograph

Maddock's and Street's famous solidification experiments revealed that melting in many situations takes place in a specific order. The experiments showed that after melting began, there was a continuous solid bed and a melt film over the inner barrel surface. Later the solid bed was completely surrounded by melt. Farther downstream, a melt pool developed between the pushing flight and the solid bed. The solid bed disappeared gradually all the way down the channel [5]. Tadmor's research group [7–9] observed similar melting phenomena. Typical results were reported and are shown in Figure 6.3.

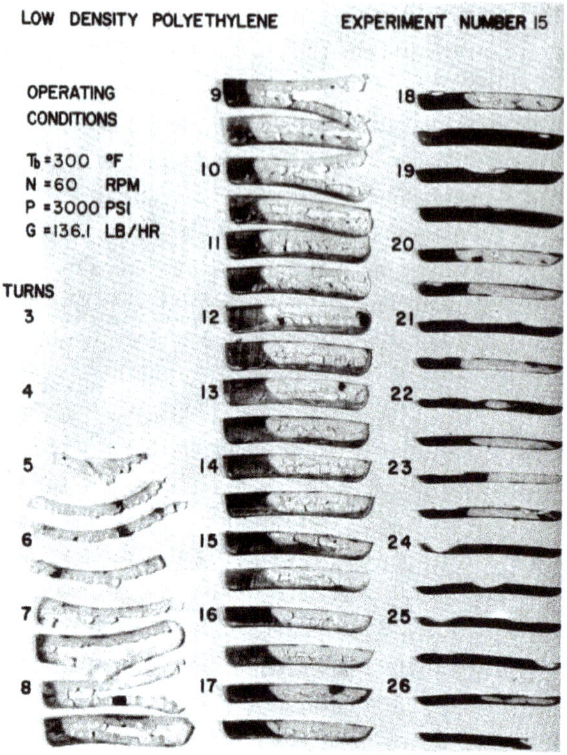

Figure 6.3 Photographs of the melting process for an LDPE resin from Tadmor and Klein [8]

Tadmor and Klein [8] studied many different materials, which included LDPE, HDPE, ABS, rigid PVC, and PP resins, using different operating conditions and extruder geometries. The results were interpreted to demonstrate that the Maddock melting mechanism prevails in quite a large range of melting processes with most resins. Other investigators' experiments led to other melting phenomena, such as the acceleration of the solid bed, breakup of the solid bed, and a melt pool existing by the trailing flight [10, 11].

6.2.1 The Melting Process as a Function of Screw Geometry

The melting process for a resin is complex and depends on many parameters, including screw speed, screw geometry, barrel temperatures, and channel pressures. Moreover, the compression ratio and compression rate also affect the pressure in the channel. The melting flux is known to increase with increasing pressure in the channel [1, 12]. A series of Maddock solidification experiments were performed at Dow using the 63.5 mm diameter highly instrumented extruder (see Figure 1.1) described in Section 8.4.1 and an ABS resin [13, 14]. Most experiments were performed using the same barrel temperature settings and at a screw speed of 60 rpm. The first series of screws had feed sections that were 6 diameters in length, 8 diameters of transition, and 7 diameters of constant-depth metering sections with a depth of 3.18 mm. The feed depths were changed such that the compression ratios were 2.0, 2.4, 2.8, and 3.2, respectively. Since the transition lengths were constant for these screws, the compression rates also changed. Photographs of cross sections parallel to the axis of the screw are shown in Figure 6.4. These experiments were performed using a small amount of black colored ABS pellets. As before, the white colored region is the solid bed while the black regions were molten just prior to stopping the screw.

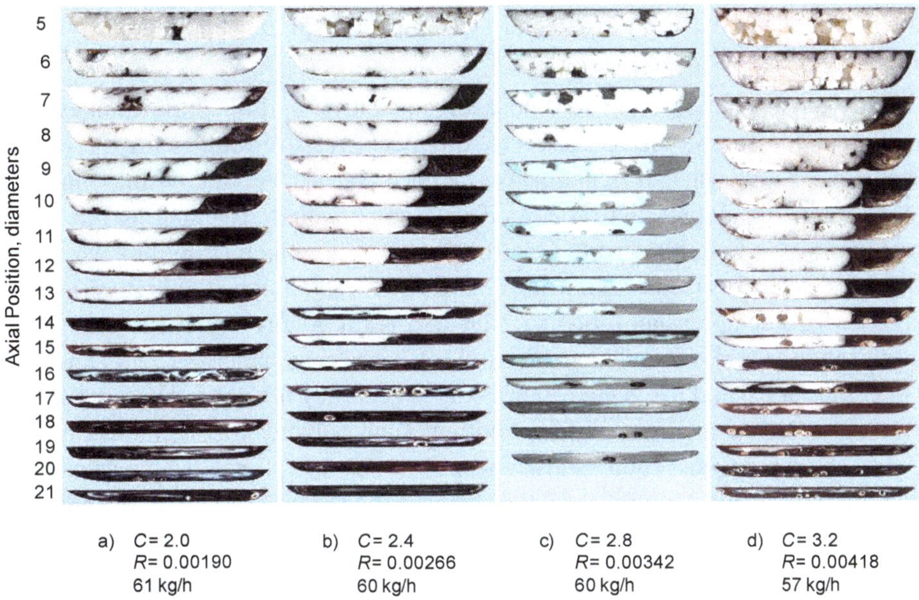

Figure 6.4 Melting profiles for a 63.5 mm diameter extruder running an ABS resin at 60 rpm for screws with a 3.18 mm deep metering channel, 6 diameters of feed section, 8 diameters of transition, and 7 diameters of metering section: (a) compression ratio of 2.0, (b) compression ratio of 2.4, (c) compression ratio of 2.8, and (d) compression ratio of 3.2 [13, 14]. The pushing flights are on the right side of the section photographs. The void marks on (c) at diameters 16 through 20 were caused by the resin shrinking slightly as it cooled

Several key observations can be made from studying the photographs in Figure 6.4. First, the pellets at 5 diameters from the start of the screw are loosely packed in the channel and clearly visible. The pellets upstream of this position would not hold the shape of the channel when the screw was removed from the extruder; that is, the pellets fell out of the machine as individual pellets. As the pellets flowed downstream, they were compacted, the entrained air was pushed out through the hopper, and the individual pellets in the solid bed could not be identified. The melting performance of the different screws appears very similar as shown by the black colored regions, but with a critical view several differences are evident. For example, qualitatively the melting rate appears to be the highest for the screws with compression ratios of 2.0 and 2.4 as indicated by the axial position where essentially all the white solid material is gone. Solid bed breakup is clearly occurring for the screws with compression ratios of 2.8 and 3.2 at diameters 15 and 16, respectively. Under select conditions, these solid bed fragments can flow downstream and lodge in another section of the screw such as a mixer, causing reduced rates and flow surging. An example of this type of problem is presented in Section 12.7.6. For this application with ABS resin, the ideal compression ratio was 2.5. Maddock solidification experiments preserve the conditions at the time when the screw was stopped. Thus, the experiment captures the conditions for a single moment. Any unsteady-state activity, such as solid bed breakup, will appear differently if the experiment is replicated.

The axial pressure profiles were measured just prior to the solidification experiment. For the screws discussed above, the axial pressure profiles are shown by Figure 6.5. As expected, the channel pressures increased with increasing compression ratio (and compression rate). As discussed previously, the melting flux for a resin increases with increasing pressure [1, 12]. Thus, the melting flux would be expected to increase with increasing compression ratio and be the highest for the screw with a compression ratio of 3.2 for ABS resin. As previously discussed and as shown in Figure 6.4, solid bed breakup was occurring for screws with compression ratios of 2.8 and 3.2. Solid bed breakup is generally not desired because it can create process instabilities and intermittent solid polymer fragments in the discharge. This pressure profile is also consistent with an optimal compression ratio of 2.4 to 2.5 for this ABS resin pellet feedstock. The higher pressure at the entry to the metering section caused a negative pressure gradient to occur in the metering channel for this case. The negative pressure gradient will result in a rate that is slightly higher than the rotational rate. The low pressures at diameter 5 are consistent with the low level of compaction observed at that location and as shown in Figure 6.4. The discharge pressure was essentially the same for all four screws and is consistent with all screws operating at essentially the same rate.

A second set of Maddock solidification experiments was performed using two screws that had compression ratios of 2.8, 6 diameter long feed sections with depths of 8.89 mm, and metering sections with depths of 3.18 mm. The first screw was used in the first set of experiments and had 8 diameters of transition section for a compression rate of 0.00342. The second screw had a transition section that was 4 diameters long and a compression rate of 0.00684. Photographs of cross sections parallel to the axis of the screw are shown in Figure 6.6.

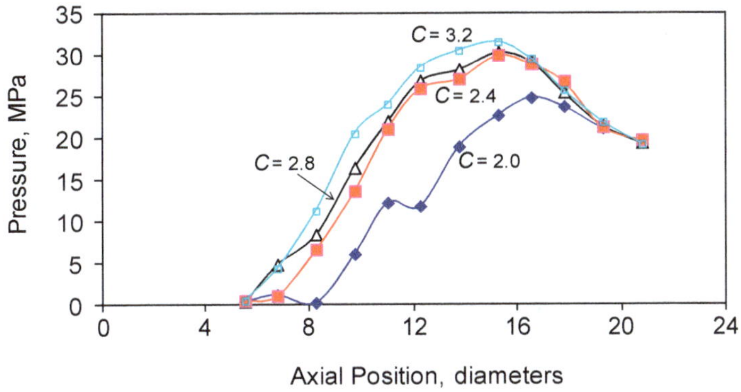

Figure 6.5 Axial pressure profiles measured for the screws used to make the cross sectional photographs in Figure 6.4

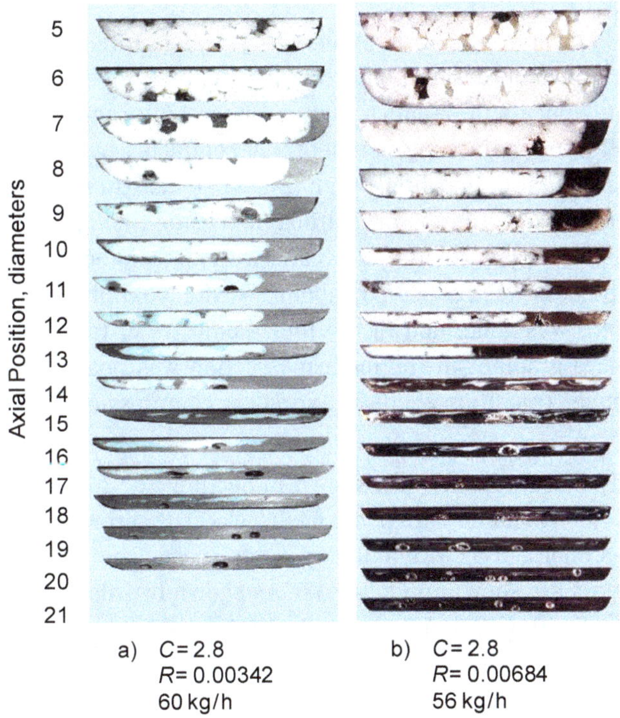

Figure 6.6 Melting profiles for a 63.5 mm diameter extruder running an ABS resin at 60 rpm for screws with a 8.89 mm deep feed channel, 6 diameters of feed section, and a metering channel depth of 3.18 mm (C = 2.8): (a) 8 diameters of transition section for R = 0.00342, and (b) 4 diameters of transition section for R = 0.00684 [13, 14]. The pushing flights are on the right side of the section photographs

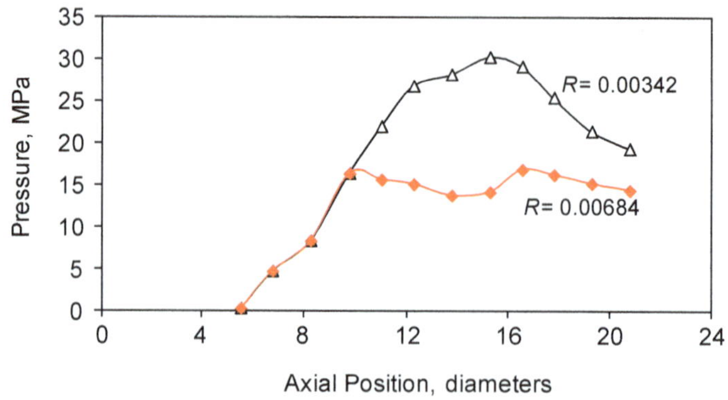

Figure 6.7 Axial pressure profiles measured for a 63.5 mm diameter extruder running an ABS resin at 60 rpm for screws with a 8.89 mm deep feed channel, 6 diameters of feed section, and a metering channel depth of 3.18 mm (C = 2.8) for the photographs of Figure 6.6

Like Figure 6.4, several key observations can be made from studying the photographs for the two compression rate screws in Figure 6.6. The main difference is the melting rate appears to be much higher for the high compression rate screw. As shown by Figure 6.6, melting is essentially completed by diameter 15 for the high compression rate screw while a high level of solids still exists at diameter 17 for the low compression rate screw. Although the high compression rate screw seems to perform at higher melting rates and without solid bed breakup, the overall rate of the process was reduced by about 7%. The axial pressure profile shows that the pressure was generated in the four diameters of transition section and then the pressure gradient in the metering section was nearly zero. In general, screws with longer transition sections and lower compression rates tend to have higher rates.

As previously discussed and also shown in Section 12.7.6, a compression ratio of about 2.4 to 2.5 is ideal for ABS pellet feedstocks. The screw with a compression ratio of 2.4 was operated as a function of screw speed. The pressure profiles for these experiments are shown in Figure 6.8. As shown by this figure, the pressure at the end of the melting section increased with increasing screw speed. This type of result is typical for most resins as solids are pushed further downstream with increasing screw speed. The pressure was the lowest at the start of the transition section at diameter 6 at the higher screw speeds. These low channel pressures were likely caused by a reduction in solids-conveying forces due to frictional heating by the high rotation rates. The higher level of frictional heating may have caused the surface temperature of the barrel wall in the feed section to be higher than optimal. These experiments were performed at the same barrel temperature settings. If the feed zone temperature was decreased while running at 120 rpm, the pressure at the entry to the transition section would likely have been higher than that shown in Figure 6.8.

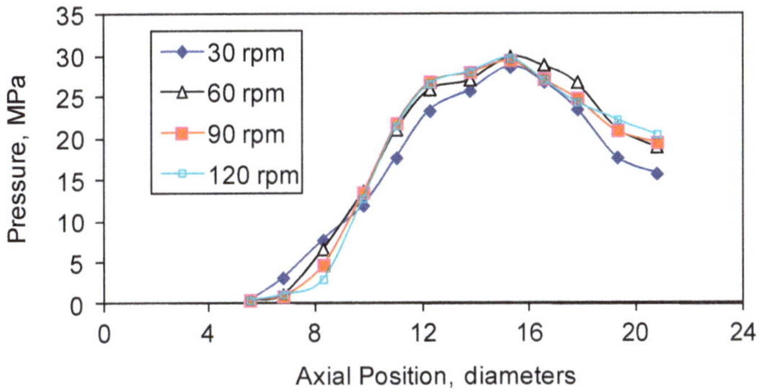

Figure 6.8 Axial pressure profiles for the screw with a compression ratio of 2.4 running ABS resin as a function of screw speed

6.2.2 Review of the Classical Literature

The extrusion process has been studied since the 1920s. Maddock [5] made the first physical description of the melting process in 1959. In his description, the heat in the melting process comes from two sources: heat conduction from the barrel, and viscous dissipation from the film at the barrel inner surface. In the melting process, a melt pool developed adjacent to the pushing flight to accumulate the melt. Maddock [5] and Tadmor's [7] solidification experiments provide very good support for this description. The Maddock melting mechanism was widely recognized and became the fundamental theory of polymer extrusion. Later researchers considered the films that existed over the screw core and trailing flight. The importance of the dissipation in these films, however, was not recognized.

Beginning from 1966 when Tadmor [9] set up the first mathematical model based on the Maddock melting mechanism and reorganizing solid bed, many other people continued working on improvements of the melting model in later years. Tadmor, Duvdevani, and Klein [7] improved the first Tadmor model by considering the melt as a non-Newtonian fluid in 1967. Chung [15] developed another model that gave analytical solutions in 1968 and improved it later by considering a melt existing at the screw surface in 1982 [16]. Donovan [17] improved the Tadmor model by considering down-channel convection in the solid bed in 1971. Mondvai, Halasz, and Molnar [18] improved the model by including films at the screw surfaces in 1973. Following that improvement, investigators began to consider the pressure term and the melt pool in their model. Edmonson and Fenner [19] included these changes and also considered the solid bed to be freely deformable in 1975. They also considered the acceleration of the solid bed. Shapiro, Halmos, and Pearson [20] set up a five-zone melting model based on a heat transfer mechanism using a radial mean temperature distribution in the melting process in 1976. A new analysis of the melting process was developed by Meijer [21]. He used this new approach for injection-molding screws [22] and for multichannel screws and barrier screws [23]. With the development of modern computation technology, investigators began to consider complex fluid mechanisms in both the melt pool and the films. Fukasi, Kunio, Shinya, and Nomura [24] considered the pressure term and rigid solid bed in their model in 1982. At about the same time, Lindt and his research group investigated and reported a body of work trying to set up a series of complete mathematical models for the melting process. Previously Lindt [25] developed a model by considering the solid bed in the center of the cross section in 1976. Lindt focused his later work on the five-zone melting model [26–29]. Lindt, Elbirli, Gottgatreu, and Baba [26] set up a model by considering all situations that previous people considered separately. Lindt and Elbirli [27] considered the cross-section circulation in the model in 1985. Also in 1985, Lindt claimed that "the development of the melting theory based on the Maddock Mechanism has been virtually completed" [28]. Altinkaynak et al. [30] performed full three-dimensional modeling of the melting process using a wrapped channel.

From the early analytical model to later complex numerical models, the understandings of the process are all based on the Maddock melting mechanism. Generally speaking, in all of these models, although researchers made more and more restrictive assumptions, there were two assumptions that were always made: (1) treating the screw channel as a rectangular channel (except Altinkaynak et al. [30]), and (2) assuming that the screw is fixed in space and the barrel is rotating in the opposite direction. The first assumption is acceptable depending on the extruder geometry. As for the second assumption, although in a single-screw extruder the screw is the rotating element, no one has developed a model based on screw rotation. Thus all of the literature models neglect the contribution from viscous dissipation at the screw core and flights. Only recently, since 1999, starting with Tang et al. [31–33], did a new model to describe the melting process begin to be developed based on screw rotation. This approach interprets the actual melting phenomena consistent with the y-direction melting dominating the process. This screw rotation model will be presented in Section 6.3.

6.2.3 Reevaluation of the Tadmor and Klein Melting Data

The classical melting data originally presented by Tadmor and Klein [8] for PE resin was examined using digital dimensional analysis [31–33], and typical results are presented in Figure 6.9 and Figure 6.10. In order to acquire these data, video capture was used on all of the melting experiments found in [8] as represented by figures like Figure 6.3. The dimensions of the screw in the feed section were then used to provide an internal standard to relate pixels to length and height. Bed width and height data from a video capture experiment are presented in Figure 6.9, which quantifies the Tadmor and Klein [8] Experiments 9, 13, and 15 for LDPE resin.

The pixel data were converted to length data in both of the cross-channel solids remaining, referred to as the x direction, and the bed thickness, referred to as the y direction. The bed width and height were then plotted as a function of the downstream distance in diameters. The results were unexpected given the literature description of the melting mechanism. The width of the bed (x direction) did not go to zero in a smooth manner. Moreover, the height of the bed (y direction) did go to zero in a smooth manner and reached zero well before the x direction was completely melted. Furthermore, at about 26 turns the x dimension went to zero in an apparent step function. These observations were repeated for the other samples captured from the literature [8] and several examples are presented in Figure 6.10. In all cases the thickness dimension went to zero smoothly and the x dimension appeared to have a step function at the end of melting. Near the end of melting, the literature and the Tadmor melting model have the last remaining portion of the solid bed positioned at the trailing flight and across the full depth of the channel, as shown by Figure 6.11(a). The observation and pixel data here have a thin plate of solids positioned as shown in Figure 6.11(b) near the end of melting. The melting results for the LDPE experiments [8] are shown in Figure 6.9. At a downstream position of 26 turns, the solids were completely melted, and thus the bed width and bed height were zero. The bed width for these experiments decreased slowly. The change in width, however, was considerably more complex for the ABS experiment shown in Figure 6.10.

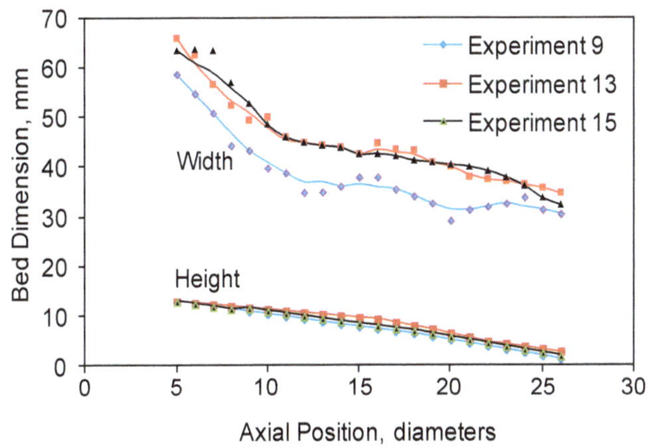

Figure 6.9 Reevaluation of the melting data from Tadmor and Klein for PE resins [8] as analyzed by Tang [31] and Campbell et al. [32, 33]

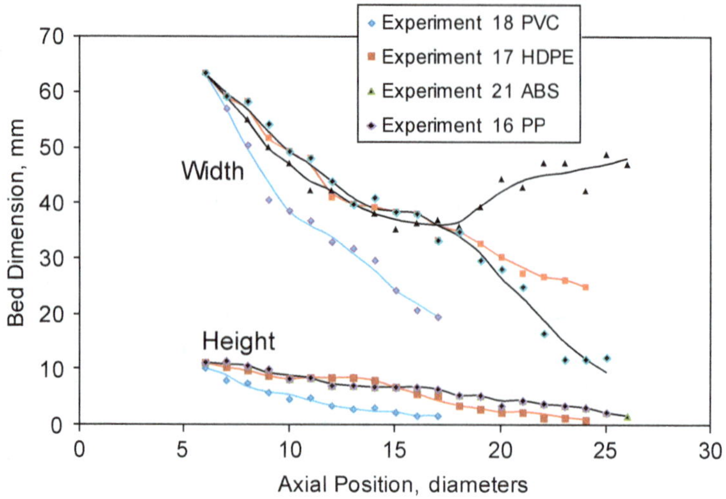

Figure 6.10 Reevaluation of the melting data for select resins from Tadmor and Klein [8] by Tang [31]

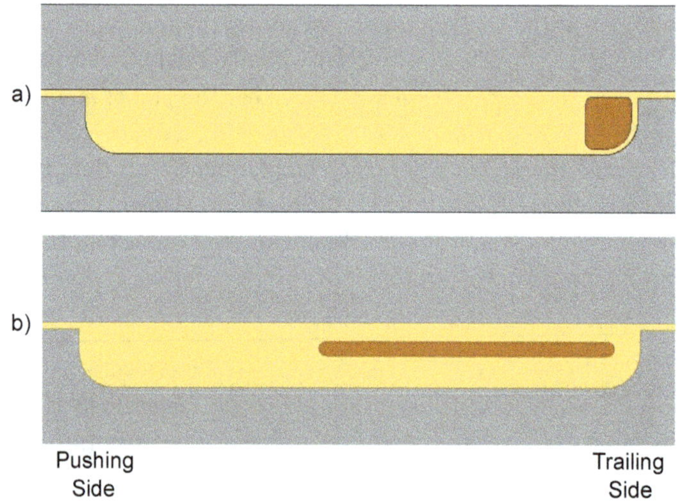

Figure 6.11 Schematics of the solid bed just prior to complete melting: (a) the solid bed is pushed to the trailing flight with the Tadmor melting model and barrel rotation physics, and (b) the solid bed is a thin plate and positioned as in the diagram (screw rotation and observation). The cream color represents molten resin

For essentially all of the data sets found in Tadmor and Klein [8] the melting results resemble the data shown in Figure 6.9 and Figure 6.10. The data analysis is of course difficult near the end of melting since the bed often breaks into segments. In essentially all cases, however, the fraction of the bed remaining in the cross-channel direction (width) is much greater than in the thickness direction. Also, in general the last third of the *x*-direction material appears to melt in the last turn of the channel, creating nearly a step function in the melting rate with respect to the axial distance, as shown in Figure 6.10. Step functions, however, do not in general occur in processes that are dominated by heat conduction or viscous dissipation. This suggests that the rapid loss of material in the x direction is an artifact of another melting mechanism. To reiterate, the digital analysis conclusively demonstrates that the y-direction thickness always goes to zero before the x-direction width [31–33]. It is interesting to note that Lindt experimentally observed this melting behavior for PP pellets and powder in 1976 [25] but continued to model the melting with the classical analysis. Lindt observed this melting process using two different screws and screw speeds ranging between 60 and 140 rpm. He believed that this nonstandard melting phenomenon was caused by a low cross-channel pressure gradient due to leakage flow across the flights. The new observations shown by Figure 6.9 and Figure 6.10 led to the development of a new physical model for single-screw melting that is based on these new observations [31, 34–36].

Several recent papers have been published that also support the observation that melting is dominated by the loss of solid in the channel-thickness direction. As was found by Lindt [25] and by Campbell et al. [32, 33], it is apparent from Figure 6.6 and Figure 6.7 that the same physics is occurring in these more recently reported experiments [23, 24, 37]. The solids melt in both the x and y directions with the y-direction loss of solids controlling the ultimate disappearance of the solid bed. In each of these cases there appears to be more than 40% of the solid left in the x direction two diameters before the entire solid has disappeared. Again, near step changes are not generally consistent with heat transfer dominated processes. The conclusion is thus that the y (thickness) direction dominated the melting process in these experiments.

6.3 Theory Development for Melting Using Screw Rotation Physics

Beginning from 1966 when Tadmor [7] set up the first mathematical model based on the Maddock melting mechanism, he proposed a reorganizing solid bed that led to a focus on the x dimension as rate controlling in the melting process. That is, the rate of melting was defined as the ratio of the solid bed width to the width of the channel. The model is based on barrel rotation and an unwrapped channel. Many other investigators continued to work on improving the melting model in later years. Numerous papers have since been written that refine the Tadmor model as developed in the previous section; all are based on the hypothesis that the solid bed reorganizes as it melts, leading to the bed being consumed in the cross-channel direction until it is completely melted. These Tadmor-based models force all of the energy dissipation to occur in the melt film between the solid bed and the barrel surface.

In this section, three models will be presented that don't force the reorganization of the solid bed and use screw rotation physics. These screw rotation models cause a significant portion of the energy dissipation to occur in the melt film between the solid bed and screw root. These models are for a conventional transition section, for a barrier melting section, and for a special case referred to as one-dimensional melting.

6.3.1 Melting Model for a Conventional Transition Section Using Screw Rotation Physics

This section will describe a melting model for a conventional transition section using screw rotation. The channel here has a tapered depth and is single flighted. Recently Campbell, Spalding, and Tang [34] have reexamined the assumptions in the literature models in order to address the reanalysis of the classical melting data discussed previously. This single-screw melting concept was developed based on recognizing that the rate-limiting melting dynamics for solid bed consumption was in the channel height and not the channel width direction. A diagram was developed for the melting process that focused on the four melt films surrounding the solid bed to set the foundation for the development of the physics of melting. These zones were originally defined by Shapiro et al. [20], and they are shown in Figure 6.12: Zone A is the solid bed, Zone B is the film (or melt pool) on the pushing side of the screw flight, Zone C is the film between the solid bed and the barrel wall, Zone D is the film between the screw root (or core) and solid polymer interface, and Zone E is the film at the trailing flight of the screw. The energy dissipation in these melt film zones will be analyzed separately.

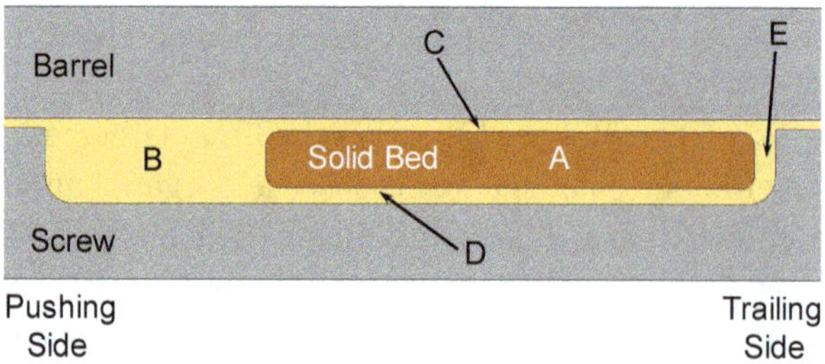

Figure 6.12 Schematic for the zones of the new melting concept: Zone A is the solid bed, Zone B is the melt pool, Zone C is the melt film located between the solid bed and the barrel wall, Zone D is the melt film between the solid bed and the screw root, and Zone E is the melt film between the solid bed and the trailing flight. The cream color represents molten resin

Previous analysis of dissipation in extrusion by Campbell and Dontula [38] showed that barrel rotation models over-predict the melt temperature. That is, the energy dissipation for barrel rotation is always higher than that for screw rotation. Barrel and screw rotation physics is clearly presented in Chapter 7 for the metering section. The melting model development here uses screw rotation velocities because the process is dissipation dominated [39–44] and dissipation is not frame indifferent. The new model here is based on the Tadmor and Klein [8] model but with boundary conditions based on a stationary barrel and a rotating screw. The velocity diagram of this adaptation is shown in Figure 6.13. The analysis begins by unwrapping the screw into the x, y, and z coordinates. The barrel is hot as depicted by the red color and its velocity is zero; that is, V_{bx} and V_{bz} are 0, as shown by Figure 6.13(a). Here the observer is in the laboratory (Eulerian) reference frame. The velocity gradient for Film C in the z direction goes from V_{sz}, the solid bed velocity in the downstream direction (z direction), to zero at the barrel interface. The x-direction velocity goes from V_{sx}, the velocity of the solid bed in the x direction, at the solid bed to zero at the barrel wall. A similar diagram was developed for the other three fluid interfaces around the solid bed, remembering that for these liquid interfaces the metal velocity is the screw velocity. Figure 6.13(b) shows the velocity diagram for the Zone D film. Here, the screw surface is depicted in a pink color to indicate that it is hot but not as hot as the barrel. The solid bed at the melt film interface is moving at a velocity V_{sz} in the downstream direction (into the page) while the velocity of the screw core is moving at V_{cz} in the opposite direction (out of the page), as shown in Figure 6.13(b). The solid bed is moving in the x direction at a velocity of V_{sx} while the velocity of the screw in the x direction is V_{cx}. The magnitude of velocity V_{sx} is only slightly higher than that for V_{cx} due to the slight increase in the radial distance, and for simplicity the values are assumed to be equal.

In the classical literature analysis the system equations were manipulated to eliminate V_{sy}, the velocity of the solid bed consumption in the thickness direction (y direction), from the analysis by using the assumption that the solid bed reorganizes. This allowed a straightforward differential analysis and a closed form solution in the cross-channel x direction for solids melting. In this analysis, the y-direction velocity V_{sy} is retained as a variable because this facilitates the calculation of the change in bed thickness, which was found to be very important in the reevaluation of the literature data, as shown in Figure 6.9 and Figure 6.10.

The nomenclature used here follows that used by Tadmor and Klein [8] as much as possible. Where deviations occur, a full explanation will be presented. A nomenclature list is presented at the end of the chapter.

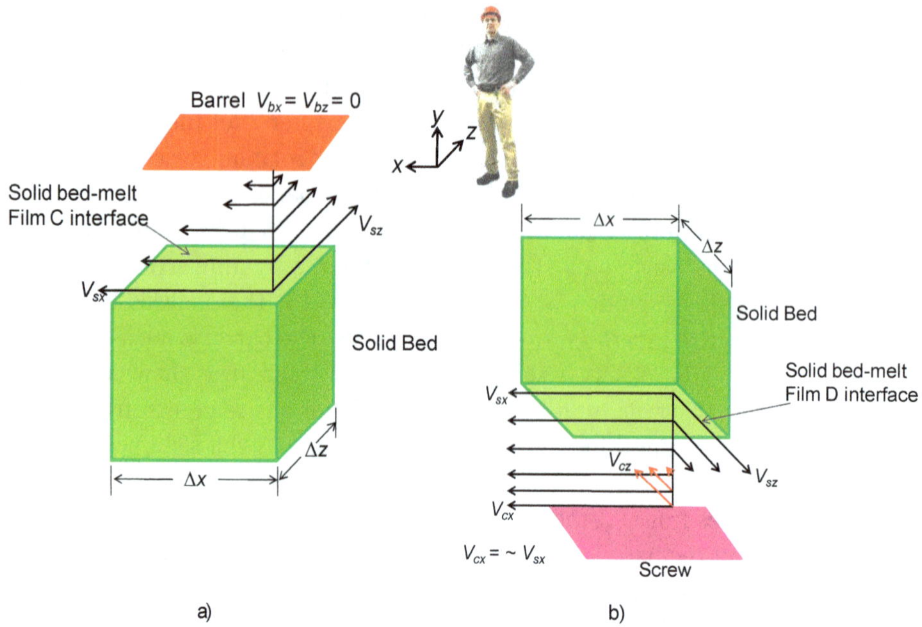

Figure 6.13 Melting differential elements for a rotating screw: a) melt film in Zone C (barrel velocity is zero), and b) melt film in Zone D with the screw moving with a velocity of V_{cz} and V_{cx}. The red colored vectors are moving out of the plane of the page towards the upstream direction while the black vectors in the *z* direction are moving into the page and in the downstream direction

The velocity of the solid bed in the *z* direction, V_{sz}, was defined by Tadmor and Klein. This velocity is set using the mass flow rate and the cross-sectional area at the entry to the transition section.

$$V_{sz} = \frac{Q_m}{\rho_s W H_1} \tag{6.5}$$

where Q_m is the mass rate of the extruder, ρ_s is the density of the solid bed, W is the average width of the channel, and H_1 is the depth of the channel at the entry to the transition section. The density of the solid bed typically increases due to the compaction of the solids as shown previously in Figure 4.1, but here the bed is assumed to be fully compacted at the entry. For conventional screws, the channel depth at the entry to the transition section is equivalent to the depth of the feed channel; that is, $H_1 = H_f$. The average width of the channel is typically used rather than the width at the barrel wall.

The velocity of the solid bed in the *x* direction, V_{sx}, due to screw rotation is as follows:

$$V_{sx} = \pi N D_f \sin\theta_f \tag{6.6}$$

$$\tan\theta_f = \frac{\pi L}{D_f} \tag{6.7}$$

where N is the screw rotation speed in revolutions per second, D_f is the diameter at the surface of the solid bed at the interface with melt Film C, θ_f is the helix angle at diameter D_f, and L is the lead length of the flight. Since this film can be very thin at the start of melting, $D_f \approx D_b$.

The thickness of the melt film δ_C for Zone C was calculated using similar assumptions used by Tadmor and Klein [8], and thus we solve their Equation 5.23 for $V_{sy,C}$, the velocity of the loss of the solid bed due to melting at δ_C:

$$V_{sy,C} = \frac{\left[\frac{k_m}{\delta_C}\left(T_b - T_m\right) + \frac{\mu V_{j,C}^2}{2\delta_C}\right]}{\rho_s \lambda + \rho_s C_v \left(T_m - T_s\right)} \tag{6.8}$$

where k_m is the thermal conductivity of the melt, T_b is the barrel temperature, T_m is the melting temperature of the resin, μ is the shear viscosity, $V_{j,C}$ is the vectorial velocity in the C film (discussed later), ρ_s is the density of the compacted solid bed, λ is the heat of fusion, C_v is the specific heat of the solid resin, and T_s is the temperature of the solid bed. The term $V_{sy,C}$ here is equivalent to Tadmor and Klein's V_{sy}. Equation 6.8 is simply an energy balance and material balance for the melting process. Melting velocity equations at the other three interfaces $V_{sy,D}$, $V_{sx,B}$, and $V_{sx,E}$ are essentially identical to that for Equation 6.8 except that the vectorial velocity used corresponds to the specific film. The derivations of these equations are presented in Appendix A6.

$$V_{sy,D} = \frac{\left[\frac{k_m}{\delta_D}\left(T_{screw} - T_m\right) + \frac{\mu V_{j,D}^2}{2\delta_D}\right]}{\rho_s \lambda + \rho_s C_v \left(T_m - T_s\right)} \tag{6.9}$$

$$V_{sx,B} = \frac{\left[\frac{k_m}{\delta_B}\left(T_b - T_m\right) + \frac{4\mu V_{j,B}^2}{2\delta_B}\right]}{\rho_s \lambda + \rho_s C_v \left(T_m - T_s\right)} \tag{6.10}$$

$$V_{sx,E} = \frac{\left[\frac{k_m}{\delta_E}\left(T_b - T_m\right) + \frac{4\mu V_{j,E}^2}{2\delta_E}\right]}{\rho_s \lambda + \rho_s C_v \left(T_m - T_s\right)} \tag{6.11}$$

where δ_D, δ_B, and δ_E are the film thickness for Zones D, B, and E, respectively; and $V_{j,D}$, $V_{j,B}$, and $V_{j,E}$ are the vectorial velocities for the films.

The melting velocities of the solid bed interfaces are the foundation of this new approach. The mathematical equality to calculate the rate of loss of the solid at the interface next to the barrel is found in Equation 6.12.

$$-\left.\frac{dY}{dt}\right|_{\delta c} = V_{sy,C} \tag{6.12}$$

where Y is the local thickness of the solid bed and t is the independent time variable. Similar balances are performed at the other melting interfaces.

The melt film thickness δ_c for Zone C is determined using Equation 6.13. Note that V_{sx} is essentially numerically equal to the traditional V_{bx} as specified by Tadmor and Klein because of the small value of the film thickness.

$$\delta_C = \left\{ \frac{\left[\left[2\, k_m \left(T_b - T_m\right) + \mu\ V_{j,C}^2\right] \dfrac{X}{\alpha}\right]}{V_{sx} \rho_m \left[C_v \left(T_m - T_s\right) + \lambda\right]} \right\}^{1/2} \tag{6.13}$$

where X is the local width of the solid bed and α is a fitting factor. Equation 6.13 is a mass balance for the Zone C film and is essentially identical to Tadmor and Klein's Equation 531 [8]. The flow equation is a first-order approximation since it neglects flows due to pressure gradients. Ideally, the value of α is set to 2 to provide an average thickness for the film in the cross-channel direction. The value of α was set to 0.5 for the simulation results reported here to force the melting lengths to be equivalent to those in the literature example [7].

The film thickness δ_E for Zone E was determined using a similar mass balance equation as follows:

$$\delta_E = \left\{ \frac{\left[3k_m \left(T_b - T_m\right) + 6\mu V_{j,E}^2\right] Z}{\rho_m V_{Ez} \left[C_v \left(T_m - T_s\right) + \lambda\right]} \right\}^{1/2} \tag{6.14}$$

where Z is the solid bed length in the down-channel direction from the point of the initiation of melting. The first-order evaluation of the average out-flow velocity in Film E in the z direction V_{Ez} is derived in Appendix A6 and is estimated as follows:

$$V_{Ez} = \frac{2}{9} V_{j,E} \tag{6.15}$$

This functional form for the thickness next to the trailing flight can only be considered a first approximation. From an engineering point of view it does appear to provide results that are consistent with literature observations; see Figure 6.5,

Figure 6.6, and Figure 6.7. The melt film thicknesses, δ_B and δ_D, for Zones B and D were found by difference:

$$\delta_B = W - X - \delta_E \tag{6.16}$$

$$\delta_D = H(z) - Y - \delta_C \tag{6.17}$$

where $H(z)$ is the depth of the channel at position z. The value of H is decreasing in the downstream direction due to the tapering channel depth in the transition section. In special transition section designs, the value of W can also decrease with the downstream position.

The x- and z-component velocities of the screw are calculated as follows, and they are used as the boundary conditions at the screw surface:

$$V_{cx} = \pi N D_c \sin\theta_c \tag{6.18}$$

$$V_{cz} = -\pi N D_c \cos\theta_c \tag{6.19}$$

$$\tan\theta_c = \frac{\pi L}{D_c} \tag{6.20}$$

here V_{cx} is the local velocity of the screw core in the x direction, V_{cz} is the velocity of the screw core in the z direction, D_c is the diameter at the screw root (core), and θ_c is the helix angle at the screw root. The direction of V_{cz} is in the opposite direction of V_{sz}, as shown in Figure 6.13, and thus it is negative. Since the transition section is tapered, V_{cx}, V_{cz}, D_c, and θ_c all depend on the downstream position and the local depth of the channel.

The scalar values for the vectorial velocities for the four polymer films that account for the heat transfer and dissipative melting may be calculated as follows:

$$V_{j,C} = \sqrt{V_{sx}^2 + V_{sz}^2} \tag{6.21}$$

$$V_{j,D} = |V_{cz}| + V_{sz} \tag{6.22}$$

$$V_{j,E} = |\overline{V}_{fz}| + V_{sz} \tag{6.23}$$

$$V_{j,B} = |\overline{V}_{fz}| + V_{sz} \tag{6.24}$$

$$\overline{V}_{fz} = \left(\frac{V_{cz} + V_{bz}}{2}\right) \tag{6.25}$$

$$V_{bz} = -\pi N D_b \cos\theta_b \tag{6.26}$$

here $\overline{V}_{fz}$ is the average velocity of the flight in the z direction. As discussed in Chapter 7, the velocity of the flight in the z direction is towards the hopper and thus in the negative z direction. The magnitude is the lowest at the screw root and the highest at the flight tip. The average velocity of the flight $\overline{V}_{fz}$ in the z direction is the same for both the leading and trailing flights, and it is negative.

The new concepts presented here remove the literature assumption that the solid bed reorganizes, and it allows melting at all solid bed interfaces. The Tadmor model allows melting at only a single interface, that is, as specified by the melting velocity $V_{sy,C}$ in Film C. The model presented here predicts melting at all four interfaces: two in the y direction and two in the x direction.

The thickness and width of the solid bed as a function of the helical downstream position z are calculated from the melting velocities at the interfaces over a small Δz increment. The calculation is progressed down the transition section until the Y_{i+1} value approaches zero. The balances for the solid in the x and y directions for an increment in the z direction are as follows:

$$Y_{i+1} = Y_i - \left[V_{sy,C} + V_{sy,D}\right]\frac{\Delta z}{V_{sz}} \tag{6.27}$$

$$X_{i+1} = X_i - \left[V_{sx,B} + V_{sx,E}\right]\frac{\Delta z}{V_{sz}} \tag{6.28}$$

$$Z_{i+1} = Z_i + \Delta z \tag{6.29}$$

where Δz is the increment in the z direction for the simulation, and Z_i is the helical position at increment i. A schematic representation of the solid bed in the channel is shown in Figure 6.14. This representation depicts the four films and how the solid bed is consumed in the melting section of the screw.

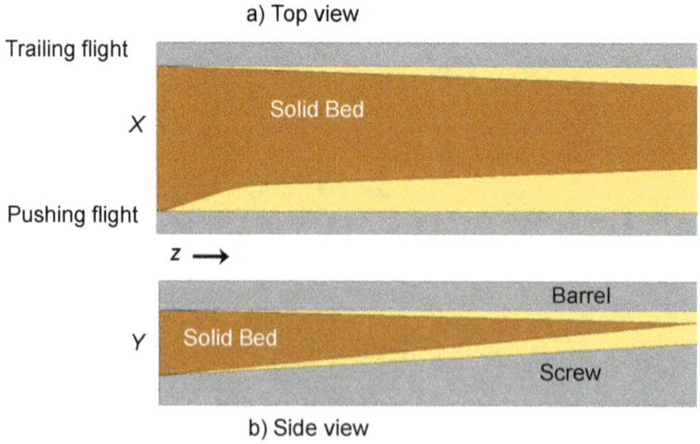

Figure 6.14 Qualitative shape of X and Y bed dimensions and melt film thickness for melting in a conventional transition section: a) top view, and b) side view. The cream color represents molten resin

6.3.1.1 Sample Simulations for Melting in a Conventional Screw Channel

The results of the proposed theory will now be compared with the observations to determine if they are consistent with the reanalysis of the literature data. For this comparison, the screw geometry and processing conditions from Example 5.1 of Tadmor and Klein [8] will be used. The screw used was 63.5 mm in diameter, square pitched ($L = D$), and had a feed section depth of 12.7 mm and a length of 3.2 diameters. The transition section was 12 diameters long, and the metering channel depth was 3.175 mm and 12 diameters long. The extruder was running a 2 MI LDPE (solid density of 0.92 g/cm^3) resin at 82 rpm for a rate of 54.4 kg/h. Other processing details can be obtained from Tadmor and Klein [8]. Simulation results are presented in Figure 6.15. As shown by this figure, the Y thickness goes to zero at a down-channel position of 250 cm (or about 12 diameters) into the melting section. It is seen that the change in the thickness (x direction) leaves about 60% of the original bed width remaining just at the end of melting. The end of melting is defined as when the y-direction thickness of the solid bed goes to zero. This simulation was performed using a PE resin with a shear viscosity of 880 Pa·s. As previously discussed, the fitting factor α was set at 0.5 in order to make the melting length agree with Tadmor and Klein's Example 5.1. Qualitatively, the new melting model with screw rotation physics and eliminating the reorganization of the solid bed matches their data very well.

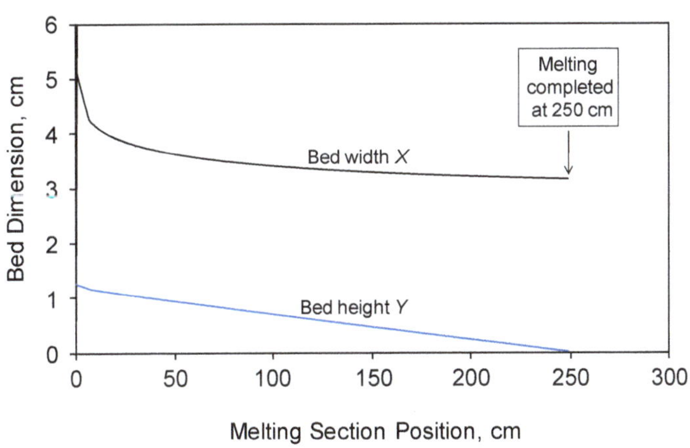

Figure 6.15 Simulation results using Example 5.1 [8] data. Melting was completed at 250 cm downstream from the start of melting for a resin with a shear viscosity of 880 Pa·s

It is now useful to examine the melting energy at the four solid interfaces for this new model and for the historical Tadmor model [8]. The dissipation data from the simulations are summarized in Table 6.1 for a PE resin with a viscosity of 880 Pa·s. Examination of the table points out that the vectorial velocities (V_j) for Zones C and D are very different for the assumption of barrel and screw rotation, as presented previously and as shown in Table 6.1. For the historic model, all energy is dissipated in the Zone C melt film, and the cumulative energy for melting was calculated at 1980 J/s. The dissipation energy is assumed to be zero for the other films; that is, the vectorial velocities ($V_{j,B}$, $V_{j,D}$, and $V_{j,E}$) for Zones B, D, and E were ignored and thus assumed to be zero. Here the values for V_{bx}, V_{bz}, and V_{sz} are 8.4, -25.9, and 2.4 cm/s, respectively. The vectorial velocity for the historical analysis for Zone D was calculated at 2.4 cm/s, but this model did not dissipate energy in the film. For the new model here with screw rotation boundary conditions, the vectorial velocity in the Zone C film is about 30% of that for the historical analysis (barrel rotation boundary conditions). For the new model the values for V_{sx}, V_{sz}, and V_{cz} are 8.3, 2.4, and -22.7 cm/s, respectively. This reduced vectorial velocity leads to the energy dissipation in Zone C to be about 60% of that using the historical analysis. The remaining 40% of the energy dissipation occurred in the remaining films, as shown in Table 6.1. Almost 25% of the total energy dissipation occurred in Film D.

Table 6.1 Calculated Vectorial Velocities for Zones C and D and the Melting Energies Developed in the Four Melt Films. The Simulations Were for a PE Resin with a Shear Viscosity of 880 Pa·s

Model Parameter	New Analysis (Screw Rotation)	Historical Analysis [8] (Barrel Rotation)
Vectorial velocity ($V_{j,c}$) in Zone C, cm/s	8.6	25.0
Vectorial velocity ($V_{j,D}$) in Zone D, cm/s	25.1	2.4
Melting energy developed in Zone C, J/s	1140	1980
Melting energy developed in Zone E, J/s	480	0
Melting energy developed in Zone D, J/s	340	0
Melting energy developed in Zone B, J/s	20	0
Total melting energy, J/s	1980	1980

Resins with a lower viscosity would intuitively dissipate less energy in the melt films and thus melt at a slower rate. Previous research, however, has shown that the melting rates of similar resins increase with decreasing viscosity [45–47]. To test the new melting model, the viscosity was decreased to 220 Pa·s. The results are presented in Figure 6.16. The position where the solid bed was completely consumed was 225 cm downstream (or about 11 diameters) from the start of the melting zone compared to 250 cm in the previous simulation for a viscosity of 880 Pa·s, as shown in Figure 6.15. This reduction in the melting length with decreasing viscosity is consistent with data published earlier [45–47].

These initial results demonstrate that this model predicts the overall dynamics found for melting of solid polymer beds, as shown in Figure 6.4 and Figure 6.9. This melting model predicts the melting rate as the summation of the loss of solid bed in both the thickness direction (y direction) and the cross-channel direction (x direction). In this model the rate-limiting process is the melting rate in the y

direction. Also, the simulated model results predict that the melting length is decreased if the viscosity is decreased for the screw analyzed. This finding from the model is consistent with previous extrusions and resin melting flux data measured using a Screw Simulator [45–47]. The Screw Simulator was discussed in Sections 4.3.1 and 4.4.

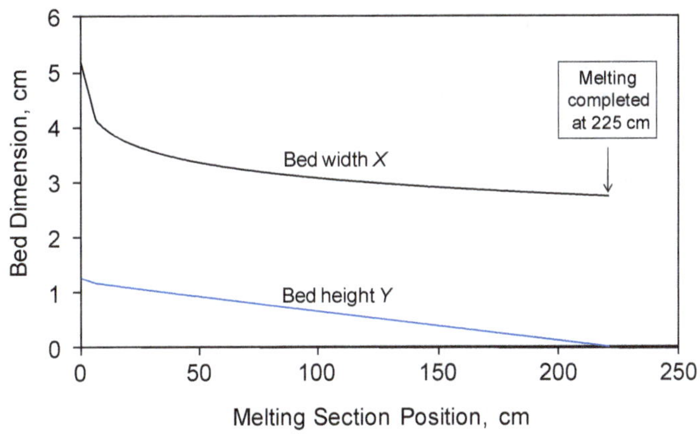

Figure 6.16 Simulation results for a lower viscosity resin. Melting was completed at 225 cm downstream from the start of melting. The shear viscosity used was 220 Pa·s

It is observed that the melt film thickness at the barrel interface is predicted by the theory to decrease when the only simulation parameter changed is the reduction of the polymer viscosity, as shown in Figure 6.17. The reduction in film thickness is about 20% at the beginning of the melting section, and it increases slightly as melting progresses downstream. This reduction in film thickness will increase conductive heat transfer and will increase the velocity gradient. Both of these processes will increase the melting rate.

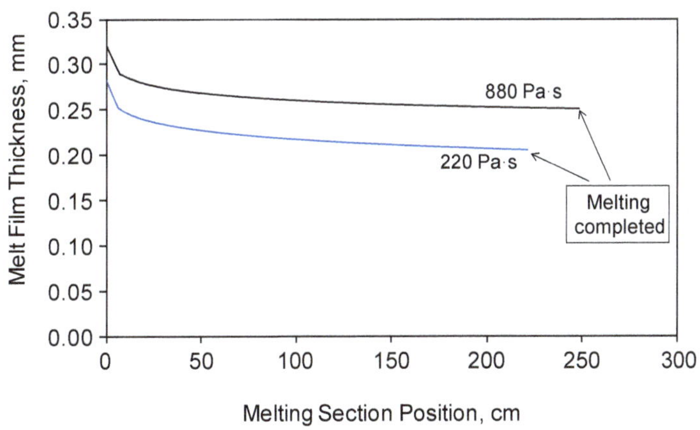

Figure 6.17 Melt film thickness at Zone C (δ_c) for the simulations using resins with shear viscosities of 220 and 880 Pa·s

The model is also consistent with observations for flow surging due to solids blockage in a channel [12]. That is, as a blockage occurs the local pressure upstream of the blockage increases, causing the local melting rate to increase. As the melting rate increases, enough of the blockage will melt such that solid polymer fragments will flow downstream. With the blockage removed, the pressure is relatively low in the region and melting rates are low, causing more solids to accumulate and increasing the size of the next blockage. Now the local pressure starts to increase and the cycle is repeated. Although pressure terms were not included in the new model, the additional pressure would likely cause the melt film thicknesses to decrease and thus cause the melting rate to increase.

The results presented here are encouraging but only qualitative and have been produced using this first-order model. Current limitations of the model are the use of a constant-viscosity function independent of temperature and shear rate. Also, the dynamic local temperature of the barrel and screw (Section 10.10) must be incorporated into the model; they are currently set as constants. An enhanced model for the film thickness at both the barrel and screw surfaces should be added to the current model along with flows induced by pressure gradients.

As previously discussed, the melting velocities and the film thicknesses are the foundation for this new melting model. The film thicknesses for Zones C, D, and E for the simulation using a shear viscosity of 220 Pa·s are shown in Figure 6.18; the Zone B film thickness was omitted due to its larger magnitude. At the barrel wall, the Zone C melt film thickness increases quickly to 0.5 mm and then decreases slowly to about 0.34 mm at the end of melting. This melt film thickness is in the range observed experimentally. The film in Zone E between the solid bed and the trailing flight increases from 0 at the start of melting to about 0.41 mm at the end of melting. Zone D for the film between the screw root and the solid bed increased from 0 to 4.1 mm at the end of melting.

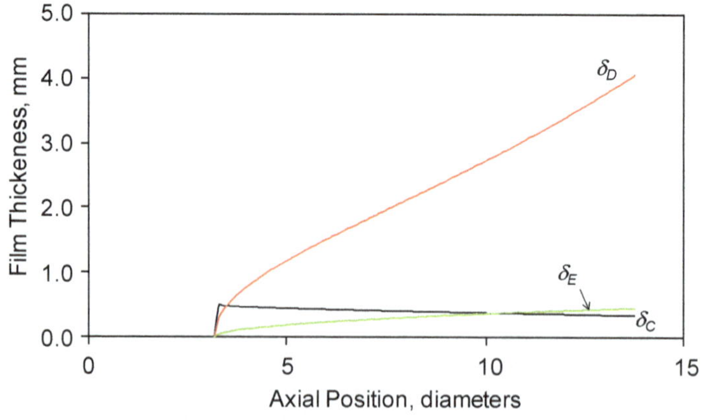

Figure 6.18 Melt thicknesses for Zones C, D, and E around a solid bed for the melting simulation using a shear viscosity of 220 Pa·s. Melting started at the entry to the transition section at 6 diameters from the start of the screws and was completed at 13.7 diameters

The thickness indicated by the red line in Figure 6.18 is the gap between the solid bed and the screw root. The screw root is moving in the minus z direction while the solid bed is moving in the positive z direction. Melted polymer will thus be dragged into the gap, and there will be a negative pressure gradient $\partial P/\partial z$ in the film. This topic will be presented in Section 6.3.1.3.

6.3.1.2 Pressure Generation in the Melt Pool

As discussed in Section 6.2.1, the meting process has the potential to generate pressure in the downstream direction. The local pressure affects the melting rate and the compaction of the solid bed and provides a pressure at the entry to the metering section. Here, the dominating pressure is assumed to be the pressure in the melt pool. Cross-channel pressures have been measured as shown in Figure 12.3, but it is not as important as the pressure in the z direction. This section will develop the equations for the pressure profile in the z direction based solely on the flow in the melt pool.

First the melted material is calculated at each increment up the taper where H_1 is the channel depth at the start of the transition section ($H_1 = H_f$). The volumetric rate of solid material remaining at increment k is Y_k in the y direction and X_k in the x direction at a position Z_k. The volumetric rate of melted material $Q_{melt,k}$ at increment k is determined by difference as follows:

$$Q_{melt,k} = p\left[WH_1 - Y_k X_k\right]V_{sz} \tag{6.30}$$

where p is the number of flight starts in the melting section. For a conventional melting section, p is set to 1. The pressure change over an incremental length Δz is based on Tadmor and Klein's Equation 6-97. This equation is based on rotational flow and pressure-induced flow for the local width and depth of the melt pool. Solving this equation for the pressure change in the increment, ΔP_k is as follows:

$$\Delta P_k = -\left[\frac{Q_{melt,k} - \frac{1}{2} p V_{bz}\left(W - X_k\right)\dfrac{H_k H_{k+1}}{\frac{1}{2}\left(H_k + H_{k+1}\right)}}{\dfrac{p\left(W - X_k\right)\left(H_k H_{k+1}\right)^2}{12\mu \frac{1}{2}\left(H_k + H_{k+1}\right)}}\right]\Delta z \tag{6.31}$$

The pressure is thus calculated as a function of position Z_k as follows:

$$P_{k+1} = P_k + \Delta P_k \tag{6.32}$$

A simulation was carried out with the melting model using the simulation conditions described earlier for the PE resin with a Newtonian shear viscosity of 220 Pa·s. For this simulation, the pressure at the end of the solids-conveying section (3.2 diameters from the start) was assumed to be 3 MPa. The model predicts that a pressure of 15 MPa will exist in the melt pool at the end of melting at about 13.7 diameters from the start of the screw. The axial pressure profile is shown by Figure 6.19. For this plot, the helical length Z was converted to diameters using Equation 1.11.

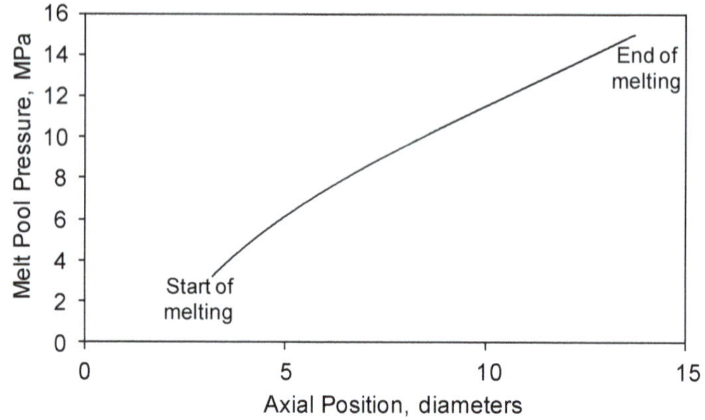

Figure 6.19 Simulated axial pressure profile for a 63.5 mm diameter screw. The pressure at the entry to the transition section was assumed to be 3 MPa. Melting was completed by diameter 13.7

6.3.1.3 Pressure in Film D and Flow from Film D to Film B

The flow in melt Film D between the solid bed and the screw root is complex due to a negative pressure gradient ($\partial P/\partial z < 0$) that must exist in the film. A schematic of the V_z velocity for the film is shown by the blue curve in Figure 6.20. As discussed earlier in the chapter, the solid bed is moving downstream in the positive z direction while the core of the screw and the flights are moving in the negative z direction, dragging fluid towards the hopper of the extruder. In general, the magnitude of the core velocity is larger than the magnitude of the solid bed velocity. If the pressure gradient in the z direction ($\partial P/\partial z = 0$) is zero, then an unsustainable situation occurs. For this case, the motion of the screw core drags material in the negative z direction, adding material to the film in the upstream positions and thus making the film thickness larger upstream. This situation, however, does not occur as shown by the Maddock solidification experiments shown in Figure 6.3, Figure 6.4, and Figure 6.6. Moreover, a zero pressure gradient would cause the material that enters the film to stay in the film. A negative pressure gradient, however, causes some additional material to flow in the positive z direction. Near the completion of the melting process, the pressure in Film D must be essentially equal to the pressure in the melt pool. Upstream of this point, the pressure in Film D must be higher than that near the end of melting due to the negative $\partial P/\partial z$. This higher pressure locally will have at least two ramifications, including the location of the bed in the channel and flow of material from Film D into the melt pool.

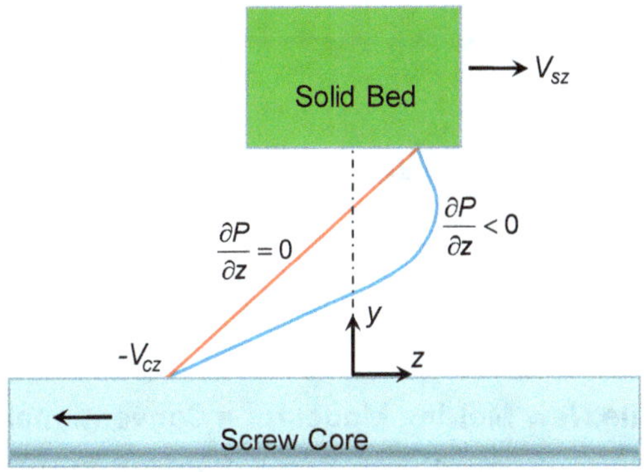

Figure 6.20 Schematic of the V_z velocity in the Film D between the screw core and the solid bed. The red and blue velocity lines are for pressure gradients that are zero and negative, respectively. The dotted vertical line is for $V_z = 0$

The locally higher pressure underneath the solid bed in Film D is postulated to exist and forces the solid bed up against the barrel surface. The pressure in Film C has been experimentally measured by many researchers, and a typical pressure profile showing that $\partial P/\partial x$ is negative is discussed in Section 12.4; here the x origin is positioned at the pushing flight as is done in all chapters. Thus, the local pressure in Film C is always less than the pressure in the melt pool. With the local pressure in Film C being less than that in the melt pool and the pressure in Film D being higher than the melt pool, it follows that difference in the pressures creates a force that pushes the solid bed against the barrel wall.

The locally high pressure underneath the solid bed and the positive $\partial P/\partial x$ in Film D causes some flow of resin from Film D to the melt pool. Thus, for a local Δz increment for Film D, there is material entering the element from the melting process and from the drag motion of the screw core, and there is material leaving the increment from the motion of the screw core and from the flow of material into the melt pool due to a positive $\partial P/\partial x$. These complex flows are consistent with observations from Maddock solidification experiments. This flow is shown in detail for the Maddock experiment shown in Figure 6.35.

At the start of the melting process, the pressure in the channel is relatively low and the solid bed may not be fully compacted. In this case, molten resin from all films has the ability to flow into the voids between the individual pellets. This process is often referred to as melt infiltration. A photograph of a cross section of a Maddock solidification experiment at the start of the melting process is shown by Figure 6.21. For this figure, the molten material prior to the solidification was black. Melt infiltration is shown by the black resin that has flowed from the films and in between the pellets. The flow of resin into the solid bed will likely cause the pressure in the films to decrease.

Figure 6.21 Photograph of a cross section from a Maddock solidification experiment at the start of the melting process. The black material shows the melt films and the regions where the melt infiltrated the loosely packed solid bed

6.3.1.4 Notes for the New Melting Model for a Conventional Channel

The new melting model presented in this section qualitatively fits the experimental data observed by many previous researchers. Like the Tadmor and Klein model [8], this model is based on simplistic assumptions and linear mathematics for the melt films. The new model, however, does not require the reorganization of the solid bed like the Tadmor and Klein model. Furthermore, the new model allows viscous dissipation and melting in all four melt films, and does not restrict all melting to the Zone C film. Melting in the Zone D melt film becomes highly important when the boundary conditions are switched from barrel rotation to the actual conditions of screw rotation.

The model developed here uses a fitting parameter to obtain melting lengths that are consistent with those observed experimentally. This fitted model is utilized to adjust for some of the non-linearities in the model. The model is not meant as a screw design tool. An improved model could be written based on the original Lindt and Elbirli model [27]. The improved model would set the boundary conditions for screw rotation rather than barrel rotation as used by Lindt and Elbirli.

6.3.2 Melting Models for Barrier Screw Sections

The focus of this section is to develop a model for the melting dynamics of the solid bed for barrier melting sections in single-screw extruders. This analysis is an extension of the screw rotation melting model presented earlier in Section 6.3.1 for conventional screws. Barrier screws have been used commercially since they were first developed in the 1960s [48, 49]. Since then, the design of the section has varied in many ways, all with the goal of increasing the melting capacity of the screw and decreasing the extrudate temperature. Details for barrier screw designs are presented in Chapter 14.

Barrier melting sections are constructed by positioning a second flight (or barrier flight) in the transition section such that the solids are maintained on the trailing side and the molten resin on the pushing side. A schematic of a cross section of a barrier melting section is shown in Figure 6.22. The resin is melted as discussed in Section 6.3.1 in the solids channel of the device. The resin that is melted near the barrel wall is conveyed across the barrier flight and collected in the melt-conveying channel. The key design parameters include the position of the barrier flight, the depths of the channels, and the undercut clearance on the barrier flight. The undercut clearance is measured by positioning a segment of straight bar stock across the two main flights and then measuring the gap between the bar and the barrier flight land. For most designs, the barrier flight undercut is constant for the entire length of the section. As a very general rule, the undercut is typically about 0.01 times the diameter of screw. Undercuts that are smaller than this rule, however, are often used. The position of the barrier flight sets the width of both channels. Many styles of barrier melting screws are commercially available, and many different variations of the channel widths and depths are practiced.

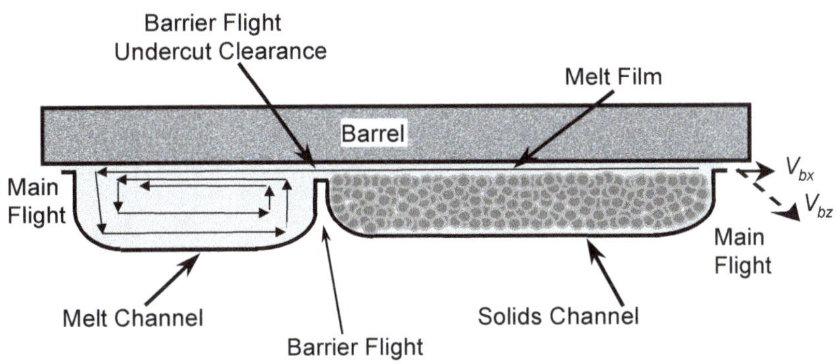

Figure 6.22 Cross-sectional view of a barrier melting section

For most barrier sections, the depth of the solids channel decreases while the depth of the melt channel increases with the downstream position. Many variations, however, are practiced. The widths of the channels are very important, and several styles are shown in Figure 6.23. Thus, the average width of the solids channel, $W_s(z)$, can be a function of z. The depth of the solids channel, $H_s(z)$, is also decreasing with increasing z. At the exit of the solids channel, the depth must increase to match the depth of the downstream metering section. The depth of the melt-conveying channel, $H_m(z)$, can increase with the increasing z. Designs exist, however, where the depth of the melt-conveying channel is essentially constant for the entire length; see Figure 6.23(c). The entry to the melt channel is typically small in order to limit the amount of solids that potentially could flow into it. Like the solids channel, the depth at the exit of the melt channel must match the depth of the metering channel.

As shown by Figure 6.23, the channel geometry is relatively complicated for barrier flight melting sections. The geometric constraint on the channel widths are provided by Equation 6.33. For most commercial designs, W is constant because the lead length of the primary flight is also constant.

$$W = W_s(z) + W_m(z) + e_b \tag{6.33}$$

a)

b)

c)

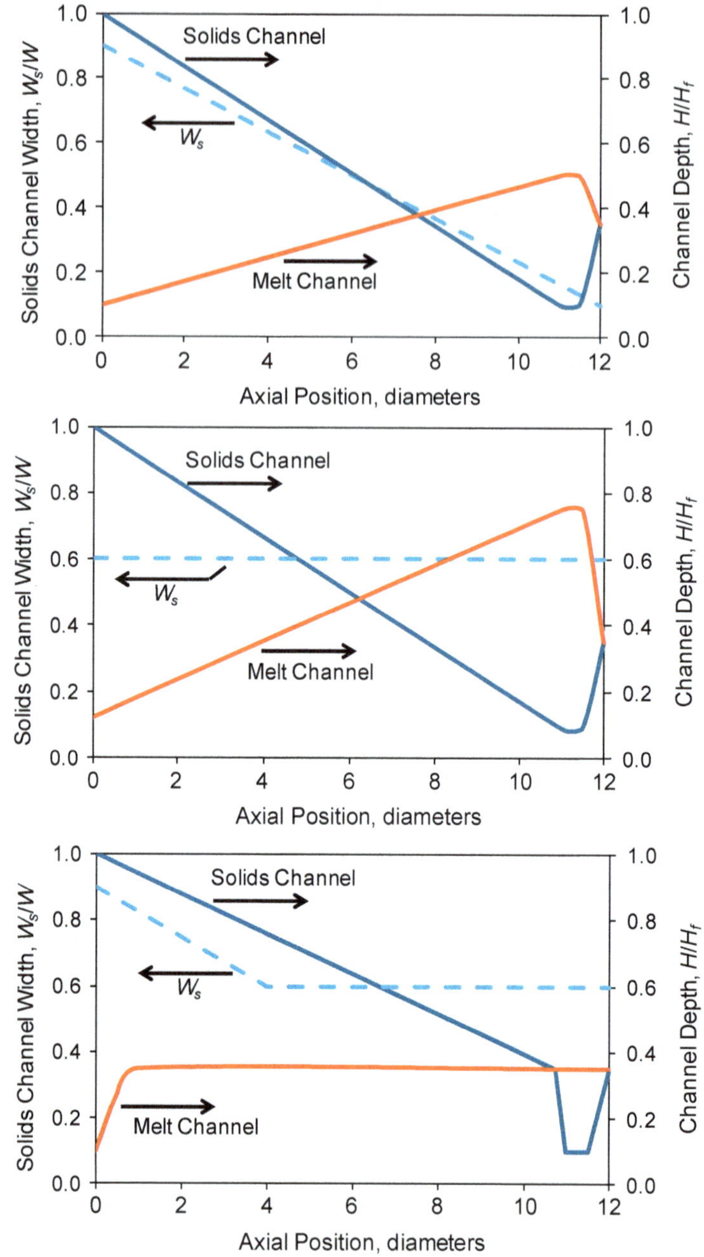

Figure 6.23 Schematic for the position of the barrier flight via the width of the solids channel (W_s) and the depths of the solids and melt channels: a) a barrier section design with a continuously decreasing W_s, b) constant-width solids channel, and c) a hybrid design. The designs all utilize a constant lead length for the primary flight

where $W_s(z)$ and $W_m(z)$ are the average width of the solids channel and average width of the melt-conveying channels at z, respectively, and e_b is the average width of the barrier flight. All widths are defined as perpendicular to the flight.

The barrier section melting concept presented here was developed based on recognizing that the rate-limiting melting dynamics for solid bed consumption was more complex than for melting in a conventional screw. This process was described by a diagram that focused on the four melt films surrounding the solid bed in the solids-conveying channel to set the foundation for the development of the physics of melting in a barrier melting section, as shown in Figure 6.24. The flow in the melt-conveying channel will be described later.

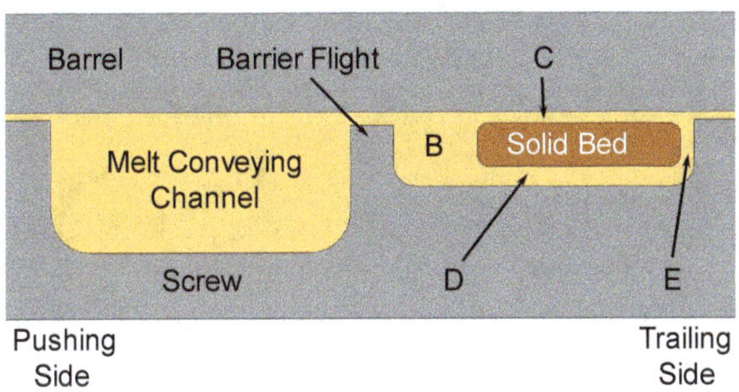

Figure 6.24 Schematic for the zones of the new barrier screw melting concept. The cream color represents molten resin

Consistent with the melting results presented in Figure 6.9, separate melting dynamics were defined again for the four melt films as originally defined by Shapiro et al. [20] and as depicted in Figure 6.24 for the solids channel; B is the film on the pushing side of the barrier flight, C is the film between the solid bed and the barrel wall, D is the film at the screw root (or core) polymer interface, and E is the film at the trailing side of the primary flight. The melting model for the solids channel is essentially identical to that for the conventional melting model presented in Section 6.3.1. The melted polymer flows over the barrier flight and into the melt-conveying channel, forming the melt pool.

Two constraints are placed on the solids-conveying channel. That is, the sum of the widths and thicknesses for the components must equal the width and depth of the solids-conveying channel. Here, the melt film thicknesses δ_b and δ_d for Zones B and D were found by difference:

$$\delta_b = W_s(z) - X - \delta_e \tag{6.34}$$

$$\delta_d = H_s(z) - Y - \delta_c \tag{6.35}$$

As stated above, both $W_s(z)$ and $H_s(z)$ are fixed by the geometry of the screw. A schematic representation of the solid bed in the channel is shown in Figure 6.25 for a barrier section geometry similar to that of Figure 6.23(a). This representation depicts how the four films change in dimension as the solid bed is consumed in the melting section of the screw.

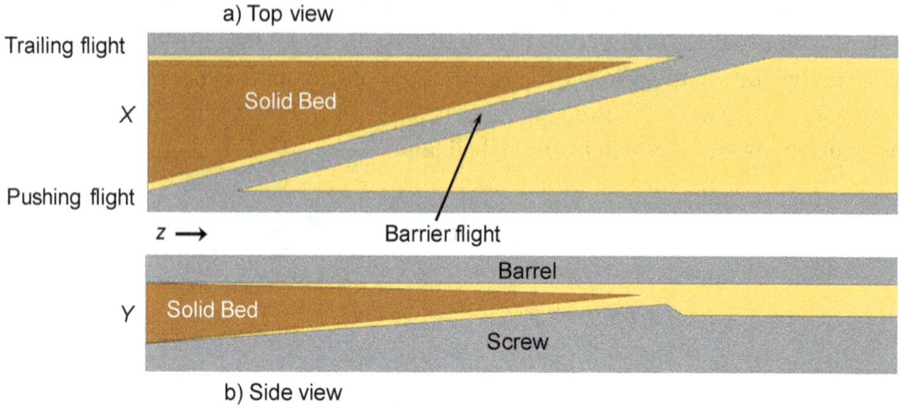

Figure 6.25 Qualitative shapes of *X* and *Y* bed dimensions and melt film thicknesses for a barrier screw model with a transverse barrier flight: (a) top view of the solids channel and the entry to the metering section, and (b) side view of the solids channel and the entry to the metering section. The cream color represents molten resin

Like melting in a conventional transition zone, screw rotation velocities are used in this analysis because the process is dominated by energy dissipation [39–44], and dissipation is not frame indifferent. The analysis was developed based on the melting model originally developed by Tadmor and Klein [8]. One major difference is that the barrel is not moving, so the velocities that are used in the simulation are adjusted to account for screw rotation as was done for melting in a conventional transition channel (Section 6.3.1). As with conventional melting discussed previously, another major difference in this analysis is that V_{sy} is retained as a variable because this leads to calculating the change in bed thickness in this reanalysis of the literature data, as shown in Figure 6.13. This provides the ability to evaluate the effect of the channel geometry on the rate of solid loss in both the cross-channel direction (x direction) and the solid thickness direction (y direction). The calculations are performed just like those outlined for a conventional melting process except with the additional constraints of Equation 6.34 and Equation 6.35.

The axial pressure and temperature distributions for the molten resin in the melt-conveying channel are calculated using the control volume method outlined in Section 7.7.5. For this method, the change in pressure and temperature are calculated using the local channel dimensions, $H_m(z)$ and $W_m(z)$, and the mass flow rate in the channel using Equation 7.54 for flow and the methods in Section 7.7.5.1 for energy dissipation and temperature. The amount of mass added to the melt channel for each increment is estimated via the rate of flow across the flight due to just the motion of the screw and ignoring the pressure-induced flow:

$$Q_{m,k+1} = Q_{m,k} + \frac{\rho_m V_{bx}\left(\lambda_b + \lambda_f\right)}{2}\Delta z \tag{6.36}$$

where $Q_{m,k}$ is the mass flow rate of material into the control volume k, $Q_{m,k+1}$ is the mass flow rate out of the control volume k, λ_b is the undercut distance between the main flight and the barrier flight, and λ_f is the mechanical clearance between the top of the main flight and the barrel wall.

6.3.2.1 Barrier Screw Melting Predictions

Two different styles of barrier section geometries were studied here. Both geometries use a melting model that is described above. For the first geometry, a barrier flight was positioned at the start of the section at the pushing flight, and it was positioned with a lead length that was larger than that for the primary flight, that is, a transverse barrier flight similar to that shown in Figure 6.23(a). This design caused the solids channel to decrease evenly in width in the downstream direction. The solids channel depth also decreased with the downstream direction. The transverse barrier flight merged with the trailing flight at the exit of the barrier section. This type of barrier section, known as a Maillefer barrier section [48], can have a stagnant region at the entry in the melt-filled section, a stagnant region at the exit of the solids channel, and reduced melting capacity as compared to other designs. The stagnant regions can lead to degraded resin if designed improperly. Most modern barrier designs do not use this placement of the barrier flight. Instead a modification shown by Figure 6.23(c) is used, mitigating these stagnant regions and increasing the width of the solids channel available for melting. The second geometry used a barrier flight that had the same lead length as the primary flight, maintaining a solids channel with a constant width; for example, see Figure 6.23(b). The depth of the solids channel decreased in the downstream direction.

Melting for resin in a barrier screw is evaluated using the model based on the physics described in Section 6.3.2. The focus of this first-order simulation is on the physics that occur on the solid bed side of the barrier section. It is assumed that once the solids melt that the molten polymer is carried across the barrier flight to the melt-conveying channel. The melting dynamics for a barrier section with a transverse barrier flight are presented in Figure 6.26 along with the melting dynamics for a conventional melting screw. The rate and screw speed were identical for these simulations for an LDPE resin with a Newtonian viscosity of 880 Pa·s. These initial results demonstrate interesting differences during melting in a conventional screw with a constant taper for the depth and a constant channel width melting section, and a barrier section with similar characteristics except the channel width is decreasing linearly as the solid is transported from the entrance of the melting section to the end of the barrier section. The most dramatic observation is that the *x*-direction bed thickness decreased much more rapidly in the transverse barrier case when compared with the conventional screw, as shown in Figure 6.26. There is also a more rapid decrease in solid thickness in the *y* direction at the end of melting in the barrier screw. Another observation for this set of simulation experiments is that it takes less down-channel distance to complete the melting for the barrier section than for the conventional transition section for the same rate and screw speed. For these similar cases, melting was completed at 180 cm downstream for the transverse barrier section while the complete melting occurred at 200 cm for the conventional screw. In practice, a rate increase would be designed into the process such that the completion of melting occurred at about the same downstream position.

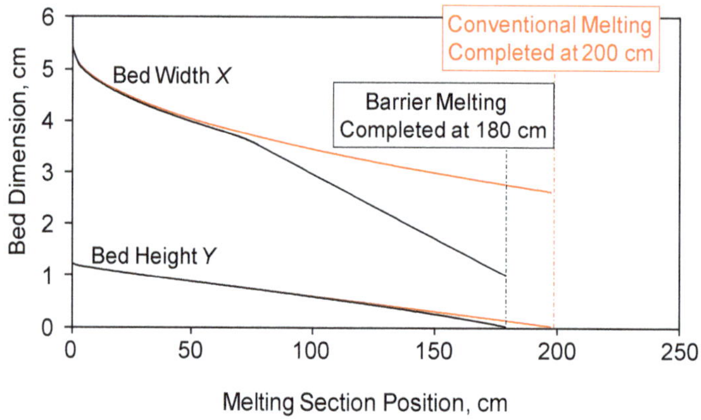

Figure 6.26 Comparison of melting dynamics for a conventional melting channel and a transverse barrier melting channel for an LDPE resin at identical rates and screw speeds. The conventional channel is in red while the barrier melting section is in black

A similar simulation was performed using a barrier melting channel with a barrier flight that is positioned parallel to the primary flight; that is, the width of the melting channel is constant with the downstream direction as shown in Figure 6.23(b). The melting dynamics for this case are shown in Figure 6.27. For this geometry, the solids channel at the entry to the barrier section is about 60% the width of the conventional melting channel. The smaller width of this parallel-flighted barrier section is caused by the abrupt addition of the constant width melt-conveying channel; that is, the melt channel occupies the remaining 40% of the screw volume. This smaller width is apparent in Figure 6.27 at a melting section position of 0 cm. The velocity of the solid bed for the barrier case is 40% higher than that for the conventional screw in order to maintain the same flow rate. The melting dynamics presented in Figure 6.27 are similar to those for the conventional screw. That is, at the end of melting the thickness in the *y* direction goes to zero while the width of the solid bed is about 40% of the original width. Moreover, this barrier section geometry required a melting length of 235 cm to complete as compared to 200 cm for the conventional screw.

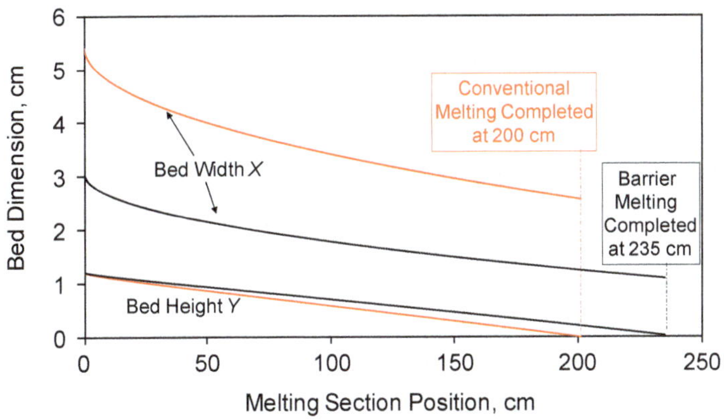

Figure 6.27 Comparison of melting dynamics for a conventional melting channel and a barrier section with the barrier flight parallel to the primary flight [50]. The conventional channel is in red while the barrier melting section is in black

The simulations suggest that conventional screws may have a higher melting capacity, as measured by the distance the solids traveled down the screw, than barrier-flighted melting sections, depending on the barrier design. However, in both designs the melting time is less than the conventional screw. Commercial practice, however, shows that barrier screws can operate at 20 to 30% higher rates than conventional screws. The combined data suggest that barrier screws can be operated at higher rates via higher specific rates by maintaining the solid bed intact. The higher specific rates also allow the extruder to discharge at lower temperatures. Conventional screws will discharge solid polymer fragments into the extrudate when the melting capacity is exceeded due to solid bed breakup, especially for high screw speeds [51]. The barrier section entry position and design are also known to be problematic [52, 53]. That is, if the design of the entry section and position downstream for the screw start are not correct, the entry section can cause a restriction and limit the rate of the extruder.

The results presented here are encouraging but only qualitative and have been produced using this first-order model. Current limitations of the model are the use of a constant viscosity function independent of temperature and shear rate. Also, the dynamic local temperature of the barrel and screw (see Section 10.10) should be incorporated into the model; they are currently set as constants.

Finally, a couple of qualitative melting effects will be presented in brief. The effects have been modeled with very simple adaptations of the models presented above. There is not really enough data in the literature to compare with the model results to determine if these models accurately represent the effect. However, the results are encouraging because they do qualitatively coincide with the few data that are present in the literature. This new model is proposed for the melting of the solid polymer bed for screws with barrier melting sections. The model alters the heat transfer and melting velocities originally proposed by Tadmor and Klein [8]. The model releases three of the original literature assumptions: (1) elimination of the V_{sy} melting velocity in the y direction of the bed, (2) the concept that the bed rearranges, and (3) barrel rotation dissipation dynamics. This required recasting the heat conduction and energy dissipation analysis using screw velocities and extending the material balances to all four solid bed interfaces at the fluid films. This model for barrier melting sections suggests that by placing the barrier flight in the transition section the rate of melting increases, leading to a shorter time to complete melting for both barrier types. This is consistent with both the experiment and theory presented where the rate-controlling step is the loss of the bed in the y direction. The model predicts the dynamics of melting previously reported as determined by digital dimensional analysis of the solids remaining in the bed, as shown in Figure 6.9 for a conventional screw. The model predicts that it may take a longer melting section to melt the entire polymer using a barrier screw due to the increased bed velocity, but since solids cannot, however, get over the barrier flight, no solids would be expected in the melt exiting the extruder at normal screw speeds.

6.3.2.2 Notes for Barrier Melting Sections

The performance of a barrier melting section is dependent on many design features, including section length, channel widths and depths, design and position of the entry and exit regions, lead lengths, and the undercut of the barrier flight. These design features are presented in more detail in Section 14.1. Poor barrier melting section designs can lead to reduced rates, resin degradation, and solid polymer fragments in the extrudate. Examples of these types of problems are presented in Sections 11.10.1 and 11.10.3 (degradation), 11.10.2 (gels in the extrudate), and 13.3.1 (solid polymer fragments).

Like the conventional melting model described in Section 6.3.1, the barrier section melting model developed here uses a fit parameter to obtain melting lengths that are consistent with those observed experimentally. This fitted model is utilized to adjust for some of the non-linearities in the model. The model is not meant as a screw design tool. An improved model could be written based on the original Elbirli, Lindt, Gottgetreu, and Baba model [54] for barrier section geometries. The original model was developed with barrel rotation boundary conditions. The improved model would set the boundary conditions for screw rotation.

6.4 Effect of Pressure on Melting Rate

Melting rates typically increase when the local operating pressure is increased. This was discussed by Klein [12] in 1972 in relation to flow surging. Flow surging is defined as a change in rate, discharge pressure, or discharge temperature when no apparent set point change on the machine was made. As an example of this mode of operation, a slow-moving solid plug flows downstream and becomes lodged in the compression section and eventually forms a plug in the channel. The pressure upstream of the plug will increase and the pressure downstream will decrease. The decrease in the downstream pressure will cause the discharge pressure and rate to decrease. The increase in local pressure at the plug will cause the melting rate to increase and eventually allow the plug to flow downstream. This will cause the downstream pressure to increase to the normal operational pressure. A flow surging problem due to this type of blockage is presented in Section 12.7.6.

The melting data [55] and models [56] developed by Chung's research group show an increase in the melting rate with increasing pressure for materials with low shear viscosities as observed using a Screw Simulator. The effect of pressure on melting rate is not prominent for the high viscosity materials. Chung explains that the melting rate increase is due to a lower melt film thickness between the solid and metal rubbing surface. The higher pressure causes some of the molten resin to flow out of the film, causing the film thickness to decrease. The higher melting rates are attributed to the thinner melt film and the higher rate of viscous energy dissipation. A similar melting rate increase with pressure was observed using the Screw Simulator at Dow; see Figure 4.11. For example, the melting flux for an LDPE resin was 0.7 kg/(m²·s) at a pressure of 0.69 MPa, and it increased to 0.9 kg/(m²·s) at a pressure of 3.45 MPa [1]. These measurements were performed under specialized lab conditions where the outside edge of the sliding solid was at atmospheric pressure (or zero pressure). At high measurement pressures, the molten material flows out of the melt film at a higher rate than what would be expected in an extrusion process. As stated by Chung [56], the local pressure gradient needs to be studied for flow of material out of the films rather than the pressure.

Energy dissipation in the melt film is slightly dependent on the cross-channel pressure gradient in the melting section. The cross-channel pressure gradient occurs due to the rotation of the screw, and it is shown in Figures 12.2 and 12.3. As shown by these figures, the pressure gradient $\partial P/\partial x$ is negative (origin at the pushing flight edge). This dissipation level has a magnitude that is typically about 10% of the viscous energy dissipation due to the motion of the screw $\left(\mu V_j^2/2\delta_c\right)$. When the dissipation due to the cross-channel pressure gradient is included, the rate of loss in the Y function is modified as follows:

$$-\frac{dY}{dt}\big|_{\delta_C} = V_{sy,C} = \frac{\left[\frac{k_m}{\delta_C}\left(T_b - T_m\right) + \frac{\mu V_{j,C}^2}{2\delta_C} + \frac{V_{j,C}\delta_C\left|\Delta P_x(z)\right|}{2X}\right]}{\rho_s\lambda + \left[\rho_s C_v\left(T_m - T_s\right)\right]} \tag{6.37}$$

where $\Delta P_x(z)$ is the pressure change across (x direction) the solid bed of width X, and as expected it is a function of the downstream position z. The third term in the numerator in Equation 6.37 is provided in Equation 6.38 below, and is the energy dissipation due to the pressure gradient over the melt Film C.

$$\text{Term 3} = \frac{V_{j,C}\delta_C\left|\Delta P_x(z)\right|}{2X} \tag{6.38}$$

represents the pressure gradient introduced into the melting rate function. The material being removed in the x direction has to be swept against this pressure gradient into the melt pool B in Figure 6.12, and thus more dissipation would be expected to occur.

6.5 One-Dimensional Melting

In 1997, there was a paper presented at the International Polymer Processing annual meeting in New York that described the melting of a low-viscosity PP resin [57]. The experimental observation was that the powder melted in a nonstandard manner. That is, no melt pool formed. Instead, the polymer melted in a one-dimensional manner (or "top down" melting) near the barrel surface. This same type of melting was observed by one of the authors for PVDC powder resins and by Dekker for PP powder [58]. Tang [31] observed that, even for commercial high molecular weight PP pellets, one-dimensional melting could occur if the screw speed is extremely low. In general, one-dimensional melting is not desirable for commercial applications since it only occurs at extremely low rates. Later in this section a model will be developed for one-dimensional melting.

Recalling the discussion earlier in this chapter, in most cases melting in the channel typically occurs at all four edges of the solid bed, with the majority of the melting occurring at the solid bed-melt film interface located between the solid bed and the barrel wall, as shown in Figure 6.2. The newly molten resin from this location is then conveyed by the motion of the screw to a melt pool located at the pushing side of the channel. For very special and sometimes unpredictable conditions, the melting process can occur by a different mechanism. In these cases, the melting is occurring predominantly at the solid bed-melt film interface located between the solid bed and barrel wall. The newly molten resin infiltrates the solid bed by flowing between the solid resin particles. A photograph of a cross section is shown by Figure 6.28. This cross section was for a PVDC powder resin at a location about 3 diameters downstream from the start of the transition section. The pressure in the screw channels was very low and the screw rotation rate was 30 rpm. The melting rate and extrusion rate were very low. Because of the low rates, a Maddock solidification experiment was performed. The melting cross sections showed that the melt film between the solid bed and barrel wall was considerably thicker than that for a conventional melting process. The grainy cross section shown in Figure 6.28 is due to a mixture of molten and solid particles at the time of the Maddock solidification experiment. As melting progresses in the downstream direction, the bulk temperature of the channel increases, the level of solids decreases, and the cross-sectional view became less grainy in appearance. The surface of the solids that were in contact with the screw showed a very thin melt film due to the conduction of the energy from the hot screw to the solids, as shown in Figure 6.28 and Figure 6.29. Although these films exist, the experimental evidence suggests that they are not significant contributors to the melting process described here.

Figure 6.28 Photograph of a PVDC powder resin that is melting via the one-dimensional melting mechanism. Even at this early location in the melting process, the thickness of the melt film adjacent to the barrel wall was thicker than normal

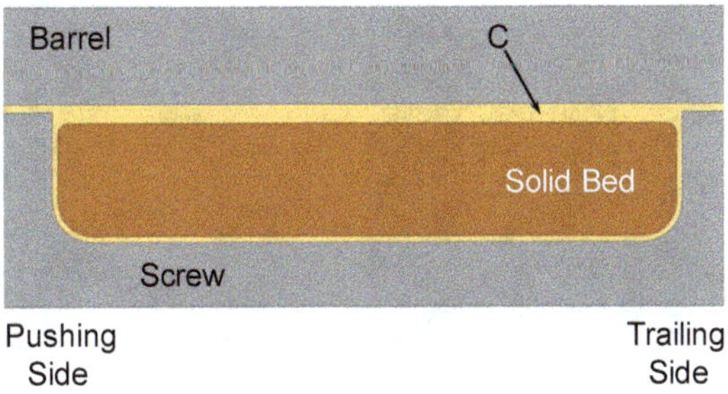

Figure 6.29 Conceptual melting physics for one-dimensional melting. The cream color represents molten resin. The dark color is a mixture of solid polymer particles with molten resin in the void regions

This is consistent with the observations of Tang [31] when evaluating PP melting. In his experiments, the melting mechanism could be switched from the conventional mechanism as shown by the photographs in Figure 6.30 to a one-dimensional melting mechanism shown in Figure 6.31. For the photographs in Figure 6.30, melt pools developed next to the pushing flights as would be expected in a conventional melting mode. This experiment was conducted with a commercial high molecular weight PP resin at a screw speed of 15 rpm and a barrel temperature of 90 °C. However, when the screw was dramatically slowed to 7 rpm and the barrel temperature was increased slightly to 100 °C, Tang observed one-dimensional melting as shown by the photographs in Figure 6.31; that is, no melt pool was observed. In this case the pellets remain next to the pushing flight, and as the polymer melts it infiltrated the solid bed.

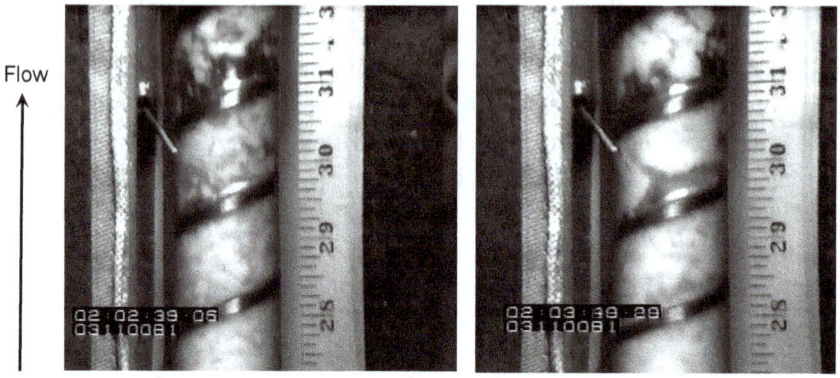

Figure 6.30 Photographs of the conventional melting of a commercial PP resin in an experimental extruder with a glass barrel. In both photographs a melt pool forms near the pushing side of the channel. The screw was rotated at 15 rpm and the barrel temperature was 90 °C. The ruler is in inches

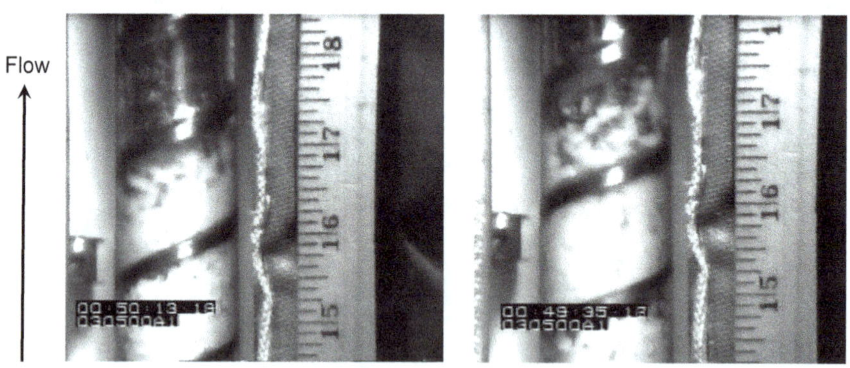

Figure 6.31 Photographs of one-dimensional melting of the same commercial PP resin as used in Figure 6.30. Melt pools do not form under these conditions. The screw was rotated at 7 rpm and the barrel temperature was 100 °C

The potential cause of these differences in melting mechanisms may be found in the pressure development in the delay zone as described by Kacir and Tadmor [59] and the source of the energy needed for melting. For the delay zone, it is generally accepted that the solid polymer is heated and melts initially close to the hot barrel wall at the end of the solids-conveying zone. Then the melted polymer continues to increase in thickness until conventional melting surrounds the solid bed. The cross-channel geometry resembles a Reynolds bearing [60]. In a Reynolds bearing the pressure at the pushing flight would decrease as the film thickness increases. It will take a certain volume of melt to encapsulate the solid bed, which would result in conventional melting. The pressure generated is a positive function of both viscosity and velocity and an inverse function of the melt film thickness. In the case of one-dimensional melting, it is thought that the pressure at the critical film thickness for encapsulation is not sufficient to compress the bed and allow the melt to encapsulate the solid bed. This pressure and source of the energy needed for melting are the key reasons for the one-dimensional melting mechanism. If the channel pressure is very low or even near atmospheric, the energy dissipation due to rubbing in the delay zone will be very low and the solids in the channel will not compact. As discussed in Section 4.1, an applied pressure is required to compress the voids out from between the solid particles. If the pressure is not available to compact the solids, molten resin from Film C can flow into the bed via the voids. The source for the energy needed to melt the resin is also very important and can come from three sources: (1) conduction from the heated barrel, (2) viscous energy dissipation from the shear flows in Film C, and (3) cross-channel flow against a pressure gradient as discussed in Section 6.4. Conduction from the heated barrel as the primary energy source occurs at elevated barrel temperatures and when dissipation in the Film C is very low. Low Film C dissipation rates occur at very low screw speeds and at higher screw speeds when the viscosity is very low. The low screw speed for the Tang experiments was likely the reason for the one-dimensional melting in Figure 6.31. If the pressure is high enough to compact the solid bed, then the melt will not infiltrate the solids and the melting mechanism will be conventional. Low pressures in the melting section can occur for a variety of reasons with the most likely reason being poor solids conveying in the upstream sections. For example, the photograph in Figure 6.28 was due to poor solids-conveying conditions for the PVDC resin. Poor solids conveying is known to occur for PVDC resins at less than optimal conditions [61]. Cross-channel pressure gradients are not high enough to deform the solid bed, and thus cross-channel energy dissipation and the deformation of the solid bed due to P_x are not significant. Screws with very low compression rates can prevent pressure from generating in the melting section and thus can contribute to one-dimensional melting.

This ends the speculative examination of the cause of one-dimensional melting. In the following section the physics of melting with only one film, leading to the phenomenon of poor melting due to one-dimensional melting (or top down melting) as shown by Figure 6.28, is developed. The theoretical differences between the conventional melting theory and one-dimensional melting are discussed in the following paragraphs. The model presented here complements previously defined models for melting in conventional channels provided in Section 6.3.1 and melting in barrier screw sections provided in Section 6.3.2.

6.5.1 One-Dimensional Melting Model

The one-dimensional melting model presented here follows the same methods as described for the conventional melting in Section 6.3.1 except that all melting is restricted to Film C. The melting is conceptualized by Figure 6.29. Comparing Figure 6.29 to Figure 6.12 it is observed that melting Zones B, D, and E are no longer available for producing dissipation to aid the melting process. Thus the expressions for δ_D, δ_B, and δ_E are reduced to zero. In practice these films are present due to energy conducting from the screw surface and the melting of very small amounts of resin. Here the melting rate for Film C is modeled using Equation 6.8. Then as the solid bed melts at the barrel interface, δ_c becomes a function of the downstream position z, as provided by Equation 6.39. The molten resin produced during the process flows into the void regions between the resin particles, displacing air and causing the bulk density of the bed to increase.

$$\delta_C = H(z) - Y(z) \tag{6.39}$$

The film at the start of the melting process has an initial thickness estimated from Equation 6.13. As the solids continue to melt, the film thickness at the barrel solid bed interface increases slightly and decreases the dissipation because the local shear rate decreased. Since the dissipation decreases, the melting rate also decreases. This theory requires the solution of only two coupled differential equations:

$$-\left.\frac{dY}{dt}\right|_{\delta c\_melt} = V_{sy,C} \tag{6.40}$$

$$\frac{dZ}{dt} = V_{sz} \tag{6.41}$$

In order to demonstrate the one-dimensional melting model, two simulations were performed at very low rates and screw speeds using an LDPE resin with a shear viscosity of 880 Pa·s. The first simulation was for the case where a substantial level of solids conveying produced a pressure in the channel that compacted the solids into a bed without voids. For this process, the melting would occur via the conventional melting model. The simulation for this case is shown in Figure 6.32 where melting occurs in all four melt films. As shown by this figure, the Y thickness goes to zero at a down-channel position of 220 cm into the melting section. It is seen that the change in the thickness (x direction) leaves about 60% of the original bed width remaining just at the end of melting. The end of melting is defined as when the y direction thickness of the solid bed goes to zero, and thus the x direction thickness also goes to zero because there is no more material to melt.

The second simulation was performed at the same conditions except that solids conveying was very poor. Here the pressure in the transition section is very low, and thus the solids are not well compacted, causing a highly porous solid bed. Due to the low channel pressure and the high porosity of the solid bed, the melting mechanism switches to the one-dimensional melting mode. The simulation results are presented in Figure 6.33. It is observed that the bed width remains constant because there is no dissipation at the screw helix wall (Zones B, D, and E), and a melt pool at the pushing flight does not form due to the infiltration of the molten resin into the porous solid bed. The bed height *Y* in this case decreases essentially at the same rate that the channel depth decreases in the transition section. As shown in Figure 6.33, at a down-channel position of 220 cm, about 75% of the resin is melted, and thus 25% of solids in the original bed remain. The molten resin has filled the voids between the resin particles, and thus the bulk density of the bed has increased.

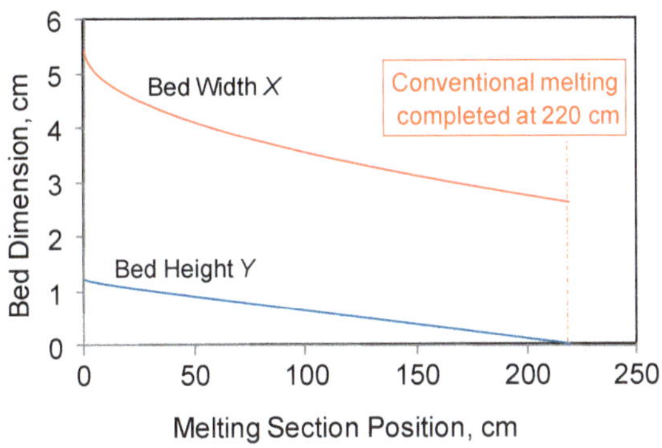

Figure 6.32 Simulation results using the melting model for the conventional melting process; that is, melting is occurring in all four melt films

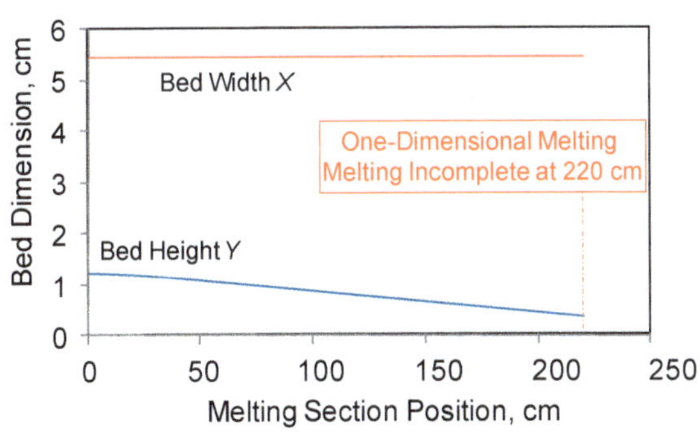

Figure 6.33 Simulation results for melting only from the barrel surface, that is, one-dimensional melting. About 75% of the resin was melted at the end of the transition section. The bed height *Y* in this case decreases essentially at the same rate that the channel depth decreases

In this simulation the model assumption is that the melt film is transported into the pores of the bed. Thus the melt film is relatively thin even for this mechanism due to the reducing depth of the channel. The model as developed does not account for any conductive heat transfer into the solid particles from the liquid infiltrate. For most extrusions, the center of the solid bed may be porous while the edges exposed to the melt film are not. The sealed edges prevent the melt from infiltrating the solid bed and the melting process occurs via conventional melting. Conventional melting was observed for the case with the measured bulk density data in Figure 4.1.

In summary, one-dimensional melting and melt infiltration into the solid bed will reduce the melting capacity of the process, as shown by Figure 6.33. Moreover, one-dimensional melting is most likely to occur for processes with very low pressures in the transition section and where conduction is the dominant energy source for melting. As such one-dimensional melting is typically not desirable in a commercial process due to the low processing rates. Instead, when one-dimensional melting occurs in a commercial process, troubleshooting procedures are needed to determine the root cause so that conventional melting can be restored.

6.6 Solid Bed Breakup

The melting mechanisms described earlier for a conventional transition section and a barrier melting section assume that the solid bed remains continuous throughout the entire melting process. For most extrusion processes, however, this is not the case. Instead the solid bed can break near the completion of melting where the bed is relatively thin and its strength is low. The strength of the bed is low at this location because the solid bed temperature is approaching the melting temperature or devitrification temperature such that the material modulus is low and because the bed is thin in the y direction. Evidence of solid bed breakup is apparent in Figure 6.3, Figure 6.4, Figure 6.6, and Figure 8.19. In these photos, there are regions were the molten resin is evident across the entire channel and solids are evident in downstream cross sections. For many screw designs and commercial processes, solid bed breakup is highly undesirable since it reduces the melting capacity of the extruder and can allow solid polymer fragments to flow into the extrudate, creating defects in the product. Moreover, solid bed breakup can cause thermal gradients in the extrudate and can lead to flow surging [12]. A complicated flow surging case where solid bed breakup is part of the root cause is presented in Section 12.7.6. Solid bed breakup is likely the mechanism for the solid fragments in the extrudates shown in Figure 8.18. Some of these fragments from a breakup can be eliminated by positioning dispersive-type mixers in the metering section of the screw. Dispersive mixers are presented in Section 8.5.

For processes with relatively low compression rates such that the air entrained between the pellets is not readily pushed back out of the hopper, solid bed breakup will eliminate a pathway back to the hopper. In this case the entrained air will discharge with the extrudate and often create defects in the product. This type of problem is presented in Section 10.2.2.

Solid bed breakup can be mitigated using screw design and process conditions for a conventional screw design. For example, the largest single contributor to bed breakup is high screw speeds, as shown by Figure 8.18. The extrudates in this figure are relatively free of solid polymer fragments at low screw speeds. But at screw speeds above about 75 rpm, solid bed breakup is occurring, and high levels of solids are discharged with the extrudates. Although not desirable for a commercial process, decreasing the rate of the line is often a short-term fix for eliminating solids in the discharge. Placing a finer screen in the screen pack is also an acceptable short-term remedy. The best long-term fix is to add an acceptable dispersive mixer to the screw or to install a high-performance screw.

Several high-performance screws take advantage of solid bed breakup by using the small solid fragments as a cooling method for the extrudate. For these designs, the solid polymer fragments are reduced in size using dispersive-type dams. The small fragments are then melted primarily by heat conduction from the hot molten resin to the cooler solid fragments, decreasing the temperature of the molten stream. Common commercially available high-performance screws that employ this technology include wave screws, Energy Transfer screws, Fusion screws, and DM2 screws. These screws and their operation are discussed in detail in Section 14.2.

Literature references discuss solid bed breakup as a phenomenon where strong forces due to pressure gradients break up the low-strength solid bed [8, 56]. The source of the pressure gradients is not evident. Zhu et al. developed an extruder with glass windows for viewing the internal processes during extrusion [62] and injection molding [63]. Their observations indicated that solid bed breakup was more likely to occur at high screw speeds, consistent with the observations of Figure 8.18. For injection molding, bed breakup is more likely to occur for long injection strokes, high discharge pressure (back pressure), and longer dwell times. They observed small cracks that occurred across the solid bed and perpendicular to the flights in the latter stages of the melting process. These cracks would then increase in width as they were filled with molten polymer. Their observations are consistent with previous researchers.

It is obvious that the strength of the solid bed would be low near the end of melting since the bed is very thin and its temperature is approaching the melting or devitrification temperature. Based on the historical melting models using barrel rotation physics, the pressure gradients postulated for breaking the bed are difficult to produce. However, screw rotation theory leads to an alternative mechanism to those based on barrel rotation theory. Screw rotation theory is presented in great detail in Chapter 7 for the metering section. For screw rotation analysis, the barrel has zero velocity and the solid bed is moving in the positive z direction at a velocity of V_{sz}. The screw has a velocity of V_c and a component in the z direction of V_{cz}. Here, the backward motion of the screw V_{cz} in the z direction has a magnitude larger than the forward velocity V_{sz}, causing a negative pressure gradient $(\partial P/\partial z)$ in Film D between the solid bed and the screw root. The velocities and pressure gradient are shown in Figure 6.20 and described in Section 6.3.1.3. That is, the backwards motion of the screw is dragging molten polymer backwards at the screw root and generating a significant level of pressure in Film D. When the strength of the solid bed is relatively high, the high pressure in Film D causes material to flow out to the melt pool B and to melt Film E. If the strength of the solid bed is relatively low, then the bed will form a small crack due to the fluid flow induced stresses, and then the crack will fill with fluid as observed by Zhu et al. [62, 63]. The solid bed breakup process is qualitatively described in Figure 6.34.

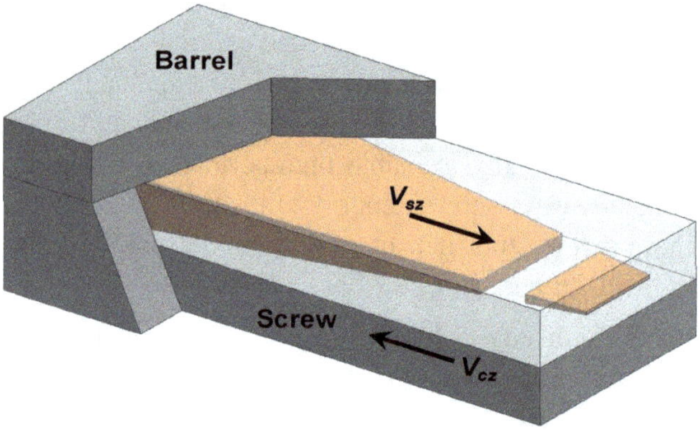

Figure 6.34 Schematic for solid bed breakup

The flow from the film between the solid bed and the screw root (Film D) can be observed in most Maddock solidification experiments, as shown in Figure 6.35. The section in this photograph was for a location early in the melting process where the strength of the solid bed is high and can withstand the pressure gradient created by the backwards motion of the screw. The high strength of the bed prevents the bed from breaking up. Flow lines due to poor mixing of the colorant into the white resin show the flow from the film into the melt pool. The flow is substantial as it is pushing the recirculating flow of the pool away from the solid bed. These flow patterns are obvious in Figure 6.3, Figure 6.4, and Figure 6.5. This flow pattern would not be possible for a barrel rotation model.

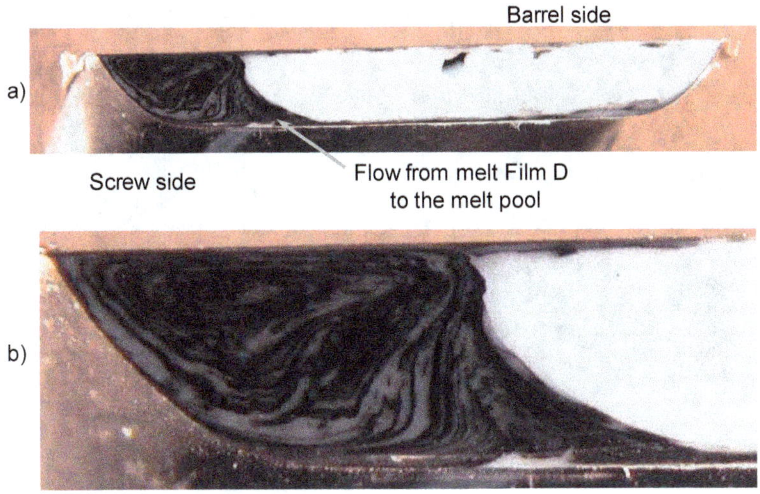

Figure 6.35 Photograph of resin solidified in the transition section after a Maddock solidification experiment for an ABS resin: (a) the arrow is pointing at the flow lines created by the flow of material from Film D out to the melt pool, and (b) an enlargement of the flow area

The recirculating flow in the melt pool and the flow entering the pool from Film D creates a region at the screw root where the flows are likely very low or form eddies. This low-flow region where the recirculating and entering flow merge is shown in Figure 6.36. For most extrusion processes, this low-flow region is not a problem and is likely very difficult to detect. For thermally sensitive materials such as PVDC resin, the long residence time of the region can cause a ribbon of degraded material to form on the root of the screw where the flows merge. This ribbon typically starts when the melt pool first forms and ends when the melting process is about 70% complete. A photograph of this type of degradation is shown in Figure 6.37. Processing changes can mitigate this type of degradation [64].

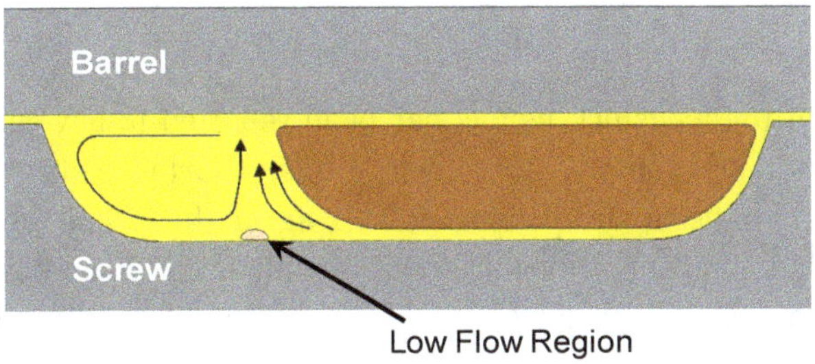

Figure 6.36 Schematic showing the recirculating flow in the melt pool and the flow entering the melt pool from Film D. A low-flow region exists where the flow streams merge

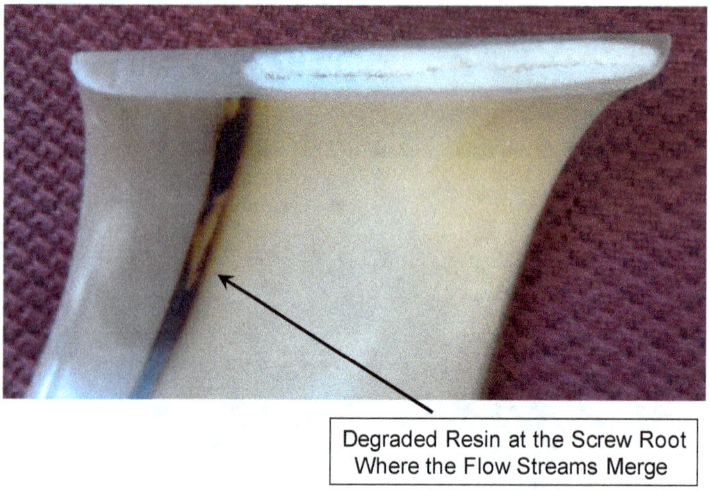

Figure 6.37 Photograph of a segment from a Maddock solidification experiment for a PVDC resin extrusion. The dark band is resin degraded due to a long residence time at the location

The degradation ribbon at the merger of the flows occurs because of the cross-channel flow of material from the region between the solid bed and the screw root to the melt pool. As shown by Figure 6.35, this flow from underneath the solid bed is relatively large. As previously stated, the flow occurs because of pressure-induced flow and the dragging of fresh material under the solid bed by the backwards motion of the screw root. This process is consistent with the physics presented for screw rotation. The flow fields developed for a barrel rotation system would not create the low-flow region such as shown in Figure 6.37.

■ 6.7 Melting Section Characteristics

The geometry of the feedstock pellets has been reported to have an effect on the melting process. For example, Bozzelli [65] has reported melting is influenced by the size and shape of the pellets, pelletizing method, and regrind particles, although details and observations are not presented. The hypothesis here is that the smaller particles melt faster since they do not have as long of a conduction pathway. This effect is often observed with PS that contains a high level of fine particles. These fine particles can originate from the resin manufacturer, occur during air conveying, or come from regrind streams. These particles can melt on the barrel wall in the solids-conveying section and have the potential to cause the rate to decrease. Once the solids are fully compacted in the melting section, however, the geometry of the pellets is typically not important. That is, as the solids are compacted into a solid bed, the individual particles lose their characteristic shape prior to melting, as shown in Figure 6.4. As discussed in the section above, regrind streams have the potential to reduce the bulk density of the feedstock, and adjustments to the screw are required for optimal solids-conveying and melting performance. The extrusion of powder resins that have a level of liquid additive on their surface will show different extrusion performances if the powder sizes separate prior to extrusion. This phenomenon is caused by the larger surface area to volume ratio of the small particles; that is, there are more surface additives on a unit mass of small particles. These small particles can be difficult to melt because of poor solids conveying and pressure generation early in the melting section.

The solids-conveying and melting performances were measured for a specialty HIPS compound using spherical pellets made using an underwater pelletizer and cylindrical pellets made using a strand cutting process. The pellets were similar in size. The pellets were then extruded using a screw with a compression ratio of 3.4 using the instrumented extruder described in Section 8.4.1. The rates and discharge temperatures were essentially identical. The axial pressure profiles are shown in Figure 6.38. For this application, the axial pressures are essentially identical. These data indicate that for a properly designed screw for an application, the geometry of the pellet may not be critical to the solids-conveying and melting processes.

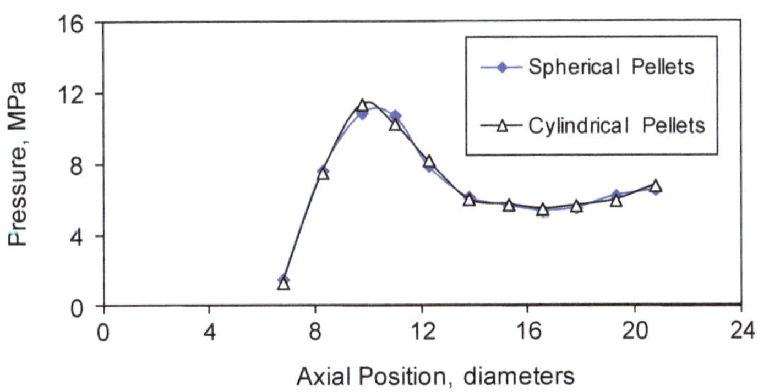

Figure 6.38 Axial pressure profiles for a 63.5 mm diameter instrumented extruder running two different pellet geometries at a screw speed of 90 rpm

Plasticating single-screw extruders will typical range in size from 15 mm up to 250 mm in diameter. Larger diameter plasticating machines are very rare and were likely built years ago. Plasticating rates above about 4000 kg/h are more economically processed using twin-screw machines.

Nomenclature

C compression ratio

C_f compression ratio that should be used for a lower density feedstock resin

$C_{pellets}$ compression ratio used for a pellet feedstock

C_v specific heat of the solid resin

D_b diameter at the barrel wall

D_c local diameter at the screw core

D_f local diameter at the surface of the solid bed at the interface with melt Film C

e_b width of the barrier flight perpendicular to the flight for a barrier melting section

H local channel depth or metering channel depth

H_f depth of the feed channel

H_1 depth of the channel at the start of the transition section; $H_f = H_1$

H_k local channel depth at the start of increment k

H_{k+1} local channel depth at the start of increment $k + 1$

H_m local channel depth of the melt-conveying channel for a barrier melting section

H_s local channel depth of the solids-conveying channel for a barrier melting section

k_m thermal conductivity of the melt

L lead length of the flight

M number of helical turns in the transition section

N screw rotation speed in revolutions/s

p number of flight starts in the melting section

P pressure

Q_m mass rate of the extruder

$Q_{melt,k}$ volumetric rate for molten resin at the entry to increment k

R compression rate of the transition section

R_f compression rate that should be used for a lower density feedstock resin

$R_{pellets}$ compression rate used for a pellet feedstock

t independent time variable

T_b barrel temperature at the inner surface

T_m	melting temperature of the resin
T_s	temperature of the solid bed
$V_{j,B}$	scalar values for the vectorial velocity in Film B
$V_{j,C}$	scalar values for the vectorial velocity in Film C
$V_{j,D}$	scalar values for the vectorial velocity in Film D
$V_{j,E}$	scalar values for the vectorial velocity in Film E
V_{bx}	velocity of the barrel in the x direction for the barrel ($V_{bx} = 0$ for screw rotation)
V_{bz}	velocity of the barrel in the z direction for the barrel ($V_{bz} = 0$ for screw rotation)
V_{cx}	velocity of the screw root in the x direction for screw rotation
V_{cz}	velocity of the screw root in the z direction for screw rotation
V_{Ez}	average velocity in Film E in the z direction
$\overline{V}_{fz}$	average velocity of the flights in the z direction for screw rotation
$V_{sx,B}$	velocity of the solid bed consumption (melting) in the x direction at the Film B interface
$V_{sx,E}$	velocity of the solid bed consumption (melting) in the x direction at the Film E interface
V_{sy}	velocity of the solid bed consumption (melting) in the y direction at the Zone C film (Tadmor)
$V_{sy,C}$	velocity of the solid bed consumption (melting) in the y direction at the Film C interface
$V_{sy,D}$	velocity of the solid bed consumption (melting) in the y direction at the Film D interface
V_{sx}	velocity of the solid bed in the x direction
V_{sz}	velocity of the solid bed in the z direction
W	average width of the channel perpendicular to the flight
W_s	average width of the solids channel of a barrier melting section perpendicular to the flight at the barrel
W_m	average width of the melt channel of a barrier melting section perpendicular to the flight at the barrel
x	independent variable in the cross-channel direction
X	local width of the solid bed
y	independent variable in the channel depth direction
Y	local thickness of the solid bed in the y direction

z	independent variable in the down-channel direction
Z	downstream position of the solid bed
α	fitting factor for Equation 6.13
δ_b	thickness of melt Film B
δ_c	thickness of melt Film C
δ_d	thickness of melt Film D
δ_e	thickness of melt Film E
$\Delta P_x(z)$	change in pressure in the x direction at the position z
θ_b	helix angle of the flight at the barrel wall
θ_c	local helix angle at the screw root
θ_f	helix angle at diameter D_f
λ	heat of fusion
λ_b	flight clearance for the main flight
λ_f	barrier flight undercut relative to the main flight
μ	shear viscosity
$\rho_{pellets}$	bulk density of the pellets at ambient conditions
ρ_f	bulk density of the feedstock mixture at ambient conditions
ρ_m	density of the molten polymer
ρ_s	density of the compacted solid bed

References

1. Spalding, M.A., Hyun, K.S., and Cohen, B.R., "An Experimental Study of the Frictional and Viscous Forces Responsible for Solids Conveying and Melting in Single-Screw Plasticating Extruders," *SPE ANTEC Tech. Papers*, **43**, 202 (1997)
2. Giles, H.F., Wagner, J.R., Jr., and Mount, E.M., III, "Extrusion: The Definitive Processing Guide and Handbook," William Andrew Publishing, Norwich, NY (2005)
3. Butler, T.I., "Low-Density Polyethylene," Chapter 27 in "Film Extrusion Manual, Process Materials, Properties," 2nd ed., Butler, T.I. (Ed.) TAPPI Press, Atlanta, GA (2005)
4. Butler, T.I., "Linear Low-Density Polyethylene," Chapter 28 in "Film Extrusion Manual, Process Materials, Properties," 2nd ed., Butler, T.I. (Ed.) TAPPI Press, Atlanta, GA (2005)
5. Maddock, B.H., "A Visual Analysis of Flow and Mixing in Extruder Screws," *SPE J.*, **15**, 383 (1959)

6. Street, L. F., "Plastifying Extrusion," *Int. Plast. Eng.*, **1**, 289 (1961)
7. Tadmor, Z., Duvdevani, I. J., and Klein, I., "Melting in Plasticating Extruders, Theory and Experiments," *Polym. Eng. Sci.*, **7**, 198 (1967)
8. Tadmor, Z. and Klein, I., "Engineering Principles of Plasticating Extrusion," Van Nostrand Reinhold Company, New York (1970)
9. Tadmor, Z., "Fundamentals of Plasticating Extrusion," *Polym. Eng. Sci.*, **6**, 185 (1966)
10. Potente, H., "Single-Screw Extruder Analysis and Design," Chapter 5 in "Screw Extrusion," White, J. L. and Pontente, H. (Eds.), Hanser, Munich (2003)
11. Menges, G. and Klenk, P., *Kunststoffe*, **57**, 598, 677 (1967)
12. Klein, I., "The Melting Factor in Extruder Performance," *SPE J.*, **28**, 47 (1972)
13. Altinkaynak, A., "Three-Dimensional Finite Element Simulation of Polymer Melting and Flow in a Single-Screw Extruder: Optimization of Screw Channel Geometry," Ph. D. Thesis, Michigan Technological University, Houghton, MI (2010)
14. Altinkaynak, A., Gupta, M., Spalding, M. A., and Crabtree, S. L., "The Investigation of the Effect of the Screw Geometry on Melting in a Single-Screw Extruder," *SPE ANTEC Tech. Papers*, **56**, 426 (2010)
15. Chung, C. I., "A New Theory for Single-Screw Extrusion, Part 1," *Mod. Plast.*, **45**, 178 (1968)
16. Chung, C. I. and Barr, R. A., "Energy Efficient Extruder Screw," US Patent 4,405,239 (1983)
17. Donovan, R. C., "A Theoretical Melting Model For Plasticating Extruders," *Polym. Eng. Sci.*, **11**, 247 (1971)
18. Mondcai, I., Halasz, L., and Molnar, I., "Extrudieren von Thermoplasten I.," *Plaste Kautschuk*, **20**, 630 (1973)
19. Edmondson, I. R. and Fenner, R. T., "Melting of Thermoplastics in Single-Screw Extruders," *Polymer*, **16**, 49 (1975)
20. Shapiro, J., Halmos, A. L., and Pearson, J. R. A., "Melting in Single-Screw Extruders," *Polymer*, **17**, 905 (1976)
21. Meijer, H. E. H., "Melting in Single-Screw Extruders: Models, Calculations, and Screw Design," Ph. D. Thesis, Technische Hogeschool Twente, Netherlands (1980)
22. Verbraak, C. P. J. M. and Meijer, H. E. H., "Screw Design in Injection Molding," *Polym. Eng. Sci.*, **29**, 479 (1989)
23. Ingen Housz, J. F. and Meijer, H. E. H, "The Melting Performance of Single-Screw Extruders," *Polym. Eng. Sci.*, **21**, 352 (1981)
24. Fukase, H., Kunio, T., Shinya, S., and Nomura, A., "A Plasticating Model for Single-Screw Extruders," *Polym. Eng. Sci.*, **22**, 578 (1982)
25. Lindt, J. T., "A Dynamic Melting Model for a Single-Screw Extruder," *Polym. Eng. Sci.*, **16**, 284 (1976)
26. Elbirli, B., Lindt, J. T., Gottgetreu, S. R., and Baba, S. M., "Mathematical Modeling of Melting of Polymers in a Single-Screw Extruder," *Polym. Eng. Sci.*, **24**, 988 (1984)

27. Lindt, J.T. and Elbirli, B., "Effect of the Cross-Channel Flow on the Melting Performance of a Single-Screw Extruder," *Polym. Eng. Sci.*, **25**, 412 (1985)

28. Lindt, J.T., "Mathematical Modeling of Melting of Polymers in a Single-Screw Extruder: A Critical Review," *Polym. Eng. Sci.*, **25**, 585 (1985)

29. Lindt, J.T., "Pressure Development in the Melting Zone of a Single-Screw Extruder," *Polym. Eng. Sci.*, **21**, 1162 (1981)

30. Altinkaynak, A., Gupta, M., Spalding, M.A., and Crabtree, S.L., "Melting in a Single-Screw Extruder: Experiments and 3D Finite Element Simulations," *Int. Polym. Process.*, **26**, 182 (2011)

31. Tang, Z., "Experimental Study and Modeling of Melting in a Single-Screw Extruder," M.S. Thesis, Clarkson University, Chemical Engineering Department, Potsdam, NY (1999)

32. Campbell, G.A., Tang, Z., Wang, C., and Bullwinkel, M., "Some New Observations Relative to Melting in Single-Screw Extruders," *SPE ANTEC Tech. Papers*, **49**, 213 (2003)

33. Campbell, G.A. and Tang, Z., "Solid Bed Melting in Single Screw Extruders, An Alternative First Order Mechanism," *SPE ANTEC Tech. Papers*, **50**, 162 (2004)

34. Campbell, G.A., Spalding, M.A., and Tang, Z., "An Alternative Analysis of Single-Screw Melting," *SPE ANTEC Tech. Papers*, **55**, 147 (2009)

35. Campbell, G.A. and Spalding, M.A., "Numerical Analysis of the Melting Process for Barrier-Flighted Single-Screw Extruders Using Screw Rotation Physics," *SPE ANTEC Tech. Papers*, **56**, 418 (2010)

36. Campbell, G.A. "One-Dimensional Melting in Single-Screw Extruders," *SPE ANTEC Tech. Papers*, **57**, 1367 (2011)

37. Somers, S.A., Spalding, M.A., Dooley, J., and Hyun, K.S., "An Experimental Study of the Flows in an Energy Transfer Screw," *SPE ANTEC Tech. Papers*, **48**, 307 (2002)

38. Campbell, G.A. and Dontula, N., "Solids Transport in Extruders," *Int. Polym. Process.*, **10**, 30 (1995)

39. Campbell, G.A., Spalding, M.A., and Carlson, F., "Prediction of Screw Temperature Rise in Single Screw-Pump Extruders," *SPE ANTEC Tech. Papers*, **54**, 267 (2008)

40. Campbell, G.A., Sweeney, P.A., and Felton J.N., "Experimental Investigation of the Drag Flow Assumption in Extruder Analysis," *Polym. Eng. Sci.*, **32**, 1765 (1992)

41. Campbell, G.A., Sweeney, P.A., and Felton, J.N., "Analysis of an Alternative Extruder Screw Pump Design," *Int. Polym. Process.*, **7**, 320 (1992)

42. Campbell, G.A., Sweeney, P.A., Dontula, N., and Wang, Ch., "Frame Indifference: Fluid Flow in Single-Screw Pumps and Extruders," *Int. Polym. Process.*, **11**, 199 (1996)

43. Campbell, G.A., Cheng, H., Wang, C., Bullwinkel, M., and te-Riele, M.A., "Temperature Rise in a Single Screw Pump," *SPE ANTEC Tech. Papers*, **47**, 152 (2001)

44. Campbell, G.A., Wang, C., Cheng, H., Bullwinkel, M., te-Riele, M.A., "Investigation of Flow Rate and Viscous Dissipation in a Single Screw Pump-Extruder," *Int. Polym. Process.*, **16**, 323 (2001)

45. Hong, B.K., Kim, H.S., and Chung, C.I., "Dependence of Melting Behavior on Melt Index," *SPE ANTEC Tech. Papers*, **47**, 272 (2001)

46. Hogan, T.A., Spalding, M.A., Cho, K.S., and Chung, C.I., "The Melting Characteristics of Polycarbonate Resins," *SPE ANTEC Tech. Papers*, **48**, 384 (2002)

47. Altinkaynak, A., Gupta, M., Spalding, M.A., and Crabtree, S.L., "Shear Stress and Melting Flux for Polycarbonate Resins," *SPE ANTEC Tech. Papers*, **54**, 1405 (2008)

48. Maillefer, C., Swiss Patent 363,149 (1962)

49. Lacher, F.K., "Extruder," US Patent 3,271,819 (1966)

50. Barr, R.A., "Extruder, or Extruder-Like Melting Apparatus," US Patent 3,698,541 (1972)

51. Spalding, M.A. and Hyun, K.S., "Troubleshooting Mixing Problems in Single-Screw Extruders," *SPE ANTEC Tech. Papers*, **49**, 229 (2003)

52. Hyun, K.S., Spalding, M.A., and Powers, J.R., "Elimination of a Restriction at the Entrance of Barrier Flighted Extruder Screw Sections," *SPE ANTEC Tech. Papers*, **41**, 293 (1995)

53. Spalding, M.A., "Metering Channel Flows and Troubleshooting Single-Screw Extruders," *SPE ANTEC Tech. Papers*, **50**, 329 (2004)

54. Elbirli, B., Lindt, J.T., Gottgetreu, S.R., and Baba, S.M., "Mathematical Modeling of Melting of Polymers in Barrier-Screw Extruders," *Polym. Eng. Sci.*, **23**, 86 (1983)

55. Mount, E.M., III and Chung, C.I., "Melting Behavior of Solid Polymers on a Metal Surface at Processing Conditions," *Polym. Eng. Sci.*, **18**, 711 (1978)

56. Chung, C.I., "Extrusion of Polymers, Theory and Practice," 2nd ed., Hanser, Munich (2011)

57. Lee, B.K., Gogos, C.G., Sebastian, D.H., and Avgousti, M., "Experimental Investigation and Modeling of the Top Down Melting Phenomenon in Single-Screw Extrusion," Proceedings of the Polymer Processing Society 13th Annual Meeting, Abstract 1-N, PPS-13, New York City, (1997)

58. Dekker, J., "Improved Screw Design for the Extrusion of Polypropylene", *Kunststoffe*, **66**, 5 (1976)

59. Kacir, L. and Tadmor, Z., "Solids Conveying in Screw Extruders Part III: The Delay Zone," *Polym. Eng. Sci.*, **12**, 387 (1972)

60. Szeri, A.Z., "Fluid Film Lubrication, Theory & Design," Cambridge University Press, New York (1998)

61. Spalding, M.A., Hyun, K.S., Jenkins, S.R., and Kirkpatrick, D.E., "Coefficients of Dynamic Friction and the Mechanical Melting Mechanism for Vinylidene Copolymers," *Polym. Eng. Sci.*, **35**, 1907 (1995)

62. Zhu, F., Wong, A.C.-Y., Liu, R., and Liu, T., "Breakup of Solid Bed in Melting Zone of Single Screw Extruder, Part 2, Strength of Solid Bed and Experimental Results," *Plast., Rubber Compos. Process. Appl.*, **26**, 8, 343 (1997)

63. Jin, Z., Gao, F., and Zhu, F., "An Experimental Study of Solid-Bed Breakup in Plasticization of a Reciprocating-Screw Injection Molding," *Polym. Eng. Sci.*, **44**, 1313 (2004)

64. Jenkins, S.R., Powers, J.R., Hyun, K.S., and Naumovitz, J.A., "High-Rate Extrusion of PVDC Polymers," *J. Plast. Film Sheeting*, **6**, 90 (1990)

65. Bozzelli, J., "Screws and Pellets: One Size Does Not Fit All," *Plast. Technol.*, **57**, 6, 21 (2011)

7 Fluid Flow in Metering Channels

This chapter describes the processes that occur in flighted screw sections that are completely filled with molten polymer. These processes include pumping, pressurization, and energy dissipation. For simple plasticating machines, the metering section of the screw is typically the only section completely filled with molten resin; the other section channels will have high levels of solid resin. In Chapter 1, the concept that the metering section of the screw must be the rate controlling step for a properly operating smooth bore extruder was introduced. If the metering section is not rate controlling, then the extruder will operate at reduced rates and has the potential to flow surge and cause resin degradation products to occur in the discharge. If the extruder has multiple stages, then the first-stage metering section must control the rate. If a stage downstream from the first-stage metering section controls rate, then flow of resin into a vent opening will occur. Sections upstream of the metering section can, however, influence the rate, such as affecting the pressure at the entry to the metering section, as presented in Chapter 1. This entry pressure will affect the pressure gradient and the pressure-driven flow component in the channel. Moreover, the metering channel depth will highly influence the temperature of the extrudate. Since the rate and discharge temperature of an extruder are critical to the economic operation of a process, a full understanding of the physical operation of the metering channel is very important. This chapter will explore the characteristics of the single-screw extruder with the goal of elucidating the mechanism that moves the fluid through the metering section as the screw is rotated. Next, the mathematical theory and design equations for single-screw fluid flow and energy dissipation will be developed.

7.1 Introduction to the Reference Frame

In order to understand the mechanisms for fluid flow in the metering channel, it is important to understand the reference frames used in the mathematical analysis of the section. As discussed in many chapters leading up to this point, the extruder is a cylindrical structure, as shown in Figure 7.1(a), that has a helical channel formed by the screw core (or root) and the helical flights. The geometry and mathematics associated with helical structures are extremely difficult. In order to avoid these difficulties, the helical channel is unwrapped from the screw to produce a trapezoidal channel in the Cartesian coordinate system, as shown in Figure 7.1(b). This is mathematically a much easier problem to address. There are theoretical difficulties, however, that have been associated with this transformation, because it results in a nonorthogonal coordinate transformation. This nonorthogonal coordinate transformation has been known for at least 40 years [1], and it has to do with the fact that the transformation is nonorthogonal from a theoretical mechanics of a continuous medium point of view. To quote the abstract from this paper: "This note presents a coordinate frame representation for the conservation laws governing fluid motion in a helical configuration as suggested by a single-screw extruder with deep flights and by the Kenics static mixer. This example demonstrates the care necessary in dealing with nonorthogonal coordinate systems." This difficulty has also been referenced regarding helical pipe flow [2]. Because of these difficulties the results of frame changes in the nonorthogonal transformed coordinates must be carefully evaluated. Here the resulting flow and energy characteristics of the extruder will be carefully examined. Most importantly the predictions of both flow rate and energy dissipation related to the polymer temperature increase will be compared with experimental data. In this development of the design equations for extruder rate and energy dissipation, the same unwrapping of the screw channel that has been used since the 1920s will be used in this analysis. It is this unwrapping of the channel that has resulted in the nonorthogonal transformation from helical to Cartesian coordinates. Moreover, the solution of the fluid mechanical equations is also very difficult because the screw channel is a moving boundary problem. Furthermore, the channel is not rectangular because of the radius of the screw; that is, the width of the channel is wider at the surface near the barrel than at the screw root. The channel is usually converted to a true rectangle for fluid flow calculations by considering the channel width to be the width in the unwrapped z direction halfway from the screw root to the barrel. The helical length (z direction) is also longer at the barrel surface than at the screw root, as shown in Figure 7.1(b). In the theory and design equation development that follows, two other important transformations are made that are different from the classical literature. First, the velocities at the flights are considered to be a function of the height in the y direction; this converts the velocities in the Cartesian coordinate system into a representation of the actual physics in the cylindrical geometry. That is, the velocity increases as the radius increases for the flight as shown by the arrows labeled $V_z(y)$ in Figure 7.1(b). This results in a linear increase in the boundary-condition velocity, as shown in Figure 7.2(b). Second, the barrel is stationary. To deal with the moving channel the reference frame is changed to a moving frame of reference attached to the screw surface, as shown in Figure 7.2(b). These concepts will be further developed later in the chapter. Therefore, in this analysis the barrel is considered to have zero velocity just as in an actual extruder. With the geometry and screw boundary conditions established for model development, it is desirable to first determine how each element of the extruder contributes to the screw pumping rate so appropriate assumptions can be made during the development of the model.

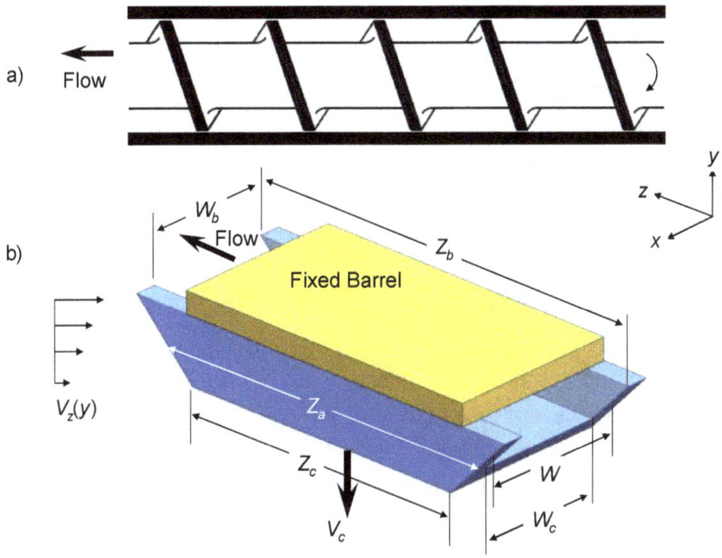

Figure 7.1 Extruder screw transformation: a) schematic of a screw inside a barrel, and b) an unwrapped channel showing the transformation with the helical length Z and the channel width W. The screw is shown moving and the barrel is stationary

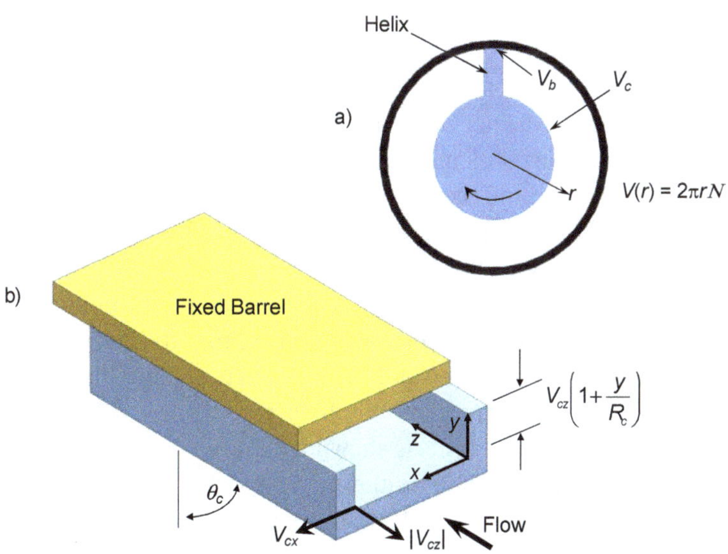

Figure 7.2 Transforming the Cartesian reference frame: a) cylindrical cross section of the screw and barrel with flow out of the surface of the page, and b) the unwound rectangular channel with a stationary barrel and the Cartesian coordinate frame positioned on the screw. V_{cz} is the velocity of the screw core in the z direction and it is negative

The next section of this chapter provides a wide range of information that elucidates the mechanism of fluid flow in channels such as metering channels. Included in the chapter are the special extrusion devices designed and built by researchers in Clarkson's Polymer Processing Laboratory. These extrusion devices allowed the visualization of dyes in fluid flows under extrusion using video photography while the screw was rotated inside of glass or plastic barrels. From these devices a mechanistic understanding was developed of how the different elements of the extruder contribute to the rate and energy dissipation in the device. A discussion of the single-screw modeling literature then follows the laboratory observation section. Using the laboratory data and the literature, a solution to the Newtonian fluid flow and dissipation was developed by modeling the screw rotation physics found in the experimental observations. With modern processes operating at higher and higher rates, designers have increased the channel depth of the metering section, causing the channels to become "boxy," that is, channels with H/Ws greater than 0.1. These boxy channels have created additional challenges for the calculation of rate and viscous energy dissipation. The polymer temperature increase for boxy pumping channels and the shear-thinning nature of many commercial polymers in the pumping section of the extruder is presented in this chapter. Both screw-rotation-based theoretical analysis and data will be presented to compare theory with the numerical solutions.

7.2 Laboratory Observations

Laboratory extrusion observations were performed using free helix extruders equipped with transparent barrels. Several free helix extruders were built. For these devices, the helix was not an integral component of the screw. Instead, the helix, screw core, and barrel elements could be rotated independently. The flows could be observed visually and the rate could be measured. The initial experiments using the first device were reported in 1992 [3, 4]. In these first experiments, it was demonstrated that the primary element in the extruder that pumps the fluid toward the die is the screw helix. The free helix extruder used in the initial investigation had some limitations when operated with a pressure at the discharge, so a new laboratory device was designed and constructed to explore the mechanisms of single-screw extruder fluid transport. A photograph of one of these free helix extruders used in these experiments is shown in Figure 7.3. Screw dimensions can be found elsewhere [5]. The configuration in Figure 7.3 was constructed with a very high surface area reservoir in order to conduct some of the mechanism studies discussed below. The glass barrel allowed video analysis to monitor the screw element rotation, the fluid movement, and the movement of dyes when investigating residence-time distributions.

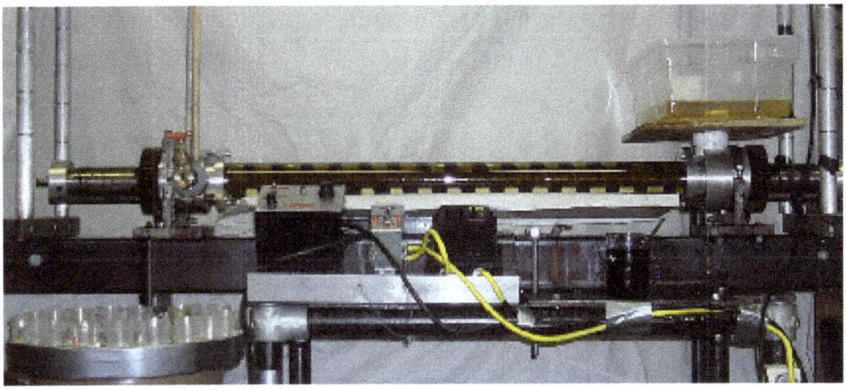

Figure 7.3 Photograph of the 49.7 mm diameter free helix extruder developed at Clarkson University

A version of this free helix device was also constructed that had a recycle system such that there was essentially no pressure change from the exit to the entrance of the extruder, as shown in Figure 7.4. That device had a barrel diameter of 58 mm and it was used to measure the effect of the different parts of the screw on the extruder performance where pressure gradients could be imposed [5]. The devices shown in Figure 7.3 and Figure 7.4 could be operated as a conventional extruder if both core and helix elements were driven at the same angular velocity with the barrel stationary. In another mode, only the helix was rotated in the same direction as the screw had been while the core and barrel were stationary. And finally the screw core was rotated in the same or different direction as the screw and the helix and barrel were stationary.

To demonstrate the operational functions for the extruder parts, a series of experiments were performed using a free helix extruder similar to that shown in Figure 7.4. The results of this set of experiments are presented in Figure 7.5. The helix when rotated alone had the highest rate for this deep channel screw, as indicated by the diamond data in Figure 7.5. When the core was rotated in the same direction and angular velocity and the helix was stationary the rate was negative. That is, the fluid was dragged back toward the screw inlet by the action of the screw core. The line in Figure 7.5 with squares represents this core rotation flow data. The device was then operated as a traditional extruder; that is, both helix and core were rotated in the same direction and angular velocity. The data collected for screw rotation is represented by the data set on the graph in Figure 7.5 by triangles.

All of the data in Figure 7.5 were analyzed using linear regression. The summation of the helix and core-regressed flow rates are plotted in Figure 7.5 as the red dotted line. The experimental superposition for the flows induced by the screw elements essentially overlaid the regression line for the screw configuration rate.

Thus, it was concluded that the helix is the pump in the single-screw extruder, and core rotation reduces the flow by dragging the fluid back toward the extruder inlet.

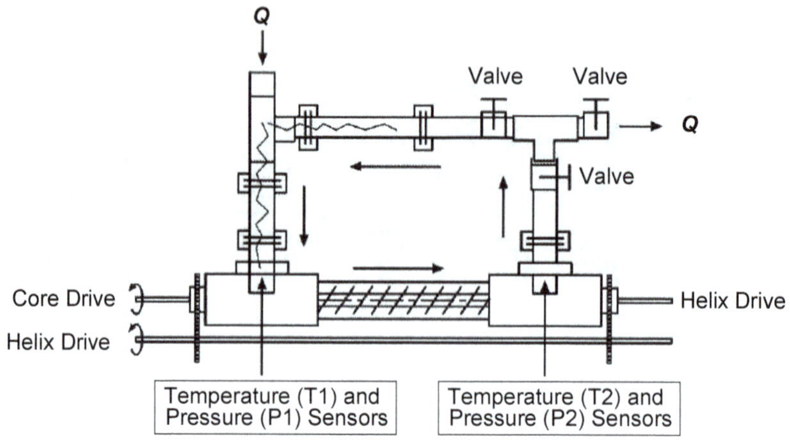

Figure 7.4 Schematic of the modified free helix extruder with a barrel diameter of 58 mm and a flight clearance of 0.1 mm [5]

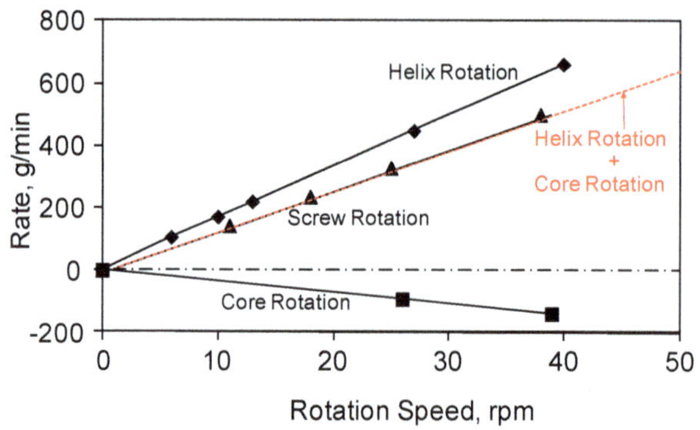

Figure 7.5 Screw element's contribution to extruder rate [5]. Data were measured using the free helix with a diameter of 52 mm using a square-pitched screw with a channel depth of 11.1 mm

The glass barrel extrusion device as configured in Figure 7.3 was utilized in a second set of experiments to demonstrate the effect of screw element rotation on the device's ability to pump against a constant discharge (or head) pressure. The essentially constant discharge pressure desired for this series of experiments was the result of the very high surface-to-volume ratio of the reservoir at the discharge of the extruder. With moderate changes in flow there would be very little change in fluid height; thus the screw discharge pressure was essentially constant. The discharge pressure in this experiment is the fluid head height maintained in the reservoir. This was necessary to assess visually the screw fluid-pumping ability due to the rotation of the different screw elements. The feed reservoir to the extruder was emptied and then the extruder was operated. In these experiments, the elements that were rotated were at identical angular speeds. After a short transient time of pumping the fluid toward the reservoir, the fluid maintained the constant fill (fluid-air interface) positions in the screw channel as shown in Figure 7.6. The feed end of the extruder was open to the atmosphere. The distance from the screw position directly under the reservoir to the fill points downstream are indicated by the arrows on the screw in Figure 7.6, and it represents the length of the screw where the pressure-driven flow (Q_p) and the element-driven flow (Q_d) are exactly equal. Thus, the net flow was zero for this low-viscosity fluid. If a screw element is more efficient at pumping, then the filled length of the screw with the fluid will be shorter. In all of the photos in Figure 7.6, the arrows indicate the position of the fluid-air interface relative to the reservoir's application of head pressure at the end of the screw.

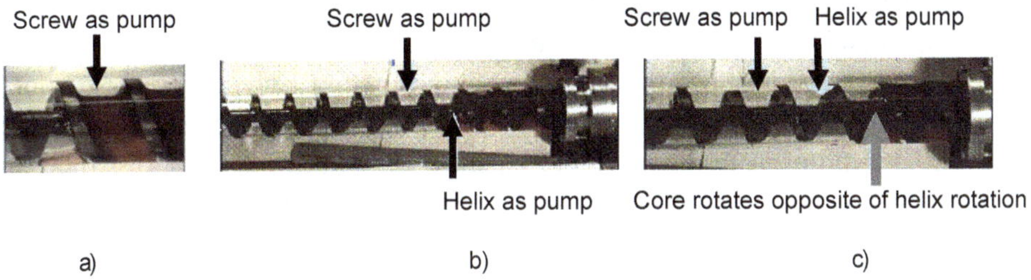

Figure 7.6 Effect of screw elements to pump against a constant head pressure [5]: a) the screw as the pump with the fill position 5 diameters from the exit, b) the helix only as the pump with the fill position 4 diameters from the exit, and c) the core rotating in the opposite direction (forwarding) and the helix moving in the standard operation with the fill position at 3 diameters from the exit

As shown in Figure 7.6(a), the active length of the screw to balance the flow driven by screw rotation and the pressure flow was about five turns. Next, core rotation was stopped such that only the helix was rotating. After steady state was achieved, the air-fluid interface moved downstream such that only about 4 diameters of screw length were needed to develop the constant pressure of the head, as shown by the arrows in Figure 7.6(b), a pumping efficiency of about 20 percent higher than screw rotation. The position of the fluid-air interface moving closer to the reservoir demonstrated again that the helix is a better pump than the screw, and thus less screw length is needed to offset the pressure-driven flow from the reservoir. Thus, it was visually demonstrated, consistent with the deduction from Figure 7.5, that the screw core rotation reduced the ability of the screw to pump the fluid toward the extruder discharge. A third experiment was conducted where the screw core was rotated at the same angular velocity as the helix but in the opposite direction. The steady state position of the air-fluid interface moved still closer to the reservoir, as shown in Figure 7.6(c). The decrease in screw length for this experiment is essentially the same as the increase in length for the screw rotation relative to the helix-driven flow. When the screw core was rotated in the direction opposite of the helix rotation, the device was a more effective pump because the helix and core were now both pumping fluid against the pressure-driven flow from the reservoir. This experiment visually confirmed the observations as to the role of the different screw elements previously presented in Figure 7.5.

In the following sections this mechanistic understanding of the pumping characteristics of the single-screw extruder will be used to develop a mathematical set of operational equations based on these observations. The mathematical analysis should reflect the mechanisms that occur in the extruder. It will be shown that the flight-driven flow (or rotational flow), Q_d, is numerically identical to the historic barrel drag flow, Q_d, when the channel depth is zero, but composed of different mechanistic components. Unlike the traditional analysis that does not predict flows for deep channels, screw rotation analysis is equally good at prediction of flow for both shallow and deep channel screws. Also, to predict the flow in a die at the extruder exit, it is important to be able to predict accurately the discharge temperature from the pumping section due to energy dissipation. The energy dissipation, both experimentally observed and predicted from the screw rotation analysis, is substantially different from that reported in the literature using barrel rotation physics. Before the detailed screw-rotation-based model development is undertaken, a literature review will be presented because the concepts developed in this book have been developed through a fundamental understanding of the contributions from the authors who developed the classical extrusion literature.

7.3 Literature Survey

The metering section of the extruder has been analyzed and investigated since the 1920s. The traditional extrusion model was first proposed by Rowell and Finlayson [6, 7] in 1922. In the 1950s, Carley, Mallouk, and McKelvey [8] developed a model based on the stipulation of barrel rotation for a rectangular channel produced by unwinding the screw channel. It was assumed that the equivalent flow rate will be obtained if the barrel is rotated in the direction opposite to that of the screw as long as the linear velocity of the unwound barrel is numerically equal to the linear velocity of the screw root or core. The flow in the extruder was then represented by two components. The x-direction flow was across the channel and produced no net down-channel contribution to the flow. The z-direction flow was down the channel and represents the entire extruder rate. This was summarized and the analysis was expanded by Tadmor and Klein [9] in the 1970s. In these historic developments, it was assumed that the flow across the channel makes no contribution to the rate, and the flights (helix) make no contribution to the screw's pumping ability. The velocity profile across the channel was originally proposed by Mohr, Saxton, and Jepson [10]. The flow down the channel was considered to be the combination of the pure drag flow and pure pressure-flow components. The pure pressure flow for an isothermal Newtonian fluid in a rectangular channel was first solved by Boussinesq [11]. The pure drag flow and pressure flow for an isothermal Newtonian fluid was first solved by Rowell and Finlayson [6, 7]; they also proposed shape factors for drag flow and pressure flow to account for the influence of the flights [6, 7]. McKelvey derived an analytical model for Newtonian fluids under isothermal and adiabatic extrusion [12, 13].

All the previous analyses were based on the assumption of an incompressible fully developed Newtonian fluid flow in a rectangular channel. Also, inertial and gravitational effects were assumed to be negligible. A Newtonian fluid gives rise to linear differential equations. Thus, the principle of superposition was applied to separate the down-channel flow into pure drag flow and pure pressure flow. For a non-Newtonian fluid, the pressure and drag flows are coupled due to the shear rate dependence of the viscosity. Therefore, the flow velocities are more complicated than merely superimposing the pressure and drag velocities. Weeks and Allen [14] proposed an apparent viscosity based on the non-Newtonian behavior to predict the rate for isothermal non-Newtonian flow. Middleman [15] derived shape factors for pressure flow and drag flow based on isothermal power law fluids in rectangular ducts. It should be noted that errors were introduced because the pressure flow, the drag flow, and the transverse flow are all coupled by the shear-rate dependence of the viscosity.

The first analysis of the effect of flight clearance on flow rates was made by Rowell and Finlayson [6]. They assumed that the clearance between the flight tip and the barrel can be approximated by two parallel plates. The leakage flow was considered a pressure flow in the two plates. Therefore the net flow in the extruder was considered to be a combination of the drag flow, the pressure flow, and the leakage flow through the flight clearance. Later Gore and McKelvey [16] added the effect of the clearance on drag flow. A more accurate analysis to account for the effect of clearance was derived by Mohr and Mallouk [17]. In their analysis, both the pressure gradient in the cross channel and the effect of the clearance on the drag flow were considered. Pittman and Rashid [18] studied the nonisothermal non-Newtonian model with leakage flow. Although most of the authors included leakage flow in their derivations, Agassant et al. [19] indicated that leakage flow is negligible with normal clearances and relatively large pressure gradients, but can be significant for highly worn screws. Barr and Chung [20] examined the effects of varying clearance on extruder performance and also concluded that at least for small clearances there is essentially no significant flow over the flights.

Due to the complicated helical screw geometry and the assumption that the down-channel drag flow was a result of matching the screw core velocity to the modeled barrel velocity, the literature assumption that the flow occurs in a rectangular channel is reasonable only if the ratio of channel depth to width is small, that is, a channel with a small aspect ratio (*H*/*W*). A schematic of the channel depth to the width perpendicular to the flight is shown in Figure 7.7. In many cases, the metering zone of a typical extruder satisfies this requirement. However, for some screws with deep channels, a significant error was thought to be introduced if the effect of curvature was neglected. To account for the effect of screw curvature in deep channel screws, Booy [21] derived numerical solutions for isothermal Newtonian fluids in deep curved channels. Booy's model consists of a screw channel bounded by an outer cylinder representing the barrel and by an inner cylinder representing the screw and two sides of a helical flight. The flow in a deep curved channel is separated into the pure tangential drag flow and the axial pressure flow between two stationary concentric cylinders [21]. Similar to the plate theory, two correction shape factors corresponding to the drag flow and pressure flow were also tabulated by Booy [21]. Booy's shape factors included the effect of channel curvature and helix angle. The analysis, however, neglected the effect of the flights. Booy's work was later improved by Pinto and Tadmor [22].

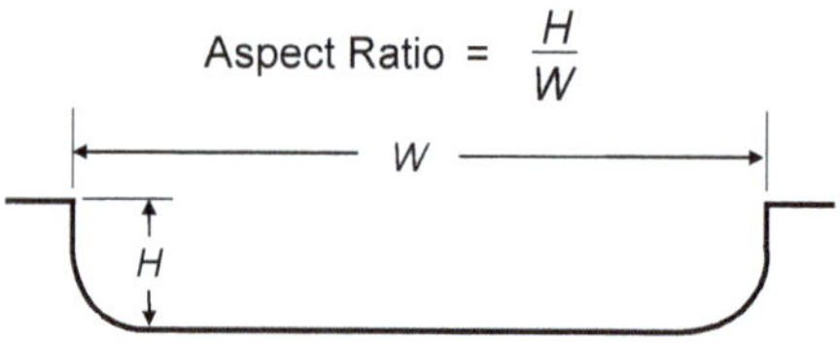

Figure 7.7 Schematic of a screw channel perpendicular to the flight edge showing the width of the channel and the depth of the channel

The first attempt to calculate the energy dissipation in the channel was made by Rowell and Finlayson [7]. Later Mohr and Mallouk [17] and Gore and McKelvey [16] included the previously neglected transverse flow into their derivations using an isothermal Newtonian model. Tadmor and Klein [9] added the influence of the flight clearance to the energy dissipation. Fenner [23] performed experiments using silicone polymers and demonstrated that the measured power was often closer to the calculated power for the screw channel alone than that including the viscous dissipation from the leakage flow through the flight clearance. Carley and Strub [24] introduced an energy shape factor to account for the effect of flights and the shape of the channel, which was neglected by the previous derivations. Booy [21] also derived numerical solutions for the energy dissipation in deep channels. Griffith [25] obtained viscous dissipation for power law fluids. Fenner [26, 27] also summarized energy dissipation in screw channels of finite width for both Newtonian and non-Newtonian isothermal flows. He highlighted the difficulties of treating the regions of the flow close to the clearance. All energy models described here evolved using flows from the barrel rotation models originally developed by Rowell and Finlayson [6, 7].

Extensive experiments were carried out to investigate the extrusion process. The first attempt to verify experimentally the flow rate equation for Newtonian fluids was made by Rowell and Finlayson [7]. The first experiments to verify the previously discussed velocity profiles for Newtonian fluids were performed by Eccher and Valentinotti [28]. Their extruder had a rotating barrel and contained a hollow screw. The existence of the backflow in the down channel was verified by their experiments. They also observed a *y*-direction velocity component on both sides of the channel in the opposing direction. Their results demonstrated the independence of the cross-channel flow, the tangential flow, and the down-channel flow. Mohr et al. [29, 30] performed experiments to measure the Newtonian and non-Newtonian flow patterns in extruders. In summary, the experimental evidence agreed with the theoretical analysis based on Newtonian fluids in a rectangular channel as long as the channel was shallow and the barrel was rotated. Although a great number of experiments were carried out to verify the flow in the extruder, few experimental data were found to verify the energy dissipation equations [9].

With the development of modern computation techniques, more and more numerical simulations occur in the literature to predict the velocity profiles, pressure distribution, and the temperature distribution inside the extruder. Rotem and Shinnar [31] obtained numerical solutions for one-dimensional isothermal power law fluid flows. Griffith [25], Zamodits and Pearson [32], and Fenner [26] derived numerical solutions for two-dimensional fully developed, nonisothermal, and non-Newtonian flow in an infinitely wide rectangular screw channel. Karwe and Jaluria [33] completed a numerical solution for non-Newtonian fluids in a curved channel. The characteristic curves of the screw and residence time distributions were obtained.

As far as heat transfer is considered, Fenner [27] made a detailed comparison of the thermally fully developed flow and thermally developing flow. He indicated that the thermally developed flow will not be achieved when heat conduction effects become significant [34]. Bruker et al. [35] experimentally verified that the thermally developing flow analysis provided a more accurate description of the flow in the extruder.

The finite element method (FEM) and finite difference method (FDM) techniques were employed independently or in combination to obtain numerical simulations of the flow and heat transfer in extruders. Dyer [36] obtained velocities, pressure differences, and flow rates in cylindrical polar coordinates using the FDM. Spalding et al. [37] used three-dimensional FEM to simulate the polymer melt flow in the screw channel and the flight lands. A comparative study of results from three-dimensional FEM and the two-dimensional FDM indicated that similar results occur if the ratio of the channel depth to the width (H/W) is smaller than 0.1 [38]. Sastrohartono et al. [39] simulated the three-dimensional flow in a curved channel using the FEM. Chang and Lin [40] utilized a solid element to approximate the screw channel. Hence the gap-wise pressure gradient and the cross-channel flow effect can be predicted. Matsuoka [41] applied a hybrid FEM/FDM to simulate the isothermal, non-Newtonian flow in the steady extrusion coating die process. The energy equation was solved by finite difference and the general Reynolds equation was solved by the finite element method. Fraser et al. [42] performed FEM analysis on a metering screw channel that had slots in the flights. The slots increased the mixing ability of the screw by permitting flow between adjacent channels.

When the numerical simulation is run for such cases as high pressure at the die or very low rate, the conventional marching schemes may result in "pressure back flow" (V_z is negative). Elbirli and Lindt [43] obtained a stable solution under an appreciable pressure backflow by transforming the coordinate from a Eulerian frame to a Lagrangian frame. In their solution, the heat transfer and residence time characteristics of the thermally developing melt flow in the extruder were coupled. Chiruvella et al. [44] employed two different approaches to solve pressure backflow problems. The first approach utilized the down-channel formulation and included the diffusion of energy in this direction. The parabolic energy equation was transformed into an elliptical equation. By doing so, the numerical difficulty that occurred using the marching schemes was eliminated. The second approach used the axial formulation. The governing equations of motion and energy were solved based on the axial and tangential coordinate frames of reference. The two approaches yielded similar results.

Because it is more complicated to solve the moving boundary problem for the rotation of the screw, the barrel rotation models described above have been extensively adopted and investigated. In practice the screw is rotated and not the barrel. The barrel rotation theory has several limitations when describing the real extrusion process, so correct interpretation of the calculated results based on barrel rotation becomes necessary. Most screw design practitioners, with substantial previous design experience, make major adjustments in design specifications to obtain effective correlations.

In order to investigate this mechanistic difference between screw and barrel rotation, Campbell, Sweeney, and Felton [3, 4] proposed a screw rotation theory, which completed the analysis of the extrusion process by converting the solution obtained from the transformed boundary conditions, really an analog description of the Lagrangian reference frame, back to the real laboratory reference frame of screw rotation. The previously discussed experiments demonstrated that the rate of the extruder is dominated by the x-direction velocity due to the helical motion of the screw flight rotation pushing the fluid from the entrance to the exit of the extruder. The z-direction core drag flow, which carries the fluid back to the inlet of the extruder, decreases the maximum potential rate by up to 50% [3] in a typical extruder. The x-direction flow and the z-direction drag flow based on screw rotation were first solved by Campbell et al. [45]. By transferring the obtained velocities into the axial formulation, the calculated rate for the screw rotation theory yields the same function for rate in both the transformed and laboratory frames. This is expected since fluid flow is reference-frame independent.

The Clarkson free helix experiments [3, 4, 45] were in agreement with the screw rotation theory. The free helix experiments provided information about the contribution of each of the elements to the fluid mechanics in the extrusion process. In the experiments, the helix and the core were separated and rotated independently, in contrast to the barrel rotation theory, which specifies the role of the helix as just constraining the fluid in the channel without contributing to the rate. The Clarkson experiments have indicated that the helix plays an important and dominant role in determining the rate. Hunt [46] showed experimentally that the "drag" flow calculated using the Lagrangian frame is identical to the down-channel components of the helix-driven flow and the core drag flow in the Eulerian frame. Experimental and theoretical comparisons for the flow and dissipation in single-screw pumps were published by Campbell et al. [5]. The analysis confirmed that screw rotation theory works equally well for shallow and deep channel screws for isothermal, Newtonian fluids. Based on this screw rotation theory, the energy dissipation originates from the combination of the core drag flow, the pressure flow, and the helical motion of the screw flights. The follow-up experimental study of barrel rotation and screw rotation then demonstrated that there is a difference in temperature profiles caused when rotating the barrel or the screw. The data were reported for seven screw-barrel cases, and the experimental barrel rotation fluid temperature increase was greater than that for screw rotation for all of the reported screw pump configurations [47]. These data will be presented in Section 7.8.

7.4 Development of Linearized Flow Analysis

In this development the traditional nomenclature of the laboratory (Eulerian) frame and the transformed (Lagrangian) frame will be utilized. Because they are being applied to a nonorthogonal coordinate transformation, the resulting design equations will be compared to determine where they differ from the expectation from an orthogonal axis transformation. The analysis begins here with a metering section channel segment schematically represented in Figure 7.8. The models developed follow the analysis by Campbell et al. [5, 45], which incorporates the effect of the helix (flight) on the down-channel flow, as shown in Figure 7.9. The helical channel in Figure 7.8 is unwound and is represented in a Cartesian coordinate system in Figure 7.9 in the same manner as in the classical literature. In the laboratory frame of reference in Figure 7.9, the motion of the screw channel under the infinitely wide fixed barrel is a moving boundary problem. The boundary of the screw channel is moving relative to the fixed laboratory coordinates where the observer is standing. With the positive direction of the coordinates indicated by the arrows, the x direction of the screw core velocity V_{cx} is positive, and the z

direction of the screw velocity V_{cz} is negative. These axis positive directions are arbitrary and were chosen to be consistent with the classical literature. By recognizing that the velocity at the flight is a function of the distance in the y direction from the screw root, the ultimate solution will be adaptable to both shallow and deep channel screws [5]. The solution will be based on the velocity at the screw core (or root) as the reference point for all velocities. The velocity up the flight is a linear function of dimension y and thus can be used in the Fourier analysis to obtain the positional velocities in the z direction and the shape factors for the down-channel flight-driven flow. As just described, this is a difficult problem to solve mathematically because it is a moving boundary problem. That is, the fluid boundaries are moving relative to the coordinate system. To overcome this mathematical difficulty, the frame of reference was changed from the laboratory reference frame shown in Figure 7.9 to a frame of reference attached to the moving channel shown by the frame in Figure 7.10.

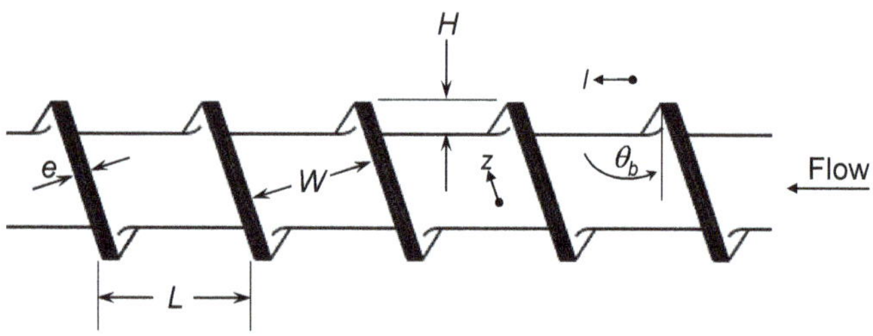

Figure 7.8 Screw pumping section with parameters labeled

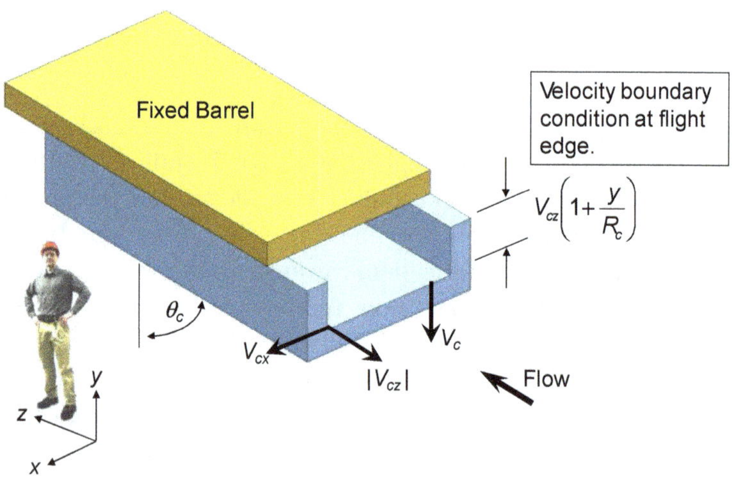

Figure 7.9 Laboratory (Eulerian) frame boundary condition for the analysis of extruder fluid flow. The observer is away from the channel. V_{cz} is the velocity of the screw core in the z direction and it is negative

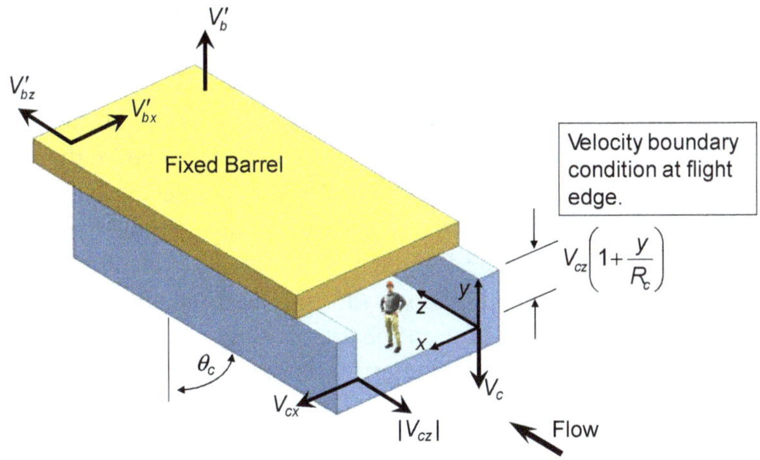

Figure 7.10 Transformed (Lagrangian) frame for the analysis of extruder fluid flow. Here the reference frame is positioned on the bottom of the screw channel. The observer on the frame would see the barrel move with the component velocities of V_{bx}' and V_{bz}'

Much of the literature assumes that the barrel rotates to obtain a solution. By assuming barrel rotation, a change in the problem physics is introduced. In this analysis the physics of the problem is not changed. This is an extremely important difference separating the classical literature analysis from the analysis here. It should be emphasized that the physics of the extruder flow has not been changed in the reference frame transformation used in this analysis. That is, the barrel is stationary and the screw channel moves in this analysis as it does in a real extruder. Only the reference frame has been changed in a manner as described by Malvern [48]. This is particularly clear in Malvern's example regarding the apparent velocities of two sailboats from the perspective of an observer on the beach and an observer on one of the boats. This is also similar to the discussion in Bird, Stewart, and Lightfoot (BSL) [49] regarding the development of different derivatives when analyzing flow and the concentration for a school of fish observed in a river [49]. In this BSL example, the discussion revolves around observations of a fish-filled steam from a bridge or from a boat moving with the average velocity of the stream. Similar to the transformation in frame for this example, in BSL's Chapter 3 for the bridge-to-boat transformation, the analysis here changes the frame so the observer is moved from the laboratory frame, a stationary position relative to the screw velocities, to a frame moving at the velocity of the screw. The observer has moved from the laboratory coordinate system to the coordinate system embedded on the screw, as shown in Figure 7.10. The screw and barrel velocities are the same as in Figure 7.9. However, the observer is now moving with the screw as indicated in Figure 7.10. The barrel velocities for the observer appear to have changed to the component velocities of V_{bx}' and V_{bz}' in Figure 7.10 because his frame of reference has changed and he is moving at the velocity of the screw channel.

The moving boundary problem shown in Figure 7.9 is now a fixed boundary problem to the observer in Figure 7.10. It was transformed mathematically because of the transformed reference frame. This type of fixed boundary problem can readily be solved using conventional mathematical techniques. The transformation can be made because the fluid mechanics of linear systems is frame indifferent. However, in order to capture the mechanisms in the original moving boundary problem, the solution obtained in the transformed frame must be converted back to the original laboratory frame. Because of frame indifference, the calculated rate will be the same in both frames. The mechanisms for the fluid motion, however, will be different. This is why those working in the area of solid mechanics always transform their solutions back to the laboratory frame in order to capture the actual deformation mechanism. As a historic note, Paul Sweeney, after completing four solid mechanics courses, was the first to recognize that the classical extruder analysis needed to be finished by transforming the solution from what is in effect the Lagrangian transformed frame of reference to the Eulerian frame (the laboratory frame). This occurred in the late 1980s when he was a graduate student in the Clarkson Polymer Processing Laboratory. Also, both experimentally and theoretically the temperature increase due to viscous dissipation will be different in the two frames, because in general dissipation is not frame indifferent. Viscous dissipation will be presented in detail in Section 7.8.

The solution of the momentum equations for high-viscosity materials may often be obtained without the inertial terms. The resulting equations are normally described as a linear model for the fluid system. The Cartesian coordinate-based Equation 7.1 to Equation 7.3 are mathematically linear as long as the viscosity is constant; that is, the viscosity is not a function of either shear rate or temperature. The analysis developed in this chapter for the temperature is a first-order nonlinear approximation because the viscosity is allowed to be a function of temperature and thus is not constant. This approximation relies on an assumption that the viscosity changes at a slow rate, and thus the velocity streamlines are not substantially altered locally due to this slow change in fluid temperature. Therefore, the linearity and independence of the equations are assumed to hold even under this assumption. Equation 7.1 describes the cross-channel velocity and pressure gradient normal to the flight in the x direction, as depicted in Figure 7.10. This equation will be simplified in the current analysis since the velocity in the y direction for the cross-channel flow is assumed to be zero. This last assumption forces Equation 7.2 to have a value of zero for the pressure gradient in the y direction. All of the terms in Equation 7.3 will be retained, and as a result the solution obtained will model the velocity in the z direction as a function of both x and y. The details of the analysis are presented in Appendix A7.

$$\frac{\partial P}{\partial x} = \mu \left[\frac{\partial^2 V_x}{\partial x^2} + \frac{\partial^2 V_x}{\partial y^2} \right] \tag{7.1}$$

$$\frac{\partial P}{\partial y} = \mu\left[\frac{\partial^2 V_y}{\partial x^2} + \frac{\partial^2 V_y}{\partial y^2}\right] \tag{7.2}$$

$$\frac{\partial P}{\partial z} = \mu\left[\frac{\partial^2 V_z}{\partial x^2} + \frac{\partial^2 V_z}{\partial y^2}\right] \tag{7.3}$$

where P is the pressure in the channel, μ is the Newtonian viscosity, and V_x, V_y, and V_z are the velocities in the x, y, and z directions, respectively. The Cartesian coordinate system used here has x in the cross-channel direction, y in the channel-depth direction, and z in the down-channel (downstream) direction. A lumped average temperature assumption is made for estimating the viscosity in the channel.

The physical transformation from the real wrapped helix on the screw to the proposed deep-channel approximation for screw rotation is illustrated in Figure 7.10 and accounts for the velocity increase at the flights as y/R_c increases, where R_c is the radius of the screw core. It is acknowledged that the screw is distorted when converted into rectangular coordinates such that it would be longer at the barrel surface than the screw core surface, and the width increases as the radius increases. Thus, the screw was represented as a channel of depth H and with the width W and length Z as functions of the local radius. This is discussed in detail by Campbell et al. [5]. Here it is proposed that the screw be represented by an average length in the z direction and by an average width in the x direction for fluid flow simulation. Taking advantage of the principle of superposition and considering that the velocity of the flights increases linearly with the channel depth [45], the physical model based on screw rotation can be now developed. The z-direction velocity of Figure 7.10 can be represented by drag and pressure terms as in Equation 7.16. The assumption made was that the plate on the top of the channel is stationary (the barrel does not move) while the channel is moving with an x and z velocity of the screw. The velocities on the channel surfaces are considered to be functions of the local radius. The governing equations of motion are the same as Equation 7.1 and Equation 7.3, but with different boundary conditions at the screw surfaces and barrel from the classical literature in order to incorporate the influence of the flight helix on the velocities and thus retain the screw moving boundary conditions of the actual screw (laboratory reference frame):

$$x\text{-direction flow} \quad y = 0 \quad V_x = V_{cx} \quad \text{screw root} \tag{7.4}$$

$$y = H \quad V_x = 0 \quad \text{barrel} \tag{7.5}$$

$$x = 0 \quad V_x = V_{cx}\left(1 + \frac{y}{R_c}\right) \quad \text{pushing flight} \tag{7.6}$$

$$x = W \quad V_x = V_{cx}\left(1+\frac{y}{R_c}\right) \quad \text{trailing flight} \tag{7.7}$$

$$z\text{-direction flow} \quad y = 0 \quad V_z = V_{cz} \quad \text{screw root} \tag{7.8}$$

$$y = H \quad V_z = 0 \quad \text{barrel} \tag{7.9}$$

$$x = 0 \quad V_z = V_{cz}\left(1+\frac{y}{R_c}\right) \quad \text{pushing flight} \tag{7.10}$$

$$x = W \quad V_z = V_{cz}\left(1+\frac{y}{R_c}\right) \quad \text{trailing flight} \tag{7.11}$$

$$\text{where} \quad V_{cx} = V_c \sin\theta_c = \pi N D_c \sin\theta_c \tag{7.12}$$

$$V_{cz} = -V_c \cos\theta_c = -\pi N D_c \cos\theta_c \tag{7.13}$$

$$V_{bx} = V_b \sin\theta_b = \pi N D_b \sin\theta_b \tag{7.14}$$

$$V_{bz} = -V_b \cos\theta_b = -\pi N D_b \cos\theta_b \tag{7.15}$$

$$\theta_c = \arctan\left(\frac{L}{\pi D_c}\right) \tag{7.16}$$

$$\theta_b = \arctan\left(\frac{L}{\pi D_b}\right) \tag{7.17}$$

where V_c is the velocity of the core, V_{cx} is the velocity of the screw core in the x direction, V_{cz} is the velocity of the screw core in the z direction, N is the screw speed in revolutions per second, D_c is the diameter of the screw core, L is the lead length, and θ_c is the helix angle at the screw core. The velocity V_{cz} is in the negative z direction. The flight clearance is ignored for this analysis. The principle of superposition is used to break the z-direction flow of Equation 7.3 into a combination of pure pressure flow $V_{pz}(x, y)$ and pure screw rotational (or flight-driven) flow $V_{dz}(x, y)$ as shown in Equation 7.18.

$$V_z = V_{pz}(x,y) + V_{dz}(x,y) \tag{7.18}$$

Substituting Equation 7.18 into Equation 7.3 and solving Equation 7.1 and Equation 7.3 for V_x, V_z, and V_p, the solution for the transformed boundary condition problem is obtained, and the equations are shown by Equation 7.21, Equation 7.23, and Equation 7.26. These equations physically represent the flow due to rotation and pressure in the transformed frame of reference in Figure 7.10. Equation 7.21 is the velocity equation for the x-direction recirculatory cross-channel flow for the observer attached to the screw, and Equation 7.23 is the apparent velocity in the z direction for the observer attached to the moving screw. In order to obtain the laboratory design equations, the appropriate boundary conditions for both the x and z directions are added back into the solutions. The velocity relationship for the frame change is as follows:

$$V_{xE} = V_{xL} + V_{cx}\left(1 + \frac{y}{R_c}\right) \tag{7.19}$$

$$V_{dzE} = V_{dzL} - \left|V_{cz}\right|\left(1 + \frac{y}{R_c}\right) \tag{7.20}$$

where V_{xL} and V_{dzL} are the cross-channel flow and the down-channel flow due to screw rotation in the Lagrangian reference frame, respectively, and V_{xE} and V_{dzE} are the cross-channel flow and the down-channel flow due to screw rotation in the Eulerian (laboratory) reference frame. The laboratory frame velocity equations obtained using Equation 7.21 and Equation 7.23 are Equation 7.22 and Equation 7.24 for the x and z components, respectively. As stated earlier, due to the assumption that the pressure gradient in the y direction is zero, no velocity for this direction is evaluated. As previously discussed, V_{cz} (and V_{bz}) are negative terms, and thus the velocity is in the negative z direction. The absolute value of these terms will be used in many of the equations to follow to ensure that the negative sign is used properly.

Equation 7.22 and Equation 7.24 represent the velocities due to screw rotation for the observer in Figure 7.9, which corresponds to the laboratory observation. Equation 7.25 is equivalent to Equation 7.24 for a solution that does not incorporate the effect of channel width on the z-direction velocity. For a wide channel it is the z velocity expected at the center of the channel where $x = W/2$ and is generally considered to hold across the whole channel. The laboratory and transformed velocities will predict very different shear rates in the channel, as will be shown in the section below relating to energy dissipation and temperature estimation. Finally, it is emphasized that as a consequence of this simplified screw rotation theory, the rotation-induced flow in the channel is reduced to two components: x-direction flow, which pushes the fluid toward the outlet, and z-direction flow, which tends to carry the fluid back to the inlet. Equation 7.26 and Equation 7.27 are the velocities for pressure-driven flow and are only a function of the screw geometry, viscosity, and pressure gradient.

Cross-channel velocity V_{xL} in the transformed (Lagrangian) frame:

$$V_{xL} = V_{cx}\left(1+\frac{H}{R_c}\right)\frac{y}{H}\left(2-3\frac{y}{H}\right) \tag{7.21}$$

Equation 7.21 is the literature expression for motion in the x direction for barrel rotation physics. The boundary conditions here are $V_{xL} = 0$ at $y = 0$ (screw root) and $V_{xL} = -V_{cx}$ at $y = H$ (flight tip). Cross-channel velocity V_{xE} in the laboratory (Eulerian) frame is obtained by adding on the x component of the moving frame as provided by Equation 7.19 as follows:

$$V_{xE} = V_{cx}\left(1+\frac{H}{R_c}\right)\frac{y}{H}\left(2-3\frac{y}{H}\right)+V_{cx}\left(1+\frac{y}{R_c}\right) \tag{7.22}$$

Down-channel velocity V_{dzL} in the transformed (Lagrangian) frame is provided by Equation 7.23. This equation provides the velocity in the z direction due to the rotation of the screw.

$$V_{dzL} = |V_{cz}|\left(1+\frac{H}{R_c}\right)\frac{4}{\pi}\sum_{i=1,3,5,...}^{\infty}\frac{\sinh\left(i\pi\frac{y}{W}\right)\sin\left(i\pi\frac{x}{W}\right)}{i\sinh\left(i\pi\frac{H}{W}\right)} \tag{7.23}$$

Equation 7.23 is the literature expression for the down-channel velocity in the z direction for barrel rotation physics. This down-channel velocity V_{dzL} in the transformed reference frame must be converted back to the laboratory frame by adjusting for the motion of the frame using Equation 7.20. The down-channel velocity due to the rotation of the screw V_{dzE} in the laboratory frame is as follows:

$$V_{dzE} = |V_{cz}|\left(1+\frac{H}{R_c}\right)\frac{4}{\pi}\sum_{i=1,3,5,...}^{\infty}\frac{\sinh\left(i\pi\frac{y}{W}\right)\sin\left(i\pi\frac{x}{W}\right)}{i\sinh\left(i\pi\frac{H}{W}\right)}-|V_{cz}|\left(1+\frac{y}{R_c}\right) \tag{7.24}$$

The down-channel velocity due to screw rotation in the laboratory frame for a channel with a small aspect ratio is provided by Equation 7.25; that is, H/W is small and less than 0.1.

$$V_{dzE} = \frac{y}{H}|V_{cz}|\left(1+\frac{H}{R_c}\right)-|V_{cz}|\left(1+\frac{y}{R_c}\right) \tag{7.25}$$

The down-channel velocity in the transformed frame for H/W less than 0.1 is shown by Equation 7.26. This equation is the same as the z-direction flow in the literature.

$$V_{dzL} = \frac{y}{H}\left|V_{cz}\right|\left(1+\frac{H}{R_c}\right) = \frac{y}{H}\left|V_{bz}\right| \tag{7.26}$$

Pressure flow velocity V_{pz} in the z direction in the channel is provided in Equation 7.27. This equation was solved in the laboratory frame and thus does not need to be transformed.

$$V_{pz} = \frac{H^2}{2\mu}\frac{\partial P}{\partial z}\left[\left(\frac{y}{H}\right)^2 - \frac{y}{H} + \frac{8}{\pi^3}\sum_{i=1,3,5,\ldots}^{\infty}\frac{\cosh\left(i\pi\frac{W}{H}\left(\frac{x}{W}-\frac{1}{2}\right)\right)\sin\left(i\pi\frac{y}{H}\right)}{i^3\cosh\left(i\pi\frac{W}{2H}\right)}\right] \tag{7.27}$$

The pressure flow velocity in the z direction for a channel with a small aspect ratio is provided by Equation 7.28; that is, H/W is small and less than 0.1:

$$V_{pz} = \frac{H^2}{2\mu}\frac{\partial P}{\partial z}\left[\left(\frac{y}{H}\right)^2 - \frac{y}{H}\right] \tag{7.28}$$

The transformed frame solutions provided by Equation 7.21 and Equation 7.23 predict that the x-y-direction flow due to the rotating screw is circulating in the channel, and the z-direction flow is drag flow down the channel toward the extruder exit, as shown in Figure 7.11. To the observer in this frame, it appears while looking at a slice of fluid across the channel that a particle in the x-y plane will only recirculate around the channel between the two flights. In this frame the particles in the slice of fluid in the z direction of the channel will appear to be "dragged" toward the exit in the positive z direction. It is perhaps unfortunate that these are the boundary conditions that result when the extruder physics is changed from screw rotation to barrel rotation. Even though the barrel does not rotate, historically this solution was assumed to represent the flow in the extruder [6, 7]. This historic mechanistic prediction, however, is not consistent with the screw component data presented at the beginning of this chapter. These particle trajectories are consistent with the predicted velocities using the transformed boundary conditions when the observer is attached to the moving frame of reference in Figure 7.10. In order to understand the flow mechanism of the extruder in the laboratory, these transformed velocity solutions must therefore be returned to the laboratory reference frame.

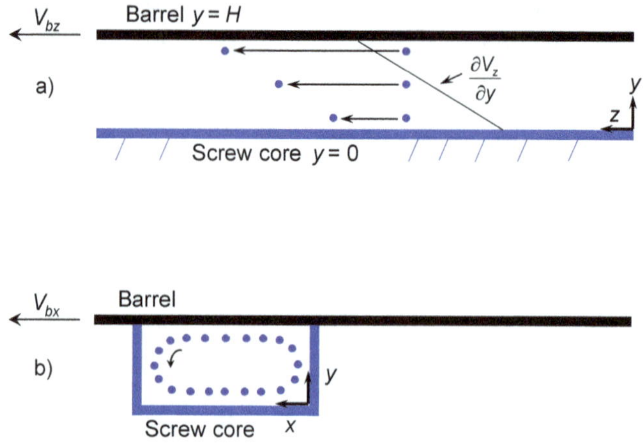

Figure 7.11 Qualitative predicted particle trajectories in the transformed frame (Lagrangian): a) down-channel flows induced by the drag motion of the moving barrel, and b) recirculation flow in the *x*-*y* plane

The experimental data presented earlier in this chapter demonstrated clearly that the helix is the forwarding or pumping component of the screw. For this simple problem, the procedure of adding back the boundary conditions changes the solution from the transformed frame to the laboratory frame. The laboratory reference frame solution is thus obtained by reversing the boundary condition manipulation used to transform the problem from a moving boundary problem to a fixed boundary problem, resulting in Equation 7.22 and Equation 7.24 These equations represent the laboratory-based solution for screw rotation pumping for the screw. Using this laboratory frame solution, the mechanistic solution can be compared to the rotating screw data. The laboratory frame solution predicts that the fluid particle starting at the centerline of the channel will now be observed to be translated in the positive x direction by the screw flights, the second term in Equation 7.22, and diagrammatically represented by the particles in Figure 7.12(b). The z-direction velocity in the channel drags material back toward the entrance of the extruder, in the direction of negative z, as shown in Figure 7.12(a). This is the theoretical solution using the laboratory reference frame for the screw velocity components, and thus the predicted fluid motion is now consistent with the data presented in Figure 7.5 and Figure 7.6. If the centerline of a fluid slice is moving at a constant velocity V_{cx} the particle now appears to move up and down in the y direction, and the forward motion accelerates and decelerates due to the positive and negative velocities from the recirculation flow in the channel, as represented by the first term in Equation 7.22.

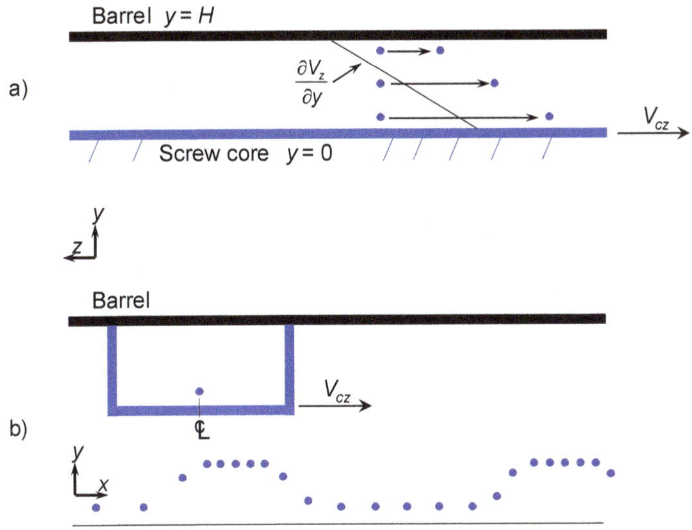

Figure 7.12 Qualitatively predicted particle trajectories in the laboratory frame (screw rotation): a) flow in the *z* direction (towards the inlet) due to the motion of the core, b) positive displacement flow in the *x*-*y* plane. The spacing between the particles is at equal time intervals. The relative positions indicate the speed of the particles. That is, the particles are moving faster near the root of the screw

A further difficulty found with only considering the transformed solution is that this is equivalent to changing the extruder physics to barrel rotation, as shown in Figure 7.13 [45], and further nonphysically based assumptions must be made. Since the assumption in the barrel drag analysis is that the flights have no effect on the flow analysis, the barrel velocity magnitude classically has to match the numerical value of the screw core velocity when calculating the velocity gradient, a necessary condition for a one-dimensional transformation. Thus, the predicted rate is based on the reduced velocity. The reduction of the barrel *z* velocity was necessitated in order to maintain the shear rates in the *z*-*y* plane at the same value regardless of which surface was moving. The one-dimensional transformation requirement is to keep the velocity-dependent shear rates consistent during the transformation from screw rotation to barrel rotation. The results of this necessary manipulation based on the assumption the flights make no contribution to the extruder screw rate can be found in Figure 7.13 as the "Literature Theory" line. For this analysis for a screw with a constant lead length, the channel depth increases with increasing H/W.

As shown in Figure 7.13, the slope of the rate with respect to H/W for the literature theory is always less than that for the new theory. Moreover, at H/Ws greater than about 0.7, the rate for barrel rotation decreases as the channel gets deeper. This is not consistent with the experimental data presented for three screws with increasing channel depth, represented by the triangles and squares. When experimentally the barrel and screw were rotated at the same angular velocity, it is expected that the pumping rate will be the same and both are positive. Several places in the historic literature state that only the *z*-direction drag needs to be considered in calculating the rate of the extruder. Further examination of the historic predictions based on assuming only the *z* velocity is involved in the pumping of the extruder leads to the following set of equations assuming that the screw and barrel surface velocities have the same magnitude. Here the velocity is defined in Equation 7.29 as is done in the literature: having a numeric value of the screw core velocity and the negative in Equation 7.30 when converting the solution to the laboratory frame to be consistent with only considering the *z*-direction drag in the calculation of the extruder flow.

$$Q_L = W\int_0^H \left[\frac{y}{H}\left|V_{bz}\right|\right] dy = \frac{1}{2}\left|V_{bz}\right| WH \quad \text{flow towards the discharge} \tag{7.29}$$

$$Q_E = W\int_0^H \left[-\frac{y}{H}\left|V_{cz}\right|\right] dy = -\frac{1}{2}\left|V_{cz}\right| WH \quad \text{flow towards the inlet} \tag{7.30}$$

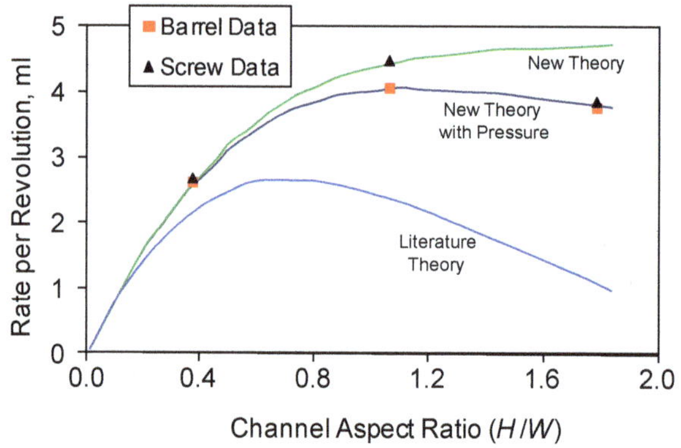

Figure 7.13 Comparison of literature drag and screw rotation for deep channels [45]. The experimental data for screw rotation and barrel rotation and the theory lines were for screws with a 7° helix angle

Thus traditional analysis predicts that when only the *z*-direction velocity is converted to the laboratory frame, the laboratory flow solution is toward the inlet of the extruder. Thus to be absolutely correct, the Literature Theory line of Figure 7.13 should be below the *x* axis and predict a negative flow for all screw channel depths.

As mentioned in the introduction to this chapter this is a necessary condition when approximating the cylindrical screw in the Cartesian coordinate system. The screw rotation theory, "New Theory" line, predicts that the rate should constantly increase as the channel gets deeper. When a fixed positive pressure occurs for the screw rotation model, the "New Theory with Pressure" line, the predictions fits the data very well for all H/Ws. Thus for modern screw designs with deeper channels, reduced energy dissipation, and lower discharge temperatures, the screw rotation model would be expected to provide a good first estimation of the performance of the extruder regardless of the channel depth for Newtonian polymers.

The screw rotation analysis leads to the model equation for the extruder discharge rate. There are now two screw-rotation-driven velocities, V_{xE} and V_{dzE}, and a pressure-driven velocity, V_{pz}, that affect the rate. V_{xE} and V_{dzE} transport the polymer fluid at right angles to one another. In order to calculate the net flow from screw rotation it is necessary to resolve the two screw-rotation-driven velocities into one velocity, V_{pL}, that can be used to calculate the screw rotation-driven flow down the screw parallel to the screw axis (or centerline) as discussed in Chapter 1 and as depicted in Figure 7.14. The resolved velocity will then be integrated over the screw channel area normal to the axis of the screw.

It is easily shown that both the velocities V_{dzE} (Equation 7.24) and V_{xE} (Equation 7.22) have a component in the "l" direction (parallel to the screw axis), and that those components can be vector summed through an area normal to the screw core:

$$V_{pl} = V_{xE} \cos\theta_c + V_{dzE} \sin\theta_c \tag{7.31}$$

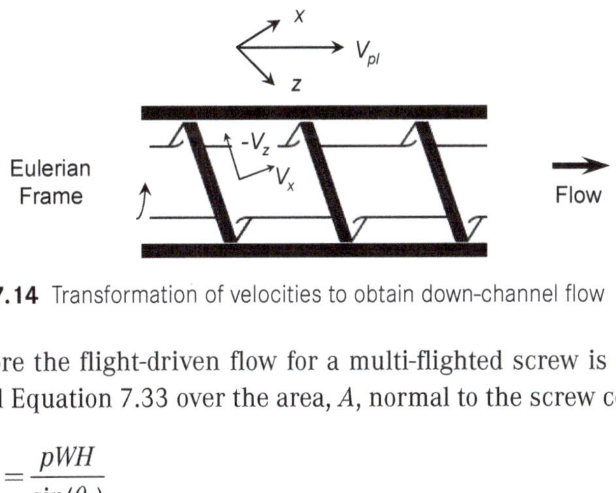

Figure 7.14 Transformation of velocities to obtain down-channel flow

Therefore the flight-driven flow for a multi-flighted screw is calculated using the integral Equation 7.33 over the area, A, normal to the screw core:

$$A = \frac{pWH}{\sin(\theta_c)} \tag{7.32}$$

where p is the number of flight starts. This leads to the following relationship for the volumetric flow rate:

$$Q_d = \frac{pW}{\sin(\theta_c)} \int_0^H V_{pl}\, dy \tag{7.33}$$

The derivation here will be for a screw with a shallow channel such that H/W is less than 0.1 so that the F_d and F_p corrections need not be made. The generalized equation derivations using Equation 7.22 and 7.24 are available in Appendix A7. Using Equation 7.22 and Equation 7.25 and substituting into Equation 7.33:

$$Q_d = \frac{pW}{\sin(\theta_c)} \int_0^H \left[\begin{array}{l} \left(V_{cx}\left(1+\frac{H}{R_c}\right)\frac{y}{H}\left(2-3\frac{y}{H}\right) + V_{cx}\left(1+\frac{y}{R_c}\right)\right)\cos(\theta_c) \\ +\left(\frac{y}{H}\left|V_{cz}\right|\left(1+\frac{H}{R_c}\right) - \left|V_{cz}\right|\left(1+\frac{y}{R_c}\right)\right)\sin(\theta_c) \end{array} \right] dy \tag{7.34}$$

and integrating:

$$Q_d = \frac{pW}{\sin(\theta_c)}\left[\left(\frac{V_{cx}}{H}\left(1+\frac{H}{2R_c}\right)\right)\cos(\theta_c) + \left(\frac{H}{2}\left|V_{cz}\right|\left(1+\frac{H}{R_c}\right) - \frac{\left|V_{cz}\right|}{H}\left(1+\frac{H}{2R_c}\right)\right)\sin(\theta_c)\right] \tag{7.35}$$

simplifying:

$$V_{cx} = V_c \sin(\theta_c) \tag{7.12}$$

$$Q_d = pW\left[\left(\frac{V_c}{H}\left(1+\frac{H}{2R_c}\right)\right)\cos(\theta_c) + \left(\frac{H}{2}\left|V_{cz}\right|\left(1+\frac{H}{R_c}\right) - \frac{\left|V_{cz}\right|}{H}\left(1+\frac{H}{2R_c}\right)\right)\right] \tag{7.36}$$

remembering that

$$\left|V_{cz}\right| = V_c \cos(\theta_c) \tag{7.37}$$

and that

$$\left|V_{bz}\right| = \left|V_{cz}\right|\left(1+\frac{H}{R_c}\right) \tag{7.38}$$

substituting:

$$Q_d = pW\left[\frac{\left|V_{cz}\right|}{H}\left(1+\frac{H}{2R_c}\right) + \frac{H\left|V_{bz}\right|}{2} - \frac{\left|V_{cz}\right|}{H}\left(1+\frac{H}{2R_c}\right)\right] \tag{7.39}$$

Simplifying Equation 7.39 results in the screw-rotation-driven volumetric flow rate for a channel with a small aspect (H/W) ratio:

$$Q_d = \frac{pWH\left|V_{bz}\right|}{2} \tag{7.40}$$

where the units for Q_d are m^3/s. As previously discussed, the classic drag flow model [8] assumed that the equivalent flow rate will be obtained if the barrel is rotated in the direction opposite to that of the screw as long as the linear velocity of the unwound barrel is numerically equal to the linear velocity of the screw core. That is, V_{cz} is used instead of V_{bz} in Equation 7.40 for barrel rotation. For screw rotation, the mathematical analysis is clear that Equation 7.40 represents the physics for screw rotation and that V_{bz} is the proper velocity component when calculating the screw pumping capacity. It is also the proper velocity when calculating the barrel rotation rate as seen from Equation 7.29, since the Lagrangian velocities for this analysis are equal in magnitude and direction to the velocities for rotating the barrel. As previously discussed, the absolute value of V_{bz} is used so as not to lose track that the screw at the flight tips is moving in the negative z direction.

The more general rotational flow is obtained by using Equation 7.24 instead of Equation 7.25 as done above, which holds for all screw geometries at constant pressure and if the material is Newtonian. The solution is converted to mass flow by multiplying the volumetric flow by the density and is provided by Equation 1.13:

$$Q_{md} = \frac{p\rho\left|V_{bz}\right|WHF_d}{2} \tag{1.13}$$

The mass generalized pressure-driven flow for screws with multiple flight starts from Equation 7.27 is as follows:

$$Q_{mp} = \frac{p\rho WH^3 F_p}{12\eta}\left[\frac{\partial P}{\partial z}\right] \tag{1.22}$$

The combined mass flow for rotation-driven and pressure-driven flow is given by Equation 1.28 and is the expected rate of the process. The average shear viscosity is calculated using the average shear rate in the channel for screw rotation and the bulk temperature. This method is also known as the generalized Newtonian method.

$$Q_m = Q_{md} - Q_{mp} \tag{1.12}$$

$$Q_m = \frac{p\rho\left|V_{bz}\right|WHF_d}{2} - \frac{p\rho WH^3 F_p}{12\eta}\left[\frac{\partial P}{\partial z}\right] \tag{1.29}$$

with

$$F_d = \frac{16W}{\pi^3 H}\sum_{i=1,3,5,\dots}^{\infty}\frac{1}{i^3}\tanh\left(\frac{i\pi H}{2W}\right) \tag{1.27}$$

and

$$F_p = 1 - \frac{192H}{\pi^5 W} \sum_{i=1,3,5,\ldots}^{\infty} \frac{1}{i^5} \tanh\left(\frac{i\pi W}{2H}\right) \tag{1.28}$$

The correction factors (or shape factors) F_d and F_p are related to the reduced flow in the z direction due to the influence of the flights, as stated in Chapter 1. For channels with small aspect ratios, these terms are both essentially 1. As the aspect ratio of the channel increases, the channel gets "boxy," and these corrections become quite important, and they can be much smaller than 1. The correction factors as a function of H/W are shown in Figure 7.15. As will be shown in Section 7.5.3, F_d and F_p do not fully correct the flow calculation. A second empirical correction function will be presented to correct the rotational flow rate.

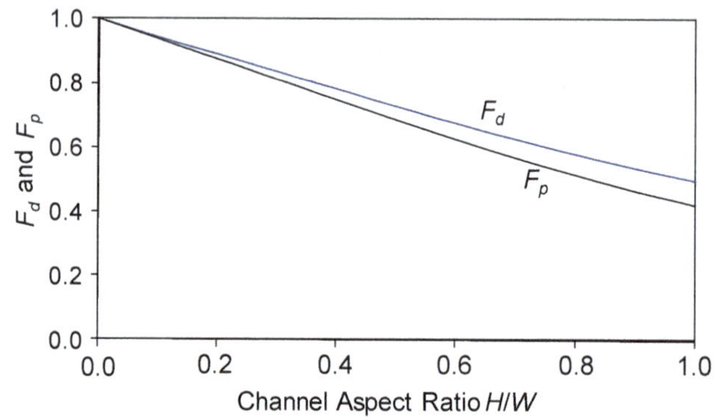

Figure 7.15 F_d and F_p factors as a function of the aspect ratio of the channel

All engineers who troubleshoot single-screw extruders should be familiar with the use of these equations to evaluate the rate characteristics of the machine. This is where the analysis should begin in almost all extruder troubleshooting problems. These equations are very useful in assessing the basic flow in many extruders, particularly those that meet the assumptions in developing this analysis: Newtonian viscosity, relatively constant temperature, and H/W of about 0.1 or less. When doing troubleshooting as described in Chapters 11 through 13, it is strongly recommended that the analyst first undertake these important flow calculations using Equation 7.29.

For non-Newtonian fluids, the average shear viscosity of the resin in the channel must be calculated using the average shear rate based on screw rotation. The average shear rate for a channel that does not have a pressure gradient ($\partial P/\partial z = 0$) can be calculated using Equation 7.22 and Equation 7.25. The motion of the screw root can be obtained by setting $y = 0$. The average shear rate is then calculated as follows:

$$\dot{\gamma}_s = \frac{\sqrt{\left(V_{xE}\big|_{y=0}\right)^2 + \left(V_{dzE}\big|_{y=0}\right)^2}}{H} = \frac{\sqrt{(V_{cx})^2 + (V_{cz})^2}}{H} = \frac{V_c}{H} = \frac{\pi N D_c}{H} \tag{7.41}$$

where $\dot{\gamma}_s$ is the average shear rate in the channel based on screw rotation. Since the majority of the flow for most cases is from the rotational component of Equation 7.18, ignoring the contribution to the shear rate from the pressure flow term may not cause an appreciable error to the calculation. If the shear rate is calculated based on barrel rotation using D_b, the shear rate can be considerably higher than that experienced in the channel, creating an error.

7.4.1 Example Flow Calculation

Two new LDPE resins are proposed to be processed on a similar extruder as described in the example in Section 1.5.1. The extruder here is configured with a downstream gear pump such that the discharge pressure for the extruder is targeted at 7 MPa. The resin manufacturer has provided viscosity data in the form of a viscosity relationship:

$$\eta = \eta_0 e^{-b(T-T_0)} \left|\dot{\gamma}\right|^{(n-1)} \tag{7.42}$$

Note that the relationship contains an absolute value of the shear rate since it can be either positive or negative. The manufacturer has supplied parameters for the proposed 0.8 dg/min (190 °C, 2.16 kg) MI resin and a lower viscosity resin with an MI of 2 dg/min. The viscosity equation parameters are provided in Table 7.1.

Table 7.1 Shear Viscosity Model Parameters for Two LDPE Resins

	0.8 MI Resin	2 MI Resin
η_0, Pa·s	9720	3300
b, 1/°C	0.00913	0.0183
T_0, °C	210	190
N	0.41	0.68

The target discharge temperature for the extrusion is 240 °C, and the maximum screw speed is 130 rpm (N = 2.167 rev/s). What will be the expected production rates for both resins at 130 rpm? At 240 °C the melt density of the resins is 735 kg/m^3. As presented in Section 1.4and Section 7.4, the net flow in the extruder is the difference between the rotational flow Q_d and the flow induced by the pressure gradient Q_p. The data in Table 7.2 was calculated from the example in Section 1.5.1 and Table 7.1:

Table 7.2 Geometric Parameter Values for the Screw Used in the Example in Section 1.5.1 and Calculated Values for a Screw Speed of 130 rpm

Parameter	Value	Equation
Barrel diameter, D_b	88.9 mm	
Lead length, L	107 mm	
Meter channel depth, H	5.08 mm	
Flight width, e	9.0 mm	
Flight starts, p	1	
Helix angle at the barrel wall, θ_b	21.0°	1.1
Average channel width, W	90.0 mm	1.10
Channel aspect ratio, H/W	0.056	
Velocity of the screw at the barrel, V_{bz}	-0.577 m/s	1.20
Unwrapped channel length for one turn, z_b	299 mm	1.11
Total helical length of the metering section, Z_b	2.23 m	1.11
Shape factor for rotational flow, F_d	0.966	1.27
Shape factor for pressure flow, F_p	0.965	1.28

The rotational flow rate is calculated using the data in Table 7.2 and Equation 1.13 as follows:

$$Q_{md} = \frac{p\rho |V_{bz}| WHF_d}{2} = \frac{(735\ \text{kg/m}^3)(0.577\ \text{m/s})(0.090\ \text{m})(5.08x10^{-3}\ \text{m})(0.966)}{2}$$
$$= 0.0937\ \text{kg/s} => 337\ \text{kg/h}$$

This is the rate that the screw can produce at 130 rpm with no pressure gradient in the channel. Because both resins are LDPE and at the same temperature of 240 °C, the densities and the rotational flow rates are the same for both the 0.8 and 2 MI resins. The specific rotational rate should always be calculated by dividing the rotational flow rate by the screw speed in rpm. The specific rotational flow rate here is 2.59 kg/(h·rpm). Next, the rate needs to be adjusted for the pressure gradient that may exist in the channel. If the upstream sections of the screw can barely deliver the rate, the pressure at the entry to the metering section will be zero as a limiting case. For this limiting case, the metering section will need to generate the 7 MPa pressure required at the inlet of the gear pump. The pressure gradient for this case, $\partial P/\partial z$, is 3.14 MPa/m, found by dividing the maximum pressure change in the meter by the helical length of the metering section. If the upstream sections of the extruder are capable of generating a pressure at the entry to the metering section, then the pressure gradient could be less and could also be negative. In this second case, the rate of the extruder will be higher than that calculated for the base case with an entry pressure of zero. The base rate case with an entry pressure of zero will be calculated here, that is, a $\partial P/\partial z$ of 3.14 MPa/m.

The base case pressure flow calculation and the above viscosity function require that the shear rate be calculated from the screw rotation equations using Equation 7.41:

$$\dot{\gamma}_s = \frac{\pi N D_c}{H} = \frac{\pi(2.167\ \text{rev/s})(0.07874\ \text{m})}{(5.08x10^{-3}\ \text{m})} = 106\ 1/\text{s}$$

The viscosity η is then calculated at 106 1/s and 240 °C using Equation 7.42 and the data in Table 7.1 as 472 and 297 Pa·s for the 0.8 and 2 MI resins, respectively. The rate due to the imposed pressure gradient is then calculated using Equation 1.22 as follows for the 0.8 MI resin:

$$Q_{mp} = \frac{p\rho W H^3 F_p}{12\eta}\left[\frac{\partial P}{\partial z}\right] = \frac{(735\ \text{kg}/\text{m}^3)(0.090\ \text{m})(5.08\times10^{-3}\ \text{m})^3\ (0.965)}{12\ (472\ \text{Pa}\cdot\text{s})}\left[\frac{7.0\times10^6\ \text{Pa}}{2.23\ \text{m}}\right]$$

$$= 0.00646\ \text{kg/s} => 17\ \text{kg/h}$$

The overall rate expected is then calculated using Equation 1.12:

$$Q_m = Q_{md} - Q_{mp} = 337 - 17 = 320\ \text{kg/h}$$

Thus, the expected maximum rate of the extruder using the analytical analysis for the 0.8 MI resin is 320 kg/h at the maximum screw speed of 130 rpm. A similar calculation was performed for the 2 MI resin, and the rate due to the imposed pressure gradient is 27 kg/h, and the expected maximum rate of the extruder is 310 kg/h. The specific rate should always be calculated by dividing the rate by the screw speed in rpm. For the 0.8 and 2 MI resins, the specific rate is calculated at 2.46 and 2.38 kg/(h·rpm). The pressure gradient here caused the rate to decrease by about 5 to 8% from the rotational rate. This type of decrease is typical for a properly operating extruder.

If the upstream sections of the screw were optimally designed, the pressure at the entry to the metering section could have been as high as 15 MPa. For this case, the pressure gradient in the metering section would have been negative, that is, a $\partial P/\partial z$ of -3.59 MPa/m. For this pressure gradient, the pressure-driven flow components are -19 and -31 kg/h for the 0.8 and 2 MI resins, respectively, and the overall rates would be 356 and 368 kg/h. In practice, the overall rate that the plant would observe is between the base pressure case and the rates calculated here with a 15 MPa pressure at the entry to the metering section.

7.5 Numerical Flow Evaluation

The calculation method and equations presented in the previous sections are for Newtonian fluids such that the flow due to screw rotation and the downstream pressure gradient can be solved independently, that is, via the principle of superposition. Since most resins are highly non-Newtonian, the rotational flow and pressure-driven flow in principle cannot be separated using superposition. That is, the shear dependency of the viscosity couples the equations such that they cannot be solved independently. Potente [50] states that the flows and pressure gradients should only be calculated using three-dimensional (3-D) numerical methods because of the limitations of the Newtonian model.

The next four sections discuss the limitations for the flow calculations using the simplified analysis of Equation 1.29. Using screw geometry to predict the overall flow from the screw-rotation-based equations developed in the previous section results in useful numeric values because the concept of frame indifference can be applied to fluid flow at constant temperature and viscosity. Unfortunately, this simplified analysis does not always provide accurate results for real extruders when extruding commercial polymers. Screws with "boxy" channels and multiple flight starts often exhibit large operating pressure deviations when compared to these simple predictions when Non-Newtonian resins are processed. The actual pressures are analytically found by simulations that solve the complete momentum balance equation simultaneously with the energy equation. These equations are traditionally solved using three-dimensional simulations using FEM or FDM techniques. The momentum equations in Cartesian coordinates are represented as Equation 7.43, Equation 7.44, and Equation 7.45. The solution of these equations is a complex mathematical and computational problem, and the detailed techniques are beyond the scope of this work. However, several examples will be presented below, using published results from these advanced numerical techniques to evaluate the complete solution. These complete solutions will then be compared with the predicted results from Equation 1.29.

$$\rho\left(V_x\frac{\partial V_x}{\partial x}+V_y\frac{\partial V_x}{\partial y}+V_z\frac{\partial V_x}{\partial z}\right)=-\left(\frac{\partial \tau_{xx}}{\partial x}+\frac{\partial \tau_{yx}}{\partial y}+\frac{\partial \tau_{zx}}{\partial z}\right)-\frac{\partial P}{\partial x} \tag{7.43}$$

$$\rho\left(V_x\frac{\partial V_y}{\partial x}+V_y\frac{\partial V_y}{\partial y}+V_z\frac{\partial V_y}{\partial z}\right)=-\left(\frac{\partial \tau_{xy}}{\partial x}+\frac{\partial \tau_{yy}}{\partial y}+\frac{\partial \tau_{zy}}{\partial z}\right)-\frac{\partial P}{\partial y} \tag{7.44}$$

$$\rho\left(V_x\frac{\partial V_z}{\partial x}+V_y\frac{\partial V_z}{\partial y}+V_z\frac{\partial V_z}{\partial z}\right)=-\left(\frac{\partial \tau_{xz}}{\partial x}+\frac{\partial \tau_{yz}}{\partial y}+\frac{\partial \tau_{zz}}{\partial z}\right)-\frac{\partial P}{\partial z} \tag{7.45}$$

The components of the stress tensor in Cartesian coordinates are as follows:

$$\tau_{xx} = -\mu \left[2\frac{\partial V_x}{\partial x} - \frac{2}{3}(\nabla \cdot \boldsymbol{v}) \right] \tag{7.46}$$

$$\tau_{yy} = -\mu \left[2\frac{\partial V_y}{\partial y} - \frac{2}{3}(\nabla \cdot \boldsymbol{v}) \right] \tag{7.47}$$

$$\tau_{zz} = -\mu \left[2\frac{\partial V_z}{\partial z} - \frac{2}{3}(\nabla \cdot \boldsymbol{v}) \right] \tag{7.48}$$

$$\tau_{xy} = \tau_{yx} = -\mu \left[\frac{\partial V_x}{\partial y} + \frac{\partial V_y}{\partial x} \right] \tag{7.49}$$

$$\tau_{yz} = \tau_{zy} = -\mu \left[\frac{\partial V_y}{\partial z} + \frac{\partial V_z}{\partial y} \right] \tag{7.50}$$

$$\tau_{zx} = \tau_{xz} = -\mu \left[\frac{\partial V_z}{\partial x} + \frac{\partial V_x}{\partial z} \right] \tag{7.51}$$

As discussed earlier, often the troubleshooter is confronted with a deep channel screw that is designed to minimize energy dissipation and discharge temperature or maximize rate for a low-pressure die. As discussed previously, the generalized Newtonian method was developed using the classical theories and thus employs barrel rotation physics. This calculation method is known to work reasonably well with small-diameter extruders with shallow channels, and it has been used extensively and successfully to design and troubleshoot extrusion processes. The extrusion troubleshooter should be aware that many commercially available computer codes use versions of the generalized Newtonian model for metering channel calculations. Recently, both FEM and FDM were used to solve the complete set of momentum equations using Equation 7.43, Equation 7.44, and Equation 7.45; see Spalding, Campbell, Carlson, and Nazrisdoust [51]. For large-diameter extruders with multiple flight starts, the generalized Newtonian method was shown to be unacceptable for predicting pressure profiles [52]. In this example, a large deep channel screw that was pumping a highly shear-thinning resin was evaluated using Equations 1.13 and 1.22. These simplified calculations were then compared with the results from numerical methods and experimental data. Other models that have been developed using different techniques include a three-dimensional FEM [37], solving Equation 7.43, Equation 7.44, and Equation 7.45 numerically with the inertial terms and the energy balance equation, a two-dimensional FDM with a rotating barrel [53], and a model with a rotating screw [45]. For the most part, these models were verified using extrusion data from relatively small diameter machines with simple screws.

Returning to the discussion comparing numerical and analytical solutions, large-diameter extrusion data sets are not readily available, and thus they are generally not used for model development. A further limitation in using these FEA and FDM codes is that they are typically difficult to use, requiring special numerical skills, and the solution takes a long time to run on a fast computer.

7.5.1 Simulation of a 500 mm Diameter Melt-Fed Extruder

Large diameter, melt-fed extruders are commonly used for the final devolatilization and pelletization of LDPE and PE copolymers in resin manufacturing plants. A full description of this type of extruder and process is provided in Section 15.3. Simulation of these processes is complicated by the multiple flights used in the design and the high H/W aspect ratios of the channels. The processes can be simulated from the feed hopper to discharge, however, since they are not required to convey solids and melt resin. This section will show the requirements and difficulties for simulating these processes.

A 500 mm diameter extruder was melt-fed with a 0.8 dg/min MI LDPE resin at a rate of 11,800 kg/h and a screw speed of 46 rpm [51, 54]. The feed temperature and pressure of the resin were measured at 225 °C and 0.1 MPa, respectively. The shear viscosity was presented previously in Section 7.4.1 and Table 7.1. The melt density was 740 kg/cm^3 at 230 °C. The extruder was relatively long, but only the first 5.6 diameters needed to be simulated to demonstrate the accuracy of the methods. For this extruder, the vertical feed pipe was 500 mm in diameter and thus forced the first diameter of the extruder to be at a temperature and pressure of 225 °C and 0.1 MPa. The screw started with a 1.5 diameter long, double-flighted feed section with a depth of 60 mm and a lead length of 750 mm. The next section (transition section) of the screw was 1 diameter long, double-flighted with a lead length of 625 mm. The last section was 3.1 diameters long, double-flighted, and was a constant-depth meter section with a depth of 27 mm and a lead length of 500 mm. The flight widths for all sections were 20 mm and the flight clearance was 0.5 mm. The H/W for this double-flighted screw was about 0.12 in the metering channel. A pressure transducer was positioned in the barrel at 5.6 diameters downstream of the start of the screw. For these conditions and the 0.8 MI LDPE resin, the pressure at this location was experimentally measured at 6.4 MPa. The barrel temperature for this data set was at 190 °C.

A three-dimensional simulation method was used to simulate this extrusion process and others presented in this book. For this method, an FDM technique was used to solve the momentum equations Equation 7.43 to Equation 7.45. The channel geometry used for this method was essentially identical to that of the unwound channel. That is, the width of the channel at the screw root was smaller than that at the barrel wall as forced by geometric constraints provided by Figure 7.1. The Lagrangian reference frame transformation was used for all calculations, and thermal effects were included. The thermal effects were based on screw rotation. This three-dimensional simulation method was previously proven to predict accurately the simulation of pressures, temperatures, and rates for extruders of different diameters, screw designs, and resin types.

The process was simulated using the generalized Newtonian method and the three-dimensional FDM technique discussed above. The simulation results were then compared to the experimentally measured pressure at 5.6 diameters from the screw start. As shown in Figure 7.16, the simulated pressure from the generalized Newtonian method was about 1.7 times higher than the actual value measured; that is, the simulated pressure was 10.8 MPa. This level of error is unacceptable for designing and troubleshooting extrusion processes. As discussed later in this chapter, the error was caused by the non-Newtonian nature of the LDPE resin and a relatively high *H/W* of 0.12. The relatively high *H/W* was the result of the double-flighted screw and deep channels, designs that are typical for large-diameter, melt-fed extruders. Large-diameter melt-fed extruders are discussed in detail in Chapter 15.

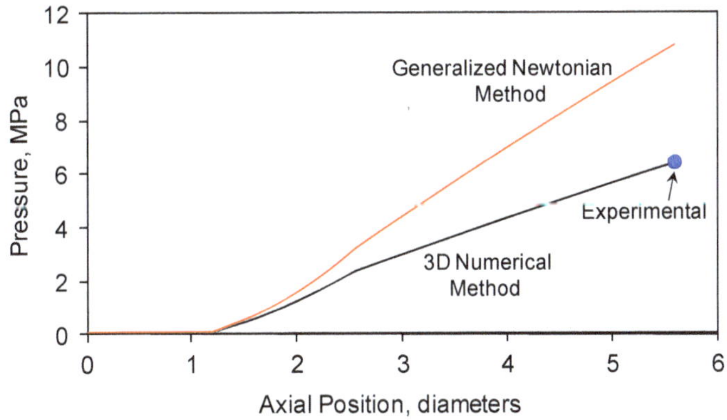

Figure 7.16 Simulated axial pressure profile for a 500 mm diameter extruder running 11,800 kg/h at 46 rpm for the 0.8 MI LDPE resin. The experimentally determined pressure at 5.6 diameters was 6.4 MPa

The three-dimensional FDM technique provided an excellent prediction of the pressure at 5.6 diameters from the start of the screw, as shown in Figure 7.16. The method, however, is difficult to use and requires relatively long computational times on a fast computer. This example is an excellent test case for determining the acceptability of a simulation code.

7.5.2 Extrusion Variables and Errors

As shown in Section 7.5.1, the prediction of the pressure profile using the generalized Newtonian method can have a high level of error for some processes. It is therefore instructive to examine the effects of several extrusion variables on the potential error when using the generalized Newtonian flow analysis. The following discussion provides a guideline as to the expected errors when using the standard and simple-to-use generalized Newtonian method of Equation 1.29 and shear viscosity at a nominal shear rate and average temperature. If a large error is expected then care should be taken when predicting the flows in a melt-filled channel. In general an error in the pressure of 10% or less would be considered acceptable for commercial screw design.

The focus of this evaluation is on the results that were reported using four different resins [52]: PC resin, LLDPE resin, EAA copolymer, and an LDPE resin. The shear viscosities for the resins at selected processing temperatures are shown in Figure 7.17 and were modeled using the power law model provided by Equation 7.42. The parameters for the model are given in Table 7.3. As shown in Figure 7.17 and the *n* values in Table 7.3, the PC resin shear-thinned the least while the LDPE resin shear-thinned the most. The LLDPE and EAA resins have *n* values between those for the PC and LDPE resins. The melt density for the LDPE and LLDPE resins at 240 °C is 735 kg/m^3. The melt density of the EAA resin at 220 °C was 785 kg/m^3, and the melt density of the PC resin at 280 °C was 1073 kg/m^3.

Table 7.3 Shear Viscosity Model Parameters for PC, LLDPE, EAA, and LDPE resins

	PC	LLDPE	EAA	LDPE
η_o, Pa·s	1650	3320	2080	9720
b, 1/°C	0.0436	0.0141	0.0210	0.00913
T_o, °C	280	240	190	210
N	0.815	0.741	0.646	0.410

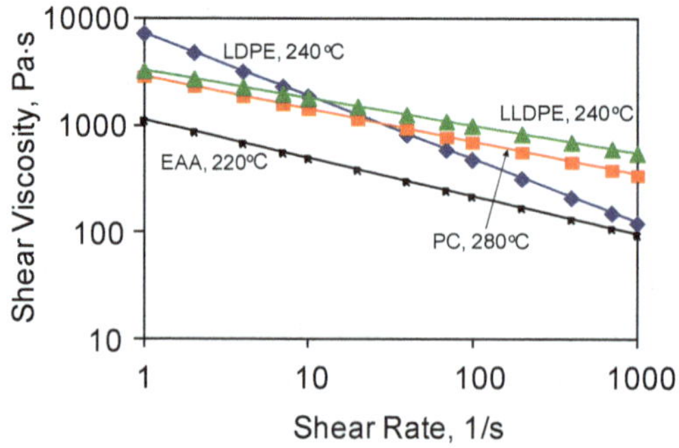

Figure 7.17 Shear viscosities for PC, LLDPE, EAA, and LDPE resins at their operating temperatures

Two different calculation methods were used for the simulations: (1) the generalized Newtonian method as developed above, and (2) the three-dimensional numerical method presented in Section 7.6.1. The generalized Newtonian method used a shear viscosity value that was based on the average barrel rotation shear rate and temperature in the channel. The average shear rate based on barrel rotation ($\dot{\gamma}_b$) is provided by Equation 7.52. Barrel rotation shear rate and the generalized Newtonian method are used by many commercial codes, and that is why it was used for this study.

$$\dot{\gamma}_b = \frac{\pi N D_b}{H} \quad \text{barrel rotation physics} \tag{7.52}$$

To understand the effect of barrel diameter on the rotational flow and the pressure gradient for an extruder, a series of numerical experiments were performed. The first simulations were performed using single-flighted screws with screw diameters ranging from 25 mm up to 700 mm. Single-flighted screws were modeled with the following dimensions: (i) the lead length was equal to the barrel diameter (square pitched), (ii) the flight width was equal to 10% of the diameter, and (iii) the flight depth was set at 5% of the diameter. For this geometry, the aspect ratio of the channel (H/W) was 0.059. The simulations were performed at a screw speed of 100 rpm and at starting temperatures of 240 °C for the LDPE and LLDPE resins, at 220 °C for the EAA resin, and at 280 °C for the PC resin. The rotational flow rates were calculated by setting the pressure gradient to zero. The ratios of the rotational flow rate calculated using the generalized Newtonian method to the rotational flow calculated using the numerical method as a function of screw diameter are shown in Figure 7.18. As previously stated, the numerical method is proven to calculate accurately the flows in the channel. Thus, if the rotational flow ratio is 1.0, then the generalized Newtonian method provides an accurate calculation of the rotational flow rate. As shown in Figure 7.19, the rotational flow ratio was near 1.0 at all screw diameters for all but the LDPE resin, indicating that the generalized Newtonian method is an acceptable method for calculating rotational flows for these resins and geometries. The rotational flow ratio for the LDPE resin had the highest ratio of 1.035. Thus, the rotational flow rate calculated using the generalized Newtonian method will be about 3.5% too high. This level of error may be acceptable for calculations and designs, but it can lead to poor predictions of pressure as will be discussed next.

The ratio of the rotational flows indicated that the diameter of the extruder is not a factor for the deviation shown in Figure 7.16. That is, for this channel with a small H/W the simplified analysis produces less than a 10% error when compared to the exact numerical solution. Thus, the rotational flow rate can be calculated quite reliably using the simple generalized Newtonian method at these conditions.

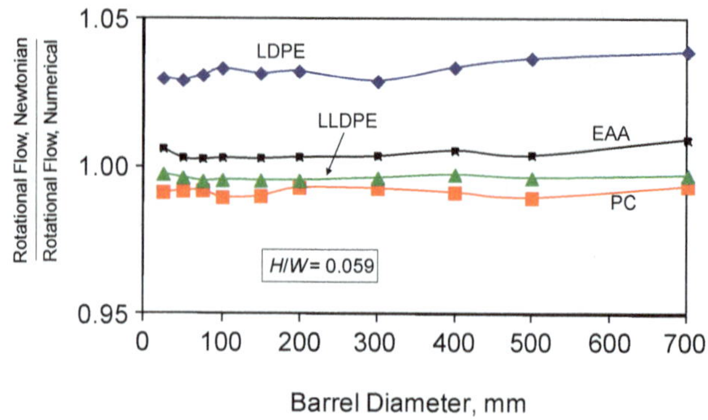

Figure 7.18 Ratio of the rotational flow rate calculated using the generalized Newtonian method to the rotational flow calculated using the numerical method as a function of screw diameter

The screw geometries described above were simulated at rates equivalent to the rotational flow rate at 100 rpm. The screws, however, were rotated at a speed of 110 rpm such that positive pressure gradients exist in the channels. The ratios of the pressure gradient calculated using the generalized Newtonian method to the pressure gradient calculated using the numerical method are shown in Figure 7.19. A ratio of 1.0 indicates that the generalized Newtonian method provides the correct pressure gradient for the extrusion process. As shown in Figure 7.19, the nearly Newtonian PC resin had a ratio of about 1.0 at all barrel diameters, indicating that the pressure gradient can be calculated with high accuracy using the generalized Newtonian method. As the n value decreased, indicating that the polymer is more shear thinning, the pressure gradient ratio increased. For the LDPE resin, the pressure gradient ratio was constant at about 2.1 for these screws. Thus, the pressure gradient calculated using the generalized Newtonian method was 2.1 times higher than the actual gradient. For most screw designs, this level of error is unacceptable. The LLDPE and the EAA resins with n values of 0.741 and 0.646 had intermediate pressure gradient ratios of 1.13 and 1.32, respectively.

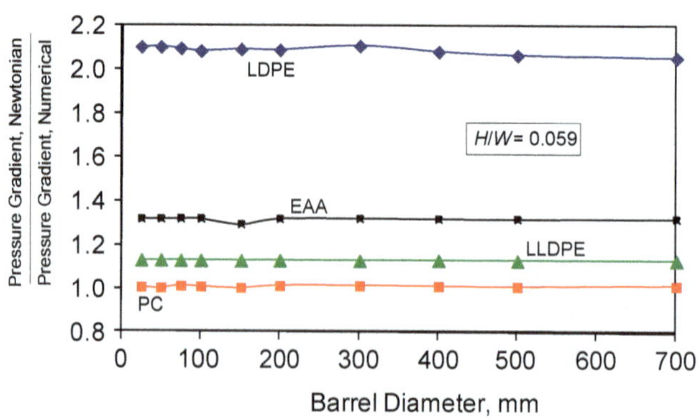

Figure 7.19 Ratio of the pressure gradient calculated using the generalized Newtonian method to the pressure gradient calculated using the numerical method as a function of barrel diameter

A second set of numerical experiments were performed using 100 mm diameter screws with a lead length of 100 mm and *H*/*W* ratios ranging from 0.023 to 0.45. These channels were created by varying the channel depth from 2 to 12 mm and using single-, double-, and triple-flighted screws. The flight widths for the single-, double-, and triple-flighted screws were 10, 6, and 5 mm, respectively. The average width of the channel decreases as the number of flight starts increases. The rotational flow rates for these screws were determined at a screw speed of 100 rpm and with the pressure gradient set to zero. The ratios of the rotational flow rate calculated using the generalized Newtonian method to the rotational flow rate calculated using the numerical method as a function of *H*/*W* are presented in Figure 7.20. The nearly Newtonian PC resin had a ratio that was about 0.99 at a low *H*/*W*, and then it increased to about 1.02 at an *H*/*W* of 0.45. For this *H*/*W* range, the rotational flow rate calculated using the generalized Newtonian method was acceptable. For the highly non-Newtonian LDPE resin, the rotational flow ratio ranged from 1.03 for channels with a low *H*/*W* to nearly 1.2 at an *H*/*W* of 0.45. As will be presented next, this high ratio of 1.2 will force the pressure gradient calculation to be in error. As before, the rotational flow ratios for the LLDPE and EAA resins were intermediate, suggesting a trend based on the non-Newtonian characteristic of the resin.

Consistent with previous computational experiments, the screw geometries described above were simulated at rates equivalent to the rotational flow rate at 100 rpm, and the screws were rotated in the simulation at a speed of 110 rpm such that positive pressure gradients exist in the metering channels. The ratio of the pressure gradient calculated using the generalized Newtonian method to the pressure gradient calculated using the numerical method is shown in Figure 7.21.

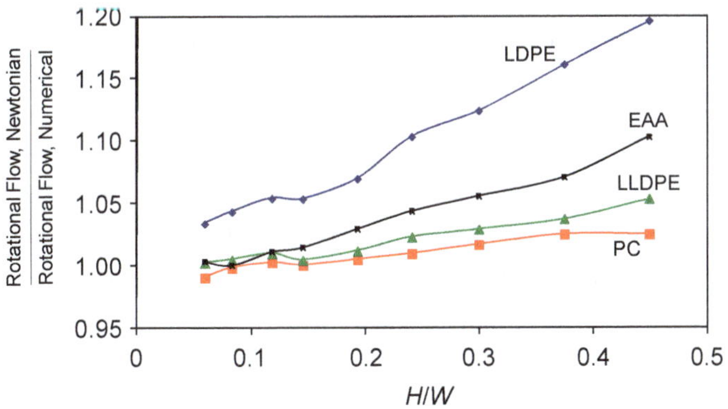

Figure 7.20 Ratio of the rotational flow rate calculated using the generalized Newtonian method to the rotational flow rate calculated using the numerical method as a function of *H*/*W*

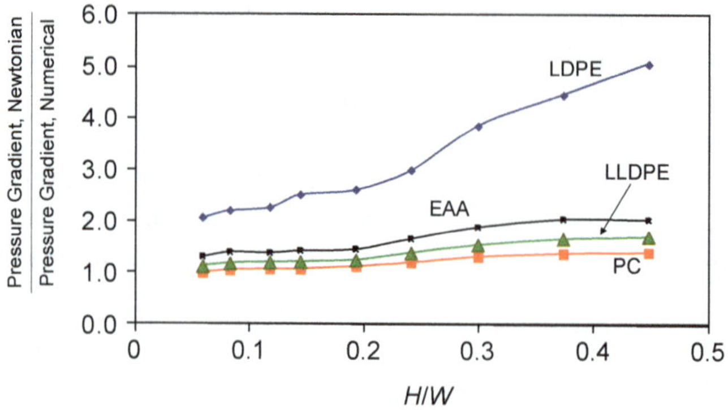

Figure 7.21 Ratio of the pressure gradient calculated using the generalized Newtonian method to the pressure gradient calculated using the numerical method as a function of *H/W*

The pressure gradient ratio for the PC resin was about 1.00 at an *H/W* of 0.059, and it increased to 1.08 at an *H/W* of 0.14. The ratio continued to increase to 1.4 at an *H/W* of 0.45. These data indicate that the generalized Newtonian method is acceptable for calculating the pressure gradient in the *H/W* range of about 0.05 to 0.19. At *H/W*s outside of this range, a level of error greater than 10% will occur. For the LDPE resin, the generalized Newtonian method is not an acceptable method for calculating pressure gradients. That is, the generalized Newtonian method will over-predict the pressure gradient by a factor of 2 to 5 times for these conditions. As expected, from the results presented above, the pressure gradient ratios for the LLDPE and EAA resins were intermediate.

The previous calculations were performed using a range of commercial resins with different levels of shear thinning behavior. In order to show the deviation for the generalized Newtonian method, the data presented in Figure 7.20 and Figure 7.21 were plotted as a function of n for several *H/W* values, as shown in Figure 7.22 and Figure 7.23. As the n value was reduced from 1.0 towards 0, and as the *H/W* ratio increases, the deviation for the generalized Newtonian method increased. These figures can be used to determine if the generalized Newtonian method is acceptable for a predetermined level of error. For example, if the level of accepted error for calculating the pressure gradient is 15%, then the acceptable range for n and *H/W* is a pressure gradient ratio of less than 1.15 in Figure 7.23.

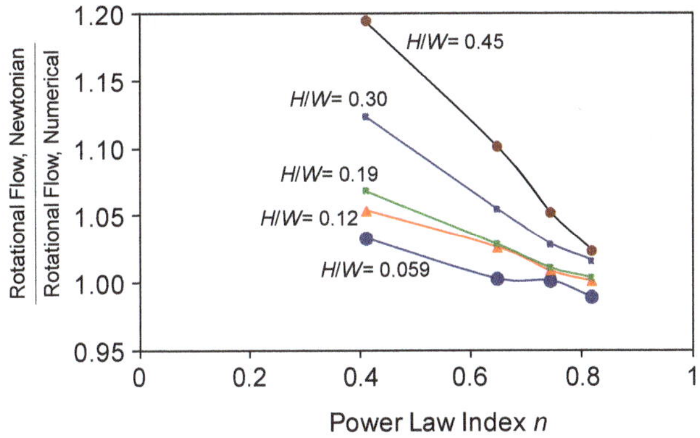

Figure 7.22 Ratio of the rotational flow rate calculated using the generalized Newtonian method to the rotational flow rate calculated using the numerical method as a function of the power law index *n*

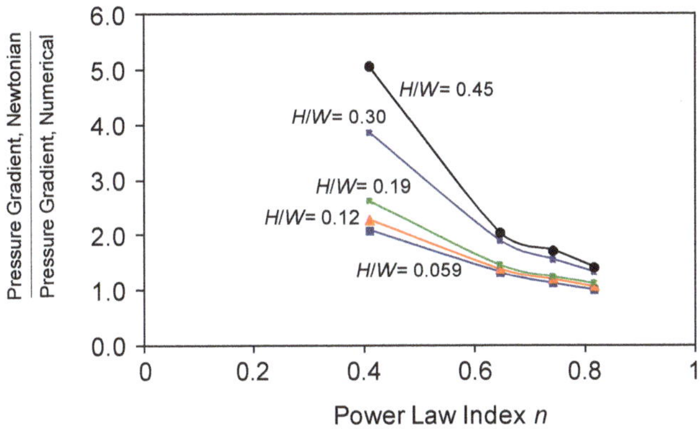

Figure 7.23 Ratio of the pressure gradient calculated using the generalized Newtonian method to the pressure gradient calculated using the numerical method as a function of the power law index *n*

The generalized Newtonian model over-predicted the rotational flow rates and pressure gradients for the channel for most conditions. This over-prediction was caused in part by the utilization of drag flow shape factors (F_d) that were too large. Then in order for the sum of the rotational and pressure flows to match the actual flow in the channel, the pressure gradient was forced to be higher than actually required by the process. It has been known for a long time [9] that the power law index has an effect on both pressure flow and drag flow. That is, the shape factors for the channel for drag flow (F_d) and pressure flow (F_p) are dependent on the shear-thinning nature of the resin. These shape factors represent the reducing effect of the flow due to the presence of the flights relative to flow between two parallel plates for Newtonian fluids. For the three-dimensional numerical method, the reduced flow at the flight edges is accounted for through the no-slip boundary conditions at the flight edge and the shear-thinning nature of the resin. That is why the numerical method predicts the experimental pressures well at all conditions.

The generalized Newtonian model has been used successfully for many years for the design of small diameter screws with relatively shallow channels. Its success has brought considerable value to those designers who understood the method and its limitations. An improved method for calculating flows is presented in the next section.

7.5.3 Corrections to Rotational Flow

Simulation of single-screw extruder screws using the standard generalized Newtonian method is known to deviate from measured performance as presented in Section 7.5.2. Part of this deviation is caused by the calculation of the rotational flow rate. As discussed above, the calculation of the rotational flow rate using this method is higher than that in the actual channel, causing the pressure gradient to be incorrectly adjusted to compensate for the error in the rotational flow term. This section provides correction factors for rotational flow such that axial pressure gradients can be quickly and accurately calculated [54].

As discussed in Section 7.4 and using the principle of superposition, the flow components were separated into rotational flow and pressure flow. The equation for the total flow and the components are as follows:

$$Q_m = Q_{md} - Q_{mp} = Q_{md} - \beta \frac{\partial P}{\partial z} \tag{7.53}$$

where Q_m is the mass rate of the extruder and β is a constant. Calculation of the term β can be obtained from Equation 1.22. It has been known for a long time [9] that the power law index (n) has an effect on both pressure flow and rotational flow shape factors. That is, F_d and F_p are dependent on the shear-thinning nature of the resin. Several empirical models were developed that focused on H/W and the power law factor n [9, 55–57]. These relationships, however, do not predict the experimental data found in Figure 7.16.

To obtain the correction factors, the rotational flow rate must be calculated using the generalized Newtonian method and the three-dimensional numerical method presented in Section 7.6. The calculations presented here were performed using five different power law resins with the form shown in Equation 7.54. Each resin had a K value of 1000. The power law index (n) values were 1.0, 0.8, 0.6, 0.4, and 0.2. Thus at a shear rate based on screw rotation ($\dot{\gamma}$) of 1 1/s, all resins had the same shear viscosity (η) of 1000 Pa·s, as shown in Figure 7.24. The melt density for the model resins was set at 1000 kg/m^3.

$$\eta = K\dot{\gamma}^{n-1} \tag{7.54}$$

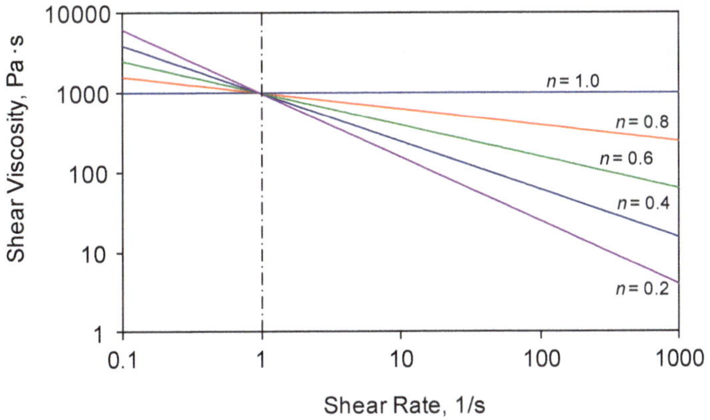

Figure 7.24 Shear viscosity for the power law fluids used for this study

A series of numerical experiments were performed using 100 mm diameter screws with lead lengths of 70 to 130 mm, and H/W ratios ranging from 0.023 to 0.18. The metering section channels were created by varying the depth from 2 to 15 mm and using single-, double-, and triple-flighted screws. The flight widths for all screws were 10 mm. The flight radii at the pushing and trailing sides were set to zero, creating a perfect rectangular channel for the generalized Newtonian method. The channel geometry used for the numerical method was essentially identical to that of the unwound channel. That is, the width of the channel at the screw root was smaller than that at the barrel wall as forced by geometric constraints.

The rotational flow rates were calculated by setting the pressure gradient ($\partial P/\partial z$) to zero. The ratio of the rotational flow rate calculated using the numerical method to that of the generalized Newtonian method is defined as F_c as follows:

$$F_c = \frac{Q_{md,numerical}}{Q_{md,Newtonian}} \tag{7.55}$$

Since the generalized Newtonian method was based on a constant viscosity for the entire channel, any factor that would cause the viscosity to vary in the channel may cause F_c to be different from 1. The factors known to cause the viscosity to vary include n and the average shear rate in the channel. To be consistent with the physical process, the average shear rate in the channel was calculated based on screw rotation using Equation 7.41. Calculation of the F_c correction factor for numerous screws at different screw speeds and thus different average shear rates indicated that the F_c value does not depend on shear rate. All data that follows were calculated at a screw speed of 100 rpm.

The F_c correction factors as a function of H/W for single-flighted, square-pitched screw geometries are shown in Figure 7.25. For these cases, only the channel depth was varied.

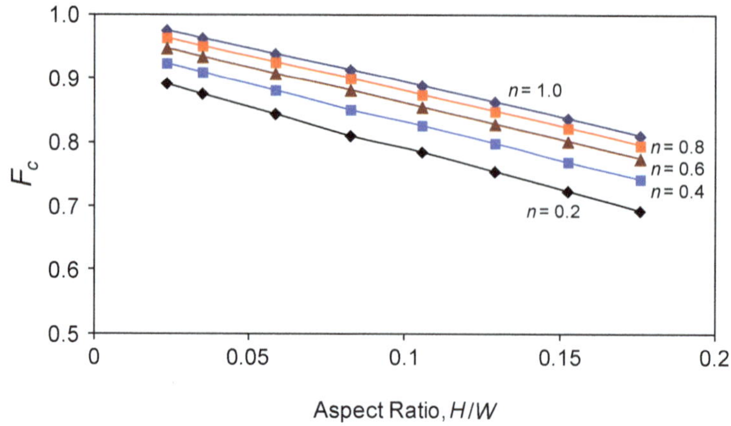

Figure 7.25 F_c correction factors as a function of H/W for single-flighted screws with square-pitched lead lengths (W/D_b = 0.85)

For a Newtonian fluid, the F_c correction factor approaches 1 as the aspect ratio approaches 0, as shown in Figure 7.25. As H/W approaches 0, the channel becomes very shallow with respect to the width, and it performs with minimal velocity reduction due to the flights. At $H/W = 0$, the channel has a depth of 0 and the rate goes to zero. As the H/W ratio increases, the F_c value decreases nearly linearly. The rate of the decrease is higher for a more shear-thinning fluid. For a screw with an H/W of 0.08 and operating with a power law index value of 0.4, the generalized Newtonian method would calculate a rotational flow rate that would be too high by about 15%; that is, the F_c for this case is 0.85.

Two additional factors needed to be included to make the correlation between the channel geometry and the correction factor. These geometric factors were required since a correlation with different lead lengths and the number of flight starts could not be predicted by H/W alone. The parameters added here were the ratio of the average channel width to the diameter of the barrel (W/D_b) and the number of flight starts (p). A series of numerical experiments were performed with constant aspect ratios H/W of 0.12 and 0.16. The width ratio W/D_b was changed by using different combinations of lead lengths and number of flight starts. For these experiments, the lead length was varied between 72 and 156 mm using single-flighted, double-flighted, and triple-flighted geometries. The combination allowed W/D_b to vary from 0.2 to 1.1. The correction factors for these geometries are shown in Figure 7.26 and Figure 7.27 for H/Ws of 0.12 and 0.16, respectively. The channel depth was varied to maintain the H/W constant. Screws with W/D_b ratios greater than 0.6 were constructed using single-flighted geometries and varying the lead length from 72 to 130 mm. Double-flighted screws were used to produce W/D_b ratios of 0.3 and 0.6 while triple-flighted screws were used to produce W/D_b ratios of 0.2 and 0.3.

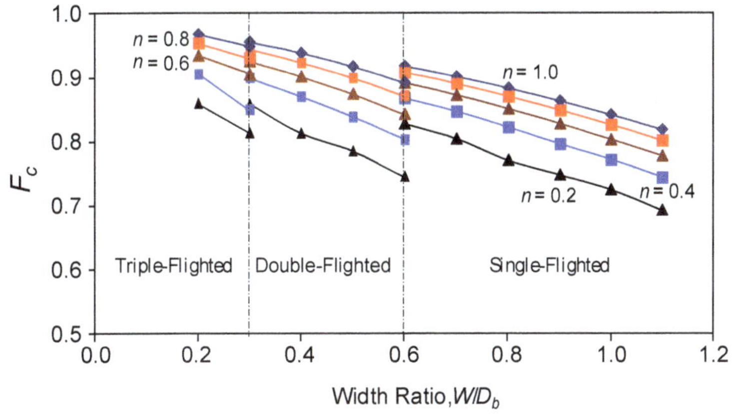

Figure 7.26 F_c factors as a function of W/D_b for a constant H/W of 0.12

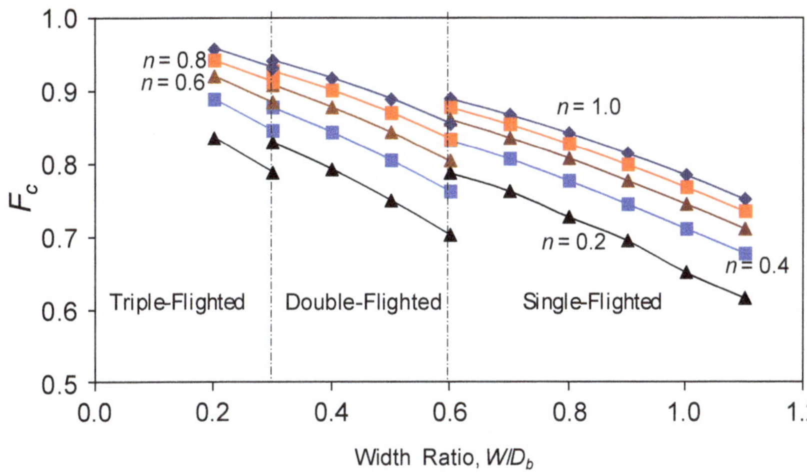

Figure 7.27 F_c factors as a function of W/D_b for a constant H/W of 0.16

As shown by Figure 7.26 and Figure 7.27, the F_c correction factor decreases as the width ratio W/D_b increases. Discontinuities occur between the number of flight starts.

In order to use the correction factors in a generalized Newtonian code, the factors need to be functionalized using an empirical equation. A total of 160 numerical experiments were performed to determine the effect of the design parameters on the correction factors. A random sample of 95 numerical data points were used to evaluate the correction factor fitting function. The equation for the correction factors is as follows:

$$F_c = 0.99755 - 0.08771\left(\frac{1}{n}\left[\frac{H}{W}\right]^{1/2}\right)^{1/2} - 8.79017\times10^{-3}\left(n\frac{W}{D_b}\right)^{1/2} + 0.14249\, n^{1/3}$$
$$- 0.58444\left[\frac{1}{n}\left(\frac{H}{W}\right)^{1/2}\left(n\frac{W}{D_b}\right)\right] - 0.45816\left[\frac{1}{n}\left(\frac{H}{W}\right)^{1/2}\left(n\frac{W}{D_b}\right)\right]^3\left[n^{1/3}\right]^{1/2} - 0.02835p \tag{7.56}$$

The quality of the F_c correlation from Equation 7.56 is shown by plotting it as a function of the F_c determined from the numerical experiments. The quality is shown by Figure 7.29 using a parity plot.

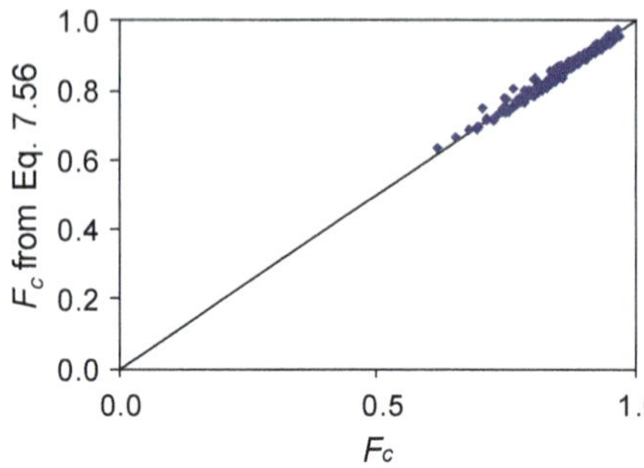

Figure 7.28 Parity plot showing the quality of the F_c correlation function (Equation 7.56) with the numerical experiments

The standard deviation for the 95 numerical data vs. the predicted fit for F_c was 0.0084. When the fitting function was applied to the complete data set, as shown by Figure 7.28, the standard deviation was 0.0095. This indicates that the function fits the total numerical population of F_c with the same goodness of fit as it does with the data used to generate the function.

Once F_c is determined using Equation 7.56, the corrected rotational flow rate is calculated using Equation 7.57. This corrected rotational flow rate should be very close to the actual rotational flow for the channel geometry. The calculation of the pressure gradient using Equation 7.58 is now much more accurate since the pressure-induced flow rate is now more accurate; that is, $Q_{mp} = Q_{md} - Q_m$.

$$Q_{md} = \frac{p\rho\left|V_{bz}\right|WHF_dF_c}{2} \tag{7.57}$$

$$Q_m = \frac{p\rho\left|V_{bz}\right|WHF_dF_c}{2} - \frac{p\rho WH^3F_p}{12\eta}\left[\frac{\partial P}{\partial z}\right] \tag{7.58}$$

Equation 7.57 and Equation 7.58 that are developed above use the V_{bz} as the velocity component as shown for screw rotation physics. As previously discussed, the classic drag flow model [8] assumed that the equivalent flow rate will be obtained if the barrel is rotated in the direction opposite to that of the screw as long as the linear velocity of the unwound barrel is numerically equal to the linear velocity of the screw core. For this classic barrel rotation model, V_{cz} is used as the velocity component instead of V_{bz}. Since V_{cz} is less than V_{bz}, the drag flow rate would be reduced. It is interesting to note here that the classical model using V_{cz} reduced the drag flow rate such that it provided a better estimate of the actual rotational flow rate, but for the wrong reason.

7.5.4 Simulation of the 500 mm Diameter Extruder Using F_c

The 500 mm diameter extrusion problem presented in Section 7.5.1 was simulated using the generalized Newtonian method with and without the F_c correction factor. The simulations and the experimentally determined pressure at a position 5.6 diameters downstream are shown in Figure 7.29. Recall that the generalized Newtonian method over-predicts the pressure by a factor of 1.7 (a 70% over-prediction) at 10.8 MPa. The method with the F_c correction factor predicts a pressure of 5.9 MPa, a pressure that is slightly lower than the experimentally determined pressure of 6.4 MPa, that is, an error less than 8%. This corrected method provides a pressure prediction that is acceptable for process troubleshooting and design.

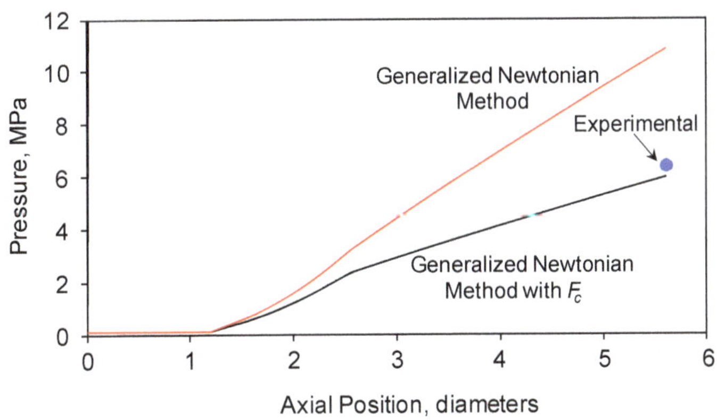

Figure 7.29 Simulated axial pressure profile for a 500 mm diameter extruder running 11,800 kg/h at 46 rpm for the 0.8 MI LDPE resin using the generalized Newtonian method with and without the F_c correction

The numerical and experimental results presented here show that the standard generalized Newtonian model must be used with caution when predicting the pressure profiles in extrusion screws. As previously stated, many commercially available codes use this method. The rotational flow calculated using the generalized Newtonian method is always higher than that experienced in an actual channel. Since the rotational flow is calculated too high, the pressure gradient must be incorrectly adjusted to compensate for the error in the rotational flow term. The F_c factors developed here allow a better calculation of the rotational flow and thus allow a better estimate of the pressure gradient. The addition of the F_c factor to the standard generalized Newtonian method was successfully demonstrated using this difficult extrusion case study. The engineering approach presented here allows the quick calculation of the rotational flow rate for non-Newtonian resins and provides an improved estimate of the pressure gradient in the channel.

The F_c factors correct for the non-Newtonian shear rheology effects that occur in the channel. The parameters that are used in the F_c correction correlation include rheological and geometric factors: power law index (n), aspect ratio of the channel (H/W), the ratio of the channel width to the screw diameter (W/D_b), and the number of flight starts (p).

Although the analysis here was performed using a power law viscosity model, other models could be used. For other viscosity models, the power law value n would be calculated using two reference shear rates, one higher and one lower than the shear rate calculated using Equation 7.41. These high and low shear rates and viscosity data would be used to determine a local n value as follows:

$$n-1=\frac{\log\eta_1-\log\eta_2}{\log\dot{\gamma}_1-\log\dot{\gamma}_2} \tag{7.59}$$

7.6 Frame Dependent Variables

The analysis so far in this chapter has concentrated on the calculation of flow rates in single-screw extruders and the application of a nonorthogonal frame change to convert the helical flow channel to a Cartesian channel, and by converting the moving boundary problem resulting from the rotating screw to a fixed boundary problem. The new transformed frame is relatively easy to evaluate. Then the solution in the transformed frame was brought back into the laboratory frame, and it became apparent that the extruder rate was dominated by the x-direction rotation of the screw flights and mitigated by a negative flow (z direction) from the screw core and helix. When using orthogonal converted axes, transformations are possible because the deformation rate tensor is frame indifferent [48, p. 400]. Chapter 6.7 of Malvern [48] describes the general concept of frame indifference and evaluates which variables of interest in evaluating flow and energy dissipation are frame indifferent under a frame change for orthogonal axis transformation. The definitions for frame change are presented in Equations 6.7.5 through 6.7.8 by Malvern. On page 387 [48], it is shown that velocity is not frame indifferent (Equation 6.7.11a), and on page 388 this is discussed in some detail. Furthermore, it is pointed out that acceleration is also not frame indifferent. On page 402 it is shown that stress is frame indifferent, but the time derivative of stress is not frame indifferent. Finally, on page 401 it is pointed out that Newtonian viscosity is also frame indifferent. As pointed out in the introduction for helix axis transformation, these certainties in orthogonal systems need to be checked in helical non-orthogonal systems.

In the next section viscous energy dissipation and the temperature increase of a polymeric fluid are evaluated for screw pumps and extruders. In order to check the dissipation, the flow characteristics for screw and barrel rotation also have to be verified. The traditional manner for the calculation of dissipation is to use the deformation rate tensor and the stress tensor. For those who are not interested in the tensor representation of the Lagrangian and Eulerian deformation rate the reader can skip to Section 7.7. The deformation rate tensor in generalized coordinates can be found in many references; for example, see Tadmor and Gogos [58].

In the Cartesian coordinate system the deformation rate tensor, $\boldsymbol{D}$, is defined as:

$$\boldsymbol{D} = \begin{pmatrix} 2\frac{\partial V_x}{\partial x} & \frac{\partial V_y}{\partial x}+\frac{\partial V_x}{\partial y} & \frac{\partial V_x}{\partial z}+\frac{\partial V_z}{\partial x} \\ \frac{\partial V_y}{\partial x}+\frac{\partial V_x}{\partial y} & 2\frac{\partial V_y}{\partial y} & \frac{\partial V_y}{\partial z}+\frac{\partial V_z}{\partial y} \\ \frac{\partial V_x}{\partial z}+\frac{\partial V_z}{\partial x} & \frac{\partial V_y}{\partial z}+\frac{\partial V_z}{\partial y} & 2\frac{\partial V_z}{\partial z} \end{pmatrix} \tag{7.60}$$

There is not an analytical velocity function for the y-direction velocity at the flights, so the wide channel approximation is used for demonstration purposes with a pressure gradient of zero. Using the equation developed previously for screw rotation for a very wide shallow channel, the transformed Lagrangian form of $\boldsymbol{D}_L$ is the same as the laboratory form for barrel rotation and is as follows:

$$\boldsymbol{D}_L = \begin{pmatrix} 0 & \frac{V_{bx}}{H}\left(2-3\frac{y}{H}\right)-3V_{bx}\frac{y}{H^2} & 0 \\ \frac{V_{bx}}{H}\left(2-3\frac{y}{H}\right)-3V_{bx}\frac{y}{H^2} & 0 & \frac{\left|V_{bz}\right|}{H} \\ 0 & \frac{\left|V_{bz}\right|}{H} & 0 \end{pmatrix} \tag{7.61}$$

The laboratory form of the deformation rate for screw rotation is defined as follows from the appropriate equations previously developed:

$$\boldsymbol{D}_E = \begin{pmatrix} 0 & \frac{V_{bx}}{H}\left(2-3\frac{y}{H}\right)-3V_{bx}\frac{y}{H^2}+\frac{V_{cx}}{R_c} & 0 \\ \frac{V_{bx}}{H}\left(2-3\frac{y}{H}\right)-3V_{bx}\frac{y}{H^2}+\frac{V_{cx}}{R_c} & 0 & \frac{|V_{bz}|}{H}-\frac{|V_{cz}|}{R_c} \\ 0 & \frac{|V_{bz}|}{H}-\frac{|V_{cz}|}{R_c} & 0 \end{pmatrix} \quad (7.62)$$

It is clear that:

$$\boldsymbol{D}_L \neq \boldsymbol{D}_E \quad (7.63)$$

Although Malvern [48] states that the deformation rate tensor is frame indifferent for orthogonal converted axes, the deformation tensor here is clearly frame dependent. It is believed that the frame dependency here is due to the nonorthogonal transformation from a helical system to the Cartesian system. Because of these difficulties with the results of the frame change [1, 2] in the nonorthogonal transformed coordinates, the resulting solutions must be carefully evaluated. Previously, the resulting flow equations were experimentally verified. In the next sections, the energy dissipation characteristics of the extruder will be carefully examined.

The shear stress tensor $\boldsymbol{\tau}$ for a Newtonian fluid is the product of the deformation rate tensor and the viscosity.

$$\boldsymbol{\tau}_{i,j} = -\mu \boldsymbol{D}_{i,j} \quad (7.64)$$

The dissipation is the scalar product of the shear stress and the deformation rate tensors.

$$\dot{\omega} = \sum_i \sum_j \boldsymbol{\tau}_{i,j} \boldsymbol{D}_{i,j} \quad (7.65)$$

Therefore this is an example of the result of using a nonorthogonal helical transformation. Since the shear rate is defined as:

$$\dot{\gamma}_L = \sqrt{\frac{1}{2}\boldsymbol{D}_L : \boldsymbol{D}_L} \quad \text{and} \quad \dot{\gamma}_E = \sqrt{\frac{1}{2}\boldsymbol{D}_E : \boldsymbol{D}_E} \quad (7.66)$$

It follows that the dissipation rate should be different for screw rotation in the laboratory frame from the expected dissipation in the Lagrangian frame for screw rotation and the laboratory frame for the barrel rotation for the same pumping rate of a particular screw.

7.6.1 Example Calculation of Energy Dissipation

In order to show the difference between screw and barrel rotation, a simple energy dissipation example is provided here. A complete analysis of energy dissipation will be presented in Section 7.7. For this example, a 50.8 mm diameter screw is rotated at a speed of 50 rpm and with a pressure gradient of zero, producing a rate of 6.8 cm^3/s. The geometric data and physical property data are provided in Table 7.4.

Table 7.4 Geometric and Physical Property Data for the Example Problem

Geometric Parameter		Physical Property	
Barrel diameter (D_b)	50.8 mm	Viscosity (μ)	1000 Pa·s
Channel depth (H)	2.54 mm	Density (ρ)	900 kg/m^3
Root diameter of screw (D_s)	45.72 mm	Heat capacity (C_v)	2500 J/(kg °C)
Channel width (W)	42.1 mm		
Helical length (Z)	83.7 mm		
Lead length (L)	50.8 mm		
Rotation rate	0.83 rev/s		
V_{cx}	0.036 m/s		
V_{cz}	-0.114 m/s		
V_{bx}	0.04 m/s		
V_{bz}	-0.127 m/s		

Evaluating the process at $y = 0.5H$, one obtains for the Lagrangian frame or barrel rotation the dissipation rate:

$$\dot{\omega}_L = \sum_i \sum_i \tau_{i,j} D_{Li,j} = 5.48 \times 10^6 \frac{J}{m^3 s} \quad \text{It follows:}$$

$$\dot{\omega}\, t_L = ZW \int_0^H \dot{\omega}_L dy = 49.0 \text{ W}$$

Next, the temperature increase for the process is calculated via an energy balance:

$$\Delta T_L = \frac{\dot{\omega}\, t_L}{\rho C_v Q} = 3.2\ °\text{C}$$

Evaluating the dissipation rate for the Eulerian or screw rotation:

$$\dot{\omega}_E = \sum_i \sum_i \tau_{i,j} D_{Ei,j} = 4.44 \times 10^6 \frac{J}{m^3 s} \quad \text{It follows:}$$

$$\dot{\omega}\, t_E = ZW \int_0^H \dot{\omega}_E dy = 39.7 \text{ W}$$

$$\Delta T_E = \frac{\dot{\omega}\, t_E}{\rho C_v Q} = 2.6\ °\text{C}$$

This analysis indicates that screw and barrel rotation should have different temperature increases even though they have the same rate predicted for the same rotation rate. In the following section the predictions will be checked experimentally.

7.7 Viscous Energy Dissipation and Temperature of the Resin in the Channel

Since historically the dissipation is evaluated using the local velocity at the boundary and the shear stress is evaluated as the product of the viscosity and the shear rate at the boundary, it follows that if the velocity is not frame indifferent then the dissipation will not be frame indifferent. As discussed previously in this chapter, rotation of the barrel at the same angular velocity as the screw are the conditions that produce the same theoretical flow rate as the rotating screw. Because the flow rate is the same and the dissipation is different, it follows that the temperature increase for barrel and screw rotation is different. This section will demonstrate this difference from both experimental data and a theoretical analysis.

The discharge temperature of the resin from the extruder is often just as important as the rate of the process. In many cases the discharge temperature limits the maximum rate at which an extruder can operate. For example, the discharge temperature will typically increase as the rate increases. Thus the maximum rate may be set by the maximum safe operating temperature for the process. The temperature of the resin in the channel and the discharge temperature are controlled by a combination of viscous energy dissipation and energy transfer through the barrel wall. Some energy can be transferred out the length of the screw, but for most processes this transfer rate is not significant and can be ignored. This section describes the viscous energy dissipation, heat transfer, and temperature of the resin in the screw channel. Numerous case studies are presented in later chapters that are limited by the discharge temperature.

Experimental and simulation results presented below will demonstrate that barrel rotation, the physics used in most texts and the classical extrusion literature, is not equivalent to screw rotation, the physics involved in actual extruders and used as the basis for modeling and simulation in this book. By changing the physics of the problem the dissipation and thus adiabatic temperature increase can be 50% in error for Newtonian fluids. For example, the temperature increase for screw and barrel rotation experiments for a polypropylene glycol fluid is shown in Figure 7.30. As shown in this figure, the barrel rotation experiments caused the temperature to increase to a higher level as compared to the screw rotation experiments.

The analysis presented here focuses on screw rotation analysis, in contrast to the historical analysis using barrel rotation [15–17]. It was pointed out recently by Campbell et al. [59] that the theory for barrel and screw rotation predicts different adiabatic melt temperature increases.

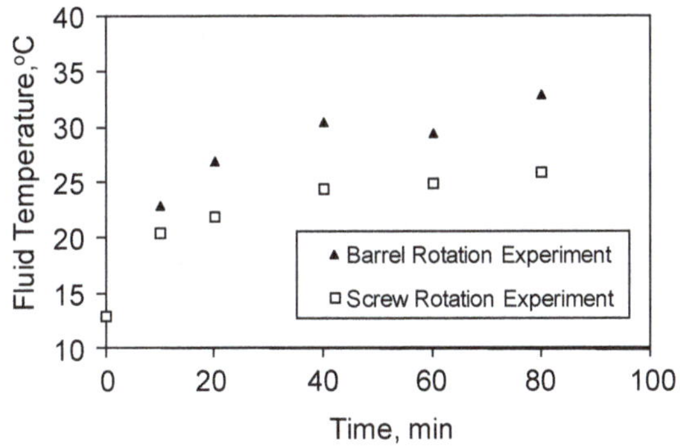

Figure 7.30 Experimental temperature increases for a Newtonian fluid (polypropylene glycol) as a function of the shearing time [47]. The data are for a square-pitched screw with a channel depth of 3.78 mm, a barrel diameter of 49.7 mm, and rotations at 55 rpm

The realization that the energy dissipation was different for screw and barrel rotation was "discovered" in Clarkson University's Polymer Processing Laboratory when Dianna Hunt [46] was evaluating if screw and barrel rotation of a screw pump produced the same rate regardless of the flight depth at several helix angles [46]. The pump stopped pumping for the deepest channel with a 30° pitch screw when the barrel was rotated but it had pumped when the screw was rotated. The very small discharge pressure caused by a 15 cm difference between the reservoir height and the discharge tube height needed to collect samples could not be overcome in the barrel rotation mode. Preliminary measurement of the fluid temperature at the exit of the screw pump appeared to indicate that the temperature of the working fluid was higher when the barrel was rotated than when the screw was rotated. The higher temperature for barrel rotation reduced the fluid viscosity and caused more pressure-induced flow due to the positive pressure gradient in the channel. This led to a two-year effort to quantify this difference in dissipation between barrel and screw rotations.

The experimental system used for this set of experiments consisted of an extruder-screw pump, fluid reservoir, temperature cooling coil, pressure and temperature measurement sensors, drive system, and a valve at the discharge. The device was very similar to that shown in Figure 7.4 except that the extruder-screw pump was designed in a manner that allowed the barrel and screw to be rotated independently. The barrel was made from cast acrylic resin and had an effective rotational length of 203.2 mm and an inside diameter of 49.7 mm. Screws were machined from steel with a flighted length of 254 mm. Thus, the flighted length of the screw was 50.8 mm longer than the effective length of the rotating barrel surface. The data reported here focuses on square-pitched screws with channel depths ranging from 4.76 to 11.0 mm [5]. As described earlier, the working fluid used in all of the visualization experiments and the experiments here was a polypropylene glycol fluid [46]. The viscosity was found to be Newtonian and to have a log-linear relationship with the reciprocal of the absolute temperature, as shown in Figure 7.31. To maintain a constant inlet temperature, the cooling coil was placed in the reservoir, and the fluid temperature was controlled by regulating the flow of coolant to the coil. The pressure difference between the inlet and outlet of the extruder was measured using pressure transducers. The temperature differences across the length of the extruder were measured using two thermocouples, one at the inlet and one at the outlet. A tachometer was used to monitor the rotational speed of the screw and barrel.

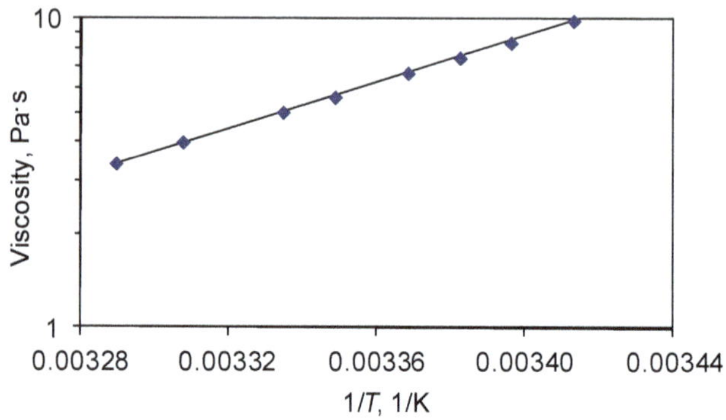

Figure 7.31 Viscosity of the polypropylene glycol fluid as a function of the absolute temperature [5]

The observation of the different rates of temperature increases during the flow experiments led to an extensive investigation into the mechanism of single-screw pump flow and energy dissipation in the Clarkson University Polymer Processing Laboratory [60]. It was decided to evaluate the temperature increase with this screw pump with no discharge; that is, $Q_{md} = Q_{mp}$. This required experimentally demonstrating that the barrel and screw rotation produced the same rate regardless of the screw channel depth. The first experiment was carried out with the outlet valve fully open. In this configuration the pressure gradient across the extruder was highly influenced by the height of the fluid discharge above the surface of the reservoir and the pressure across the open valve and the discharge pipe. The extruder was operated at several rotation speeds. Fluid was collected at a fixed time for determination of the rate. The indication of steady state was that no significant change in differential pressure occurred over a 20 minute time period. The inlet temperature was controlled to ±0.2 °C using the reservoir temperature controller.

The measured rates for both screw and barrel rotations as a function of rotation speed are shown in Figure 7.32 by the blue triangles and red squares, respectively. For this device, the measured rate for screw rotation was always higher than that for barrel rotation. As discussed previously, the flighted length of the screw was 50.8 mm longer than the rotating section of the barrel. Since the discharge pressure was the same for both rotations, the pressure gradient was higher for the barrel rotation case because the active length for the rotating barrel was shorter than that for screw rotation. To understand the effect of pressure, the pressure flows were calculated using Equation 1.22 using the pressure gradients known to exist in the channel, and they are shown by the dashed lines in Figure 7.32. The pressure flow for barrel rotation was more negative than that for screw rotation. Therefore, it follows that the measured rate for screw rotation exceeds the measured rate for barrel rotation for this device and operation.

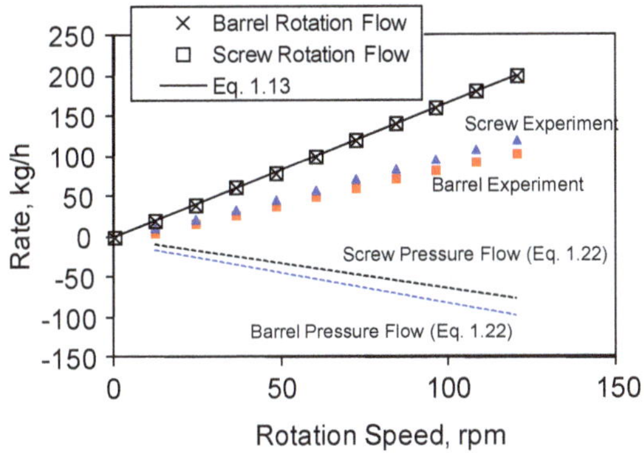

Figure 7.32 Screw rotation and barrel rotation flow rate comparison for a deep channel screw [5] running a polypropylene glycol fluid

The measured rate is the superposed sum of the positive rotational flow and the negative pressure flow. The rotational flows can be obtained by subtracting the negative pressure flow from the measured rate as shown in Figure 7.32 by the open square [□] and x-markers [×] for screw and barrel rotation, respectively. As shown in this figure, screw rotation and barrel rotation flows were the same within experimental error. The solid black line is the rotational flow predicted using Equation 1.13. Thus and as expected, barrel rotation and the screw rotation produce the same rotational rates.

Having established that the barrel and screw rotation pumped the same quantity of fluid when rotated at the same angular velocity, the effect of screw and barrel rotation on dissipation and thus temperature increase can be evaluated. A second set of experiments was designed to determine the effect of barrel and screw rotation on the temperature increase [59]. For these experiments, the extruder was operated at open discharge until it reached steady state at a screw speed of 55 rpm.

Then the valve was fully closed and the screw or barrel was rotated at 55 rpm for ten- to sixty-minute intervals. With the valve closed, the extruder was operated such that $Q_{md} = Q_{mp}$. At the end of the interval, the valve was opened and the fluid was pumped from the extruder to the reservoir. The temperature difference between the inlet and outlet of the extruder was then recorded. The temperature increase observed was always substantially more than the experimental measurement error. The experiment was stopped for about two hours before a new run was started to be certain that the screw pump had returned to its base-state laboratory temperature.

Temperature increases for the screw and barrel rotation experiments are shown by Figure 7.33 and Figure 7.34 for a deep and shallow screw, respectively. In both cases the screw was square pitched and the rotation speed of the barrel or screw was 55 rpm. It was observed that the temperature increase was higher for barrel rotation as compared to screw rotation at every rotation time. Thus, viscous energy dissipation is higher for barrel rotation. For example, the difference in temperature increase (ΔT) for the shallow screw was about twice as high for barrel rotation. For the deep channel screw, the difference in the temperature increase was about 50% higher for barrel rotation. As expected, the shallow channel dissipated a higher level of energy due to the narrower passages, and thus caused the temperature to be higher than that for the deep screw. For example, the screw with the shallow channel (H = 4.7 mm) had a temperature increase as the steady state approached that was about twice as high as that for the screw with the deep channel (H = 11 mm). For more details, the reader is referred to the complete study [47] where seven screws with varying helix angles and channel depths were studied.

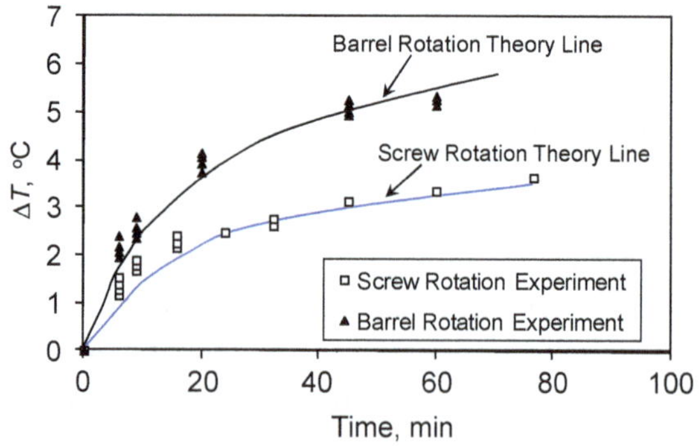

Figure 7.33 Screw and barrel temperature increase observed and predicted for an 11.1 mm deep channel screw with a square pitch (17.9°) for rotation rates of 55 rpm [5]

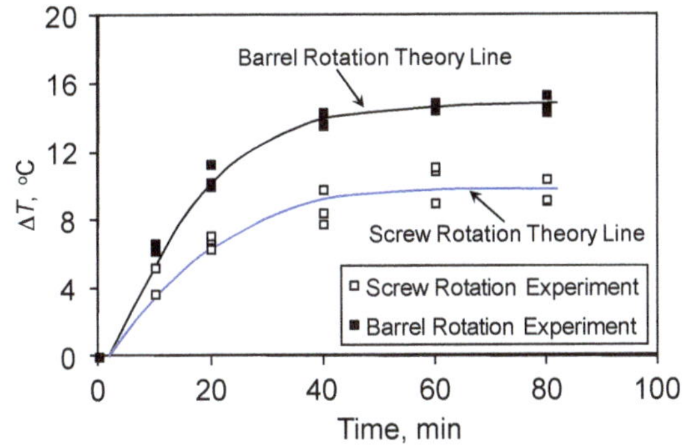

Figure 7.34 Screw and barrel temperature increase observed and predicted for a 4.7 mm deep channel screw with a square pitch (17.9°) for rotation rates of 55 rpm [5]

Since the goal in most screw design projects is to provide a molten resin at the proper temperature to a downstream operation such as a die, it is important to predict accurately the discharge temperature at the end of the extruder. Because the viscous dissipation is always higher for barrel rotation and since the screw rotates in commercial extruders, the dissipation model should be based on screw rotation physics. It is also apparent that the simulations based on barrel rotation assumptions will not provide a proper prediction of energy dissipation and discharge temperature. The theory used for these low-viscosity, closed-valve experiments must be modified to account for the flow in the extruder, since the preceding analysis was developed for a closed discharge extruder where $Q_{md} = Q_{mp}$, and thus was a transient analysis on a captured fluid. When dealing with commercial extruders the analysis must reflect the steady-state temperature increase as the resin flows down the metering section of the screw. The next sections will describe the energy balances.

Ideally, one would like to use the tensorial approach used in the Section 7.7 to calculate the energy dissipation. The generalized energy equation has a number of terms [58]:

$$\rho C_v \frac{DT}{Dt} = -(\nabla \cdot \boldsymbol{q}) + (-\boldsymbol{\tau} : \nabla \boldsymbol{v}) \tag{7.67}$$

where C_v is the heat capacity, $\boldsymbol{q}$ is the heat flux vector, and $\boldsymbol{v}$ is the velocity vector. Quoting from Tadmor and Gogos [58]: "We need a complete velocity and temperature field inside the machine. Because we do not have these we cannot calculate $\dot{\omega} = \int_V (-\boldsymbol{\tau} : \nabla \boldsymbol{v}) dV$ the total dissipation." Therefore, in this analysis the power input will be calculated as has been done by previous investigators [58, 61].

The temperature increase prediction lines in Figure 7.33 and Figure 7.34 were developed using the screw and barrel dissipation theory developed in Appendix

A7, which follows the analysis found in Middleman [61] and Tadmor and Gogos [58]. As shown in these figures, a very good fit exists between the measured data and the theory developed by Cheng [60]. As discussed previously, the analysis here is for $Q_{md} = Q_{mp}$. The theoretical equation to make this prediction is as follows:

$$\Delta T = \frac{\omega_i - \omega_0}{\rho\, C_v\, V} \tag{7.68}$$

where ΔT is the temperature increase, ω_i and ω_0 are the energy dissipation for the rotating screw and the energy transferred to the surroundings, respectively, and V is the volume of fluid in the screw pump. The temperature increase is dependent on the energy dissipation for the time of rotation with the valve closed minus the energy transferred to the environment through the barrel and down the core of the screw. The energy inputted by the rotation via viscous dissipation as a function of time is as follows:

$$\omega_i = t \iint_s n_i \tau_{ij} v_j \, ds \tag{7.69}$$

where $\boldsymbol{n}$ here is the normal unit vector, $\boldsymbol{\tau}$ is the shear stress tensor, $\boldsymbol{v}$ is the velocity vector, t is the time duration for the dissipation, and the s is the moving surface. The energy dissipation terms will be discussed in detail in the next section. The energy conducted out of the screw pump as a function of time through the barrel wall is:

$$\omega_0 = tK_b A\left(T_i - T_0\right) \tag{7.70}$$

where K_b is the heat transfer coefficient at the barrel wall, A is the heat transfer area, T_i is the time-dependent temperature of the fluid in the screw pump, and T_0 is the environmental temperature.

7.7.1 Energy Dissipation and Channel Temperature for Screw Rotation

In the following paragraphs a method is developed that models and predicts the temperature effects in an extruder. It is then followed by an example to demonstrate the use of the new dissipation model. This model is then extended for use as a control volume calculation method that allows the prediction of fluid temperatures as a function of the axial direction.

In this approach the energy input into the polymer fluid is modeled as the work done by the moving surface on the fluid. The rate of work ($\dot{\omega}_i$) is modeled as the product of the force (F) times the velocity (U).

$$\dot{\omega}_i = \sum FU \tag{7.71}$$

In the Cartesian system used for an unwrapped screw, the rate of work can be represented in terms of the viscosity, the local velocities, and the screw geometry.

$$\dot{\omega}_{total} = p\int_0^Z \int_0^W \left[\left(\tau_{xy}\big|_s\right)\left(U_x\big|_s\right) + \left(\tau_{zy}\big|_s\right)\left(U_z\big|_s\right)\right] dxdz + \dot{\omega}_\lambda \tag{7.72}$$

The double integral represents the nonzero terms of the dissipation rate tensor as adapted by Middleman [61] and Bernhardt and McKelvey for adiabatic extrusion [62]. The nontensorial approach was adopted by Tadmor and Klein in their classical text on extrusion [9]. In essence these are the nonzero terms of the dissipation rate tensor when it is applied to the boundary of the fluid at the solid-fluid interface. In the following development this historic analysis was adopted for energy dissipation for a rotating screw. In this case the velocities U_i are evaluated at the screw surface s and calculated in relation to screw rotation theory. The work in the flight clearance $\dot{\omega}_\lambda$ was previously described in the literature [9]. The shear stresses τ_{xy} and τ_{zy} can be calculated using either the Newtonian viscosity or a power law viscosity. This equation is developed for multiple flight starts with p being the number of starts. In both cases the viscosity should be temperature dependent, and the shear rate should be evaluated at the moving surface when used to evaluate the viscosity such as Equation 7.73.

$$\mu = K_o e^{-b\Delta T} \left|\dot{\gamma}_{total}\right|^{n-1} \tag{7.73}$$

If $n = 1$ then this is a Newtonian fluid and K_o is the Newtonian viscosity. The shear stress, τ, is the viscosity times the shear rate. The reader is encouraged to go to references [4–6, 45] to become familiar with the literature development of the screw surface velocities.

The viscous energy dissipation is calculated using velocities in the laboratory (Eulerian) reference frame. As previously stated here and by Malvern [48], velocities are not frame indifferent. The Eulerian velocities V_{xE}, V_{dzE}, and V_{pz} provided by Equation 7.22, Equation 7.24, and Equation 7.27 are the starting points for these calculations. Viscous dissipation is obtained by multiplying the shear stress on the screw surface by the velocity on the screw surface and integrating over the screw surface area. As a consequence of screw rotation, the viscous dissipation in the extruder arises from six components: (1) z-direction drag flow at the screw core, (2) x-direction drag flow at the core, (3) z-direction driven flow at the flight, (4) pressure flow at the screw core, (5) pressure flow at the screw flights, and (6) leakage flow over the flight lands. Since the velocity in the x direction is assumed not to be dependent on x, the dissipation component for the x-direction driven flow at the flight is assumed to be zero. These six dissipation components are calculated independently by using the principle of superposition. The full derivations for the energy dissipation components are provided in Appendix A7. The rate of work from screw rotation to the polymer $\dot{\omega}_s$ is expressed as the above components, and they have units of J/s. As previously stated, the derivation here is for the experiment using the screw pump where $Q_{md} = Q_{mp}$. The dissipation for a standard metering channel will be described later.

The rate of work inputted for flow in the z direction due to motion of the core, $\dot{\omega}_{core\_z}$, is as follows:

$$\dot{\omega}_{core\_z} = p\int_0^{W_c}\int_0^{Z_c}\left(\tau_{yz}\,|_{y=0}\right)\left(V_{dzE}\,|_{y=0}\right)dz\,dx \tag{7.74}$$

$$\dot{\omega}_{core\_z} = p\mu Z_c V_{cz}^2\left(1+\frac{H}{R_c}\right)\frac{8}{\pi}\left[\sum_{i=1,3,5,\ldots}^{\infty}\frac{1}{i\sinh\left(i\pi\frac{H}{W}\right)}\right] - \frac{p\mu V_{cz}^2 Z_c W_c}{R_c} \tag{7.75}$$

The rate of work for flow in the x direction due to the motion of the core, $\dot{\omega}_{core\_x}$, is as follows:

$$\dot{\omega}_{core\_x} = p\int_0^{W_c}\int_0^{Z_c}\left(\tau_{yx}\,|_{y=0}\right)\left(V_{xE}\,|_{y=0}\right)dz\,dx \tag{7.76}$$

$$\dot{\omega}_{core\_x} = p\mu\left(\frac{2V_{cx}^2}{H}+\frac{3V_{cx}^2}{R_c}\right)Z_c W_c \tag{7.77}$$

The rate of work for flow in the z direction due to the motion of the flights, $\dot{\omega}_{flight\_z}$, is as follows:

$$\dot{\omega}_{flight\_z} = 2pZ_c\int_0^H\left(\mu\left|V_{cz}\right|\left(1+\frac{H}{R_c}\right)\frac{4}{\pi}\sum_{i=1,3,5,\ldots}^{\infty}\frac{\sinh\left(i\pi\frac{y}{W}\right)}{\sinh\left(i\pi\frac{H}{W}\right)}\frac{\pi}{W}\right)\left|V_{cz}\right|\left(1+\frac{y}{R_c}\right)dy \tag{7.78}$$

The integration shown in Equation 7.78 is complex and is performed numerically.

The viscous dissipation that occurs due to pressure-induced flow is calculated in the same manner and it has two components due to the motion at the core and at the flights. For pressure-induced flow in the z direction and the motion of the screw core $\dot{\omega}_{core\_p}$:

$$\dot{\omega}_{core\_p} = \frac{pH^2 Z_c\left|V_{cz}\right|}{2}\left|\frac{\partial P}{\partial z}\right|\left[\frac{W_c}{H}-\frac{8}{\pi^3}\sum_{i=1,3,5,\ldots}^{\infty}\frac{\sinh\left[i\pi\frac{W}{H}\left(\frac{W_c}{W}-\frac{1}{2}\right)\right]+\sinh\left[\frac{i\pi W}{2H}\right]}{i^3\cosh\left(i\pi\frac{W}{2H}\right)}\right] \tag{7.79}$$

When $\frac{W_c}{W}$ approaches 1, then the above equation reduces to:

$$\dot{\omega}_{core\_p} = \frac{pH^2 Z_c |V_{cz}|}{2} \left|\frac{\partial P}{\partial z}\right| \left(\frac{W_c}{H} - \frac{16}{\pi^3} \sum_{i=1,3,5,...}^{\infty} \frac{\tanh\left[i\pi \frac{W}{2H} \right]}{i^3} \right) \tag{7.80}$$

The absolute value of the pressure gradient is used for all dissipation calculations. Although the pressure-induced flow is in different directions for a positive- and negative-valued pressure gradient operation, the dissipation level, however, is identical. Viscous energy dissipation is always positive definite.

For pressure-induced flow in the z direction, the motion of the screw flights $\dot{\omega}_{flight\_p}$ is as follows:

$$\dot{\omega}_{flight\_p} = 2pZ_c \int_0^H \left(\frac{H^2}{2} \left|\frac{\partial P}{\partial z}\right| \frac{8}{\pi^3} \left(\sum_{i=1,3,5,...}^{\infty} \frac{\tanh\left(i\pi \frac{W}{2H} \right) \sin\left(i\pi \frac{y}{H} \right)}{i^2} \frac{\pi}{H} \right) \right) |V_{cz}| \left(1 + \frac{y}{H} \right) dy \tag{7.81}$$

The integration shown in Equation 7.81 is complex and is performed numerically.

Rowell and Finlayson described leakage flow through the flight clearance as flow between two parallel plates [7]. The rate of energy dissipation between the flight lands and the barrel wall is given by Equation 7.82. This dissipation is considered to be the same as that for barrel rotation, and it is provided in the literature [9].

$$\dot{\omega}_\lambda = \frac{p\mu_f \pi^2 N^2 D_b^2 e Z_b}{\lambda} \tag{7.82}$$

The heat transferred through the barrel wall over the channel, E_T, is given by Equation 7.83. A heat transfer expression could also be written for energy transferred from the molten polymer to the barrel over the flight land and also from the melt to the screw surface.

$$E_T = pK_b W_b Z_b (T_m - T_b) \tag{7.83}$$

where T_m is the melt temperature of the resin in the channel and T_b is the temperature of the barrel wall.

The total dissipation is the sum of the component dissipations for the screw and the energy rate transferred through the barrel wall. The temperature change ΔT is computed as shown in Equation 7.84 for the experiments detailed in Figure 7.32 to Figure 7.34.

$$\Delta T = \frac{\left(\dot{\omega}_{core\_z} + \dot{\omega}_{core\_x} + \dot{\omega}_{flight\_z} + \dot{\omega}_{core\_p} + \dot{\omega}_{flight\_p} - E_T\right)}{Q_m C_v} \tag{7.84}$$

Equation 7.84 assumes that the viscosity is not a function of temperature. It is presented here to describe the method. In Section 7.7.5, a control volume method will be described that allows the temperature to vary in the down-channel direction.

A simple form for the dissipation can be calculated for channels with small aspect ratios (H/W less than 0.1). For these cases, Equation 7.22, Equation 7.25, and Equation 7.28 are the velocity equations used for the calculations. The derivations of the dissipation equations are presented in Appendix A7.

7.7.2 Energy Dissipation and Channel Temperature for Barrel Rotation

The dissipation is calculated for barrel rotation in the screw pump device in a similar manner. For this case, the velocities V_{xL}, V_{dzL}, and V_{pz} are used for the calculations, and they are provided by Equation 7.21, Equation 7.23, and Equation 7.27, respectively. The rate of work input for barrel rotation $\dot{\omega}_b$ is obtained by multiplying the normal shear stress for the barrel surface by the velocity of the barrel surface integrated over the barrel surface area. The traditional method for calculation of dissipation for barrel rotation is as follows:

$$\dot{\omega}_b = p\int_0^{W_b}\int_0^{Z_b}\left(\tau_{yz}\big|_{y=H}\right)\left(V_z\big|_{y=H}\right)dzdx + p\int_0^{W_b}\int_0^{Z_b}\left(\tau_{yx}\big|_{y=H}\right)\left(V_x\big|_{y=H}\right)dzdx + \dot{\omega}_\lambda \tag{7.85}$$

For the x component of flow:

$$V_x = V_{cx}\left(1+\frac{H}{R_c}\right)\frac{y}{H}\left(2-3\frac{y}{H}\right) = V_{bx}\frac{y}{H}\left(2-3\frac{y}{H}\right) \tag{7.21}$$

The velocity component in the z direction for drag flow and pressure flow are provided in Equation 7.86 and Equation 7.87. This equation uses Equation 7.23 and Equation 7.27 for H/Ws that are less than 0.1. If the channel aspect ratio H/W is greater than 0.1, then Equation 7.23 and Equation 7.27 should be used. The generalized solution using Equation 7.23 and Equation 7.27 is provided in Appendix A7.

For the z component of drag flow for H/W less than 0.1:

$$V_z = \left|V_{bz}\right|\frac{y}{H} \tag{7.86}$$

Here V_{bz} is a positive value for barrel rotation. For the z component of pressure flow for H/W less than 0.1:

$$V_{pz} = \frac{H^2}{2\mu}\frac{\partial P}{\partial z}\left[\left(\frac{y}{H}\right)^2 - \frac{y}{H}\right] \tag{7.87}$$

Substituting the shear stresses and velocities and then integrating provides the rate of work for the channel as Equation 7.88 for barrel rotation. As before for screw rotation, dissipation is a positive definite quantity, and thus the absolute value of the pressure gradient is used. The dissipation between the flight lands and barrel wall is the same for barrel and screw rotation.

$$\dot{\omega}_b = \frac{p\mu V_{bz}^2 W_b Z_b}{H} + \frac{p\left|V_{bz}\right| H W_b Z_b}{2}\left|\frac{\partial P}{\partial z}\right| + \frac{4p\mu V_{bx}^2 W_b Z_b}{H} + \frac{p\mu_f \pi^2 N^2 D_b^2 e Z_b}{\lambda} \tag{7.88}$$

The overall energy balance for the metering channel for barrel rotation is given by Equation 7.89:

$$\Delta T = \frac{\left(\dot{\omega}_b - E_T\right)}{Q_m C_v} \tag{7.89}$$

7.7.3 Temperature Increase Calculation Example for a Screw Pump

Experiments were run on a laboratory screw pump to evaluate the fluid temperature increase of an enclosed fluid as a function of element rotation time. For this case, the device was similar to that shown in Figure 7.4, and the discharge was blocked using a valve. The blocked flow caused the rate to be zero and $Q_{md} = Q_{mp}$. Both the device and the fluid were at room temperature at the start of each experiment. The task at hand is to use the information below and the equations presented in Chapter 7 to calculate the temperature increase for the fluid for a total time of 30 seconds in three-second increments. The dimensions of the single-flighted extrusion device are provided in Table 7.5.

Table 7.5 Laboratory Screw Pump Dimensions for the Temperature Increase Problem

	Parameter
Barrel diameter, mm	49.66
Lead length (single-flighted), mm	50.8
Channel depth, mm	11.1
Flight clearance, mm	0.76
Flight width, mm	7.2
Effective screw length, mm	254
Fluid volume for screw rotation, m^3	2.66×10^{-4}

The viscosity for the Newtonian polypropylene glycol fluid used for the experiments was measured and then modeled using Equation 7.90. The temperature response of this fluid was very similar to the polypropylene glycol fluid shown in Figure 7.31 except that this fluid was considerably more viscous. The density and heat capacity were 1015 kg/m^3 and 2000 J/(kg °C), respectively.

$$\mu = \mu_0 \exp\left(\frac{\Delta E}{R_g}\left(\frac{1}{T} - \frac{1}{T_0}\right)\right) \tag{7.90}$$

where μ is the viscosity, μ_0 is the viscosity at the reference temperature and equal to 908.5 Pa·s, ΔE is the activation energy at 78,010 J/mol, R_g is the ideal gas constant (8.314 J/[mol K]), T is the operating temperature in Kelvin, and T_0 is the reference temperature (295 K). The activation energy is high, but it is typical of polypropylene glycol fluids. At the start of the process, the device and fluid are at the reference temperature. The environment around the device is also at the temperature T_0.

The screw pump was operated at a screw speed of 85 rpm (N = 1.417 revolutions/s). The calculated results are shown in Table 7.6, which was generated using the method found earlier in this chapter and in Appendix A7.

Table 7.6 Temperature of the Fluid as a Function of Time and the Element Rotating

Time, s	ΔT for barrel rotation case, °C	Viscosity for barrel rotation case, Pa·s	ΔT for barrel rotation case neglecting dissipation in the clearance, °C	ΔT for screw rotation case, °C	Viscosity for screw rotation case, Pa·s	ΔT for screw rotation case neglecting dissipation in the flight clearance, °C	ΔT for screw rotation using the abbreviated method of Equation 7.90 ($H/W < 0.1$), °C
0	0	908.5	0	0	908.5	0	0
3	4.0	596.6	2.4	3.4	631.0	2.0	2.7
6	6.6	455.1	4.2	5.8	492.6	3.6	4.7
9	8.5	371.0	5.8	7.6	407.4	5.0	6.4
12	10.2	314.6	7.1	9.2	348.9	6.1	7.8
15	11.6	273.7	8.2	10.4	306.1	7.2	9.0
18	12.8	242.7	9.3	11.6	273.2	8.1	10.0
21	13.9	218.1	10.2	12.6	247.2	9.0	11.0
24	14.8	198.3	11.0	13.5	225.9	9.7	11.9
27	15.7	181.8	11.8	14.3	208.3	10.4	12.7
30	16.5	167.9	12.5	15.1	193.4	11.1	13.4

The governing equations for screw rotation were modified from the equations in the chapter to provide a simple and illustrative example. A straightforward calculation of the total energy generation for the time increment is performed by multiplying the summation of these terms by the viscosity of the fluid at the beginning of the increment and the time for heating during the increment. A more accurate solution would be obtained either by using many more increments or by iterating at each time using a new temperature for the viscosity based on some temperature between the inlet temperature and the exit temperature. It is instructive to examine each energy element contribution in the format presented here. The velocity and geometric components are calculated as follows:

$V_{cx} = \pi N D_c \sin\theta_c = 0.062$ m/s $\qquad V_{cz} = -\pi N D_c \cos\theta_c = -0.105$ m/s

$V_{bz} = -\pi N D_b \cos\theta_b = -0.210$ m/s $\quad R_c = (D_b - 2H)/2 = 0.014$ m

$$\theta_c = \arctan\left(\frac{L}{\pi D_c}\right) = 30.5° \quad \theta_b = \arctan\left(\frac{L}{\pi D_b}\right) = 18.0°$$

$W_c = L\cos(\theta_c) - e = 0.037$ m $\qquad W_b = L\cos(\theta_b) - e = 0.041$ m

The discharge of the screw is closed by the valve, $Q_{md} = Q_{mp}$, which leads to the following pressure gradient term in the device using Equation 1.28:

$$\frac{1}{\mu}\frac{\partial P}{\partial z} = \frac{\Delta P}{\mu Z_s} = \frac{6V_{cz}\left(1+\frac{H}{R_c}\right)F_d}{H^2 F_p} = \frac{6\left|V_{bz}\right|F_d}{H^2 F_p} = \frac{6(0.210\ \text{m/s})(0.845)}{(0.0111\ \text{m})^2(0.821)}$$

$$= 10{,}500\ \text{m}^{-1}\text{s}^{-1} = 10{,}500\ \frac{\text{Pa}}{(\text{Pa}\cdot\text{s})(\text{m})}$$

Here the viscosity is combined with the pressure gradient to make the calculations simple. This combined method is acceptable only for Newtonian fluids. The rate of work divided by the viscosity for the components is next.

Dissipation ratio for flow in the z direction due to the motion of the core:

$$\frac{\dot{\omega}_{core\_z}}{\mu} = \frac{pZ_c\left|V_{cz}V_{bz}\right|}{2}\left(\frac{8}{\pi}\sum_{i=1,3,5,\ldots}^{\infty}\frac{1}{i^3\sinh\left(\frac{i\pi H}{W_c}\right)}\right) - \frac{pV_{sz}^{\ 2}Z_c W_c}{R_c} = 0.012\ \text{J/(Pa}\cdot\text{s}^2)$$

Dissipation ratio for flow in the x direction due to the motion of the core:

$$\frac{\dot{\omega}_{core\_x}}{\mu} = p\left(\frac{2V_{cx}^{\ 2}}{H} + \frac{3V_{cx}^{\ 2}}{R_c}\right)Z_c W_c = 0.028\ \text{J/(Pa}\cdot\text{s}^2)$$

Dissipation ratio for flow in the z direction due to the motion of the flights:

$$\frac{\dot{\omega}_{flight\_z}}{\mu} = 2Z_c \int_0^H V_{sz}\left(1+\frac{y}{R_c}\right)\left|V_{bz}\right|\frac{4}{\pi}\sum_{i=1,3,5,...}^{\infty}\left(\frac{\sinh\left(\frac{i\pi y}{W_c}\right)}{\sinh\left(\frac{i\pi H}{W_c}\right)}\frac{\pi}{W_c}\right)dy = 0.172\ \text{J/(Pa}\cdot\text{s}^2)$$

Dissipation ratio for flow in the z direction induced by the pressure gradient:

$$\frac{\dot{\omega}_{flight\_p}}{\mu} = 2p\Delta Z_c \int_0^H \left|V_{cz}\right|\frac{H}{2\mu}\left|\frac{\partial P}{\partial z}\right|\left(1+\frac{y}{R_c}\right)\frac{8}{\pi^3}\sum_{i=1,3,5,...}^{\infty}\frac{\tanh\left(\frac{i\pi W_c}{2H}\right)\pi}{i^2 H}\sin\left(i\pi\frac{y}{H}\right)dy$$
$$= 0.064\ \text{J/(Pa}\cdot\text{s}^2)$$

The dissipation due to the pressure flow and the screw core rotation is then:

$$\frac{\dot{\omega}_{core\_p}}{\mu} = \frac{p\left|V_{cz}\right|Z_c H^2}{2\mu}\left|\frac{\partial P}{\partial z}\right|\left(\frac{W_c}{H}-\frac{16}{\pi^3}\sum_{i=1,3,5,...}^{\infty}\frac{\tanh\left(\frac{i\pi W_c}{2H}\right)}{i^3}\right) = 0.093\ \text{J/(Pa}\cdot\text{s}^2)$$

Dissipation ratio for flow through the flight clearance (λ = 0.00075 m):

$$\frac{\dot{\omega}_\lambda}{\mu_f} = \frac{p\pi^2 N^2 D_b^2 e Z_b}{\lambda} = 0.309\ \text{J/(Pa}\cdot\text{s}^2)$$

The k + 1 temperature increment is calculated using this equation where Δt is the increment of time (3 s) that the fluid is being heated. The temperature increase for a time increment is calculated using Equation 7.92. The heat transfer coefficient was experimentally determined for the screw pump device by operating the device until the temperature remained constant. It was determined to be 5.5 J/(m²·s·°C). The subscript k+1 indicates the current temperature calculation and k is the previously calculated temperature.

$$\Delta T_k = \left(T_{k+1} - T_k\right) = \frac{\left(\frac{\dot{\omega}_{core\_x}}{\mu}+\frac{\dot{\omega}_{core\_z}}{\mu}+\frac{\dot{\omega}_{flight\_p}}{\mu}+\frac{\dot{\omega}_{flight\_z}}{\mu}+\frac{\dot{\omega}_{core\_p}}{\mu}+\frac{\dot{\omega}_\lambda}{\mu}\right)\mu(T)\Delta t - K_b W_b Z_b\left(T_k - T_0\right)\Delta t}{\rho C_v V} \quad (7.91)$$

$$\Delta T_k = \left(T_{k+1} - T_k\right) = \frac{\left(0.678\ \frac{\text{J}}{\text{Pa}\cdot\text{s}^2}\right)\mu(T)\Delta t - K_b W_b Z_b\left(T_k - T_0\right)\Delta t}{\rho C_v V} \quad (7.92)$$

Equation 7.92 was used to construct the screw rotation data in Table 7.6.

The fluid temperature increase was calculated using the equation for barrel rotation using Equation 7.93. Like the equation for screw rotation, this equation was developed for a screw pump with a closed discharge.

$$\Delta T_k = \left(T_{k+1} - T_k\right) = \frac{\left(\frac{V_{bz}^2}{H} + \frac{3V_{bz}^2 F_d}{HF_p} + \frac{4V_{bx}^2}{H} + \frac{\pi^2 N^2 D_b^2 e}{\lambda W_b}\right)\mu(T) W_b Z_b \Delta t - K_b W_b Z_b \left(T_k - T_0\right)\Delta t}{\rho C_v V} \quad (7.93)$$

Equation 7.93 was used to calculate the barrel rotation channel temperature increases for Table 7.6. As shown by the data in Table 7.6, the temperature increase for barrel rotation is higher than that for screw rotation, consistent with the experimental data presented previously.

Examination of the energy contributions for the dissipation elements above indicates that a large portion of the energy is dissipated in the flight clearance. The fourth and seventh columns in the table were generated neglecting the dissipation source in the flight clearance. It is observed that the difference in temperature increase is considerably higher when this source is included. Neglecting the dissipation source in the flight clearance is not recommended.

The temperature increase for screw rotation was calculated using the same channel geometry and conditions as discussed above, except that the channel depth was changed from 11.1 mm to 3.8 and 18.4 mm. As expected, the temperature increase for the process was the highest for the shallowest screw and the lowest for the deepest screw. The temperature increase as a function of time for the three channel depths is shown in Figure 7.35. The channel with a depth of 3.8 mm is in the commercially important range for a machine of this diameter.

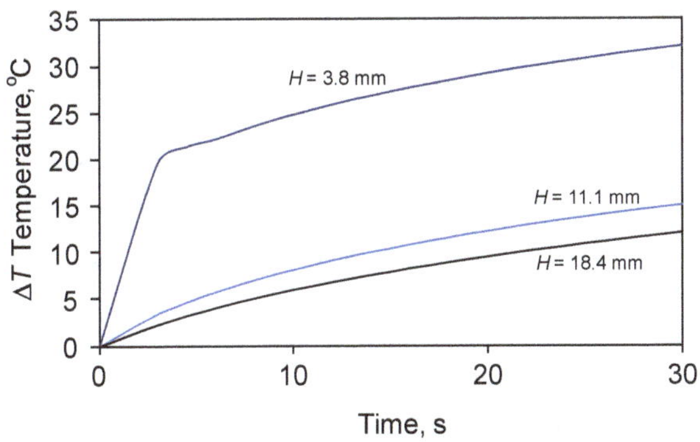

Figure 7.35 Temperature increase for present example as a function of channel depth for screw rotation

Calculation of the energy dissipation using Equation 7.92 is the proper method for this closed-discharge screw pump problem. The resulting energy dissipation terms are relatively complex due to the summations and the integrations. For screws with shallow channels such that H/W is less than 0.1, the energy calculation can be simplified to Equation 7.94. The flight drag and pressure dissipation were not used because the contributions from the flights were neglected; that is, H/W is less than 0.1. The calculation of the example with a channel depth of 11.1 mm in the above example is shown in the last column of Table 7.6. As shown by the data in this table, calculations produced temperature increases of 15.1 and 13.4 °C for the full calculation and the simplified calculation, respectively. This error here is likely unacceptable for this method due to the H/W of 0.3. For a channel with a depth of 3.8 mm and an H/W of 0.10, the calculation methods provide temperature increases of 32.0 and 30.3 °C for the full and simplified calculations, respectively. As expected, the simplified calculation under-predicts the energy dissipation. The simplified method for calculation of the temperature increase for a closed-discharge screw pump for screw rotation as in the present example is as follows:

$$\Delta T_k = \left(T_{k+1} - T_k\right) = \tag{7.94}$$
$$\frac{\left(\frac{2V_{cx}^2}{H} + \frac{3V_{cx}^2}{R_c} + \frac{V_{cz}^2}{H} + \frac{3\left|V_{cz}V_{bz}\right|F_d}{HF_p} + \frac{\pi^2 N^2 D_b^2 e Z_b}{\lambda Z_c}\right)\mu(T) W_c Z_c \Delta t - K_b W_b Z_b \left(T_k - T_0\right)\Delta t}{\rho C_v V}$$

Equation 7.94 was used to calculate the simplified screw rotation channel temperature increases for Table 7.6.

7.7.4 Heat Transfer Coefficients

As discussed in the earlier sections, the heat transfer at the barrel wall is an important method of adding or removing energy from the extrusion process. The heat transfer coefficient at the barrel wall has been studied by many groups. Jepson [63] developed a simple wiping model for heat transfer where the heat transfer coefficient depended on the wiping rate of the flights and the clearance between the flight tip and the barrel wall. That is, the heat transfer coefficient increased with increasing screw speed and number of flight starts and decreasing flight clearance. Jepson's method placed the heat transfer coefficient between 100 and 850 J/(m^2·s·°C). For example, a 200 mm diameter extruder equipped with a single-flight screw with a standard flight clearance of 0.2 mm operating at 100 rpm would have a heat transfer coefficient at the barrel wall of about 400 J/(m^2·s·°C).

Todd [64] applied heat transfer technology from batch mixer tanks using close-fitting agitator blades [65] to model heat transfer data from twin-screw extruders. The model that best fit his data is as follows:

$$\left(\frac{K_b D_b}{k}\right) = 0.94\left(\frac{D_b^2 N\rho}{\mu}\right)^{0.28}\left(\frac{C_v\mu}{k}\right)^{0.33}\left(\frac{\mu}{\mu_w}\right)^{0.14} \tag{7.95}$$

where k is the thermal conductivity of the resin, μ is the shear viscosity of the resin in the channel, and μ_w is the shear viscosity at the wall. The viscosity for the resin in the channel can be estimated at the average shear rate based on the depth of the channel and the bulk temperature. The last term of Equation 7.95 is used to correct for the viscosity at the wall due to a different temperature and shear rate condition. The quantities in the brackets are all dimensionless. For the 200 mm diameter screw presented above and for a 2 MI LDPE resin, the heat transfer coefficient estimated using Equation 7.95 is about 100 J/(m²·s·°C). This value is considerably less than that estimated by Jepson [63]. Unless otherwise stated, heat transfer coefficients provided by Equation 7.95 were used for the simulations provided in this book. A more complete version of this type of analysis was developed by White et al. [66].

7.7.5 Temperature Calculation Using a Control Volume Technique

The temperature increase calculation in Section 7.7.1 and Section 7.7.2 was based on the viscosity using the temperature at the entry to the metering section. Because the temperature of the resin increases as it flows downstream, the shear viscosity continuously decreases. A better method to calculate the temperature of the resin in the channel is to divide the channel into many Δz_c or Δz_b increments, and then for each increment, perform an energy balance on each control volume [67]. A schematic of the control volume is shown in Figure 7.36. The energy balance includes convection into and out of the volume, dissipation due to rotation and pressure flows, and energy conduction through the barrel wall and the root of the screw. This section will describe a control volume method for temperature calculation for both screw rotation and barrel rotation.

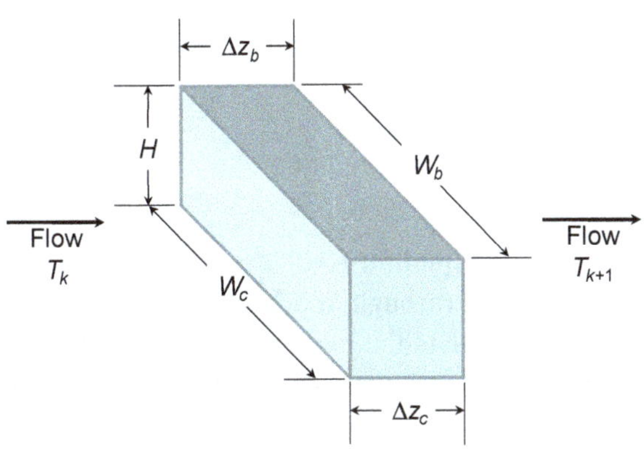

Figure 7.36 Control volume slice for temperature calculations

7.7.5.1 Temperature Calculation for Screw Rotation

The rate of work inputted by the rotating screw, energy conducted through the barrel wall, and energy conducted to the screw can be accounted for in each volume. The energies dissipated $\dot{\omega}'$ here are based on the flow in a single channel and a unit length for the Z_c surface. The calculation is as follows for the kth volume for screw rotation:

The energy dissipations for unit length of Z_c with units of J/(m·s) are as follows:

$$\dot{\omega}'_{core\_z} = \mu V_{cz}^2 \left(1+\frac{H}{R_c}\right)\frac{8}{\pi}\left[\sum_{i=1,3,5,\ldots}^{\infty} \frac{1}{i\sinh\left(i\pi \frac{H}{W}\right)}\right] - \frac{\mu V_{cz}^2 W_c}{R_c} \tag{7.96}$$

$$\dot{\omega}'_{core\_x} = \mu\left(\frac{2V_{cx}^2}{H} + \frac{3V_{cx}^2}{R_c}\right)W_c \tag{7.97}$$

$$\dot{\omega}'_{core\_p} = \frac{H^2 \left|V_{cz}\right|}{2}\left|\frac{\partial P}{\partial z}\right|_k \left(\frac{W_c}{H} - \frac{16}{\pi^3}\sum_{i=1,3,5,\ldots}^{\infty} \frac{\tanh\left[i\pi \frac{W}{2H}\right]}{i^3}\right) \tag{7.98}$$

where $\partial P/\partial z$ is the pressure gradient at the start of the increment at k.

$$\dot{\omega}'_{flight\_z} = 2\int_0^H \left(\mu\left|V_{cz}\right|\left(1+\frac{H}{R_c}\right)\frac{4}{\pi}\sum_{i=1,3,5,\ldots}^{\infty} \frac{\sinh\left(i\pi \frac{y}{W}\right)}{\sinh\left(i\pi\frac{H}{W}\right)}\frac{\pi}{W}\right)\left|V_{cz}\right|\left(1+\frac{y}{R_c}\right)dy \tag{7.99}$$

$$\dot{\omega}'_{flight\_p} = 2\int_0^H \left(\frac{H^2}{2}\left|\frac{\partial P}{\partial z}\right|_k \frac{8}{\pi^3}\left(\sum_{i=1,3,5,\ldots}^{\infty} \frac{\tanh\left(i\pi\frac{W}{2H}\right)\sin\left(i\pi\frac{y}{H}\right)}{i^2}\frac{\pi}{H}\right)\right)\left|V_{cz}\right|\left(1+\frac{y}{H}\right)dy \tag{7.100}$$

$$\dot{\omega}'_{\lambda} = \frac{\mu_f \pi^2 N^2 D_b^2 e Z_b}{\lambda Z_c} \tag{7.101}$$

The integrations shown in Equation 7.99 and Equation 7.100 are complex and are performed numerically. The heat transferred through the barrel wall over the channel for unit length Z_c is E_T', and it is calculated using Equation 7.102. A heat transfer expression could also be written for energy transferred from the molten polymer to the barrel wall above the flight land and transferred from the melt to the screw.

$$E_T' = \frac{K_b W_b Z_b}{Z_c}(T_m - T_b) \tag{7.102}$$

The energy balance for the $k + 1$ increment in the Δz_c direction is as follows:

$$T_{k+1} = T_k + \frac{\left(\dot{\omega}'_{core\_x} + \dot{\omega}'_{core\_z} + \dot{\omega}'_{flight\_p} + \dot{\omega}'_{flight\_z} + \dot{\omega}'_{core\_p} + \dot{\omega}'_{\lambda} - E_T'\right)}{Q_m\, C_v} p\Delta z_c \tag{7.103}$$

For this technique, the viscosity is calculated at T_k, and T_{k+1} is unknown. Since the increment Δz_c is relatively small, the temperature does not change enough over the control volume to cause a severe error in the temperature calculation. The value of T_{k+1} is calculated after the energy input terms and heat conduction terms of Equation 7.96 to Equation 7.102 are calculated. The theory line temperatures in Figure 7.37 and Figure 7.38 were calculated using this technique.

Calculation of the temperature profile using the control volume method and Equation 7.103 is the preferred method. The resulting energy dissipation terms are relatively complex due to the summations and the integrations. These terms, however, are easily calculated using computer simulation routines. For screws with shallow channels such that H/W is less than 0.1, the flights can be ignored and the temperature profile calculation can be simplified to Equation 7.104. The flight drag and pressure dissipation were not used because the contributions from the flights were neglected; that is, H/W is less than 0.1. The simplified calculation under-predicts the energy dissipation. The simplified method for calculation of the temperature profile using the control volume method and screw rotation is as follows:

$$T_{k+1} = T_k + \frac{\left(\frac{2\mu W_c V_{cx}^2}{H} + \frac{3\mu W_c V_{cx}^2}{R_c} + \frac{\mu W_c V_{cz}^2}{H} + \frac{|V_{cz}| H W_c}{2}\left|\frac{\partial P}{\partial z}\right|_k + \frac{\pi^2 \mu_f N^2 D_b^2 e Z_b}{\lambda Z_c} - \frac{K_b W_b Z_b}{Z_c}(T_k - T_0)\right)}{Q_m C_v} p\Delta z \tag{7.104}$$

Now that the temperature is known at the $k+1$ position, the pressure gradient at the end of the volume can be estimated using Equation 7.105 as follows by evaluating the viscosity at temperature T_{k+1} and the average shear rate in the channel using Equation 7.41:

$$\left.\frac{\partial P}{\partial z}\right|_{k+1} = \left(Q_m - \frac{p\rho |V_{bz}| W H F_d F_c}{2}\right)\frac{12\mu(T_{k+1}, \dot{\gamma}_s)}{p\rho W H^3 F_p} \tag{7.105}$$

The pressure at the end of the increment ($k+1$) is as follows:

$$P_{k+1} = P_k + \left(\left. \frac{\partial P}{\partial z} \right|_{k+1} \right) \Delta z_c \tag{7.106}$$

For these types of simulations, the rate Q_m is set at a specific screw speed and entry pressure P_0. The simulation is then progressed until the discharge pressure at the end of the metering section is obtained. If the calculated discharge pressure is not correct, the screw speed is adjusted until the correct discharge pressure is obtained. In many cases, P_0 is known based on experiments or from baseline simulations. The baseline technique is presented in detail in Chapter 9.

7.7.5.2 Temperature Calculation for Barrel Rotation

The temperature calculation for the barrel rotation case is similar to that provided in Equation 7.93 for a control volume and an applied pressure gradient. The calculation is as follows:

$$T_{k+1} = T_k + \frac{\left(\frac{\mu V_{bz}^2}{H} + \frac{|V_{bz}| H W_b}{2} \left. \frac{\partial P}{\partial z} \right|_k + \frac{4 \mu V_{bx}^2 W_b}{H} + \frac{\pi^2 \mu_f N^2 D_b^2 e}{\lambda} - K_b W_b \left(T_{k-1} - T_0 \right) \right)}{Q_m C_v} p \Delta z_b \tag{7.107}$$

The pressure is calculated in a method similar to that used in Equation 7.105 and Equation 7.106.

7.7.6 Numerical Comparison of Temperatures for Screw and Barrel Rotations

In this section, the calculation of the discharge temperature using the control volume method described in Section 7.7.5 was used to demonstrate the effect of pressure gradient and rheology on the discharge temperature [68]. For this study, an 88.9 mm diameter screw with a length-to-diameter ratio (L/D) of 5 was used along with PC and PS resins. The screw was square pitched and had a channel depth of 6.35 mm. The calculations were performed at a rotation rate of 100 rpm. The increase in melt temperature for a Newtonian-like PC resin is presented in Figure 7.37 for a flow that was at 60% of the rotational flow for this screw. The fraction of the rotational rate is conveniently set using the parameter F_r as shown in Equation 7.108 and Equation 7.109. F_r has a value of 0 when the extruder is operating with a rate equal to the rotational rate, that is, no imposed pressure gradient. When F_r has a value of 1, the pressure gradient is such that the flow of the machine is 0;

that is, $Q_{mp} = Q_{md}$. The pressure gradient provided by Equation 7.109 can then be used to estimate the temperature increase using Equation 7.104 for screw rotation and Equation 7.107 for barrel rotation.

$$Q_{mp} = F_r Q_{md} \tag{7.108}$$

$$\frac{\partial P}{\partial z} = \frac{\Delta P}{Z} = \frac{6\mu |V_{bz}| F_d F_r}{H^2 F_p} \tag{7.109}$$

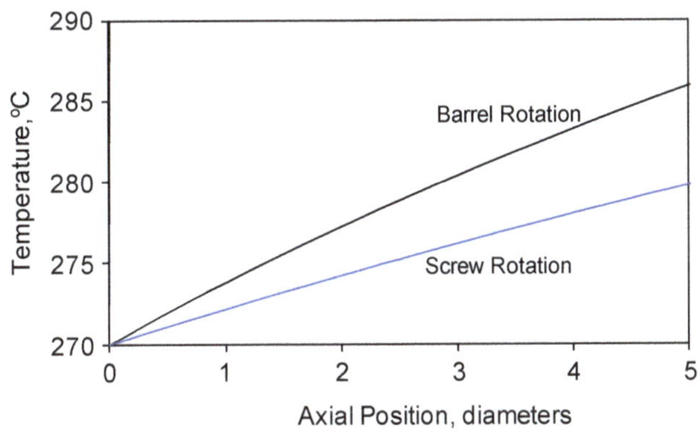

Figure 7.37 Calculated axial temperature for the PC resin for barrel and screw rotation for a discharge pressure that caused the machine to operate at 60% of maximum flow (F_r = 0.4) [68]. The difference in the discharge temperatures for the two cases was about 6 °C

As indicated by Figure 7.37 and Table 7.7 and as expected, the increase in melt temperature for a PC resin was always higher for the barrel rotation case as compared to screw rotation. If a very high die pressure was needed and the rate is reduced to 10% of the rotational rate (F_r = 0.9), the difference between discharge temperatures for the rotation cases was predicted at about 18 °C. In other terms, the temperature increase (47 °C) for screw rotation was about 72% of the temperature increase (65 °C) for barrel rotation. The discharge temperature difference for the two rotation cases decreases to about 4 °C for a low die pressure case where the rate is 90% of the rotational flow rate (F_r = 0.1). For this case, the melt temperature increase (6 °C) for screw rotation was about 60% of the temperature increase (10 °C) for barrel rotation.

The predictions for a typical highly shear-thinning PS resin are shown in Figure 7.38. The difference in predicted discharge temperature was not as dramatic for the different rotation cases. Like the PC resin and as expected, the melt temperature increase for the PS resin was always higher for the barrel rotation case. As shown by Figure 7.38, the melt temperature increased by 22 °C for barrel rotation while it increased 19 °C for the screw rotation case. Thus, the melt temperature increase for screw rotation was about 86% of the temperature increase for barrel rotation. For shear-thinning materials it appears that the higher shear rate for barrel rotation lowers the effective viscosity enough to overcome the larger surface area and thus mitigating the effects of energy dissipation. This may be why for many shear-thinning polymers the prediction of melt temperature increase using barrel rotation theory is in the correct range.

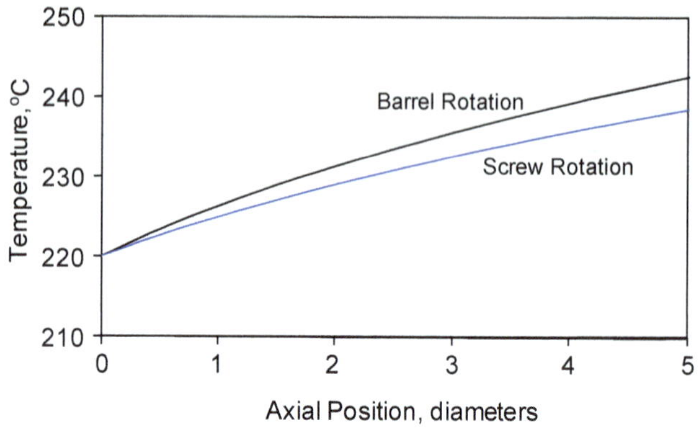

Figure 7.38 Calculated axial temperature for the PS resin for barrel and screw rotation for a discharge pressure that caused the machine to operate at 90% of maximum flow (F_r = 0.1) [68]. The difference in the discharge temperatures for the two cases was about 4 °C

Table 7.7 Comparison of Barrel and Screw Temperature Increase for PC and PS Resin

Resin	F_r	% of Maximum Flow	Barrel Rotation, °C	Screw Rotation, °C	Difference, °C
PC	0.9	10	65	47	18
PC	0.4	60	15	9	6
PC	0.1	90	10	6	4
PS	0.1	90	22	19	3

7.8 Metering Section Characteristics

The previous sections of this chapter described the pumping and energy dissipation mechanisms for melt-filled channels such as the metering channel. Clearly, the performance of the extruder such as rate, pressure generation, and discharge temperature are highly dependent upon the design and operation of the metering section channel. This section will describe the strategies for specifying the metering channel geometry so that the performance targeted by the design can be achieved in practice.

As an example of process design, LLDPE resin is commonly used for blown film, cast film, and extrusion coating processes. Even though the resin grades are similar (MIs do vary) for these three processes, the extrusion equipment is significantly different due to the requirements of the die and downstream equipment. The blown-film process requires an extrudate that is relatively low in temperature and typically in the range of 200 to 220 °C. In order to plasticate and produce an extrudate in this temperature range the metering channel is relatively deep and the screw would be designed to rotate at speeds less than 100 rpm. For a 150 mm diameter screw with a square-pitch lead length, the metering channel depth would be between 9 and 12 mm. The cast film process requires an extrudate that is slightly higher in temperature and typically in the range of 240 to 260 °C. The same 150 mm diameter screw would have a metering channel depth of 6 to 9 mm and the screw would rotate at higher speeds. The extrusion coating process requires an extrudate that is very high in temperature and often approaching 310 °C. The metering channel depth would typically be about 3 to 4 mm for the 150 mm diameter screw, and the screw would be designed to rotate at very high speeds up to 230 rpm. These examples clearly show that the extrudate temperatures are set in part by the geometry of the metering section channel and the conditions of the process.

The specific rate is often a good measure of the relative discharge temperature for a process. The specific rate is simply the rate divided by screw speed. For the examples above, the specific rate for the metering channels would be the highest for the blown-film screw with the deep metering channel and the lowest for the extrusion coating process screw with the very shallow channel. Thus as a general guideline, the extrudate temperature decreases when the specific rate of the screw increases. The specific rates for the screws in these examples increased because the channel depth increased. The specific rate could also be changed by adjusting the lead length. The calculated specific rotation rate can be also used as a similar guideline for discharge temperature. This guideline, however, can be violated if the channel is extremely deep and a large positive pressure gradient exists. In Section 13.4.3, an example is shown where a process was rate limited by high discharge temperatures because the screw was too deep. In this case, the specific rotation rate for the screw was very high because the metering channel was too deep, but the pressure gradient and deep channel caused the pressure flow term to be very large, causing the specific rate during operation to be about 20% of the calculated specific rotation rate. For this example, an optimal channel depth was determined that maximized the specific rate and minimized the extrudate temperature.

For processes that have a positive pressure gradient in the metering channel, the observed rate can be reduced considerably as indicated above for deep channels. This rate loss is caused by the H^3 term in the pressure flow equation; for example, see Equations 1.22 and 1.29. Thus as the channel depth increases, the rotational rate increases linearly with H and the rate loss increases to the H^3. If the H dependence on F_p is included, the rate loss is even higher. For some situations, the specific rate of a metering channel can be increased by increasing the lead length of the screw. As the lead length increases, the average channel width W increases. Because the pressure-driven flow and rotational flow are directly dependent on W, the loss of rate due to a positive pressure gradient is less for changing the lead length as compared to increasing the channel depth. As the lead length is increased, however, the total helical length Z to generate pressure decreases for a fixed axial length. In general, the optimal metering channel geometry for an application will be obtained by optimization of the channel depth and lead length.

Mixing effects in the metering channel are relatively minor. Moreover, the residence time distribution and the tendency of the channel to degrade resin depend on the size of the flight radii. This long residence time region in the flight radii are caused by the formation of Moffat eddies, and they can cause the residence time to be long enough to degrade the resin. Mixing performance of metering channels, residence time distributions, and Moffat eddies will be discussed in detail in Chapter 8.

As shown by the example problem of Section 7.7.3, the dissipation of energy at the flight lands can be a large component of the overall dissipation. Flight land widths are typically specified as 10% of the diameter. Thus for a 100 mm diameter screw, the width of the flight perpendicular to the edge (e) is 10 mm. For many applications where minimizing the temperature is important, decreasing the width from 10% to about 8% can decrease the dissipation level and temperature while not affecting the mechanical integrity of the flight.

Nomenclature

A	channel cross section area perpendicular to axis of the screw or area for heat transfer
b	temperature coefficient for the shear viscosity model of Equation 7.42
C_v	specific heat of the molten polymer
D	local diameter
D_b	diameter of the barrel
D_c	diameter of the screw core
$D_{i,j}$	deformation rate tensor component
$\mathbf{D}_E$	Eulerian frame deformation rate tensor

$\boldsymbol{D}_L$	Lagrangian frame deformation rate tensor
e	flight width at the tip of the screw and perpendicular to the flight edge
E_T	heat transferred through the barrel wall over the channel
F_c	rotational flow correction factor
F_d	shape factor for rotational flow
F_p	shape factor for pressure flow
F_r	parameter for setting mass flow rate induced by pressure via Equation 7.108
H	local distance from the screw root to the barrel
k	the index for the control volume method when used as a subscript
k	thermal conductivity of the molten resin
K	constant for Equation 7.44
K_b	heat transfer coefficient at the barrel wall
K_o	constant for Equation 7.73
l	axial distance
L	lead length
n	power law index
N	screw rotation speed in revolutions/s
p	number of flight starts
P_{dis}	discharge pressure
P	pressure in the channel
$\boldsymbol{q}$	heat flux vector
Q	volumetric flow rate
Q_d	volumetric rotational flow rate
Q_m	mass flow rate
Q_{md}	mass rotational flow rate
Q_{mp}	pressure-induced mass flow rate
Q_p	volumetric pressure flow rate
r	independent variable for the radial direction of a screw
R_c	radius of the screw core
R_g	ideal gas constant, 8.314 J/(mol K)
t	independent variable time
T	temperature

T_0	reference temperature or initial temperature
T_m	temperature of the molten resin in the channel
T_b	temperature of the barrel wall
$\boldsymbol{v}$	velocity vector
V	fluid volume for the screw pump for the example in Section 7.7.3
$V(r)$	velocity of the screw surface as a function of the radial position r
V_c	velocity of the screw core
V_{cx}	x component of velocity of the screw core
V_{cz}	z component of velocity of the screw core
V_b	velocity at tip of the flight land for screw rotation or velocity of the barrel surface for barrel rotation
V_{bx}	x component of velocity of the screw flight at the barrel wall
V_{bz}	z component of velocity of the screw flight at the barrel wall
V_b'	velocity of barrel as observed in the Lagrangian frame
V_{bx}'	x component of velocity of the barrel as observed in the Lagrangian frame
V_{bz}'	z component of velocity of the barrel as observed in the Lagrangian frame
V_x	velocity component in the x direction
V_{xE}	velocity component in the x direction for the Eulerian (laboratory) frame
V_{xL}	velocity component in the x direction for the Lagrangian (transformed) frame
V_y	velocity component in the y direction
V_z	velocity component in the z direction
V_{pl}	transformation velocity for the velocity in the axial direction due to screw rotation
V_{dz}	down-channel velocity (z direction) due to rotational flow as a function of y
V_{dzE}	down-channel velocity (z direction) due to rotational flow as a function of y for the Eulerian frame
V_{dzL}	down-channel velocity (z direction) due to rotational flow as a function of y for the Lagrangian frame
V_{pz}	z component of the velocity due to a pressure gradient
W	average width of the channel
W_b	channel width perpendicular to flight at the barrel wall
W_c	channel width perpendicular to flight at the screw core
x	independent variable for the cross-channel direction perpendicular to the flight edge

y	independent variable for the direction normal to the barrel surface
z	independent variable in the down-channel direction (or helical direction)
z_b	helical position of the channel at the barrel wall
$z(r)$	helical position of the channel at radial position r
Z_a	average helical length of the metering channel
Z_b	helical length of the metering channel at the barrel wall
Z_c	helical length of the metering channel at the screw core
β	constant for Equation 7.53
$\dot{\gamma}$	average shear rate in the channel
$\dot{\gamma}_b$	average shear rate in the channel based on barrel rotation
$\dot{\gamma}_s$	average shear rate in the channel based on screw rotation
ΔE	activation energy for Equation 7.90
η	shear viscosity of the polymer at the average channel temperature and average shear rate
η_0	pre-exponential factor for the shear viscosity model using Equation 7.42
θ_c	helix angle at the screw core
$\theta(r)$	helix angle at radial position r
θ_b	helix angle at the barrel
λ	mechanical clearance between the top of the screw flight and the barrel wall
μ	Newtonian viscosity
μ_o	constant for Equation 7.90
μ_f	Newtonian viscosity in the flight land
μ_w	Newtonian viscosity at the barrel wall for the heat transfer Equation 7.95
ρ	melt density of the fluid
τ	shear stress
ω_i	energy dissipated for component i
ω_0	energy transferred through the barrel wall
$\dot{\omega}_{core\_z}$	energy dissipation rate for screw core movement and flow in the z direction
$\dot{\omega}_{core\_x}$	energy dissipation rate for screw core movement and flow in the x direction
$\dot{\omega}_{core\_p}$	energy dissipation rate for screw core movement and flow induced by pressure

$\dot{\omega}_{flight\_z}$ energy dissipation rate for flight movement and flow in the z direction

$\dot{\omega}_{flight\_p}$ energy dissipation rate for flight movement and flow induced by pressure

$\dot{\omega}_{\lambda}$ energy dissipation rate in the flight land

$\dot{\omega}_{total}$ total energy dissipation rate

$\dot{\omega}'_{core\_z}$ energy dissipation rate for screw core movement and flow in the z direction per unit length Z_c

$\dot{\omega}'_{core\_x}$ energy dissipation rate for screw core movement and flow in the x direction per unit length Z_c

$\dot{\omega}'_{core\_p}$ energy dissipation rate for screw core movement and flow induced by pressure per unit length

$\dot{\omega}'_{flight\_z}$ energy dissipation rate for flight movement and flow in the z direction per unit length Z_c

$\dot{\omega}'_{flight\_p}$ energy dissipation rate for flight movement and flow induced by pressure per unit length Z_c

$\dot{\omega}'_{\lambda}$ energy dissipation rate in the flight land per unit length Z_c

References

1. Tung, T.T. and Laurence, R.L., "A Coordinate Frame for Helical Flows," *Polym. Eng. Sci.*, **15**, 401 (1975)
2. Germano, M., "The Dean Equations Extended to a Helical Pipe Flow," *J. Fluid Mech.*, **203**, 289 (1989)
3. Campbell, G.A., Sweeney, P.A., and Felton, J.N., "Experimental Investigation of the Drag Flow Assumption in Extruder Analysis," *Polym. Eng. Sci.*, **32**, 1765 (1992)
4. Campbell, G.A., Sweeney, P.A., and Felton, J.N., "Analysis of an Alternative Extruder Screw Pump Design," *Int. Polym. Process.*, **7**, 320 (1992)
5. Campbell, G.A., Wang, Ch., Cheng, H., Bullwinkel, M., and te-Riele, M.A., "Investigation of Flow Rate and Viscous Dissipation in a Single Screw Pump-Extruder," *Int. Polym. Process.*, **16**, 323 (2001)
6. Rowell, H.S. and Finlayson, D., "Screw Viscosity Pumps," *Engineering*, **114**, 606 (1922)
7. Rowell, H.S. and Finlayson, D., "Screw Viscosity Pumps," *Engineering*, **126**, 249 (1928)
8. Carley, J.F., Mallouk, R.S., and McKelvey, J.M., "Simplified Flow Theory for Screw Extruders," *Ind. Eng. Chem.*, **45**, 974 (1953)
9. Tadmor, Z. and Klein, I., "Engineering Principles of Plasticating Extrusion," Van Nostrand Reinhold Company, New York (1970)

10. Mohr, W.D., Saxton R.L., and Jepson, C.H., "Theory of Mixing in the Single-Screw Extruder," *Ind. Eng. Chem.*, **49**, 1857 (1957)
11. Boussinesq, M.J., *J. Mathematique Pures et Appliquees*, serials 2.13, 377 (1868)
12. Carley, J.F. and McKelvey, J.M., "Extruder Scale-Up Theory and Experiments," *Ind. Eng. Chem.*, **45**, 989 (1953)
13. McKelvey, J.M., "Experimental Studies of Melt Extrusion," *Ind. Eng. Chem.*, **45,** 982 (1953)
14. Weeks, D.J. and Allen, W.J., "Screw Extrusion of Plastics," *J. Mech. Eng. Sci.*, **4**, 380 (1962)
15. Middleman, S., "Flow of Power Law Fluids in Rectangular Ducts," *Trans. Soc. Rheol.*, **9**, 83 (1965)
16. Gore, W.L. and McKelvey, J.M., "Theory of Screw Extruders," in "Rheology: Theory and Applications," Eirich, F.R. (Ed.), vol. 3, New York, Academic Press (1959)
17. Mohr, W.D. and Mallouk, R.S., "Flow, Power Requirement, and Pressure Distribution of Fluid in a Screw Extruder," *Ind. Eng. Chem.*, **51**, 765 (1959)
18. Pittman, J.F.T., and Rashid, K., "Heat Transfer in Recirculating Extruder Channel Flow," *J. Polym. Eng.*, **5**, 1 (1985)
19. Agassant, J.-F., Avenas, P., Sergent, J.-Ph., and Carreau, P.J., "Polymer Processing: Principles and Modeling," Hanser Publications, Munich (1991)
20. Barr, R.A. and Chung, C.I., "Effects of Radial Screw Clearance on Extruder Performance," *SPE J.*, **23**, 71 (1966)
21. Booy, M.L., "Influence of Channel Curvature on Flow, Pressure Distribution, and Power Requirements of Screw Pumps and Melt Extruders," *SPE Trans.*, **3**, 176 (1963)
22. Pinto, G. and Tadmor, Z., "Mixing and Residence Time Distribution in Melt Screw Extruders," *Polym. Eng. Sci.*, **10**, 279 (1970)
23. Fenner, R.T., "Extruder Screw Design," Iliffe, London (1970)
24. Carley, J.F. and Strub, R.A., "Basic Concepts of Extrusion," *Ind. Eng. Chem.*, **45**, 970 (1953)
25. Griffith, R.M., "Fully Developed Flow in Screw Extruders," *Ind. Eng. Chem. Fundam.*, **1**, 180 (1962)
26. Fenner, R.T., "The Design of Large Hot Melt Extruders," *Polymer*, **16**, 298 (1975)
27. Fenner, R.T., "Developments in the Analysis of Steady Screw Extrusion of Polymers," *Polymer*, **18**, 617 (1977)
28. Eccher, S. and Valentinotti, A., "Experimental Determination of Velocity Profiles in an Extruder Screw," *Ind. Eng. Chem.*, **50**, 829 (1958)
29. Mohr, W.D., Squires, P.H., and Starr, F.C., "Flow Patterns in a Single-Screw Extruder," *SPE J.*, **16**, 1015 (1960)
30. Mohr, W.D., Clap, J.B., and Starr, F.C., *SPE Trans.*, **1**, 113 (1961)
31. Rotem, Z. and Shinnar, R., "Non-Newtonian Flow Between Parallel Boundaries in Linear Movement," *Chem. Eng. Sci.*, **15**, 130 (1961)

32. Zamodits, H. and Pearson, J. R. A., "Flow of Polymer Melts in Extruders. Part 1. The Effect of Transverse Flow and of Superposed Steady Temperature Profile," *Trans. Soc. Rheol.*, **13**, 357 (1969)
33. Karwe, M. V. and Jaluria, Y., "Numerical Simulation of Fluid Flow and Heat Transfer in a Single-Screw Extruder for Non-Newtonian Fluids," *Numer. Heat Transfer, Part A*, **17**, 167 (1990)
34. Fenner, R. T., "Principles of Polymer Processing," Macmillan, London (1979)
35. Bruker, I., Miaw, C., Hasson, A., and Balch, G., "Numerical Analysis of the Temperature Profile in the Melt Conveying Section of a Single Screw Extruder: Comparison with Experimental Data," *Polym. Eng. Sci.*, **27**, 504 (1987)
36. Dyer, D. F., "A Numerical Solution for the Single Screw Extrusion of a Polymer Melt," *AIChE J.*, **15**, 823 (1969)
37. Spalding, M. A., Dooley J., Hyun, K. S., and Strand, S. R., "Three-Dimensional Numerical Analysis of a Single-Screw Extruder," *SPE ANTEC Tech. Papers*, **39**, 1533 (1993)
38. Bruin, S., Van Zuilichem, D. J., and Stolp W., "A Review of Fundamental and Engineering Aspects of Extrusion of Biopolymers in a Single-Screw Extruder," *J. Food Process Eng.*, **2**, 1 (1978)
39. Sastrohartono, T., Jaluria, Y., Esseghir, M., and Sernas, V., "Transport Phenomena in Materials Processing and Manufacturing," ASME-HTD-Vol. 196, 39 (1994)
40. Chang, R. and Lin, K., "The Hybrid FEM/FDM Computer Model for Analysis of the Metering Section of a Single-Screw Extruder," *Polym. Eng. Sci.*, **35**, 1748 (1995)
41. Matsuoka, T. and Takahashi, H., "Finite Element Analysis of Polymer Melt Flow in Profile Extrusion Coating Die," *Int. Polym. Process.*, **3**, 183 (1991)
42. Fraser, K., Coyle, D. J., and Bruker, I., "Evaluation of an Injection Molding Mixing Screw," *SPE ANTEC Tech. Papers*, **35**, 214 (1989)
43. Elbirli, B. and Lindt, J. T., "A Note on the Numerical Treatment of the Thermally Developing Flow in Screw Extruders," *Polym. Eng. Sci.*, **24**, 482 (1984)
44. Chiruvella, R. V., Jaluria, Y., Sernas, V., and Esseghir, M., "Extrusion of Non-Newtonian Fluids in a Single-Screw Extruder with Pressure Back Flow," *Polym. Eng. Sci.*, **36**, 358 (1996)
45. Campbell, G. A., Sweeney, P. A., Dontula, N., and Wang, Ch., "Frame Indifference: Fluid Flow in Single Screw Pumps," *Int. Polym. Process.*, **11**, 199 (1996)
46. Hunt, D. L., M. S. Thesis, Clarkson University (1995)
47. Campbell, G. A., Cheng, H., Wang, C., Bullwinkel, M., and te-Riele, M. A., "An Experimental Investigation of Screw Configuration on Temperature Rise in Single Screw-Pump Extruders," *SPE ANTEC Tech. Papers*, **48**, 189 (2002)
48. Malvern, L. E., "Introduction to the Mechanics of a Continuous Medium," Prentice-Hall (1969)
49. Bird, R. B., Stewart, W. E., and Lightfoot, E. N., "Transport Phenomena," John Wiley and Sons, Inc., New York (1960)

50. Potente, H., "Fundamental Mechanisms in Polymer Processing–Extrusion," *SPE ANTEC Tech. Papers*, **44**, 3604 (1998)

51. Spalding, M. A., Campbell, G. A., Carlson, F., and Nazrisdoust, K., "The Effect of Barrel Diameter and Rotation on the Numerical Simulation of Single-Screw Extruders," *SPE ANTEC Tech. Papers*, **52**, 792 (2006)

52. Spalding, M. A. and Campbell, G. A., "The Accuracy of Standard Drag Flow and Pressure Gradient Calculations for Single-Screw Extruders," *SPE ANTEC Tech. Papers*, **54**, 262 (2008)

53. Agur, E. E. and Vlachopoulos, J., "Numerical Simulation of a Single-Screw Plasticating Extruder," *Polym. Eng. Sci.*, **22**, 1084 (1982)

54. Spalding, M. A. and Campbell, G. A., "An Engineering Approach to the Correction of Drag Flow Calculations for Single-Screw Extruders - Equation Correction," *SPE ANTEC Tech. Papers*, **57**, 1211 (2011)

55. Robinson, W. D. B, Doctoral Thesis, Univ. of Delaware (1960)

56. Metzner, V. L., "Handbook of Fluid Dynamics," V. L Streeter (Ed.), Section 7, 28, McGraw-Hill, (1961)

57. Wheeler, J. A. and Whissler, E. H., "The Friction Factor–Reynolds Number Relation for the Steady Flow of Pseudoplastic Fluids through Rectangular Ducts. Part I. Theory," *Am. Inst. Chem. Eng. J.*, **11**, 207 (1965)

58. Tadmor, Z. and Gogos, C. G., "Principles of Polymer Processing," 2nd ed., Chapter 2, John Wiley & Sons, Inc., Hoboken, NJ (2006)

59. Campbell, G. A., Cheng, H., Wang, C., Bullwinkel, M., and te-Riele, M. A., "Temperature Rise in a Single Screw Pump," *SPE ANTEC Tech. Papers*, **47**, 152 (2001)

60. Cheng, H., "An Investigation of Flow Rates and Viscous Dissipation in Extrusion," Chemical Engineering M. S. Thesis, Clarkson University (1997)

61. Middleman, S. "Fundamentals of Polymer Processing," Chapter 6, McGraw-Hill, New York (1977)

62. Bernhardt, E. C. and McKelvey, J. M., "Analysis of Adiabatic Plastics Extrusion," *SPE J.*, **10**, 419 (1954)

63. Jepson, C. H., "Future Extrusion Studies," *Ind. Eng. Chem.*, **45**, 992 (1953)

64. Todd, D., "Heat Transfer in Twin-Screw Extruders," *SPE ANTEC Tech. Papers*, **34**, 54 (1988)

65. Sandall, O. C. and Patal, K. G., "Heat Transfer to Non-Newtonian Pseudoplastic Fluids in Agitated Vessels," *Ind. Eng. Chem. Process Des. Dev.*, **9**, 139 (1970)

66. White, J. L., Kim, E. K., Keum, J. M., and Jung, H. C., "Modeling Heat Transfer in Screw Extrusion with Special Application to Modular Self-Wiping Co-Rotating Twin-Screw Extrusion," *Polym. Eng. Sci.*, **41**, 1448 (2001)

67. Tucker, C. L., "Computer Modeling for Polymer Processing-Fundamentals," Hanser Publications, Munich (1989)

68. Campbell, G. A., Spalding, M. A., and Carlson, F., "Prediction of Screw Temperature Rise in Single Screw-Pump Extruders," *SPE ANTEC Tech. Papers*, **54**, 267 (2008)

8 Mixing Processes for Single-Screw Extruders

Single-screw extruders are expected to provide a homogeneous extrudate at high rates and at the proper discharge temperature and pressure. If the extrudate is not homogeneous, then defects in the final product are highly likely. Poorly mixed extrudates are much more likely to occur at high process rates and thus high screw speeds. This chapter presents the fundamental concepts of mixing, the primary mixing that occurs during the melting process, and secondary mixing devices. High-performance screws also provide a high level of mixing. High-performance screws will be presented in Chapter 14.

A considerable amount of literature exists on mixing mechanisms, processes, and devices. Most of the devices developed utilize fundamental principles to provide mixing to a polymer system. Some devices, however, have been placed on the market with good intention but provide lower mixing performances. The number of devices on the market is considerably more than what could be covered here. Instead only the most-used devices and their performances will be covered in this chapter. The reader is directed to other sources for detailed mixing mechanisms, devices, and applications [1, 2].

In some single-screw processes that perform compounding operations or use masterbatch colorants, degradation products can be incorrectly diagnosed as poorly mixed components. In these cases, the technical solution is to identify the source of the degradation and eliminate it. If a technical solution is employed to improve the mixing performance of the process, then the modified operation will simply disperse and distribute the degradation products. Resin degradation is not discussed in this chapter. It is presented in Chapter 11 along with numerous case studies.

8.1 Common Mixing Operations for Single-Screw Extruders

The mixing performance of an extrusion process starts with the feed delivery system upstream of the extruder. In many cases, a blending operation is used to mix two or more different types of pellet feedstocks. The system must provide an acceptable level of uniform solids to the feed hopper. If the mixing in the feed delivery section is not adequate or the feed composition varies with time, then the quality of the discharged material can show signs of a poorly mixed system with variations in the extrudate composition. For example, if a processor desires to use a 35:1 letdown ratio of a color concentrate for a plasticator on an injection-molding machine, but instead the feed delivery system provides momentary color letdown ratios varying from 20:1 to 100:1, color gradients are likely to appear across the part and also as variations between parts. On some single-screw compounding operations, the feedstocks are combined using a high-intensity batch mixer. If the feedstocks have varying levels of toughness, then the high-intensity mixer will selectively fragment the more brittle components, creating dust and very small particles known as fines. These fines will have a higher concentration of the brittle component. The fines often segregate from the pellets and cause a compositional variation in the feedstock to the extruder. This variation in the feedstock composition will likely cause a variation in the discharge material from the extruder. An example of this problem is shown in Figure 5.35. High-intensity mixers can be made to perform adequately for these applications if the process and material limitations are fully understood.

Although solids conveying in the feed channel provides little to no mixing of the components, an improperly designed system can cause the feed section to operate unstably and thus cause the extruder to flow surge. The flow-surging process can cause oscillatory levels of energy to be dissipated into the resin, causing thermal gradients and temperature transients in the discharge [3]. Moreover, if a machine is operating improperly and with a color concentrate and a natural resin, some of the resin can flow to the discharge end of the screw as solids or unmixed molten material. Natural-colored resin solids or pigmented masterbatch solids can cause a color (or lack of color) streak in the final product. Thus, it is extremely important to verify that the extruder is operating properly before mixing processes are assessed. Here proper operation is defined as the metering section controlling the specific rate as outlined in Section 1.4 for a smooth-bore machine.

The melting section of the screw is where the primary mixing occurs. The thin melt film between the solid bed and the barrel surface where much of the melting is occurring provides a high level of elongational stresses that create a high level of dispersive mixing between the components. This mixing mechanism will be described in detail in Section 8.4.

A level of mixing occurs in the melt-filled channels of metering sections but to a lesser degree. The mixing here occurs due to the motion of the advancing flight and backward dragging of the screw root. This type of mixing will be discussed in Section 8.3.2.

The most common distributive and dispersive secondary mixers are discussed in Section 8.5. These mixers include devices such as Maddock mixers, blister mixers, and gear mixers. They are used to finish the mixing process that was not completed during the primary mixing step, to mitigate thermal gradients, and to distribute and disperse an additive that is added downstream of the feed hopper. Processes that utilize downstream injection include the addition of a physical blowing agent through the barrel wall and into molten polymer to produce a foamed product. Some specialty processes could add a liquid additive to the resin in a similar manner. For large diameter devolatilizing extruders, a stripping agent can be added through the barrel wall to aid in the removal of unreacted monomer from the resin. In all cases, one or more secondary mixing devices are used to combine the second stream with the molten resin. The unreacted monomer and stripping agent are removed in a downstream vent opening. This type of process is discussed in Section 15.3.

8.1.1 Common Mixing Applications

Many processors use single-screw extruders for a variety of mixing applications. The most common application is the addition of a pigment to a natural resin, producing a product with the correct color. For this process, the pigments are almost always precompounded into a base resin using a twin-screw extruder to form color concentrate pellets. The processor generally purchases the concentrate from a color-compounding vendor. The color concentrate is then added to the natural resin by the processor at a ratio of between 25 and 75 parts natural resin to one part color concentrate, that is, a 25:1 to 75:1 letdown ratio. This type of process is fairly common for extrusion, injection-molding, and blow-molding processes. In all cases, the final product must be uniform in color. Thus the process requires that adequate mixing occur in the single-screw machine. The process allows the processor to dedicate a large silo to the natural resin and maintain only small quantities of the color concentrate. The system is considerably less expensive than maintaining an inventory of each resin in a factory color, and it eliminates the possibility of having a large quantity of a precompounded factory resin in an obsolete color. In all cases, the discharge from the machine is expected to be homogeneous in composition, color, and temperature. If the discharge is not homogeneous, then part defects will occur and the cost to manufacture will increase.

For blown-film processes, the processor will often blend several different types of polyolefin resins to achieve a unique balance of physical properties such as impact strength and modulus [4, 5], and they also typically add fluoropolymers to aid in processing [6]. The processor may blend up to five resins, add them to the hopper of the extruder, and produce a uniform and high-value product at high rates. Improper mixing in the extruder, however, can result in gels in the discharge and a film with poor physical properties. The gels here are most likely solid particles of the resin with the highest melting temperature in a matrix of partially mixed material.

Eliminating temperature gradients in the discharge is also extremely important in extrusion and injection-molding processes. If the discharge from the plasticator of an injection-molding machine is not uniform, then cosmetic surface defects are likely to occur [7]. For PS foam sheet production using a tandem extrusion line [8], the resin is plasticated in a primary extruder and discharged at temperatures near 230 °C. This discharge is too hot to produce acceptable foam. Instead, the discharge is sent to a secondary extruder where the resin-blowing agent mixture is cooled to about 140 °C; the actual temperature depends on the blowing agent that is used. The secondary cooling extruder utilizes a slowly rotating screw with relatively deep channels, and either cold or chilled water is flowed through the barrel casings. As the cold water removes energy through the barrel wall, thermal gradients develop in the resin in the deep channels of the screw. These thermal gradients impede further energy removal and will cause poor quality foam if discharged to the die. Thus, the minimization of temperature gradients in the resin in the channels and in the discharge is extremely important in maintaining high foam quality and high rates. Tandem foam extrusion is discussed in detail in Section 15.4.

Single-screw extruders have also been used to compound resins from either concentrates or directly from the feedstock materials, producing new compounds that can be pelletized [9, 10] or delivered to the final downstream shaping process (in-line processes). The most common types of products include resin colorization, flame-resistant resins from natural feedstocks, and specialty resin blends. Compounding pigments directly with a single-screw extruder is possible [11], but the pigments can be either difficult to disperse or can agglomerate, creating large pigment particles that appear in the final pellets. This type of compounding process is being used more often with in-line production operations. For these special cases, the compounding step at the twin-screw extruder is being eliminated, and the melting, compounding, and pressurization are performed using a single-screw extruder upstream of the fabrication equipment. In-line production operations have economic advantages since the original compounding is eliminated and the processor has opportunities for optimization of the product composition.

Most converters and processors do not have the ability to quantify the degree of mixing for the material discharged. Instead, they typically produce either acceptable articles or articles with poorly mixed components or with thermal gradients. Poor mixing is defined here as a process where either a composition or thermal gradient causes the article to be unacceptable. Many operations can cause the mixing to be poor, including a poor screw design, improper operation of the equipment, and poor selection of the components to be mixed and extruded. These topics will be discussed in detail in this chapter.

Clearly, single-screw extruders are not as efficient as twin-screw compounding extruders due to their different intensities in the mixing process [12]. For example, compounding carbon black powders into most resins is virtually impossible with a single-screw extruder, yet it is routinely performed using twin-screw machines. For select applications, however, an optimized and properly operating single-screw extruder can be used to perform compounding processes [10]. As will be discussed and shown later, the primary mixing that occurs in a single-screw, plasticating machine occurs during the melting (or devitrification) process [13, 14]. The mixers that are traditionally used, such as a Maddock-style mixer, are performing secondary mixing and are usually needed due to less-than-optimal primary mixing during melting. In general, the best mixing performance occurs when the proper combination of primary mixing through the melting process is followed by one or more secondary mixers.

8.2 Dispersive and Distributive Mixing Processes

The two types of mixing processes that occur in polymer processing are dispersive and distributive mixing. These processes are used to melt solid polymer fragments in a molten matrix, mix soluble materials including blowing agents, mix immiscible polymer blend systems, mix solid particulates into a polymer matrix, and mitigate thermal gradients. Dispersive mixing is the mixing process that causes fluid elements to stretch, via shear flows and elongational flows, and then break into smaller size fluid elements. The stretching rate is linear for constant shear fields while it is exponential for constant elongation rate flow fields. Many of the common dispersive mixers rely primarily on shear flows. For example, if a molten polymer stream flows through a channel, especially a channel with a small clearance, the shear stress in the process will cause the elements in the flow to stretch. Common dispersive mixers include Maddock, blister, and spiral dam mixers. The macroscopic effect is to reduce the size of locally high concentrations of a component. For example, solid polymer fragments that are flowing with a stream of molten resin will be reduced in size as they pass through the shear stress field created by the mixing flight of a Maddock mixer. For an immiscible polymer blend, dispersive mixers are used to reduce the size of the minor component. As the minor component enters the shear stress field as a large droplet, the droplet becomes stretched until the interfacial forces cause the stretched fluid to break into smaller size droplets. The droplet size of the minor component after leaving the shear stress field will depend on a complicated balance between the stress applied to the system, the interfacial forces, and the ability of the droplets to coalesce.

Dispersive mixers are also used to reduce the size of solid components in a molten polymer matrix. Solid components include pigments, flame-retardant additives, and fillers. Many of these feedstock materials are in the form of loosely bounded agglomerates of smaller primary particles. As the agglomerate enters the shear stress field in the mixer, the agglomerate will be decreased in size if the applied shear stress is higher than the cohesive forces bonding the primary particles together.

Dispersive mixers are excellent at reducing the size of a minor component, but they may not distribute the minor component evenly through the matrix material. Distributive mixing processes work by dividing the flow stream, reorienting the flows, and then recombining the flow streams. Consecutive dividing, reorientation, and recombining of the flow streams can increase the surface area between components very quickly. The process distributes the minor components throughout the flow domain. Distributive mixers typically will not reduce the size of the minor component. Common distributive mixers designed into screws include gear, pin, and knob mixers. Static mixers are distributive mixers that are placed downstream of the screw and typically just upstream of the die. An in-depth discussion on dispersive and distributive mixing can be obtained elsewhere [15–17].

Elongational flow fields have the capability of stretching and reducing the size of minor components. The dispersive mixing occurs due to elongational stresses in the elongational flow fields, and it is considerably more effective than the dispersive mixing that occurs in a shear flow field. Elongational flow fields typically occur when the flow velocity is either accelerating in a contracting channel or is decelerating in an expanding channel.

Although mixers tend to be classified as either dispersive or distributive mixers, almost all mixers have a combination of the mixing types. For example, a Maddock mixer is labeled a dispersive mixer. This mixer will divide the flow into several channels (distributive mixing), accelerate the flow as it passes from the in-flow flute to the mixing flight land (elongational dispersive mixing), disperse the minor component by flowing across the mixing flight land, and then recombine the flow at the end of the mixer (distributive).

8.3 Fundamentals of Mixing

The fundamentals and numerical treatment of mixing of two or more liquid components are described in this section. The simplest fundamental explanation is for two miscible liquid components. For example, the spontaneous mixing of two liquid materials is fundamentally controlled by the free energy of mixing:

$$\Delta G_m = \Delta H_m + T\Delta S_m \tag{8.1}$$

where G_m is the free energy of mixing, H_m is the heat of mixing, S_m is the entropy of mixing, and T is the temperature in Kelvin. For example, spontaneous mixing occurs when an alcoholic beverage is added to water or a soft drink and the two liquids spontaneously form a homogeneous solution. This process is driven by two phenomena. First, the fluids have similar chemistry, so their solubility parameters are very close. Second, they have a very high entropy of mixing, because they are small molecules and thus have a large number of potential ensembles as they are placed in the space provided. The last process that helps this low-viscosity mixing process by getting the two materials in close proximity is the turbulence that occurs when the flow is driven by a pressure gradient or a mixing blade. In the beverage mixing case, a spoon was used to agitate the low-viscosity liquids in the glass. Turbulence is accomplished when the Reynolds number is greater than about 4000:

$$\mathrm{Re} = \frac{V\rho D}{\mu} \tag{8.2}$$

where Re is the Reynolds number, V is the velocity of the fluid, ρ is the density, D is the hydrodynamic diameter, and μ is the fluid viscosity. For fluids that have a viscosity similar to water it is easy to get velocities high enough to get turbulence and thus produce microscopic eddies that bring the two liquids in very close physical contact such that diffusion can complete the mixing.

For polymers the mixing situation is very different. First, each polymer molecule takes up much more space because of its molecular weight and limited configurational motion due to restricted bond angles. Thus the entropy of mixing, S_m, is very much smaller than that for small molecules. Therefore, the spontaneous mixing of polymers is controlled by the heat of mixing H_m. As a result, only a few true polymer-polymer solutions exist, and the first to be discovered was a solution of PPO and PS resins. Next, the viscosity of polymers is thousands of times higher than that of common liquids, so the Re is almost always less than 10. This Re is at least two orders of magnitude too small for turbulence, regardless of the pressure or the velocity of a mixing device. Thus turbulent eddies do not exist to bring the two polymeric fluids together. Therefore, polymer-polymer mixing is brought about by thinning the striations of the two polymer materials being mixed such that the thickness of the striations are small enough to provide a "homogeneous" appearance and the physical properties of the mixed materials. The most effective method for creating a large increase in surface area is the so-called "baker's fold" where the surface area between the dough and the butter of "puff paste" increases exponentially each time the dough is rolled out and then refolded before being rolled out again. The baker's fold process is shown schematically in Figure 8.1 for the stretch and fold process and also the stretch, cut, and stack process. For polymeric materials, this exponential increase in surface area can be obtained by using devices that produce chaotic mixing. Several screw configurations have been patented that produce chaotic mixing of polymers. The figure clearly shows that repeated stretching and stacking of the elements provides finer and finer striations and thus a rapid increase in interstriation surface area. These concepts can be extended to distributive mixing of fillers in the polymer, and in this case the filler will be dispersed in a truly random manner in the polymer matrix.

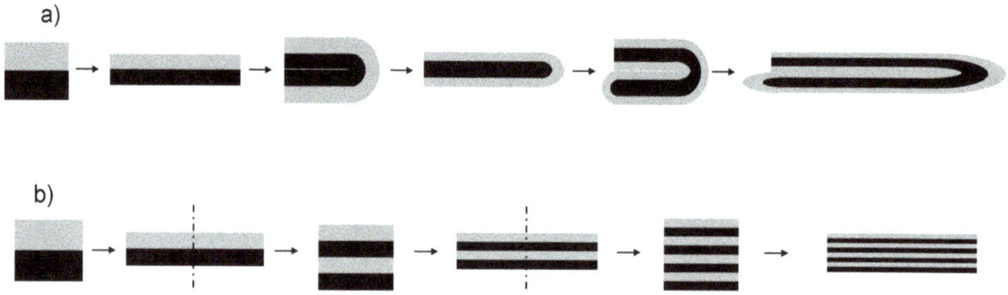

Figure 8.1 Distributive mixing mechanism for the baker's fold process: a) stretch and fold method, and b) stretch, cut, and stack method

Other entropy types have been developed to determine the mixing level in systems, including Shannon entropy and Renyi entropy. Shannon entropy has been used to show distributive mixing levels in color mixing experiments [18, 19], while Renyi entropy has been used to show a more rigorous analysis of distributive mixing [20, 21]. The reader is referred to the references as these types of entropies are beyond the scope of this chapter.

8.3.1 Measures of Mixing

Two calculations are often used to evaluate the mixing in polymer processes. The first is the generation of surface area as simulated in two dimensions by the stretching of a line. The second is the standard deviation of the concentration of the minor component in a residence-time evaluation of the mixing process. This section will present both calculations.

Stretching of a line of polymer is a commonly used measure of mixing that represents the generation of surface area [22, 23]. This increase in interfacial area between two materials is depicted in Figure 8.2.

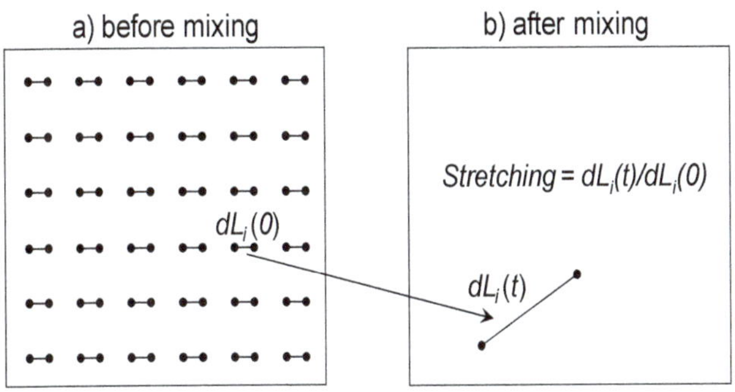

Figure 8.2 Physical representation of the line-length concept for prediction of surface area generation: a) an array of line segments before mixing, and b) the length of the *i*th segment increases to $dL_i(t)$ after a mixing process

Overall stretching $dE(t)$ is defined as:

$$dE(t) = \frac{\sum_{i=1}^{n} dL_i(t)}{n} \tag{8.3}$$

where n is the total number of line segments considered in the cross section and $L_i(t)$ is the length of the segment at time t.

A statistical measure for mixing can be obtained by calculating the standard deviation of a particle concentration, after numerically tracking a compact matrix of particles [24]. For example, a numerical matrix of particles containing 10,000 tracer particles is introduced at the inlet of a rectangular channel. It is tracked for 150 seconds in the flow field. The entire rectangle is divided into a 40 by 40 grid. The number of particles in each small grid is counted after tracking the particles in the matrix. The number of particles is equivalent to the concentration of a tracer. If there was uniform mixing, the particles in the matrix would have become uniformly distributed throughout the cross section. The standard deviation is defined as follows. Let A be the area of the cross section, N_0 the total number of particles tracked, and N_g the number of divisions into which the area is divided. For the case specified here N_0 = 10,000 and N_g = 1600. The area of each rectangular element and the concentration is as follows:

$$\Delta A = \frac{A}{N_g} \tag{8.4}$$

$$C_n = \frac{N_n}{\Delta A} \tag{8.5}$$

where C_n is the concentration in the nth rectangular element and N_n denotes the number of particles in the nth element. The concentration in each of the 1600 elements is obtained after tracking all the particles and counting the particles reaching each of the rectangles. The average of the concentrations $\bar{C}$ is as follows:

$$\bar{C} = \frac{1}{N_g}\sum_{n=1}^{Ng} C_n = \frac{N_0}{A} \tag{8.6}$$

The standard deviation, σ, can be calculated from Equation 8.7:

$$\sigma = \left[\frac{1}{N_g}\sum_{n=1}^{Ng}\left(C_n - \frac{N_0}{A}\right)^2\right]^{1/2} \tag{8.7}$$

A well-mixed system would have a standard deviation that would approach zero. If the mixture is poorly mixed, the standard deviation will be relatively high.

8.3.2 Experimental Demonstration of Mixing

The fundamentals of mixing using standard deviation and overall stretching will be demonstrated using a free helix extruder. This demonstration will show why there has been a strong interest in introducing chaotic mixing sections into single-screw extruders. The free helix extruder was previously described and shown in Figure 7.3. The device has a reservoir that feeds a screw that can be operated in several modes. This device can be operated as either a conventional single-screw pump or a single-screw chaotic mixer [25]. The screw consists of a helix that is ground such that the flight width perpendicular to the flight, W, is essentially constant as a function of depth. The screw core (or root) is machined such that it slides into the center of the helix. The helix and core can be rotated in either direction independently. This allows the device to produce a so-called deterministic chaos. The deterministic chaos is produced by the periodic programming of the velocities at the core and helix.

In order to demonstrate visually this difference in mixing that occurs in a conventional screw channel while it is rotating and the result of chaotic mixing, two experiments were performed using the free helix extruder and a color dye such that stretching could be observed. The first problem was how to accurately place the dye drop in the channel. A technique was developed where a quantity of a polyol fluid was placed in an aluminum foil cavity of the exact dimensions of the screw channel cross section. Then a drop of concentrated crystal violet dye was added and the composite was frozen using liquid nitrogen. The frozen composite was then physically positioned in the channel through the throat of the extrusion device using forceps. The composite was allowed to melt prior to the start of the extrusion experiment. The device was either operated as a conventional extruder (extrusion mode) or using a chaotic mixing program (chaotic mixing mode). The chaotic program used here was as follows: the helix and core were moved as a screw for one period at a speed of five rpm; the helix was then stopped and the core rotation was reversed and maintained at five rpm for one period. Now the core was turning in such a direction as to continue pushing fluid out of the device. This sequence was then repeated until all the dye had visibly discharged from the extruder.

The stretching of the dye for operation in extrusion mode is shown in Figure 8.3. As shown in Figure 8.3(a), the dyed drop stretched in a nearly constant and helical trajectory down the extruder. Closer to the exit the drop continued to be stretched as indicated by the smaller cross section of the tinted fluid filament, as shown in Figure 8.3(b). The typical mixing in this mode of operation results in a dye monofilament that remains contiguous. Although not shown here, the trajectory and stretching depends on the starting position of the dye drop.

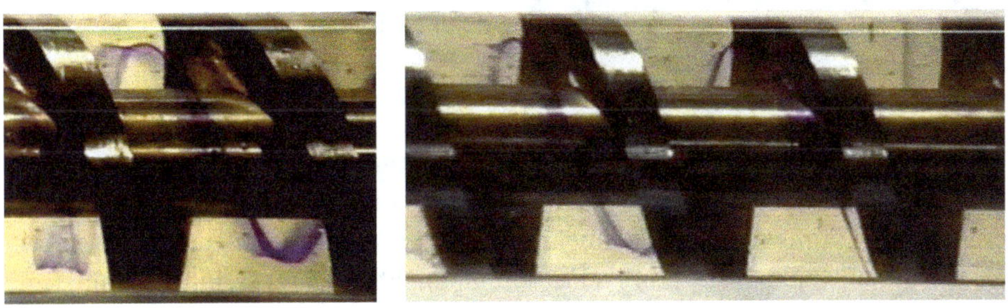

a) b)

Figure 8.3 Photographs of the deformation of dye for operation of the free helix extruder in extrusion mode: a) photograph of the early stretching of the drop near the start of the experiment, and b) photograph of the stretching near the exit of the extruder

The free helix extruder was then operated in the chaotic mixing mode as described above with a period of 4 s. The trajectories of the dye drop and stretching are shown in Figure 8.4 and Figure 8.5. As shown in Figure 8.4(a) in the first few periods of the program, fluid stretching was already giving an improved mixing as indicated by the dye spreading across the channel much more than in the extrusion mode shown in Figure 8.3(a). In Figure 8.4(b) the dye continued to be stretched into the primary fluid in the midsection of the device. Near the discharge end of the device the dye is so well dispersed that it is difficult to see with the naked eye, as shown in Figure 8.5.

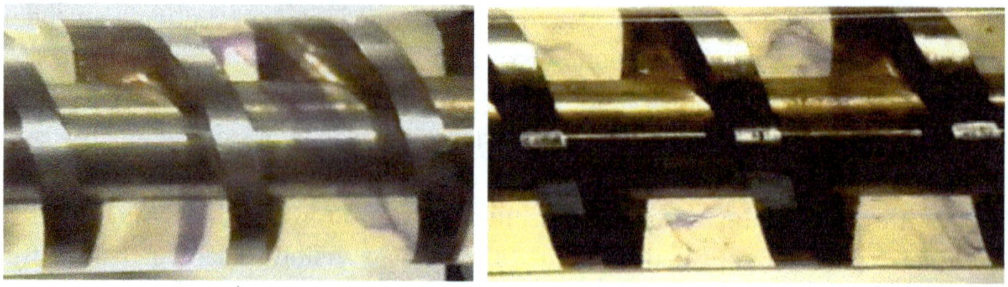

a) b)

Figure 8.4 Photographs of the deformation of dye with the chaotic mixing program: a) photograph of the trajectory and stretching of the dye for the first several periods of flow, and b) photograph of a downstream midsection of the device after several periods of flow

Figure 8.5 Photograph of the deformation of the dye with chaotic mixing near the discharge of the free helix extruder

The measure of mixing for these experiments can be studied using an overall stretching rate, calculated by averaging the stretching over the entire rectangular cross section of the extruder channel. Numerical results presented here are shown for two-dimensional flow. The V_z velocity component is not taken into account. Only the cross-sectional mixing is considered here. Ten thousand equally spaced particles are introduced, filling the entire rectangular cross section. A particle is introduced very near to each of the previous particles. The initial distance between the pairs of particles is very small ($dL_i(0) = 10^{-7}$ length unit, Figure 8.2). The length of the line ($dL_i(t)$) between the pair of particles is watched with respect to time, while tracking the particles. A stretching parameter S_i is defined for the ith particle as follows:

$$S_i = \frac{dL_i(t)}{dL_i(0)} \tag{8.8}$$

To get an overall measure of stretching at any time t, $dL_i(t)$ for all the 10,000 points is calculated and an average $dE(t)$ is calculated. The overall stretching $[dE(t)/dE(0)]$ obtained for 100 seconds is plotted in Figure 8.6. For the extruder operational mode, the plot of the stretching S_i as a function of time is a straight line with a value near one. For the chaotic mixing program, stretching increases exponentially as a function of time. Results obtained show that the regular chaotic pattern gives much better mixing as indicated by the almost two orders of magnitude increase in stretching rates.

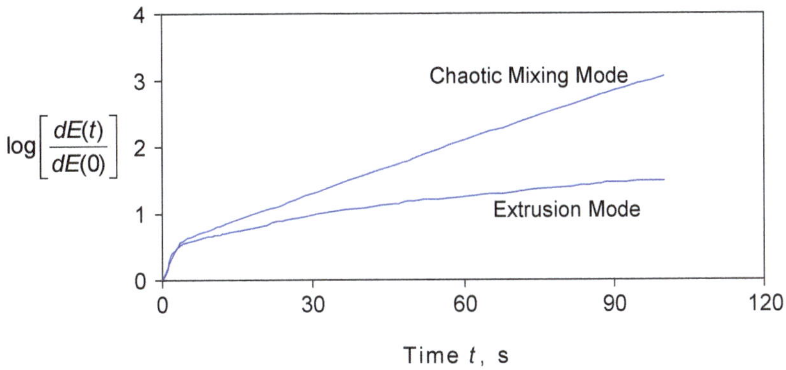

Figure 8.6 Average stretching (S_i) for operational modes of a conventional extrusion and the chaotic mixing. The *y*-axis is logarithmic. The top wall moving curve corresponds to a conventional single-screw extruder with a rotating screw. The results are shown for an h/W aspect ratio of one and a period of 4 s

The free helix experiments presented above were simulated numerically using the same geometry and mixing modes. The simulations were performed by placing randomly four small matrices of particles into the square cross section at $z = 0$. The starting positions are shown in Figure 8.7(a). Each of the matrices shown in Figure 8.7(a) contains 1600 closely spaced particles. They are placed in the shape of a square with 0.01 unit long edges. Next, the boundary conditions are set to that of extrusion mode; that is, the core and helix are rotated at the same speed and same direction as used for extrusion. For this case, each matrix gets stretched into a single long streak of particles, as shown by Figure 8.7(b) and Figure 8.7(c). These patterns are similar to the experimentally determined dye experiments shown in Figure 8.3. As expected, the patterns produced are highly dependent on the starting positions of the matrices.

The numerical experiment was also performed using the chaotic mixing program presented above. The compact matrices of particles were positioned in the channel in the same position as shown in Figure 8.7(a). In this case, the particles in the matrices get scattered throughout the cross section of the extruder as shown in Figure 8.8. Clearly, the overall stretching rates and the deformation of matrices show that the flow pattern of the chaotic mixing program gives more mixing than does the conventional operation of an extruder. The theory shows that the matrices are stretched into continuous filaments in screw rotation and that they are spread over the total volume when the extruder is operated in the chaotic mixing mode.

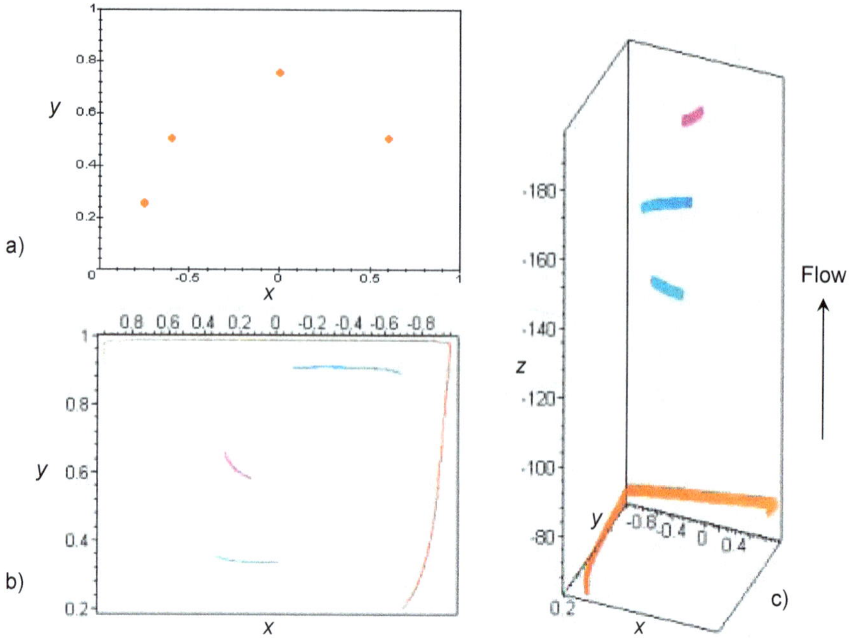

Figure 8.7 Numerical simulation of the stretching of four randomly placed compact matrices of particles into a square channel for the free helix geometry operated in extrusion mode: a) initial position of the matrices at $z = 0$, b) end view of the particles at 100 s of rotation, and c) three-dimensional view of the particles after 100 s of rotation

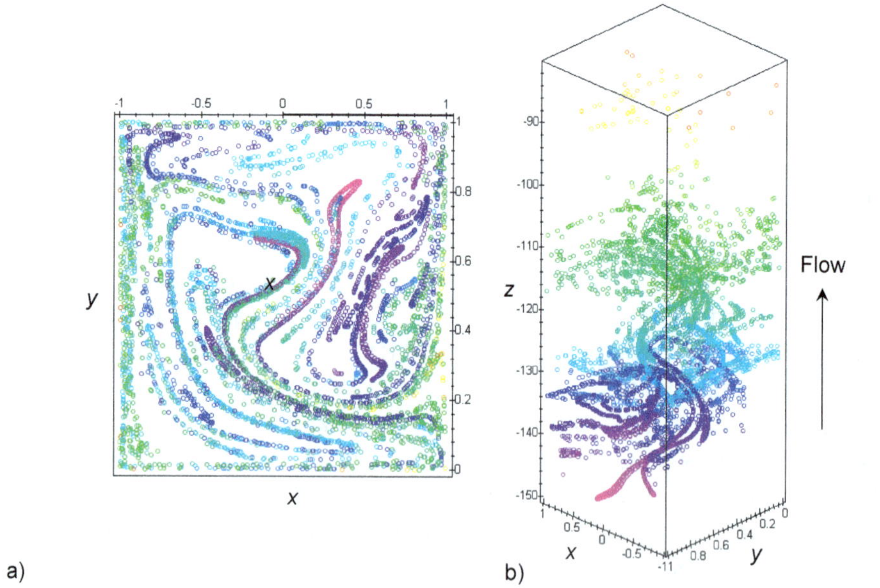

Figure 8.8 Numerical simulation of the stretching of four randomly placed compact matrices of particles into a square channel for the free helix geometry operated using the chaotic mixing program: a) end view of the particles at 100 s of rotation, and b) three-dimensional view of the particles after 100 s of rotation

The residence-time distribution (RTD) for a process can be easily observed using dye and it can be predicted numerically. For example, three drops of dye were positioned in the feed region of the free helix extruder, and the screw was rotated in extrusion mode at a screw speed of five rpm, and then the RTD was observed by measuring the concentration of the dye in the discharge as a function of time. The concentration was measured using a spectrophotometer. The extruder was operated with a zero pressure gradient. For the free helix operating in extrusion mode the RTD is shown by Figure 8.9. As expected, three concentration peaks were observed, one originating from each drop. The three peaks indicated that very little mixing occurred in the process. The concentration profile depends on where the dye was positioned in the channel. For the cases here, the three dye drops were evenly spaced across a line perpendicular to the flight and at the mid-depth position. The maximum relative concentration was intermediate for this run, and the duration for dye in the discharge was about 25 s for the experiment. The numerical simulation indicated that small trace amounts of dye will persist longer in the discharge.

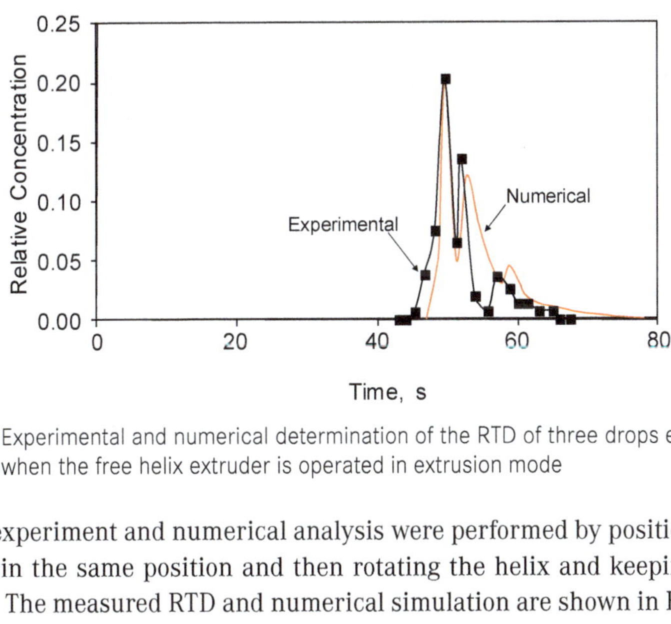

Figure 8.9 Experimental and numerical determination of the RTD of three drops extruded when the free helix extruder is operated in extrusion mode

A similar experiment and numerical analysis were performed by positioning three dye drops in the same position and then rotating the helix and keeping the core stationary. The measured RTD and numerical simulation are shown in Figure 8.10. As shown by this figure, the three drops were better mixed than in extrusion mode, but the mixing was still poor as indicated by the magnitude of the first concentration peak. The long concentration tail was caused by the dye near the walls of the screw.

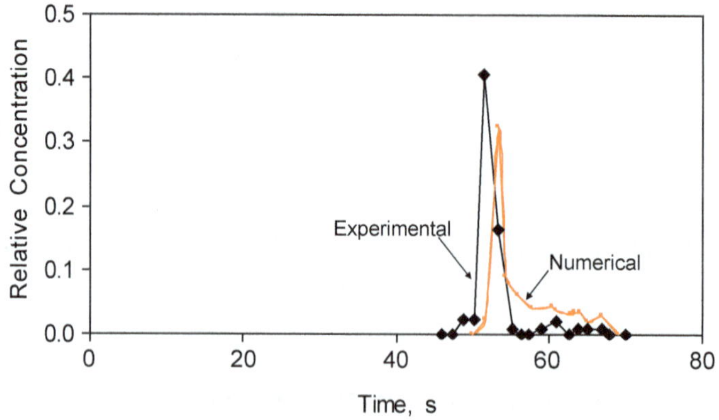

Figure 8.10 Experimental and numerical determination of the RTD of three drops extruded when the free helix extruder is operated with just the helix rotating (core is stationary)

The free helix extruder was also operated using the chaotic program discussed above using three dye drops positioned as before. The experimental and numerical results are shown in Figure 8.11. The experimental data for this run shows the increased time that the dye takes to exit the extruder and the broadened peak of the dye. The mixing level is also indicated by the low relative concentration of the peak. Moreover, the individual peaks of the three drops cannot be observed in the RTD. The relative concentration decreases very quickly at the end of the experiment, indicating that the dye is moved away from the walls and out of the Moffat eddies (discussed later) by the reorientation of fluid caused by the chaotic program. This mixing level and program are not possible in commercial extruders due to the mechanical limitations of building a helix that can withstand the forces that occur with highly viscous resins, but it does show how deterministic chaos can improve the mixing that occurs in screw channels.

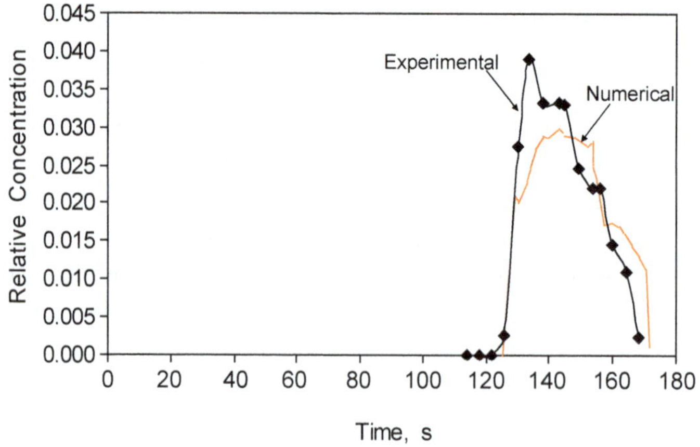

Figure 8.11 Experimental and numerical determination of the RTD of three drops extruded when the free helix extruder is operated in chaotic mode

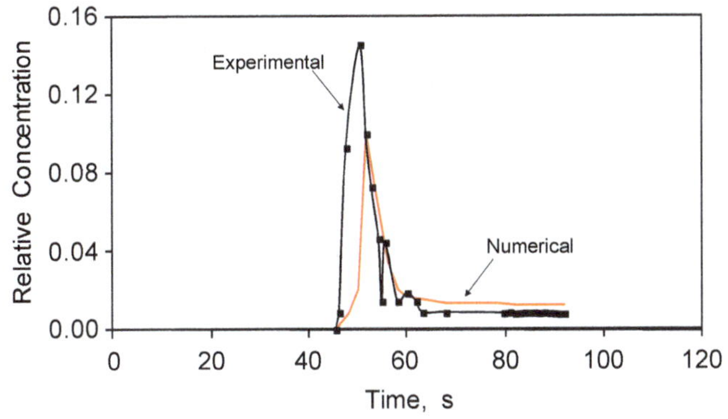

Figure 8.12 Experimental and numerical determination of the RTD for a sheet of dye positioned across the channel in the feed and perpendicular to the flight. The screw was operated in extrusion mode

In a final RTD experiment, a sheet of dye was frozen as before and positioned in the feed channel perpendicular to the flight tip. The sheet positioned the dye evenly across the entire cross section. After the dye thawed, the extruder was operated at five rpm in extrusion mode. The experimental and numerical RTDs for this experiment are shown in Figure 8.12, and they show the characteristic residence-time distribution for a single-screw extruder. The long peak indicates that most of the dye exits at one time. The shallow decay function indicates wall effects pulling the fluid back up the channel of the extruder, while the extended tail describes dye trapped in the Moffat eddies that greatly impede the down-channel movement of the dye at the flight corners. Moffat eddies will be discussed more next. Due to the physical limitations of the process, sampling was stopped before the tail had completely decreased to zero concentration.

As shown in Figure 8.9 though Figure 8.11 the residence time distribution for three drops placed at the feed area of the free helix extruder changes dramatically depending on the manner in which the device was operated. The theory and the data fits are quite good in all cases. As indicated by Figure 8.10, the RTD for helix rotation is almost a Dirac function, consistent with the concept that the helix is almost a positive displacement pump. The dispersion of the RTD was due to the flights providing some drag flow back towards the extruder entrance. In Figure 8.9 the screw rotation flow for the three drops has a wider dispersion and a smaller maximum value, which would lead to a smaller standard deviation. The RTD for the chaotic mixing program had even smaller maximum concentrations and much broader time of measurable output. These combine to give much smaller standard deviations and thus much better mixing than either the screw rotation or the helix rotation, as shown in Figure 8.11.

This difference at the end of the RTD curves is due to the flow characteristic of Moffat eddies. Moffat eddies [26] are recirculation or vortices that occur in sharp corners as shown in Figure 8.13 and Figure 8.14. When fluid is put in motion with top-driven cavity flow, the main circulation is shown in Figure 8.13. A secondary circulation is set up in the stationary corners of the channel, creating a low-velocity helical eddy that is outside the high-velocity flows of the main part of the channel. This flow region is commonly called a Moffat eddy. The long residence times that Moffat eddies create can lead to the degradation of the resin and difficulties in color changes and purging. These problems will be discussed in detail in Chapter 11.

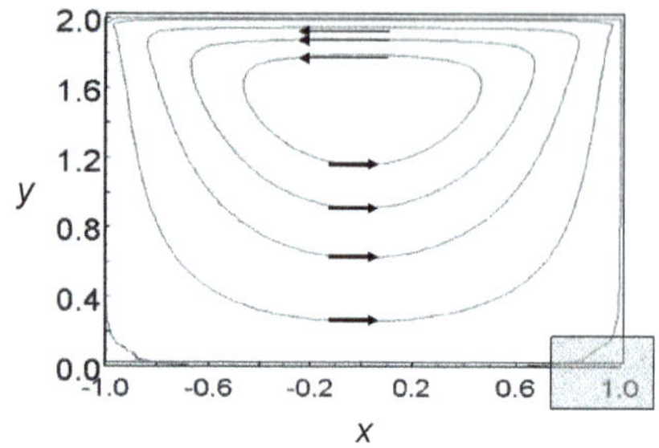

Figure 8.13 Two-dimensional flows in a screw channel with $h/W = 1$ and operating in extrusion mode. The arrows show the recirculation flows. The shaded area in the lower right corner is expanded in Figure 8.14 to show the Moffat eddy

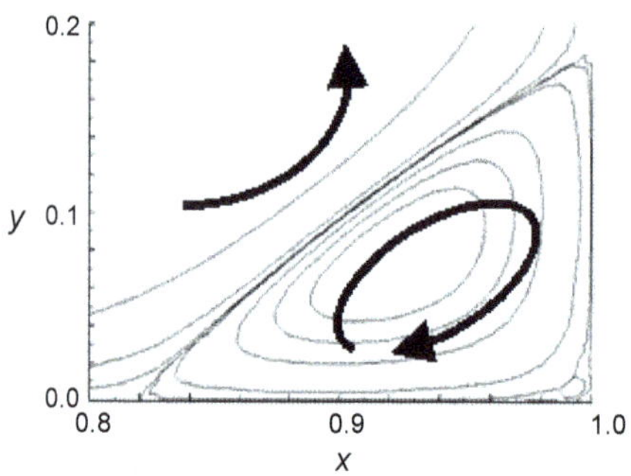

Figure 8.14 Two-dimensional flows in a screw channel with $h/W = 1$ and operating in extrusion mode. The region shown is the shaded area in Figure 8.13. Cross-channel flow lines and the corner circulation of the Moffat eddy are visible in the schematic

8.4 The Melting Process as the Primary Mechanism for Mixing

The first and sometimes the highest level of mixing that occurs during the plasticating extrusion process is during the melting process, as shown by Benkreira et al. [13]. This concept is fairly obvious from the photograph contained in Figure 8.15. For single-flighted, conventional screws, Benkreira showed evidence that most of the mixing occurs during the melting process with only small levels of additional mixing occurring in the downstream channels of a conventional screw. As will be shown later, if melting is not completed inside the extruder, significant defects can occur due to poor mixing. As melting progresses downstream, the thickness and width of the solid bed become smaller until all the resin is fully melted. For conventional single-flighted screws, the melting (or transition) section channel starts at the depth of the feed section and becomes shallower until it ends at the entry of the metering section. Recall that the metering channel depth is typically one-half to one-sixth the depth of the feed channel. The melting process, however, is not limited to the transition section of the screw. In most cases, the melting process extends several diameters downstream and into the metering section. Most researchers have ignored the importance of the melting process on mixing performance.

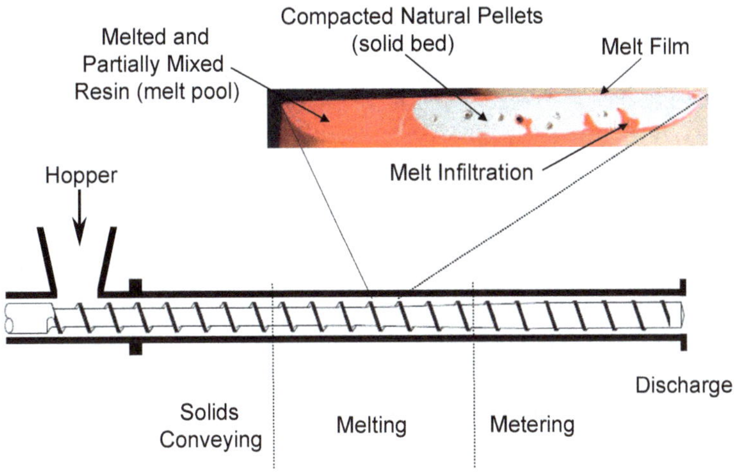

Figure 8.15 Schematic of the melting and mixing process that occurs during the plastication of natural ABS (white) resin with a red color concentrate resin using a conventional, single-flighted screw

The mechanism for mixing during the melting process was clearly demonstrated by Benkreira et al. [13] in 1992. For this mechanism, the solid particles are moving down the channel in the z direction at the velocity of the solid bed. As melting occurs, the pellets exposed to the solid bed–melt film interface undergo a phase change (or devitrify) and are transferred to the melt film, as shown by Figure 8.16. This newly molten material is then accelerated and is forced to undergo elongation. The elongation occurs because the motion of the screw relative to the barrel is considerably higher than that for the solid bed and is in a different direction. That is, as the fluid element moves away from the bed due to the introduction of newly molten material, the cross-channel (V_{bx}) and the downstream (V_{sz}) components of the flow accelerate and stretch the element. The elongation causes the material to form a thin striation of material as shown in Figure 8.15 and Figure 8.16. As indicated in Figure 8.16, the thickness of the melt film increases in the cross-channel flow direction. This is due to the melting flux that is occurring across the entire width of the bed.

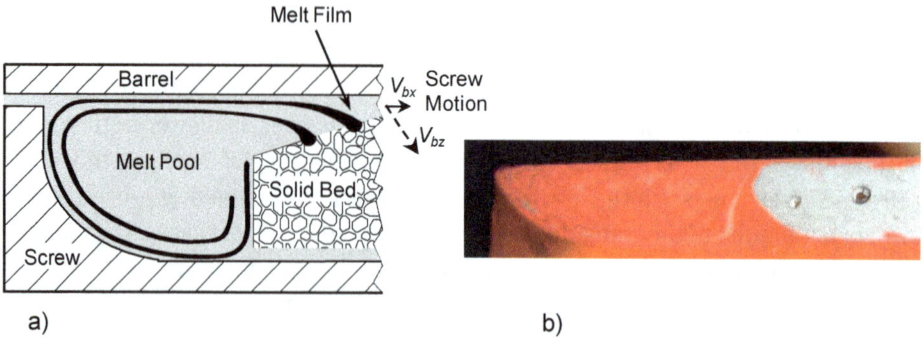

Figure 8.16 The mixing mechanism during the melting process: a) schematic that shows the mixing mechanism during the melting process. Several of the pellets were colored black to show the flow paths and the elongation of the fluid elements. The melt film thickness is enlarged to show the elongation of the dark-colored pellets. b) a photograph of a cross section from a Maddock solidification experiment showing the mixing process of a color concentrate into a natural ABS resin

Several features are illustrated in Figure 8.16. First, the melt film thickness is shown increasing in the cross-channel direction. This occurs due to the addition of newly molten material migrating from the solid bed–melt film interface into the film. At the trailing flight side of the channel, the thickness of the melt film is the smallest and about the size of the flight clearance, and as material is melted and accumulated in the film its thickness increases. The thickest location occurs at the edge nearest the melt pool. The film thickness is enlarged in Figure 8.16 to show the details of the melting and mixing mechanism. As the newly molten material is added to the film, it pushes the older material away from the interface and into a location where the velocity is higher, causing the fluid element to elongate. This elongation is shown in Figure 8.16(a) by the reduction in width of the melting streak in the melt film. The material then flows to the melt pool where the individual streaks create a pattern with many laminar striations. These streaks are shown here as a color mixing process, but they could be compositional or thermal differences.

The striations that occur in the melt pool are caused by streaks from many different pellets. Some of these streaks are occurring due to the pellets at the viewing cross section, but most are created by pellets that have melted upstream of the view. For color mixing processes, the best mixing that will occur during the melting process will be for very low color concentrate letdown ratios. With low color letdown ratios, such as 25:1 natural resin to a colored resin, the amount of colored pellets to natural pellets is extremely high; for this case it is four colored pellets per 100 natural pellets. A larger number of colored pellets will create a larger number of color streaks during melting and provide a higher level of striations or interfacial area between the colored and natural materials.

The melting process described above provides the first and likely the best level of mixing. This mechanism, however, is sometimes difficult to maintain in a plasticating extruder. The most common problem encountered is the breakup of the solid bed. If the solid bed breaks up, then a solid bed fragment can flow with the molten material downstream and not be subjected to the high stresses that occur during the melting process. Solid bed breakup is a complicated process [27, 28] and is poorly understood. In general, solid bed breakup is most likely to occur at high screw speeds. Solid bed breakup was discussed in Section 6.6 and will be discussed in Section 8.4.1.

8.4.1 Experimental Analysis of the Melting and Mixing Capacity of a Screw

The melting and mixing capacity of a screw can be assessed by performing an extrusion study using pigmented resins. Several of these melting studies were performed using a conventional screw and for select high-performance screws. The results for the high-performance screws will be presented in Chapter 14. In all cases, the experiments were performed using a 63.5 mm diameter, 21 *L/D* highly instrumented extruder (see Figure 1.1) and an ABS resin. Twelve pressure transducers were placed in the barrel wall to measure the axial pressure profile. The ABS resin was precompounded with TiO_2 such that the final composition was 2%. A black color concentrate resin (30% pigment) was made from the same base resin. The white-colored resin and black color concentrate were dry blended at a ratio of 100 parts white resin to one part black resin, and they were added to the hopper. The extruder was equipped with three barrel temperature control zones, and they were set at 200, 230, and 250 °C for zones one (feed), two, and three, respectively. The extrudate was forced through an 11 mm diameter circular die, producing a solid rod sample. After steady state was achieved at a specific screw speed, an extrudate sample was collected and allowed to solidify, as shown by Figure 8.17. This sample was then cut perpendicular to its axis to reveal the quality of the mixing and to determine if solid polymer fragments are present in the extrudate.

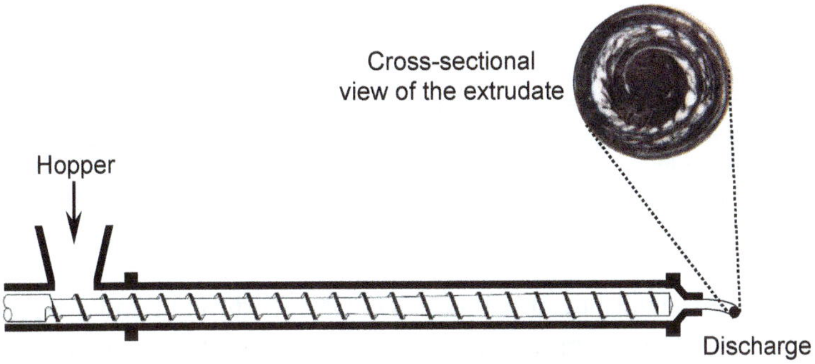

Figure 8.17 Extruder schematic showing location of the extrudate cross-sectional views for the melting capacity and mixing experiments

The baseline melting-mixing performance for this experiment was performed using a conventional single-flighted screw. This screw had a square pitch lead length and the following geometry: 1) a 6 diameter long feed section with a constant channel depth of 8.89 mm, 2) an 8 diameter long transition section, and 3) a 7 diameter long metering section with a constant depth of 3.18 mm. The extruder was operated at screw speeds ranging from 30 to 150 rpm, and extrudate samples were collected and then visually examined for solid polymer fragments (melting capacity) and mixing, as shown by Figure 8.18. As shown by this figure, the extrudate at screw speeds of 30 and 60 rpm contained essentially no solid fragments (white regions) and had relatively small amounts of poorly mixed material. As the screw speed was increased further, however, some solid fragments were obvious. Solids were evident by the nonuniform diameter of the extrudate stream and by the cross-sectional views. Moreover, the solids level in the extrudate increased as the screw speed increased beyond 90 rpm; some trace amount of solids were first observed at about 80 rpm. Thus, operation of an extruder at speeds higher than the operational limits of the screw can cause solids to appear in the discharge. Since there are considerably more white pellets than black pellets, the fragments will most likely cause a white streak to appear in the final article. The highest possible rate that provided an acceptable quality discharge for this conventional screw was at a screw speed of about 75 rpm and a rate of about 70 kg/h. The cross-sectional views shown in Figure 8.18 will be used as a baseline for several high-performance screws presented in Chapter 14.

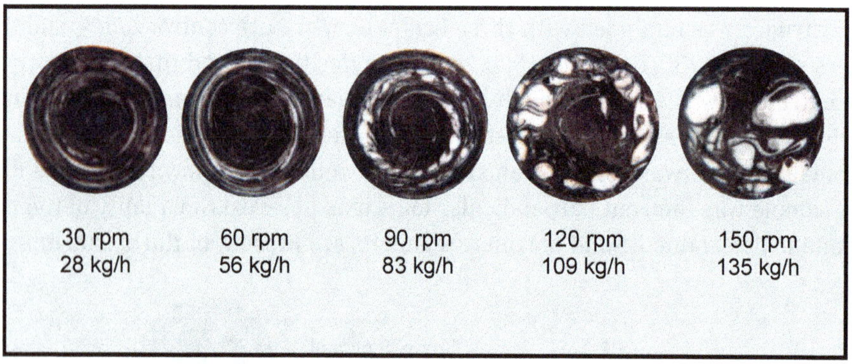

Figure 8.18 Cross-sectional views of extrudate samples at a letdown ratio of 100 to 1 of a white pigmented ABS resin with a black color concentrate for a melting-mixing experiment. The samples were produced at screw speeds ranging from 30 to 150 rpm. The large round white regions for the cross sections at screw speeds of 90 rpm and higher are solid particles discharged with the extrudate

The mixing requirement presented here is extremely difficult and noncommercial due to the very high letdown ratio used and the use of TiO_2 in the main resin. For experiments where natural ABS resin (no added TiO_2) was used at a 100:1 letdown ratio with the black color concentrate, the very small white streaks at screw speeds of 30 and 60 rpm were not visible. Moreover, commercial operations typically use letdown ratios ranging from 35:1 to 50:1. At these commercial letdown ratios and using a translucent main resin, the fine streaks in the extrudates caused by the lack of color concentrate would not be visible. The mixing assessment here is also complicated by the relatively short barrel length of only 21 *L/D*. Most commercial extruders have barrel lengths that are longer than this laboratory machine. A longer barrel length would allow improved melting and mixing capabilities.

Melting capabilities for a screw can also be assessed by performing a Maddock solidification experiment as outlined in Section 10.3.1. The solidification experiments presented here were performed using the white- and black-colored ABS resins and conditions identical to those used for the mixing experiments presented above. The baseline melting performance was performed using the conventional single-flighted screw described above and a screw speed of 74 rpm [29]. The cross-sectional views are shown in Figure 8.19. For each cross section, the barrel surface is along the top, the screw root is along the bottom, and the pushing and trailing flights are along the left and right edges, respectively. As shown by this figure, the solid bed spans the entire channel width at 6 diameters from the start of the screw. By 8.5 diameters, about 85% of the channel was occupied by the solids bed and about 15% by the melt pool. At 8.5 diameters, the depth of the channel has decreased and the pressure has increased, causing the solid pellets to compact into a bed. By 12 and 16 diameters, about 80% and 65% of the channel width is occupied by solids, and the channel depth has decreased further, as indicated by the thickness of the cross section. From these three cross sections, it is obvious that a considerable level of mixing has taken place via the melting process. At 17 diameters, the solid bed is not present in the view. Instead the melt pool is occupying the entire width of the channel. But at 18 diameters, there appears to be a relatively high level of white pigmented material in the center and right side of the channel. This high level of white resin was caused by a breakup of the solid bed, and it represents solid material or material that is at a temperature near the T_g but not mixed with the black pigment from the color concentrate pellets. The cross section at diameter 19 shows even a higher level of solids or unmixed material. At diameter 20, at a position one diameter from the discharge end of the screw, there are many striations between the white and dark areas, indicating that a significant level of mixing occurred during the melting process and the downstream sections of the screw. The cross sections shown in Figure 8.19, however, are just a view of the operation at one moment in time. Unsteady-state processes such as solid bed breakup can cause some solids or unmixed white material to flow very close to the tip and discharge with the extrudate. Solid bed breakup was discussed in Section 6.6.

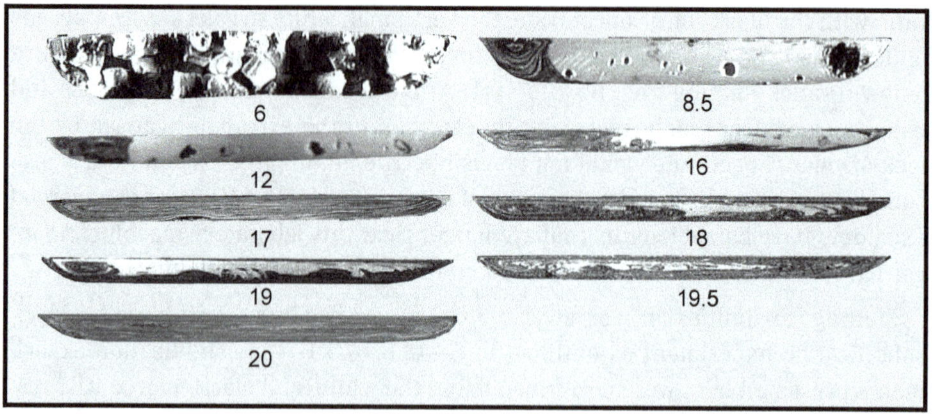

Figure 8.19 Cross-sectional views for the conventional, single-flighted screw [29]. A black-colored concentrate was added at a letdown ratio of 220:1 to a white-tinted ABS resin. Labels indicate the axial distances in screw diameters from the start of the screw

The extrudate samples and the solidification experiment clearly show the importance of the melting process on the mixing quality of the extruder discharge. The cases shown here are for color masterbatches in natural or white resins, but the concept applies to any solid mixture or blend added to the hopper of an extruder. Color masterbatches were used here because it is much easier to visualize than other compositional variations or thermal gradients.

8.4.2 Mixing and Barrier-Flighted Melting Sections

The mixing and melting performance of a process can be increased by replacing a standard conventional transition section with a barrier-flighted melting section. Barrier-flighted melting sections are discussed in detail in Sections 6.3.2 and 14.1. For these processes, the melting rates are increased by keeping the solid bed intact, thus mitigating solid bed breakup. The strategic positioning of the barrier flight maintains the continuity of the bed. The undercut distance between the primary flight and the barrier flight also contributes to the mixing and melting performance of the process. The undercut distance is typically between 0.2 and 1.5 mm, or about 1% of the screw diameter. A smaller undercut will trap and melt small solid fragments and increase the mixing performance of the screw. The smaller undercut, however, can cause higher discharge temperatures and higher internal channel pressures. A case study where a barrier-flighted section increased rates and improved the mixing performance for a TPU resin is presented in Section 13.4.2.

8.5 Secondary Mixing Processes and Devices

The metering section of a flood-fed, smooth bore extruder must be the rate-controlling step for a properly operating machine. As discussed in the previous sections, this section can also provide a level of secondary mixing to the molten material coming from the upstream melting section. The term secondary mixing is used here to denote mixing that occurs between flights of the main channel section for a conventional flighted screw or for the mixing processes that occur due to a mixing device positioned in the metering section of the screw. Due to the high viscosities of the resins, all flows in the extruder are laminar. Mixing devices such as Maddock-style mixers, blisters, and spiral dams are generally positioned in the metering section and provide a process for trapping and melting solid particles that did not melt in the upstream sections of the screw. Numerous high-performance mixing sections have been developed that totally replace the metering section of the screw. These high-performance devices provide a high degree of secondary mixing performance while having the capability of generating pressure for the downstream equipment. Several of these devices will be discussed in Chapter 14.

As presented earlier, secondary mixing performance can be experimentally determined by studying the growth rate of interfacial area between two components such as colored and natural resins. Once a steady state is achieved in the mixing device, the materials are solidified via cooling as in a Maddock solidification experiment or by chemically crosslinking the material. The resulting solidified mass is removed from the mixer and sliced to view the striations of the different colored materials. The level of interfacial growth is directly proportional to the increase in the number of striations [30]. Secondary mixing occurs by a combination of shear strain and reorientation of the interfacial area. Reorientation can occur due to the presence of a partial restriction in the flow field such as a mixing device like a Maddock mixer, a pin positioned in the flow channel, or simply a flight edge.

Traditional mixing devices [31, 32] are commonly designed into screws downstream of the melting section for plasticating screws, or near liquid additive injection ports for melt-fed sections. For plasticating machines, these mixers are effective at trapping and melting solid fragments, homogenizing molten streams, and mitigating thermal gradients. As previously discussed, the first main mixing operation is occurring during the melting process upstream of these mixers. Since these mixers were developed and are implemented as a secondary process, they will be referred to here as secondary mixers. For liquid additive injection through the barrel wall and into the molten polymer stream, the correct combination of distributive and dispersive mixing is required to produce an optimal extrudate.

These mixers can be extremely important especially for those processes operating at very high rates or for processes that demand a very high-quality extrudate.

Secondary mixers will mitigate the color streaks shown in the melt pool of Figure 8.19 or the spiral patterns shown in Figure 8.18, and trap and melt any solid fragments entrained in the melt. There are many secondary mixer designs on the market, all having advantages and disadvantages with their use. Some of the more commonly used secondary mixers for dispersive mixing include Maddock-style mixers, blister rings, and spiral dams. These devices work well when it is highly desired to trap and melt solid polymer fragments, impart a high shear stress to the polymer, or create interfacial area between components. But, since these devices all work by flowing the molten stream through a narrow passage, energy dissipation can be high, causing the temperature of the material exiting the device to be higher than desired. If the passageway is too large (not restrictive enough), then the mixing effect will be reduced and thus may not be acceptable for the application. Moreover, if the entering stream has too high a level of polymer solids, blockage of the device is possible and the rate of the process will be reduced. Some of the common distributive mixers on the market include pin mixers, gear mixers, knob mixers, and pineapple mixers. These mixers all work by the multiple dividing, reorientation, and then recombining of the flow. This reorientation of the flows provides an excellent way to create interfacial mixing at an exponential rate. Since the channels on these devices are relatively large, there are typically only low levels of viscous dissipation and only minor increases in the temperature of the resin. But since the channels are open (not restrictive) they are unable to trap and melt polymer solids that may have come from the upstream sections of the screw. These solids have the potential to flow through a distributive mixer unaffected, causing a defect in the final product, especially at high rates. Although a mixer may be classified as either a dispersive or distributive mixer, many mixing devices provide a combination of the two types of mixing.

Numerous mixing sections are available commercially from many manufacturers. Only the most commonly used mixers will be discussed in this section.

8.5.1 Maddock-Style Mixers

Gene LeRoy at the Union Carbide Corporation (UCC) Research Center in South Charleston, West Virginia received a patent in 1969 for one of the most widely used mixing devices ever invented [33]. The device is constructed by cutting several pairs of fluted channels into the screw in the axial direction, as shown in Figure 8.20. Each pair was designed with an in-flow flute and an out-flow flute. Between these flutes is a mixing flight that is undercut from the main flight. The mixing flight traps solid particle fragments and disperses them into the molten resin matrix. The number of flute pairs depends on the diameter of the screw. The device is positioned in the metering channel of the screw.

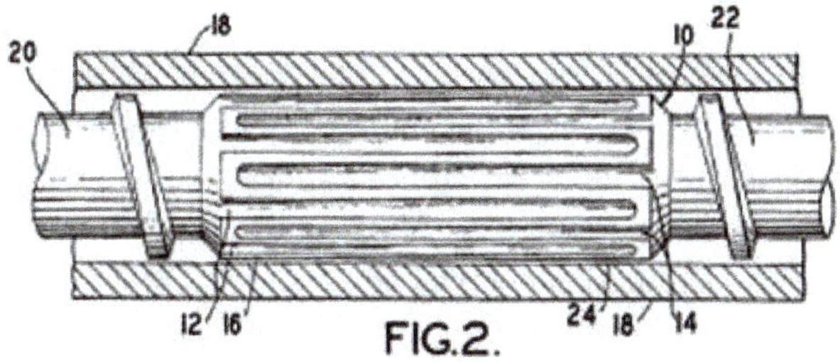

FIG.2.

Figure 8.20 Schematic of LeRoy's mixer [33]

Very few people remember Gene LeRoy for being the inventor of this very unique device, but almost everyone knows the name of the device as the Maddock Mixer or the UCC Mixer. Maddock was the individual who perfected the design and commercialized this very important breakthrough invention. The interesting trivia of this story is that LeRoy invented this mixing device for an entirely different purpose years before Maddock received his 1973 patent [34, 35]. While it didn't prove of further value for the gel reduction project LeRoy was working on, it did provide Maddock with an idea as to how it might be modified to become a useful addition to an extruder screw to help ensure that the polymer is fully melted and mixed into a homogeneous mass. This mixing device had such a large impact on improving the melt quality for polyethylene blown film applications that UCC released their patent rights to the public domain. Numerous spin-off devices were later patented including the Dray mixer [36] and Gregory's spiral mixer [37] in 1974. The mixers will be referred to here as Maddock or Maddock-style mixers.

Maddock-style mixers [38] are very commonly used due to their low cost to manufacture and their ability to disperse solid fragments, trap and melt polymer solids, and mitigate color and compositional gradients. Many styles are on the market under two basic types: 1) flutes parallel to the screw axis, and 2) flutes in a spiral pattern in the same direction as the flights. Schematics for these devices are shown by Figure 8.21 and Figure 8.22. For small-diameter screws, the mixer is generally constructed with four in-flow flutes (or channels) and four out-flow flutes. Larger diameter screws will have more paired flutes due to the larger available area at the screw circumference. For a Maddock mixer with the flutes parallel to the axis of the screw, molten polymer flows into the in-flow flutes via a pressure gradient and then either continues to flow downstream in the flute or is passed through a small clearance between the mixing flight and the barrel wall. This small clearance is responsible for providing the dispersive mixing characteristics of the device. Screw manufacturers typical specify the mixer flight height position relative to the main flight as an undercut. The undercut for a 63.5 mm diameter screw is typically about

0.5 to 1.2 mm, although for some applications and designs the clearance can be smaller. For this size screw with an undercut of 0.50 mm and a flight clearance of 0.07 mm, the clearance between the mixing flight and the barrel wall is 0.57 mm. The material that flowed across the mixing flight is accumulated in the out-flow flute and is then flowed via pressure to the discharge end of the mixer. The wiper flight shown in Figure 8.21 is set at the same height as the flight in the metering section. For mixers with the flutes in a spiral pattern, some of the forwarding flow in the flutes is due to the rotational movement of the flute relative to the barrel wall. Performance and simulation details can be found in the references [39–41].

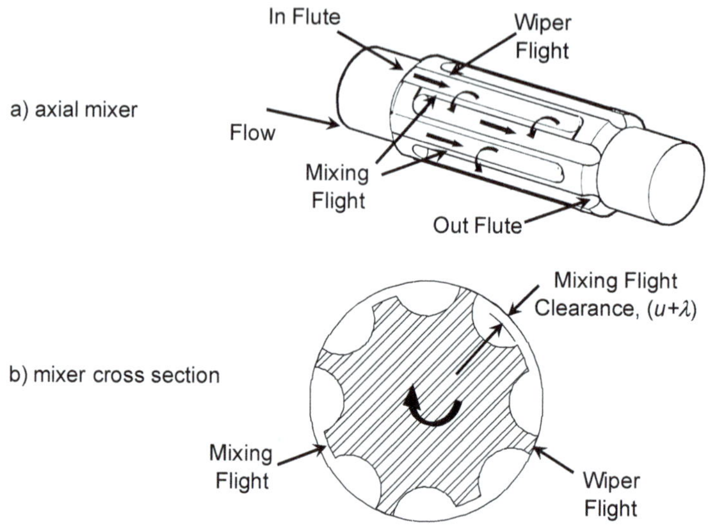

Figure 8.21 Schematic for Maddock-style mixers: a) a mixer with the flutes aligned in the axial direction, and b) a cross-sectional view perpendicular to the screw axis showing the clearance for the mixing flight (courtesy of Jeff A. Myers of Robert Barr, Inc.)

The specification of the undercut on the mixing flight for Maddock-style mixers is critical to its performance. As previously stated, all material must flow through the clearance provided by the sum of the undercut and flight clearance. If the clearance is too large, some medium- and small-size solid polymer fragments will not be trapped and melted by the device. If the clearance is too small, then a high-pressure gradient can occur and there exists the possibility of increasing the temperature of the resin beyond its thermal capabilities, that is, causing degradation. As a general rule, the clearance of the mixing flight of a Maddock mixer with the flutes parallel to the screw axis should be no smaller than the point where the pressure gradient across the flight is zero:

$$(u+\lambda)=\frac{pWH\cos\theta_b}{n_f L_m} \tag{8.9}$$

where u is the undercut of the mixing flight past the metering channel flight, n_f is the number of in-flow flutes (or out-flow flutes), and L_m is the axial length of the mixing flight. The geometric dimensions for the metering section of the screw are as follows: p is the number of flight starts, H is the metering channel depth, W is the average channel width, λ is the flight clearance, and θ_b is the helix angle at the barrel wall. This calculation sets the rotational flow rate across the mixing flight to the rotational flow rate in the metering channel. As an example, the minimum undercut that would be used for a 63.5 mm diameter (D_b) extruder with a single-flighted (p = 1), square-pitched lead length (L = 63.5 mm) metering section with a flight clearance (λ) of 0.06 mm would be about 0.33 mm (u). The metering section average width (W) and depth (H) for this case are 54 and 3 mm, respectively. The Maddock mixer has four in-flow flutes and four out-flow flutes (n_f = 4), and it has an axial mixing length of 100 mm (L_m). Thus, if the extruder is discharging a low level of solids like that shown in Figure 8.18 at 90 rpm, then the mixing flight undercut can be decreased to no smaller than about 0.33 mm. For most applications the undercut clearance will likely be larger than this. The screw would be removed from service and a weld would be applied to the mixing flight, and then the flight would be ground down to an undercut of 0.33 mm. This type of procedure is designed for typical low- to medium-viscosity materials with relatively low levels of solids in the stream. The actual application, however, may require a larger clearance; the clearance calculation is a guideline only.

The shear stress that the material experiences for flow across the mixing flight of the mixer can be estimated using Equation 8.10 and Equation 8.11. The shear stress level is responsible for breaking up agglomerates and dispersing solid polymer fragments. A higher shear stress level will improve the ability of the mixer to disperse smaller size fragments. Like all calculations in this book, this calculation is based on screw rotation physics.

$$\dot{\gamma}_M = \frac{\pi(D_b - 2u - 2\lambda)N}{(u + \lambda)} \tag{8.10}$$

$$\tau_M = \eta \dot{\gamma}_M \tag{8.11}$$

where $\dot{\gamma}_M$ is the average shear rate for flow over the mixing flight in 1/s, N is the screw rotation rate in revolutions/s, η is the shear viscosity at the temperature of the mixing process and at shear rate $\dot{\gamma}_M$, and τ_M is the shear stress that the material will experience for flow over the mixing flight.

Several other design factors are important for the correct operation of Maddock-style mixers. These include the positioning of the mixer downstream from the melting section, the distance between where the meter flight ends and the mixer starts, and the elimination of polymer stagnation regions. The mixer must be positioned on the screw downstream far enough that only low levels of solid polymer fragments exist. If the level of solids is too high in the stream, then the fragments may be melted and dispersed at a rate slower than the rate of the entering solids, causing the mixer to become plugged with solids and reducing the rate of the machine. As shown in Figure 8.22, the mixer should be positioned about 0.3 to 0.5 diameters away from the end of the upstream metering section flight. This creates an annular gap where the material is allowed to flow evenly into all in-flow flutes of the mixer. The annular gap is often undercut as shown by Figure 8.22(b). If the flights extend close to the mixer entry, then it is possible that the in-flow flute near the trailing side of the flight will not operate completely full of resin and thus may cause the resin to stagnate and degrade. Moreover, flute depths should be streamlined and shallower at the entry end of the out-flow flute and the exit end of the in-flow flute. A common design error is to make these regions too deep, creating stagnation regions and causing polymer degradation. In summary, it is highly important that the undercut on the mixing flight be set correctly to create the level of mixing and shear stress required for the application and that the device is positioned properly on the screw and is streamlined such that it does not cause material to have long residence times and thus cause resin degradation.

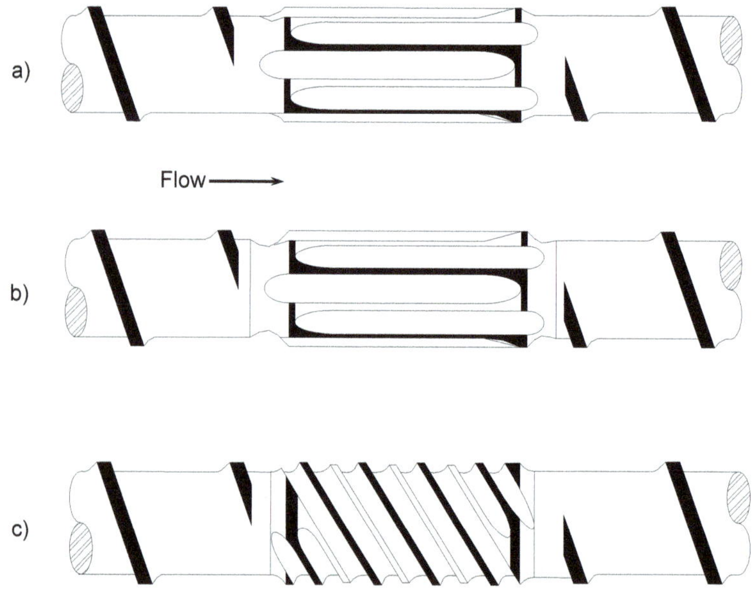

Figure 8.22 Schematic for Maddock-style mixers: a) a mixer with the flutes aligned in the axial direction, b) an axial mixer with pressure-relief zones at the entry and exits, and c) a mixer with spiral flutes (courtesy of Jeff A. Myers of Robert Barr, Inc.)

Other versions of Maddock-style mixers are available commercially that include undercuts on all flights (mixer and wiper) and designs where the in-flow and out-flow flutes are open to the upstream and downstream metering channels. For this latter case, it is possible that some resin can bypass the dispersive mixing process by bypassing the mixing flight. If solid polymer fragments exist in the flow, this design can allow a level of solids to pass downstream.

8.5.1.1 Best Practices for Maddock Mixer Designs

Poorly designed Maddock mixers are ubiquitous in the plastics industry. Poor designs can reduce the mixing performance required for a process and cause resin degradation. Two major design flaws include the shape of the flutes and the size of the mixing flight undercut. Poor mixing can result from the undercut clearance of the mixing flight being too large. A large undercut will reduce the shear stress imposed on the resin as it is forced over the undercut flight. Unmixed gels in PE resins are common in film products. This type of gel will be discussed in detail in Section 11.10.2. These gels are composed of highly entangled polymer chains that when cooled after the die solidify first, causing a gel that looks like a small solid polymer fragment. Unmixed gels can be mitigated by passing the molten resin through a Maddock mixer with a small mixer undercut. For most PE resins, the undercut should provide about 200 kPa of shear stress. Typically, this will set the undercut to about 0.5% of the screw diameter. Most screws in the industry have undercuts that are too large with clearances between 1.0 and 1.5% of the diameter.

Resin degradation is very common for poorly designed Maddock mixers, as shown in Figure 8.23 and Figure 8.24. The resin degradation in Figure 8.23 for this PE extrusion is obvious at the leading and trailing edges of the mixer flutes. The resin at these locations was essentially stagnant [84, 85], creating a region with very long residence times. The high temperature, presence of oxygen, and the long residence time caused the resin to crosslink. Small process disturbances can cause the degraded material to separate from the flute and flow with the extrudate into the product. For delicate film processes, the gel can be large enough to cause the film to break, shutting down the line.

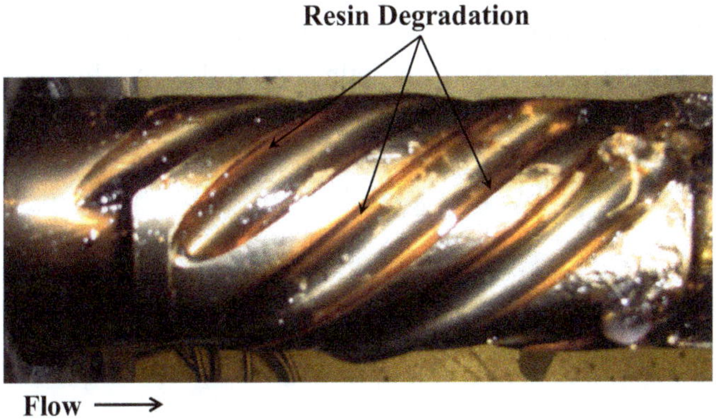

Figure 8.23 Photograph of a Maddock mixer. The flutes are too deep, causing resin degradation at the leading and trailing sides of the flutes

Figure 8.24 A Maddock mixer with shallow flutes: a) photograph of a screw with flutes that are too shallow, and b) Maddock solidification experiment showing resin degradation at the root of the flutes (courtesy of Libo Du of Dow, Inc.)

The long residence time and resin degradation are caused by flutes that are improperly fabricated. For Figure 8.23, the flutes were cut too deep such that the edges become parallel to the radial direction, i.e., a vertical edge. The flute shape and vertical edge are obvious on the photograph in Figure 8.25(a). As a comparison, well-designed mixer flutes are shown in Figure 8.25(c). A well-designed flute is fabricated using a ball end mill with a tool diameter equal to the desired width of the flute. Metal is removed from the screw until a perfect semicircular channel cross-section results, as shown in Figure 8.25(d). In many cases, the ball end mill is used to cut the flute deeper, creating vertical edges in the cross-section as shown in Figure 8.25(b). Some screw designers cut the flutes deep because they are trying to minimize pressure consumption by the device while not understanding the ramifications of resin degradation.

For Figure 8.24, the flutes were too shallow, causing very high temperatures in the flutes and resin degradation at the core. The flute depth was about 25% of the channel width, or half as deep as an optimal flute design. This design will also require a high pressure gradient for operation, adding a high level of undesired pressure drop.

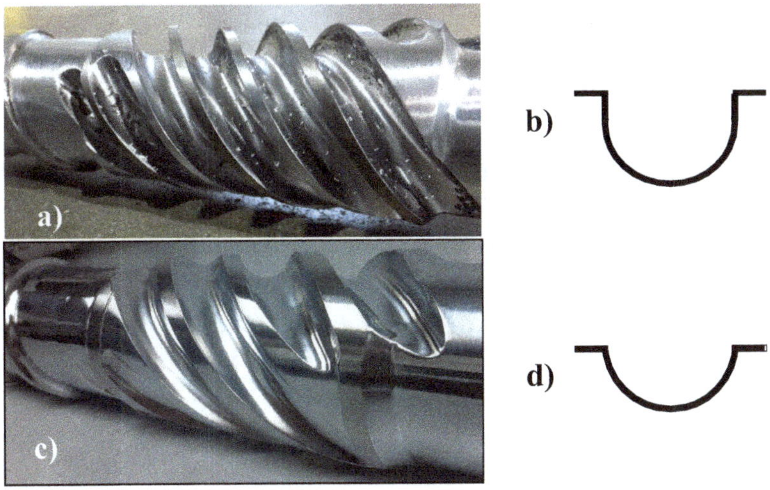

Figure 8.25 Effective and poor Maddock mixer designs: a) a photograph of a poor design with very deep flutes, b) a schematic of a deep flute (poor design) where the depth is greater than half of the width, c) photograph of an effective design with well-designed flutes, and d) a schematic of an effectively designed flute where the depth is equal to half of the width

Another common error is to design too many flute pairs into a mixer. The barrel circumference and the diameter of the ball end mill used for fabrication fixes the number of flute pairs for a well-designed Maddock mixer. Moreover, customer trials and field experience has indicated that spiral Maddock mixers are better at mitigating the degradation of resin in the flutes when compared to standard axial Maddock mixers where the flutes are parallel to the axis of the screw. The helix angle on the mixing flutes is ideally set at 60°. This angle has proven to aid in the movement of resin through the flutes while not allowing stagnation or resin degradation. The lead length of the flutes is related to the helix angle as follows:

$$\tan\theta_b = \frac{L}{\pi D_b} \tag{8.12}$$

where θ_b is the helix angle of the flutes, L is the lead length, and D_b is the inside barrel diameter. With the helix angle set at 60°, the width of the flute in the circumferential direction at the barrel can be calculated from the diameter of the ball end mill using Equation 8.13.

$$W_b = \frac{D_m}{\cos(90 - \theta_b)} \tag{8.13}$$

where D_m is the diameter of the ball end mill that will be specified to cut the flute into the metal stock and W_b is the width of the flute in the circumferential direction as shown in Figure 8.26. For North American produced screws, the ball end mill should be specified in inches to the nearest 1/16 inch such that low-cost, off-the-shelf mills can be used for the fabrication process.

The width of the mixing flight and the wiping flight should be set equal and not less than 6.4 mm (0.25 inch) for screws 2.5 inches in diameter and larger. If the widths are made smaller, then the flight may not have enough strength and failure could occur. Next the number of flute pairs and the ball end mill size are successively chosen until the width of the wiping and mixing flights are about 6.4 to 10 mm (0.25 to 0.4 inch). The calculation of the width of the flights in the circumferential direction is as follows:

$$W_f = \frac{\pi(D_b - 2\lambda) - 2N_f W_b}{2N_f} \tag{8.14}$$

where W_f is the width of the mixing and wiping flights in the circumferential direction, λ is the flight clearance on the wiping flight, and N_f is the number of flute pairs. The flight clearance of the wiping flight should be identical to the flight clearance on the main flights in the metering sections. The widths of the mixing and wiping flights are typically specified in the direction perpendicular to the flight edge. The conversion of the flight width in the circumferential direction to the perpendicular direction (W_p) is as follows:

$$W_p = W_f \cos(90 - \theta_b) \tag{8.15}$$

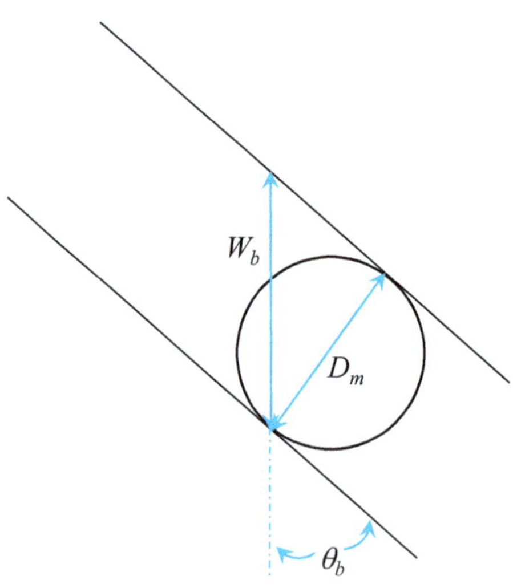

Figure 8.26 Schematic for the design of the flutes showing angles and key dimensions

Typical dimensions for spiral Maddock mixers using the criteria above are provided in Table 8.1 for common screw diameters for North America and metric diameters. Numerous screws have been designed and tested with these criteria, all with acceptable mixing performance and without degrading the resin.

Table 8.1 Typical Maddock Mixer Design Parameters for North American and Metric Diameters. All Designs Have a 60° Helix Angle. The Flute Width is Perpendicular to the Flight Edge

Diameter, inch	Lead Length, inch	Number of Flute Pairs	Ball Mill Size, inch	Flute Depth, inch	Flight Width, inch	Mixing Flight Undercut, inch
2.5	13.6	3	0.875	0.438	0.255	0.015-0.025
3.5	19.0	3	1.250	0.625	0.334	0.020-0.035
4.5	24.5	4	1.250	0.625	0.277	0.025-0.045
6	32.6	5	1.375	0.688	0.254	0.030-0.060
8	43.5	6	1.375	0.688	0.435	0.040-0.080
Diameter, mm	**Lead Length, mm**	**Number of Flute Pairs**	**Ball Mill Size, mm**	**Flute Depth, mm**	**Flight Width, mm**	**Mixing Flight Undercut, mm**
60	326	3	20	10.0	7.15	0.3-0.6
80	435	4	20	10.0	7.15	0.4-0.8
100	544	4	25	12.5	8.94	0.5-1.0
150	816	5	32	16.0	8.72	0.7-1.5

8.5.2 Blister Ring Mixers

Blister rings are another common type of dispersive mixer used on single-screw extruders. These devices are constructed by positioning the root of the screw near the barrel, creating an annular flow path between the screw and barrel, as shown in Figure 8.27. For a 63.5 mm diameter screw, the clearance between the barrel wall and screw is typically 1 mm and the axial length is between 0.3 to 1 diameter. Screw manufacturers typically specify the diameter and length of the blister land. Like the Maddock-style mixers, blister ring mixers are positioned well downstream from the melting section, where only low levels of solids are present. Often the mixer is positioned at the tip of the screw, as shown in Figure 8.27(b). The relative motion between the screw and the barrel creates shear stress that can break some solid fragments into smaller particles, melt low levels of polymer solids, and mitigate thermal gradients. Short flightless sections are typically designed just upstream and downstream of the blister so that the flow is even across the circumference of the mixer. A similar flightless section was positioned upstream and downstream of Maddock mixers. Flow in the axial direction is driven by a pressure gradient. Because of their simple design they are relatively inexpensive to fabricate.

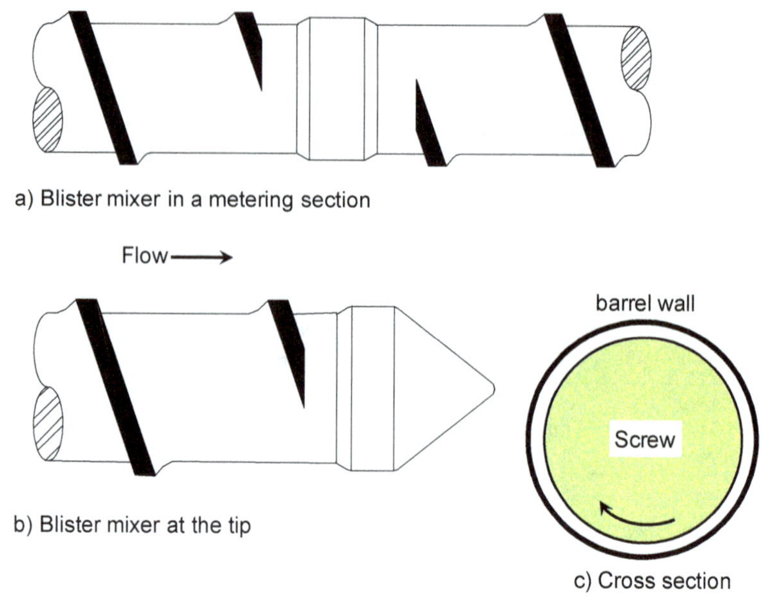

Figure 8.27 Schematic of blister ring mixers: a) a blister mixer positioned in a metering section, b) blister section positioned at the discharge end of the screw near the tip, and c) the cross-sectional view showing the region of annular flow between the screw and barrel wall (courtesy of Jeff A. Myers of Robert Barr, Inc.)

Blister mixers provide a level of dispersive mixing to the flow stream. The mixer typically has a shorter land length perpendicular to the main flow direction as compared to a Maddock mixer. For example, for a 114.3 mm diameter screw, the length of the gap for flow-through is simply the circumference (πD_b) at the barrel wall for a blister mixer. For a 114.3 mm diameter screw with a two-diameter-long axial Maddock mixer with four in-flow flutes and four out-flow flutes, the length of the gap is $8D_b$, a distance that is 2.5 times higher than that for the blister mixer. The short gap length for the blister mixer will cause higher pressure gradients, especially for clearances that are very small. For this reason, Maddock mixers are often favored over blister mixers.

Blister mixers are commonly used as seals on screws. They are often positioned at the very end of the first stage of a two-stage plasticating screw such that a vacuum pulled on the vent opening does not pull air all the way through to the hopper. For a properly operating two-stage extruder, the seal can be provided by a full and pressurized metering section, and then the sealing ability of the blister is not necessary. Blister mixers are also used upstream and sometimes downstream of a liquid-injection point while a distributive type mixer such as a gear mixer is underneath the injection point. For this case, the upstream blister provides the seal, the gear mixer then distributes the liquid into the polymer, and the downstream blister provides a level of dispersive mixing. An example of liquid injection into an extruder is provided in Section 15.3 and Figure 15.9.

The shear stress that the material experiences for flow across a blister mixer can be estimated using Equation 8.16 and Equation 8.17. For this calculation, the flow in the axial direction is ignored to simplify the mathematics, and thus only the angular flow is considered. Ignoring the axial flow causes the calculated shear rate to be lower than in the actual channel.

$$\dot{\gamma}_B = \frac{2\pi D_B N}{(D_b - D_B)} \tag{8.16}$$

$$\tau_B = \eta \dot{\gamma}_B \tag{8.17}$$

where $\dot{\gamma}_B$ is the average shear rate in the blister section in 1/s, D_B is the diameter of the blister, η is the shear viscosity at the temperature of the mixing process and at shear rate $\dot{\gamma}_B$, and τ_B is the shear stress applied to the material flowing through the mixer. This calculation is based on screw rotation physics.

8.5.3 Spiral Dam Mixers

Spiral dams are dispersive mixers that are easily designed and constructed and work well as a trap for solid polymer particles. The mixers are typically two to five diameters in length and are designed with a mixing flight starting at the pushing side of a channel and ending at the trailing side, as shown in Figure 8.28. The design is based on Maillefer's 1962 patent [42]. They are generally positioned at the end of metering sections, although many variants are practiced commercially. Material enters the section on the trailing side, is forced to flow through a small

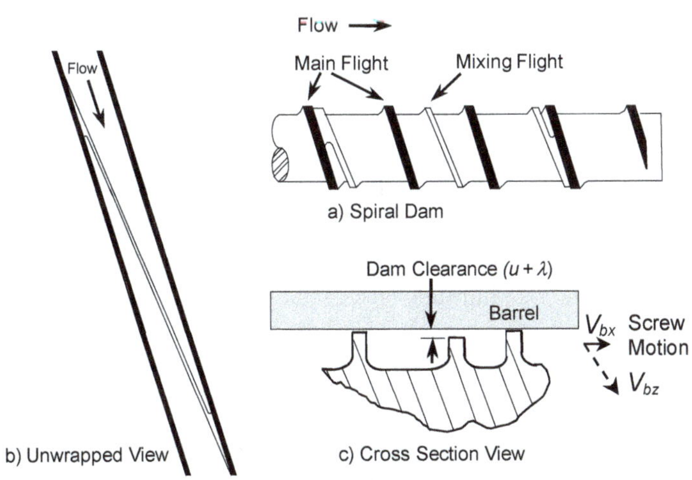

Figure 8.28 Schematic of a spiral dam: a) side view, b) unwrapped view, and c) a cross-sectional view perpendicular to the flight edge showing the clearance between the dam and the barrel wall (courtesy of Jeff A. Myers of Robert Barr, Inc.)

clearance between the mixing flight tip of the secondary flight and the barrel wall, and is then flowed out of the device at the pushing side of the channel. Screw manufacturers typical specify the height of the mixing flight as an undercut distance between the tip of the mixing flight and top of the primary flight. The mixing flight undercut for a 63.5 mm diameter screw is typically between 0.5 and 1 mm. A mixer designed with a smaller clearance may mix and trap solids better and disperse other fragments, but it will also cause the temperature of the material to increase and may reduce the rate of the machine. Like the Maddock-style mixers, the mixer channel depths should be streamlined and shallower at the entry end of the out-flow channel and the exit end of the in-flow channel, minimizing stagnation regions and thus polymer degradation. A case study where a spiral mixer was incorrectly specified for an application is presented in Section 11.12.4.

The shear rate and the stress the polymer experiences over the mixing clearance for spiral dam mixers are approximated by Equation 8.18 and Equation 8.19. For the shear rate calculation, the contribution due to pressure flow is relatively small compared to the component due to rotation and thus it was neglected.

$$\dot{\gamma}_S = \frac{\pi(D_b - 2u - 2\lambda)N}{(u + \lambda)} \tag{8.18}$$

$$\tau_S = \eta\dot{\gamma}_S \tag{8.19}$$

where $\dot{\gamma}_S$ is the shear rate for flow across the mixing clearance in 1/s, u is the undercut distance between the primary flight and the mixing flight, and τ_S is the shear stress.

8.5.4 Pin-Type Mixers

Pin mixers are distributive mixers that work by dividing, reorienting, and recombining the flow streams. For most devices, this process is performed multiple times. As previously stated, the flow channels on these devices are typically large, creating high distributive mixing while contributing very little to the pressure gradient and energy dissipation in the channel. These devices work well for streams that are completely molten, but they lack the ability to trap and melt solid polymer fragments. Pin mixers are commonly seen commercially due to their effectiveness at reorienting the flow, their low cost, and the ability to install them as a modification to an existing screw.

Pin-type mixers perform well for reorientation of the flow fields and improving the mixing for fully molten streams [43]. These devices are constructed by installing a series of strategically positioned short metal cylinders at the root of the screw, as shown in Figure 8.29. The configuration shown in Figure 8.29 is for a retrofit modification where the pins were added after the screw was determined to have less-than-acceptable mixing performance. Another and likely preferred design is to place the pins in a region with no flights. For this case, the axial distance between the pins should be large enough to allow the polymer to reestablish flow in the absence of the pins [43]. Positioning the pins too close together provides a lower level of mixing since the flows tend to channel through a path of least resistance; the pin rows act as interrupted flights rather than individual pins. Regions with very long residence times can occur at the downstream side of the pin, leading to polymer degradation. In an effort to minimize this type of degradation, a pin mixer design with a hole bored from the barrel side of the pin on the upstream edge down to near the screw root on the downstream edge was developed [44, 45]. The hole diverts some of the resin from an area of high flow to the region where flow is very low and residence times are very high. The design has not had a strong commercial interest.

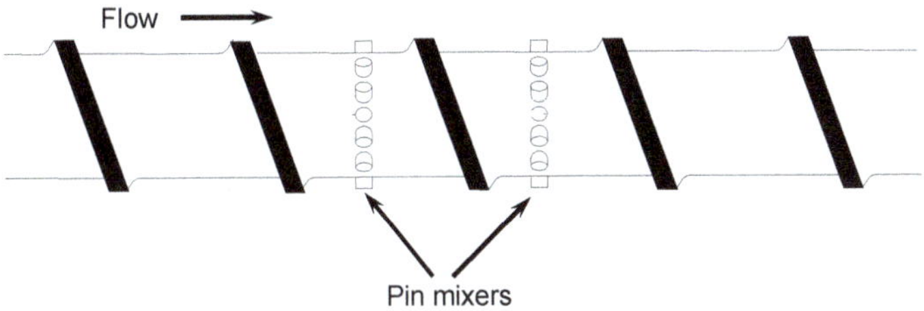

Figure 8.29 Schematic of a conventional single-flighted metering section where the screw was retrofitted with two rows of pins (courtesy of Jeff A. Myers of Robert Barr, Inc.)

8.5.5 Knob Mixers

Knob-type mixers are essentially distributive mixers, but the clearance between the lands of the knob peaks can be undercut beyond the flight clearance in the metering section to provide a small level of dispersive mixing. The knobs can be arranged with all peaks in the same angular plane (straight knob mixer) or they can be configured in a spiral pattern (pineapple-type knob mixer). Schematics for these mixers are shown in Figure 8.30. Many versions and proprietary designs are available commercially.

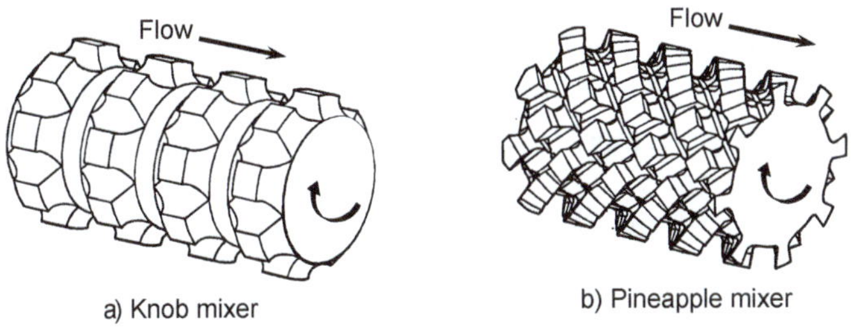

Figure 8.30 Schematic of two common knob-type mixers: a) a straight knob mixer with the knobs in the same angular position, and b) a pineapple mixer with the knobs positioned in a spiral pattern in the same direction as the main flights of the screw channel

Knob mixers work well to eliminate fine striations in color uniformity and thermal gradients. Since knob mixers are primarily distributive mixers, they are not suited for mitigating solid polymer fragments from an incomplete melting process.

8.5.6 Gear Mixers

Gear-type mixers are distributive mixers that perform well for the addition of liquid additives to molten polymer streams. For this application the liquid additives are injected through a hole in the barrel wall and typically upstream of the mixer. Applications include foam production where a blowing agent such as butane is added, or a devolatilization process where a stripping agent such as water or nitrogen is added. For the devolatilization process, the stripping agent and volatile components are removed from the polymer using a downstream vent opening. Gear mixers are also used for the addition of liquid dyes or plasticizers. Many styles of gear mixers are available and they are typically designed using a multiflighted channel design and then slitting the flights to allow material to flow into the adjacent channel, as shown in Figure 8.31. In general, a well-designed gear mixer utilizes a pressure gradient to flow material through the device.

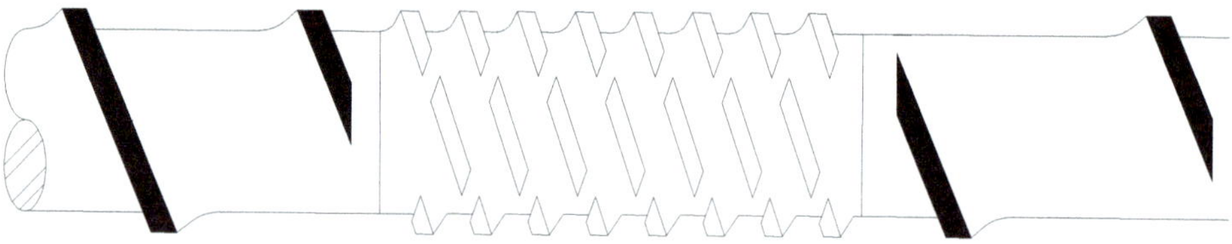

Figure 8.31 Schematic of a gear-type distributive mixer (courtesy of Jeff A. Myers of Robert Barr, Inc.)

The open channels of distributive mixers do not have the ability to trap and melt small solid polymer fragments from incomplete melting in upstream processes. Moreover, combination mixers that perform both dispersive and distributive mixing often only force a fraction of the resin to flow across a restrictive flight, and thus some of the material can bypass without being subjected to a high shear stress. As previously indicated, devices that do not have the ability to trap and melt solid polymer particles can cause poorly mixed extrudates, especially at high screw speeds. Mixers that do not have the capacity to trap and melt solids should not be the sole mixer if solid fragments are likely to occur in the melt stream.

8.5.7 Dynamic Mixers

Dynamic mixers are a class of secondary mixers where part of the mixing device is allowed to move relative to the barrel and screw surfaces or has an active channel positioned in the barrel wall. In general, these mixers provide improved mixing over traditional secondary mixers, but they are more complicated and costly to produce, require higher maintenance levels, and tend to have higher levels of wear. These devices are typically located at the tip of the screw and are often incorporated as part of a nonreturn valve for injection-molding screws. Like other secondary mixers, there are many of these mixers available on the market. The most common of these mixers include the Cavity Transfer Mixer (CTM), the Twente Mixing Ring (TMR), and the Barr Fluxion ring mixer. The CTM mixer [46–48] is a device that is attached to the end of an existing extruder. It has both a barrel section and a screw section. Photographs of a CTM are provided in Figure 8.32. Both sections have concave pockets bored into them, allowing the flowing material to be transferred between the rotor cavities on the screw and the stationary cavities on the barrel wall. The device combines a unique blend of elongational flows and dispersive mixing in the tight clearance regions and flow reorientation in the cavities. The disadvantages of the device include cost and the need for a barrel extension. Additional mixing performance data for the device is in the references [46–48].

a) b)

Figure 8.32 Photographs of a CTM: a) cavities on the end of a screw, and b) cavities on a ring that is inserted into an oversize barrel segment. Eight cavity rings were used to complete the mixer on the barrel segment

The Twente Mixing Ring [49–51] and the Barr Fluxion ring mixer [52–54] are adaptations to the CTM design, allowing the addition of the mixer to the screw without requiring a barrel extension with inset cavities. Schematics for these mixers are shown by Figure 8.33 and Figure 8.34. Both devices are designed with a floating sleeve ring that is allowed to rotate freely from both the screw and barrel. For example at a screw speed of 80 rpm, the speed of the sleeve rings were measured using high-speed data acquisition from pressure transducers at about 10 to 13 rpm [53]. The ring of the TMR is perforated with cylindrical holes, as shown in Figure 8.33. The material in the ring is temporarily suspended in the holes until it is allowed to transfer to the hemispherical cavities in the root of the screw. The floating mixing ring for the Fluxion mixer consists of a series of perpendicular holes that extend through the ring, as shown in Figure 8.34. Each series of holes is divided by a dam that prevents material from flowing directly into the adjacent channel. Material is only allowed to transfer to the adjacent channel by entering the grooves in the root of the screw. As expected, the performance of these mixers exceeds those of typical secondary mixers, but because of the floating sleeve the potential for wear is high. Moreover, small amounts of foreign material that does not melt can get trapped in the holes of the sleeve, creating a restriction to flow and ultimately requiring a shutdown for cleaning. For injection-molding applications, the sleeve can be designed to slide in the axial direction and thus perform as a nonreturn valve [54].

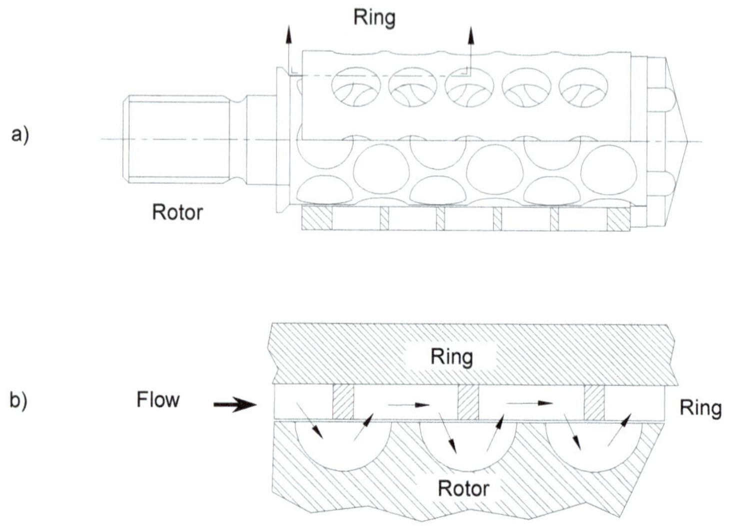

Figure 8.33 Schematic for a Twente Mixing Ring: a) mixer with a cutaway section of the ring exposing the surface of the rotor, and b) flow path for the mixer

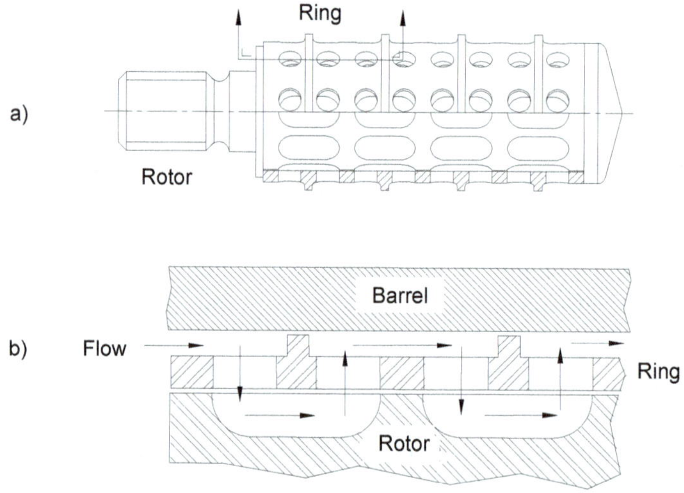

Figure 8.34 Schematic for a Barr Fluxion mixing ring: a) mixer with a cutaway section of the ring exposing the surface of the rotor, and b) the flow path for the mixer

The mixing performance for a TMR and a Fluxion ring mixer was measured using a 63.5 mm diameter, 21 *L/D* extruder and a conventional screw with a removable tip [53] by collecting extrudate samples as described in Section 8.4.1. The tip was 2.4 diameters in length and could be exchanged between a conventional flighted section, a TMR mixer, or a Fluxion mixer. The conventional flight tip had a channel depth of 3.56 mm. The experiments were performed with the same ABS resin used in the studies presented previously. For these runs, the extruder was operated at a screw speed of 80 rpm with a 220:1 ratio of white ABS to a black color concentrate ABS resin. The extrudate samples as collected via the schematic shown by Figure 8.17 are shown in Figure 8.35. As shown by this figure, considerable levels of solid polymer fragments and unmixed material were discharged using the conventional flighted screw tip, with no mixer present. This result was consistent with the data shown by Figure 8.18 for a screw with a slightly shallower metering channel depth. As indicated by Figure 8.35, similar mixing levels were obtained from the TMR and the Fluxion mixer. This level of mixing for the extrudate was extremely remarkable considering that the materials entering the devices were essentially 60% unmixed (white pigmented ABS), and considering that the mixing criteria here exceeds the current demands of the commercial community. Other performance characteristics for these mixers were previously presented in the references [50–54].

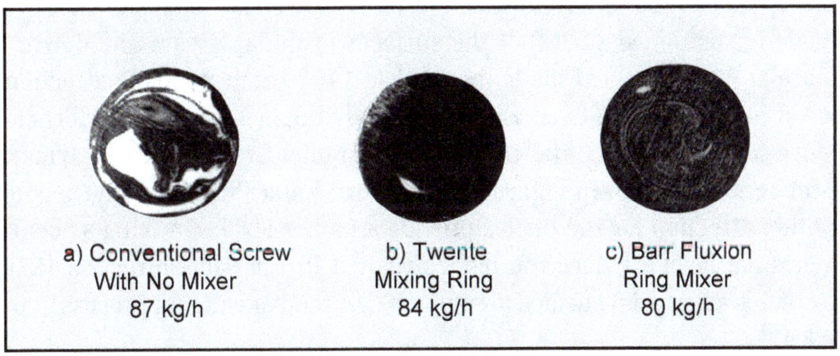

Figure 8.35 Cross-sectional views [53] of the extrudate samples using a conventional screw with segmented tip for the testing of different types of dynamic mixers: a) conventional screw with no mixer, b) Twente Mixing Ring, and c) Barr Fluxion ring mixer. The extruder was operated at a screw speed of 80 rpm and a letdown ratio of 220:1 of a white pigmented ABS resin to a black color concentrate

8.5.8 Static Mixers

Static mixers are secondary mixers and are used commercially to increase the interfacial surface area of polymer materials and thus improve the mixing by decreasing the striation thickness of the components. They can improve the mixing of color masterbatches into natural resins. They are also used to reduce the thermal gradients in the melt stream. The purpose of distributive mixing using static mixers in most systems is to produce a physical state where the dissimilar components have small enough physical striations, or "thickness," such that diffusion can complete the chemical or thermal mixing process for making the system homogeneous or such that the striations are small enough such that they appear to be homogeneous to the observer. These devices are typically positioned in transfer pipes and injection-molding nozzles. Their installation is simple and they are relatively inexpensive. As the name implies, the mixers are stationary in the pipe and all flow is due to a pressure gradient. There are many types and styles of static mixers available commercially. The most widely used mixers include Kenics, SMX, KMX, and Ross mixers. These mixers will be discussed in this section. In some extrusion systems, static mixers are used to compensate for unresolved issues in the extruder, often as a result of non-optimum screw or extruder design.

The Kenics mixer was one of the early static mixers. It was constructed by suspending helical elements or blades in a tube [55]. This early mixer divided the flow stream and changed the direction of the stream lines. The blades were all positioned with the same rotation direction. Modern Kenics [56] mixers, however, divide the flow and then reverse the flow field for a number of times thus producing smaller and smaller striation thicknesses. For this device the flow field is reversed by blades that alternate rotation direction. These blades twist 140°. The next blade is positioned such that the surfaces of the upstream and downstream blades are at 90°. This next blade then twists 140° in the opposite direction. The number of blades and the twist angle are highly important to the performance of the device for both mixing and the pressure gradient required for polymer flow. Kenics mixers are shown in Figure 8.36. Modern Kenics mixers with the alternating rotation direction for the blades provide a higher level of mixing as compared to the original design where the blades rotated in the same direction [57]. The device creates interfacial surface area much like the baker's fold process shown in Figure 8.1.

Simulations of the mixing performance for Kenics mixers were performed by Galaktionov et al. [57, 58] to show the effect of blade design. For these simulations, the mixing and pressure performance were determined as a function of the twist angle and the number of alternating blades. The numerical results are shown in Figure 8.37 and Figure 8.38. As shown in Figure 8.37, the best mixing occurs at a twist angle of about 140° to 150°. Commercially available Kenics mixers are designed with a 140° twist angle. As expected, the mixing performance of the mixer increases with more blades, as shown in Figure 8.38. As shown by this figure, as the number of blades increase from 0 to 20, the mixing increases substantially. At the end of a 20 blade device, the striations get so thin that the eye cannot discern any heterogeneity. Log (trA) and Log (I) are mathematical systems to in effect count the striations and mixing level. Details of the mathematical system can be found elsewhere [57, 58]. The addition of blades to a mixer will require a large upstream pressure to drive the flow. References [57] through [63] have excellent lists of additional references for those who wish to explore this subject more completely.

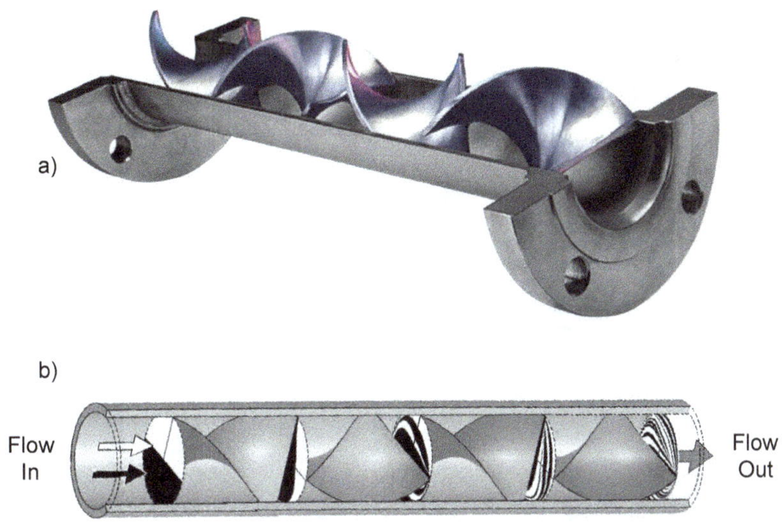

Figure 8.36 Modern Kenics mixers with alternating twist directions: a) cutaway view, and b) schematic with mixing striations (courtesy of Kevin G. Walsh of Chemineer, Incorporated)

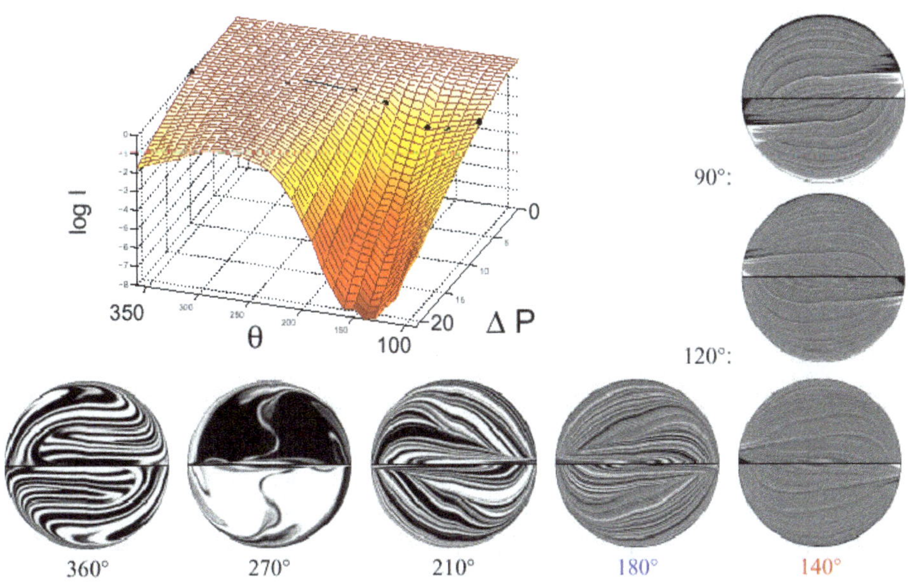

Figure 8.37 Kenics mixer and mixing quality as a function of the helix angle [58] (courtesy of Han Meijer of Technische Universiteit Eindhoven)

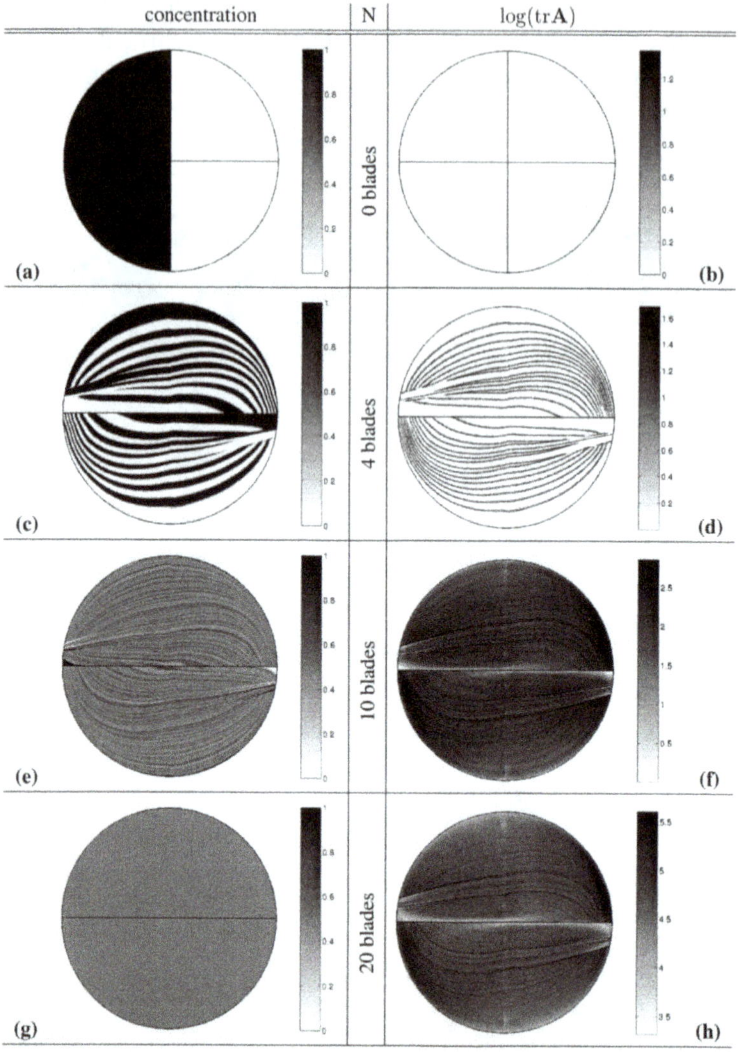

Figure 8.38 Mixing efficiency as a function of the number of blades for a twist angle of 140° [57] (courtesy of Han Meijer of Technische Universiteit Eindhoven)

KMX and SMX mixers are commercially available and widely used static mixers. Schematics of the mixers are shown in Figure 8.39, and they are similar in physical construction. The SMX [64] has straight rods in its elements and the KMX [65] has curved rods. The SMX mixer is designed with two elements, as shown in Figure 8.39(b), with the second element turned 90° in the tangential direction. The element is constructed with eight cross bars (N_x = 8) shaped as four X-shaped pairs, and with three bars that are parallel (N_p = 3) to the flow direction. The angle between the bars in the X-shape is 90°. The flow through the device is driven by a pressure gradient, and it divides and stretches the flow eight times per element.

When the material flows into the second element, the interfacial area between the components increases exponentially. It has been found that the KMX mixer would likely perform better for materials with relatively large viscosity differences [61].

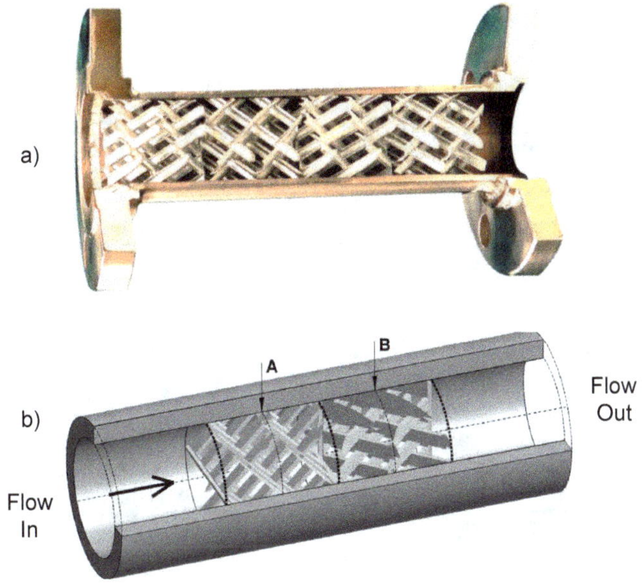

Figure 8.39 Static mixers based on rod flow dividers: a) KMX static mixer with four elements (courtesy of Kevin G. Walsh of Chemineer, Incorporated), and b) SMX static mixer with two elements [64]

The number of cross bars, parallel bars, and elements will affect the performance of an SMX static mixer. Singh at al. [62] performed simulations of the two-component flow through an SMX mixer. The number of cross bars and the number of elements were varied, and the results are shown in Figure 8.40. Contrary to intuition, which would tend toward the most bars to get optimum mixing, the optimum was shown to be six cross bars in the channel. Moreover, after only four elements the number of striations and thus the interfacial surface area was extremely high such that the black and white patterns are becoming difficult to see.

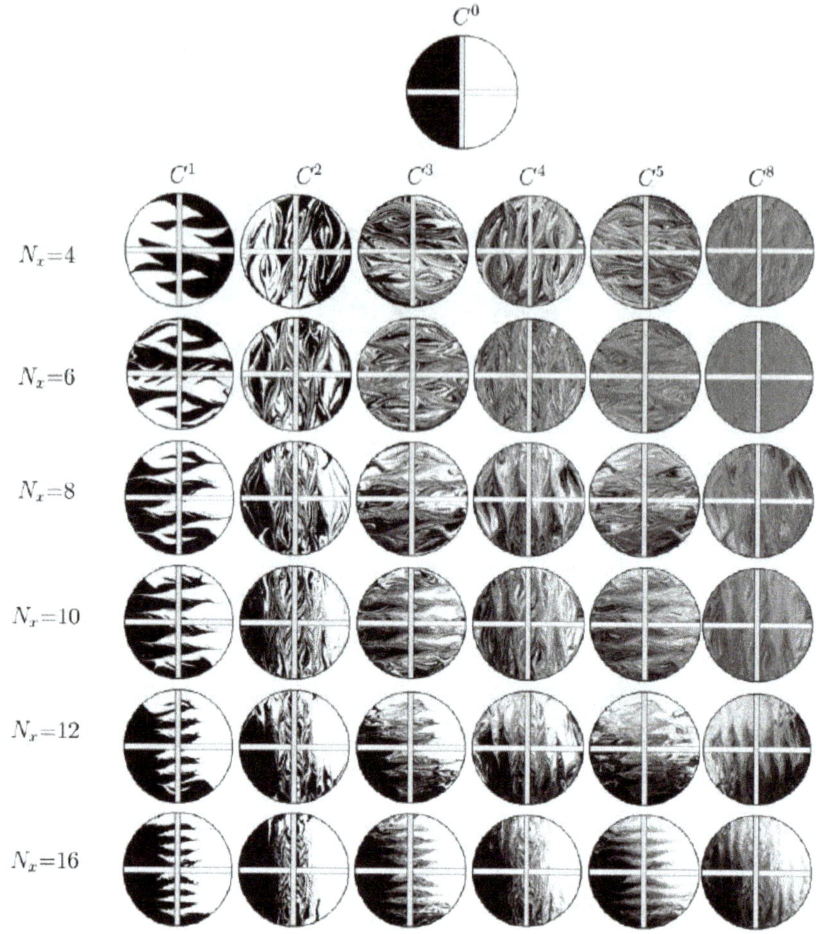

Figure 8.40 Mixing performance of an SMX static mixer as a function of the number of cross bars (N_x) and the number of elements (C^n) [62]. The angle between the cross bars was held constant at 90°, and the number of parallel bars (N_p) was three (courtesy of Han Meijer of Technische Universiteit Eindhoven)

The Ross LPD and LLPD static mixers are also commercially available and widely used. The Ross mixer consists of a series of semielliptical plates that are discriminately positioned in a tubular housing [66]. A typical single element consists of two plates perpendicular to each other and at a specified angle, as shown in Figure 8.41. Most Ross mixers are constructed with four support rods to provide maximum rigidity. The mixer in Figure 8.41 shows five elements. Each element is positioned at 90° tangentially from the upstream element.

Figure 8.41 Ross static mixer mixing components [66]

The mixing performance of the Ross mixer as a function of the number of elements and the angle between the plate pairs was calculated by Singh at al. [60]. The mixing effectiveness for the Ross mixer with alternating right and left elements is shown for elements with different angles from 90° to 160° and up to twelve elements by Figure 8.42. It is easily seen that the mixing improves up to an angle of about 140° to 160°.

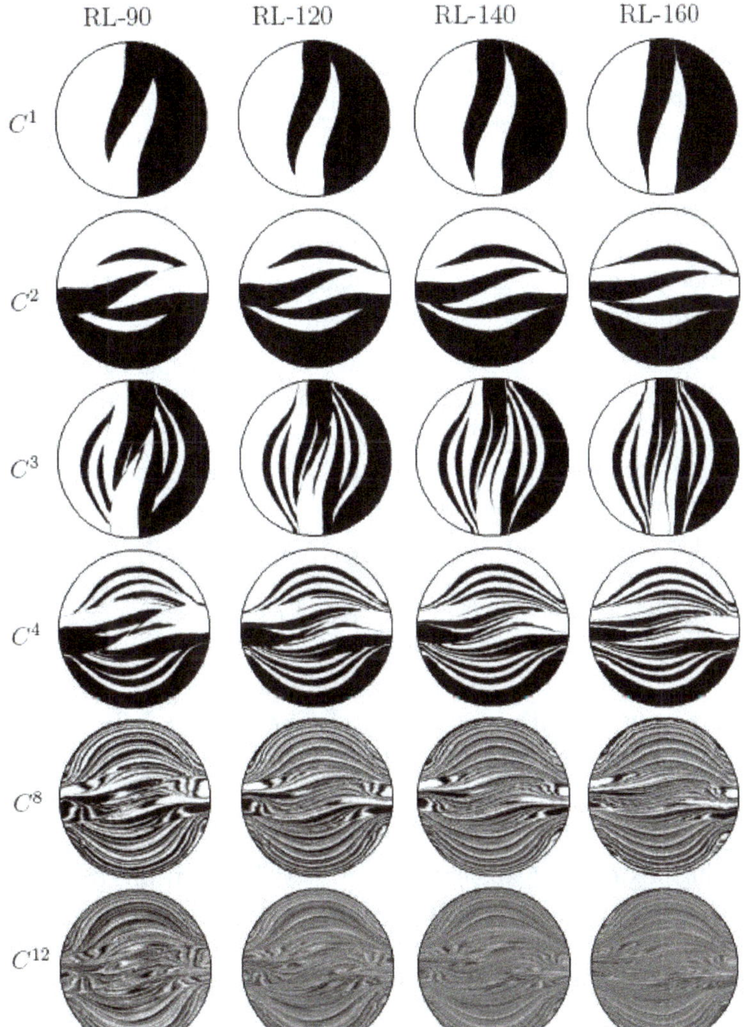

Figure 8.42 Mixing quality in a Ross LPD mixer with right-left elements [60]. The term RL-90 means that the plates alternate directions right and then left, and the C^n is the number of n elements in the mixer (courtesy of Han Meijer of Technische Universiteit Eindhoven)

8.6 Mixing Using Natural Resins and Masterbatches

Color masterbatches or concentrates are commonly used to color natural resins during the final fabrication step. This technique is sometimes referred to as natural-plus, referring to the extrusion or molding of natural resin plus a color concentrate [67, 68]. The natural-plus method is cost effective since only a small amount of color concentrate is produced using a compounding operation rather than the entire resin feedstock. Moreover, it eliminates the need to store large quantities of factory-colored resin. Instead only the natural resin and relatively small quantities of the color concentrates are stored on-site. Several disadvantages to the system, however, create challenges for its implementation, including the proper selection of pigments, the method of compounding, and the selection of the base resin for the color concentrate. Troubleshooting color problems associated with the injection-molding process was provided by Angel [68].

Color streaks can occur in the product due to poorly dispersed pigment in the color masterbatch [69]. For this case, the pigment particles in the color concentrate are either agglomerated from a poorly designed twin-screw compounding process or are not dispersed due to a low-viscosity carrier resin [69–71]. Most concentrate manufacturers use a low-pressure compounding process with a dispersion aid added to the pigment. This dispersion aid allows the pigments to shear and fracture during the melting process for the masterbatch product. If the dispersion aid is not added properly or is incorrect for the compounding process, the local high pressures that can occur during the resin melting process can cause the pigment to form agglomerates. These agglomerates can be very difficult to break up later in the compounding process, especially for carrier resins with very low viscosities. The agglomerates in the masterbatch material show up later during the molding or extrusion process as color streaks. A case study that had color streaks due to pigment agglomerates is provided in Section 11.12.2.

The melting-mixing mechanism described above is clearly illustrated with color masterbatch systems. Furthermore, the composition and melting characteristics of the masterbatch relative to the natural resin can affect the mixing that occurs during the melting process. Benkreira and Britton [69] have experimentally shown the effect of melt viscosity of the masterbatch on color mixing for these systems. In general, the masterbatch should be less viscous than the natural resin that will be colored. This viscosity difference will preferentially decrease the stresses of the colorant resin such that the elongation of the colorant is effective during the melting process, as shown in Figure 8.16. Their mixing experiments indicated that the viscosity ratio at the processing conditions of the natural resin to that of the masterbatch resin should be as high as possible. Masterbatches with very low viscosities, however, can be difficult to produce since the stresses during the compounding operation may not be high enough to disperse the pigments. In general and as a compromise, the viscosity at processing conditions of the masterbatch should be about one half that of the natural resin. If the viscosity of the masterbatch is higher than that of the natural resin, then color streaking is very likely to occur. The best option for this kind of color streaking is to switch to a masterbatch with a lower viscosity, that is, a viscosity less than that of the natural resin. In most cases, the same resin type as the natural resin is used for the masterbatch resin.

In many cases, the masterbatch melt flow rate (MFR) or melt index (MI) are compared to that of the natural resin. Since these indices are measured at shear rates considerably less than those that occur during processing, they are often not the best indicators for masterbatch selection. The viscosity at the temperature and shear rates for processing should be used and not the MFR or MI. For example, a PS resin with an MFR of 10 dg/min will have experienced a shear rate of 20 s^{-1} during the flow rate measurement using Equation 3.51. During the melting process in a single-screw extruder, however, the shear rate is in the range of 1000 to 3000 s^{-1}.

8.7 Mixing and Melting Performance as a Function of Flight Clearance

The clearance between the flight tip and the barrel wall is an extremely important parameter for mixing and the performance of an extruder. For a typical machine this clearance is about $D_b/1000$. Thus for a 140 mm diameter machine, the flight clearance is about 0.14 mm. This allows for enough mechanical clearance to permit rotation without seizing, yet provides a well-wiped barrel surface for maintaining acceptable heat transfer. This clearance will increase as the screw wears during service. Increased flight clearance in the melting section will cause the melt film between the solid bed and barrel wall to increase. This increased melt film thickness will reduce the melting capacity of the screw by reducing the level of viscous energy dissipation, requiring a longer distance in the barrel for melting [31, 72, 73]. If the melting capacity decreases too much, solids will discharge with the extrudate. Thus an increased flight clearance in the melting section will reduce the mixing and melting performance of the screw. Increased clearance in the metering section with only molten polymer present, however, will cause the mixing performance to increase. Increased flight clearance will reduce the specific rate of the machine, causing the screw to rotate faster to produce at the same rate. The decreased specific rate will improve the mixing performance, but at the expense of higher discharge temperatures.

8.8 High Pressures During Melting and Agglomerates

High pressures in compacted solids during melting are known to agglomerate some additives such as pigments [11]. For these pigments and compounding on single-screw extruders, agglomerates can be created and discharged with the extrudate when high pressures are generated in the melting section of the screw, that is, at high flood-fed specific rates. In general, once these agglomerates are formed there is not enough stress generated in the downstream sections of the screw to break them into small and acceptable-sized particles. At low levels of fill during starve-fed operation, however, the pressure in the melting section can be reduced dramatically, reducing the agglomeration of the additives to essentially zero. For the starve-feeding method, a feeding device above the hopper is controlling the rate of the process, causing the early sections of the screw channels to operate partially filled and with essentially zero pressure. Pressure in the melting section, however, is required in single-screw extruders [74] to maintain high melting rates, as discussed in Chapter 6. If the pressure is low enough to minimize agglomerate formation, then the melting rate and thus overall rate may be too low to be commercially attractive.

8.9 Effect of Discharge Pressure on Mixing

Operation of the extruder at high discharge pressures is a method of improving the mixing quality of the extrudate. For a single-stage screw, increasing the discharge pressure will increase the pressure in the melting zone and improve the melting rate [74, 75] while reducing the tendency for solid bed breakup. Moreover, increasing the discharge pressure will improve mixing in the fully molten sections but to a lesser extent. Increasing the discharge pressure, however, can be difficult for many processes. Common ways of increasing the discharge pressure include using a control valve or positioning static mixers in the transfer line. Increasing the discharge pressure, however, can cause the discharge temperature to increase beyond the operating limits of the process or resin.

8.10 Shear Refinement

Shear refinement (more appropriately called stress refinement) is a process where a molten stream of resin is forced through a dispersive mixer and the resulting resin has improved optical properties. It is a well-known process for improving the haze and gloss properties for LDPE blown film grade resins [76–81]. Haze is a measure of light transmittance while gloss is a measure of light reflectance from the surface. Haze in LDPE thin film comes mainly from surface haze caused by the scattering of light due to surface roughness. Surface roughness results from melt flow disturbances at the die exit due to melt elasticity and from crystallization close to the film surface. Shear refining temporarily reduces melt elasticity, resulting in less surface roughness and thus lower haze and higher gloss. The applied stress reorients some of the polymer chains, causing temporary changes in some of the physical properties and melt rheology of the resin. This process is a physical change and is reversible; the relaxation time is relatively long.

Shear refinement was studied for several LDPE resins using small-diameter single-screw extruders [77, 78, 80]. For these experiments, the resins were extruded five times using extruders equipped with underwater pelletizers. At the end of the experiments, both the haze and the gloss of the blown films produced from the resins were improved, that is, the gloss increased and the haze decreased. The researchers showed evidence that the primary molecular structures were unchanged. The amount and level of shearing, however, was not reported. Moreover, it is not clear whether high shear levels or high stress levels were the cause of the improvement in the optical properties.

Chung, Powell, and Werling [81] developed a shear-refining device that was capable of separating the effects of temperature, geometry, and the melting process on shear refinement of LDPE resins. For this apparatus shown in Figure 8.43, a 2 MI (190 °C, 2.16 kg) LDPE resin was melted in a single-screw extruder and then fed to the shear-refining module. The shear-refining module consisted of a short-segment screw with a Maddock-style mixer with in-flow and out-flow flutes. Several rotors were constructed that varied in geometry, allowing the investigation of the effect of mixer clearances, mixer land length, and the number of passes over mixing flights on resin properties. In all cases, the resin discharged from the module was compared to the resin entering the module for both haze and gloss. The feed temperature was studied for the range of 200 to 250 °C, and it was determined to have the largest effect on the optical properties of the material discharged. As the feed temperature was increased the optical properties of the discharged resin decreased. Moreover, at a temperature of 250 °C (and higher) shear refinement of the LDPE resin was not significant. The researchers attributed this to the relatively low viscosity of the resin at 250 °C and thus the low shear stress levels applied in the process. The optical properties also increased with increasing rotor speed and thus the shear rate of the material in the device. The data suggested that the number of mixing flights is not important and that only one properly designed mixing flight is needed.

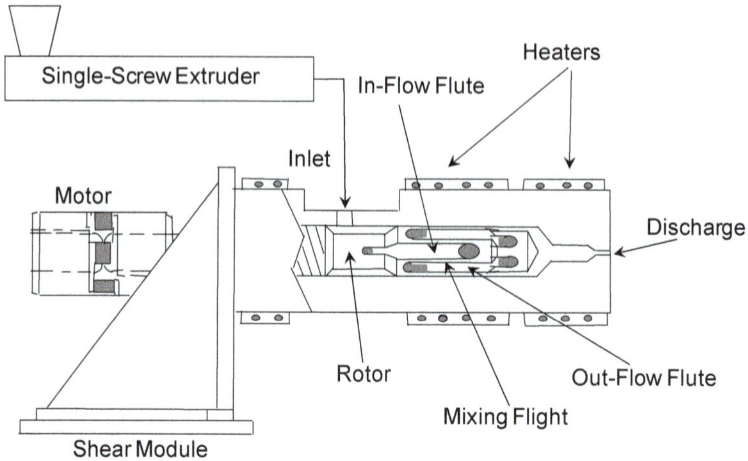

Figure 8.43 Schematic for the shear refinement apparatus built by Chung et al. [81]

The gloss and haze levels measured by Chung et al. [81] correlated very well with the applied stress in the shear refiner experiments. For these experiments the haze decreased and gloss increased nearly linearly with the applied shear stress, as shown in Figure 8.44. Thus, application of a shear stress of about 200 kPa to a 2 dg/min MI LDPE resin can improve the haze and gloss levels of the product film. The shear stress for the shear refiner was calculated using Equation 8.10 and Equation 8.11.

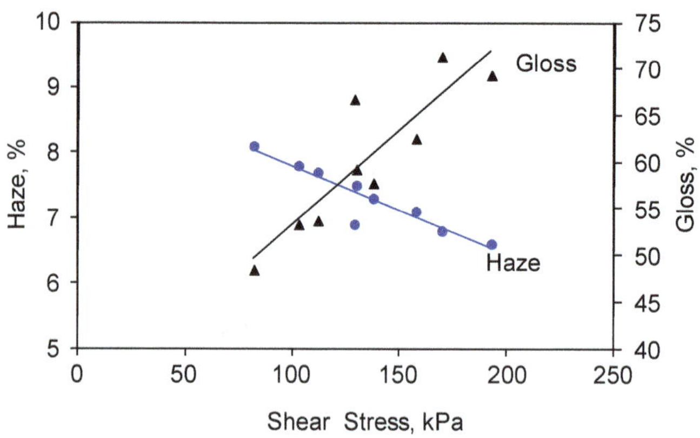

Figure 8.44 Film haze and gloss as a function of the shear stress for shear-refined 2 dg/min MI (190 °C, 2.16 kg) LDPE resin [81]

Mixers with relatively small mixing flight clearances can provide shear stress levels near 200 kPa. For example, a 150 mm diameter extruder operating at a screw speed of 80 rpm with a mixing flight clearance of 0.6 mm ($u + \lambda$) will typically have a shear stress of 210 kPa over the mixing flight. This calculation is for a temperature of 230 °C at the entry to the mixing flight and a viscosity of 200 Pa·s and a shear rate of 1050 1/s. Smaller mixing flight clearances will provide higher shear stresses but may cause unacceptable pressure gradients and temperature increases across the flight.

8.11 Direct Compounding Using Single-Screw Extruders

Direct compounding is a process where a mixture containing multiple feedstocks is preblended and then fed directly to the hopper of a single-screw extruder. Here the compounding of the components is performed in the single-screw extruder rather than in a separate compounding and pelletization process. For many feedstock systems, a single-screw extruder can be effective for compounding, eliminating a costly precompounding process. The process is very common for polyolefin resins [5, 82, 83] and is used occasionally in specialty applications. The success of direct compounding of dissimilar resins depends on the differences between the melt rheologies and the ability of the screw to melt, disperse, and distribute the components. If the rheologies are too different or the screw is incapable of dispersing the components, gels can occur in the final product. Moreover, the physical properties of the final product will likely be unacceptable. Proper clearances on the screw features designed to disperse and melt the resin, such as barrier flight clearances, mixing flight clearances, and clearances in mixing regions of high-performance screws, are keys to producing extrudates with acceptable qualities.

Direct compounding was proposed for a blend of HIPS resin and an elastomeric resin for a specialty product. For this example the composition of the final product was about 50% HIPS resin and 50% of the elastomer. On pilot line equipment, the resin pellet blend was extruded and the product had a high and unacceptable level of defects due to gels. In this case the gels were undispersed fragments of the elastomeric component and had diameters of 1 mm or less. In order to determine the feasibility of direct compounding using an enhanced screw design, the components were studied using the shear refiner shown in Figure 8.43. The shear-refined product was then examined for visual gels and normalized per unit mass of product. The gel level as a function of the shear stress imposed on the compound is shown in Figure 8.45. As indicated by this figure and as expected, the level of gels decreased as the shear stress was increased. If the final product had an acceptable gel level of 500 gels/kg, the shear-refining experiments indicated that the level of stress needed by the process is about 130 kPa. Thus a dispersive mixing feature in the screw would need to be designed at the proper location and clearance to provide at least 130 kPa of stress to the resin blend while not severely restricting the flow path and causing overheating of the resin.

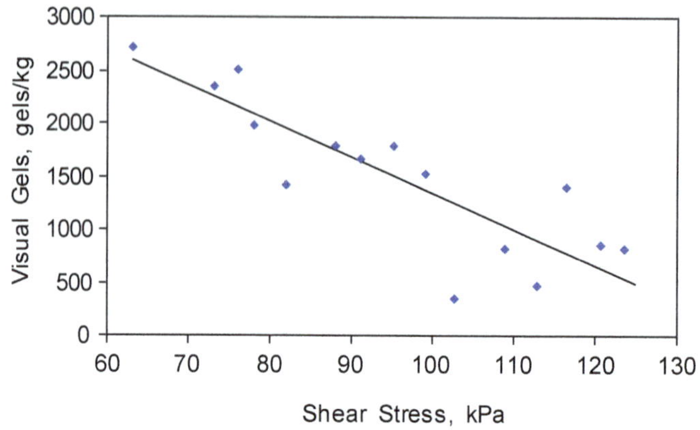

Figure 8.45 Visual gels in a film produced from a mixture containing 50% HIPS resin and 50% of an impact-modifying elastomer. The components were dispersed using the shear refiner shown in Figure 8.43 as a function of shear stress

Nomenclature

A cross-sectional area for a mixing region

C_n concentration of a minor component in element N_n

C^n designation for n number of elements in an SMX static mixer

$\overline{C}$ average concentration in area A

D hydrodynamic diameter for fluid flow for calculating the Reynolds number

D_b inside barrel diameter

D_B diameter of the land of a blister mixer

D_m diameter of the ball end mill that will be specified to cut the Maddock mixer flute into the metal stock

$E(t)$ overall stretching function with time

G_m free energy of mixing

H depth of the metering channel

H_m heat of mixing

i	summation index
l	the axial direction, m
l_m	axial length of the metering section of the screw, m
L	lead length
L_i	length of segment i
L_m	axial length of the mixing flight for a Maddock-style mixer, m
n_f	number of in-flow flutes (or out-flow flutes) for a Maddock-style mixer
N	screw rotation rate in revolutions/s
N_f	number of Maddock mixer flute pairs
N_g	number of divisions into which the area A is divided
N_n	nth rectangular element in area A
N_p	number of parallel bars in an SMX mixer
N_x	number of cross bars in an SMX static mixer
N_0	number of tracer particles introduced in area A
p	number of flight starts, dimensionless
Re	Reynolds number
S_i	stretching parameter for the ith particle
S_m	entropy of mixing
T	temperature
t	time
T_g	glass transition temperature
u	undercut for the mixing flight of a Maddock mixer or the mixing flight of a spiral dam section
V_z	velocity of the fluid in the z direction
V_{bx}	velocity of the screw in the x direction (cross channel and perpendicular to the flight edge)
V_{bz}	velocity of the screw in the z direction relative (downstream direction and parallel to the flight edge)
W	width of the metering channel
W_b	width of the Maddock mixer flute in the circumferential direction
W_f	width of the mixing and wiping flights of a Maddock mixer in the circumferential direction
W_p	width of the mixing and wiping flights of a Maddock mixer perpendicular to the flute edge

x cross-channel direction for an unwound channel system

y depth direction for an unwound channel system

z downstream direction for an unwound channel system

$\dot{\gamma}_B$ shear rate for flow across a blister mixer

$\dot{\gamma}_M$ shear rate for flow across the mixing undercut clearance for a Maddock mixer

$\dot{\gamma}_S$ shear rate for flow across the mixing undercut clearance for a spiral mixer

η shear viscosity at an average shear rate and temperature

θ_b helix angle of the flights at the barrel wall

λ clearance for the primary flight in the metering channel

μ Newtonian fluid viscosity

σ standard deviation

τ_B shear stress for the flow across a blister mixer

τ_M shear stress for the flow over the mixing undercut clearance for a Maddock mixer

τ_S shear stress for the flow over the mixing undercut clearance for a spiral mixer

References

1. Manas-Zloczower, I. and Tadmor, Z., "Mixing and Compounding of Polymers, Theory and Practice," Hanser Publishers, Munich (1994)
2. Gale, M., "Mixing in Single-Screw Extrusion," Smithers Rapra, Shawbury, United Kingdom (2009)
3. Spalding, M.A., Powers, J.R., Wagner, P.A., and Hyun, K.S., "Flow Surging in Single-Screw, Plasticating Extruders," *SPE ANTEC Tech. Papers*, **46**, 254 (2000)
4. Beagan, C.M., McNally, G.M., and Murphy, W.R., "Metallocene Catalysed Polyethylene in Blown Film Applications - A Comparison Between Monoextruded Blended Films and Coextruded Films," *SPE ANTEC Tech. Papers*, **44**, 128 (1998)
5. Schwank, D., "Film Applications," Chapter 3 in "Film Extrusion Manual," Butler, T.I. (Ed.), TAPPI Press, Atlanta, GA (2005)
6. Woods, S.S. and Pocius, A.V., "The Influence of Polymer Processing Additives (PPAS) on the Surface and Optical Properties of Polyolefin Plastomer Blown Film," *SPE ANTEC Tech. Papers*, **46**, 191 (2000)
7. Koppi, K.A., Ceraso, J.M., Cleven, J.A., and Salamon, B.A., "Gloss Modeling of Injection Molded Rubber-Modified Styrenic Polymers," *SPE ANTEC Tech. Papers*, **48**, 390 (2002)
8. Moore, E.R., "Styrene Polymers" from "Encyclopedia of Polymer Science and Engineering," **16**, 2nd ed., John Wiley & Sons (1989)

9. Gale, M., "Compounding with Single-Screw Extruders," *Adv. Polym. Technol.*, **16**, 251 (1997)

10. Black, T., "Single-Screw Compounding," in "Plastics Compounding Equipment and Processing," Todd, D. B. (Ed.), Hanser Publishers, Munich (1998)

11. Elemans, P. H. M. and van Wunnik, J. M., "The Effect of Feeding Mode on Dispersive Mixing Efficiency in Single-Screw Extrusion," *Polym. Eng. Sci.*, **41**, 1099 (2001)

12. Dennis, H. R., Hunter, D. L., Chang, D., Kim, S., White, J. L., Cho, J. W., and Paul, D. R., "Effect of Melt Processing Conditions on the Extent of Exfoliation in Organoclay-Based Nanocomposites," *Polymer*, **42**, 9513 (2001)

13. Benkreira, H., Shales, R. W., and Edwards, M. F., "Mixing on Melting in Single-Screw Extrusion," *Int. Polym. Process.*, **7**, 126 (1992)

14. Lindt, J. T. and Ghosh, A. K., "Fluid Mechanics of the Formation of Polymer Blends. Part I: Formation of Lamellar Structures," *Polym. Eng. Sci.*, **32**, 1802 (1992)

15. Meijer, H. E. H. and Janssen, J. M. H., "Mixing of Immiscible Liquids," Chapter 4 in "Mixing and Compounding of Polymers, Theory and Practice," Manas-Zloczower, I. and Tadmor, Z. (Ed.), Hanser Publishers, Munich (1994)

16. Janssen, J. M. H. and Meijer, H. E. H., "Dynamics of Liquid-Liquid Mixing: A 2-Zone Model," *Polym. Eng. Sci.*, **23**, 357 (1983)

17. Ottino, J. M. and Chella, R., "Laminar Mixing of Polymeric Liquids; A Brief Review and Recent Theoretical Developments," *Polym. Eng. Sci.*, **35**, 1766 (1995)

18. Alemanskin, K., Camesasca, M., Manus-Zloczower, I., Kaufman, M., Kim, E. K., Spalding, M. A., Trumbull, W. A., and Swain, R. D., "Entropic Mixing Characterization in a Single-Screw Extruder," *SPE ANTEC Tech. Papers*, **50**, 167 (2004)

19. Alemanskin, K., Manus-Zloczower, I., and Kaufman, M., "Entropic Analysis of Color Homogeneity," *Polym. Eng. Sci.*, **45**, 1031 (2005)

20. Wang, W., Manas-Zloczower, I., and Kaufman, M., "Characterization of Distributive Mixing in Polymer Processing Equipment using Renyi Entropies," *Int. Polym. Process.*, **16**, 315 (2001)

21. Wang, W., Manas-Zloczower, I., and Kaufman, M., "Entropic Characterization of Distributive Mixing in Polymer Processing Equipment," *AIChE J.*, **49**, 1637 (2003)

22. Erwin, L. and Mokhtarian, F., "Analysis of Mixing in Modified Single-Screw Extruders," *Polym. Eng. Sci.*, **23**, 49 (1983)

23. Bigio, D. I. and Connor, J. H., "Principal Directions as a Basis for the Evaluation of Mixing," *Polym. Eng. Sci.*, **35**, 1527 (1995)

24. Ling, F. H. and Zhang, X. "Mixing of a Generalized Newtonian Fluid in a Cavity," *J. Fluids Eng.*, **117**, 75 (1995)

25. Campbell, G. A., Bomma, S., St. John, S., and Chempath, S., "Chaotic Mixing in a Free Helix Single Screw Device," *SPE ANTEC Tech. Papers*, **48**, 1441 (2002)

26. Moffat, H. K., "Viscous and Resistive Eddies Near a Sharp Corner," *J. Fluid Mech.*, **18**, 1 (1964)

27. Zhu, F. and Chen, L., "Studies on the Theory of Single-Screw Plasticating Extrusion. Part I: A New Experimental Method for Extrusion," *Polym. Eng. Sci.*, **31**, 1113 (1991)

28. Han, C.D., Lee, K.Y., and Wheeler, N.C., "A Study on the Performance of Barrier-Screw Extruders," *Polym. Eng. Sci.*, **31**, 831 (1991)

29. Somers, S.A., Spalding, M.A., Dooley, J., and Hyun, K.S., "An Experimental Study of the Flows in an Energy Transfer Screw," *SPE ANTEC Tech. Papers*, **48**, 307 (2002)

30. Bigio, D.I., Boyd, J.D., Erwin, L., and Gailus, D.W., "Mixing Studies in the Single-Screw Extruder," *Polym. Eng. Sci.*, **25**, 305 (1985)

31. Chung, C.I., "Extrusion of Polymers: Theory and Practice," Hanser Publishers, Munich (2000)

32. Schopper, V., "Screw Designs," Chapter 3.3 in "Screw Extrusion," White, J.L. and Potente, H. (Ed.), Hanser Publishers, Munich (2003)

33. LeRoy, G., "Apparatus for Extrusion of Thermoplastics," US Patent 3,486,192 (1969)

34. Maddock, B.H., "Mixing of Thermoplastic Materials," US Patent 3,730,492 (1973)

35. Maddock, B.H., "Thermoplastic Material Mixing," US Patent 3,756,574 (1973)

36. Dray, R.F., "Mixing Element for Extruder Screw," US Patent 3,788,612 (1974)

37. Gregory, R.B., "Mixing Section for Extruder Feed Screw," US Patent 3,788,614 (1974)

38. Maddock, B.H., "An Improved Mixing-Screw Design," *SPE J.*, **23**, 23 (1967)

39. Tadmor, Z. and Klein, I., "Design of Certain Fluted Mixing Sections of Extruder Screws," *Polym. Eng. Sci.*, **13**, 382 (1973)

40. Han, C.D., Lee, K.Y., and Wheeler, N.C., "A Study on the Performance of the Maddock Mixing Head in Plasticating Single-Screw Extrusion," *Polym. Eng. Sci.*, **31**, 818 (1991)

41. Zitzenbacher, G., Karlbauer, R., and Thiel, H., "A New Calculation Model and Optimization Method for Maddock Mixers in Single-Screw Plasticating Technology," *Int. Polym. Process.*, **22**, 73 (2007)

42. Maillefer, C., Swiss Patent 363,149 (1962)

43. Yao, W.G., Tanifuji, S., Takahashi, K., and Koyama, K., "Mixing Efficiency in a Pin Mixing Section for Single-Screw Extruders," *Polym. Eng. Sci.*, **41**, 908 (2001)

44. Yao, W.G., Takahashi, K., Koyama, K., and Dai, C.G., "Experimental Study on Distributive Mixing Characteristics of a New Type of Pin Mixing Section," *Polym. Eng. Sci.*, **37**, 615 (1997)

45. Yao, W.G., Takahashi, K., Koyama, K., and Yamashita, Y., "The Effect of a New Type of Pin Mixing Section on the Performance of a Single-Screw Extruder," *Polym. Eng. Sci.*, **38**, 1623 (1998)

46. Gale, G.M. and Hindmarch, R.S., "RAPRA's Glittering Prize," *Eur. Rubber J.*, July-August, 29 (1982)

47. Gale, G.M., "Extruder Mixer," US Patent 4,419,014 (1983)

48. Wang, C. and Manus-Zloczower, I., "Flow Field Analysis of a Cavity Transfer Mixer," *Polym. Eng. Sci.*, **34**, 1224 (1994)

49. Semmekrot, G.J.M., "Distributive Mixer Device," US Patent 5,013,233 (1991)

50. Esseghir, M., Gogos, C.G., Yu, D., Todd, D.B., and David, B., "A Comparative Study on the Performance of Three Single-Screw Elements in Melt-Melt Mixing of Immiscible Blends," *Adv. Polym. Technol.*, **17**, 1 (1998)

51. Ingen-Housz, A.J. and Norden, S.A., "Mixing Ring," *Int. Polym. Process.*, **10**, 120 (1995)

52. Barr, R.A., "Floating Sleeve Mixer and Method," US Patent 5,988,866 (1999)

53. Myers, J.A., Barr, R.A., Spalding, M.A., and Hughes, K.R., "Experimental Comparison of Floating Ring Mixing Devices," *SPE ANTEC Tech. Papers*, **45**, 157 (1999)

54. Salamon, B., Spalding, M.A., Powers, J.R., Serrano, M., Sumner, W.C., Somers, S.A., and Peters, R.B., "Color Mixing Performance in Injection Molding: Comparing a Conventional Screw and Non-Return Valve to Those Designed for Improved Mixing," *SPE ANTEC Tech. Papers*, **46**, 479 (2000)

55. Armeniades, C.D., Johnson, W.C., and Raphael, T., "Mixing Device," US Patent 3,286,992 (1966)

56. Grout, K.M. and Devellian, R.D., "Dispersion Producing Method," US Patent 3,704,006 (1972)

57. Galaktionov, O.S., Anderson, P.D., Peters, G.W.M., and Meijer, H.E.H., "Morphology Development in Kenics Static Mixers (Application of the Extended Mapping Method)," *Can. J. Chem. Eng.*, **80**, 604 (2002)

58. Galaktionov, O.S., Anderson, P.D., Peters, G.W.M., and Meijer, H.E.H., "Analysis and Optimization of Kenics Mixers," *Int. Polym. Process.*, **18**, 138, (2003)

59. Anderson, P.D. and Meijer, H.E.H., "Chaotic Mixing Analyses by Distribution Matrices," *Appl. Rheol.*, **10**, 119 (2000)

60. Singh, M.K., Kang, T.G., Anderson, P.D., Meijer, H.E.H., and Hrymak, A.N., "Analysis and Optimization of Low-Pressure Drop Static Mixers," *AIChE J.*, **55**, 2208 (2009)

61. Heniche, M., Tanguy, P.A., Reeder, M.F., and Fasano, J.B., "Numerical Investigation of Blade Shape in Static Mixing," *AIChE J.*, **51**, 44 (2005)

62. Singh, M.K., Anderson, P.D., Meijer, H.E.H., "Understanding and Optimizing the SMX Mixer," *Macromol. Rapid Commun.*, **30**, 362 (2009)

63. Van Der Hoeven, J.C., Wimberger-Friedl, R., and Meijer, H.E.H., "Homogeneity of Multilayers Produced with a Static Mixer," *Polym. Eng. Sci.*, **41**, 32 (2001)

64. Sulzer Chemtech website; *http://www.sulzerchemtech.com*

65. Chemineer Incorporated website; *http://www.chemineer.com*

66. Charles Ross & Son Company website; *http://www.staticmixers.com*

67. Furches, B. and Bozzelli, J., "Screw Design Efficiency in Color Distributive Mixing with ABS Resins," *SPE ANTEC Tech. Papers*, **33**, 6 (1987)

68. Angel, G., "Injection Molding and Color: A Basic Look at Design, Processing, and Troubleshooting," *SPE ANTEC Tech. Papers*, **48**, 2508 (2002)

69. Benkreira, H. and Brittin, R.N., "The Optimization of Masterbatch Formulations for Use in Single-Screw Machines," *Int. Polym. Process.*, **9**, 205 (1994)

70. Mack, M., "Corotating Intermeshing Twin-Screw Extruder: Berstorff's System," Chapter 5 in "Plastics Compounding Equipment and Processing," Todd, D.B. (Ed.), Hanser Publishers, Munich (1998)

71. Canedo, E.L. and Valsamis, L.N., "Selecting Continuous Compounding Equipment Based on Process Considerations," *Int. Polym. Process.*, **9**, 225 (1994)

72. Tadmor, Z. and Klein, I., "Engineering Principles of Plasticating Extrusion," Van Nostrand Reinhold Co., New York (1970)

73. Klein, I., "The Effect of Flight Clearance on the Performance of Plasticating Extruders," *SPE ANTEC Tech. Papers*, **20**, 237 (1974)

74. Spalding, M.A., Hyun, K.S., and Cohen, B.R., "An Experimental Study of the Frictional and Viscous Forces Responsible for Solids Conveying and Melting in Single-Screw Plasticating Extruders," *SPE ANTEC Tech. Papers*, **43**, 202 (1997)

75. Klein, I., "The Melting Factor in Extruder Performance," *SPE J.*, **28**, 47 (1972)

76. Huck, N.D. and Clegg, P.L., "The Effect of Extrusion Variables on the Fundamental Properties of Tubular Polythene Film," *SPE Trans.*, 121 (July 1961)

77. Fujiki, T., "Concept of Secondary Heterogeneous Structure of Long-Chain Branched Polyethylene," *J. Appl. Polym. Sci.*, **15**, 47 (1971)

78. Rokudai, M., Mihara, S., and Fujiki, T., "Influence of Shearing History on the Rheological Properties and Processability of Branched Polymers. II. Optical Properties of Low-Density Polyethylene Blown Films," *J. Appl. Polym. Sci.*, **23**, 3289 (1979)

79. Baker, W.E., Rudin, A., Schreiber, H.E., and El-Kindi, M., "The Effect of Processing on Rheological and Molecular Characteristics of a Low Density Polyethylene," *Polym. Eng. Sci.*, **33**, 377 (1993)

80. Van Prooyen, M., Bremner, T., and Rudin, A., "Mechanism of Shear Modification of Low Density Polyethylene," *Polym. Eng. Sci.*, **34**, 570 (1994)

81. Chung, C.I., Powell, T.M., and Werling, C.L., "Shear Refining of Branched Low Density Polyethylene," *SPE ANTEC Tech. Papers*, **44**, 76 (1998)

82. Heaney, J., Billham, M., Clarke, A.H., McNally, G.M., and Murphy, W.R., "The Effect of Blending LLDPE Comonomers with Conventional LDPE on the Thermal and Mechanical Properties of Packaging Films," *SPE ANTEC Tech. Papers*, **51**, 2955 (2005)

83. Walia, P., "INSPIRE At-Press TPO Technology," *SPE ANTEC Tech. Papers*, **55**, 1443 (2009)

84. Sun, X., Gou, Q., Spalding, M.A., Womer, T.W., and Uzelac, N., "Optimization of Maddock-Style Mixers for Single-Screw Extrusion," *SPE ANTEC Tech. Papers*, **62**, 898 (2016)

85. Sun, X., Spalding, M.A., Womer, T.W., and Uzelac, N., "Design Optimization of Maddock-Style Mixers for Single-Screw Extrusion Using Numerical Simulations," *SPE ANTEC Tech. Papers*, **63**, 1017 (2017)

9 Scaling of Single-Screw Extrusion Processes

Commercial screws for single-screw extruders are designed using a wide range of methods that include: (1) prior designs using similar resins, (2) simple rotational flow and pressure gradient calculations for the metering section and practical experience for the upstream sections of the screw, (3) trial-and-error designs where the actual screw is tested in production or a lab, (4) scaling rules and ratios, and (5) extrapolation of existing processing data. In all cases, the goal of the designer is to develop a screw and process that operates at the target performance. The target performance is defined here as providing an extrudate at a specified rate, discharge pressure, and discharge temperature. Moreover, the extrudate must be homogenous, with no entrained solid polymer particles and no entrained air. In most cases, the designer achieves the target rate and discharge pressure, but the temperature of the extrudate can be more difficult to obtain. Quality laboratory data or production data for the same resin using a screw with known dimensions can be extremely valuable for designing a new screw. Without these data or prior experience with the resin, designing a screw can be very complex, and the risk of not achieving the target performance can be very high.

All screw manufacturers have a database of prior screw designs that are known to work for a specific resin and application. That is, the database contains screw geometries that are known to provide a specified rate at a known discharge pressure and temperature. When a new screw is ordered, the design would likely come from the database and be adjusted to accommodate the overall length of the extruder and the customer requirements for rate and discharge temperature and pressure. If the customer requirements are not extremely different from the database screw performance, then the risk of not meeting the customer's target performance is minimal. Obviously, screw designs for new resins and processes cannot be done using this method. For these cases, experimental extrusions are necessary to provide an acceptable design.

Scaling rules and ratios have also been developed to guide the screw designer. These rules have many different foundations, but they all attempt to maintain a design parameter constant with the scaling. For example, it might be desirable to maintain a constant discharge temperature and scale from a 100 mm diameter extruder to a 150 mm machine. In this case, the ratio of the metering section channels would be related via a rule to the diameter ratio of the machines such that specific energy input is constant. Scaling rules have been used to design successfully screws in limited ranges of the scaling features.

This chapter will present an introduction to scaling rules and an engineering design method for screw design. The engineering method combines experimental data from a process with numerical methods for metering channel calculations. This design method has been used successfully to perform scale-up designs, rate expansions, and process optimization. Two case studies are provided to show the method.

9.1 Scaling Rules

Scaling rules were developed for quick design and scale-up of extrusion processes. As such, the rules can provide screw channel dimensions and process conditions that are acceptable for rate and power consumption. The rules, however, may not provide the best estimate for maintaining a similar discharge temperature. Moreover, the rules will not allow the designer to adjust the discharge temperature of a design. This section will introduce these classical design techniques.

The first scale-up method was developed by Carley and McKelvey [1] in 1953. In this original work, the rate and power consumption at a fixed speed and for a geometrically similar extruder was simply a ratio of the extruder diameters as follows:

$$Q_2 = Q_1 \left(\frac{D_2}{D_1} \right)^3 \tag{9.1}$$

$$P_{e2} = P_{e1} \left(\frac{D_2}{D_1} \right)^3 \tag{9.2}$$

where P_e, Q, and D are the power consumed, the rate, and the barrel diameter, respectively, and subscript 1 is for the base-case machine and 2 is for the scaled machine. Thus, if the diameter of the barrel is increased by a factor of two, the rate and power are predicted to increase by a factor of eight. Here, geometrically similar means that all dimensions of the larger machine are a factor of two greater than that for the smaller diameter machine. Both extruders are operated at the same screw speed and discharge pressure while the discharge temperature is expected to be nearly the same. The scale-up rule is very simple and is only meant to provide a guide to the rate and power expected for a process. Since this writing by Carley and McKelvey, many other investigators have expanded the method [2–7].

These scaling rules were used successfully by designers who were highly experienced. In most cases, the designer will also make contingency plans if the design does not meet the target performance. These rules were developed during a period when computations were slow and expensive. Moreover, many of the rules were based on barrel rotation physics. It is interesting to note that while most of the scale-up rules worked relatively well for flow, they were often in error for predicting discharge temperature. With modern computational power and the development of screw rotational physics, flow and temperature simulations should be performed using screw rotational models presented in this book instead of scaling rules. Scaling rules and ratios can be found in the references [8, 9].

9.2 Engineering Design Method for Plasticating Screws

The engineering design approach presented here uses experimental data and experience to set the channel dimensions in the solids conveying and melting sections, while numerical methods are used to design the metering channel to obtain the desired performance. This combination allows the designer to meet the design targets of rate, discharge pressure, and discharge temperature. The approach has been used successfully to design a large number of screws in design modes that include scale-up, scale-down, rate increases for a machine, and resin changes. The case studies presented later in this chapter demonstrate this technique for a successful scale-up design and a rate increase for an existing extruder.

The design approach starts with the collection of baseline process data for the resin from a lab extruder or from a production line. At a minimum, the baseline data that are needed include the rate, screw speed, motor current, barrel temperatures, discharge pressure, and discharge temperature. The rate, barrel temperatures, and discharge pressure are typically easy to collect. The other values need to be studied to get the best values. For example, the screw speed measured and reported by the extruder control panel can be inaccurate to the point of creating errors in the baseline calculations. To eliminate this problem, the screw speed should always be measured by counting revolutions and using a stopwatch. If the screw is rotating too fast to count, then a strobe light-type tachometer or other suitable device should be used.

The discharge temperature should always be measured with a handheld temperature probe because it provides the most accurate measurement. Thermocouple sensors positioned through transfer lines and exposed to the melt stream do not provide an acceptable measurement of the discharge temperature due to the high thermal conductivity of the metal components (transfer line and thermocouple sheath) and the very low thermal conductivity of the polymer stream [10]. For some processes, the discharge temperature cannot be safely measured using a handheld probe, and another method must be used. The motor current during operation is typically shown on the control panel. This value should be verified by a qualified electrician.

At this point, the process data needed to make the new design have been collected: rate, screw speed, motor current, barrel temperatures, discharge pressure, and discharge temperature. The next step is to obtain a copy of the screw print and the motor and gear box details. If a print of the screw is not available, the manufacturer will often provide a schematic, or the screw can be removed and measured. The nameplate data on the motor and gear box are needed so that the energy required for the extrusion can be estimated. The data for the motor include power rating, base speed and the maximum current (at full power). The maximum screw speed is then calculated by dividing the base speed by the gear box reduction and the belt sheave reduction (if used). Additional information for motors and drive systems can be obtained in Section 10.6. The energy inputted by the screw to the resin is estimated as follows:

$$P_m = P_{\max}\left(\frac{A}{A_{\max}}\right)\left(\frac{N}{N_{\max}}\right) \quad \text{speeds} \leq \text{base motor speed} \tag{9.3}$$

$$P_m = P_{\max}\left(\frac{A}{A_{\max}}\right) \quad \text{overspeeding} \tag{9.4}$$

$$E = \frac{(3600 \cdot J \cdot kg)}{(kW \cdot g \cdot h)} \frac{P_m}{Q} \tag{9.5}$$

where P_m is the power that is dissipated in kW, P_{max} is the nameplate power (kW) for the motor, A is the motor current observed during the extrusion, A_{max} is the nameplate motor current at full load, N is the screw speed during extrusion, and N_{max} is the maximum screw speed that the extruder is capable of running at the base speed. Overspeeding techniques can be used to increase the speed of the motor beyond the base speed rating. Overspeeding is discussed in Section 10.6. After the power is computed, the specific energy inputted to the resin from the screw, E, in J/g is calculated using Equation 9.5 and the extrusion rate, Q, in kg/h. The calculation neglects the inefficiencies in converting electrical power to mechanical power and the power factor. Motor efficiencies and power factors are discussed in Section 10.6.1.

Physical properties that need to be obtained for the resins include the shear rheology, melt density, thermal conductivity, and heat capacity. The shear rheology of the resin can be measured using a capillary rheometer or obtained from the resin supplier. Melt densities of most unfilled resins are available [11] and data for select resins are provided in Section 4.7. If the resin has a high level of fillers or is a blend with other resins, then the melt density should be measured using a melt indexer (Section 4.7). Heat capacity and thermal conductivities are available for select resins in Sections 4.5 and 4.6, respectively.

Next, the baseline extrusion process is numerically simulated for just the metering section. For this step, the resin temperature and pressure at the inlet of the metering section are iteratively guessed, and the metering section simulation is performed at the baseline rate and screw speed. The discharge pressure and discharge temperature are the outputs from the calculation. If the calculated discharge pressure and temperature match the baseline discharge pressure and discharge temperature, then the simulation process is finished and the entry conditions to the metering section are established. If the baseline process data do not match well with the simulation, then the guessed inlet pressure and temperature to the metering section are changed, and the process is continued. If an acceptable match of the baseline and calculated discharge pressure and temperature cannot be obtained, then several problems could exist, including: (1) the screw used is different from the screw print, (2) the shear rheology of the resin is not correct, (3) the screw speed used for the calculations is not correct, (4) the rate is not correct, or (5) the existing process is not working properly and is not completely full of resin [12]. The design cannot progress to the next step until the simulation provides an acceptable estimate of the pressure and temperature at the entry to the metering section. This process assumes that the simulation code used is verified using a wide range of resins and channel aspect ratios. The simulation methods acceptable for use are discussed in Section 9.2.1.

The target performance needs to be specified for the new process. In most cases, the end user is requesting that the rate of the line be increased while maintaining the same discharge temperature for the same resin. For a case such as this, the discharge temperature would be held constant, the rate would increase, the discharge pressure would increase, and the specific energy would likely remain unchanged. Thus, a new screw with different channel depths that meets the target performance would need to be built for the existing extruder.

The discharge pressure at the new and higher rate must be estimated. In many cases, the higher rate will require a higher discharge pressure. The new discharge pressure will be the sum of the pressures needed to run the die, transfer line, and the screen pack system (if installed). The pressure change with rate for a screen pack system was detailed previously [13, 14] and is discussed in Section 11.1.1. The pressure increase at the higher rate for the die can be obtained using several methods, including experimentally, by numerical simulation of the die, or by a simple calculation for a circular restriction. The pressure change for a rate increase for flow through a circular restriction [15] and a power law fluid is as follows:

$$\Delta Q = (Q_1 - Q_2) / Q_1 = 1 - (1 - \Delta P)^{1/n} \tag{9.6}$$

$$\Delta P = (P_1 - P_2) / P_1 \tag{9.7}$$

where n is the power law index for the resin, Q_1 and P_1 are the rate and discharge pressure at the baseline (condition 1), and Q_2 and P_2 are the rate and pressure at the target (condition 2). If a gear pump is attached downstream of the extruder, then the pressure will likely not change when the rate is increased. For this case, the gear pump would provide the pressure needed to operate the die at the higher rate condition.

Finally, the simulation is performed using the geometry of the proposed metering section design and the target performance. For these simulations, the entry pressure and temperature for the metering section are those calculated from the baseline simulations. The design of the metering section is changed (channel depth, lead length, number of flights) along with the screw speed until the target performance is achieved. After a candidate design is identified, the designer must check if the motor and gear box can supply the required power to the new process. For this calculation, it is assumed that the specific energy of the new process is the same as the baseline process. Equation 9.3 through Equation 9.5 are used to verify that the motor can supply the same specific energy for the target process at the screw speed and rate determined using the simulations.

The feed and melting sections upstream of the metering section are set using experience and existing experimental data. For example, if the base case worked well with a compression ratio and compression rate of 3.0 and 0.0035, respectively, then the target screw would be designed with the same upstream specifications. The compression ratio and compression rate are as follows:

$$C = \frac{H_f}{H} \tag{9.8}$$

$$R = \frac{(H_f - H)\sin\theta_b}{ML} \tag{9.9}$$

$$\tan\theta_b = \frac{L}{\pi D_b} \tag{9.10}$$

where C is the compression ratio, H_f is the channel depth of the feed section, H is the depth of the metering channel, R is the compression rate in the transition section, M is the number of turns in the transition section, θ_b is the helix angle at the barrel wall, L is the lead length, and D_b is the diameter of the barrel. The compression rate for the transition section of the screw describes the rate that the channel depth decreases as the resin is transported through the section. Compression rate and ratios are discussed in Sections 6.1 and 10.8.2.

A sensitivity analysis of the input variables is advised to establish contingency plans if the target performance is not obtained. For example, if the discharge temperature for the target extruder is too high, then the barrel temperatures in the metering section could be decreased. A numerical analysis for the sensitivity of the barrel temperatures would guide the designer on how much the discharge temperature could be decreased using barrel temperature set points. The sensitivity of the assumed entry pressure should also be studied.

The design approach has been used with two-stage vented extruders and for designs that incorporate mixing sections such as blister rings and Maddock-style mixers [16]. As before, the experimental data are used to establish a baseline performance and provide information on the solids conveying and melting sections. The sections that are operating with only molten resin would be designed using the numerical simulation technique.

9.2.1 Process Analysis and Simulations

All simulations and process analyses that appear in the next chapters were performed using the engineering design method described in Section 7.5.1 for flow and Section 7.7.5.1 for thermal effects. Since solving the full 3-D equation set in Section 7.5.1 is difficult and time consuming, the generalized Newtonian method with the correction factor F_c described in Section 7.5.3 can be used here for flow. In this chapter and later chapters, the simulated sections of the process are shown using solid lines while the estimated sections are shown with dotted lines. The 3-D simulation method was previously proven to predict accurately the simulation of pressures and temperatures for extruders of different diameters, screw designs, and resin types.

9.3 Scale-Up from a 40 mm Diameter Extruder to an 80 mm Diameter Machine for a PE Resin

A 40 mm diameter laboratory extrusion line was used to make demonstration parts using a specialty PE resin. The desire was to build a larger line with considerably higher extrusion rates. The shear viscosity of the resin is shown in Figure 9.1. For the temperature range of the process, the heat capacity and thermal conductivity were assumed to be constant at 2500 J/(kg·°C) and 0.25 J/(m·s·°C), respectively. Since the resin was not filled, the melt density was comparable to that of an LDPE resin.

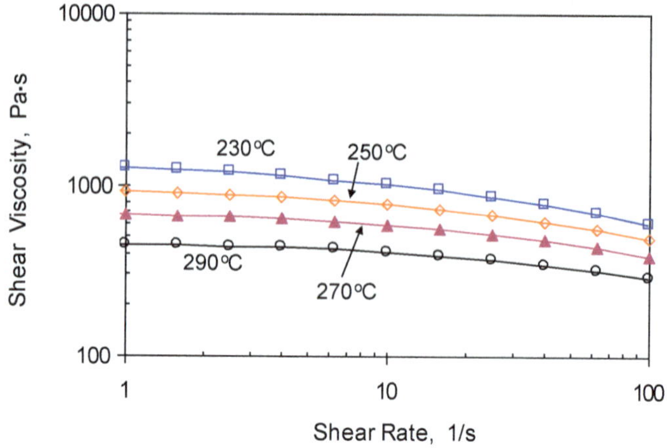

Figure 9.1 Shear viscosity for the PE resin used in the scale-up case study

The extruder that provided the baseline performance data was 40 mm in diameter and had an *L/D* ratio of 28. The screw was single flighted and constructed with three sections: (1) an 8 diameter long feed section with a constant depth of 8.89 mm, (2) a 7 diameter long transition section, and (3) a 13 diameter long metering section with a constant depth of 3.30 mm. The lead length and flight width perpendicular to the flight edge were 40 and 4 mm, respectively. The extruder was equipped with a 16 kW motor. The motor and gearbox combination was capable of a maximum screw speed of 105 rpm. The baseline performance for the lab screw is shown in Table 9.1. At these conditions, the motor was operating at 50% of its maximum current load (A/A_{max} = 0.5). The barrel temperatures were 160, 220, and 275 °C for the feed section through the discharge section, respectively. The screw had a compression ratio of 2.7 and compression rate of 0.0060.

Table 9.1 Baseline Performance for the 40 mm Diameter Laboratory Extruder

	Performance
Rate, kg/h	10.3
Screw speed, rpm	28
Specific rate, kg/(h·rpm)	0.37
Discharge pressure, MPa	6
Discharge temperature, °C	275
Specific energy, J/g	750

The baseline process was simulated using the data in Table 9.1 to determine the pressure and temperature at the entry to the metering section. After several iterations, the pressure and temperature were determined to be 22 MPa and 190 °C, respectively. The simulated axial pressure and temperature are shown in Figure 9.2.

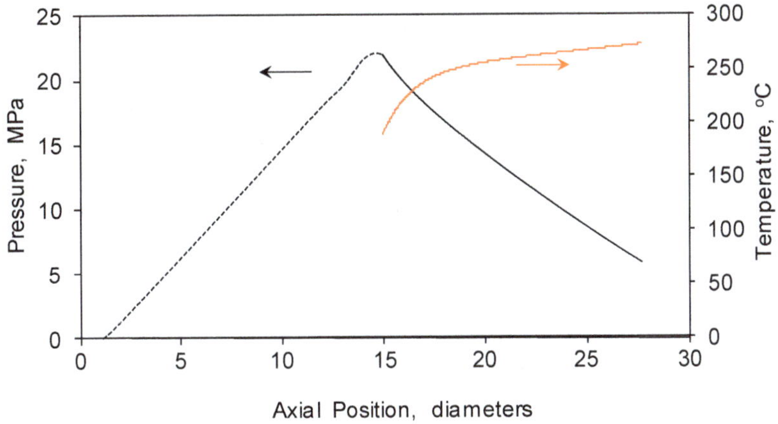

Figure 9.2 Simulated axial pressure and temperature for the baseline process at 10.3 kg/h and a screw speed of 28 rpm. The solid lines are for the simulated profiles. The dashed line is the estimated pressure. The simulation predicts a discharge pressure and temperature of 5.8 MPa and 273 °C, respectively

For the baseline process, the experimental data and the numerical simulations fit quite well. The entry pressure and temperature to the metering section of 22 MPa and 190 °C are very reasonable for this resin and application. These entry conditions will be used to design the new process.

The new commercial process was based on the lab process and has the target performance shown in Table 9.2. To achieve this performance, an 80 mm diameter extruder with an *L/D* ratio of 28 was specified. The extruder was built with a 100 kW motor and a gear box such that the maximum speed of the screw was 117 rpm. A gear pump was positioned downstream of the extruder such that the extruder only needed to generate 6 MPa of pressure at all rates.

Table 9.2 Target Performance for the 80 mm Diameter Process

	Performance
Rate, kg/h	120
Discharge pressure, MPa	6
Discharge temperature, °C	less than 280
Specific energy, J/g	750

The strategy now is to set the geometry of the metering channel such that the target performance can be obtained. For this design, the compression ratio was set at 2.7, the same ratio as the lab screw. The compression rate was considered high for the lab screw, and it was decreased slightly from 0.0060 to 0.0052 for the target screw. The metering section entry pressure and temperature were set at 22 MPa and 190 °C as determined by the baseline simulations.

Numerous simulations were performed before the target performance and screw design were obtained for this process. Moreover, the final design reported here is not a unique solution. Other design solutions that meet the target performance are possible. The design that was developed here had three basic sections: (1) a 7 diameter long feed channel with a constant depth of 12.6 mm, (2) a 7 diameter long transition section, and (3) a 14 diameter long metering section with a constant depth of 4.67 mm. The screw was single flighted and had a lead length of 100 mm. For each design tested the screw speed had to be iteratively determined such that the target rate of 120 kg/h was obtained.

The metering section of the screw presented above was simulated at 120 kg/h and a discharge pressure of 6 MPa. The screw had to be rotated at a speed of 56 rpm to obtain 120 kg/h. The barrel temperatures were 160, 220, and 275 °C for the feed section through the discharge section, respectively. The simulated axial pressure and temperature are shown in Figure 9.3.

At a screw speed of 56 rpm and a rate of 120 kg/h and with the motor operating at full current (full torque), the extruder motor would be inputting a specific energy of 1440 J/g to the resin via the screw. Since the 80 mm diameter extruder is expected to require only 750 J/g, the motor is expected to operate at about 52% of the maximum current. Thus, the extruder has enough power to operate the proposed screw design at 120 kg/h and a screw speed of 56 rpm.

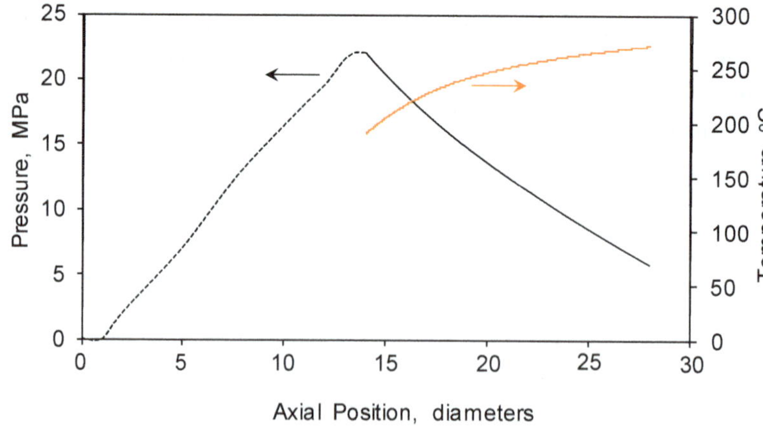

Figure 9.3 Simulated axial pressure and temperature for the target process at 120 kg/h and a screw speed of 56 rpm. The solid lines are for the simulated profiles. The simulation predicts a discharge pressure and temperature of 5.8 MPa and 272 °C, respectively

The 80 mm diameter screw was built with the channel geometry presented above, and it was installed and tested. After several days of operation, the performance of the extruder was measured at a rate of 120 kg/h. The predicted performance and the actual performance are summarized in Table 9.3.

Table 9.3 Predicted and Actual Performance of the 80 mm Diameter Extrusion Process

	Predicted	Actual
Rate, kg/h	120	120
Screw speed, rpm	56	58
Specific rate, kg/(h·rpm)	2.14	2.07
Discharge pressure, MPa	5.8	6
Discharge temperature, °C	272	272
Specific energy, J/g	750	720

As shown by the data in Table 9.3, the 80 mm diameter designed screw met the target performance and operated at essentially the predicted screw speed. The slightly higher screw speed that was used (58 rpm) was likely a result of the lower compression rate used for the 80 mm screw. The lower compression rate likely reduced the pressure at the entry to the metering section to a value less than 22 MPa. This lower entry pressure would cause the specific rate to decrease slightly and thus require the screw speed to increase to produce the target rate.

9.4 Rate Increase for an 88.9 mm Diameter Extruder Running a HIPS Resin

A profile extrusion line was required to increase its rate from the current 75 kg/h to about 230 kg/h in order to meet business demand. The extruder was 88.9 mm in diameter and was running a HIPS resin. Operation of the existing extrusion equipment, however, caused the extrudate temperature to be too high at rates higher than about 80 kg/h. The objective of this project was to increase the rate of the profile line from the current 75 kg/h to a maximum rate of 230 kg/h while maintaining the extrudate temperature below 195 °C.

The profile line runs a blend of virgin HIPS resin (2.1 dg/min at 200 °C and 5.0 kg) and between 0 and 20% of recycle material. The recycle material is from the same extruded parts and thus is composed of the same HIPS resin. The shear viscosity of the resin at 170, 190, and 210 °C is shown in Figure 9.4.

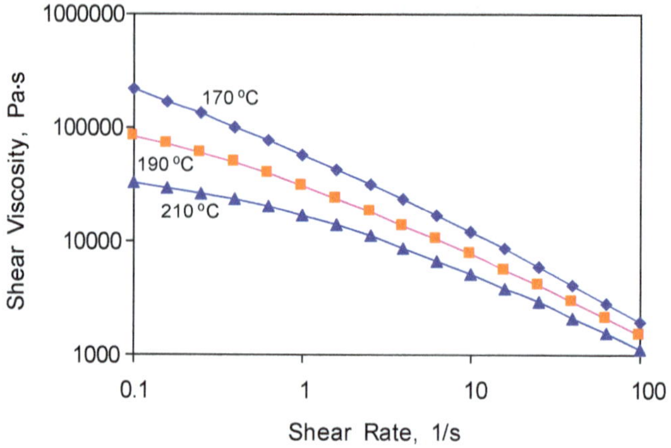

Figure 9.4 Shear viscosity data for the HIPS resin at 170, 190, and 210 °C

The bulk density of the pellets and the recycle material were measured at ambient conditions at 620 and 460 kg/m^3, respectively. The device used to measure the bulk density is shown in Figure 4.2. A mixture containing 80% pellets and 20% recycle material was estimated to have a bulk density of 590 kg/m^3.

Baseline process data were collected on the profile line and the data are summarized in Table 9.4. The specific rate for this process was 2.85 kg/(h·rpm) with a specific energy of 680 J/g. This specific rate was slightly higher than the calculated specific rotational flow rate of 2.73 kg/(h·rpm), indicating that a relatively small negative pressure gradient existed in the metering section of the screw. The extrudate temperature was measured with a handheld thermocouple at 196 °C. During data collection, the temperature controllers on the barrel zones were operating with very small levels of cooling. Additional cooling was available. The data in Table 9.4 forms a baseline for the scale-up of the current process and the design of a new process that operates at a rate of 230 kg/h.

Table 9.4 Baseline Extruder Process Conditions and Performance for the Profile Line

Performance Parameter	Value
Rate, kg/h	77
Screw speed, rpm	27
Specific rate, kg/(h·rpm)	2.85
Discharge temperature, °C	196
Extruder discharge pressure, MPa	10.6
Motor load, % full current	53
Specific energy, J/g	680
Barrel temperatures for zones 1 through 4, °C	149, 149, 182, 188
Screw fluid temperature in, °C	111
Screw fluid temperature out, °C	117

The baseline process consisted of an 88.9 mm diameter, single-screw extruder with a nominal 25 *L/D* single-stage screw. The current screw was designed with three sections as follows: (1) 5 diameters of a feed section with a constant depth of 14.2 mm, (2) 14.4 diameters of a transition section, and (3) 6 diameters of metering with a constant depth of 5.7 mm. The lead length of all sections of the screw was 88.9 mm, and the flight width perpendicular to the tip edge was 9.5 mm. The compression rate of the screw was determined to be 0.0020 and the compression ratio was 2.5. This compression ratio and rate are lower than the typical design specifications for the extrusion of PS resin pellets. For 100% PS pellets, the ideal compression rate and ratio are 0.0035 to 0.0048 and 2.7 to 3.0, respectively. Since the recycle resin decreases the bulk density of the feedstock, the depth of the feed channel should be increased, causing the compression ratio and compression rate to increase slightly. The guideline for increasing the compression ratio and compression rate due to a reduced bulk density feedstock is provided by Equations 6.3 and 6.4. With the bulk density decreasing from 620 kg/m^3 for the pellet feedstock to 590 kg/m^3 for a mixture containing 20% recycle, the ideal compression rate and compression ratio for this feedstock mixture is 0.0037 to 0.0050 and 2.8 to 3.2, respectively. The specific rotational flow rate for the metering section was calculated at 2.73 kg/(h · rpm). A 19 mm diameter cooling hole extended from the shank to 3 cm from the tip of the screw. A screw cooling lance was used to cool the screw using a heated oil system. The screw was slightly worn and had an average flight clearance of 0.20 mm, a clearance that was about twice that of a new screw. This larger flight clearance was acceptable from a flow perspective, but it will limit the level of energy transfer out through the barrel wall. The extruder was equipped with a 55 kW, 1750 maximum rpm motor, and a gear box and belt sheave ratio of 13.9:1 and 2.29:1, respectively. Thus, the gearbox and sheave belt system provided a total reduction ratio of 31.8:1 and a maximum screw speed at the base motor speed of 55 rpm. Resin flowed from the extruder tip, through a short transfer line, a screen pack, and then through the profile die. The profile was passed through several downstream calibrators and an air cooling system. The profiles were cut to length at the end of the line.

The baseline extrusion process was numerically simulated using the processing conditions in Table 9.4 and the method described in Section 9.2.1, that is, with a rate of 77 kg/h, a screw speed of 27 rpm, and a discharge pressure of 10.6 MPa. The iterative calculation process was used to estimate a bulk temperature of 160 °C and a pressure of 13.1 MPa at the entrance to the meter section. The axial pressure and temperature profile for the simulation is shown in Figure 9.5.

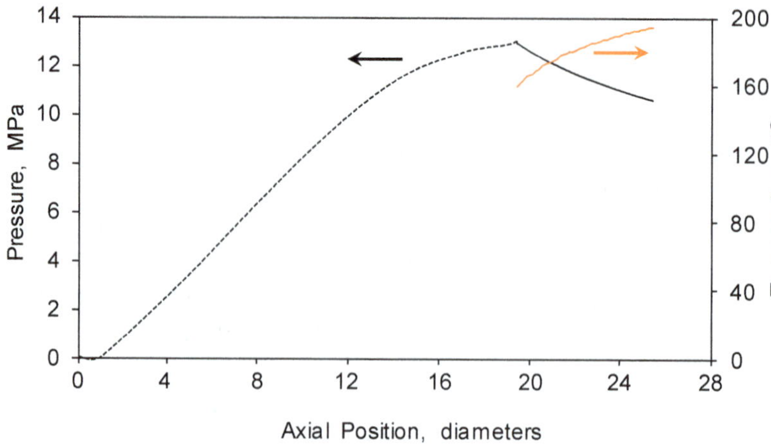

Figure 9.5 Simulation of the current process at a rate of 77 kg/h and a screw speed of 27 rpm. The specific rate was 2.85 kg/(h·rpm)

The axial pressure and temperature profiles are considered very typical for this type of process. As indicated by Figure 9.5, the discharge temperature was estimated at 195 °C, a temperature consistent with the measured temperature at the plant of 196 °C. The simulations were performed with essentially zero cooling on the screw. If cooling were applied, then the discharge temperature decreased to near 180 °C and the torque level on the screw became very high. The plant was contacted and they were instructed to remove the screw cooling lance and inspect it. A severe flow restriction was found in the cooling lance, confirming that internal screw cooling was not effective. These simulation data form a baseline for designing the high-rate screw.

Conversion of the current process to the high-rate profile process will require several modifications. These modifications include a new high-rate screw, a new die, profile calibration equipment, and a new water cooling tank to cool the profile. The die will have two profiles such that each profile will be operating at a maximum rate of 115 kg/h. The screen pack upstream of the new die, however, will be unchanged and will need to operate at a rate of 230 kg/h. Since only the pressure upstream of the screen pack for the current process is known, it is difficult to predict the discharge pressure required by the extruder to run the screen pack and new die at a total rate of 230 kg/h. For the purpose of the design work here, it is assumed that a discharge pressure of 13 MPa is adequate for the new process. An Energy Transfer (ET) screw [17] was chosen as the high-performance design style for this application. ET screws and simulations are discussed in Chapter 14. The metering channel for this screw style has two channels that vary in channel depth in the axial direction. The flights between the channels are selectively undercut to provide a level of dispersive mixing. The design targets for the screw are summarized in Table 9.5.

Table 9.5 Process Targets for the Profile Line and the New High-Rate Screw

Performance Parameter	Value
Rate, kg/h	230
Discharge temperature, °C	195
Extruder discharge pressure, MPa	13

Numerous simulations and design iterations were made to establish an acceptable screw design for the profile extruder. The ET screw design specifications that provide an extrudate at the target rate and conditions are too complicated to present here, but the average channel depth and the lead length of the ET section were 8.18 and 108 mm, respectively. The entry channel depth of the ET section was 6.35 mm, and this depth will be used to calculate the compression rate. The ET section length was 76 cm or seven turns. Downstream from the ET section will be a 21 cm long single-flighted metering section with a constant depth of 6.35 mm. The depth of this single-flight section will be used to set the compression ratio. About 142 cm of axial length were available for the feed and transition sections.

The ET screw was simulated at a rate of 230 kg/h and a discharge pressure of 13 MPa, as shown in Figure 9.6. For this operation, the barrel temperatures were 170 and 175 °C for zones 3 and 4, respectively, and the material entering the ET section was assumed to have a temperature of 160 °C and a pressure of 17 MPa. This entry pressure is slightly higher than that used for the current screw because a higher compression ratio will be specified for the ET screw. The screw was rotated at a speed of 58 rpm for a specific rate of 3.97 kg/(h·rpm) and the discharge temperature was 195 °C. The simulations were performed using slightly lower barrel temperature settings so as to adjust the discharge temperature to 195 °C. These settings are hypothesized to be acceptable since they are only 12 °C lower than the current settings. Moreover, the heat transfer rate will be much improved with the new screw since its flight clearances will be set at the standard clearance of 0.09 mm rather than the 0.20 mm for the current worn screw. At this screw speed, the motor will be overspeeded and at a motor current that is 78% of the maximum current to input 680 J/g at 58 rpm.

The ET design was numerically studied as a function of rate to determine the effect of screw speed on the discharge temperature. The discharge temperature and screw speed as a function of rate are shown in Figure 9.7. As shown by this figure, the discharge temperature increases with increasing rate, as expected. At rates less than about 230 kg/h the discharge temperature is less than the target value of 195 °C.

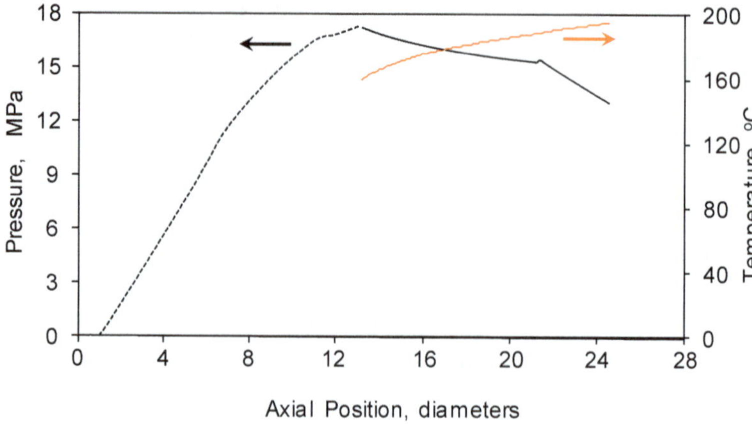

Figure 9.6 Simulation of the ET screw at a rate of 230 kg/h and a discharge pressure of 13 MPa. The rate was obtained at a screw speed of 58 rpm and a discharge temperature of 195 °C, for a specific rate of 3.97 kg/(h·rpm)

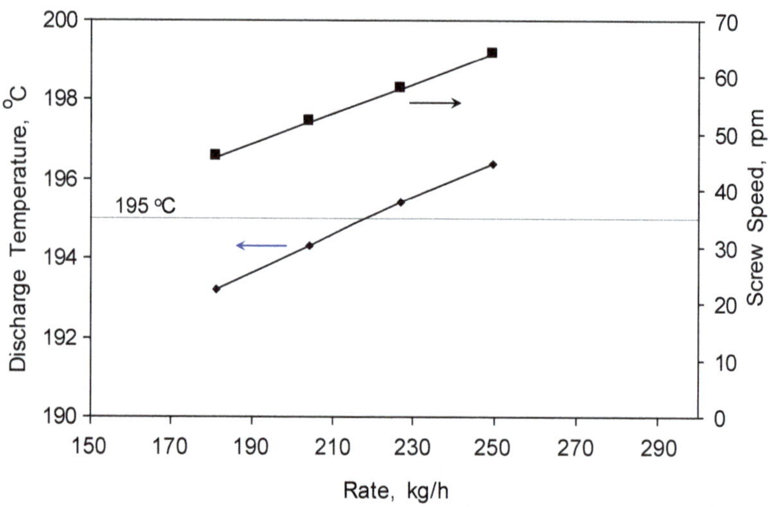

Figure 9.7 The effect of rate on the discharge temperature and screw speed for the high-rate ET screw. The calculations assumed that the discharge pressure was 13 MPa

As previously discussed, the discharge pressure required to operate the screen pack and the new die system at 230 kg/h is unknown. The discharge pressure for the design was, however, assumed to be 13 MPa. Simulations were performed at 230 kg/h for discharge pressures ranging from 11 to about 16 MPa, as shown in Figure 9.8. As shown in this figure, an error in the estimation of the discharge pressure required to operate the equipment had very little effect on the discharge temperature. Thus, changes in the pressure required to drive the downstream process will not affect the discharge temperature to a significant level for this process.

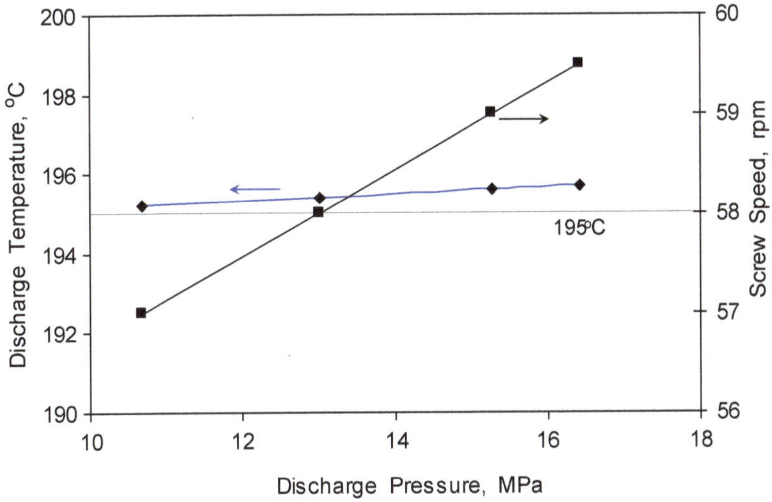

Figure 9.8 The effect of discharge pressure on the discharge temperature and screw speed for the high-rate ET screw. The calculations are for a rate of 230 kg/h

The barrel temperature settings were studied to determine their effect on the discharge temperature of the extrudate. As shown in Figure 9.9 and as expected, decreasing zones 3 and 4 barrel settings created a very significant change in the discharge temperature of the extrudate. Although this extruder is air cooled, decreasing the barrel temperatures to near 170 to 175 °C is likely an acceptable option. Changing the barrel temperatures on zones 3 and 4 is the best option for adjusting the temperature of the extrudate for this process.

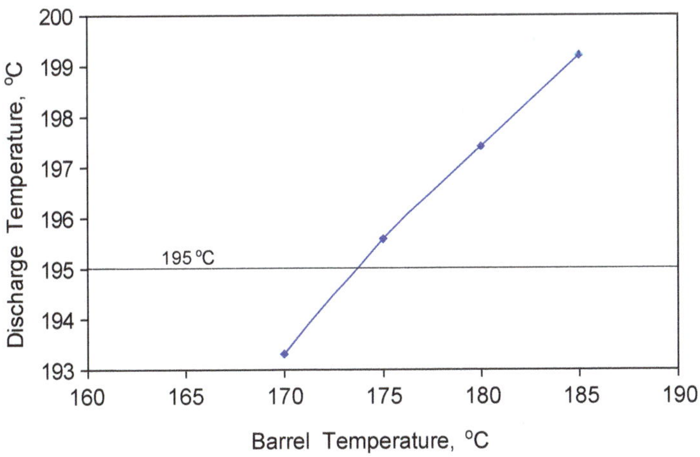

Figure 9.9 The effect of the zone 3 and 4 barrel temperature settings on the discharge temperature for the high-rate screw. The rate and screw speed were held constant at 230 kg/h and 58 rpm, respectively

Based on the baseline data and previous experience with the processing of this HIPS resin, the compression ratio and rate used for this screw will be 2.8 and 0.0046, respectively. These values are slightly higher than those used for the existing screw, but they are known to work well for HIPS resin and are well suited for the bulk density of the feedstock mixture. The compression ratio and the depth of the single-flighted metering section of 6.35 mm specify the feed channel depth at 17.8 mm. The entry depth to the ET section and the feed channel depth are used to specify the transition section length at 88.9 cm using a compression rate of 0.0046.

The material of construction for the ET screw is critical in order to provide the proper strength in the event of an improper start-up procedure and to minimize cost. The maximum stress that the screw will experience during operation will be in the feed section where the root diameter is the smallest and the torque is the greatest. The calculation of the power that a screw can transmit safely is provided in Section 10.4.5. This calculation ignores the strength contribution from the flight. The safety factor for this screw made from three different materials is provided in Table 9.6.

Table 9.6 Safety Factors for the ET Screw for Three Common Metal Stocks

Material of Construction	Tensile Strength, MPa	Safety Factor
AISI 4140 Steel	690	1.27
AISI 4340 Steel	758	1.39
Stainless Steel 17-4 PH	1206	2.22

The safety factors for the AISI 4140 and 4340 steel are too low and unacceptable for this process. In general, the safety factor should be at least 2 to prevent screw breaks during cold starts. The screw should be made from stainless steel 17-4 PH alloy, providing a safety factor of 2.22 for this application. As discussed in Section 10.4.5, the strength of the flight was ignored for this calculation. Additional strength, however, will be obtained from the flight.

Nomenclature

A	motor current during operation
A_{max}	nameplate motor current at full load
C	compression ratio of the screw
D_b	diameter of the barrel
D_1	barrel diameter for extrusion condition 1

D_2	barrel diameter for extrusion condition 2
E	specific energy input from the screw to the resin in J/g
F_s	safety factor in torsion for the screw
H	channel depth in the metering section
H_f	channel depth in the feed section
L	flight lead length
M	number of turns in the transition section
N	speed for the screw in revolutions/s
N_{max}	maximum screw speed in revolutions/s
P_{e1}	required power from the motor to the process for an extruder with a diameter of D_1
P_{e2}	required power from the motor to the process for an extruder with a diameter of D_2
P_m	power at the screw shaft in kW
P_{max}	maximum power rating for the motor in kW
P_1	pressure at the die for rate condition 1 in Equation 9.4
P_2	pressure at the die for rate condition 2 in Equation 9.4
Q	extrusion rate in kg/h
Q_1	rate for condition 1 in Equation 9.3
Q_2	rate for condition 2 in Equation 9.3
R	compression rate in the helical direction
θ_b	helix angle at the barrel wall

References

1. Carley, J.F. and McKelvey, J.M., "Extruder Scale-Up Theory and Experiments," *Ind. Eng. Chem.*, **45**, 985 (1953)
2. Fenner, R.T. and Williams, J.G., "Some Melt Flow and Mechanical Design Aspects of Large Extruders," *Polym. Eng. Sci.*, **11**, 474 (1971)
3. Chung, C.I., Nichols, R.J., and Kruder, G.A., "Screw Horsepower Dependence on Screw Speed and Size - Theory and Experiment," *Polym. Eng. Sci.*, **14**, 28 (1974)
4. Chung, C.I., "On the Scale-Up of Plasticating Extruder Screws," *Polym. Eng. Sci.*, **24**, 626 (1984)

5. Guo, Y. and Chung, C. I., "Dependence on Melt Temperature on Screw Speed and Size in Extrusion," *Polym. Eng. Sci.*, **29**, 415 (1989)
6. Potente, H., "Existing Scale-Up Rules for Single-Screw Plasticating Extruders," *Int. Polym. Process.*, **6**, 267 (1991)
7. Potente, H., Hanhart, W., and Reski, T., "Design and Processing Optimization of Extruder Screws," *Polym. Eng. Sci.*, **34**, 937 (1994)
8. Chung, C. I., "Extrusion of Polymers: Theory and Practice," Hanser Publishers, Munich (2000)
9. Potente, H., "Single-Screw Extruder Analysis and Design," Chapter 5 in "Screw Extrusion," White, J. L. and Potente, H. (Eds.), Hanser Publishers, Munich (2003)
10. McCullough, T. W. and Spalding, M. A., "Predicting Actual Temperature Distributions in a Polymer Stream Using an Adjustable-Depth, Exposed-Tip Thermocouple Assembly," *SPE ANTEC Tech. Papers*, **42**, 412 (1996)
11. Zoller, P. and Walsh, D. J., "Standard Pressure-Volume-Temperature Data for Polymers," Technomic Publishing Co., Inc., Lancaster, PA (1995)
12. Spalding, M. A., "Metering Channel Flows and Troubleshooting Single-Screw Extruders," *SPE ANTEC Tech. Papers*, **50**, 329 (2004)
13. Carley, J. F. and Smith, W. C., "Design and Operation of Screen Packs," *SPE ANTEC Tech. Papers*, **21**, 594 (1975)
14. Todd, D. B., "Determining Pressure Drop in Extrusion," *Plast. Compd.*, **17**, 23 (1994)
15. Hyun, K. S. and Spalding, M. A., "Use of Process Data Obtained from a Data Acquisition System for Optimizing and Debugging Extrusion Processes," *Adv. Polym. Technol.*, **15**, 29 (1996)
16. Zitzenbacher, G., Karlbauer, R., and Thiel, H., "A New Calculation Model and Optimization Method for Maddock Mixers in Single Screw Plasticising Technology," *Int. Polym. Process.* **22**, 73 (2007)
17. Chung, C. I. and Barr, R. A., "Energy Efficient Extruder Screw," U. S. Patent 4,405,239 (1983)

10 Introduction to Troubleshooting the Extrusion Process

Eventually every single-screw extrusion process will experience periods when the machine is operating at a performance level that is less than designed. During these periods, the cost of manufacturing will increase due to the production of off-specification products, loss of production rate, high levels of recycle, higher labor costs, and lower daily production of prime product. In extreme cases, the problem can be so severe that the line must be shut down. Obviously the plant needs to restore the operation of the line to the original performance level as soon as possible to maximize profitability. Many things can cause an extruder to malfunction, including mechanical and electrical failures, installation of new equipment, process changes, and resin changes. For example, electronic component failures can occur due to a lightning strike to a plant, or a barrel zone heater could fail causing an uneven heating of the barrel. The root cause and solution to these problems are typically easy to identify and fix. For many cases, however, the root cause of a problem is not readily apparent.

The next chapters will describe a wide variety of problems that have been observed by the authors and others. This chapter will include general remarks about the troubleshooting process, equipment needed for troubleshooting, common mechanical and electrical problems, screw manufacturing, and how to avoid costly troubleshooting operations for a new installation. Chapters that follow this one will provide actual case studies where extruders were malfunctioning, showing the symptoms and the corrections required to restore the performance of the machines. The troubleshooting chapters are grouped by the major symptom of the process. Often a single root cause can create multiple symptoms.

10.1 The Troubleshooting Process

The goal of all troubleshooting operations is to restore the process to its original performance as quickly as possible. If the process is operational and producing a high level of off-specification product, then the manufacturing costs can be very high. Restoring the line to its original performance quickly will reduce costs by eliminating some quality control operations and labor wasted in making product that is not fit for use, reducing resin consumption, eliminating recycle due to off-specification product, and decreasing energy consumption. Moreover, if the line is inoperable due to the defect, the downtime of the line can be extremely costly, especially if the line is sold out. In this latter case, the goal would be to bring the line back to production operation as quickly as possible. Excellent overviews of the troubleshooting process for extrusion systems were provided by Gould [1] and by Christie [2], and by Tobin [3] for injection molding. Procedures to troubleshoot processes in general were outlined by Mager [4] and Fogler and LeBlanc [5].

Troubleshooting a process can range from solving a very simple problem such as replacing a malfunctioning barrel heater to a very difficult flow problem that is very complicated to diagnose. Collecting the proper information on machine performance can minimize the time required to restore the machine to its original performance while reducing the cost of the troubleshooting process. The machine owner will provide details and information for the operation. Typically, the information will be a collection of facts, ideas on the root cause, and data that are not relevant to the problem. The troubleshooter must be able to listen to the information provided and then sort the important facts from the nonrelevant information. Often, several different solutions will be possible. The best solution will be based on a combination of the cost of lost production, the time and cost to implement, machine owner acceptance, and the risk associated with the modified process.

The first thing that a troubleshooter should do is talk to the plant personnel and the process operators about the defect. The operators in many cases witnessed the event that caused the problem or they can provide the recent history leading up to the failure. In some cases the operator may have inadvertently caused the problem. Interviewing the operator and having the operator assist in the diagnosis of the problem can speed up the troubleshooting process. After the interviews, the troubleshooter must verify the accuracy of the information. Verification of the information can be as simple as viewing computer fault information on a control panel to questioning events that are impossible to reproduce or verify. The information that is verified will become part of the basis for setting hypotheses on the root cause of the problem.

The troubleshooter should obtain the performance and modification history for the machine. This information is typically available from electronic data storage devices associated with the extruder, the machine owner, maintenance personnel, and the operators. A schematic for the screw is required for most troubleshooting operations. Most screw vendors will provide, as a courtesy, a diagram that shows the flow channels of the screw. If these diagrams are not provided during the original purchase, a diagram of the screw should be made by plant personnel prior to installation. Other required data include the rate, screw speed, motor current, barrel temperature settings, and discharge pressure and temperature. Additional sensor data are occasionally available on some machines, including pressure measurements in the barrel, gear pump rate via its rotation rate, and sensors specific to a process. The screw speed and motor current values displayed on the control panel should be verified. For the screw speed, counting the rotations of the screw at the back of the gearbox for a time period must be performed to verify the speed displayed on the panel. Most displays show the screw speed within a 5% error, but for some processes the screw speed display has been in error by a factor of two. Verifying the motor current is easily performed by an electrician using a simple current meter. As a general practice, all instruments and sensors should be verified for accuracy.

All mechanical and electrical components should be examined and verified that they are functioning properly. These components include the solenoid valves for water cooling systems, cooling water pumps, cooling fans, electrical heaters, thermocouples and other temperature sensors, pressure sensors, and gear pump operations.

Most extrusion processes measure the discharge temperature using a thermocouple sensor positioned in the transfer line downstream from the tip of the screw. The thermocouple is often positioned into the polymer stream about 1 cm past the inside wall of the transfer line. This equipment configuration is the best for most applications, but it can provide temperature measurements that are considerably different from the actual resin temperature [6]. The thermocouple measures the temperature at the sensor junction, and this junction is influenced by the temperature in its vicinity, including the temperature of the resin, the temperature of the sensor sheath, and the temperature of the transfer line. Since the thermal conductivity for a metal is typically 300 times higher than that for a polymer, thermal conduction is more influenced by the surrounding metal than the temperature of the resin in the transfer line. Thus, for a transfer line that is controlled at a temperature less than the bulk resin temperature, the thermocouple is going to report a temperature that is less than the bulk temperature of the resin. A better way to measure the discharge temperature is to place a handheld thermocouple in the resin discharging at the die opening. In order to mitigate conduction of energy away from the junction, the thermocouple sheath should be immersed in hot resin. If the transfer line is relatively short with no significant cooling or heating occurring in the line, then the measured value should be close to the actual discharge temperature from the extruder. For many systems where the polymer flow is always inside lines, handheld measurements are not possible, and the troubleshooter must rely solely on the temperature measurement through the transfer line. Handheld temperature sensors that measure the infrared (IR) radiation level from the resin provide an excellent measure of relative temperatures, but because of the difficulty in measuring the emissivity of the polymer these units are not as accurate as a handheld thermocouple sensor.

Next, the troubleshooter should perform a basic series of calculations for the screw and process. These calculations include the rotational flow rate and pressure flow rate as outlined in Section 1.4. These calculations will allow the troubleshooter to determine if the metering section of the screw is the rate-limiting section of the process. As previously stated for a smooth bore machine, the metering section of a single-stage screw or the first-stage metering section of a multi-stage screw must control the rate for the process. These calculations are simple and will guide the troubleshooter to develop appropriate hypotheses on the root cause of the process defect. Other calculations that should be performed are the compression rate and compression ratio (Section 6.1 and Section 10.8.2). The compression rate and ratio should be within an acceptable range for the resin processed. To aid in the calculations and the relevance of their application, the bulk density of the feedstock resin and the shear viscosity should be measured. The shear viscosity is needed for the pressure flow calculation while the bulk density is needed to assess the compression ratio and compression rate. The specific energy inputted by the motor to the resin should be calculated (Section 10.8.1) and compared to similar processes.

Often troubleshooting guides are provided by equipment manufacturers for common problems. These guides are helpful for many of the simpler problems associated with the equipment. Some resin manufacturers are an excellent resource for troubleshooting processing problems that are specific to a resin. Subject matter experts or extrusion consultants are also resources for troubleshooting an extrusion process.

Spare parts for common components such as heaters for barrels, transfer lines, and dies, thermocouples, pressure transducers, drive belts, and fuses should be kept in stock. Since the goal is to keep the line operational at all times, keeping these low-cost but necessary components in stock can reduce the amount of downtime due to simple failures. For operations where the resin is abrasive or corrosive, a spare screw should be kept in stock. As the screw wears in the extruder and the performance decreases beyond an economic limit, then the screw should be replaced with the spare screw and the worn screw should be sent to a screw manufacturer for refurbishment.

10.2 Hypothesis Setting and Problem Solving

With the plant interview information, verification of the data, and the completion of the simple calculations, an experienced troubleshooter will develop a set of hypotheses for the root cause of the defect. After the hypotheses are established, a series of experiments need to be developed that accept or reject the hypotheses. Once a hypothesis is accepted via experimentation, then the next step is to develop a technical solution to remove the defect. Often more than one technical solution is possible. The best technical solution will depend on the cost and time to implement the solution, machine owner acceptance, and the risk associated with the modified process. An accepted hypothesis must drive the technical solution. If a hypothesis is not accepted prior to developing a technical solution, then the troubleshooter may be working on the wrong problem and the defect may not be eliminated from the process.

A hypothesis is a proposed explanation for an observation. The hypothesis should be stated such that it is clear and testable. For each hypothesis, an alternative hypothesis should be stated that is accepted should the original hypothesis be proven false. Developing alternative hypotheses allows the troubleshooter to quickly arrive at the defect while moving through a complicated decision-making process. The alternative hypothesis provides a logical branch that directs the troubleshooter to the root cause. Each troubleshooting process should be developed with a set of alternative hypotheses, designing experiments that exclude one or more of the hypotheses, and then performing the experiments such that definitive results occur [7]. At the conclusion of this process, a new set of alternative hypotheses may be required to continue the decision-making process.

As an example of an alternative hypothesis, a simple case study is presented here. For this case, an extruder is discharging degradation products into the product stream. The hypothesis and alternative hypothesis statement is "the metering section of the screw is not operating full and not under pressure, creating regions where resin can degrade, or the process is operating with the metering section full and under pressure and the degradation products are coming in with resin or generated in some other section of the process." This hypothesis is referred to here as the alternative hypothesis because it allows for a decision to be made about where the root cause occurs. Once an experiment is developed to test the hypothesis, the troubleshooter can then focus the next hypothesis and experiments on only one side of the decision branch. Developing acceptable hypotheses depends on validated information, a fundamental knowledge of the process, and knowledge of the properties of the resin. The early chapters of this book provide the fundamental knowledge while the last six chapters present case studies that identify the root causes for extrusion defects. Some of the most common root causes along with the defects that they create are provided in Table 10.1. As shown in this table, several defects can occur from the same root cause, and a particular defect can be produced by several root causes.

Table 10.1 Common Root Causes and Potential Defects

Root Cause	Potential Defects	Section
Blockage at the entry of a barrier section or spiral mixer	Low specific rates, high discharge temperatures, resin degradation, flow surging	11.10.1 12.7.6
Small flight radii	Resin degradation	11.7
Low compression rate and low compression ratio	Splay and air entrapment	11.12.4
Pigment agglomerates, stagnant regions in the flow path	Black streaks or specks	11.12.2 11.12.3
Improper solids-conveying temperatures for the barrel	Flow surging and low specific rates	12.7.4
High screw temperatures	Flow surging and low specific rates	12.7.3
Process exceeded the melting capacity of the screw	Solid polymer fragments in the extrudate	13.4.2
Improperly designed vent diverter	Flow of resin out the vent opening	13.4.6
Improper pump ratio for a two-stage screw	Flow of resin out the vent opening	13.1

Once the root cause has been properly identified as the source of the defect, a technical solution must be devised that eliminates the root cause from the process. As previously discussed, often more than one solution exists. The best solution will depend on many factors, including the economic conditions for the plant and product line, the cost the defect is creating at the plant, the cost and time for implementing the technical solutions, and the risk associated with each solution. Plant personnel will ultimately decide on the best technical solution for their plant.

Three case studies are presented next that demonstrate approaches to troubleshooting problems. The first two cases were developed with poor hypotheses, while the last case study had a problem that was solved quickly using strong hypotheses and a strong experimental plan for verification.

10.2.1 Case Study for the Design of a New Resin

A new GPPS resin was tested at a customer's injection-molding plant as an improvement over an incumbent resin manufactured by a competitor. The new resin performed well in the process except that it created parts with a 5% rejection rate due to a splay defect. The competitive resin was reported to run well but with a lower defect rate. The plant manager asked that the new resin be redesigned such that it had a defect rate comparable to the competitive resin. Here the hypothesis was that the new resin had a poor performance relative to the incumbent resin, and the technical problem to be solved was that the new resin needed to be modified such that it performed as well as the incumbent resin. As will be shown later, this technical problem was the wrong problem to be solved.

Performance information for the incumbent resin was missing from the early parts of the decision-making process. The decision that the technical problem was the performance of the new resin was based on anecdotal information from plant personnel on the performance of the incumbent resin. That is, the plant personnel believed that the reject level for parts made from the incumbent resin was less than 5%. A statistical analysis of the part defect rates was not performed. This lack of information early in the process allowed the plant manager to propose a poor technical solution without understanding the root cause for the defect. Later in the troubleshooting process, a statistical analysis of the defect rate indicated that the incumbent resin had a defect rate that was statistically equivalent to the new resin.

The primary and alternative hypotheses here are that the injection-molding machine process was creating the defects in the parts, or the resin design was creating the defects. The statistical analysis on the part defects molded using the new and incumbent resins showed that the resins were not the root cause for the defects. A small set of exploratory experiments on the injection-molding machine allowed the development and acceptance of the hypothesis that the screw design used in the plasticator was not effective at melting the resin at high rates and expelling entrained air out through the hopper. The new technical problem to be solved was increasing the melting capacity of the process and eliminating the entrainment of air via process changes and screw modifications. The technical details and the modifications that were made to the screw are presented in detail in Section 11.12.4. When the modifications to the screw were finished, the splay defects were eliminated and the capacity of the plant was increased by about 14%.

This example clearly shows that developing and accepting a hypothesis based on accurate and complete information is necessary for setting an acceptable technical solution. If the plant manager could have persuaded the resin manufacturer to develop a new resin that was similar to the incumbent resin, then the defect would still be there, the cost of the troubleshooting process would have been extremely high, the supplier would have incurred unnecessary development costs, and a high level of defective parts would still have occurred because the root cause would not have been removed.

10.2.2 Case Study for a Surface Blemish

A surface blemish on a specialty sheet product was severely limiting the rate of the process. The blemish appeared as a small (2 mm diameter) hemispherical pit or crater in the surface. The level of surface defects could be minimized but not totally eliminated by reducing the rate of the process by 50%. A series of exploratory experiments were performed and the defects could be eliminated by decreasing the temperature of the last six barrel zones. The first two barrel zones were not adjusted so as to not change the solids-conveying behavior of the resin. When large temperature changes (decreases in set point temperatures of up to 50 °C) were made to these zones, the extrudate temperature was decreased by about 15 °C and the defects were totally eliminated. The hypothesis developed was that when the extrudate temperature exceeded a specified value then surface defects occurred. The technical solution was to develop a process that discharged the resin at less than the specified temperature and at a high rate. The reason the defects occurred at higher discharge temperatures, however, was unknown.

The resin was analyzed for moisture and other volatiles that might cause a gas to be evolved at higher temperatures. All analyses, however, did not indicate that a gas was evolving or that the material was degrading. These data conflict with the stated hypothesis. The technical solution for this path was to design a screw with a very deep metering section such that the extrudate is discharged at as low a temperature as possible. Since the material will discharge at increasing temperatures with increasing screw speeds, the maximum rate will be bounded when the extrudate exceeds the maximum specified temperature.

In this case, the experiment developed to test the hypothesis that high discharge temperatures create the surface defects was flawed. The screw had a very low compression ratio and compression rate. A better hypothesis is that when the barrel zone temperatures are decreased over the melting section, the temperature and bulk density of the solid bed are decreased, allowing entrained air to exit through the hopper. To test this new hypothesis, the barrel zone temperatures over the melting and metering sections were selectively changed such that the melting zones were kept low while the metering zones were increased. For these tests, the defects in the sheet only appeared when the melting zone barrel temperatures were high. The discharge temperature was not a factor in controlling the defects. The entrained air was likely enabled by solid bed breakup that eliminated a pathway for the air back to the hopper. The only path out for the air was with the extrudate. The new technical solution was to design a screw with a high compression rate and compression ratio such that entrained air can be forced backward and out through the hopper.

A new screw was designed with a higher compression rate and compression ratio. The new screw was installed and the defects were totally eliminated. This case study shows a poorly developed experiment that incorrectly validated a poor hypothesis. If the second experiment had been performed first, the original hypothesis of high discharge temperatures creating surface defects would have been invalidated. Clearly, the experimental plan must be such that it definitively validates or invalidates the hypothesis.

10.2.3 Case Study for a Profile Extrusion Process

A customer wanted to switch to a resin with a higher modulus such that a large profile part could be made with additional strength. The initial production trial was performed and limited data were collected. The information that came out of this trial was that the part profile could not be maintained in-specification, the discharge temperature was higher than normal, and the motor was operating at the maximum current load. Since the trial did not last long, rate data were not collected. The new resin was more viscous than the original resin. Based on these very limited data, a second trial was developed based on two hypotheses. The first hypothesis was that the higher modulus of the new resin created a blockage at the entry to a barrier-melting section of the screw. The blockage would cause the motor current to increase dramatically and cause the specific rate for the extruder to decrease. This type of screw defect is presented in Section 11.10.1. The alternative hypothesis is that a blockage did not occur and some other section of the process other than the barrier section was the root cause.

Preliminary work was performed prior to the trial and included the measurement of the viscosities of both resins around the discharge temperatures and the calculation of the specific rotational flow rate for the metering channel geometry. The specific rotational flow rate for the resins was calculated at 11.3 kg/(h·rpm). At the start of the trial, the extruder and line were processing the original resin at a specific rate of 11.0 kg/(h·rpm), a value that is consistent with the specific rotational rate. The part profile was acceptable. Next, the new resin with the higher modulus was added to the line. Within 30 minutes the part profile dimensions were out of specification and the motor was operating at near the maximum current limit. If the first hypothesis is correct that a blockage is occurring at the entry to the barrier-melting section, then the specific rate should be significantly less than the calculated rotational rate of 11.3 kg/(h·rpm). In this case, the specific rate was measured at 11.1 kg/(h·rpm), and thus the first hypothesis that the barrier section was causing a blockage was not valid. A second hypothesis was developed that the higher viscosity of the new resin caused too much energy to be dissipated, increasing the motor load and the discharge temperature such that an acceptable part profile could not be produced. To test this hypothesis, the metering-zone barrel temperatures were slowly decreased in 10 °C increments until the cooling ability of the zones were at the maximum capability. In order to maintain the motor torque at an acceptable level, the temperature of the zone in the solids-conveying section was increased slightly. The extruder was allowed to come to a steady-state operation. Within 45 minutes the part profile was on specification and the extruder was operating at a specific rate of 11.0 kg/(h·rpm) and discharging at a temperature that was about 15 °C less than that at the start of the trial. The data indicates that the second hypothesis is valid. The technical solution for this case was to design a process that discharges the new high-modulus resin at a lower discharge temperature. In this case, plant personnel opted for a new screw design with a deeper metering section to decrease the energy dissipation level and decrease the extrudate temperature.

This case study was developed with an alternative hypothesis and then a second hypothesis, and the experiments were designed properly to determine quickly the root cause of the defect in the part profile. If the hypotheses and experiments had not been developed properly, the time required to troubleshoot the problem would have increased or the project would have failed.

10.3 Equipment and Tools Needed for Troubleshooting

Several tools are very important for troubleshooting a process. These tools include measuring devices for determining the geometry of the screw, measuring temperature, and collecting process data. Troubleshooting a process without these tools can be very difficult.

Common measuring instruments are helpful in determining the geometry of the screw and other components. Measuring devices include a tape measure and micrometers for measuring channel depths and outside screw diameters. If the equipment owner does not have a print of the screw and does not know the diameter of the machine, a simple piece of string can be wrapped around the circumference of the screw shank where it is visible between the feed casing and the gearbox. With a single wrap of string on the shank, the troubleshooter measures the circumference by marking the string at the meeting point with a felt-tip marker. The diameter is then obtained by dividing the length of the string between the ink marks by π. The method will provide the nominal diameter of the machine even though the shank at this location is likely smaller than the barrel diameter by up to eight times the flight clearance; that is, the clearance in this section is typically four times the clearance between the flights and barrel wall in the downstream sections.

A sled device for measuring the depth of screw channels, barrier flight undercuts, and mixer flight undercuts is shown in Figure 10.1. To use this device, the sled is positioned on the top of the screw with the micrometer tip contacting the top of the flight. The micrometer is zeroed at this position. Next the sled is slid axially on the screw, and the micrometer is adjusted until the top of the micrometer probe contacts the root of the screw. A similar procedure can be used to determine the undercut depths for barrier flights or mixer flights.

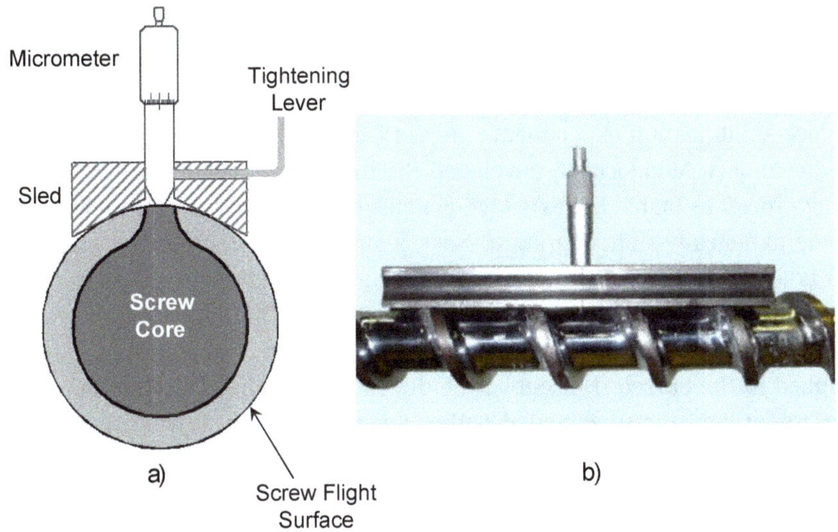

Figure 10.1 Micrometer and sled for measuring the depth of screw channels, barrier flight undercuts, and mixer flight undercuts: a) schematic view perpendicular to the screw axis showing how the micrometer is zero-positioned over the top of a flight, and b) photograph of the micrometer sled measuring the depth of a channel

A handheld thermocouple measurement device and an IR temperature sensor are very important to have for understanding the performance of the extruder. These devices have been discussed previously in this chapter. In many of the case studies presented later, the extrudate temperature was measured by immersing a handheld thermocouple probe into the extrudate. The entire probe length is immersed or draped in the extrudate to minimize heat conduction to cooler parts of the probe. The highest temperature measured is the reported value. If it is unsafe to measure the extrudate temperature using a handheld device, then an IR temperature sensor is used to estimate the discharge temperature.

A data acquisition system is an ideal device for collecting and storing process data. Sequential electronic process data can be extremely valuable for troubleshooting the most complicated process. The data should be collected at a frequency that is at least 10 times faster than the instability that is being studied. These devices are relatively simple to connect to the existing control panel. In general, thermocouple wires of the same type should be connected to the termination points of the extruder thermocouples and then to the acquisition device. Most pressure sensor display units, screw speed tachometers, and motor current sensors have output terminals with a 0 to 10 volt range that is linear with the measured values. A qualified electrician should make the connections between the panel and the acquisition device, minimizing the chance of electrical shock and damage to the acquisition device.

10.3.1 Maddock Solidification Experiment

Maddock solidification experiments are very useful for troubleshooting the flows in the channels. Maddock [8] developed the method to visualize the melting and mixing processes inside the extruder. For this experiment, the extruder is allowed to come to a steady-state operation. Next, a small amount of colored masterbatch resin is added to the natural resin at a letdown ratio of about 200 parts natural resin to 1 part masterbatch, and then it is added to the hopper of the extruder. When the color appears in the discharge, screw rotation is stopped and full cooling is applied to the barrel. If possible, the die should be immediately removed from the extruder, and the unextruded pellets should be removed from the hopper. A vacuum cleaner works well for removing the pellets from the hopper. After the resin is completely solidified in the extruder, the screw is pushed out of the extruder using either a hydraulic jack system or a threaded screw push rod. Often the barrel temperatures must be increased slightly from ambient temperature to allow the solid polymer to separate from the barrel wall. To view the melting profile and the mixing performance of the screw, the solidified polymer must be removed from the screw and then sectioned. The sections can be either perpendicular to the flight edge or parallel to the screw axis. A photograph of a solidification experiment using ABS resin is shown in Figure 10.2. Solidification experiments are relatively simple to perform for screws with diameters up to about 100 mm. Solidification experiments have been performed on larger diameter extruders, but the force required to push the screw out of the barrel can become excessive. The relative mixing of a black color concentrate in a natural resin and the melting level are shown in Figure 10.2(c) for a solidified section removed from the middle of the transition section.

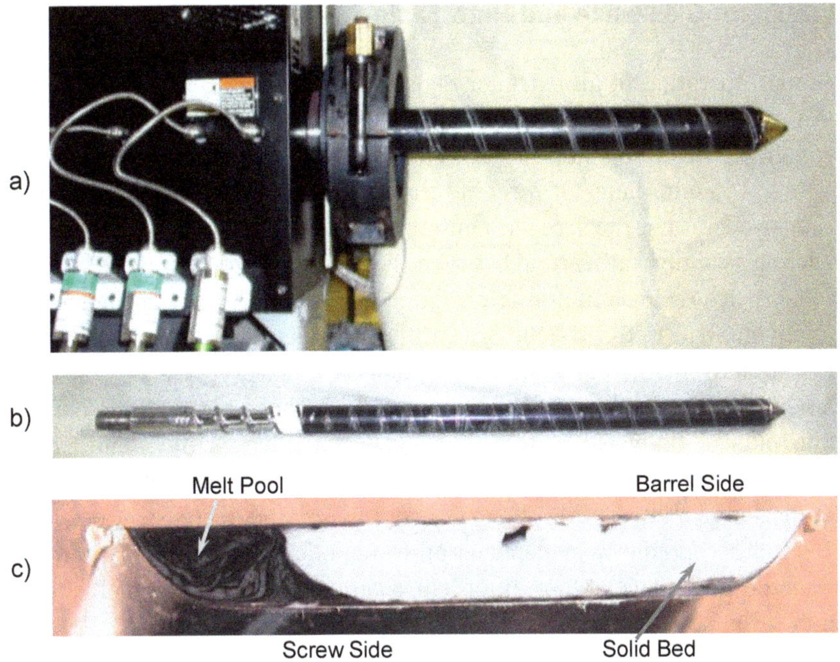

Figure 10.2 Photograph of a Maddock solidification experiment: a) the screw is being pushed out of the barrel with the solidified resin, b) the screw with resin solidified in the channels, and c) a cross-sectional view parallel to the screw axis in the melting section

10.4 Common Mechanical Problems

Several common mechanical problems can occur, leading to high barrel and screw wear, reduced extrusion performance, or a catastrophic breakdown of the equipment. Several of these mechanical issues cause the barrel to become out of alignment with the gearbox, causing a high level of machine wear and in the worst cases seizing of the screw in the barrel. Most of these problems can be eliminated by performing routine maintenance on the machine as directed by the equipment manufacturer. These mechanical problems are discussed in this section. Troubleshooting of mechanical problems is also documented in other sources [9, 10].

10.4.1 Flight Clearance and Hard Facing

The proper operation of an extruder depends on having a close fit of the screw inside the barrel. Typically, the clearance between the outside diameter of the screw and the inside diameter of the barrel is 0.001 times the barrel diameter. Thus, for a 150 mm diameter machine the clearance is 0.15 mm. During normal and proper operation, the screw is centered in the barrel by a thin film of molten resin acting as a lubricating fluid between the screw flight tips and the barrel wall. For nonabrasive resins and properly aligned extruders, it is not uncommon to see essentially no wear in the system for continued service of 10 to 20 years. But if the resin is abrasive or the barrel and feed casing are not properly aligned with the gearbox, then excessive wear will increase the clearance between the barrel wall and the outside diameter of the screw very quickly. As the flight clearance increases, the performance of the machine will typically be reduced, leading to higher discharge temperatures, process instabilities, and decreased rates [11-13]. The level of acceptable wear depends on the application, but typically the worn parts should be replaced when the flight clearance increases to more than four times the original clearance.

Screws and barrels are manufactured from materials such that the screw will wear at a rate faster than the barrel. Typically the flight tips of the screws will have a welding of a hard surface material. These surface materials have a hardness ranging from 40 to 56 as measured using the Rockwell C scale. The thickness of these hard facing materials is usually around 1.5 mm, and they can be applied with one to three layers across the full flight width as shown in Figure 10.3. The application of multiple layers can strengthen the weld and provide a harder surface. A less costly method is to apply just a single layer of the hard facing such as shown in Figure 10.3(b). Some screw manufacturers inlay the hard facing in the center of the flight, as shown in Figure 10.3(c). For this practice, the hard facing covers only about 40 to 50% of the flight width. Inlay hard facing is less expensive to apply and does not have as long of a service life as compared to full flight hard facing. Barrels are also lined with hard coatings, and they typically have a Rockwell C hardness of 48 to 66 depending on the application and the expected service life. In order to provide the proper wear surfaces, the hard facing material must be compatible with the liner material used for the barrel. If the materials are not compatible, then the wear rate on the barrel may be too high, reducing the useful life of the barrel. The screws are designed to wear at a faster rate than the barrels because screws are relatively easy to replace and refurbish.

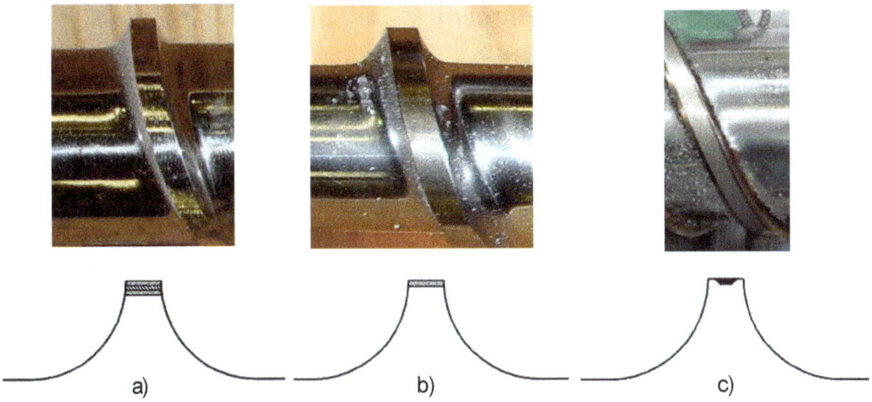

Figure 10.3 Schematic showing different hard facing styles on flight edges: a) a triple-pass hard facing where the first layer might be stainless steel 316 and the second and third layers would be a specialized hard facing material, b) single-pass hard faced flight, and c) a flight tip with a hard facing material inlaid. Photographs of the flight construction are provided above the schematic

Often the hard facing on the flight tips will have very small cracks that are perpendicular to the flight edge. A small level of these cracks is common for most screws manufactured. The level of cracking depends on the welding process used and the type of hard facing material. If a crack becomes large enough that a piece of paper can be inserted into it, then the crack is unacceptable because it may lead to the eventual failure of the weld in that area. In these cases, the hard facing flight fragment can break off the screw and will flow downstream with the resin, potentially causing additional equipment damage. For many processes, the fragments will be collected on the screen system before damaging gear pumps and dies. The performance of the screw will likely not change significantly until a large enough section of the hard facing has broken off. Metal fragments that are trapped on screens should always be examined to determine their source.

The feed casing can also show wear under extreme usage and after a long service. Often the inside diameter of the feed casing is slightly oversize in diameter, or the outside diameter of the screw is undercut in this region such that the clearance between the flight tips and the casing is about four times that used in the barrel. This large flight clearance is there to mitigate interference between the screw and the casing that would be caused by different levels of thermal expansion for the cold feed casing and the warmer screw, and for the tolerances of the radial bearing in the gearbox. The inside diameter of the feed casing should be measured each time the barrel is changed or removed from the casing. If the wear on the casing is excessive the rate of the extruder may decrease or the machine may operate unstably. An example of this instability is presented in Section 12.7.7.

10.4.2 Barrel and Screw Alignment

The alignment of the barrel and feed casing with the gearbox is extremely important for the long-term operation of the equipment. If the alignment is not within the SPI tolerance, then excessive mechanical wear will occur to both the barrel and the screw. For example, a 150 mm diameter 24 *L/D* extruder must be aligned within 0.305 mm at both the feed and discharge ends of the centerline at both room temperature and at the processing temperature. Misalignment that results in deviations greater than 0.305 mm will cause excessive barrel and screw wear. Moreover, systems with higher levels of misalignment will wear at a higher rate. If the misalignment is extreme, the screw will be forced to flex once per revolution. Depending on the flex level and the induced stress to the screw, the screw can fail after 1 million to 10 million rotations [14]. An extruder operating at a screw speed of 60 rpm and at an operation rate of 8000 h/yr will rotate 6 million times in less than three months. Typically the screw will break near the middle section. Extremely long screws such as those used in 40 *L/D* two-stage, vented machines are prone to flexural-type breaks if poorly aligned. The break will typically occur in the first-stage metering section. This type of misalignment can also cause significant wear to the radial bearing in the gearbox, reducing the useful service life of the bearing or causing the screw to seize in the barrel.

The gearbox, feed casing, and barrel should be aligned when first installed and whenever one of the components is removed or replaced. During routine maintenance or if a shift occurs in the machine, an alignment of the components should be considered. The alignment should be performed by personnel trained and skilled in the procedures. On several machines, local machinists were hired to align large-diameter extruders. These alignments were later discovered to be misaligned by more than seven times the SPI specification. Basic alignment procedures were previously outlined by Myers and Puhalla [15]. The alignments can be made using optical or laser systems. The preferred technique depends on the application. For this procedure, the screw and downstream equipment are first removed from the extruder. Next, a borescope is centered in the hole of the drive quill of the gearbox. A target is then positioned at the center point of the barrel at the discharge end of the machine. Optical micrometers on the borescope are used to determine the position of the barrel axis at the discharge relative to the position of the borescope. A more sophisticated method can check the alignment across the entire length of the barrel.

Inspection of the barrel and screw for wear can provide information on the root cause for the wear. For example, if wear occurs at an axial location on only one side of the barrel and on all sides (angular direction) of the screw, then the likely root cause is that the barrel is out of alignment at that axial location. Conversely, if the barrel is worn on all sides and the screw is worn on only one side at the same axial location, then the root cause is likely a local bend in the screw.

Several common mechanical problems can occur that will cause an extruder to move out of alignment. These problems include wear of the radial bearing in the gearbox, a shift in the concrete floor supports, improper support of the equipment attached to the discharge end of the extruder, and malfunctioning of the barrel supports. The radial bearing is designed to maintain the centerline for rotation of the screw within a small tolerance. If the radial bearing fails, then the axis for rotation will vary more than the specification, and it will be impossible to maintain the screw in the proper position relative to the gearbox centerline and the barrel. For older plants, the concrete floor or supports can shift slightly, causing sections of the extruder to shift and potentially create an alignment problem between the barrel and gearbox. Heavy downstream equipment also needs to be supported independently of the extruder and should be aligned with the extruder axis. Often the barrel can be perfectly aligned with the gearbox when the downstream equipment is disconnected. If the downstream equipment is not aligned and supported properly, the attachment process can force the discharge end of the barrel out of alignment, causing excessive wear to the components or the seizing of the screw in the barrel. For example, a worn barrel of a large-diameter, two-stage vented extruder was replaced with a new one. An expert was brought onto the site to laser-align the barrel and feed casing to the gearbox. After the alignment expert left the plant, a heavy die and screen changer system was attached to the front of the extruder. The die system was not supported well, and when it was attached to the barrel it forced the discharge end of the barrel down by about 2 mm. This caused a high wear rate of the barrel and screw surface and resulted in the discoloration of the resin with degradation products. After a few hours of service, the barrel and screw were damaged beyond repair and were removed from the machine.

10.4.3 Extruder Barrel Supports

The most common root cause for misalignment is a malfunctioning barrel support. The barrel supports must be able to support the weight of the barrel and allow the barrel to expand thermally along its axis and in the radial direction during the heating process. Extruders are typically fixed in position by the gearbox, and thus the barrel will thermally expand in the direction of the discharge. That is, as the extruder is heated from ambient to the processing temperature, the length and the outside diameter of the barrel will increase. The barrel supports are designed to allow the thermal expansion of the barrel while maintaining the alignment of the barrel with the gearbox. For example, as the barrel length increases, the barrel must slide along the barrel support, as shown in Figure 10.4.

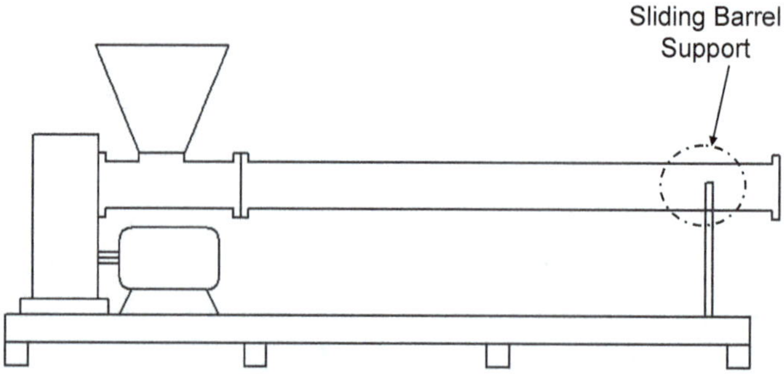

Figure 10.4 Diagram for the operation of barrel supports. The barrel slides on the supports during thermal expansion, allowing the barrel axis to maintain its elevation. If the barrel support does not allow the barrel to slide, then the discharge end of the barrel is forced down and possibly out of alignment

For a 150 mm diameter 28 L/D extruder, the barrel will increase in length by about 8.8 mm as the temperature is increased from ambient (T_0) to an average temperature (T) of 200 °C. If the barrel cannot slide on the support, the barrel will be pulled down by the support by a small amount (ΔA). If this amount is greater than the SPI specification for the extruder, then a high and unacceptable level of wear will occur to both the barrel and screw. A diagram showing the extruder support and the thermal expansion is shown in Figure 10.5, and the change in barrel elevation is calculated in Equation 10.1 to Equation 10.4. For a coefficient of thermal expansion for steel (C_e) of 1.2×10^{-5} mm/(mm·°C), a vertical support height A_0 of 700 mm, and the support positioned 3800 mm (S) downstream from the gearbox, the barrel will be forced down by 0.05 mm if the barrel cannot slide on the barrel support. This change in height is less than the SPI guideline of 0.308 mm, and thus the barrel and screw should wear at a normal level. If the alignment of the barrel, however, was such that the discharge end of the extruder was very low and in specification, the additional change due to a malfunctioning barrel support could cause the barrel and screw to wear at an excessive rate.

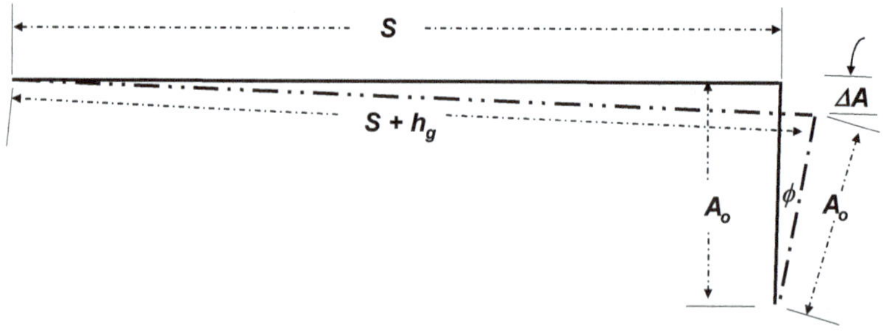

Figure 10.5 Diagram showing the change in barrel elevation at the support for a barrel support that does not allow sliding in the axial direction during thermal expansion. The original position of the barrel and support are shown with solid lines, and the thermally expanded system with the malfunctioning barrel support is shown by dotted lines

$$h_g = SC_e(T - T_0) \tag{10.1}$$

$$\varphi = \arcsin(\frac{h_g}{A_0}) \tag{10.2}$$

$$A_T = A_0 \cos(\varphi) \tag{10.3}$$

$$\Delta A = A_0 - A_T \tag{10.4}$$

where h_g is the change in length of the barrel section between the gearbox and the barrel support, φ is the angle between the support at temperature T and the vertical position at T_0, A_T is the vertical distance between the bottom of the support and the barrel at temperature T. The barrel supports need to be maintained according to the manufacturer's specifications to ensure that the barrel slides on the supports during thermal expansion.

Large-diameter machines (over 250 mm in diameter) often have barrel support columns that are water cooled to maintain their temperature near ambient. These machines have large barrel supports that can allow a significant amount of energy to conduct from the barrel into the supports, causing the support column to expand thermally and elevate the barrel. The increase in elevation can be easily calculated using the coefficient of thermal expansion, the height of the support column, and the temperature increase of the column. It is highly important that the water flow rate is high enough to maintain the support column near ambient temperature.

10.4.4 First-Time Installation of a Screw

The first time that a new screw is installed into an extruder, it is important that the screw be installed into the machine at room temperature. If the outside diameter of the screw is slightly too large or if it has an unacceptable yet small bend in it, the cold screw may be easily inserted into a hot barrel. A hot barrel will be larger in diameter than the specification at room temperature. After the cold screw is inserted into the hot barrel, the screw will thermally expand and potentially create a very tight fit with the barrel. This tight fit will also exist when both the screw and barrel are cooled back to ambient. In one instance, a manufacturing error caused a screw to be too large in diameter. The cold screw, however, readily slid into the extruder with the barrel hot. As the temperature of the screw increased over the next few hours during the heat-up period, it became wedged in the barrel. A large hydraulic jack was required to push the screw from the barrel, causing a considerable level of damage to both the screw and barrel. Once the screw is fit-checked by inserting a cold screw into a cold barrel, then subsequent installations can be performed by inserting the screw cold into a hot barrel.

10.4.5 Screw Breaks

As stated earlier, a screw can break due to repetitive flexing caused by a poorly aligned barrel. A more common break occurs in the feed section of the screw, where the cross-sectional area for strength is the lowest and where the local torque is the highest. For example during the startup of the extruder, the barrel heaters must be turned on and the machine must be allowed to remain at the processing temperature for a period of several hours, depending on size. This "heat soak" period allows the resin in the barrel to melt and obtain a viscosity that is near that observed during processing. When the screw is first rotated, the motor current and thus the torque on the screw will be near normal. If the extruder, however, is not heat soaked long enough, the resin can have an extremely high viscosity and the instantaneous torque on the screw can be very high when the motor is turned on in a "cold start." If the motor is large enough or the material of construction for the screw is not matched with the drivetrain, then the torque in the feed section of the screw can exceed the torsional strength limits of the material of construction, causing the screw to break. If the material of construction is matched with the drivetrain, then the machine controls will shut down the motor when the current exceeds a maximum value. In many cases, the screw will have gone through several cold starts before it breaks.

The material of construction for the screw is critical in order to provide the proper strength in the event of an improper startup procedure such as a cold start and to minimize cost. The choice of the material of construction depends on the motor size, the maximum screw speed at the base motor speed, and the depth of the feed section. The maximum torque that the motor and gearbox combination can supply to the screw, $T_{s,max}$, (Nm) is computed as follows:

$$T_{s,\max} = \frac{P_{\max}}{2\pi N_{\max}} \tag{10.5}$$

where N_{max} is the maximum screw speed in revolutions/s, and P_{max} is the maximum power rating for the motor in watts. The maximum screw speed is determined by dividing the base motor speed by the gearbox reduction and the reduction due to a belt sheave. Next, the maximum torque that the screw can deliver before the base metal yields is calculated using:

$$T_{\max} = \frac{\pi\left[\left(D_s - 2H_f\right)^4 - d^4\right]S_m}{16(D_s - 2H_f)} \tag{10.6}$$

where D_s is the diameter of the screw, H_f is the channel depth of the feed section where the break is most likely to occur, d is the cooling hole diameter, S_m is the shear strength of the material, and T_{max} is the maximum torque that can be transmitted by the screw before breaking. This equation should only be used as a guide because it is based on a cylinder with a hole in the center, and thus it does not account for the strength of the flight. Thus, there is an inherent safety factor built into the equation. For design purposes, it is assumed that the shear strength of the base metal is about 60% of the ultimate tensile strength. Tensile strengths and the assumed values for shear strength are provided in Table 10.2. The safety factor, F_s, for the screw is then calculated as follows:

$$F_s = \frac{T_{max}}{T_{s,max}} \tag{10.7}$$

In general, F_s should be greater than 1.5 to provide an acceptable strength for the screw. If the safety factor is less than 1.5, then the designer should consider a different base metal alloy with a higher shear strength.

Table 10.2 Tensile Strengths and Shear Strengths for Several Base Alloys Used to Manufacture Screws

Base Alloy for the Screw	Ultimate Tensile Strength, MPa	Shear Strength (S_m), MPa (assumed to be 60% of the ultimate tensile strength)
Stainless steel 304	241	145
Stainless steel 316	241	145
Hastealloy C-276	551	331
Nitralloy 135-M	586	351
AISI 4140 steel	690	414
AISI 4130 steel	758	455
Stainless steel 17-4 PH	1206	724
Stainless steel 15-6 PH	1206	724
CPM 10-V	2070	1240
D-2 Tool steel	2070	1240
H-13 Tool steel	2070	1240

For example, a 150 mm diameter extruder is configured with a 250 kW motor with a maximum speed of 1750 rpm. The gearbox reduction is 17.5:1, providing a maximum screw speed of 100 rpm. The screw is constructed from AISI 4140 steel and has a feed channel depth of 22 mm. Using Equation 10.5, the maximum torque capable at the screw is 23,900 Nm. From Equation 10.6 and Equation 10.7, the screw is capable of delivering 97,000 Nm before it fails, providing a safety factor of 4.1. Thus, the material selection is acceptable for this application. For most applications, this analysis is adequate. An improved calculation method, however, was presented previously by Myers [16] that takes into account the strength contribution from the flight.

10.4.6 Protection from High-Pressure Events

Single-screw extruders have the ability to generate high pressure at the discharge very quickly, especially for resins with a high viscosity. For most applications, extruder barrels are capable of operation at a maximum pressure of about 70 MPa while dies and other downstream equipment are rated at considerably lower pressures, 35 MPa or less. For small extrusion systems operating at rates less than 50 kg/h, the extruder and downstream equipment can be protected from high-pressure events using software controls and mechanical pressure relief devices such as rupture disks. For processes operating at rates higher than 50 kg/h, rupture disks are typically not capable of releasing the molten polymer at a high enough and safe rate. For these high-rate processes, the extruder and die must be protected from high-pressure events using pressure sensors and software controls.

10.4.6.1 Relief Rupture Disks for Processes Operating at Low Rates

Properly installed rupture disks are effective at relieving pressure at the discharge of extruders and gear pumps for process flow rates of 50 kg/h and less. At these low rates and for a high-pressure event, the rupture disk will break and allow molten polymer to flow to the floor at a rate high enough to mitigate the high pressure in the line. Rupture disks are manufactured in a wide range of pressure ratings, that is, the pressure at which the disk breaks and relieves the pressure. In general, rupture disks with the proper pressure rating should be installed on all processes operating at 50 kg/h and less. At higher process flow rates, rupture disks become less viable for protecting the process since rupture disk systems have limited capabilities of releasing molten resin. Typical locations for the rupture disks are transfer lines downstream from pressure-generating devices such as extruders and gear pumps. If a process continues to cause rupture disks to fail, then the process should be studied to determine the root cause of the problem.

Rupture disks must be installed properly and inspected periodically to maintain safe operations. Two common mistakes for the installation of these devices are shown in Figure 10.6. In Figure 10.6(a), the flow path in the stem downstream from the rupture disk is too long, and because it extends away from the transfer line it can be too cold for adequate polymer flow. If this section causes the polymer to freeze during a high-pressure event, then the device is rendered useless. Instead, the flow path for the device downstream from the frangible disk should be as short as possible, and it must be at the process temperature during operation. In a second scenario, the device in Figure 10.6(a) is activated and hot resin is vented through the cold and relatively long pipe stem. The device in this case worked as planned. The extruder would then be shut down to investigate the root cause of the overpressurization and to install a new rupture disk. If the rupture disk, however, is not replaced, the polymer in the cold discharge end of the device will freeze, forming a new plug. When the line is started back up, the machine will operate normally except that the machine is not protected from the next high-pressure event. A second common installation problem is shown in Figure 10.6(b). In this case, the device is not flush mounted to the wall of the transfer line, creating a stagnant region and a place for resin to degrade and plug the channel. If this plug is strong enough, the machine will not be protected from a high-pressure event. The device must be installed such that the tip of the relief device is flush with the wall of the transfer line.

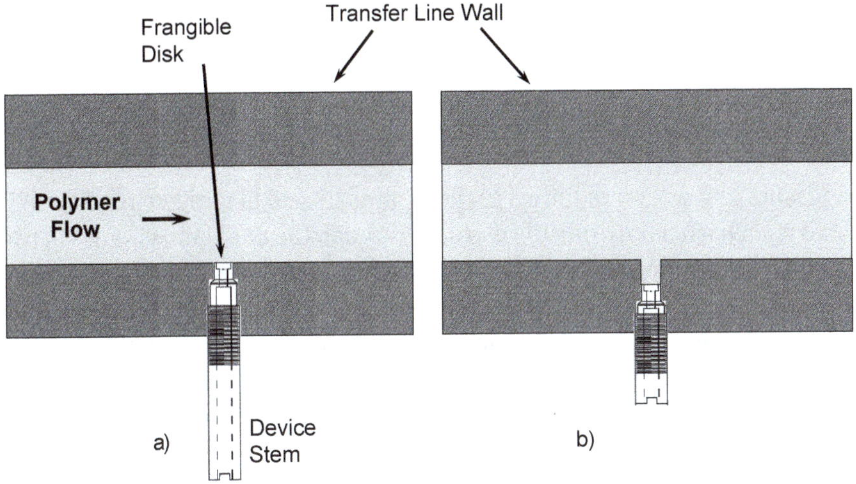

Figure 10.6 Schematic of two pressure-relief devices that are installed improperly: a) the stem on this device may be too long, causing the polymer to solidify in the stem during a rupture, and b) the rupture disk should be flush with the flow surface and not recessed as shown

10.4.6.2 Protection from Overpressurization for Processes Operating at High Rates

As discussed previously, processes operating at rates higher than about 50 kg/h should not be protected solely by a rupture disk device. Rupture disks are limited by the amount of molten resin that they can divert from the system to the floor. For example, if a line that is operating at 1000 kg/h overpressurizes the transfer line, diverting resin to the floor via a rupture disk at a rate of 50 kg/h is not going to relieve the pressure and protect the equipment. For processes operating at these high rates, the equipment should be equipped with redundant pressure sensors that are connected to the motor controls. If the pressure becomes too high in one or both of the sensors, the control logic shuts down the motor.

10.4.7 Gearbox Lubricating Oil

The lubricating oil in the gearbox should be changed periodically as directed by the manufacturer. Typically, the oil should be changed every 8000 hours of operation under normal conditions. When the oil is changed, it should be inspected for metal particles. Metal particles in the oil indicate that the gearbox is experiencing a high level of wear that could lead to a catastrophic failure, that is, failure of the internal gears, shaft bearings, thrust bearing, or the radial bearing. Identification of the composition of the metal fragments will often indicate the component in the gearbox that is failing.

10.4.8 Particle Seals and Viscoseals

Most plasticating screws require a particle (or dust) seal to mitigate the flow of fine or small resin particles exiting the extruder through the annular clearance between the shank of the screw and the barrel wall. Although they are commonly referred to as dust seals, they are actually fine threaded channels that push the solid particles back into the feed channel of the screw. If these particles are not pushed back into the feed channel, then a substantial amount of fine resin particles will exit the extruder through the clearance and accumulate on the floor in front of the gearbox. The seal is constructed by cutting a spiral channel into the shank in the same direction as the main flight, as shown in Figure 10.7. Typically, the annular clearance is about four times that between the main channel flight tips and the barrel wall. A typical particle seal channel for a 150 mm diameter screw will be single flighted with a lead length of 12 mm, a channel depth of 0.7 to 1.0 mm, and a channel width of 6 mm. Many other designs, however, have been implemented successfully. A seal that is designed properly should have very little leakage of particulate resin from the extruder.

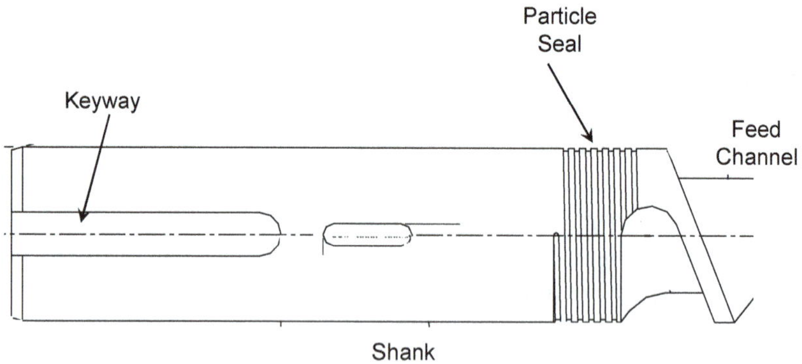

Figure 10.7 Schematic for a particle seal for a plasticating extruder

Viscoseals must be used on the shanks of all melt-fed extruders, especially those machines that operate flood-fed with the resin under pressure at the feed opening. These viscoseals are designed in essentially the same manner as the dust seal shown in Figure 10.7. In this case, the pressure in the feed causes the resin to flow out of the extruder through the annulus formed between the shank of the screw and the barrel wall and also back through the channel of the seal. The rotational motion of the screw, however, causes a high enough level of rotational flow to push the resin back into the feed channel. The required length of the seal depends on the balance between the ability of the resin to flow out and the rotational flow rate back into the feed channel. For example, large-diameter melt-fed extruders operate with a feed pressure near 0.1 MPa (see Section 15.3) and the length of the melt seal is typically less than one diameter on the shank. During operation, less than half of this length is typically occupied with molten resin. For a cooling extruder on a tandem foam sheet line, the pressure of the resin entering the cooling extruder can be as high as 30 MPa. The viscoseal design for these machines is more complicated because the pressure gradient for flow through the annulus is very high. Typically, the seal is very long and up to 3 diameters in length. Moreover, cooling is typically used at both the barrel side and screw side to increase the viscosity of the resin. A diagram showing screw cooling at the seal is shown in Figure 10.8. If the material temperature decreases, causing the viscosity to increase, the ability to flow resin out through the annulus will decrease, and the seal will become more effective.

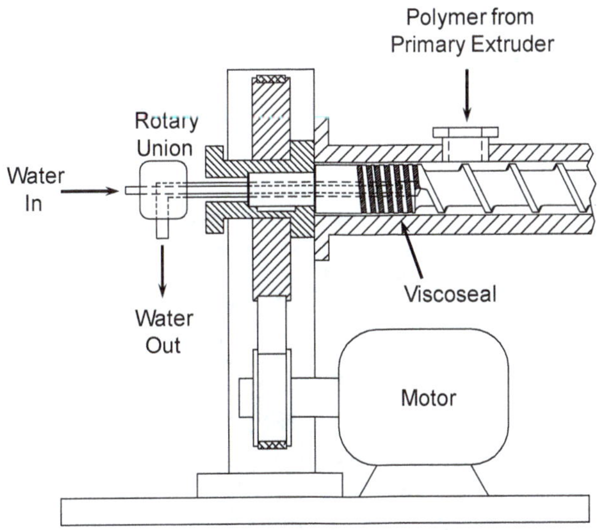

Figure 10.8 Schematic of screw cooling at a viscoseal for a tandem foam sheet line

10.4.9 Screw Cleaning

For most processes, the screw will need to be removed from the extruder, cleaned, and inspected for wear and damage. Almost all machine manufacturers remove the screws from the discharge end of the barrel so as not to disrupt the alignment of the barrel to the gearbox. For this procedure, the excess resin in the screw channels is forced out of the extruder by rotating the screw while the hopper is empty of resin. A purge compound could also be used to push out the excess resin. After the resin (or purge material) stops flowing, the downstream equipment is disconnected from the extruder, and the screw is pushed out of the barrel in small incremental lengths while the barrel is still at the processing temperature. With a segment of the screw protruding out of the barrel and still hot, the soft resin should be removed from the screw using brass brushes or brass gauze. This procedure is repeated until the entire screw is clean and out of the barrel. The screw should never be cleaned by burning the resin off of the screw using an acetylene torch as this can undo the annealing of the base steel, cause the screw to bend, damage the plating, and damage the hard facing on the flights [17].

Screws should always be cleaned using processes that do not alter the surface treatment. Typically, the screw should be cleaned using brass tools as the brass is not hard enough to damage the surface of the screw. A common practice, however, is to chip off solidified resin using steel tools that are harder than the screw. For example, the screw in Figure 10.9 was allowed to cool with resin residue. Next, the hardened resin was chipped off of the screw using a chisel and a hammer. This screw was constructed from a soft stainless steel billet, and the steel chisel was considerably harder. As shown in Figure 10.9, the cleaning process resulted in numerous dents in the surface of the screw. If the dents are deep enough, resin degradation is likely to occur.

Figure 10.9 Photograph of a large diameter screw with deep dents. The screw was constructed from stainless steel and the dents were caused by cleaning the screw with a steel chisel and hammer

10.5 Common Electrical and Sensor Problems

Electrical problems related to barrel heaters, band heaters, and instruments occur constantly in most production lines. In most cases, the problem is easily identified and fixed. For example, most barrel heaters are configured with a current meter on the control panel. If the controller is calling for heat on the zone and the meter shows that no current is flowing to the heater, then several obvious things could cause the problem. Potential problems include a barrel heater has burned out, the controller is malfunctioning, a solid state relay has failed, a wire is loose, or the meter is malfunctioning. All these root causes are relatively easy to determine by an experienced electrician. Malfunctioning heaters can be very troublesome due to the lack of process control, especially during equipment startups. Moreover, for new installations the extruder motor, gear pumps, and downstream devices must be wired for the proper rotation direction. The root cause of other electrical problems can be more difficult to diagnose. Several of these electrical problems will be discussed next.

10.5.1 Thermocouples

Thermocouples and wire leads require constant maintenance due to wire breaks and malfunctioning thermocouples. In most cases, the root cause of the problem is easy to determine. When a thermocouple loses continuity the instrument panel typically reports either an open circuit or a very negative temperature. A more difficult but also common problem occurs when the control thermocouple is not located close to the heating element on the control system. In this case, it is possible for the thermocouple to be influenced by another heated control zone while allowing its control zone to operate at temperatures that can be either very hot or very cold. Such cold conditions can cause the polymer to solidify in the transfer line while a hot condition can cause the resin to degrade. During the installation and startup of new equipment, it is important that the thermocouples are matched with the proper controller and control zone.

10.5.2 Pressure Sensors

Data from pressure sensors can be misleading if they are not calibrated properly, not sized properly for the application, are old and have a high level of strain hardening on the diaphragm, or are constructed from the wrong material. For most installations, an electronic calibration of the pressure sensor is acceptable.

However, for processes that require precise knowledge of the pressure in the system, off-line calibration should be considered. Off-line calibration is performed by inserting the sensor into a pressurized test cell, as shown in Figure 10.10. The pressure is produced using a hydraulic system and a weight pushing down on a plunger. The transducer should be sized such that the maximum pressure that the line operates at is essentially near the maximum measurement pressure of the system. For example, if the pressure at the entry to a die never exceeds 8 MPa, then a pressure transducer with a maximum pressure capability of 10 MPa should be installed. If instead a transducer with a maximum pressure rating of 70 MPa is installed, then a considerable level of measurement error is possible since the device is operating in the lower 10% of its range. Strain hardening of the metal on the transducer diaphragm is typically not a severe problem, but it can reduce the response time of the device and its accuracy in the lower range. The material of construction for the sensor diaphragm must be compatible with the resin being extruded. That is, the diaphragm must be chemically inert to the resin and its degradation products, and it must be durable enough to provide a useful service life in use with abrasive feedstocks.

Figure 10.10 Photograph of a pressurized testing cell for calibrating pressure sensors. This unit can pressurize the cell to 8 MPa

10.5.3 Electronic Filters and Noise

Electronic filters are used on most data measurement and storage systems. The filter should be set to mitigate process noise while not decaying the actual process signal. For example, if a filter is set too high for a pressure transducer located in a transfer line, it may be incapable of showing a flow surging problem due to poor solids conveying, as shown in Figure 10.11. In cases like this one, the filter should be adjusted to the appropriate level so that any pressure variation can be observed by the operator.

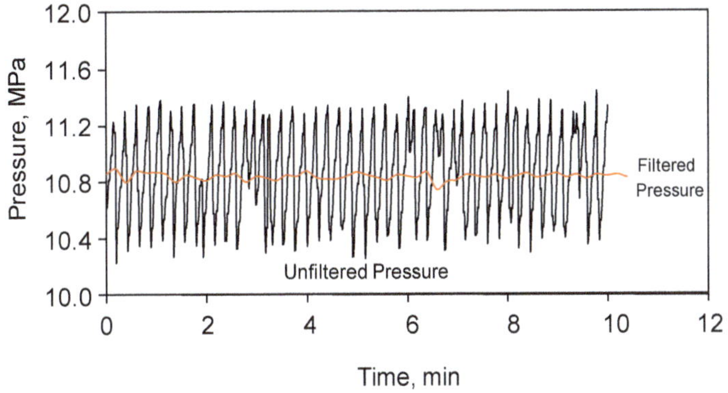

Figure 10.11 Filtered and unfiltered discharge pressure data. The unfiltered data were collected using a portable data acquisition system while the filtered data was the pressure that was observed by plant personnel. The plant personnel believed that they had a very stable process even though this extruder was flow surging. The transducer was positioned in the transfer line and the extruder was operating at a screw speed of 60 rpm

Low-voltage signal wires are very prone to noise induced from high-voltage alternating current (AC) power wires and motors. The source of the noise can sometimes be determined by sequentially turning off motors or other equipment and then watching the response of the device. Proper wire shielding, use of devices that transmit a current rather than a voltage, and placement of the wires mitigate this type of noise. As an example, an unshielded thermocouple was positioned in a plastic sheet, and the sheet with the thermocouple was passed through a lamination process. At the start of the process, the thermocouple reported an accurate measurement of the temperature with essentially no noise in the signal, as shown in Figure 10.12. At about 100 s into the process, the thermocouple wires were influenced by the electromagnetic fields produced by the motors and the electrical heaters in the process. The thermocouple response during the period with the noise was essentially useless.

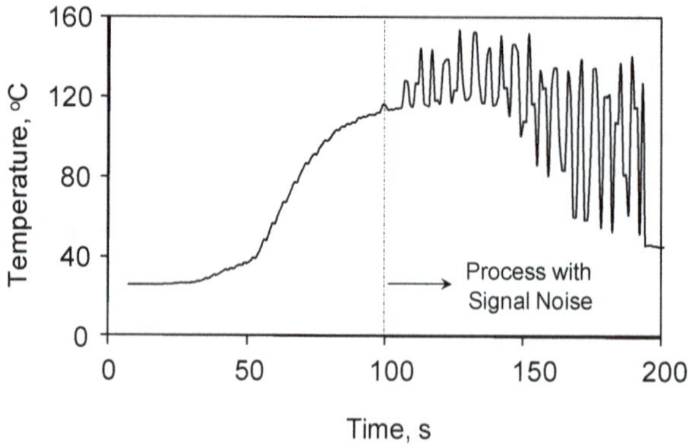

Figure 10.12 Temperature plot of a thermocouple sensor moving downstream through a process line. The process noise occurred because the unshielded sensor wires were in the proximity of a motor and heaters

10.6 Motors and Drive Systems

The availability of power and torque to the shank of the screw is extremely important to the success of an extrusion process, especially if modifications to the line are planned to increase the rate. That is, the proper level of power must be available at the proper screw speed. Thus, the motor size and speed, belt sheaving if used, and gearbox reduction must be specified to provide enough power to the screw to plasticate and pump the resin at a high rate.

The electrical current used by the motor is an important indicator of the performance of the extruder. The current as measured in amperage is directly proportional to the torque applied to the motor shaft and also to the screw. The motor current can be used to estimate the power and specific energy that are dissipated by the screw. The variation in the motor current with time is also an indicator of the stability of the extruder. Numerous examples of the motor current stability and process stabilities are presented in Chapter 12.

Small-diameter extruders with motors less than about 200 kW often use a belt drive and sheave system between the motor and gearbox. For these machines, a reduction in speed and an increase in torque are obtained from both the sheave speed reduction ratio and the gearbox ratio, as described by Equation 10.8. For example, if the sheave ratio is 1.5, the gearbox ratio is 12, and the motor has a maximum speed of 1750 rpm, the maximum speed of the screw will be 97 rpm while the torque will increase from the motor shaft to the screw by a factor of 18, as described by Equation 10.9. The gearbox and sheave reduction are selected to provide an acceptable level of torque and speed to the screw. If the scope of the extrusion line changes, such as when higher rates are required or different resins need to be extruded, then the speed reduction (or timing) of the machine may be altered by changing the sheave pulleys. For example, a small-diameter adhesive extruder was limiting the rate of a production line because it was operating at its maximum screw speed of 95 rpm. The extruder motor was currently operating at about 50% of the full current load. The sheave ratio was changed such that the screw speed would operate at a maximum speed of 120 rpm. Since the screw speed was increased for this operation, the maximum torque available at the screw decreased. Thus, the motor current required to operate the screw will increase for the new configuration. Since the motor was originally operating at 50% of the full motor load, the motor will have enough torque available to run the new configuration. Whenever timing changes are made to an extruder by changing the sheave pulleys, it is recommended that the equipment manufacturer be consulted before the change is made. Improper sheave pulleys have the potential to exceed the torque specifications for the gearbox.

The rotation speed of the screw to that of the motor is simply:

$$N = \frac{N_m}{R_{bs} R_g} \tag{10.8}$$

where N_m and N are the speeds for the motor and screw in revolutions/s, respectively, R_{bs} is the belt sheave reduction, and R_g is the speed reduction for the gearbox. The belt sheave reduction is the ratio of the pulley diameter attached to the input side of the gearbox to the pulley diameter attached to the motor shaft. If a belt sheave system is not used, then R_{bs} is set to one. As the maximum speed of the screw is reduced through the reductions, the torque is increased as follows:

$$T_{sh} = R_{bs} R_g T_{mot} \tag{10.9}$$

where T_{sh} and T_{mot} are the torques for the screw and motor, respectively. The relation between power, torque, and speed at the screw is as follows:

$$P_s = 2\pi N \tau_{sh} \tag{10.10}$$

where P_s is the power at the screw in watts.

Gearboxes are typically manufactured with ratings for torque (or power) and speed for the input shaft that attaches to the motor. Along with these ratings is a safety factor that indicates the relative durability of the gear system in the gearbox at the rated torque and speed. The safety factor is defined as the ratio of the maximum input power for the gearbox to the input power. For example, a gearbox with a rating at the input shaft of 100 kW, a speed of 1750 rpm, and a safety factor of 1.5 sets the specification of the gearbox. The safety factor of 1.5 indicates that the gearbox is capable of operation at a maximum motor size of 67 kW at a speed of 1750 rpm.

Operating with a larger motor, however, may shorten the lifespan of the gear drive system in the box. When specifying and purchasing new equipment, it is recommended that the safety factor be at least 1.5.

Numerous styles of alternating current (AC) and direct current (DC) motors are commonly used with extrusion drivetrains. The actual type of motor will depend on the size, application, and on the gearbox. Both AC and DC motors provide the maximum rated torque at the maximum motor current and at speeds less than or equal to the maximum speed. Thus at full motor current, the torque is constant up to the maximum rated speed, as shown in Figure 10.13. For this speed range, the voltage on the field of a DC motor is maintained at the maximum rated level. The speed of the motor can be increased by about 20% higher than the base speed if the field is operated with a reduced voltage (or in a field-weakened state). For example, a standard DC motor that operates at a base speed of 1750 rpm with a full voltage on the field can often be operated at a speed of up to 2100 rpm by decreasing the voltage on the field. At these field-weakened speeds, the motor provides the maximum rated power at full motor current. Since the power level is unchanged in this region, the torque at the shaft must decrease as the motor speed increases beyond the maximum standard speed, as shown in Figure 10.13 and Equation 10.10. AC motors have essentially the same torque and power response as shown in Figure 10.13. Overspeeding of an AC motor, however, is performed by altering the frequency and voltage to the motor using an AC drive system. It is recommended that the manufacture of the equipment be consulted when overspeeding motors.

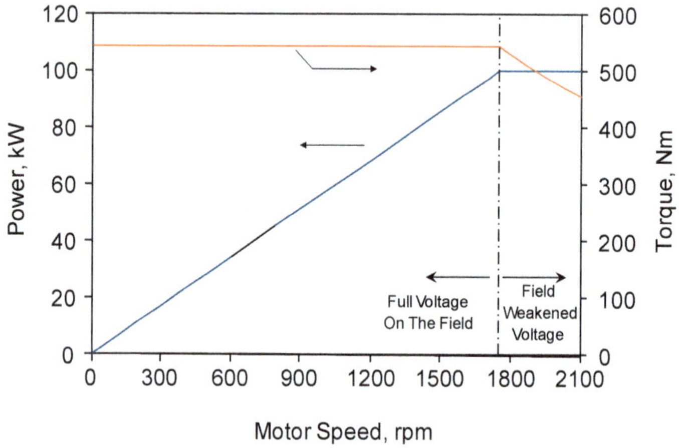

Figure 10.13 Power and torque as a function of motor speed for a 100 kW DC motor with a maximum speed of 1750 rpm with a full voltage on the field and at the maximum motor current. The motor can be operated at screw speeds up to 2100 rpm by weakening the field voltage

Operating by overspeeding the motor is a common practice to increase the maximum rate of an extrusion line. For example, if an extrusion line is operating at a rate with a maximum screw speed and base motor speed, overspeeding the motor will allow the screw to rotate at a faster speed and process at a higher rate. The success of this approach, however, is dependent on several conditions. First, there must be additional torque available from the motor based on the operation at the base motor speed. Also, the screw design used must be able to provide a high-quality discharge at the new rate, and the downstream equipment must be able to operate at the higher rate. A complete description of motors and their use can be found elsewhere [18, 19].

10.6.1 Motor Efficiencies and Power Factors

Most extruders manufactured today are built using AC vector motor systems instead of the older DC motor systems. AC motors operate at about 5% higher efficiencies, require less maintenance, and are lower in cost than a DC motor. Typical efficiencies for AC and DC motors rated at 83 kW are provided in Table 10.3. It is estimated that about a third of the electrical power consumption in a manufacturing plant using extruders is by the motor on the extruder. Thus, a 5% improvement in efficiency for an AC motor over a DC motor can reduce the energy consumption of the plant by a significant amount [20]. The difference in efficiencies between the two motors can be even greater if a new AC vector motor is compared to an older DC motor.

Table 10.3 AC and DC Motor Efficiencies for Motors Rated at 83 kW at Full Speed (Nominal 1750 rpm) and as a Function of Load [20]

Motor Type	Motor Efficiency, %		
	50% Load	75% Load	100% Load
DC	89.8	90.3	89.7
AC	95.0	95.4	95.0

The power factor of an AC circuit is the ratio of the real power flowing to the system to the apparent power. The power factor ranges from 0 to 1. If the current and voltage are in phase, then the power factor is exactly 1. However, if the current and voltage are slightly out of phase, the real power will be less than the apparent power, and more current will be flowing through the circuit to provide the same level of power. Utility companies penalize customers that consume electrical power at a low power factor. Thus, designing an extrusion system that operates at a high power factor will reduce the energy costs for the plant.

AC vector motors typically operate with a power factor in the range of 0.90 to 0.98 [20]. The power factor for a DC drive system can be considerably lower and as low as 0.5. Replacing an older DC motor with an AC vector motor can be economically attractive due to an increase in both the efficiency and the power factor.

10.7 Typical Screw Channel Dimensions

There are practical limits to the dimensions for the channels relative to the diameter of the screw. For example, the depth of the metering section is typically between 3% and 7% of the diameter but can be as deep as 12% of the diameter for very viscous materials and special applications. Metering sections with shallow channels will produce an extrudate with a higher discharge temperature than that for a deeper channel. The depth of the feed channel is in the range of 1.5 to up to 6 times the depth of the metering section, depending on the application and the bulk density of the feedstock. The lead length is typically equal to the diameter, but it has a range between 0.8 and 1.3 times the diameter. For single-flighted screws, the flight width perpendicular to the flight edge is about 10% of the diameter. If more than one flight is present, then the width of the flight will generally be smaller. The clearance between the tip of the flight and the barrel wall is typically 0.001 times the diameter. Thus for a 150 mm diameter screw, the flight clearance on a side is about 0.15 mm and the outside diameter of the screw will be 149.70 mm. Variations from this guideline do exist and depend on the machine manufacturer and application.

The section lengths and types also depend on the application, materials, and the downstream equipment. For conventional screws, the feed section is between 5 and 7 diameters in length, the transition section length is 5 to 9 diameters, and the metering section is 4 to 9 diameters. If a Maddock-style mixer is used, its section length is generally about 2 diameters. These dimensions are guidelines since many screws are designed with different values. Screw channel dimensions can be measured or obtained from the screw drawing.

10.8 Common Calculations

Several calculations should be made at the start of the troubleshooting process. These calculations include the rotational flow and pressure flow rates for the metering section, the energy inputted by the motor to the resin, the energy exiting with the extrudate, the energy flux through the barrel wall, and several indices that define the geometry of the screw. The calculations can often guide the troubleshooter to specific root causes. By far the most important calculations are the rotational flow and pressure flow rates for the metering section. Calculations for these rates were provided earlier in Section 1.4.

10.8.1 Energy Dissipated by the Screw

An energy balance on the extruder is often helpful in diagnosing extrusion problems. For example, the energy inputted to the resin from the motor via the screw is typically between 450 J/g for plasticating some amorphous resins up to about 900 J/g for some semicrystalline resins. If the level of energy input has changed or is considerably different than what is expected, then the troubleshooter can start looking at sections of the machine that dissipate energy. The power and specific energy input into the polymer from the extruder screw are estimated using Equation 10.11, Equation 10.12, and Equation 10.13 below:

$$P_m = P_{\max}\left(\frac{A}{A_{\max}}\right)\left(\frac{N}{N_{\max}}\right) \quad \text{speed} \leq \text{base motor speed} \tag{10.11}$$

$$P_m = P_{\max}\left(\frac{A}{A_{\max}}\right) \quad \text{overspeeding motor} \tag{10.12}$$

$$E = \frac{(3600 \cdot \text{J} \cdot \text{kg})}{(\text{kW} \cdot \text{g} \cdot \text{h})}\frac{P_m}{Q} \tag{10.13}$$

where P_m is the power that is dissipated in kW, P_{max} is the nameplate power (kW) for the motor, A is the motor current observed during the extrusion, A_{max} is the nameplate motor current at full load, N is the screw speed (rpm) during extrusion, and N_{max} is the maximum screw speed (rpm) that the extruder is capable of running at the base motor speed. N_{max} is determined by dividing the base motor speed by the gearbox and belt sheave ratios as defined in Equation 10.8. After the power is computed, the specific energy input to the resin from the screw, E, in J/g is calculated using Equation 10.13 and the extrusion rate, Q, in kg/h. The ratio A/A_{max} is equal to the drive load. The calculation neglects the inefficiencies for converting electrical power to mechanical power and the power factor. For example, drives and motor systems operating at low speeds will typically have power factors less than about 0.8. These systems will require higher currents (A) than those with higher power factors for the same mechanical energy inputted to the resin. Power factors improve as the motor is operated near its nameplate rating [18–20]. Motor efficiencies and power factors were discussed previously in Section 10.6.1.

The specific energy from Equation 10.13 is reported in J/g for convenience in analyzing the process. If a change in a process causes the specific energy to increase by 50 J/g, the troubleshooter can translate this to an adiabatic temperature increase of 20 °C because many unfilled resins have heat capacities near 2.5 J/(g °C). This temperature change calculation, however, is an approximation because extruders typically do not operate adiabatically.

10.8.2 Screw Geometry Indices

Several geometric indices should be calculated at the start of the troubleshooting process. These indices will help determine if the screw is suitable for the resin and the application. The major indices include the compression rate and compression ratio of the melting section and the pump ratio for a two-stage screw. The compression ratio is defined as the ratio of the cross-sectional area perpendicular to the flight tip for the feed section to that of the metering section. The compression ratio for a screw with different lead lengths in the feed and metering sections is given in Equation 10.14 and for the simple case where the lead length is constant for the sections in Equation 10.15.

$$C = \frac{\left(L_f \cos\theta_{b,f} - e_f\right)H_f}{\left(L_m \cos\theta_{b,m} - e_m\right)H} \quad \text{for different lead lengths in the feed and metering sections} \tag{10.14}$$

$$C = \frac{H_f}{H} \quad \text{for constant lead lengths in the feed and metering sections} \tag{10.15}$$

where C is the compression ratio, H_f is the channel depth of the feed section, H is the depth of the metering channel, $\theta_{b,f}$ and $\theta_{b,m}$ are the helix angles at the barrel wall for the feed and metering sections respectively, e_f and e_m are the widths of the flight tip perpendicular to the flight edge for the feed and metering sections, and L_f and L_m are the lead lengths for the feed and metering sections, respectively.

The compression rate for the transition section of the screw describes the rate of the channel depth change as the resin is transported through the transition section. The compression rate is calculated using Equation 10.16.

$$R = \frac{(H_f - H)\sin\theta_b}{ML} \tag{10.16}$$

where R is the compression rate in the helical direction for the transition section, M is the number of turns in the transition section, θ_b is the helix angle at the barrel wall, and L is the lead length in the section. For a barrier screw, the compression rate is calculated based on the solids channel.

The pump ratio, P_r, describes the ability of a two-stage screw to pump and generate pressure at the discharge of the screw. The ratio is calculated by dividing the rotational flow rate of the second-stage metering section, $Q_{d,m2}$, by the rotational flow rate of the first-stage metering section, $Q_{d,m1}$, as shown in Equation 10.17. For a screw with a constant lead length in both sections, the pump ratio is simply the ratio of the channel depths in the second-stage (H_{m2}) and the first-stage (H_{m1}) sections.

$$P_r = \frac{Q_{d,m2}}{Q_{d,m1}} \quad \text{for screws with different lead lengths in th metering sections} \quad (10.17)$$

$$P_r = \frac{H_{m2}}{H_{m1}} \quad \text{for screws with the same lead lengths in the metering sections} \quad (10.18)$$

For most two-stage vented processes, the pump ratio is typically between 1.3 and 2.0. If the pump ratio is too low, then vent flow (or vent flooding) will occur because the second-stage metering section cannot pump the resin faster than the first stage. If the pump ratio is set too high, the second-stage metering channel can become too deep to generate the pressure required by the downstream process. In this latter case, vent flow will also occur. First- and second-stage metering section lengths also affect the pumping ability of the machines. Optimizing the pump ratio can also provide stability to the process [21].

10.9 Barrel Temperature Optimization

Optimization of the barrel temperature zones can improve the performance of an extrusion process. The barrel temperature settings in the feed zone and melting section control the solids-conveying rate and affect the melting rate, respectively. The temperature zones in the metering section can affect the temperature of the extrudate. In many cases, the goal is to improve process stability, maximize rate, and decrease the discharge temperature. Barrel temperature control for the first-stage metering and second-stage metering sections can mitigate vent flow in special circumstances. This section provides several guidelines for optimizing the temperature profile of the barrel.

Several procedures exist for the optimization of barrel zone temperatures, and they all start by optimizing the temperature in the solids-conveying zone [10, 22]. The starting temperature for the solids-conveying zone can either be the recommended feed zone temperature from the resin manufacturer, the current operating temperature, or the melting (or devitrification) temperature plus about 25 °C. The barrel zones over the metering section should be set to a temperature near the expected temperature of the extrudate. The barrel zones over the melting section should be set at an intermediate temperature. Because all extruders are unique with different diameters, lengths, heater positions, and thermocouple positions, the optimal barrel zone temperatures for each extruder will be slightly different for a specific resin. Next the solids-conveying zone temperature is increased (or decreased) in 10 °C increments until a maximum rate is obtained for a specific screw speed. Once the maximum rate at the optimal solids-conveying barrel temperature is determined, the discharge temperature can be adjusted by increasing or decreasing the temperature of the barrel zones in the metering section of the screw. The melting rate can be increased by increasing the barrel temperature in the transition section. This procedure is simple and can be quickly performed on commercial extruders. A more elaborate method uses a designed experiment where numerous factors are selected at two or more levels as outlined by Himmelblau [23] and described by Wagner and Perdikoulias [24].

To illustrate the barrel optimization process [22], an extrusion process was optimized first for rate and then for discharge temperature. Each of the experiments was conducted at a screw speed of 50 rpm. The initial extruder barrel temperatures were selected based upon common practice for an HDPE resin with an MFR of 5.9 dg/min (190 °C, 2.16 kg). The extruder used in this study was a fully instrumented 63.5 mm diameter, 21 *L/D* single-screw machine. The extruder was equipped with three barrel temperature and cooling zones. The first 2.5 diameters of the screw were inside a water-cooled feed casing. Twelve pressure transducers were positioned in the barrel wall to measure the axial pressure profile of the process. The resin flowed through a short transfer line and then to a die system. The screw was a conventional, single-flighted screw with a square-pitch lead length. The starting barrel temperatures were 170, 195, and 220 °C for Zones 1 through 3, respectively. To demonstrate the effect of temperature on solids conveying, the Zone 1 temperature was increased in 10 °C increments and the Zone 3 temperature was held constant at 220 °C. For each condition, Zone 2 was set at the midpoint value between the Zone 1 and Zone 3 temperatures. This process was repeated across a range of feed zone temperatures until a maximum extrusion rate was identified. At each new condition, process data were collected, including discharge temperature, rate, motor current, and pressures along the axial length of the barrel. The discharge temperature was measured by inserting a handheld thermocouple into the extrudate.

The effects of the Zone 1 temperature on the rate, motor current, and discharge temperature are shown in Figure 10.14 and Figure 10.15. In Figure 10.14, the rate was initially at 26.1 kg/h at a screw speed of 50 rpm for a Zone 1 temperature of 170 °C. As the Zone 1 temperature was increased, the rate (and specific rate) increased to a maximum of 32 kg/h at a Zone 1 temperature of about 230 °C. This represents a rate increase of about 21%. As the Zone 1 temperature increased, the motor current also increased. The increase in motor current was a result of the higher rates and increased discharge pressures. Further optimization is possible by increasing the temperature of Zone 2. For many operations, increased rate could be simply obtained by increasing the speed of the screw rather than by barrel temperature optimization. Operating at the maximum specific rate will typically allow the process to operate at its highest stability.

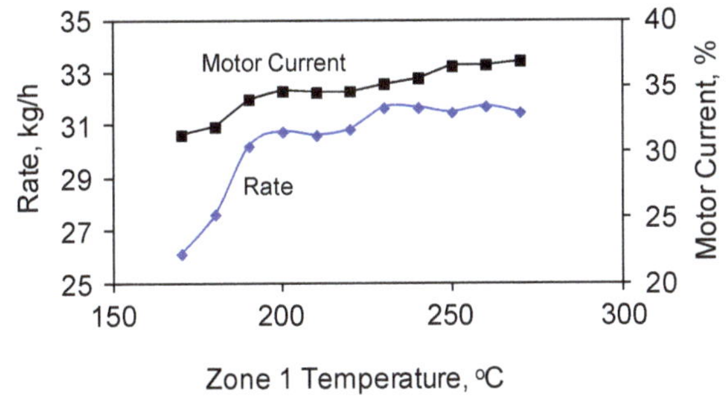

Figure 10.14 Extrusion rate and motor current for an HDPE resin at a screw speed of 50 rpm. The highest rate was achieved at a Zone 1 temperature of about 230 °C

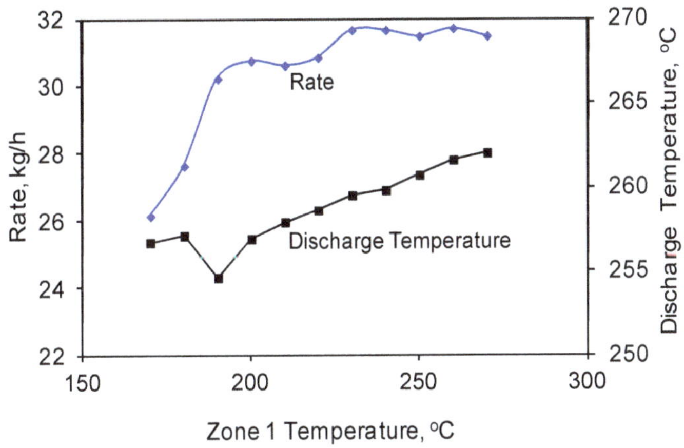

Figure 10.15 Extrusion rate and discharge temperature for an HDPE resin at a screw speed of 50 rpm. Discharge temperature increased when the Zone 1 temperature was increased

The axial pressure profiles for the HDPE resin at Zone 1 temperatures of 170 and 230 °C are shown in Figure 10.16. The axial pressures increased when the Zone 1 temperature increased from the initial temperature to the optimized temperature. The shift to higher axial pressures and higher discharge pressure with an increase in the Zone 1 barrel temperature was a result of the increased forwarding forces in the solids-conveying and melting sections and the higher rates. From this figure, it is clear that the change in the Zones 1 and 2 barrel temperatures has caused a significant change in the process. Although not shown here, the process stability as measured by pressure fluctuation at the discharge was improved for the optimized process.

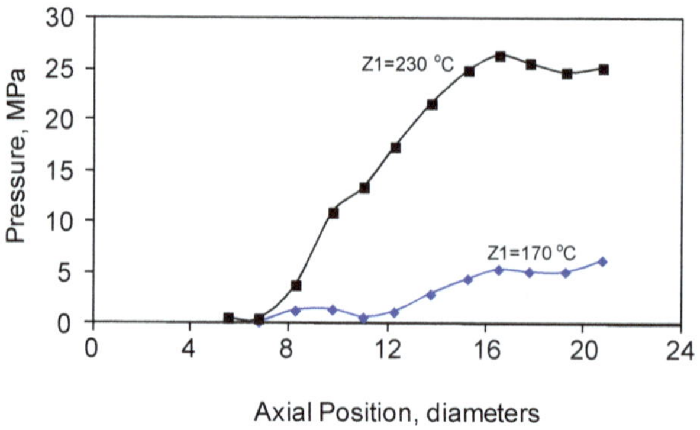

Figure 10.16 Axial pressure profiles for the HDPE resin at Zone 1 (Z1) temperatures of 170 and 230 °C

The discharge temperature for the HDPE resin increased with increasing Zone 1 temperature, as shown in Figure 10.15. For example, the discharge temperature increased from 257 to 260 °C when the Zone 1 temperature was increased from 170 to 230 °C. Although this discharge temperature increase is relatively small, it might impact the operation of the downstream equipment and processes. To demonstrate that discharge temperature can be adjusted at the higher rates, the Zone 3 temperature was decreased from 220 to 190 °C. As shown in Figure 10.17, the discharge temperature decreased from about 260 to 235 °C. As the Zone 3 temperature was decreased the rate also decreased from 31.7 kg/h to 30.2 kg/h. The ability to change the discharge temperature with the metering-zone temperatures is highly dependent on the rate and the diameter of the extruder. The extrudate temperature is much more difficult to change via barrel zone control for large-diameter extruders due to heat transfer limitations.

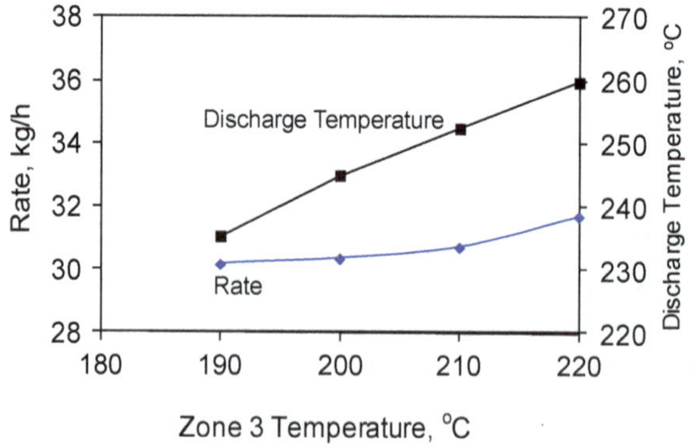

Figure 10.17 Extrusion rate and discharge temperature for an HDPE resin at a screw speed of 50 rpm. Discharge temperature decreased when the Zone 3 temperature was decreased

10.10 Screw Temperature Profile

The processes in all sections of the extruder depend on the temperature of the screw. As discussed in Chapter 5, the solids-conveying section depends on a balance between the forwarding forces at the barrel wall and the pushing flight of the screw and the retarding forces at the root of the screw. These forces are directly proportional to the coefficients of dynamic friction (or shear stress at the interface). Since the coefficients of friction are a function of temperature, optimal solids conveying depends on the temperature of the screw and barrel. If the temperature of the screw in the feed section is outside of the optimal range, then flow surging can occur. Numerical results have indicated that the melting rates in the transition section are dependent on the temperature of the screw [25, 26]. For the metering section, the screw root temperature is a boundary condition for the process. The temperature of the screw, however, is typically not controlled and is not known.

The first melting and metering channel models assumed that the screw was adiabatic [27, 28]. That is, the models assumed that energy is not transported from the polymer in the channels to the root of the screw. This assumption was made to simplify the mathematical models even though thermal gradients were known to exist in the axial length of the screw. For example, the temperature of the tip of the screw is typically near the discharge temperature of the resin while the shank of the screw is relatively cool and is sometimes at temperatures less than 50 °C. Thus, a thermal gradient exists along the axial length of the screw. This gradient along with the relatively high thermal conductivity of the metal alloy causes a substantial amount of energy to conduct down the axial length from the tip of the screw to the shank. Clearly, energy must transport from the polymer in the channel to the root of the screw.

The temperature of the screw was measured by several investigators [29-32]. The measurements were performed by mounting thermocouples in an axial hole bored in the center of the screw or by protruding the thermocouples into the melt flow. The sensor signals were then transmitted to a chart recorder using an electrical rotary union. The technology available at the time of these measurements limited the number of sensors in the screw and the quality of the data.

Transport of energy in the screws was modeled previously for single-screw extruders [30-32] and for twin-screw extruders [33]. In order to predict the axial screw temperature in a single-screw extruder, heat conduction along the screw has to be modeled. The model developed by Derezinski [32] included heat conduction from the barrel through the screw flights to the screw surface, heat conduction from the polymer to the screw root, and heat conduction in the axial direction. The model showed that the screw does not behave adiabatically and that the steady-state heat conduction in the screw depends greatly on the size of the extruder.

An empirical model that describes the axial temperature profile was developed by Cox and Fenner [30] based on the research performed by Edmondson [31]. The model is as follows:

$$T_s = T_b\left(1 - e^{\lambda l}\right) + T_{so}e^{\lambda l} \tag{10.19}$$

$$\lambda = \frac{\ln\left[\dfrac{\left(T_s(l_2) - T_b\right)}{\left(T_{so} - T_b\right)}\right]}{l_2} \tag{10.20}$$

where T_s is the axial screw temperature, l is the independent variable for the axial location on the screw, T_b is the barrel temperature in the metering section, T_{so} is the material feedstock temperature, λ is a fitting parameter defined in Equation 10.20, and l_2 is the axial location where the melt film first forms on the screw. $T_s(l_2)$ is equivalent to the melting point temperature of the polymer. With this empirical correlation, the value for l_2 is unknown, and hence it would need to be determined experimentally.

In a recent study [34], a HIPS resin (2.8 dg/min at 200 °C and 5 kg) was extruded using the fully instrumented 63.5 mm diameter, 21 *L/D* single-screw machine discussed in Section 10.10. The screw used in this study was a conventional, single-flighted screw with a square-pitch lead length. The screw was used in numerous other studies [25, 35, 36], and it was designed with three sections: (1) a 6 diameter long feed channel with a constant depth of 8.89 mm, (2) an 8 diameter long transition section, and (3) a 7 diameter long metering section with a constant depth of 3.18 mm. The screw had a detachable screw tip and was bored along the axis from the tip to the shank with a 15.9 mm hole. Seventeen resistive temperature devices (RTDs) were installed in the screw by drilling small holes in the center of the channel perpendicular to the axis. The RTDs were positioned in small metal plugs, and the plugs were then press fitted into the small hole, as shown by Figure 10.18. The RTDs were attached to the metal plugs using both a ceramic compound and a high-temperature epoxy cement. The tips of the RTDs were positioned about 1 mm from the surface of the screw root. The RTDs were nearly evenly spaced down the axial length of the screw. The axial locations of the RTDs are shown in Table 10.4. The RTD wires passed through the center bore of the screw and out through an electrical rotary union, as shown in Figure 10.19. The screw temperatures from the RTDs along with the barrel temperatures were measured and stored using a data acquisition system.

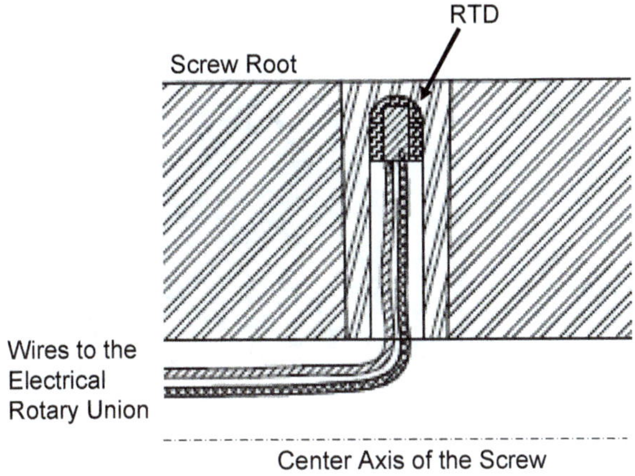

Figure 10.18 Schematic of an RTD positioned near the surface of the root of the screw

Table 10.4 RTD Labels and Axial Positions

RTD Label	Axial Position, cm	Axial Position, diameters
T1	1.6	0.3
T2	9.5	1.5
T3	17.5	2.8
T4	25.4	4.0
T5	33.3	5.3
T6	41.0	6.5
T7	49.2	7.8
T8	57.2	9.0
T9	65.2	10.3
T10	73.0	11.5
T11	81.0	12.8
T12	88.9	14.0
T13	96.8	15.3
T14	104.8	16.5
T15	112.7	17.8
T16	120.7	19.0
T17	128.6	20.2

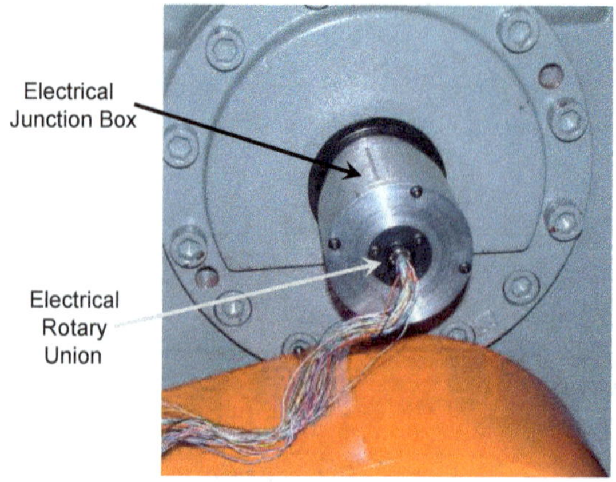

Figure 10.19 Photograph of the electrical junction box and the electrical rotary union attached to the shank end of the screw

The response of the RTDs and the temperature of the screw were tested with the screw not rotating. For this experiment, the temperatures were first measured with the extruder at ambient conditions. Next, the barrel temperature set points were increased to 200, 220, and 240 °C for Zones 1, 2, and 3, respectively. The downstream die system was heated at the same time as the barrel and at a set point temperature equal to Zone 3 (240 °C). The temperature profile of the screw as a function of axial length is shown in Figure 10.20 for select heating times. For heating times less than about 40 minutes, the temperature of the screw tip was less than that for the segments upstream. This was caused by the relatively slow heating of the downstream die system and the conduction of energy from the end of the barrel to the die. The barrel zones reached the set point temperatures in about 38 minutes, as shown in Table 10.5. The time required to achieve a steady-state temperature for the screw, however, was considerably longer at 105 minutes. Thus, the time required for the screw to reach a steady state was about two to three times longer than that for the barrel.

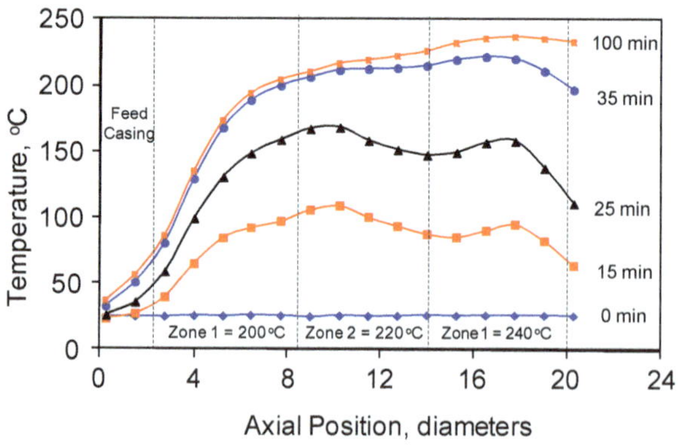

Figure 10.20 Axial screw temperature profiles during a heating cycle. The screw was not rotating

Table 10.5 Time to Steady State During the Heating Cycle

	Probe	Time to Steady State, minutes	Steady-State Temperature, °C
Zone 1	Barrel	31	200
	T4	99	135
	T5	73	173
	T6	71	194
Zone 2	Barrel	38	220
	T7	67	204
	T8	66	210
	T9	65	217
	T10	73	219
	T11	83	222
Zone 3	Barrel	38	240
	T12	80	226
	T13	82	232
	T14	77	235
	T15	76	235
	T16	92	235
	T17	105	234

The HIPS resin was extruded at screw speeds of 30, 60, and 90 rpm at barrel temperatures of 200, 220, and 240 °C for Zones 1, 2, and 3, respectively. The screw temperatures in Zone 3 as a function of time at the screw speeds are shown in Figure 10.21. Because the RTDs were positioned within 1 mm of the screw root surface, they were influenced by the temperature of the material flowing in the channels. Prior to the experiment, the screw was allowed to come to a steady-state temperature without rotation. Next, the screw speed was slowly increased to a speed of 30 rpm. The time for the screw to reach a steady state after changing the screw speed to 30 rpm was found to be about 10 minutes. The temperature of the T12 and T13 locations decreased with the introduction of the resin. This was caused by the flow of cooler solid resin that conducted energy out from the screw and into the solids. At sensor positions downstream from T13, the screw temperature increased at a screw speed of 30 rpm, indicating that the resin was mostly molten in these locations. These data suggest that the solid bed extended to somewhere between 15.3 and 16.5 diameters, that is, between T13 and T14. When the screw speed was increased to 60 rpm, the T12 and T13 sensors decreased in temperature, the T14 sensor was essentially constant, and the T15, T16, and T17 sensor temperatures increased. These data are consistent with solids moving further downstream with the increase in screw speed. For this case, the end of the solids bed was likely just upstream of the T14 sensor. If the solid bed were beyond this location, the T14 temperature would have decreased. Likewise, if the solid bed ended further upstream of the T14 sensor, the temperature would have increased. When the screw speed was increased to 90 rpm, the T12, T13, and T14 temperatures decreased while the T15, T16, and T17 temperatures increased. As before, the solids bed was conveyed further downstream with the increase in screw speed. At a screw speed of 90 rpm, the solid bed likely ended between the T14 and T15 sensor positions, that is, between 16.5 and 17.8 diameters. These RTDs were influenced by the cooler solid material because they were positioned within 1 mm of the screw root surface.

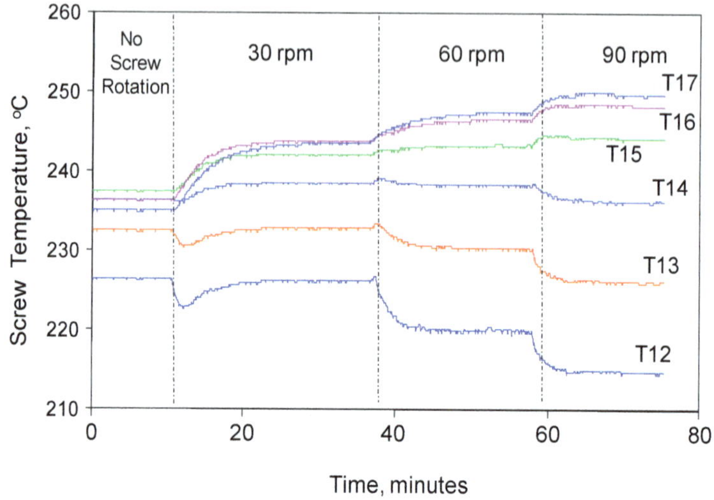

Figure 10.21 Zone 3 screw temperature response to changes in screw speed. The barrel zones were 200, 220, and 240 °C for Zones 1, 2, and 3, respectively

The screw temperature data in Figure 10.21 suggest that the discharge temperature increased as the screw speed increased, as indicated by T17. The extrusion performance for the data shown in Figure 10.21 is summarized in Table 10.6. As shown by this table, the discharge temperature measured using a handheld device increased as the screw speed was increased from 30 to 60 rpm. At a screw speed of 90 rpm, the discharge temperature decreased slightly. This decrease was likely caused by solid polymer fragments moving closer to the tip of the screw. Previous mixing studies using the same screw geometry clearly showed that solids are conveyed closer to the tip with increasing screw speed as discussed in Section 8.4.1 and elsewhere [35]. As shown in Table 10.6, the discharge temperature was nearly the same as the temperature of the screw at the discharge end (T17).

Table 10.6 Extrusion Measurements for Barrel Temperatures of 200, 220, and 240 °C for Zones 1, 2, and 3, Respectively

Screw Speed, rpm	Time to Steady State, min	T17, °C	Discharge Temperature, °C	Rate, kg/h
30	11	243	247	28
60	10	247	249	53
90	6	250	246	74

The axial screw temperature profiles for the screw speeds are shown in Figure 10.22. These profiles were constructed from the data set shown in Figure 10.21 by using the data collected at steady state. As shown in this figure, the temperature profile would approximate the model developed by Cox and Fenner [30], but the temperature distribution is more complicated than this simple model.

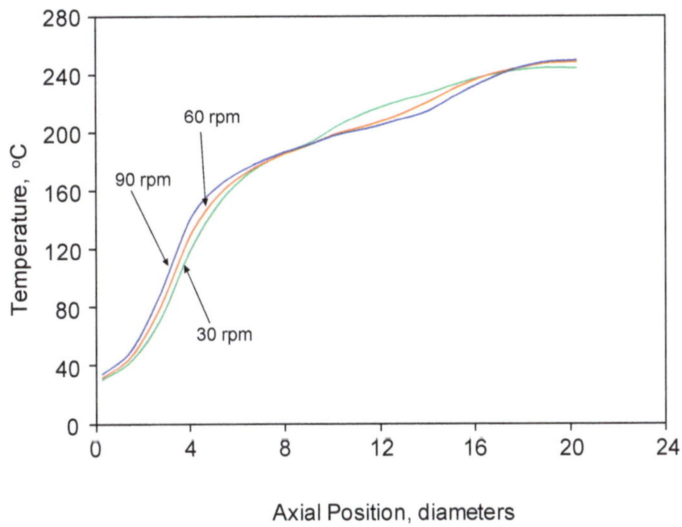

Figure 10.22 Axial screw temperature profiles as a function of screw speed for barrel temperatures of 200, 220, and 240 °C for Zones 1, 2, and 3, respectively. The data point labels for the sensors were omitted for clarity

The effect of barrel temperature on the screw was also investigated in this study. For these experiments, the extruder was operated at a constant screw speed of 60 rpm. The axial screw temperature profiles for the barrel temperatures specified in Table 10.7 are shown in Figure 10.23. The barrel temperatures were initially set at 200, 220, and 240 °C for Zones 1, 2, and 3, respectively. At these conditions the T17 and discharge temperatures were measured at 246 and 248 °C. Next, the Zone 3 temperature was decreased in two steps to 220 °C and then to 200 °C. As shown in Figure 10.23 and Table 10.7, decreasing the Zone 3 temperature caused the discharge temperature and the T17 temperature to decrease. That is, the 40 °C change in the barrel temperature caused the discharge temperature to decrease by 22 °C and the T17 temperature to decrease by 27 °C. Next, the Zone 2 barrel temperature was increased to 240 °C such that the barrel profile was 200, 240, and 200 °C for Zones 1, 2, and 3, respectively. The axial screw temperature profile is shown in Figure 10.23 and the extrusion performance in Table 10.7. For this change, the T17 and discharge temperatures were measured at 219 and 229 °C, respectively. As shown in Figure 10.23, the axial screw temperatures were nearly identical to that of the other runs from the feed location to about 9 diameters downstream. The screw temperature, however, was considerably higher at locations in Zone 2. The screw temperature was essentially the same in Zone 3 as that for the previous run with a Zone 3 temperature of 200 °C. These results clearly show how the barrel zone temperatures affect the discharge temperature and the sections of the screw in the zone. These data indicate that a significant amount of energy is conducting from the barrel wall to the root of the screw for this small-diameter extruder.

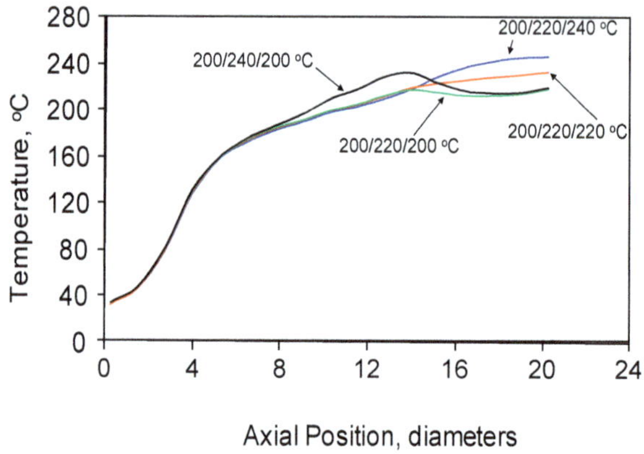

Figure 10.23 Axial screw temperature profile at a screw speed of 60 rpm and as a function of barrel temperature. The barrel temperatures for the three zones are indicated in the figure. The data point labels for the sensors were omitted for clarity

Table 10.7 Extrusion Measurements for a Screw Speed of 60 rpm and Different Barrel Temperatures

Barrel Temperatures for Zones 1, 2, and 3, °C	T17, °C	Discharge Temperature, °C	Rate, kg/h
200, 220, 240	246	248	53
220, 220, 220	233	237	51
200, 220, 200	219	226	49
200, 240, 200	219	229	48

The cooling process for the screw and barrel during a Maddock solidification experiment [8] was performed. For this experiment, the extruder was operated at barrel temperatures of 200, 240, and 200 °C for Zones 1, 2, and 3, respectively, and at a screw speed of 60 rpm. After the extruder was at a steady state, screw rotation was stopped and full barrel cooling was applied to solidify the molten polymer in the screw channel. Select barrel zones and screw temperatures are shown for the cooling cycle in Figure 10.24. The barrel zones and most of the screw cooled at similar rates and were at essentially ambient temperature after 110 minutes. The screw section containing the T15, T16, and T17 sensors, however, was the last region to cool, as shown in Figure 10.24. This result was due to the large and slow cooling of the downstream die system. The screw temperature data suggests that the screw with the solidified plastic is ready to remove from the extruder soon after the barrel temperatures reach ambient temperature. For the extruder here, the screw with the solidified plastic could be pushed from the extruder after about 150 minutes from the time the screw rotation was stopped and full cooling was applied.

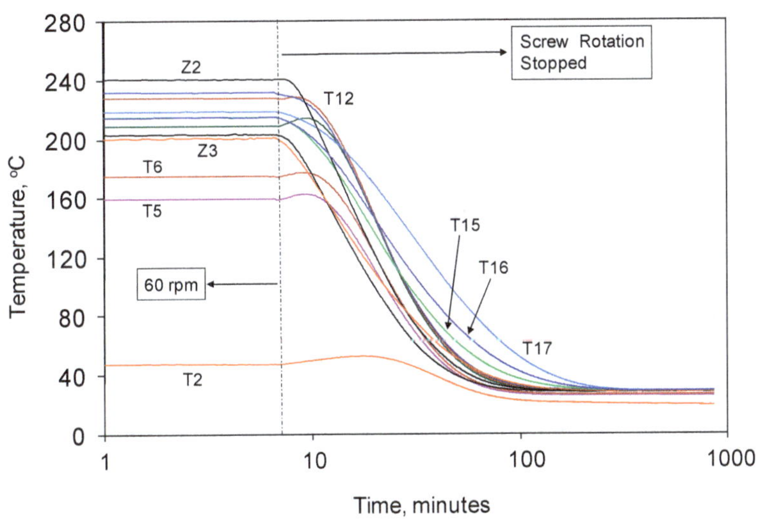

Figure 10.24 Select barrel (Z2 and Z3) and screw temperatures as a function of cooling time during a Maddock solidification experiment

Axial screw temperature profiles have been measured before, but they have been limited by the number of sensors and the quality of the data. The data presented here expand the existing knowledge of extrusion and also provide a new method for determining the ending location of the solid bed.

The transient temperature data indicated that the screw required a longer time than the barrel to achieve a steady state after a process change. In most cases, the screw required about three times longer to achieve a thermal steady state than the barrel. Additionally, it was found that the discharge temperature was typically about 2 to 10 °C higher than the screw temperature at the discharge end. Transient temperature data also suggested the ending location of the solid bed. For a change in screw speed, the temperature sensors indicated that either hot molten resin was flowing into the region or that cooler solid material was flowing into the proximity of the sensor.

The axial screw temperature profiles are consistent with the general trends that would be predicted using the Cox and Fenner [30] model, but the temperature of the screw is obviously affected by all barrel temperature zones and not just the zone over the metering channel. The data shows that heat conduction from the barrel to the screw root is highly important. This conclusion is consistent with the observations and model by Derezinski [32].

10.11 The Screw Manufacturing and Refurbishing Process

Optimal operation of an extrusion process depends on the design and geometry of the screw. All other mechanical operations downstream are important, but the success of minimizing resin usage, minimizing scrap and labor costs, and maximizing profits depends highly on the design of the screw. This section describes the process for manufacturing and refurbishing a screw. It is included here to help the troubleshooter understand the limitations of the fabrication process, screw modifications, and refurbishing. The manufacturing process presented here is standard for the industry although variations and trade secrets do exist. Small-diameter screws typically can be fabricated in less than six weeks while large-diameter screws can require six months or more. Screws are made from many different metal alloys to provide the required strength, durability, and chemical resistance. For many applications, 4140 steel is the preferred alloy for screws based on availability, strength, ease of screw fabrication, and cost. As previously discussed, other alloys can be selected for screws requiring higher torsional strength in the feed section.

The manufacturing process starts with a cylindrical billet of the base metal that is oversized in diameter, as shown in Figure 10.25(a). Prior to most steps, including at the start, the billet is straightened in a large hydraulic press. This process removes local bends and also large drifts in the axial center of the screw. A straightening press is shown in Figure 10.26. Next, the billet is placed in a lathe and its diameter is decreased to about 1.5 mm larger than the final outside diameter of the screw. A lathing operation is shown in Figure 10.27. A channel is then milled into the billet where the flights will eventually be, as shown in Figure 10.25(c). This channel for full-width flight hard facing is about 10% wider than the flight width, and it will be used to position the hard facing metal. Again, the billet is straightened in the press.

a)
b)
c)
d)
e)

Figure 10.25 Photographs of screws in different steps of the fabrication process: a) rough billet as delivered by the foundry, b) a billet that has been straightened and lathed to a diameter that is 1.5 mm larger than the outside diameter of the screw, c) a billet with channels milled into it where the full-flight hard facing will be welded, d) the billet with the hard facing welds completed, and e) billet with the hard facing ground to 1.5 mm oversize in diameter (courtesy of Jeffrey A. Kuhman of Glycon Corporation)

Figure 10.26 Photograph of a hydraulic press used to straighten screws and billets (courtesy of Jeffrey A. Kuhman of Glycon Corporation)

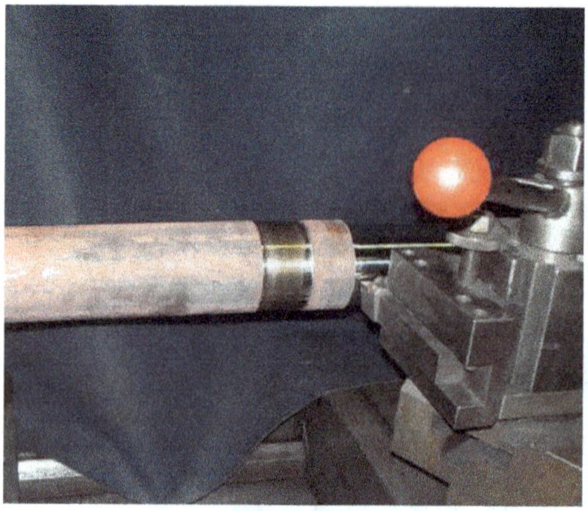

Figure 10.27 Photograph of a lathe performing the turndown from the rough billet to one that is 1.5 mm oversize in diameter (courtesy of Jeffrey A. Kuhman of Glycon Corporation)

The hard facing is welded onto the billet in the channel. Many variations of this process are typically performed, and the particular one used depends on end-use customer requirements, the screw manufacturer, and the material of construction for the billet. The preferred welding process is a technique called plasma transfer arc welding or PTA welding, as shown in Figure 10.28. During the welding process, the entire billet is maintained at a temperature near 400 to 430 °C by using numerous natural gas burners positioned underneath the billet, as shown in Figure 10.28(a). The welding material is then added as a fine powder to a hot plasma zone. One method is to weld a 0.5 mm thick layer of stainless steel (SS) 316 into the channel first. The SS 316 layer provides excellent weld strength between most metals including 4140 steel and hard facing materials. Next, the first layer of the hard facing is welded on top of the SS 316 layer. Colmonoy 56 is the most common hard facing material used in the United States, but many others are available. During the welding process, some of the SS 316 material will diffuse into the hard facing material and cause the hardness of the weld to be less than that of the pure hard facing material. To attain the highest level of hardness, a second welding of the hard facing is applied over the top of the first welding. The hardness of the final surface for Colmonoy 56 is about a 50 to 55 on the Rockwell C scale. Single-pass welding methods where just a single layer of hard facing is welded into the channel have been developed and can provide nearly the same performance. After the welding is completed, the screw is removed from the welding lathe, placed in a tank, and then covered with a mineral insulator such as vermiculite. The mineral insulator allows the screw to cool at a very slow rate of about 10 to 15 °C/h so that large cracks do not develop in the hard facing surface. A cooling tank is shown in Figure 10.29. After about one to three days depending on diameter, the screw is removed from the tank and is straightened. A billet with the hard facing applied is shown in Figure 10.26(d). Next, the outside diameter of the welded billet is ground back to 1.5 mm oversize in diameter, as shown in Figure 10.26(e).

a) b)

Figure 10.28 A photograph of a screw in a PTA welding lathe: a) natural gas burners are heating the screw billet to a temperature of about 400 to 430 °C, and b) the plasma arc of the PTA welder is visible in this photograph (courtesy of Jeffrey A. Kuhman of Glycon Corporation)

Figure 10.29 A photograph of a cooling tank containing three screws (courtesy of Jeffrey A. Kuhman of Glycon Corporation)

Next, the flow channels are cut into the billet between the hard facing layers. This cutting process can be performed using manual mills, computer-controlled mills, or by whirling machines. For the milling process, the welded billet is first centered on a lathe. A vertical milling machine is then used to cut the channels as the lathe feeds the cutting front into the cutting tool, as shown in Figure 10.30(a) for a Maddock-style mixer and Figure 10.30(b) for the keyway in the screw shank. The flow channels are cut short of the design depth by about 0.3 mm. For the whirling process, a ring tool with numerous cutting edges is positioned around the welded billet. This cutting tool is positioned at an angle to the axis of the screw and rotated at relatively high speeds, as shown in Figure 10.31. As the tool is rotated, the lathe of the machine feeds the cutting front into the tool. The whirling process is considerably faster than the milling process and subjects the billet to lower levels of stress, and thus these billets do not require as much time during the straightening process. At the end of this milling (or whirling) process, the screw is straightened.

a) b)

Figure 10.30 Photograph of screw in a vertical milling machine: a) the mill is cutting the flow channel for a Maddock-style mixer, and b) the keyway is being cut into the shank of the screw (courtesy of Jeffrey A. Kuhman of Glycon Corporation)

Figure 10.31 Photograph of screw in a whirling machine (courtesy of Jeffrey A. Kuhman of Glycon Corporation)

Mixing sections are cut into the billet next using the milling machine and lathe combination, as shown in Figure 10.30(a). Depending on the mixer design, cutting these sections tends to be more time consuming.

At this point in the process, the hard facing has been welded onto the flight tips, the flighted channels are milled such that the channel depth is about 0.3 mm less than that required by the design, the mixing sections are milled into the billet, and the screw is oversize in diameter. Next, the billet is centered in a grinding lathe, and the outside diameter of the screw is decreased to about 0.5 mm oversize. A grinding lathe showing the grinding of the hard facing to near the outside screw dimension is shown in Figure 10.32. The screw is then placed in a lathe where the surface finish of the flight channels is improved by using a belt sander, as shown in Figure 10.33. The belt sander increases the depth of the channel to the final depth, removes surface imperfections created by the cutting process, and creates a smooth finish. Again, the screw is straightened.

Figure 10.32 The surface grinding process where a grinding lathe removes excess hard facing material from the outside diameter of the screw (courtesy of Jeffrey A. Kuhman of Glycon Corporation)

Figure 10.33 A photograph of a belt sander used to perform the final finish of the channels. The side guard has been temporarily removed to show the sanding belt (courtesy of Jeffrey A. Kuhman of Glycon Corporation)

The main section of the screw is nearly finished at this point. Next the drive shank and discharge tip details are fabricated into the screw. If the screw requires a plating material such as chrome or nickel, finished regions of the screw that should not be chromed are masked off using a metal tape or paint. Regions that are not finished such as the flight lands do not need to be masked since the plating will be removed in the final grinding operation. With the screw as straight as possible, the final outside diameter of the screw is set by performing the final grinding of the flights. After the final grinding, the screw is polished and is inspected before shipping it to the end-use customer.

In most cases, the screw performs as designed at the customer's plant. However, for applications that are new, for a process scope change such as the extrusion of a different resin, or if processing problems occur, the screw may need to be modified. Several types of screw modifications are easy to perform while others are difficult or not practical to perform. For example, if the depths of any or all of the channels need to be increased the screw can be placed back into the milling lathe (or whirling machine) and the channel depths can be increased. For this type of process, the chrome plating would first need to be stripped off the screw using an electroplating process. Like the production of the original screw, the channel depths would need to be sanded to remove imperfections created by the cutting process. These processes would likely not cause the screw to bend to a high level and it can be easily straightened in the press. Each location where the screw is straightened, however, may cause an outside location of the screw to be outside the screw axis radius on some sides and inside the radius on the other side. The location on the outside will be ground to position while a slightly large flight clearance will occur on the opposite side. The screw is then replated, polished, and shipped to the customer.

If the screw modification requires a welding operation, the area for the weld must be relatively small because the welding process tends to induce a high level of stress that creates bends in the screw. Regions that are easily welded include barrier flights and mixing flights. For these cases, the designer decreases the undercut between the main flight and the barrier or mixing flights. First, the undercut flight is welded to increase its height. Next a grinding or milling operation is used to position the flight at the desired undercut. Because these operations require welding, bending in the screw will be more prevalent, causing a higher level of difficulty when the screw is straightened.

It is very difficult and not practical to decrease the depths of the flights by welding metal back into the channels. This level of welding is generally unacceptable due to the large level of bends that will occur in the screw. For this case, it is often less costly and quicker to build a new screw with shallower channels.

As previously stated, large flight clearances due to wear will cause the performance level of the screw to decrease. When the performance decreases to an unacceptable level, the screw must be either replaced or refurbished. The cost of refurbishing a screw is about 50 to 70% of the cost of a new screw, and refurbishing requires about four weeks. A schematic of the flight tip at different stages of the refurbishing process is shown in Figure 10.34. For this process, the old hard facing is first removed in the grinding lathe. Next, a single welding of SS 316 and two consecutive weld passes of the hard facing material are placed on the flight. Like before, the screw is allowed to cool slowly in the cooling pit. Next, the screw is straightened. The method for the final finishing of the flight depends on whether the screw is going to be plated with a coating and the preferred practice of the fabricator. If the screw is not going to be plated, the preferred sequence is shown in Figure 10.34(d) through Figure 10.34(f). Some of the welded material causes the screw to be oversize, and the hard facing also extends over the side of the flight, as shown in Figure 10.34(d). Next, the screw is ground to the original outside diameter. The edge material is then removed using a milling lathe. In order to produce an acceptable flight edge, some of the base metal must be removed on the sides of the flights. This process will cause the flight radii to become smaller and the width of the flight to become narrower, as shown in Figure 10.34(e). Since some of the base metal on the flight edges is removed during the process, the screw can typically be refurbished up to three times before the flight edges become too small and too stressed for practical operation. Flight edges that become too thin and stressed will tend to break away, causing damage to the downstream sections of the process. Moreover, flight radii that are too small can cause regions with long residence times that will eventually allow the resin to degrade as discussed in Section 11.7 [37]. If the screw is going to be plated, then often the flight edges are milled first, then the screw is plated, and then the screw is ground to the original outside diameter.

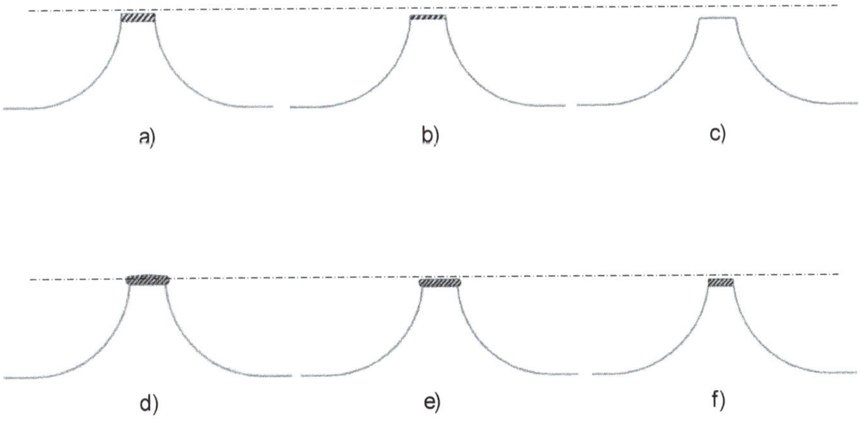

Figure 10.34 Schematics of flight tips during the refurbishing process: a) schematic of a new flight tip with a single pass of hard facing, b) a worn flight tip, c) a worn flight tip with the old hard facing removed, d) flight tip with a single-pass welding of hard facing material, e) hard facing ground to the outside diameter of the screw, and f) flight tip with the flight edges machined. The dotted line shows the relative position of the barrel wall to the flight tip

10.12 Injection-Molding Plasticators

The screw and barrel of injection-molding plasticators have similar design and function as the screws and barrels of extruders. The most distinct differences for the plasticator include: (1) the screw typically rotates intermittently, (2) the screw retracts as the injectate or shot is produced, (3) the feed section tends to be longer than those of extruder screws to accommodate the retraction length, (4) a nonreturn valve is positioned on the tip of the screw, and (5) a hydraulic cylinder is positioned behind the screw that adjusts the axial forces on the rear of the screw. The basic design, operation, and manufacturing of the screw, however, are essentially the same as screws made for extruders. That is, as the screw rotates, the resin is plasticated and pumped under pressure through the nonreturn valve and is stored between the nonreturn valve and the nozzle. To accommodate the flow of the resin the screw retracts to produce the volume of the injectate. As the screw retracts, a portion of the feed channel is pushed behind the feed opening of the casing, reducing the length of the solids-conveying section. To compensate for this motion, the solids-conveying section length is often longer than those used in extrusion applications. After the injectate is plasticated, screw rotation is stopped, the nozzle valve is opened, and the hydraulic cylinder positioned behind the shank of the screw injects the polymer into the mold tooling. To prevent the injectate from flowing back into the channels of the screw, the nonreturn valve closes with the forward motion of the screw. A schematic of a poppet-style nonreturn valve is shown in Figure 10.35. Many variations of the injection-molding process exist. A complete description of these processes can be found elsewhere [38].

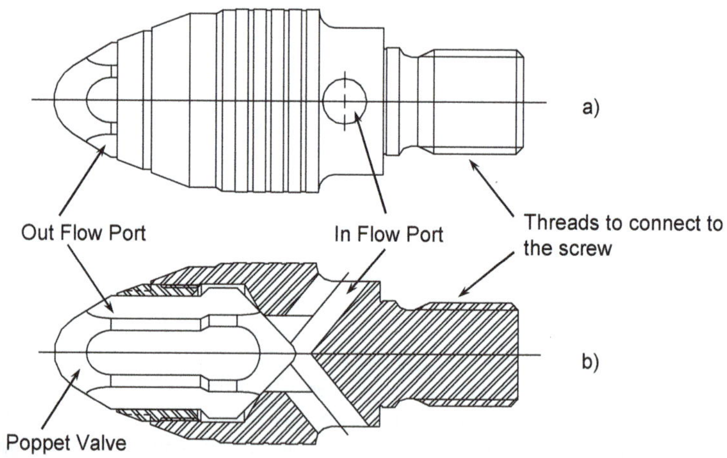

Figure 10.35 Schematics of a poppet-style nonreturn valve: a) full view, and b) a cutaway view showing the flow paths for the poppet valve in the closed position. The poppet valve is forced forward during plastication to allow flow from the screw channels to the tip. The poppet is forced back during the injection step, preventing polymer from flowing back into the screw channels (QSO valve schematic courtesy of Jeffrey A. Kuhman of Glycon Corporation)

The pressure at the tip of the nonreturn valve during plastication and the temperature of the injectate are key attributes for the performance of the plasticator. The pressure at the tip is controlled by adjusting the hydraulic pressure on the cylinder behind the shank of the screw, that is, the back pressure. The pressure at the tip is directly proportional to the hydraulic pressure as shown in Equation 10.17. The proportionality constant or the intensification factor is typically 8 to 14 for most injection-molding machines. The factor is typically stamped on a placard on the machine. The control of the hydraulic pressure has a strong effect on the plastication rate and the injectate temperature. As the hydraulic pressure is increased, the plasticating rate decreases and the injectate temperature increases, as shown in Figure 10.36. The temperature of the injectate is measured by performing an "air shot." For this procedure, the plastication process is performed normally and then the nozzle is retracted from the tooling so that the molten resin can be injected into the air and onto a piece of cardboard. The temperature of the injectate is then carefully measured using a handheld thermocouple probe.

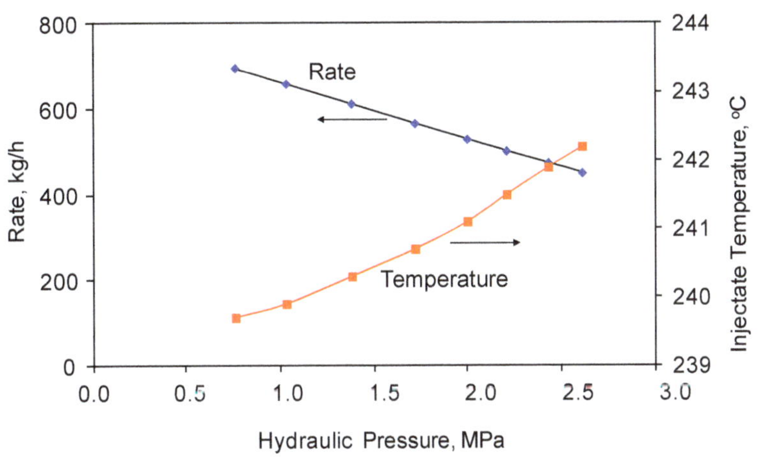

Figure 10.36 The effect of hydraulic pressure on injectate temperature and specific rate. The data are for a 125 mm diameter screw and an intensification factor of 10 for a molding machine running a HIPS resin at a screw speed of 100 rpm

Increasing the hydraulic pressure is a common practice for improving the mixing abilities of the screw or eliminating void regions in the screw that can lead to the degradation of the resin. As the hydraulic pressure is increased, the plasticating rate decreases at the same screw speed, causing the specific rate to decrease. A decrease in the specific rate typically causes the mixing quality to increase. The time for the plastication process and the injectate temperature, however, will increase. If the plasticating time is the rate-limiting step of the molding process, increasing the hydraulic pressure will increase the cycle time and the cost to mold the part. Moreover, an increase in the injectate temperature can be unacceptable for some resins and molding operations. A case study for void regions in the screw channel is presented in Chapter 11.

10.12.1 Calculations for Injection-Molding Plasticators

Rotational flow and pressure flow rate calculations for the screw geometry and process conditions are performed for the injection-molding process in the same manner as for an extrusion process. Since the plasticator of an injection-molding process is not a continuous process, the instantaneous rate must be calculated based on the time that the screw is actually rotating. The instantaneous rate is the rate that is compared to the calculated flow rates for the screw. The instantaneous specific rate is calculated as follows:

$$Q' = \frac{3600m}{tN_i} \tag{10.21}$$

where Q' is the specific rate in kg/(h·rpm), m is the total mass of the parts and runner system ejected from the press, t is the plasticating time in seconds, and N_i is the screw speed in rpm.

The pressure at the tip of the screw is controlled by setting the hydraulic back pressure at the shank end of the screw. The pressure at the tip is calculated from the hydraulic back pressure and the process intensification factor as follows:

$$P_t = IP_b \tag{10.22}$$

where P_t is the pressure at the tip of the screw, I is the process intensification factor, and P_b is the hydraulic back pressure. The tip pressure is essentially equivalent to the discharge pressure of an extruder.

10.13 New Equipment Installations

Specifying and installing the proper equipment for a process are keys to minimizing the long-term cost of producing products and minimizing costly troubleshooting operations during the startup. But often the objective to purchasing equipment is to minimize the initial capital cost. Minimizing this initial purchase cost, however, may require the purchaser to add costly modifications to the line after installation, creating higher operating costs, lengthy troubleshooting, and a delay to market entry. Principles and strategies are presented here that show how to avoid this mistake, and two case studies are provided as learning tools.

In some regions of the world labor is relatively inexpensive and it can be used advantageously by innovative producers. Labor costs, however, are typically not the largest cost in the manufacturing process. A more typical ranking of the costs of operation, from the highest cost percentage to the lowest is (1) resin and feedstocks, (2) labor, and (3) plant and equipment [39, 45]. Figure 10.37(a) shows a typical and simplified breakdown of the costs for a process that is designed and operated properly. Thus to minimize costs, the processor must reduce resin waste, increase labor efficiency, minimize downtime, maximize production rate, and get the most from production machinery. If the process equipment is improperly designed and operated, production rates will decrease, the cost of the product will increase, resin waste can increase, recycle streams will increase, and the cost structure to produce the product will change as shown in Figure 10.37 (b). Specifying improper equipment to decrease the initial purchase cost can potentially cause the equipment cost per kg of product to increase.

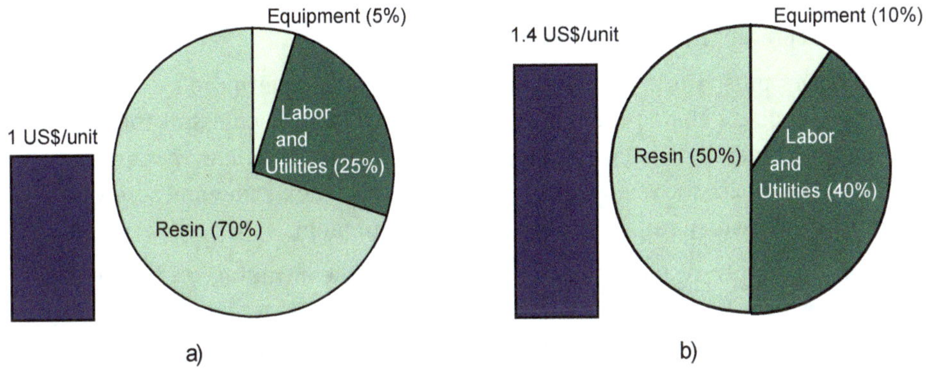

Figure 10.37 Typical cost structures for two processes: a) a process that is designed and operated properly, and b) a process that is designed for reduced initial cost. The basis cost has units of US$/kg for in-specification product

The goal for all new equipment purchases and installations is to select the proper equipment and technologies to get the maximum utilization of the resin, reduce downtime, and increase the rates to reduce the cost of labor per unit mass of finished product. Since resin costs are typically the highest component cost of an operation, it is very important to manage the resin usage via machinery components. Many times, the lack of processing knowledge of the decision makers negatively impacts the long-term profitability of the company. As companies grow larger through internal growth and through consolidation, every department is driven to reduce costs. Unfortunately, cutting costs with respect to equipment specifications can be disastrous.

New equipment is often purchased from the lowest cost provider. Motors may be sized too small, and parts supplied by the lowest bidder may not have the life expectancy of the ones that have a higher initial cost. Rotating equipment that is undersized and pushed to the limit can result in lengthy downtimes and low rates and may cause higher scrap rates and higher resin consumption per part. Additionally, excessive troubleshooting time at startups can result in delaying product deliveries, frustrating end-use customers, and delaying the entry of a new product into the market. Improperly specified equipment can become considerably more expensive than purchasing the right equipment in the first place. The equipment purchased at the lowest cost could be the most expensive long-term option and could even jeopardize the success of the company.

There are three basic principles to understand and apply in value analysis and cost reduction when selecting the proper equipment. The first is that equipment costs are typically amortized over 10 years, and it is usually a small fraction of the finished product cost. The costs are typically less than 5% of the total cost per unit mass for a product. Secondly, decreasing the amount of resin in the finished product without sacrificing product quality is a major cost reduction. And finally, increasing production efficiency by increasing line speeds and decreasing downtime will decrease the cost of production.

Processors must work with equipment suppliers to determine the capabilities and limitations for the equipment to be supplied for all sections of the line. The four basic equipment categories and design strategies that follow the three principles listed above are the resin supply system to the line, extruder and screw design, die and downstream equipment, and optional equipment.

The resin supply system should be designed to take advantage of the raw materials in the lowest cost and most effective form. Additives tend to be more expensive than the base resin. Gravimetric rather than volumetric supply of the material is more conducive to minimizing the use of the more expensive feedstock components. The ability of the equipment to utilize reliably 100% of in-plant regrind, additive concentrates, and recycled materials is one of the most important factors to be considered.

All components of the extruder and screw must be designed for the most difficult conditions. Motors should be sized with more than adequate power and designed with a built-in growth factor. The gearbox should have a maximum torque rating for the worst possible conditions plus a 20% or more rating to allow for growth and then have a service factor of at least 1.5 for reliability. The screw designer should be given complete data on all of the materials that will be processed as well as the full range of temperatures and pressures expected. Be sure to consider the additives, types, and amounts that are planned for use. If the feedstock components are abrasive, corrosive, or the degradation products are corrosive, then this information must be provided to the suppliers. A poorly designed screw can compromise rates and extrudate quality and can decrease the amount of acceptable product per kg of material used [36, 40–43].

Many options are commercially available for dies and downstream equipment. In order to optimize a particular process for an application, all available market options should be studied so as to integrate the best technologies into the process. Consider all of the materials to be processed and expected rates for now and in the future. As with the screw and barrel, the die and downstream equipment must be fabricated from the proper materials to mitigate abrasion and corrosion.

The processor may want to consider optional equipment that will enhance the performance of the line to minimize resin consumption and maximize rates. Gravimetric blending, as mentioned above, can reduce the consumption of high-cost additives and maintain a very accurate record for resin inventory. Melt pumps (gear pumps) can be very useful in the conservation of resin by reducing variations in extruder rate and allowing down-gauging and reduction in resin usage per part. Since resin is the most expensive part of the operation, even a small reduction in resin usage can provide a large cost savings. Additionally, the melt pump can facilitate higher usages of regrind and off-specification materials, can improve screw performance over a wide range of materials [44], and can extend the life of gearboxes and the time between screw and barrel replacements by reducing extruder discharge pressure. Melt pumps can be used to decrease the extrudate temperature to the die. For example, Figure 10.38 shows that if an extruder is operated at a discharge pressure of 32 MPa (no melt pump), the discharge temperature is about 260 °C. If a melt pump is installed between the extruder and the die, the discharge pressure from the extruder can be decreased to nearly 8 MPa and the discharge temperature is decreased to 248 °C. Static mixers are very effective in assisting the extruder to better distribute additives and colorants and by homogenizing the temperature gradients in the extrudate. A more uniform temperature and melt viscosity makes the die easier to adjust and reduces point-to-point gauge variation across the die.

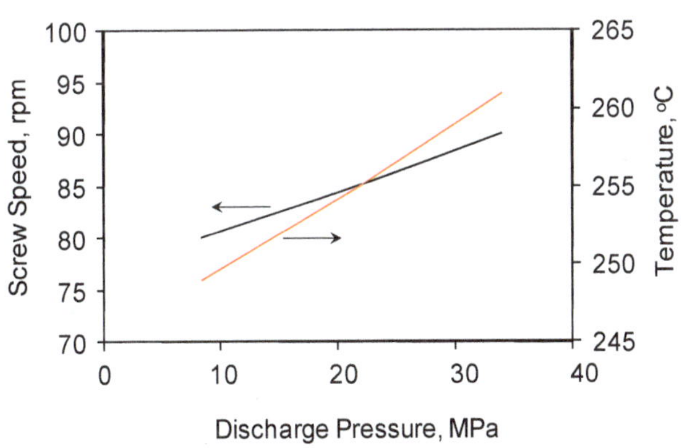

Figure 10.38 The effect of discharge pressure on screw speed and discharge temperature at constant rate. The analysis was for a 150 mm diameter extruder running an engineering resin. An extruder discharge pressure of 8 MPa would be typical for a process running a melt pump while a high pressure of 32 MPa would represent a process without a melt pump

No matter which category of equipment is involved, it is important to design the line such that the maximum anticipated rate is less than 75% of the maximum rate possible for the line. That is, if it is desired to operate a process at 750 kg/h, the line and components should be specified at a rate of at least 1000 kg/h. This allows for a reasonable capacity increase in the future and ensures that components have a longer service life.

For existing lines, the processor must understand the rate-limiting step for the process. If a rate increase is desired and the troubleshooter improves a section of a line that is not the rate-limiting step, then a rate gain will not occur. Sometimes the rate-limiting step for a process is not obvious and can be difficult to determine.

When selecting machinery components the goals are to maintain or improve product quality, have line flexibility to run a wide variety of products, have the ability to increase rates in the future, and minimize the long-term cost of the equipment. An additional goal is often to minimize installation time.

Two case studies are presented here that show equipment purchases that were performed by two different companies. The first case study followed the principles and strategy rules previously discussed, and the equipment allowed for high rates, minimal installation time, and a low-cost product. The other case study did not follow the strategies, and the installation and process were flawed by numerous problems. The consequences of not having the proper equipment will be detailed along with the additional costs associated with the mistake. Additional case studies on equipment installations can be found elsewhere [39, 45].

10.13.1 Case Study: A Large Diameter Extruder Purchase

A large diameter single-screw extruder was needed to replace an aging worn-out machine on a complicated customer line. The goal of the purchase was to optimize the design of the extruder to the current products, increase the rate by 35%, and install the unit with the lowest cost for long-term ownership. The line was sold out and thus the amount of downtime for the installation of the new extruder needed to be minimal. Although the purchase cost of the new equipment was a concern, the design team understood that it was not the largest cost for the project. The largest potential costs for the project were identified as poor operation of the machine if it was not specified properly, not achieving the rate increase, and the potential of a delay in production due to an extensive troubleshooting process during installation and startup.

A team was assembled that included experts in resins, machinery design, process controls, and plant operations. The machinery manufacturer chosen was the one that had the most knowledge and manufacturing capabilities for this size of machine and process. Again, all team members were acutely aware of the project goals and timing.

The design of the line included a motor and gearbox that were large enough for a rate that was at least 50% higher than the current operation. Moreover, the length of the extruder was increased to accommodate the higher rates and discharge pressures. If adequate power and increased capabilities were not specified correctly, then the project would not succeed due to a failure to achieve a rate increase. Other details were examined and redesigned to provide optimal performance for the process. This up-front design process delayed the fabrication of the extruder by about four months. The cost of the additional options for this line increased the purchase cost by about 35%.

During the planned shutdown at the customer's plant, the old extruder was removed and the new extruder was installed. At the end of the two-week shutdown, the line was started back up at the original rate. The extruder and line operated flawlessly. Over the next several weeks, the rate of the line was slowly increased without difficulties to a rate that was 35% higher than the original rate. Within another four weeks, the rate of the line was increased to about 75% higher than the original rate. Thus, the goals of a fast installation and rate increase were obtained and exceeded, respectively. Although the initial purchase cost of the extruder was 35% higher than the base case, this additional cost was insignificant compared to the potential of a long shutdown or troubleshooting operation and the cost reductions associated with a 75% rate increase.

10.13.2 Case Study: Extruder and Line Purchase for a New Product

A new-to-the-world extruded product was scheduled to be introduced to the market using a resin that was about twice as viscous as materials used for the competing technology. At the start of the project, a single-screw extrusion line was being designed and quoted by several manufacturing firms. The goal of the company was to obtain the lowest cost for the installation of the line and the shortest time for the delivery of commercial products to the market. The equipment manufacturers were made aware of the company goals. Thus, the team did not follow the design strategies listed above.

In order to meet the goals, the equipment manufacturers were forced to reduce the size of many of the components, including the motor size, gearbox, barrel diameter and length, and many auxiliary components of the line. It was hypothesized that the equipment manufacturers did not believe that the resin was as viscous as reported. The gearbox was, however, specified with a safety factor of 1.5 for the motor and belt sheave system installed.

During the startup of the line, the extruder was quickly identified as not having enough torque from the motor to make the product. At this point the best option was to change the belt sheave ratio between the motor and gearbox such that additional torque was available to the screw. When this was done, it decreased the safety factor on the gearbox from the delivered value of 1.5 to about 1.05 for operation. This reduction in safety factor may significantly reduce the life of the gearbox and make the long-term cost of ownership for this line higher than estimated by the company. Even with this belt sheave change, the motor was still not large enough to achieve the rate targets specified by the company. A redesign of the screw and numerous other modifications to the line were implemented before commercial production could be achieved. These modifications required about 12 months to make and allowed the line to operate at only 80% of the target capacity. Since the economics for the process were based on operation at 100% capacity with a desire to increase the rate with time, the cost to produce this product was considerably higher than planned.

A better design for this line would have included a larger gearbox and motor, a longer and larger diameter barrel, and many larger auxiliary components. These items were not specified with the original delivery due to the desire to minimize the initial cost of the purchase. The outcome of the design, however, was significantly higher costs to provide new components, a long and costly troubleshooting process for the line, and a delay of a new product to the marketplace.

10.13.3 A High-Density Foamed Sheet Product

A coextrusion foam sheet process was rate limited due to high discharge temperatures and surface defects [45]. The foam sheet was produced using two HDPE resin layers and a chemical blowing agent. The HDPE resin had a solid density of 0.96 g/cm^3. At the maximum rate, the sheet formed would have a nearly perfect surface quality, a foamed solid density of 0.84 g/cm^3, and a sheet haul-off velocity of 6.1 m/min. The width and thickness of the sheet were held constant. If the rate was increased slightly beyond a sheet velocity of 6.1 m/min, severe surface blemishes would occur, and the sheet would be out of specification and was scrapped. Moreover, if the level of the chemical blowing agent was increased, surface defects would be severe due to the collapse of foam cells. The line was sold out and the product was valued by the sheet. Moreover, additional product could be readily sold if available. Each layer was supplied by identical diameter smooth-bore, single-screw extruders. The discharge temperature of the extrudate at the die lips was measured at 247 °C.

A process assessment [46] determined that the screws were not optimized for the application, and that they produced an extrudate that was too hot. Optimized screws could increase the velocity of the sheet and lower the density, and thus the number of cut sheets produced would increase.

Two screws were designed such that the extrudate temperature could be decreased. The details of the design are beyond the scope of this section, and only the economics for the change are reported. The cost of the two screws was $85,000. The optimized screws were identical and operated at higher specific rates such that the extrudate temperature was reduced. The extrudate temperature was measured at the die lips at 238 °C. For these screws and the lower extrudate temperature, the solid density of the sheet was decreased to 0.65 g/cm^3 by adding more chemical blowing agent. The line speed was easily increased to 8.7 m/min for a 43% rate increase for producing cut sheets. Moreover, the mass rate of the extruder was increased by only 10% to obtain the 43% increase in sheets, decreasing the amount of resin in each sheet by 21%.

A cost model was established for this case to understand the cost savings from the optimized screws. The details of the model were previously reported [45]. The overall manufacturing cost of the sheet product in units of $/m was calculated and shown in Figure 10.39. It indicates that both the material cost and the conversion cost per unit length of product can be reduced significantly by using the optimized screw design: the material cost is reduced by 21% and the conversion cost is reduced by 29%. This is due to the decrease in the extrudate temperature, allowing higher mass rates through the extruder, and lower foam densities in the sheet. The lower densities reduce the amount of resin and cost in the sheet.

As shown in Figure 10.39, the optimized screws reduced the cost of the sheet by $3.47/m. Because of the high cost reduction and the high volume of this production line, the payback period for the upfront investment of $85,000 for the optimized screws was calculated to be only 2.1 days. The cost reduction as a result of the optimized screw was about $7.3 MM/yr.

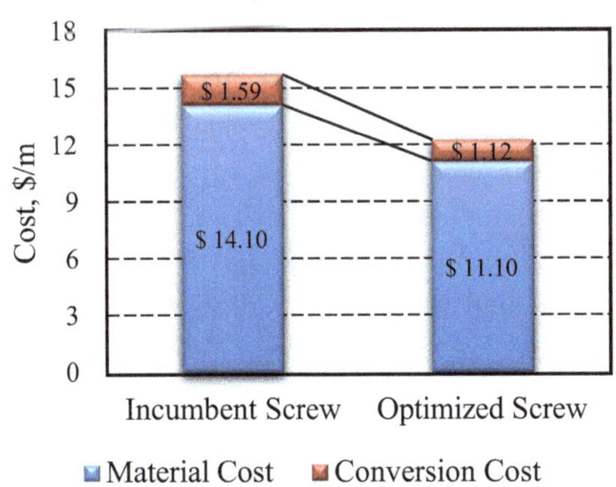

Figure 10.39 Comparison of the cost structure using the incumbent and optimized screw designs for the high-density foamed sheet product

10.13.4 Summary for New Equipment Installations

From these case studies, it is apparent that the specification and selection of equipment for any extrusion line is very critical. While being the lowest cost portion of operating an extrusion line, the specification and operation of the equipment can also influence the cost and usage of resin and labor costs per unit of finished goods produced. Additionally, extended startup times due to troubleshooting, modification, repair, or replacement of improperly specified equipment can delay introduction of a new product to a market. In some cases these issues can result in total failure of the product, or in the least, loss of a competitive edge and timing for introduction. Downtime and repair costs due to premature equipment failure can significantly increase the cost of the finished product, erode profit margins, and jeopardize the ability to fill customer contracts. The result could be the loss of customers to the competition.

Thus, a processor must have the ability to design and engineer a product in a timely and efficient manner and to produce that product at a high production rate and with high quality. These capabilities allow these producers to minimize resin waste, increase labor efficiency, minimize downtime, and get the most from production machinery. If processors fail to take advantage of these strengths, then they will lose their competitive edge. As mentioned earlier and as demonstrated with the case studies, the failure to design and install equipment that results in the highest rates, best quality, and lowest maintenance cost does not make sense in light of the pitfalls. For example if an extrusion line is designed for a rate of 1360 kg/h and the design of the equipment is marginal and causes a 1% overweight in the finished product due to operating the machine and line to its limits, the cost in excess resin alone is \$240,000/yr. This calculation is based on a resin cost of \$2.20/kg and an operation time of 8000 hours per year. This does not include the issues mentioned earlier and is far more costly than designing the equipment to easily handle the rate in the first place.

It is imperative that the processor utilize every advantage available to ensure success since the specification of equipment in an extrusion line also affects resin usage and labor. Thus, it is always more cost effective in the long run to design and install an extrusion line that (1) has a maximum rate capability of at least 25% more than the expected maximum rate, and (2) has a properly engineered line (that might have a higher capital cost) to achieve maximum profitability.

Nomenclature

A	motor current during operation
A_0	vertical support height for a barrel
A_T	vertical distance between the bottom of the support and the barrel at temperature T
A_{max}	nameplate motor current at full load
C	compression ratio of the screw
C_e	thermal expansion coefficient for steel
d	cooling hole diameter
D_s	diameter of the screw at the flight tips
e_f	flight width perpendicular to the tip in the feed section
e_m	flight width perpendicular to the tip in the metering section
E	specific energy inputted from the screw to the resin in J/g
F_s	safety factor in torsion for the screw
h_g	change in length of the barrel section between the fixed end and the barrel support due to thermal expansion
H	channel depth in the metering section
H_f	channel depth in the feed section
I	process intensification factor for an injection-molding plasticator (Equation 10.22)
l	independent variable for the axial position on the screw
l_2	axial location on the screw where the melt film first forms on the screw (Equation 10.20)
L	flight lead length
L_f	flight lead length in the feed section
L_m	flight lead length in the metering section
m	total mass of the parts and runner system ejected from an injection-molding press (Equation 10.21)
M	number of turns in the transition section
N	speed for the screw in revolutions/s
N_i	speed of the screw in rpm
N_m	speed for the motor in revolutions/s
N_{max}	maximum screw speed in revolutions/s

P_m	power at the screw shaft
P_b	hydraulic back pressure for an injection-molding plasticator (Equation 10.22)
P_{max}	maximum power rating for the motor in watts
P_r	pump ratio for a two-stage screw
P_t	pressure at the tip of the screw for an injection-molding plasticator (Equation 10.22)
Q	extrusion rate in kg/h
Q'	specific rate in kg/(h·rpm) for Equation 10.21
R	compression rate in the helical direction
R_{bs}	belt sheave reduction
R_g	speed reduction for the gearbox
S	support distance from the fixed end of the barrel
S_m	shear strength of the material used for the screw
t	plasticating time in seconds (Equation 10.21)
T	temperature
T_b	axial barrel temperature
T_{max}	maximum torque that can be transmitted by the screw before breaking
T_{mot}	torque that the motor supplies to the screw (a function of the motor load)
T_s	axial screw temperature
T_{sh}	torque at the screw shank or shaft (function of the motor load)
T_0	ambient temperature
T_{s0}	material feedstock temperature for Equation 10.19
$T_s(l_2)$	equivalent to the melting point temperature of the polymer (Equation 10.20)
$T_{s,max}$	maximum torque on the screw at maximum motor load
δ_f	mechanical clearance between the top of the screw flight and the barrel wall
ΔA	change in length for a barrel support due to a change in equipment temperature
$\theta_{b,f}$	helix angle at the barrel wall in the feed section
$\theta_{b,m}$	helix angle at the barrel wall in the metering section
φ	angle between the support at temperature T and the vertical position at T_0

References

1. Gould, R.J., "Introduction: Basics of Extrusion Troubleshooting," Chapter 1 in "The SPE Guide on Extrusion Technology and Troubleshooting," Vlachopoulos, J. and Wagner, J.R. (Eds.), Society of Plastics Engineers, Newtown, CT (2001)
2. Christie, A., "Troubleshooting the Extruder," Chapter 9 in "Film Extrusion Manual," Butler, T.I. (Ed.), TAPPI Press, Atlanta, GA (2005)
3. Tobin, W.J., "Troubleshooting Injection-Molded Parts," WJT Associates, Louisville, CO (1996)
4. Mager, R.F., "Troubleshooting the Troubleshooting Course or Debug D'Bugs," Center for Effective Performance, Atlanta, Georgia (1983)
5. Fogler, H.S. and LeBlanc, S.E., "Strategies for Creative Problem Solving," Prentice Hall PTR, Upper Saddle River, New Jersey (1995)
6. McCullough, T.W. and Spalding, M.A., "Predicting Actual Temperature Distributions in a Polymer Stream Using an Adjustable-Depth, Exposed-Tip Thermocouple Assembly," *SPE ANTEC Tech. Papers*, **42**, 412 (1996)
7. Platt, J.R., "Strong Inference," *Science*, **146**, 347 (1964)
8. Maddock, B.H., "A Visual Analysis of Flow and Mixing in Extruder Screws," *SPE J.*, **15**, 383 (1959)
9. Vlachopoulos, J. and Wagner, J.R. (Eds.), "The SPE Guide on Extrusion Technology and Troubleshooting," Society of Plastics Engineers, Newtown, CT (2001)
10. Giles, H.F., Wagner, J.R., Jr., and Mount, E.M., III, "Extrusion: The Definitive Processing Guide and Handbook," William Andrew Publishing, Norwich, NY (2005)
11. Maddock, B.H., "Effect of Wear on the Delivery Capacity of Extruder Screws," *SPE J.*, **15**, 433 (1959)
12. Barr, R.A. and Chung, C.I., "Effects of Radial Screw Clearance on Extruder Performance," *SPE J.*, **22**, 71 (1966)
13. Lounsbury, D.C., "New Studies on Screw Wear and its Effect on Output, Stock Temperature, Uniformity, and Processor Economics," *SPE ANTEC Tech. Papers*, **28**, 441 (1982)
14. Juvinall, R.C. and Marshek, K.M., "Fundamentals of Machine Component Design," 3rd ed., John Wiley & Sons, Inc., Hoboken, NJ (2000)
15. Myers, J.A. and Puhalla, M., "Barrel Alignment: A Critical Factor in Reducing Extruder Wear," *SPE ANTEC Tech. Papers*, **51**, 323 (2005)
16. Myers, J.A., "Practical Approach to Screw Breakage and How to Avoid Failures," *SPE ANTEC Tech. Papers*, **50**, 310 (2004)
17. Womer, T.W., "How to Clean Screws Nondestructively," *Plast. Technol.*, July, 23 (2009)
18. Shultz, G.P., "Transformers and Motors," SAMS, Carmel, Indiana (1992)
19. Kramer, W.A., "Motors and Drives for Extrusion Applications," *SPE ANTEC Tech. Papers*, **45**, 268 (1999)

20. Barlow, S., "Reducing Electrical Energy Costs for Extrusion Processes," *SPE ANTEC Tech. Papers*, **55**, 1157 (2009)
21. Derezinski, S.J., "Stability of Two-Stage Single-Screw Extruders," *SPE ANTEC Tech. Papers*, **48**, 312 (2002)
22. Crabtree, S.L., Spalding, M.A., and Pavlicek, C.L., "Single-Screw Extruder Zone Temperature Selection for Optimized Performance," *SPE ANTEC Tech. Papers*, **54**, 1410 (2008)
23. Himmelblau, D.M., "Process Analysis by Statistical Methods," John Wiley & Sons, Inc., New York (1970)
24. Wagner, J.R. and Perdikoulias, J., "Extrusion Process and Screw Design Investigation via Experiment Design and Computer Simulation," *SPE ANTEC Tech. Papers*, **51**, 311 (2005)
25. Altinkaynak, A., Gupta, M., Spalding, M.A., and Crabtree, S.L., "Numerical Investigation of the Effect of Screw Temperature on the Melting Profile in a Single-Screw Extruder," *SPE ANTEC Tech. Papers*, **53**, 430 (2007)
26. Campbell, G.A., Spalding, M.A., and Tang, Z., "An Alternative Analysis of Single-Screw Melting," *SPE ANTEC Tech. Papers*, **55**, 147 (2009)
27. Tadmor, Z., Duvdevani, I.J., and Klein, I., "Melting in Plasticating Extruders Theory and Experiment," *J. Polym. Eng. Sci.*, **7**, 198, (1967)
28. Tadmor, Z. and Klein, I., "Engineering Principles of Plasticating Extrusion," Van Nostrand Reinhold Company, New York (1970)
29. Marshall, D.I., Klein, I., and Uhl, R.H., "Measurement of Screw and Plastic Temperature Profiles in Extruders," *Soc. Plastics Engrs. J.*, **20**, 329 (1964)
30. Cox, A.P.D. and Fenner, R.T., "Melting Performance in the Single-Screw Extrusion of Thermoplastics," *Polymer*, **16**, 49, (1975)
31. Edmondson, I.R., Ph.D. Thesis, London University (1973)
32. Derezinski, S.J., "Heat Transfer in Extruder Screws," *SPE ANTEC Tech. Papers*, **46**, 83 (2000)
33. Keum, J. and White, J.L., "Heat Transfer Coefficients and Screw Temperature Profiles in Modular Twin Screw Extrusion Machines," *SPE ANTEC Tech. Papers*, **50**, 108 (2004)
34. Pham, H.T., Spalding, M.A., Crabtree, S.L., Pavlicek, C.L., and Altinkaynak, A., "Experimental Measurement of the Surface Temperature of a Single-Screw Extruder Screw," *SPE ANTEC Tech. Papers*, **54**, 289 (2008)
35. Spalding, M.A., Dooley, J., Hyun, K.S., and Strand, S.R., "Three Dimensional Numerical Analysis of a Single-Screw Extruder," *SPE ANTEC Tech. Papers*, **39**, 1533 (1993)
36. Spalding, M.A. and Hyun, K.S., "Troubleshooting Mixing Problems in Single-Screw Extruders," *SPE ANTEC Tech. Papers*, **49**, 229 (2003)
37. Spalding, M.A., Dooley, J., and Hyun, K.S., "The Effect of Flight Radii Size on the Performance of Single-Screw Extruders," *SPE ANTEC Tech. Papers*, **45**, 190 (1999)
38. Kamal, M.R., Isayev, A.I., and Liu, S.-J., "Injection-Molding Technology and Fundamentals," Hanser Publications, Munich (2009)

39. Smith, D., Spalding, M. A., and Gould R. J., "Selecting Equipment to Minimize Production Costs and Maximize Profitability," *SPE ANTEC Tech. Papers*, **52**, 814 (2006)

40. Spalding, M. A., Powers, J. R., Wagner, P. A., and Hyun, K. S., "Flow Surging in Single-Screw, Plasticating Extruders," *SPE ANTEC Tech. Papers*, **46**, 254 (2000)

41. Hyun, K. S. and Spalding, M. A., "Use of Process Data Obtained from a Data Acquisition System for Optimizing and Debugging Extrusion Processes," *Adv. Polym. Technol.*, **15**, 29 (1996)

42. Hyun, K. S., Spalding, M. A., and Powers, J. R., "Elimination of a Restriction at the Entrance of Barrier Flighted Extruder Screw Sections," *SPE ANTEC Tech. Papers*, **41**, 293 (1995)

43. Spalding, M. A., "Metering Channel Flows and Troubleshooting Single-Screw Extruders," *SPE ANTEC Tech. Papers*, **50**, 329 (2004)

44. Pham, H. T. and Hyun, K. S., "Gear Pump Assisted Extrusion," *Polym. Eng. Sci.*, **32**, 488 (1992)

45. Spalding, M. A., Sun, X., and Womer, T. W., "Cost Analysis for Installing New and Optimized Screws for Single-Screw Extrusion Lines," *SPE-ANTEC Tech. Papers*, **64**, (2018)

46. Spalding, M. A. and Gou, Q., "Process Optimization of Single-Screw Extrusion Systems for Polyolefin Resins," *SPE-ANTEC Tech. Papers*, **61**, 1088 (2015)

11 Contamination in the Finished Product

The quality of a product depends on many things including dimensions, physical properties, color, clarity, and the level of defects due to surface imperfections and contaminants. Contaminants can cause surface defects, color streaks, tinting, and optical defects in films and can reduce the mechanical properties of the product. Most products are allowed small and acceptable levels of defects. But if the level of defects due to contaminants is higher than the specification allowance, then the product cannot be sold as prime and must be downgraded, adding additional costs to the product and possibly reducing the capacity of the line. Minimizing the contamination level of a product is one of several keys to optimizing the profitability of a plant.

Contaminants in the extrudate can originate from many sources. Common sources include recycle streams, dust, poor housekeeping, improper labeling of resin feedstocks, air and water entrapment, and equipment failures. Contaminants can also be from decomposition products produced by improperly designed equipment, from undispersed resin components, or from partially melted resin. Melt filtration is effective at removing some types of contaminants while others such as soft gels can flow through very fine screens. This chapter will describe the use of melt filtration systems, the most common sources of foreign contaminants, common process problems that can lead to decomposed resin contaminants, and poorly dispersed resin components. Several case studies will be presented that describe the defects, the root cause of the defects, and the process changes required to eliminate the defects.

11.1 Foreign Contaminants in the Extrudate

Foreign material contamination in a resin feedstock can destroy value to the processor by the production of off-specification product and by shutting down a line due to the formation of holes in a product web or a hole in a blown film bubble. To mitigate the economic loss, the processor must be able to identify the contaminant quickly to minimize the level of off-specification product. Moreover, the off-specification product needs to be isolated before a decision can be made to either add it back as a recycle or dispose of it. Foreign contamination can originate from numerous sources including metal from failed process components, paper and incompatible plastic from shipping bags, color concentrates, wood and fabric fibers, dirt, dust, the addition of the wrong resin to the system, and a host of gel types. Most processes have unit operations and storage systems with openings to the environment, and thus there are many opportunities for contaminants to get into the feedstock system. Proper housekeeping and operation discipline can mitigate contamination from foreign materials.

Foreign contamination is typically first discovered by quality control checks of the finished product or by the loss of the web for film processes. If a melt filtration system is installed downstream of the extruder, the larger size particles will be collected. After the contaminants are collected they must be analyzed for composition. Some types of contaminants are easily identified using a microscope or hand lens and include paper and cloth fibers, dirt, and metal fragments. Other contaminants such as gels or foreign resins are not as easily identified, and their identification often requires advanced analytical procedures. Many resin manufacturers offer these types of services to their customers. After the contaminant is identified, the source must be determined and then eliminated. Elimination of the source can be simple for common contaminants but can be a challenge for contaminants that exist at a very low level.

11.1.1 Melt Filtration

Foreign material and some types of decomposed resin products can be removed using melt filtration systems located just downstream from the discharge of the extruder. In some cases, these devices are located upstream of a gear pump in order to protect the pump from damage. Typical screen packs are constructed from one or more screens positioned upstream of a support called a breaker plate. The breaker plate is a relatively thick plate of metal with holes cut in it to allow the polymer to flow to the downstream sections of the process. The size, number, and pattern of the holes in the breaker plate are critical to providing the proper strength, minimizing the pressure gradient across the plate, and providing the necessary support for the screens. Holes that are too large will minimize the pressure gradient but will not provide enough support for the screens. Conversely, holes that are too small may support the screens well but will cause the pressure gradient to be unacceptably high and thus limit the rate of the process. Breaker plates can be constructed with multiple plates to provide options for material strength and reduced pressure change. As foreign material collects on the screen system, the pressure on the upstream side of the screen pack will increase. This process is referred to as blocking. When the upstream pressure becomes too high and the rate is limited, the blocked screens need to be replaced with fresh screens.

Melt filtration systems come in many styles [1, 2], and the selection of a style will depend on the resin processed, the application, cost constraints, and the level of filtering desired. The simplest system uses a screen or multiple screens against a breaker plate positioned in a flange at the discharge end of the extruder. This system is very low cost, but the extruder must be shut down and the flange disassembled to change the screen. Numerous manufacturers produce dual-cavity screen systems where one of the screens is always in service and the other screen is either idle or being cleaned for the next use. When the screen in service becomes blocked and thus the pressure gradient across the system is too high, then the idle screen is slid or rotated into place while the blocked screen is removed. This type of screen filtering system can switch screens while the extruder is in operation, but the switch does momentarily disrupt the flow to the downstream processes. That is, during the screen switch the motion of the sliding mechanism disrupts polymer flow and the new screen may allow air to be introduced into the extrudate. For some processes such as pelletizing operations, this type of interruption might be acceptable. For more delicate processes such as blown film, the disruption will shut down the line. Several continuous screen changers are available. For these units, the pressure gradient across the device is essentially constant during operation, and they do not entrap air or disrupt the flow during changes. These systems feed the screen from a roll and index the screen as needed to maintain a constant pressure at the inlet side of the device. Heating and cooling units where the screen enters and exits the device are used to soften the resin during screen advancement and then cool the resin to make a seal.

The mesh number and the wire diameter of the screen determine the particle size level that is removed from the extrudate. For example, a screen with a 140 mesh and a wire diameter of 0.076 mm will remove hard particles with a size larger than the opening of the screen of 0.105 mm. The mesh number is the number of wires per inch of screen. Flow openings and wire sizes for common size square-weave screens are provided in Table 11.1. Different screen manufacturers, however, will use different wire diameters for a given mesh number. Thus, the average opening for flow can be different for different screen manufacturers. For many applications, multiple screens are stacked to add strength and service life. A typical screen pack may have screens with the following mesh numbers: 20/60/140/60/20. The upstream 20 and 60 mesh screens collect the larger size particles and minimize the blocking of the 140 mesh screen. These upstream screens also provide mechanical integrity (tearing and sliding) for the 140 mesh screen when it is placed into service. The 60 mesh screen directly downstream of the 140 mesh screen provides support for the 140 mesh screen and the last 20 mesh screen is positioned in front of the breaker plate and provides support for the finer mesh screens. And finally, a symmetric screen pack ensures that it cannot be installed backwards. The opening width in a square-weave mesh can be calculated as follows:

$$d_o' = \left[\frac{25.4}{M} - d_w'\right] \tag{11.1}$$

where M is the mesh number in wires per inch, d_w' is the diameter of the wire in mm, and d_o' is the width of the opening in the screen in mm.

Table 11.1 Typical Flow Opening Sizes and Wire Diameters for Common Size Square-Weave Screens

Mesh Size (*M*), wires/inch	Wire Diameter (d_w'), mm	Average Opening (d_o'), mm
12	0.889	1.23
12	0.584	1.53
20	0.711	0.559
20	0.406	0.864
40	0.305	0.330
40	0.254	0.381
50	0.229	0.279
60	0.191	0.232
80	0.140	0.178
100	0.114	0.140
150	0.066	0.103
180	0.058	0.083
200	0.053	0.074
250	0.041	0.061
325	0.036	0.043
400	0.025	0.038
450	0.025	0.031

Large-size soft polyolefin-type gels or rubber components often deform and flow through relatively smaller size openings in screen packs. Thus, fine mesh screens are sometimes ineffective at removing deformable crosslinked gels from the process stream. Screens can be very effective, however, at trapping and melting solid polymer fragments that are discharged from the extruder. In this case, the polymer fragments are melted via conduction from the hotter molten stream. If the solid fragment level from the extruder, however, is too high, solids can accumulate on the screen faster than they are melted, causing the screen differential pressure to increase and eventually block. For this case, if the extruder is shut down for a short period of time and then restarted, the solids will have time to melt and pass through the system, reducing the pressure on the upstream side of the screen. The operating conditions for the extruder, however, would need to be modified during the startup in order to mitigate the discharge of solids. Typically, the rate of the process would need to be decreased by decreasing the screw speed, or the barrel zone temperatures in the transition section would need to be increased.

The pressure change required to operate a screen changer system with fresh screens can be estimated using a method developed by Carley and Smith [3] and later modified by Todd [4] for a power law fluid.

$$\Delta P_{screen} = 2^{n+3}\left(3+\frac{1}{n}\right)^{n}(900\pi)^{-n}\, m\left(\frac{W}{\rho D_s^2}\right)^{n}\left(\frac{d_w\,(d_w+d_o)^{2n}}{d_o^{3n+1}}\right) \tag{11.2}$$

$$\eta = m\dot{\gamma}^{n-1} \tag{11.3}$$

where ΔP_{screen} is the pressure change (or pressure drop) across the screen (Pa), η is the shear viscosity (Pa·s), n is the power law index, m is the power law consistency index at a fixed temperature, $\dot{\gamma}$ is the shear rate (1/s), W is the mass flow rate (kg/h), ρ is the melt density of the extrudate (kg/m^3), D_s is the diameter of the screen surface (m), d_w is the diameter of the wires used to construct the screen (m), and d_o is the average opening of the mesh (m). Care must be exercised with the units specified when using this equation. Carley and Smith [3] have verified the method using square-woven and full-twill-woven mesh screens. For screen filtering systems with multiple screens, the total pressure change is the sum of the changes for the individual screens.

Dutch twill weave screens have a different pattern for the weaving of the wires to make the mesh. The weaves are complicated, using two different size wires; the wires are labeled weft and warp. Dutch weave screens are generally used when a high level of filtration is desired and a relatively large pressure change is acceptable. Typical wire diameters and average openings for Dutch twill weave screens are provided in Table 11.2. Equation 11.2 can be used for Dutch weave screens [3] by using the wire diameter, d_w, as follows:

$$d_w = \left(d_{wp} + 2d_{wt}\right) \tag{11.4}$$

where d_{wp} and d_{wt} are the wire diameters for the warp and weft wires, respectively.

Table 11.2 Typical Flow Opening Sizes and Wire Diameters for Common Size Dutch Twill Weave Screens

Mesh Size*	Warp Wire Diameter (d_{wp}), mm	Weft Wire Diameter (d_{wt}), mm	Average Opening (d_o), mm
20 × 250	0.249	0.201	0.110
30 × 360	0.249	0.152	0.090
40 × 560	0.178	0.102	0.070
120 × 600	0.102	0.064	0.040
80 × 700	0.102	0.076	0.035
165 × 800	0.071	0.051	0.024

*Mesh size in number of wires per inch (warp × weft)

Many processes will eventually operate at rates higher than the original design rate. For these cases, the designer must be able to estimate the pressure required at the discharge of the extruder to operate the downstream equipment. Equation 11.2 can be used to estimate the pressure change that is expected across a screen filter system at higher processing rates using a simple ratio as follows:

$$\frac{\Delta P_{screen,2}}{\Delta P_{screen,1}} = \left(\frac{W_2}{W_1}\right)^n \tag{11.5}$$

where indices 1 and 2 refer to the original design and the new design, respectively. For example, if an LLDPE resin with a power law index of 0.74 had a pressure gradient across a screen system of 6 MPa at a rate of 1000 kg/h, the pressure gradient for a rate of 1600 kg/h is expected to be about 8.5 MPa.

The pressure change across a breaker plate can also limit a process. The pressure change across the plate can be estimated by using simple equations for flow through a tube [5]. The calculation ignores end effects and is shown in Equation 11.6:

$$Q = \left(\frac{\pi \rho N_h R^3}{\frac{1}{n} + 3}\right)\left(\frac{R \Delta P_{bp}}{2mL}\right)^{1/n} \tag{11.6}$$

where N_h is the number of holes in the breaker plate, R is the radius of the holes, L is the thickness of the breaker plate, ΔP_{bp} is the pressure change across the plate, and Q is the rate in kg/s. The pressure change that is calculated for the breaker plate should be added to the sum of the pressure changes for the screens to estimate the pressure change for the device.

11.1.2 Metal Fragments in the Extrudate

Metal fragments in the feedstock that are magnetic can often be trapped upstream of the hopper using rare-earth magnetic collection devices. Nonmagnetic metal fragments and fragments not trapped by the magnetic device will pass through the extrusion system and will be trapped on screen pack systems located just downstream from the extruder. The most likely sources for metal are upstream processing unit operations, recycle grinding equipment, tools, machine hardware such as bolts, washers, and nuts, and flight hard facing edges from the screw. If the metal fragments are large enough, they can damage the screw and barrel and downstream equipment such as gear pumps and dies. In all cases, it is important to discover the source of the metal contamination. The composition of the metal will help identify the source. Once the source has been identified, the problem should be corrected so that the extrusion equipment is not at risk and so that the final part is free of defects due to metal fragments.

As an example, metal fragments were found on the screen pack system of a complex PP compounding line. The metal fragments were not magnetic and were determined to be stainless steel. Since only an upstream mixing tank was constructed from stainless steel components, the tank was opened and inspected. The inspection determined that a mixing blade had broken off, and parts of it had flowed downstream. The mixer was quickly fixed, and the line was brought back into service. The previous shift of compounded resin was quarantined and was then removed to an off-specification storage area. The quick response of the plant allowed the source of the metal fragments to be quickly identified and eliminated, minimizing the amount and cost of the off-specification PP resin.

11.1.3 Gas Bubbles in a New Sheet Line

New or refurbished chill rolls can cause gas bubbles to form in thin sheets during the initial startup of the line. A photograph of these bubble defects is shown in Figure 11.1 for a thin PP resin sheet product. Chill rolls are typically fabricated from carbon steel and are then chrome plated to provide a highly controlled surface smoothness. During the chrome plating process, the carbon steel rolls are placed into a bath where the chrome is electroplated onto the carbon steel. During the bonding of the chrome to the carbon steel, hydrogen is generated. After the rolls are chrome plated they are moved into an oven at 170 to 200 °C for about four hours. This treatment is used to release any residual hydrogen in the micropores of the chrome. If this step is not done properly, then during production operations the hydrogen gas can diffuse into the resin sheet and form small bubble defects. The bubbles will continue to appear in the sheet until the gas is completely exhausted.

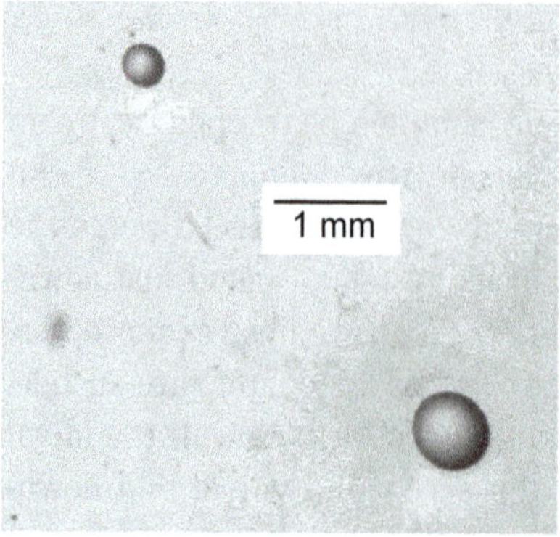

Figure 11.1 Photograph of gas bubbles in a thin PP resin sheet. The sheet for this case was about 0.7 mm thick

The bubble defects can be eliminated by simply heating the rolls to 100 °C for about 12 hours. During this period, the rolls should be rotated. After this treatment, the rolls should allow normal operation without the bubble defects.

11.2 Gels in Polyolefin Resins

A common contaminant in polyolefin products is gels. The term gel is commonly used to refer to any small defect that distorts a film product, creating an optical distortion. There are many types of gels [6, 33], and the most common include (1) highly oxidized material that appears as brittle black specks, (2) gels that are crosslinked via an oxidative process, (3) unmixed gels that are highly entangled polymer chains that are undispersed but not crosslinked, (4) unmelted resin, and (5) a different type of resin or contaminant such as wood, cloth fibers, or dirt. A crosslinked resin gel is typically formed during an oxidation process, resulting in the crosslinking of the resin chains and the generation of color bodies. Highly entangled gels are typically high molecular weight polymer chains that are highly entangled and thus difficult to disperse during the extrusion process. When analyzed using a hot-stage microscope, this gel type will melt as the stage temperature is increased. When the stage temperature is then decreased, the gel will crystallize before the surrounding material, creating the appearance of a gel as a solid polymer fragment. Since these gels are not oxidized they are not associated with color. They are commonly referred to as undispersed or unmixed gels. Unmelted resin exiting with the discharge can sometimes occur, especially at high extrusion rates. These gels will melt during heating with a hot-stage microscope, and typically they will not re-form during the cooling phase. Numerous sophisticated methods are available for analyzing gels including epifluorescence microscopy, polarized light microscopy, and electron microscopy with X-ray analysis. Photographs for several film gel examples are shown in Figure 11.2.

Gels appear in film processes with sizes from less than 0.2 mm up to very large gels with diameters greater than 1.8 mm [6]. Typically, smaller gels are found in film samples at a higher frequency than larger gels. Larger gels are obviously more problematic since they are easier to see and can cause problems in downstream processing operations including blowing, printing, and lamination. They can also reduce the mechanical properties of the film such as elongation (stretch).

Gels can be generated from many different sources and include (1) the resin manufacturer, (2) the converting process, (3) pellet blending of resins with significantly different shear viscosities, (4) pellet blending of different resin types, and (5) direct contamination. Modern resin manufacturing processes exclude oxygen from the system and are very streamlined such that process areas with long residence times do not exist. Moreover, careful quality control techniques at the manufacturing plant are typically in place to verify that the resin does not contain an unacceptable level of gels. Therefore crosslinked and oxidative gels are likely not generated by the manufacturer. Improperly designed extrusion equipment and processes, however, are common, leading to the oxidative degradation of resins and crosslinked gels. Several case studies in the next sections show how poorly designed processing equipment can lead to crosslinked and unmixed gel contamination of products.

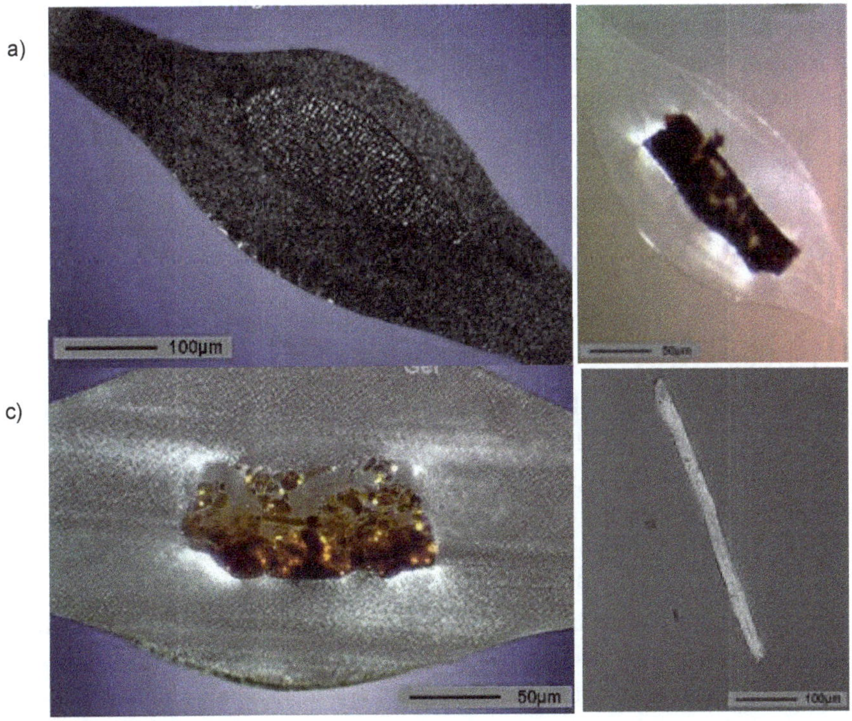

Figure 11.2 Photographs for several common gel types found in PE films: a) a highly oxidized and crosslinked gel, b) a gel with a small particle of carbonized PE, c) a foreign contaminant, and d) a gel caused by a fiber contaminant. Photographs were provided by B. Vastenhout of The Dow Chemical Company

11.2.1 Protocols for Gel Analysis

Established protocols for gel analysis in polymer films are well documented in the literature [6–8]. Typically a film with defects is visually inspected using a low power dissecting microscope. The gels can be classified based on size, color, and shape, and isolated using a razor blade or scissors. Cross sections of the gels ranging from 5 µm to 10 µm thick are collected at temperatures below the T_g of the film using a cryogenic microtome; about -80 °C to -120 °C. For optical examination, a thin section containing the gels is placed on a glass microscope slide with a drop of silicon oil and covered with a glass cover slip. Additional sections are collected for examination via hot stage microscopy and for compositional identification if needed.

After collecting the sections, the remaining polished block-face containing the remainder of the gel is retained. In many instances, gels arise from inorganic contaminants such as the metallurgy from pellet handling equipment, extruders, or components from masterbatches. Examination of these inorganic components are best performed with the block-face sample using a scanning electron microscope (SEM) equipped with an energy dispersive X-ray detector (EDX) [9, 10]. In some cases, additives or inorganic residues are present in low concentrations within the gels. A method to enrich the concentration of these materials is to expose the block-face containing the gel to oxygen plasma. Etching will preferentially remove the polymer at a much faster rate than the inorganic materials, enriching these components for elemental analyses. It must be noted that prior to SEM and EDX analyses, a thin conductive coating like carbon is typically evaporated onto the sample to render it conductive under the electron beam.

The next sections will demonstrate various methods of analyses used for common gel types.

11.2.1.1 Oxidized Gels

The most common type of gel is caused by oxidative processes that crosslink the PE chains. The best way to identify this gel type is by observation with polarized light and ultraviolet (UV) light sources. Transmitted polarized light microscopy represents an effective technique [11] that can be used to investigate structures in crystalline films. For example, black speck gels were contaminating a multilayer film product. The gels were relatively brittle when cut for analysis. The source was unknown. The detail of a gel is clearly visible using transmitted polarized light, as shown in Figure 11.3(a). Close examination of this gel using epi-fluorescence with an ultraviolet light source caused an intense fluorescence emission, as shown in Figure 11.3(b). This type of emission suggests thermal oxidation and crosslinking of the polymer. Micro-infrared analysis of the gel indicated that it contained oxidized PE and maleic anhydride, as shown by the spectrum in Figure 11.4. This material likely formed on the metal surfaces of the extruder and then flaked off during a minor process instability. The material then flowed downstream and contaminated the film as a gel.

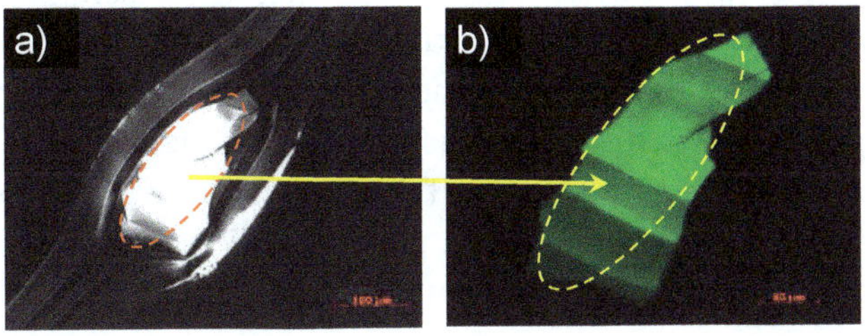

Figure 11.3 Transmitted polarized light images of a thermally oxidized and crosslinked gel in a multilayer film: a) photograph in polarized light, and b) the gel fluorescing under UV light. Photographs were provided by E. Garcia-Meitin of The Dow Chemical Company

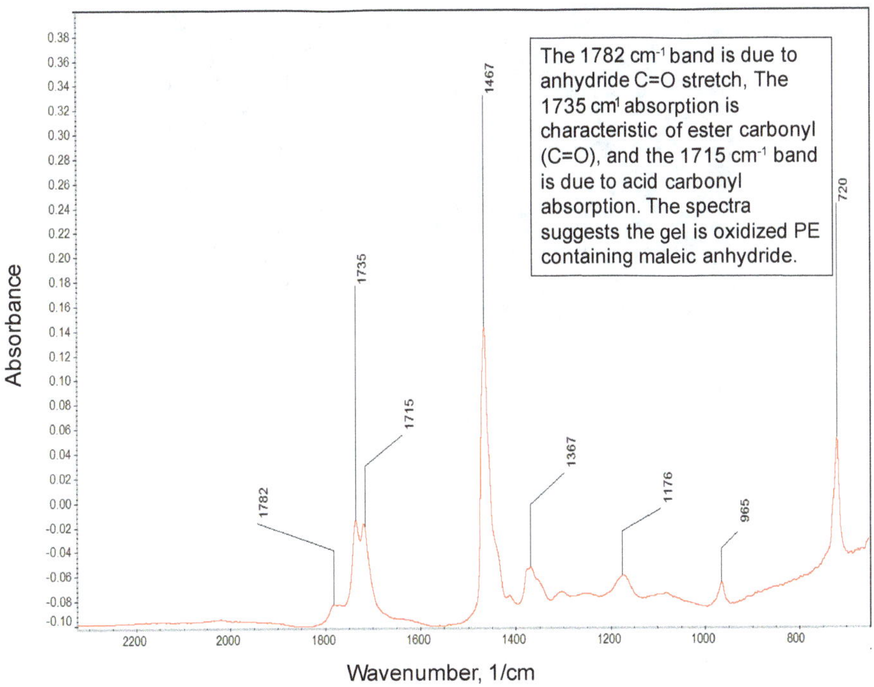

Figure 11.4 The micro-infrared spectrum of gel shown in Figure 11.3. The spectrum suggests it is an oxidized PE gel containing maleic anhydride

11.2.1.2 Crosslinked Gels

Crosslinked gels are oxidized gels, but the level of oxidation is not enough to cause them to fluoresce under UV light. The gels may have a level of crystallinity and thus be birefringent under polarized light. For example, the slightly birefringent gel shown in Figure 11.5(a) was studied using a temperature programmable hot stage, polarizing light microscope [11]. The optical melting temperature (T_m) of the gel was measured at 128 °C and consistent with the PE used to make the product, as shown in Figure 11.5(b). To determine if the gel was unmixed (highly entangled but not crosslinked), the gel was held above the melting temperature (135 °C) and then stressed. A dental tool was used to stress the top of the glass slip cover. Crosslinked gels will appear birefringent, as shown in Figure 11.5(c), in response to the anisotropy of stress distribution in the gel to polarized light. The gel dimensions and shape remained after cooling, verifying crosslinking, as shown in Figure 11.5(d). If the gel was highly entangled and not crosslinked (unmixed gel), the gel would have disappeared after the stress and cooling were applied.

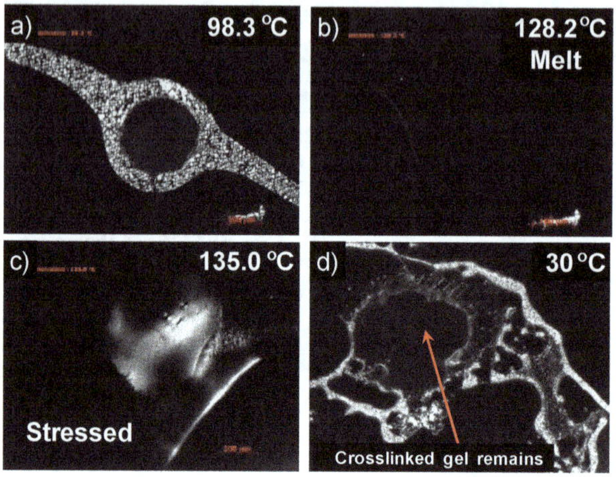

Figure 11.5 Hot stage microscopy of a crosslinked gel in a crystalline monolayer film: a) below the melting temperature, b) optical melting point at 128 °C, c) appearance of birefringence after stressing at 135 °C, and d) intact crosslinked gel after cooling to 30 °C. Photographs were provided by E. Garcia-Meitin of The Dow Chemical Company

11.2.1.3 Gels from Foreign Contamination

The origin of defects causing discoloration in polyolefin pellets can be identified using light and electron microscopy. For example, PE pellets from an in-plant recycle re-pelletizing process contained pellets that were off color and had black specks, as shown in Figure 11.6(a). One of these defects was isolated using the cross-sectioning technique, as shown in Figure 11.6(b). The cross section revealed an intense reddish particle that caused the discoloration of the pellet.

SEM and EDX microanalyses were used to determine that the defects contained primarily iron and oxygen, and it likely was iron oxide. Figure 11.7 shows a backscatter electron image (BEI) of the pellet block-face sample showing the defect causing the discoloration and the elemental spectrum. Metallic based defects can originate from process equipment, railcars used for shipment, pellet transfer lines, and poor housekeeping. The origin of the iron oxide was likely from a storage bin.

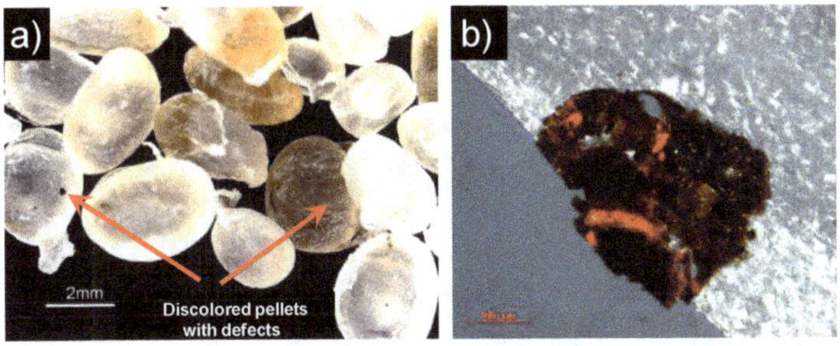

Figure 11.6 Photographs of foreign contamination in pellets of a re-pelletized reclaim stream: a) photomicrograph of discolored PE pellets containing dark defects, and b) transmitted polarized light micrograph of a pellet cross section containing a defect. Photographs were provided by E. Garcia-Meitin of The Dow Chemical Company

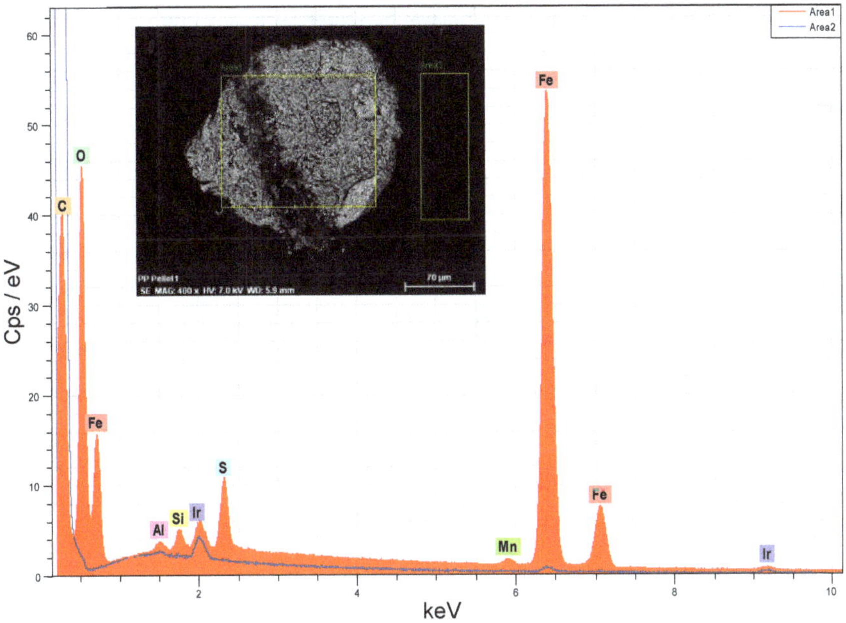

Figure 11.7 EDX microanalysis of an inclusion in a polyolefin pellet cross section of Figure 11.6(b). The analysis indicated that the particle was likely iron oxide

In another example, a multilayer film product was experiencing occasional gels. The gels were isolated and the cross sections were collected as shown in Figure 11.8(a). These gels contained highly birefringent particles that resided in the core layer. The outer film layers appeared amorphous and the core layer was slightly birefringent. The optical melting temperature of the core layer was determined to be 123 °C while the birefringent gels melted at 265 °C. The melting temperature of 123 °C was consistent with the PE resin used to produce the core layer. The higher melting temperature material and micro-infrared analyses of the defects indicate that they were foreign contaminants, and they were identified as a polyester resin.

The polyester resin was used in another process in the converting plant, and it inadvertently contaminated the PE feedstock.

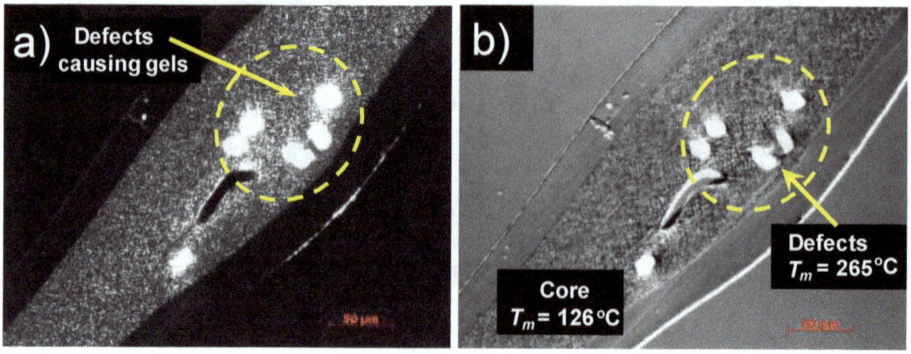

Figure 11.8 Photographs of gels in the core layer of a three layer film: a) transmitted polarized light, and b) hot stage microscopy was used to determine the melting temperatures of the core resin and defects. Photographs were provided by E. Garcia-Meitin of The Dow Chemical Company

Another common contaminant that produces gels is fiber, as shown in Figure 11.9. In many cases, these contaminants are cotton fibers from clothing and gloves or cellulosic fibers from packaging materials. Fourier transform infrared (FTIR) spectroscopy is one of the best techniques for determining the chemical functionality of organic based defects in PE films. The infrared absorbance characteristics of the defect were determined using FTIR spectroscopy, as shown in Figure 11.10. The broad absorption bands near 3600 cm^{-1} to 3100 cm^{-1} are due to hydroxyl (-OH) stretching vibrations, the C–H vibration stretch is near 2916 cm^{-1} to 2851 cm^{-1}, and the ester carbonyl group absorption is near 1734 cm^{-1}. Based on the infrared absorption characteristics, the defect in the PE film is a cellulosic fiber with degraded PE resin.

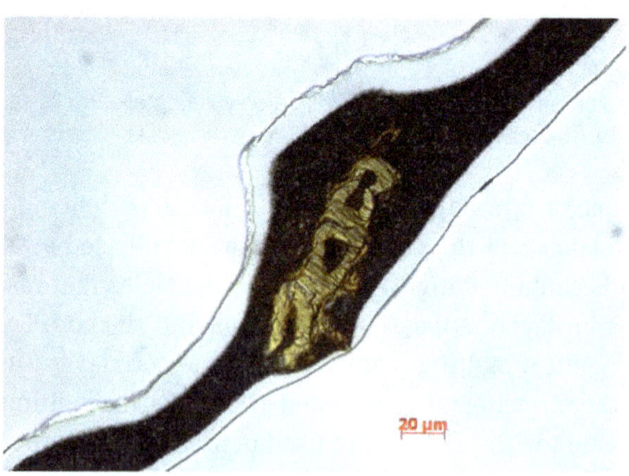

Figure 11.9 Transmitted bright-field image of a PE film containing a fibrous gel. Photographs were provided by E. Garcia-Meitin of The Dow Chemical Company

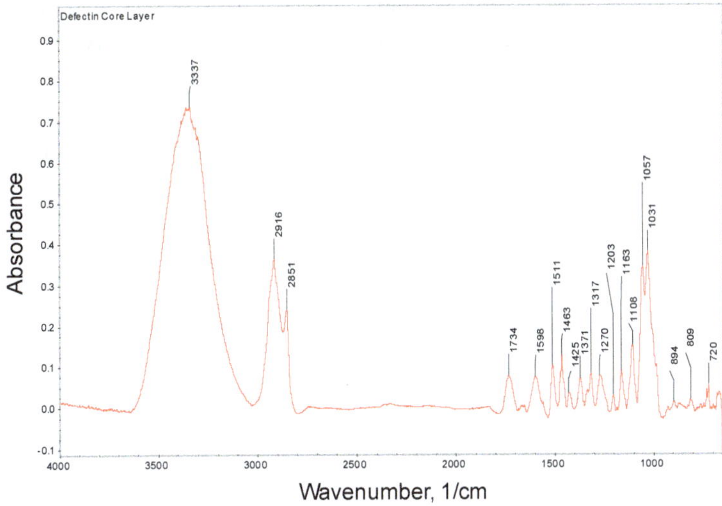

Figure 11.10 FTIR spectrum of defect in a polyolefin film. The spectrum indicates that the gel is cellulosic fiber and degraded PE resin

Once the contaminant is identified, the troubleshooter must determine how the material entered the feedstock stream. Process controls must be identified and implemented to mitigate the contaminant source.

11.3 Resin Decomposition in Stagnant Regions of a Process

Poorly designed extrusion lines often have regions in the process that are stagnant or have very low velocities. These regions can cause the resin to degrade and form a multitude of degradation products. As discussed in Chapter 2, all resins degrade at high temperatures and long durations. The durations typically range from minutes for thermally sensitive resins to 12 days or longer for LDPE resins. In all cases, the degraded resin will break away from the extruder surfaces and flow downstream, eventually contaminating the extrudate and the product. Screws, transfer lines, filtering systems, and dies must be designed such that stagnant regions and regions with very long residence times (that is, very low velocities) do not exist.

Several common problems can occur in the design of screw channels leading to stagnant regions and regions with very long residence times. Resin degradation originating in the screw channels will typically appear in all locations in the finished product. That is, as the degraded fragment breaks away from the screw channel, the downstream sections of the screw will tend to disperse and distribute the fragments to all portions of the extrudate. These fragments will typically exit the extruder as a "shower" of fragments. These common problems will be discussed in the case studies presented in this chapter. If the degraded fragments always occur in the same location in the product, then the location for the degradation is downstream from the screw (and static mixer if used) such as the transfer lines, filtering system, and die.

11.3.1 Transfer Lines

Transfer lines are essential to every extrusion line in the plastics industry. A transfer line serves as a flow passage for molten polymer between the discharge of the extruder and the downstream equipment including screen changing systems, gear pumps, and dies. Many transfer lines are designed to minimize pressure consumption between the extruder and die. It is important, however, to design the transfer line such that the residence time and pressure consumption are manageable, and the shear stress at the wall is high enough to eliminate stagnation at the wall. If stagnation at the wall occurs, then a continuous level of gels will occur in the product.

When designing a transfer line, many considerations are given to pressure consumption. The pressure consumption needs to be managed to a reasonable level for the operation of the extrusion line, as an excessively high pressure change will adversely impact the rate as well as the discharge temperature, especially for a smooth-bore extruder. The pressure change along a cylindrical transfer line for a power-law fluid can be calculated as follows [5]:

$$\Delta P = \frac{2LK}{R^{3n+1}} \cdot \left[\frac{Q\left(\frac{1}{n}+3\right)}{\pi} \right]^{n} \tag{11.7}$$

where L is the length of the transfer line, R is the radius of the transfer line, Q is the volumetric flow rate, K is the consistency factor of the power-law model, and n is the power-law coefficient.

As indicated by Equation 11.7, the pressure change through the pipe is proportional to the length of the transfer line and inversely proportional to the $3n + 1$ order of the transfer line radius. Thus, a shorter transfer line length and a larger transfer line diameter will reduce pressure consumption. While it is commonly agreed by designers that the transfer line needs to be made as short as possible, in practice its length is largely dependent upon the equipment layout design, as the transfer line needs to spatially connect the extruder with the downstream equipment. This places a constraint on the design freedom for the transfer line length. Therefore, instead of shortening the length, increasing the diameter of the transfer line is regarded by designers as a more viable approach for managing pressure consumption. This practice, however, has become a common root cause for gels and degradation for many resins.

A large-diameter pipe is very effective at reducing the pressure consumption, but it also decreases the polymer shear stress at the pipe wall. A high wall shear stress is essential for preventing resin degradation, as it provides the driving force to renew the resin layer adjacent to the wall surface. This layer of polymer moves at a very low flow velocity. Without adequate wall shear stress, this polymer layer will stay on the wall and experience an excessively long residence time inside of the transfer line. The long residence time will eventually cause the resin to degrade and appear in the final product as gels and black specks. PE resins can exhibit degradation at a time scale of 1 hour or more if continuously exposed to processing temperatures, even with antioxidants [34]. This type of degradation is often overlooked when designing a transfer line.The shear stress at the transfer line wall should be maintained at a level of 20 to 30 kPa or higher [35]. At this stress level, the resin near the pipe walls will periodically slough off the wall and thus not be in the pipe long enough to degrade. At lower stress levels, the resin near the wall will not slough off and has the potential to be in the pipe long enough to degrade. As expected, the pressure gradient required to achieve flow at these high stress levels will be considerably higher than that for a large-diameter pipe that allows gels to form. For smooth-bore extruders, the higher discharge pressure needed will cause the specific rate to decrease slightly and cause the extrudate temperature to increase. For grooved-bore extruders, the temperature effect with increasing discharge pressure is not a factor since the specific rate will be nearly constant with changing discharge pressure.

The stress at the wall and residence time distributions can be obtained using computational fluid flow techniques [36]. A simplistic method for calculating the shear stress and pressure gradient, however, using a Newtonian fluid is as follows [5]:

$$\frac{\Delta P}{L} = \frac{8Q\eta}{\pi R^4} \tag{11.8}$$

$$\dot{\gamma}_w = \frac{4Q}{\pi R^3} \tag{11.9}$$

$$\tau_w = \frac{4Q\eta}{\pi R^3} \tag{11.10}$$

where ΔP is the pressure change for a pipe of length L, Q is the volumetric flow rate in m^3/s, η is the Newtonian shear viscosity, R is the pipe radius, $\dot{\gamma}_w$ is the shear rate at the wall, and τ_w is the shear stress at the wall.

Transfer line elbows must be designed with a gentle bend such that the bend radius is at least five times greater than the diameter of the pipe. This type of design produces a streamlined flow path without regions where the velocity is extremely low. In contrast, on several occasions transfer line elbows were constructed by drilling holes into a square block of metal, as shown in Figure 11.11(b). In this case, two regions exist that will have extremely low velocities and will allow the degradation of resin: the far corner of the flow path, and the backside and downstream channel of the elbow. These positions are shown in Figure 11.11(b). A better design with a bend radius that is five times larger than the diameter of pipe is shown in Figure 11.11(a). In this case, the half channels are cut into two metal blocks using a ball end mill. The blocks are then bolted together to make the elbow. If the source for degradation is downstream from a screen changer and static mixers are not used, the degradation often appears at the same location in the product.

The design of a die for an application is a very complex process. The design must be such that regions with long residence times do not exist, and it must provide uniform flow at the exit. Moreover, coextrusion dies must be designed with layer uniformity and without process instabilities known as the wave instability and the zig-zag instability [12, 13]. These instabilities are beyond the scope of this book and thus are not discussed further. The velocities in a die should be at least 1.5 cm/s for thermally stable resins (with antioxidant systems) and at least 3 cm/s for resins that are not thermally stable.

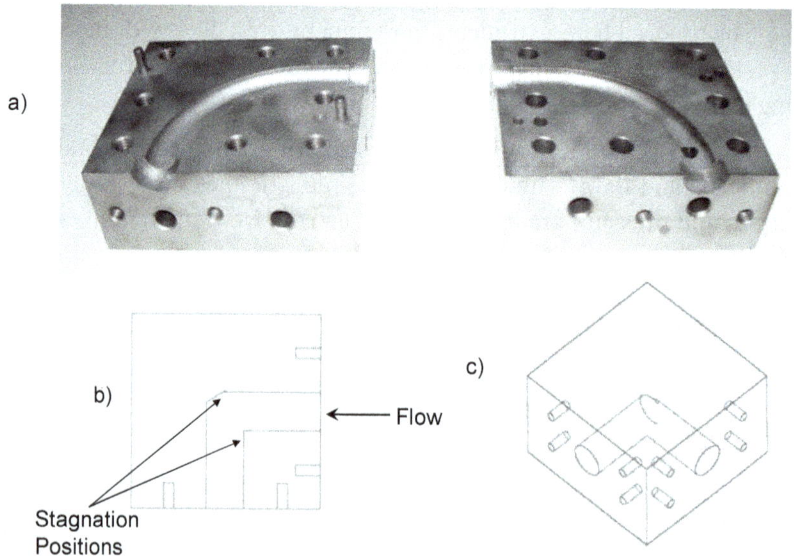

Figure 11.11 Schematic of two transfer line elbows: a) a properly designed elbow made by cutting half channels into two metal blocks using a ball end mill, b) a top view of an improperly designed elbow where a single block of metal was drilled from both corners, and c) oblique view of the improperly designed block. Resin degradation products are likely to form in the stagnation positions shown in b)

11.4 Improper Shutdown of Processing Equipment

Shutting down an extrusion line occurs for many reasons, including planned shutdowns for maintenance, shift changes, changing filtering screens, product changes, and many unplanned events. A shutdown period is defined here as a period when the screw is not rotating and thus resin is not discharging from the line. If the shutdown period is relatively short such that very little resin degradation can occur, then the extruder barrel temperatures can be maintained at the operating set point temperatures. If the length of the shutdown is long relative to the time required to create a significant level of degradation products, then the extruder should be either purged with a more thermally stable resin or the barrel set point temperatures should be decreased to considerably lower temperatures. Sparging the resin with an inert gas to exclude oxygen is also effective at mitigating gels. An extruder that is maintained at process conditions long enough to create degradation products can be very difficult to bring back online running prime product. In this case, the surfaces of the screw and all metal surfaces in contact with molten resin may become coated with degradation products, as shown in Figure 11.12. The time to purge can be extremely long and very expensive. For example, LLDPE resins can form crosslinked gels and black specks after 30 minutes of being offline at process temperatures. If the shutdown is under 30 minutes, the barrel can be held at the process temperature. If the shutdown is longer but the line will be brought back online soon, the screw could be rotated at a low speed of 5 rpm to keep resin flowing, mitigating the formation of degradation products. For longer shutdown periods, the extruder should be purged using an LDPE resin and then cooled to ambient temperature.

Figure 11.12 Photograph of a screw that had numerous shutdowns where the extruder was maintained at operating temperatures for an extended time. The extruder was purged prior to removing the screw, yet dark degraded resin covers most areas of the screw

Antioxidant chemicals are typically added at levels that stabilize the resin during normal melt processing [37]. They are not meant to protect the resin from degradation during an extended shutdown period. As discussed in Chapter 2, antioxidants are slowly consumed during the extrusion process, and thus they can be fully consumed during an extended shutdown period. When they are fully consumed, the resin system is not protected from degradation, and thus degradation reactions will occur at accelerated rates.

Extruding PC resins for optical sheet applications can be difficult since the resin forms black specks on the screw surfaces, transfer lines, filtering system, and the die during extended shutdown periods at the process temperature. These black specks can decrease the quality of the finished sheet product. Other resins such as HIPS and ABS will also create black specks. PS resin will depolymerize, causing a loss in molecular weight and thus a reduction in physical properties. If the depolymerization level is high enough, then components with a brown color can be discharged to the product stream. Decomposition products for common resins are provided in Table 2.4.

11.5 Equipment Purging

Purging an extrusion line is often necessary during resin changes and color changes and to remove resin degradation products from the screw and system. Many different purge compounds are commercially available and with different methods for their use. Purge compounds can be classified into three major categories: (1) reactive chemical purge resins, (2) abrasive purge resins, and (3) simple purge resins. The chemical purge resin is added to the extruder hopper and is extruded such that the channels and downstream flow areas are filled with the compound. Next, screw rotation is stopped and the compound is allowed to contact the metal surfaces for up to 15 to 20 minutes, and then it is purged using the next resin. The purge material is designed to chemically break the polymer backbone and loosen degradation products from metal surfaces. Many of the compounds have foaming agents that expand into regions that are void of resin. The technique is effective for processes that are very streamlined and thus do not have stagnant or void regions in the flow path. Purging degraded resin, purge material, and old resin from processes that contain stagnation (void) regions, however, can be extremely difficult, causing the unwanted resin and purge compound to contaminate the extrudate for extended periods of time after startup. Abrasive purge resin compounds contain mineral fillers, glass strands, or thermoset resins dispersed into a base resin. As expected, these compounds tend to abrade the old resin and degradation products from the metal surfaces. Many of these compounds are developed so that they have a low affinity for metals, allowing the compound to purge cleanly from the system. Lastly, a simple resin purge uses the next grade or color of resin to push out the current resin from the system.

The purging process must be as quick as possible and utilize the least amount of purge compound and resin. Longer purge times will add additional costs to the product and will decrease the profitability of the plant. The best practice for purging with resin is to vary the screw speed from a level that is relatively low to a speed that is considerably higher [14, 15]. For example, the extruder could be operated at a screw speed of 15 rpm for several minutes, and then the speed should be increased as quickly and as safely as possible to about 60 rpm or higher. The screw speed should be held at this high level for several minutes before decreasing the screw speed back to 15 rpm. The process should be repeated until the extrudate is free of contaminants and old resin.

Many resins such as LLDPE resins are thermally sensitive. This class of resins can form crosslinked gels at processing temperatures in under 30 minutes. During planned shutdowns of the extruder, these resins cannot be maintained at processing temperatures for extended periods of time. Ideally the extruder should be purged with a more thermally stable material such as an LDPE resin before the barrel temperatures are decreased. When the extruder is started back up, the LLDPE resin can be used to purge the LDPE resin. Failure to purge the LLDPE resin from the extruder can lead to the creation of crosslinked gels that adhere to the surface of the screw, transfer lines, and die components. Purging these gels from the system can be costly and time consuming.

Stagnant regions in the process stream are problematic for both production operations and system purging. For these poorly designed systems, the stagnant regions will be filled with degraded resin and hard carbonized materials. The purge will be effective at removing most of the degradation materials, but the stagnant areas will either not be purged or will retain some of the purge material. During the startup of the process after purging is complete, some of the purge material and degraded material in the stagnant areas will move from the stagnant area to the main flow stream, contaminating the extrudate. The best way to eliminate this problem is to eliminate the stagnant regions in the process. Several case studies are included in this chapter where stagnant regions occurred in the process.

11.6 Oxygen Exclusion at the Hopper

The exclusion of atmospheric oxygen can often reduce the oxidative degradation processes that occur in the extruder, especially for thermally sensitive resins such as LLDPE and EAA resins. For this process, an inert gas such as nitrogen is added inside the hopper just above the feed opening of the screw. A gas vent is positioned on the top cover of the hopper system to discharge the nitrogen purge and the air entrained with the incoming pellet stream. Engineering controls must be developed so that the added nitrogen gas is incapable of flooding the extrusion area and thus reducing the breathing oxygen content to dangerously low levels.

When nitrogen inerting of pellets entering a flood-fed extruder is required, it is recommended that a small-diameter manifold (for example, drilled tubing) be installed across the feed hopper as close as practical to the feed throat of the extruder to help ensure the uniform distribution of nitrogen. A single entry point through the side of the hopper will result in the nitrogen channeling up the side, and the opposite side of the hopper will see little to no nitrogen flow, allowing oxygen to enter the extruder. The manifold can consist of a 9.5 mm diameter tube with two to three uniformly spaced drilled openings, as shown in Figure 11.13. For proper nitrogen gas distribution, the cross-sectional area of the manifold should be at least three times the sum of the cross-sectional areas of the openings. When installing the manifold, the drilled holes should be pointed down towards the screw.

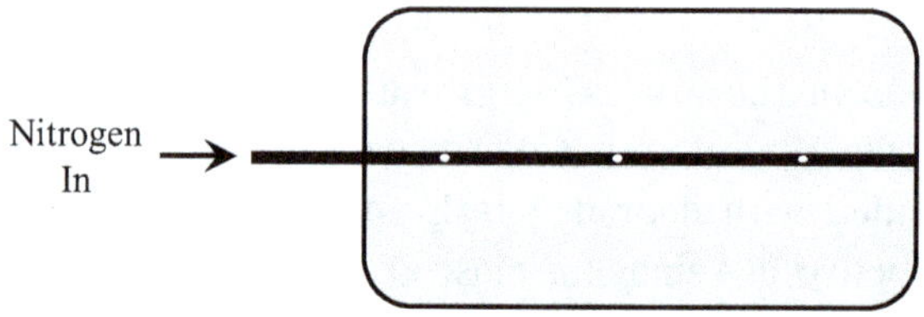

Figure 11.13 Schematic of a nitrogen purge manifold for a small diameter extruder. The white exit holes here are facing the screw

In order to ensure adequate inerting, the required nitrogen flow rate should be three times the air volume coming in with the pellet feed to the extruder. For example, if the pellet feed rate to the extruder is 200 kg/h, and if the pellet feedstock bulk density is 660 kg/m^3 and the solid density of the polymer is 920 kg/m^3, the incoming air rate is 0.086 m^3/h. Therefore, the minimum nitrogen feed rate is 0.26 m^3/h or 4.3 L/min. The calculation is as follows:

$$V_a = m\left(\frac{1}{\rho_b} - \frac{1}{\rho_s}\right) \tag{11.11}$$

where ρ_b is the bulk density of the feedstock in the hopper, ρ_s is the solid density of the resin, m is the mass rate for the extrusion in kg/h, and V_a is the air rate that is being expelled from the extruder in m^3/h. The units for ρ_b and ρ_s are kg/m^3. Based on a mass balance, the recommended amount of nitrogen purged (V_N) should be three times the volume of air expelled as follows:

$$V_N = 3V_a \tag{11.12}$$

If the extruder feed is granular or powder form, additional calculations are recommended to ensure the nitrogen flow does not result in fluidization of the feedstock. Also, whenever nitrogen is used, adequate ventilation is required to prevent asphyxiation and death. Oxygen exclusion is extremely important for large-diameter melt-fed extruders used for the production of polyolefin resins. Like discussed above, oxygen from the atmosphere can cause the formation of crosslinked gels on metal surfaces. The gels will eventually cause contamination in the pelletized product. The main regions for oxygen intrusion are at the feed section, screw shank, and vent openings for devolatilizing extruders. This topic is covered in more detail in Section 15.3.3.

11.7 Flight Radii Size

The size of flight radii on the screw channel is important for the proper performance of a single-screw extruder. The specifications of the flight radii are commonly made using a combination of personal experience and guidelines [16] from the Society of the Plastics Industry, Inc. (SPI). These guidelines state "unless otherwise specified the root radius will not be less than one-half of the flight depth up to 25 mm radius." Many screws are often designed, however, with flight radii that are very small and approach values that are between 10 and 20% of the channel depth. Moreover, during the screw fabrication process, manufacturing errors can result in flight radii that are less than that specified in the design. Improper design that results in radii that are too small for the application can create regions with long residence times, leading to material degradation. These long residence time regions are caused by Moffat eddies. Moffat eddies were discussed previously in Section 8.3.2. An example of material degradation in the flight radii of a screw channel is shown in Figure 11.14. Figure 11.14(a) is a photograph of a Maddock solidification experiment using a red color concentrate resin in a process using an LLDPE resin. The flight radii and depth of the metering channel were 5 and 18 mm, respectively, that is, radii that are about 28% of the depth of the channel. Degradation at the flight radii was observed. Figure 11.14(b) is a photograph of a screw that was removed hot from an extruder running LLDPE resin. Degradation is clearly observed at both the pushing and trailing flights. For both examples, the flight radii were too small. Larger radii would have prevented the degradation by decreasing the residence time of the material in the radii region. Once the degradation products are formed, a small process upset will eventually occur and dislodge the degradation products from the screw, and they will flow downstream and contaminate the final product.

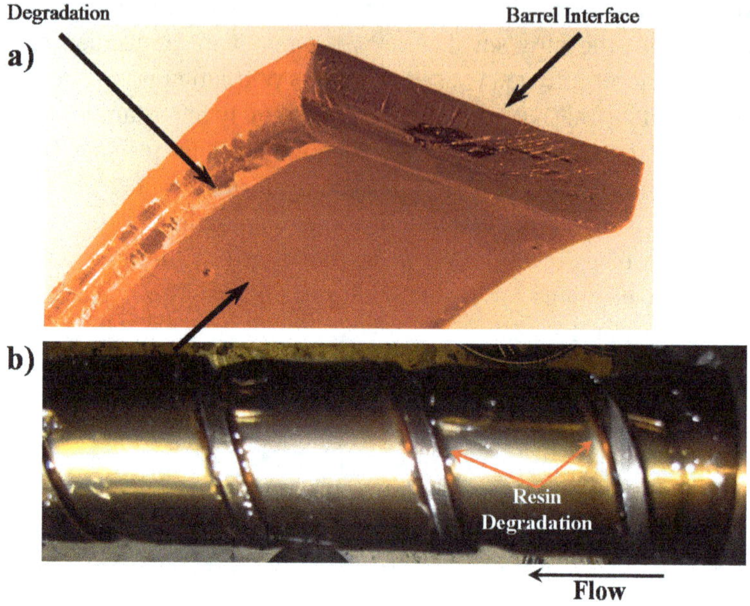

Figure 11.14 Degradation at the flight radii for LLDPE extrusions due to Moffat eddies: a) degradation at the pushing flight radius for a 152.4 mm diameter extruder with an 18 mm deep metering channel and a 5 mm flight radius, and b) degradation at the flight radii for a small diameter extruder

For existing screws, the flight radii can be measured with radius gauges, or they can be estimated by the tangent point on either the flight edge or the root of the screw, as shown in Figure 11.15. The tangent point is where the curvature of the flight radii merges into a straight edge. If the channel radius is less than the depth of the channel, the tangent points will be on both the flight edge and the screw root. The radius in this case, however, is easier to estimate from the flight edge. Thus a channel with a radius that is half the size of the depth will have a tangent point midway up the flight. For a radius that is larger than the depth of the channel, the tangent point is on the root of the screw.

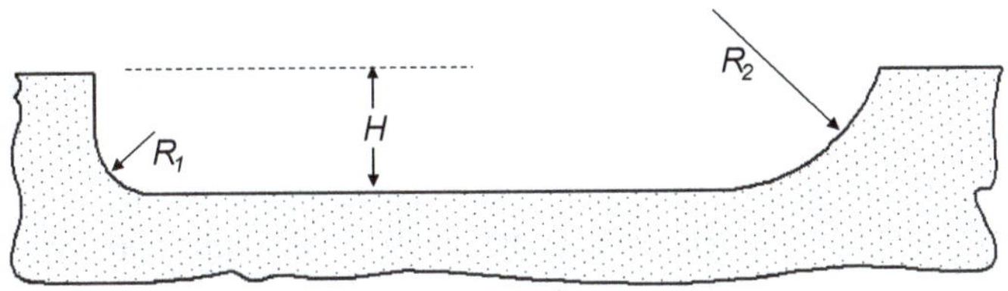

Figure 11.15 Flight radii and tangent points in a screw channel. The pushing flight R_1 in this case is estimated at 0.5 the depth of the channel (H) while the radius R_2 at the trailing side of the channel is estimated at 1.5 times the channel depth

The effect of flight radii size in the metering section has been studied previously [17–20, 38], mostly from numerical simulations. These simulations indicate that flows at the radii of a screw channel can be extremely low in both the cross-flow and downstream directions for channels with very small radii [17–19]. These low flows lead to regions with very long residence times. If the residence time is long enough, material will degrade and eventually contaminate the extrudate. The magnitude of the flows in these regions, however, is considerably higher for channels with radii larger than the depth of the channel. Some designers often decrease the radii in the metering section so that the cross-sectional area available for flow will be maximized for a given channel depth. These designers believe that the specific rate will be increased by decreasing the radii. Numerical simulations [20], however, indicate that the pressure gradient and thus the rate are essentially the same for both channels with large and small radii, indicating that the small radii have little effect on the rate. Flight radii recommendations are provided in Table 11.3. In general, the flight radii should be between 0.5 to 2.5 times the depth of the metering channel. Thus for most resins, the SPI guideline as a minimum is appropriate for this section, but larger radii are better; radii up to 2.5 times the depth are acceptable. For resins that do not adhere readily to metal surfaces, the molten resins will tend to pull away from the walls of the screw, and thus even for very small radii there will be no regions with high residence times. Conversely, thermally sensitive resins, resins that adhere to metal, and resins with low elasticities would extrude with lower levels of degradation if the flight radii were large and about 2.5 times the depth of the channel. For PE resins, radii that are 1.5 times the depth are recommended.

Table 11.3 Flight Radii Recommendations to Minimize Resin Degradation at the Channel Edges [20]

Screw Section	Flight Radii to Channel Depth Ratio
Solids conveying	0.25
Transition	Blend radii sizes from the solids-conveying section to the metering section
Metering	0.5 to 2.5

Degradation rarely occurs in the feed section of a plasticating screw due to the very low temperatures of the resin in the section. Thus, the size of the flight radii in this section is not influenced by resin degradation. Instead, the size of the flight radii is based on the ability to convey solids away from the hopper and into the melting section. As discussed in Section 5.3.2.2, experimental solids-conveying data indicated that the flight radii size in the feed section results in two competing factors that affect rate: (1) reduced area for a channel with large radii, resulting in reduced rates, and (2) increased forwarding forces for a channel with large radii, leading to higher rates. Factor 1 dominates at low solids-conveying pressures and factor 2 at high pressures. The second factor is resin dependent, and in most cases small radii are recommended; that is, radii of about one-quarter of the channel depth for the feed section are recommended.

The flight radii in the melting section should be adjusted to reflect the amount of molten material present in the channel. Like the metering section, the goal is to minimize degradation of material at the flight radii. For example, at the entry region of the melting section where only small amounts of molten material are present at the pushing side of the channel and the channel is relatively deep, the flight radii should be about that used for the solids-conveying zone. When the melt pool becomes large at the exit of the melting section, the flight radii should approach the radii used in the metering section. Thus in order to minimize degradation, flight radii should transition from the radii in the feed section to those in the metering section. This is the current practice used by most screw designers.

11.8 Drying the Resin

Many resins must be dried prior to processing in an extruder. Improper drying can allow small amounts of water to be entrained with the pellets. The water here can be adsorbed onto the surface of the pellet feedstock or absorbed into the bulk of the pellets. For most processes, the resin is air conveyed from a storage bin or silo to a desiccant drier. The drier is set at a temperature and humidity level such that the resin can be dried according to the resin manufacturer's recommendation. The humidity level is measured as the dew point temperature of the drying air. The operation of desiccant driers was presented previously [1]. The resin must stay in the drier longer than a minimum residence time. Failure to meet the residence time, temperature, and humidity specifications has the potential of allowing too much water to enter the extrusion process. For polyesters, PC resins, TPU resins, and many resins made by condensation reactions, trace amounts of water and the high temperatures in the extruder can cause the molecular weight of the resin to decrease, reducing the physical properties of the product. This degradation method was described in Chapter 2. In other cases, the water can exit with the extrudate and create bubbles in the product. During injection molding, the water vapor can migrate to the surface of the tool and create a splay defect. Splay is a common term for surface imperfections that occur on injection-molded parts, and it is discussed later in Section 11.11.1.

The drier must be sized properly to maintain the resin longer than the minimum recommended time. For example, an injection-molding process was running a HIPS resin on large size housing parts. During the startup of the shift, the parts produced were in specification and did not have splay defects. After about 50 minutes of operation, small levels of splay were observed in about 30% of the parts. During the startup phase of the process, the resin was stored in the desiccant drier for more than two hours. But during steady-state operation, the drier only provided a residence time of about 20 minutes. The recommended drying conditions for this resin were 70 °C temperature, −40 °C dew point, and a residence time of 45 minutes. For this application, the drier was too small and did not provide enough residence time to dry the resin.

11.9 Color Masterbatches

Color masterbatches or concentrates are commonly used to color natural resins during the final fabrication step. This technique is sometimes referred to as natural-plus, referring to the extrusion or molding of natural resin plus a color concentrate. The process is discussed in detail in Section 8.6. Poor mixing in the extruder or by the compounding process that is used to make the masterbatch can cause color variations and streaks in the product. For example, color streaks can occur in the product due to poorly dispersed pigment in the color masterbatch. For this case, the pigment particles in the color concentrate are either agglomerated from a poorly designed twin-screw compounding process or are not dispersed due to a low viscosity carrier resin. For example, a twin-screw compounding operation for making a specialty resin from a HIPS resin and specialty filler chemical was not designed properly. Here the filler chemical was partially agglomerated prior to the melting process in the twin-screw extruder. As shown in Figure 11.16, the resin was colored black and the filler chemical was white. These white agglomerates could not be eliminated in the final injection molding process and created defects in the product. The goal for this type of application is to produce masterbatches that are free of filler agglomerates because the final plasticating process is incapable of dispersing them.

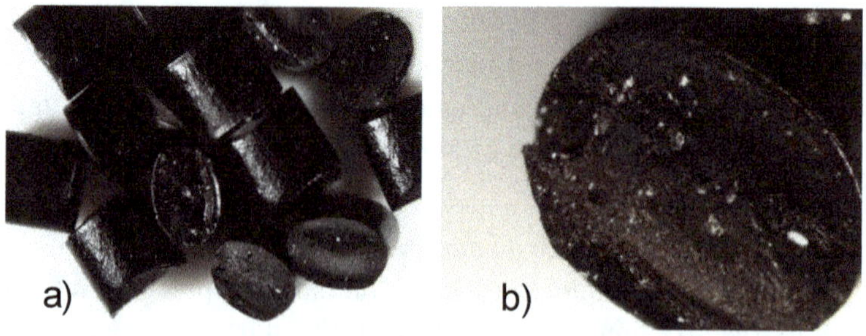

Figure 11.16 Photographs of specialty HIPS resin pellets made using a poorly designed process. The white specks are filler agglomerates: a) 1X magnification, and b) 4X magnification

The root cause of dark color streaks in gray-colored parts can sometimes be more difficult to determine since both poorly dispersed pigments and resin degradation products can be the source. In these cases, removing the color masterbatch from the process and operating with just the natural resin will allow the root cause to be identified. That is, if the dark color streaks still occur in the product after the color masterbatch is removed, then the source of the problem is related to resin degradation. Two case studies are shown in Section 11.12 to illustrate these defects.

11.10 Case Studies for Extrusion Processes with Contamination in the Product

Diagnosing and eliminating a problem that occurs due to an improperly operating extruder can be difficult and time consuming. This section contains several case studies where improperly designed processes created contamination defects in the final articles. The axial pressure profiles of the metering channels for these case studies and other studies in the next chapters were calculated using the method described in Section 9.2.1. The simulated axial pressure profiles are shown with solid lines while the estimated pressure profiles are shown by dashed lines.

11.10.1 Intermittent Crosslinked Gels in a Film Product

A film plant was extruding an LDPE resin into a specialty product using a cast process [21, 22]. Due to high demand, a new 88.9 mm diameter, single-stage, 33 *L/D* extruder was installed in the plant. Soon after startup the product was acceptable and high quality. After 12 days, the line began to experience intermittent discharges of crosslinked material (gel showers) and carbon specks. Photographs of these gels are shown in Figure 11.17. In some cases, the gel showers were observed two to three times per day and would last from one to five minutes. The gels were clearly crosslinked and were brown in color. The extrudate temperature was higher than expected for the process. The intermittent gels resulted in production downtime due to purging and in numerous customer complaints. A high and costly level of quality control was required to remove the gel-contaminated product from the prime product. Due to the high amount of downtime and the high levels of quality control needed, operation of the new line was considerably more expensive than planned.

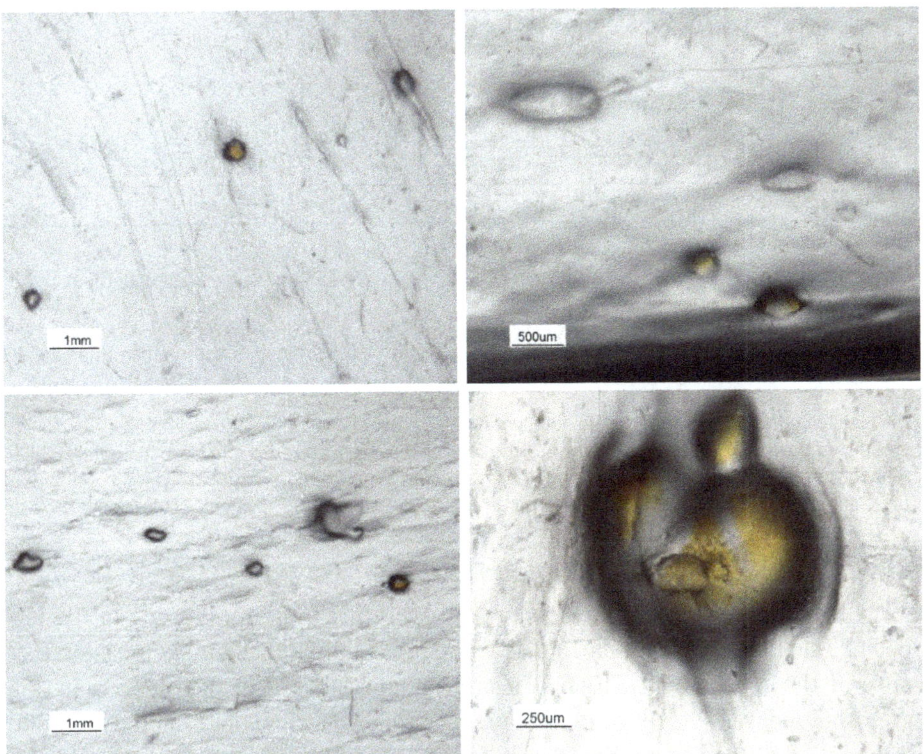

Figure 11.17 Photographs of crosslinked gels in an LDPE resin film for the case study of Section 11.10.1

At the startup of the line, the extruder was operated at 91 rpm to produce the required rate of 148 kg/h for a specific rate of 1.63 kg/(h·rpm). The temperature of the extrudate was measured through the transfer line wall at 232 °C. Due to process safety constraints the extrudate temperature could not be measured using a handheld temperature sensor. The extrusion rate was required in order to maintain the downstream take-away equipment at its maximum rate. At first the extruder appeared to be operating well except that the specific rate was lower than predicted. That is, the screw was rotated at an rpm that was higher than expected to produce the 148 kg/h. At 91 rpm, the rotational flow rate was calculated at

228 kg/h; the specific rotational flow rate was calculated at 2.51 kg/(h·rpm). Thus, the line was operating at only 65% of the rotational flow rate. A barrier design screw was used for this process, and the screw channel dimensions are given in Table 11.4. The lead length was 124 mm for the main flight of the barrier section and 88.9 mm for all other sections of the screw. The lead length for the barrier flight was larger than the main flight of the barrier section because the barrier flight started at the pushing side of the main flight at the entrance to the section and transitioned to near the trailing side at the end of the section. Both the width and the depth of the solids channel for the barrier section decreased in the flow direction. The LDPE resin had an MI of 6.4 dg/min (190 °C, 2.16 kg).

Table 11.4 Screw Channel Dimensions for an 88.9 mm Diameter, 33 *L/D* Barrier Screw for Running an LDPE Resin

	Depth, mm	Length, diameters	Notes
Feed section	15.7	4.7	
Barrier section		9.1	0.25 mm barrier flight undercut
First meter	5.84	12.6	
Maddock-style mixer	25.2	2	0.51 mm undercut on the mixing flight
Second meter	5.84	4.6	

The lead length was 124 mm for the main flight of the barrier section and 88.9 mm for all other sections of the screw. The main flight width and clearance were 9 and 0.09 mm, respectively, in all sections of the screw. The first 2.5 diameters of the screw were inside a water-cooled feed casing. The compression ratio was 2.7 and the compression rate was 0.0050. The specific rotational rate was calculated at 2.51 kg/(h·rpm).

It was hypothesized that the extruder was operating partially full due to the low specific rate during operation. To determine if partially full channels were the root cause of the reduced rate, high discharge temperature, and degraded material, screw rotation was stopped and the screw was removed while hot from the extruder. Examination of the polymer on the screw indicated that in the first meter section about half of the channel width on the trailing sides of the flights for all but the last diameter (about 11 diameters) were filled with a dark-colored, partially carbonized LDPE resin, indicating that these regions were stagnant. The reduced flow rate caused these regions to be partially filled, creating void regions on the trailing side of the channel. Some of the resin adhered to the trailing side of the screw in the void regions and stayed there for extended time periods, as shown in Figure 11.18. Figure 11.19 shows schematically a cross section of the channel in the partially filled section. The resin adhering in the void regions eventually degraded into the dark-colored, crosslinked material. Small process variations dislodged some of this material and caused the intermittent gel showers that contaminated the product. Moreover, compacted solids were found wedged in the channel at the entrance to the barrier section. The wedged material was caused by the relatively large width of the entering solid bed being forced into the continually decreasing width of the solids channel of the barrier section.

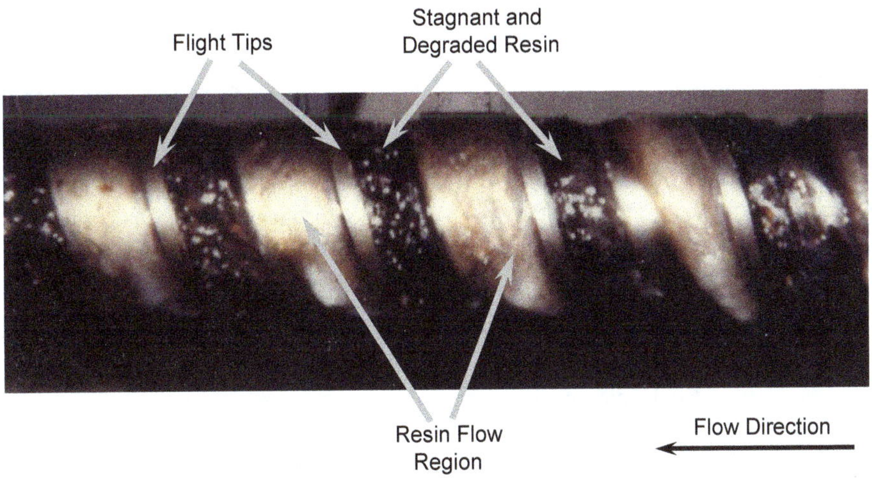

Figure 11.18 Photograph of the main flow and stagnant regions on a barrier screw running LDPE resin [22]. The white polymer is the main flow region, and the dark material is degraded resin in the stagnant region

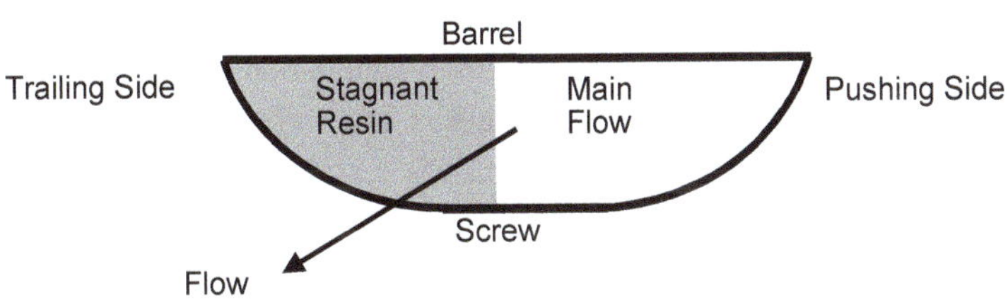

Figure 11.19 Schematic of the cross section perpendicular to the flight of the partially filled metering channel of Figure 11.18

An estimate of the pressure gradient in the screw metering channel was calculated to determine independently if the screw channels were full. The pressure gradient needed to reduce the rate from the rotational flow value to that measured for this LDPE resin was estimated at 1.6 MPa/diameter. The pressure measured at the extruder discharge was 13 MPa, and thus about 8 diameters of metering section were required to generate this pressure, suggesting that about 9 diameters in the first metering section were at a pressure of zero and thus only partially filled. The axial pressure profile for this case is shown in Figure 11.20. The solid line in this figure was calculated using the methods described previously in Section 9.3, and the dashed line represents the expected pressure profile and was not calculated. This numerical analysis is very consistent with the experimental results described above; experimentally, about 11 diameters in the first metering section were partially filled or at zero pressure. These data and computations lead to the conclusion that the barrier section flow was not matched properly to the flow requirements of the metering sections.

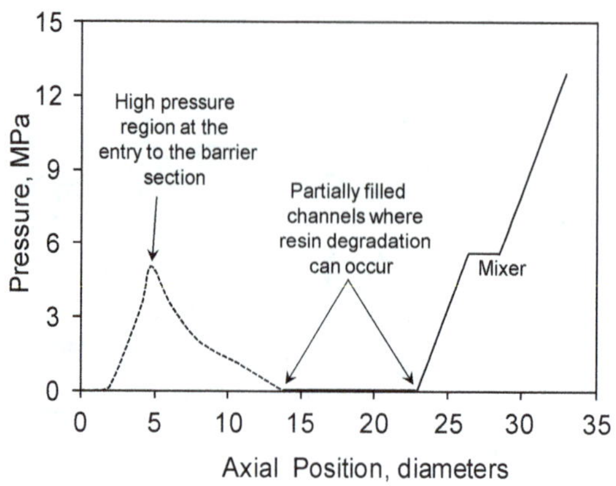

Figure 11.20 Axial pressure profile for the barrier screw. Sections of the screw at zero pressure are partially filled with resin and are a location for the decomposition of the resin. The solid line in this figure was calculated using the methods described previously, and the dashed line represents the expected pressure profile and was not calculated

The compacted solids upstream of the entry to the barrier section and the downstream void region indicated that a restriction at the entrance to the barrier section was limiting the rate. Two technical solutions were provided to the plant. The first solution was a newly designed barrier screw that did not have the restrictive entry region. The entry was positioned further upstream on the screw where the channel pressures would be lower, and thus the resin would not be as compacted at the start of the barrier section. The second solution was to mitigate this restriction on the existing screw by removing metal. The first technical option would require at least six weeks to build and to implement a new screw. The second option is low cost and could be implemented in the plant very quickly. Plant personnel opted to modify the existing screw. The modification was made to the first 3 diameters of the barrier section. For this modification, the depth of the melt channel was increased to that of the solids channel at the entrance, and it was tapered into the depth of the melt channel over 2 diameters. Then the barrier flight was removed for the first 2 diameters and blended in with the melt channel, the shallower of the two channels in this region. The barrier flight for the next diameter was blended into the original undercut. The cross section of the barrier section perpendicular to the main flight at 2 diameters into the section before and after the modification is shown in Figure 11.21. This modification was made in an attempt to allow some of the solids to pass into the melt-conveying channel, eliminating the wedging of solids against the entrance of the barrier flight by streamlining the flight and allowing an increase in the specific rate of the screw.

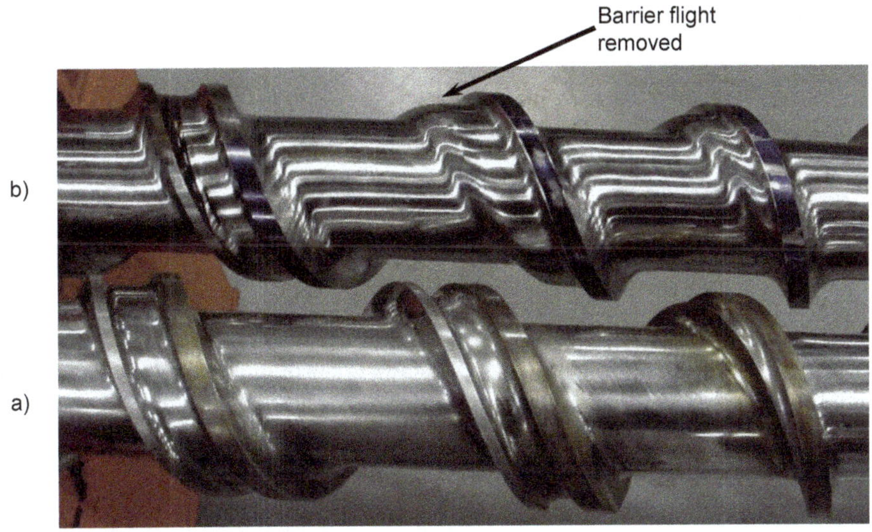

Figure 11.21 Photograph showing the modification to the barrier flight at the entry to the barrier section

After the screw modification, the 148 kg/h rate was obtained at a screw speed of about 69 rpm with an extrudate temperature of 223 °C. Thus, the specific rate increased from 1.63 kg/(h·rpm) before the modification to 2.14 kg/(h·rpm) after the modification, a specific rate increase of about 30%. At a screw speed of 69 rpm, the rotational flow rate was calculated at 173 kg/h; now the extruder was operating at about 86% of the rotational flow rate. The calculated axial pressure gradient required to maintain the flow of the extruder at the reported flow rate showed that pressures in the screw never decreased to zero, indicating that the channels were full as shown in Figure 11.22. No adverse effects were experienced with the reduced discharge temperature (8 °C lower), no unmelted material was observed in the extrudate, and no gel showers occurred after the modification. A summary of the extrusion performance before and after the modification is shown in Table 11.5.

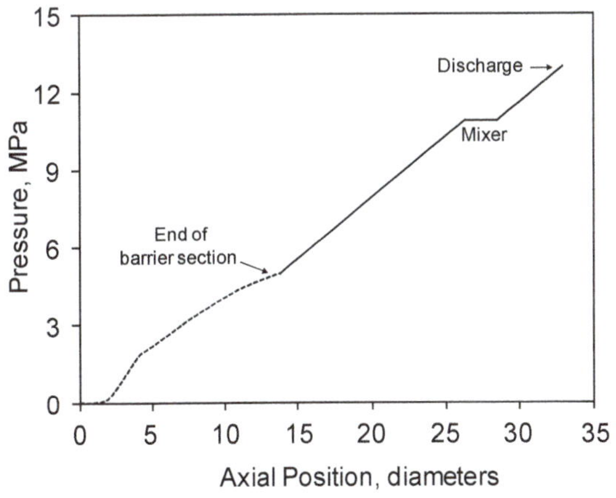

Figure 11.22 Axial pressure profile for the barrier screw after the screw modification. All sections of the screw are filled and operating under pressure. The solid line in this figure was calculated using the methods described previously, and the dashed line represents the expected pressure profile and was not calculated

Table 11.5 Extrusion Rates and Performance Before and After Modifications to the Entrance of the Barrier Section

	Before Modification	After Modification
Rate, kg/h	148	148
Screw speed, rpm	91	69
Specific rate, kg/(h·rpm)	1.63	2.14
Discharge temperature, °C	232	223
Gel showers	Yes	No

Barrier-flighted screws for plasticating extruders are economically important to commercial processors. The barrier section permits extrusion at higher rates, reduces the occurrence of solid bed breakup, and improves the quality of the extrudate. The entrance to the section, however, can cause a restriction in the flow path that can reduce the rate of the extruder. This restriction occurs due to an abrupt change in the cross-sectional area at the exit of the feed section and entrance to the barrier section, and it is accentuated by polymer properties. Modification to the barrier flight and melt channel at the entrance can eliminate this restriction without decreasing the performance advantages of the barrier section. Although not observed during this case study, flow surging may occur with this type of restriction. This type of restriction has been observed on many different resins including LLDPE, HDPE, PP, and PS resins. The restriction typically becomes more acute for resins with a higher modulus and for smaller diameter screws. If this restriction is likely to occur, the screw designer will often position the start of the barrier section farther upstream so that the solids do not have time to compact prior to entering the barrier section.

11.10.2 Small Gels in an LLDPE Film Product

The stretch performance of an LLDPE resin film product manufactured using a cast film process was less than desired at a customer's plant. That is, the film was breaking under the tension of the process. Film breaks are typically caused by flaws or imperfections in the film such as gels. At this plant, several of the lines produced films with acceptable stretch performance while other lines were having chronic breaks of the film. The same LLDPE resin was used on all lines. The resin had an MI of 2.3 dg/min (190 °C, 2.16 kg) and a solid density of 0.917 g/cm^3. The lines with the chronic film breaks were purchased at the same time and were essentially identical. The extruders were 114.3 mm in diameter and had *L*/*D* ratios of 32. Barrier melting screws were used for the process, and the channel dimensions are shown in Table 11.6. The key features for these screws, the barrier flight undercut and the mixing flight undercut on the Maddock mixer, were both 1.5 mm, and the depth of the solids channel at the end of the barrier section was 3.2 mm.

Table 11.6 Screw Channel Dimensions for a 114.3 mm Diameter, 32 *L/D* Barrier Screw for Running an LLDPE Resin on a Cast Film Process with Unmixed Gels in the Extrudate

	Depth, mm	Length, diameters	Notes
Feed section	19.1	7.5	
Barrier section		15	1.5 mm barrier flight undercut
First meter	7.62	5.5	
Maddock-style mixer		2	1.5 mm undercut on the mixing flight
Second meter	7.62	2	

The lead length was 172 mm for the main flight of the barrier section and 114.3 mm for all other sections of the screw. The main flight width and clearance were 11 and 0.11 mm, respectively, in all sections of the screw. The first 2.7 diameters of the screw were inside a water-cooled feed casing. The compression ratio was 2.5 and the compression rate of the barrier section was 0.0030. The depth of the solids channel at the end of the barrier section was 3.2 mm. The specific rotational flow rate was calculated at 5.2 kg/(h·rpm).

The specific rotational flow rate for the screw was calculated at 5.2 kg/(h·rpm). The line typically ran at a screw speed of 100 rpm and a rate of 590 kg/h for a specific rate of 5.9 kg/(h·rpm). Since the operating specific rate was higher than the specific rotational flow rate, a negative pressure gradient must exist in the metering channel. At these conditions, the extrudate had an unacceptable level of gel imperfections in the film. These gels occurred continuously, and they could not be removed from the extrudate using a screen pack system with a 200 mesh screen. The discharge temperature of the extrudate was measured with a handheld sensor at 290 °C.

The gels from film samples were typically less than 1 mm in diameter. A typical gel is shown in Figure 11.23. This gel was analyzed using a hot-stage microscope at temperatures ranging from ambient up to about 135 °C. As shown by the series of photographs for the same gel, the gel melted completely at a temperature of about 135 °C. When the temperature of the stage was decreased the gel reappeared. But when the gel was increased to 135 °C and a small amount of pressure and movement (smearing) was applied to the gel to impart a small amount of stress, the gel disappeared and would not re-form during a second cooling. Thus, the stress due to the motion was enough to disentangle the polymer chains such that when the temperature was decreased the gel did not reappear. The gels did not have a color tint and thus were likely not oxidized resin. The analysis suggested that these gels were composed of higher molecular weight materials that melt at slightly higher temperatures and require additional stress to disentangle the chains. If the chains are not disentangled in the melt state, they will recrystallize first during the cooling process, creating a gel in the film product. As discussed previously, this type of gel is classified as an unmixed gel. These data suggest that a screw that applies a higher shear stress to the resin during the melting process and at the mixer may mitigate the gel problem.

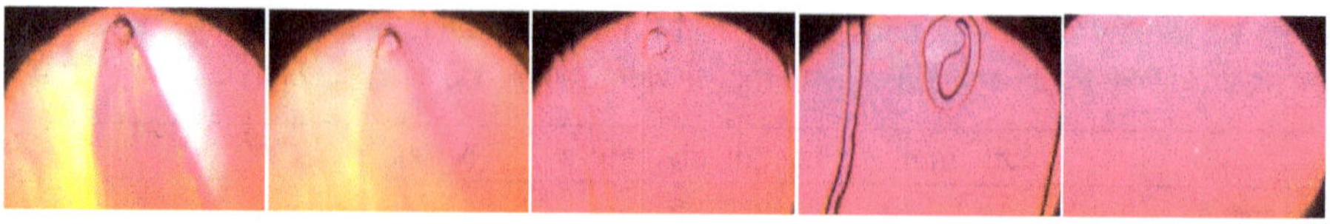

100 °C 125 °C 126 °C 135 °C 135 °C (pressure smeared)

Figure 11.23 Photographs of an unmixed gel at select temperatures using a hot-stage microscope. The circular segment of the gel in the photograph at 100 °C had a diameter of about 0.8 mm. As shown by this sequence of photographs, the unmixed gel melted at about 135 °C. When the gel was smeared by moving the glass cover plate, the stress was enough to disentangle the polymer chains such that the gel would not reappear upon cooling

The data indicated that a similar screw that imparted slightly higher shear stresses would eliminate the unmixed gels from the extrudate. Two technical solutions were proposed to the plant. The first solution was to remove the screw from service and rebuild the barrier flight, mixing flight, and the exit of the solids channel of the barrier section to smaller clearances. The second option was to build a new screw with tighter clearances. The first technical solution would shut the line down for about four weeks and was unacceptable to plant personnel. The second solution would allow the plant to operate but at reduced rates and higher levels of quality control while a new screw was fabricated. The new screw that was designed was very similar to the original screw. Only three major changes were made to the screw. First, the barrier flight clearance was decreased from 1.5 mm for the original screw to 1.0 mm for the new screw. This modification was made to increase the shear stress of the resin soon after it melts. Second, the depth of the solids channel at the exit of the barrier section was decreased from 3.2 to 1.5 mm. This change was made to limit the level of solids discharging from the barrier section and flowing downstream. Lastly, the clearance on the mixing flight on the Maddock mixer was decreased from 1.5 to 1.0 mm to impart a higher level of shear stress near the end of the pumping process. A summary of the screw geometry for this new screw is shown in Table 11.7. Since the metering channel dimensions were not changed, the specific rotational flow rate was the same as the original screw at 5.2 kg/(h·rpm).

The new screw was installed and operated at a screw speed of 100 rpm for a rate of 580 kg/h and a specific rate of 5.8 kg/(h·rpm). Thus, the smaller clearance on the barrier flight, Maddock mixer clearance, and the solids channel depth at the end of the barrier section caused the specific rate to decrease slightly as compared to the original screw. The discharge temperature of the extrudate was measured with a handheld sensor at 293 °C, a temperature that was 3 °C higher than the original screw. These tighter clearances were enough to eliminate essentially all of the unmixed gels that were occurring with the original screw. The process worked well with essentially zero film breaks in the process due to unmixed gels. Moreover, the modification did not change significantly the extrudate temperature.

Table 11.7 Screw Channel Dimensions for a 114.3 mm Diameter, 32 L/D Barrier Screw for Running an LLDPE Resin on a Cast Film Process with an Improved Screw Design

	Depth, mm	Length, diameters	Notes
Feed section	19.1	6.5	
Barrier section		15	1.0 mm barrier flight undercut
First meter	7.62	6.5	
Maddock-style mixer		2	1.0 mm undercut on the mixing flight
Second meter	7.62	2	

The lead length was 172 mm for the main flight of the barrier section and 114.3 mm for all other sections of the screw. The main flight width and clearance were 9 and 0.09 mm, respectively, in all sections of the screw. The first 2.7 diameters of the screw were inside a water-cooled feed casing. The compression ratio was 2.5 and the compression rate in the solids channel of the barrier section was 0.0033. The depth of the solids channel at the end of the barrier section was 1.5 mm.

11.10.3 Degassing Holes in Blow-Molded Bottles

A high-speed blow molding operation using an HDPE resin was experiencing defects that limited the rate of the process due to a 5% loss in yield, and it created additional costs for quality control inspection. The defects occurred sporadically as either voids (holes) or locations where the wall was extremely thin. Photographs of three typical defects in the parts are shown in Figure 11.24. The defect shown in Figure 11.24(a) occurred due to the entrainment of gas in the extrudate. When the part was blown in the tooling, the small amount of gas caused the part to become extremely thin and eventually created a large void. The level of gas entrainment defects was increasing slowly with the age of the screw. Defects shown in Figure 11.24(b) and Figure 11.24(c) were caused by small particles of contaminants and entrained gas. The particle contaminants were later identified as degraded HDPE resin. The HDPE resin had a solid density of 0.949 g/cm^3 and an MI of 0.25 dg/min (190 °C, 2.16 kg). Since the problem only occurred sporadically and led to a 5% loss of production, the root cause for the defects was very difficult for the blow molder to determine.

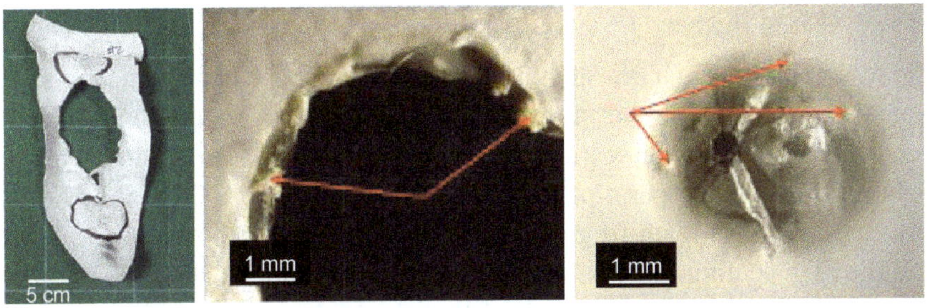

a) b) c)

Figure 11.24 Photographs of the defects for the high-speed blow-molding parts: a) a large cutout from the part showing a large hole, b) a small hole defect, and c) a location with a very thin wall. The arrows show locations that have degraded resin particles

Table 11.8 Screw Geometry for a 152.4 mm Diameter Screw for a High-Speed Blow-Molding Process Running an HDPE resin (Original Design). The Screw had an L/D of 33 and a Barrier Melting Section

	Depth, mm	Length, diameters	Notes
Feed section	24.2	8	
Barrier section		18	1.9 mm barrier flight undercut
First meter	10.1	3	
Maddock mixer		2	1.9 mm undercut on the mixing flight
Second meter	10.1	2	

The lead length was 152.4 mm for the feed and metering sections, and it was 191 mm for the main flight and mixing flight of the barrier section. The main flight width and original clearance were 11 and 0.11 mm, respectively, in the feed and barrier sections. The first 2.5 diameters of the screw were inside a water-cooled feed casing. The compression ratio was 2.4 and the compression rate was 0.0034. The solids channel depth at the exit of the barrier section was 2.5 mm. The specific rotational flow rate for the screw was calculated at 12.8 kg/(h·rpm).

The resin was plasticated using a 152.4 mm diameter, 33 L/D extruder that was equipped with a barrier screw. A summary of the screw geometry is provided in Table 11.8. The screw was typical for screws used for this application, but it did have a relatively long feed section and a very restrictive entry to the solids channel of the barrier section. For example, the width of the solids channel perpendicular to the flight and just upstream of the barrier section was about 12.7 cm. Within 1 diameter into the barrier section, the width of the solids channel was reduced to 8.6 cm. Thus, the cross-sectional area available for solids flow was decreased by about 30%. It was hypothesized that the rate was restricted by this entry section, creating downstream sections of the screw that were partially filled. Like the case study of Section 11.10.1, these partially filled sections could lead to degraded resin. The specific rotational flow rate for the screw was calculated at 12.8 kg/(h·rpm). The barrel was also found to be highly worn such that the flight clearance was up to 0.6 mm in some axial locations. Further investigation determined that an incorrect hard facing material was placed on the flight tips. The hard facing used was incompatible with the bimetallic liner used in the barrel, causing extreme and excessive wear to the barrel.

The screw operated at a rate of 495 kg/h at a screw speed of 50 rpm for a specific rate of 9.9 kg/(h·rpm), discharging at 26 MPa and 220 °C. Barrel Zone 3 was near the entrance to the barrier section and it was in control at the set point temperature, but the zone was operating at 80% of the maximum cooling rate. The zones upstream and downstream from this location were in control and operating normally. The high cooling rate for Zone 3 indicates that a large energy flux was dissipated in this region. It is hypothesized that the high energy flux was caused by solids forced into the barrier entry restriction, causing a high pressure in the region. The pressure gradient calculation for the metering section of the screw and the estimation of the pressure in the early sections of the screw (dotted line) are shown in Figure 11.25. The calculations indicate that much of the melt channel of the barrier melting section is operating partially filled. Like the case study in Section 11.10.1, these partially filled channels will cause some of the resin to have extremely long residence times and will cause the resin to degrade. When the degradation products break away from the stagnant region, they cause the defects in the blow-molded parts. Some of the degradation products also occurred as a gas, causing hole defects in the parts.

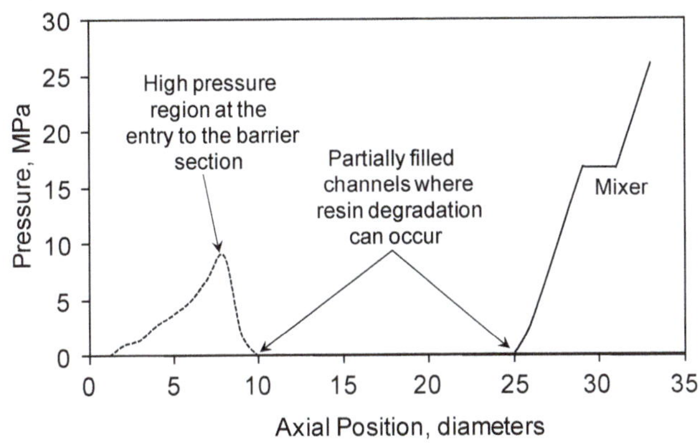

Figure 11.25 Simulated axial pressure profile for the 152.4 mm diameter screw running HDPE resin in a high-speed blow-molding application

The design defect for this case is the abrupt reduction in the cross-sectional area near the entry to the barrier section. This abrupt reduction caused the restriction and became the rate-limiting step of the process. The restriction was compounded by the long length of the feed section. Here the feed section was 8 diameters in length, allowing the solid bed to become highly compacted prior to entering the barrier melting section. A shorter solids-conveying section would not have compacted the resin to as high a level and might allow the bed to deform while flowing into the restricted region. As previously discussed, the remaining design defect was the use of an incompatible hard facing on the screw. The hard facing caused excessive wear to the barrel. The next barrels that were installed were compatible with the hard facing, eliminating the high wear defect.

The restriction was mitigated by modifying the screw as outlined in the case study in Section 11.10.1. That is, the depth of the melt channel of the barrier section was increased to that of the solids channel at the entrance, and it was tapered into the depth of the melt channel over 2 diameters. The barrier flight was removed for the first 2 diameters and blended in with the melt channel, the shallower of the two channels in this region. The barrier flight for the next diameter was blended into the original undercut. With this modification the restriction still existed but it was spread over a three-diameter length of the screw instead of over half of a diameter.

The modified screw was installed into the plasticator and was tested for performance. The screw was able to run at a rate of 625 kg/h at a screw speed of 47 rpm for a specific rate of 13.3 kg/(h·rpm). This specific rate is higher than the calculated rotational flow rate of 12.8 kg/(h·rpm) and indicates that a negative pressure gradient exists in the metering sections. The simulated axial pressure profile is shown in Figure 11.26. As shown by this figure, the pressure in the channels is positive in all sections downstream from the hopper, and thus the channels are completely filled with resin. Since the stagnant regions in the barrier section do not exist, degradation of the resin will not occur. The defects in the blow-molded parts were eliminated.

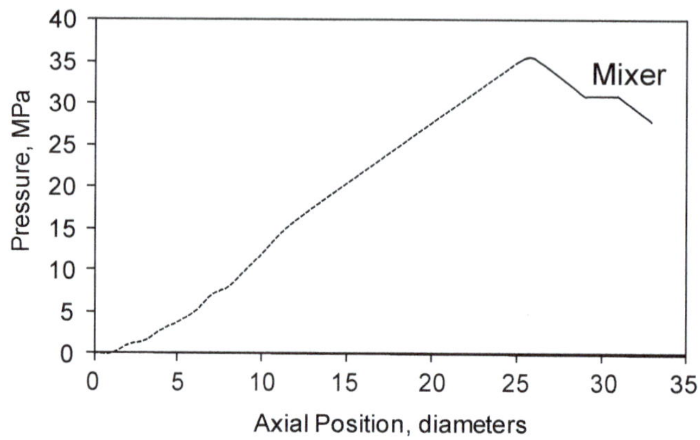

Figure 11.26 Simulated axial pressure profile for the modified 152.4 mm diameter screw. The entry to the barrier section was still restrictive, but it did not control the rate

11.11 Contamination in Injection-Molded Parts

Many of the contamination defects that typically occur in extrusion processes can also occur in injection-molded parts. The most obvious defects are caused by foreign material contamination, resin degradation, and surface defects known as splay. The next sections provide case studies where contamination caused defects in injection-molded parts.

11.11.1 Splay Defects for Injection-Molded Parts

Splay is a common term for surface imperfections that occur on injection-molded parts. Like the term gels for films, splay can occur from several root causes. Many of the root causes originate from the plasticating system and are related to moisture in the resin, degradation, entrained air, and thermal gradients. Splay defects can also be caused by sections of the process downstream from the plasticator. These types of defects are beyond the scope of this book and are discussed elsewhere [23].

The most common splay defect is caused by gases that are entrained with the molten polymer and become trapped near the surface of the part. Photographs of molded parts with splay defects are shown in Figure 11.27 and Figure 11.28. Frequently, the gas causes a cosmetic change in the surface gloss. In severe cases, the defect can be just below the surface of the part as shown in Figure 11.28. In most cases, the parts are unacceptable and must be rejected as off-specification product. The gas can originate from the resin feedstock as entrained air or moisture, or it can come from degradation of the resin within the process. For example, engineering resins that are not dried properly prior to molding can show splay defects. An improperly designed screw can allow air that is entrained between the pellets to discharge into the tooling, creating a splay defect. Typically a screw with a higher compression ratio is effective at forcing the air back out through the hopper rather than allowing it to exit with the injectate.

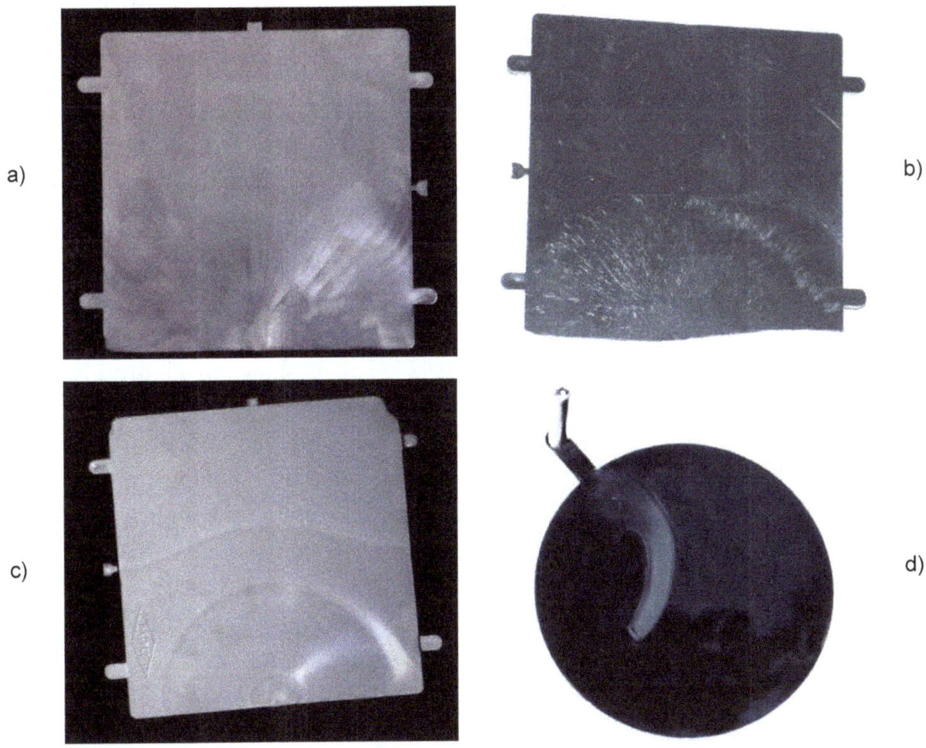

Figure 11.27 Photographs of injection-molded parts with splay defects: a) splay due to water and other volatiles, occurring in a jetting pattern from the gate, b) splay due to volatiles and occurring in a halo-like pattern, c) halo splay due to gloss gradients in the part surface, and d) blush defect at the gate (parts courtesy of Jack Little of The Dow Chemical Company and photographs by Aleah Z. Pavlicek)

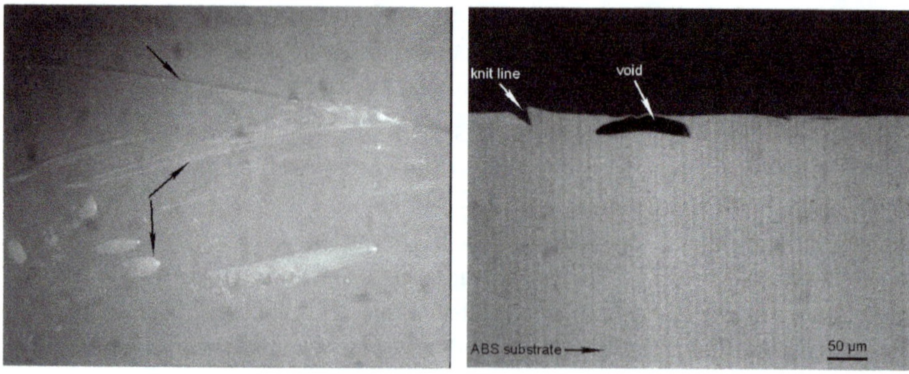

a) b)

Figure 11.28 Photographs for surface defects in an ABS injection-molded part: a) surface photograph, and b) cross-sectional view showing a void just under the surface (photographs courtesy of James T. Seliskar of The Dow Chemical Company)

Like the extrusion processes discussed earlier in this chapter, stagnant regions and partially filled regions in the screw channels of injection-molding plasticators can cause the resin to degrade, producing splay defects. Processors will attempt to mitigate the defect by increasing the back pressure on the screw, causing the partially filled regions to become completely filled with resin. If the error in the screw design is not too severe, then this technique may be effective at eliminating the splay defect. Increasing the back pressure, however, will cause the temperature of the injectate to increase and will often increase the cycle time for the process. The best way to remove the defect is by eliminating the root cause in the screw design. Several cases studies address splay defects in the next sections.

Halo surface defects are a form of splay (or blush) that occurs as concentric rings that are typically in center-gated parts [24, 25], as shown by Figure 11.27(c). The defect is caused by gloss gradients in the surface of the parts. The gloss gradients can occur due to a thermal gradient in the polymer feed to the part, particularly when the temperature of the feed to cavity increases during the injection process. The thermal gradient of the feed causes different levels of gloss or roughness on the surface that appear as concentric rings about the gate, especially for engineering resins that have multiple phases, such as blends and rubber-modified resins. Inconsistent injection speed of the screw and poor temperature control of the tool can cause halo defects. Other troubleshooting methods for mitigating surface defects on injection-molding parts are presented elsewhere [23, 26].

11.12 Injection-Molding Case Studies

Several case studies are presented in the next sections that show some common root causes of contamination in injection-molded parts. In these case studies, the problem is presented in a manner that the troubleshooter would encounter during a trial or information-gathering session. In each case study, the modifications required to fix the process are detailed along with supporting fundamental information. Two of the case studies used Energy Transfer (ET) screws to eliminate the defects. ET screws and other high-performance screws will be discussed in Chapter 14.

11.12.1 Injection-Molded Parts with Splay and Poor Resin Color Purge

An injection molder was producing housing parts using an ignition-resistant polystyrene (IRPS) resin. The molder complained that some of the parts occasionally had splay and the plasticator was very difficult to purge during color changes. When the press was changed from a black-colored resin to a gray resin, the black resin would create streaks in the gray parts for several days. The splay, however, would occur with both IRPS colored resins, as shown in Figure 11.29 for a gray-colored part. The splay and defective gray parts with black streaks caused the scrap rate to be high, increasing the costs and decreasing the production capacity of the press.

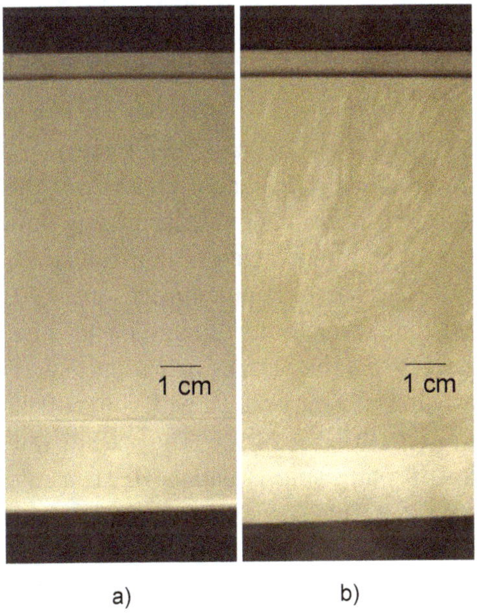

Figure 11.29 Photographs of two IRPS resin part sections: a) a part with an acceptable surface, and b) a part with splay

Table 11.9 Screw Channel Dimensions for the 120 mm Diameter, 18 *L/D* Conventional Screw Running an IRPS Resin in an Injection-Molding Operation

	Depth, mm	Length, diameters	Notes
Feed section	14.6	8	
Transition section		10	The undercut of the barrier flight was 1.0 mm
End of transition	9.47		

The lead length was 120 mm for the main flight of the screw. The main flight width and clearance were 13 and 0.12 mm, respectively, in all sections of the screw. The compression ratio was 1.5 and the compression rate was 0.0013. The ratio of the flight radii in the meter section to the meter depth was 0.2. The specific rotational rate was calculated at 9.3 kg/(h·rpm).

The injection-molding press was producing a part and runner system that had a mass of 2.15 kg. The mass was plasticated using a 120 mm diameter, 18 *L/D* screw. The screw used for the process had a barrier melting section that extended to the end of the screw, as shown by the specifications in Table 11.9. That is, the screw did not have a metering channel. Instead, the last sections of the barrier section were required to produce the pressure that was needed to flow the resin through the nonreturn valve and into the front of the screw. The specific rotational flow rate for the screw for the IRPS resin was calculated at 9.3 kg/(h·rpm) based on the depth of the channel at the end of the transition section. The screw was built with an extremely low compression ratio and compression rate of 1.5 and 0.0013, respectively. For IRPS resins and other PS resins, screws with low compression ratios and compression rates tend to operate partially filled. The compression ratio and compression rate for the screw are preferred to be around 3.0 and 0.0035, respectively. The flight radii on the screw were extremely small at about 0.2 times the channel depth. For IRPS resin, the ratio of the radii to the channel depth should be about 1.

The part and runner system had a mass of 2.15 kg. The screw was rotated at a speed of 48 rpm for an average plasticating time of 27.5 s and a pressure at the screw tip of 16 MPa. Over a series of six injections (or shots), the plasticating time ranged from 26.1 to 28.4 s. This relatively large range suggests that the screw channels were not running completely full. The barrel temperatures were 175, 225, 230, 235, 235, and 240 °C for the feed zone through the discharge end zone, respectively. The temperature of the discharge from the plasticator was measured by ejecting it onto a board and then measuring its temperature with a handheld probe. The temperature of the discharge was measured at 246 °C. The specific rate for the operation was calculated at 6.3 kg/(h·rpm) using Equation 10.21, a rate considerably less than the calculated specific rotational flow rate of 9.3 kg/(h·rpm). Thus, the screw was operating at a specific rate that was about 65% of the specific rotational flow rate. Since the pressure at the tip of the screw during plastication (16 MPa) was relatively high and the screw did not have a conventional metering section, the screw was expected to operate at a low specific rate. Due to the complicated nature of this screw geometry, the only way to be sure of the degree of fill is to shut down the machine, remove the screw, and study the residual polymer in the channels. It was hypothesized that some of the screw channels were partially filled, causing the resin to degrade, creating gas and black specks. The gas would cause the surface splay. These regions would also be difficult to purge during color changes.

At the start of the molding trial, the parts were produced using the black-colored IRPS resin. When the switch was made to the gray-colored resin, the first four parts were essentially all black, and the next 10 parts had increasing levels of gray color. The next 10 parts were essentially all gray except for small black patches on the top and sides. These observations were consistent with the hypothesis that the black resin was adhering to the screw at the flight radii and also in the stagnant regions of the screw. After 25 parts were molded, the resin flow was shut off to the screw and the screw was allowed to rotate and pump the excess resin from the barrel. Next, the screw was removed while hot and examined for regions with black resin. Black resin was found on the nonreturn valve, at the pushing and trailing flights of the barrier section, and in a small stagnant zone early in the barrier section. The older black resin found in the trailing and pushing sides of the channel at the entry to the barrier section is shown in Figure 11.30. A partially filled zone was located between the completely filled zone that creates the pressure at the screw tip and just downstream from the entry to the barrier section. In this partially filled zone, a hard and stable coating of a black carbonized material that did not contain molten polymer was found across most sections of the screw. These results confirm the hypothesis that the screw was operating improperly with partially filled channels. Moreover, the small flight radii created additional stagnant regions, contributing to the long times required for color purges and to the splay.

The barrier screw design specified in Table 11.9 was unacceptable for the plastication of IRPS resins, and thus a modification to the screw was not a solution. Instead, an ET screw [27] was designed and built specifically for this press and resin. The specification for the screw is summarized in Table 11.10. The screw was designed with a much higher compression rate and compression ratio, and the flight radii were equal to the channel depth in all sections. The higher compression ratio will maintain the metering channel full of resin. The compression rate was higher than desired due to the extremely short barrel length of 18 L/D. The ET section had minimum and maximum depths of 2.0 and 11.2 mm, respectively; the distance-averaged depth of the ET section was 7.5 mm. The compression ratio based on the distance-averaged depth was calculated at 2.7. The ET flight undercut, primary flight clearance, and flight width were 2.0, 0.12, and 7.0 mm, respectively. The lead length was 133 mm for all sections of the screw. The specific rotational flow rate for this screw and the IRPS resin was calculated at 8.5 kg/(h·rpm).

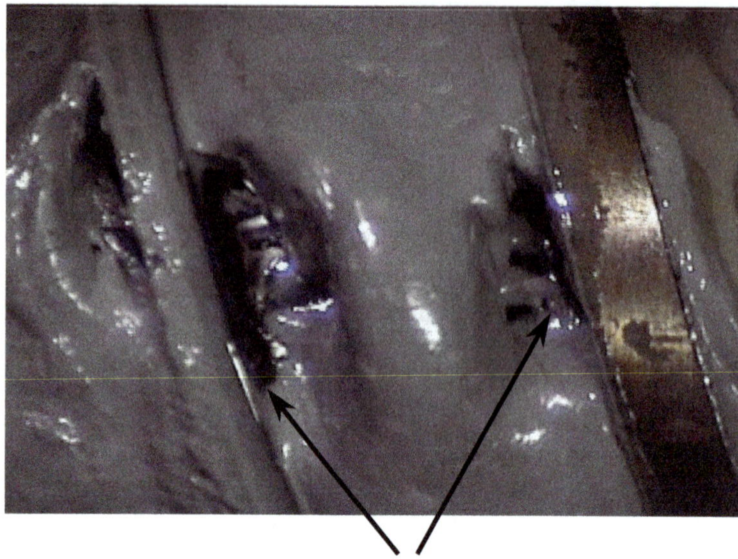

Figure 11.30 Photograph of the barrier screw after running 25 injections of a gray-colored IRPS resin after a black resin. The black-colored resin shows regions that did not purge well and thus were stagnant

Table 11.10 Screw Channel Dimensions for a 120 mm Diameter, 18 *L/D* ET Screw Designed for Plasticating an IRPS Resin in an Injection-Molding Operation

	Depth, mm	Length, diameters	Notes
Feed section	20.3	6	
Transition section		4.5	
High-performance metering section	7.57	7.5	Double-flighted and average depth reported

The lead length was 133 mm for the main flight of the screw. The main flight width and clearance were 12 and 0.12 mm, respectively. The compression ratio was 2.7 and the compression rate was 0.0090. The ratio of the flight radii to the local channel depth was 1.0 in all sections. The specific rotational rate was calculated at 8.5 kg/(h·rpm).

The high-performance screw was installed in the press and was operated with barrel temperatures of 150, 185, 230, 235, 235, and 240 °C for the feed zone through the discharge end zone, respectively, and a tip pressure of 9 MPa. At these conditions, the 2.15 kg part and runner system was plasticated at a screw speed of 44 rpm in 27.1 s for a measured specific rate of 6.5 kg/(h·rpm). The plasticating time was essentially constant at 27.1 s between injections. This measured specific rate was about 80% of the specific rotational flow rate, and the decrease was due to the pressure gradient created by a tip pressure of 9 MPa. Based on this information the screw was operating full and under pressure. Like before, the discharge temperature was measured using a handheld probe. For the ET screw, the temperature of the discharge was measured at 233 °C, a temperature that was about 13 °C less than that for the original screw.

The parts produced were in specification and did not have the splay defect. When the press was switched between different colored resins, the purge time was very short and typically required less than 20 parts. The higher compression ratio and compression rate screw combined with channels that had large radii provided a constant plastication rate and no detectable degradation of the resin.

11.12.2 Black Color Streaks in Molded Parts: Case One

An injection-molding plant was producing large parts from a natural PP resin. The natural resin was blended with a 35 to 1 letdown ratio of a light gray color concentrate, and this mixture was plasticated, mixed, and then injected into the tooling to produce the part. The molder was experiencing problems with black streaks on the surface of the parts, causing a high scrap rate. A photograph of the black streaks is shown in Figure 11.31. When the back pressure on the screw was increased from 0.7 MPa to 2.5 MPa, the fraction of parts with black streaks decreased from 50% down to less than 10%. The material used for these parts was a natural PP resin with an MFR of 18 dg/min (230 °C, 2.16 kg).

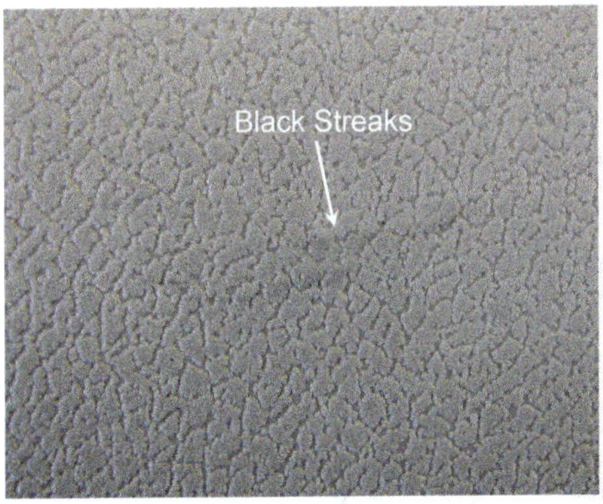

Figure 11.31 Photograph of black-colored streaks in a PP molded part

The injection-molding press was equipped with a 140 mm diameter, 20 *L*/*D* plasticator. The press was designed with a pressure intensification factor of 10. That is, for a back pressure setting of 2.5 MPa the pressure at the discharge of the screw during rotation was 25 MPa. The basic channel dimensions of the barrier screw are provided in Table 11.11. The specific rotational flow rate for the metering section of the screw was calculated at 9.5 kg/(h·rpm). Due to the very short metering section length (2 diameters), the specific rate that the screw will operate at will be highly dependent on the discharge pressure during operation, that is, the back pressure setting. The screw was capable of operating at a maximum speed of 99 rpm.

Table 11.11 Screw Channel Dimensions for a 140 mm Diameter, 20 *L*/*D* Barrier Screw Running a PP Resin in an Injection-Molding Operation

	Depth, mm	Length, diameters	Notes
Feed section	22.9	8	
Barrier section		7.5	1.4 mm barrier flight undercut
Meter section	7.37	2	
Spiral-type mixer		2.5	Mixer provided distributive and dispersive mixing with a clearance between the flight and barrel wall of 3.5 mm
The lead length was 157 mm for the main flight of the screw. The main flight width and clearance were 14 and 0.14 mm, respectively, in all sections of the screw. The compression ratio was 3.1 and the compression rate was 0.0049. The specific rotational flow rate was calculated at 9.5 kg/(h·rpm).			

This press had a new barrel, barrier screw, and nonreturn valve, and the press had been thoroughly inspected for operation. The molder had spent considerable time working with the current process to achieve the best results. They concluded that high temperatures helped with the recovery time and minimized the level of defects. The main concern was that the defects were coming from the degradation of the PP resin or were related to the masterbatch colorant. Limitations were imposed by the molder in changing process conditions as the plant could not afford to shut down to run experiments. That is, they needed to continue to make parts to meet their production demands. Some adjustments were made to increase the back pressure and decrease the barrel temperatures, resulting in a constant barrel or "flat" temperature profile. The molder determined that a constant barrel temperature profile at 235 °C coupled with 2.5 MPa back pressure allowed a more consistent operation of the barrel heating zones and more consistent recovery time while minimizing the level of color streaks. The cycle time, however, increased to an unacceptable level of 85 s with the plasticating process limiting the rate.

Dark-colored streaks in a gray-colored part can be difficult to diagnose properly since the color streak could be originating from several different sources. The sources include but are not limited to the following root causes: (1) degradation of the resin in the screw channels, (2) degradation of the resin in the nonreturn valve or runner system of the mold tooling due to a local high temperature with long residence times, (3) degraded material or contamination entering with the resin feedstock, and (4) poorly dispersed pigments in the color masterbatch. Degradation in the screw channels could be caused by a stagnant region or a section of the screw that is operating partially filled and at zero pressure. These partially filled sections would allow the decomposition of the resin, forming a dark-colored material. Occasional process upsets would dislodge some of this degraded material, causing it to contaminate the injectate and thus the final part.

At the start of the trial, the target for the plant was to run a 65 s cycle time with a 1% scrap rate or less due to all defects. The plasticator was running at 235 °C for all four barrel zones, a screw speed of 99 rpm, and with a back pressure of 2.5 MPa. The barrel zone located over the metering section of the screw was measured at 262 °C, a temperature that was more than 25 °C over the set point temperature. For this case, the cooling capability of the zone was unable to remove the dissipated energy fast enough. This process gave a cycle time of 82 s but kept the scrap rate the lowest at about 10%. The plasticating time was 36 s, and the instantaneous specific rate for this process was measured at 3.45 kg/(h·rpm). This instantaneous rate is the rate that is compared to the calculated flow rate for the screw. This plasticating rate of 3.45 kg/(h·rpm) was considerably less than the calculated rotational flow rate for the screw; the rotational flow rate for the screw was previously calculated at 9.5 kg/(h·rpm). This press was rate-limited by the current 36 s plasticating time. Observations showed slight inconsistencies for the screw plasticating times. Although these inconsistencies were minor, they led to a concern that the screw and plasticator were not functioning properly. The injectate temperature was measured by producing an "air shot" and then measuring the temperature using a handheld temperature sensor. The injectate temperature was measured at 259 °C, a temperature that is considered relatively high for this type of process.

Several mechanisms could cause the specific rate of the screw to be considerably less than the calculated specific rotational flow rate for the screw. These mechanisms include (1) normal operation for a screw with a very short metering section and a low-viscosity resin, (2) the screw is rate-limited by solids conveying, causing the downstream sections of the screw to operate partially filled, and (3) the entry to the barrier section is restricting flow (see Section 11.10.1) to the downstream sections of the screw and causing the downstream sections to operate partially filled. The goal was to determine which of the above mechanisms was responsible for the low specific rates for the plasticator.

In order to understand why the plasticator operated at a low specific rate, several changes were made to the process. First the back pressure was decreased to 1.4 MPa for a discharge pressure of 14 MPa. This allowed the plasticating time to decrease to 21 s and the specific rate to increase to 5.9 kg/(h·rpm). The parts produced at these conditions were 100% scrap due to black streaks. Next, the color concentrate flow was turned off and allowed to run out. The color was removed from the system to determine whether the color streaks were caused by degraded PP resin or from poorly compounded pigments in the color concentrate. If the streaks were caused by degraded resin, the streaks should still be present after the color concentrate is removed. Once the parts were completely natural in color, there were no black streaks present in the parts. The screw speed and back pressure were varied between 50 and 90 rpm and between 0.7 and 2.5 MPa, respectively, in an attempt to disrupt the process. The hypothesis was that if any degraded resin had accumulated on the screw due to a poor screw design, then the variations in screw speed and back pressure would cause the degraded resin to exit with the injectate, creating streaks in the parts. In all cases, the parts did not contain streaks, indicating that the screw was operating properly with the natural PP resin material. The axial pressure profile for the screw operating at a discharge pressure of 25 MPa is shown in Figure 11.32. The process data indicated that the molder could operate the screw at 99 rpm and a back pressure of 0.7 MPa to produce high-quality parts with a cycle time of 65 s or less. The very low specific rates were caused by a short metering section and the low viscosity of the PP resin.

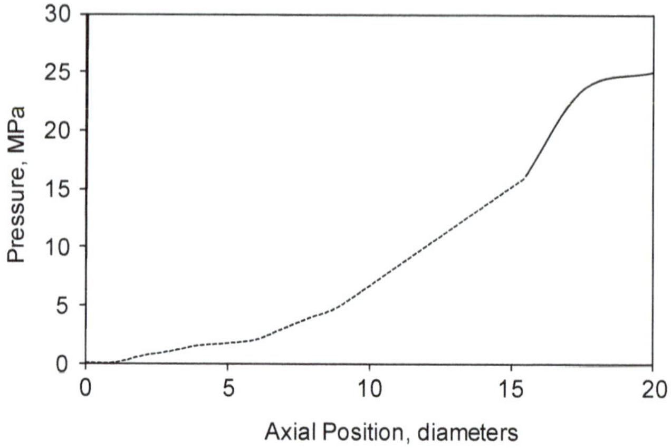

Figure 11.32 Axial pressure profile for the injection-molding screw described in Table 11.11 at a rate of 340 kg/h, a screw speed of 99 rpm, and a discharge pressure of 25 MPa

The molding conditions were returned to the original settings of a screw speed of 99 rpm and a back pressure of 2.5 MPa, and then the masterbatch colorant was added back into the process at a letdown ratio of 35 to 1. As soon as the colorant was observed in the parts, the black streaks reappeared. Based on the data here, the root cause for the black streaks was the color concentrate. Although not completely evaluated, it is likely that pigment agglomerates were created during the compounding operation. These agglomerates cannot be effectively removed using this barrier screw. When the back pressure was increased to 2.5 MPa, the mixing abilities of the screw increased slightly and the level of streaks in the parts decreased. The obvious goal was to obtain a color concentrate free of agglomerated pigment particles, allowing the press to operate using a back pressure of 0.7 MPa and a minimum cycle time.

The MFR for the PP resin and the masterbatch colorant were measured to see if the masterbatch material met the criteria defined by Benkreira and Brittin [28] and as discussed in Section 8.6. The MFRs were 20 and 116 dg/min (230 °C, 2.16 kg) for the natural PP resin and the masterbatch colorant, respectively. Obviously, the color concentrate masterbatch was not well matched for the natural PP resin according to the guidelines developed by Benkreira and Britton. That is, the carrier resin used to make the masterbatch was too low in viscosity to break up pigment agglomerates during the twin-screw compounding process. If the pigment, however, could have been dispersed in this carrier resin, the masterbatch would have been acceptable. As previously stated, comparing the shear viscosity of the natural resin and the masterbatch is preferred over comparing the MFRs. For these materials, the shear viscosity of the natural PP resin and the color masterbatch were 160 and 40 Pa·s, respectively. As expected the ratio of the shear viscosity for the natural resin to that of the concentrate was about four, a value larger than the guideline value of two. A better masterbatch could be made with a more viscous carrier PP resin such that additional stress can be applied during compounding to aid in the dispersion of pigment agglomerates.

The technical solution of changing the color concentrate masterbatch was proposed to the machine owner. Due to preset specifications, the color concentrate masterbatch could not be changed. The pigment agglomerates in the masterbatch would need to be dispersed in the plasticator of the injection-molding machines. That is, a new screw would need to be designed and fabricated that is capable of dispersing the agglomerates while meeting the cycle time of 65 s. A DM2 high-performance screw was chosen for the application with the channel specifications shown in Table 11.12. A description of the DM2 screw is provided in Section 14.2.4. The distance-averaged depth of the DM2 metering section was 8.9 mm, providing a specific rotational flow rate of 8.94 kg/(h·rpm). Multiple dispersing dams in the high-performance section provided a gap between the peak of the dam and the barrel wall of 1.1 mm. The clearance between the mixing flights of the spiral mixer and the barrel wall was 0.89 mm. The dispersing ability of this screw was considerably higher than the barrier screw specified in Table 11.11 as shown by comparing the dispersion gaps. For the barrier screw, the barrier undercut was 1.4 mm and the spiral mixer undercut was 3.5 mm while the high-performance screw had dispersion gaps of 1.1 and 0.89 mm for the DM2 and mixer sections, respectively.

Table 11.12 Screw Channel Dimensions for a 140 mm Diameter, 20 *L/D* DM2 High-Performance Screw Running a PP Resin in an Injection-Molding Operation

	Depth, mm	Length, diameters	Notes
Feed section	24.1	6	
Transition section		5	
DM2 high-performance metering section	8.9	7	Double-flighted and average depth reported. The dispersion gap was 1.1 mm
Spiral-type mixer		2	Mixer provided distributive and dispersive mixing with a clearance between the flight and barrel wall of 0.89 mm
The lead length was 152 mm for the main flight of the screw. The main flight width and clearance were 14 and 0.14 mm, respectively, in all sections of the screw. The compression ratio was 3.3 and the compression rate was 0.0078. The specific rotational flow rate was calculated at 8.94 kg/(h·rpm).			

The high-performance screw was installed and tested. The screw immediately produced high-quality parts without black streaks and with a cycle time of 65 s. The plasticating time was 15 s at a screw speed of 99 rpm for a specific rate of 8.3 kg/(h·rpm). This specific rate is just slightly less than the specific rotational rate of 8.94 kg/(h·rpm) calculated for the screw. The back pressure used for this plastication was 0.7 MPa. These results indicate that the dispersion gaps on the high-performance screw were small enough to provide a high enough stress level to disperse the pigment agglomerates that were in the color concentrate masterbatch.

11.12.3 Black Streaks in Molded Parts: Case Two

An injection-molding plant was producing large electronic housing parts from a HIPS resin. The molded parts were gray in color using a natural HIPS resin and a gray color concentrate masterbatch. Many of the molded parts had to be scrapped due to black color streaks, as shown in Figure 11.33. The molder was claiming that the color concentrate was not adequately mixed into the resin and that a pre-color resin would be required to solve the problem. A pre-color resin is a resin that has the color compounded into it, increasing the cost of the resin. The HIPS resin used was a natural resin with an MFR of 6 dg/min (200 °C, 5 kg).

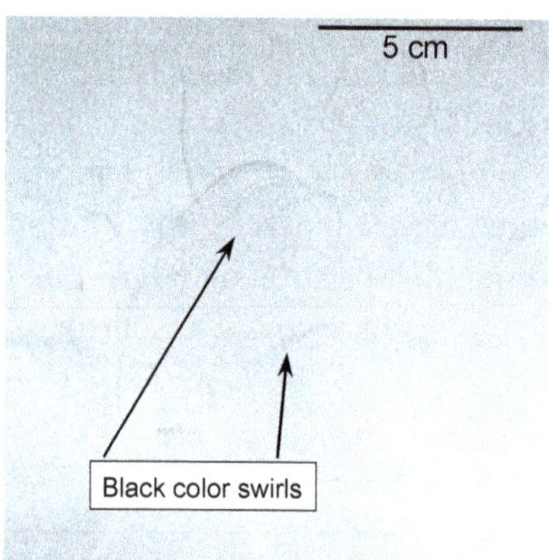

Figure 11.33 Black color streaks on the surface of an injection-molded electronics housing made from HIPS and a gray color concentrate masterbatch

The parts were molded using a 2500 ton injection-molding press with a 125 mm diameter, 21 *L/D* plasticator. Several debugging operations were tried and reported by plant personnel, and all either failed or provided an unacceptable solution. These included increasing the back pressure, increasing the color concentrate loading, and positioning a static mixer inside the nozzle. Black color streaks were present with increased back pressure and when the color concentrate loading was increased from about 2 to 5%. The color swirls were, however, mitigated with the addition of static mixers in the nozzle. This technique was not accepted due to problems with filling the part and an increase in the cycle time. The static mixers required that the injection pressure be increased to unobtainable levels to maintain cycle time.

The screw used for this process was a conventional, single-flighted screw with a spiral dam positioned in the metering section of the screw. The specifications for the screw are provided in Table 11.13. The spiral dam started at the entry to the metering section at the pushing side of the channel. The dam ended at the end of the metering section and at the trailing side of the flight. The undercut of the dam was 0.89 mm relative to the main flight. The specific rotational flow rate for the screw was calculated at 6.4 kg/(h·rpm). The flight radii sizes were estimated from the radii tangent points on the flight edge. For this case, the flight radii were about 20% of the depth of the channel. The flight radii for this screw were extremely small and were likely a region where degradation of the resin was occurring.

Table 11.13 Screw Channel Dimensions for a 125 mm Diameter, 21 *L/D* Conventional Screw Running a HIPS Resin in an Injection-Molding Operation

	Depth, mm	Length, diameters	Notes
Feed section	15.1	11	
Transition section		6	
Meter section with a spiral dam	6.35	4	The undercut of the spiral flight was 0.89 mm

The lead length was 140 mm for the main flight of the screw. The main flight width and clearance were 12 and 0.13 mm, respectively, in all sections of the screw. The compression ratio was 2.4 and the compression rate was 0.0039. The ratio of the flight radii in the meter section to the meter depth was 0.2. The specific rotational flow rate for the screw was calculated at 6.4 kg/(h·rpm).

The injection press was producing a 2.69 kg part and gating from natural HIPS resin with 2% of a color concentrate. The barrel temperatures were 215, 230, 230, 230, 240, 250, and 250 °C for the feed section through the nozzle, respectively. The 2.69 kg part and gating were plasticated in 34.6 s at a screw speed of 52 rpm and 9.5 MPa pressure at the screw tip. Screw retraction was fairly steady, indicating a constant plasticating rate. For this machine and screw, the measured specific rate was 5.4 kg/(h·rpm). As previously stated, the specific rotational flow rate of the screw was calculated at 6.4 kg/(h·rpm), and a sufficient pressure gradient existed to reduce the specific rate to 5.4 kg/(h·rpm). Thus, this screw was operating properly and hydraulically full.

As discussed in Section 11.12.2, several root causes exist for black streak contamination in molded parts. These root causes include degradation of the resin in the screw channels and nonreturn valve or runner system, degraded material entering with the resin feedstock, and poorly dispersed pigments in the color masterbatch. To determine if the color concentrate masterbatch was the source of the black streaks, the color concentrate was removed from the feedstock and only natural HIPS resin was used. After about 10 parts, the housings were relatively free of pigment. Many particles of black material were, however, present in the parts, as shown in Figure 11.34. Most of the particles were between 1 and 5 mm in diameter and were positioned in the interior of the part, away from the surface. These particles would not be visible if the color concentrate were used. Some of the smaller particles occasionally contacted the surface of the tool and created a black color streak. These streaks were likely those that were observed during the molding of the cabinets with the color concentrate. It was hypothesized that the black particles were degraded resin that was coming from the stagnant regions of the screw and nonreturn valve. The next step was to remove the screw and look for stagnant regions.

Figure 11.34 Backlit photograph of a part section of an electronic housing with the gray color concentrate removed. The black particles observed in the interior of the part were identified as degraded HIPS resin

To locate the regions where the material was degrading in the screw, pellet flow to the screw was stopped and rotation of the screw was continued until all natural HIPS resin emptied from the plasticator. Next, the screw was removed from the barrel and examined for black and degraded resin. There was a relatively large amount of severely degraded resin at the entry to the spiral dam on the collection side of the channel, as shown in Figure 11.35. This region is known to be a place for long residence times, that is, a location for resin degradation. The degraded resin was soft, leathery, and black in color. There was also a considerable amount of this degraded material at both the pushing and trailing flight radii, starting in the middle of the transition section and extending the rest of the length of the screw. The nonreturn valve also had considerable levels of black degraded resin adhering to its surface. These leathery, black specks were the root cause of the black colored swirls in the parts. It is not known whether the screw or the nonreturn valve was the major contributor of the black, degraded resin. These procedures convinced the plant personnel that it was their process equipment that was creating the black color streaks and not the resin or the color concentrate masterbatch. The only permanent solution to this problem was to eliminate the regions where the polymer was degrading. That is, the existing screw and nonreturn valve needed to be replaced with streamlined equipment, that is, with a screw and nonreturn nozzle that do not have regions with long residence times for the resin.

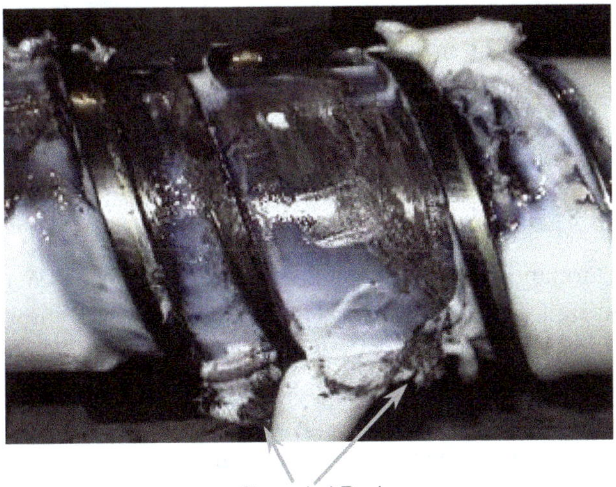

Figure 11.35 Photograph of the entry to the spiral dam of the screw. All pigment was purged out of the screw prior to shutting down. The dark material in the photograph is the degraded resin due to long residence times

A high-performance screw was designed for the injection-molding press. This screw was designed with a much higher compression ratio and with large flight radii in all sections of the screw. The details of the screw are provided in Table 11.14.

Table 11.14 Screw Channel Dimensions for a 125 mm Diameter, 21 *L/D* High-Performance Screw Running a HIPS resin in an Injection-Molding Operation

	Depth, mm	Length, diameters	Notes
Feed section	21.0	6	
Transition section		5	
High-performance meter section	6.86	7.3	Double-flighted and average depth reported
Spiral-type mixer		0.7	Mixer provided distributive and dispersive mixing with a clearance between the flight and barrel wall of 0.89 mm

The lead length was 140 mm for the main flight of the screw. The main flight width and clearance were 12 and 0.13 mm, respectively, in all sections of the screw. The compression ratio was 3.1 and the compression rate was 0.0076. The ratio of the flight radii to the local channel depth was 1.0 in all regions that contained molten resin. The specific rotational flow rate of the screw was calculated at 7.1 kg/(h·rpm).

The press was started back up using the high-performance screw and natural HIPS resin with 2% of a light gray color concentrate masterbatch. For this startup the same 2.69 kg part was produced and the barrel set point temperatures were the same as before. After steady operation was obtained (about 10 parts), the 2.69 kg part and gating were plasticated in 33.5 s at a screw speed of 52 rpm and 9.5 MPa pressure at the screw tip. The screw retraction rate was steady, indicating a constant plasticating rate. The measured specific rate was 5.6 kg/(h·rpm), a specific rate that was about 3% higher than that for the original screw. The specific rotational flow rate of the screw was calculated at 7.1 kg/(h·rpm), and a sufficient pressure gradient existed to reduce the specific rate to 5.6 kg/(h·rpm). Thus, this screw was operating properly and hydraulically full. During the remainder of the trial, black color streaks were never observed.

The high-performance screw was monitored closely for about one month after its installation. During this period, black color streaks were never observed, and plant personnel indicated that the problem was resolved. One month of operation should have been a long enough time period to generate degraded materials in stagnant regions of the screw if they existed. Since the degraded material has not been observed in the parts, it is concluded that the stagnant regions of the original screw were the main cause of the black color streak problem. The nonreturn valve, although not very streamlined, does not appear to be a large contributor of degraded resin.

As previously mentioned, plant personnel were successful at mitigating the black streaks with the use of static mixers positioned inside the injection nozzle. Since the degraded resin was soft and leathery, it was easily broken into smaller particles as they passed through the mixer. Thus, the particles were not eliminated; they were just reduced in size such that they did not create as many black streaks. The additional pressure and time that was required to inject the resin through the static mixers and into the mold, however, increased the cycle time to an unacceptable level. Although static mixers can be used to mitigate these types of problems, they do not eliminate the root cause of the degradation problem.

11.12.4 Silver Streaks in a Clear GPPS Resin Injection-Molded Packaging Part

Visual defects occurred in a clear PS packaging part when new injection-molding presses were installed in a commercial application and run at increasing rates. These defects accounted for about 5% of the parts molded, and the defects could be minimized by running the process at a slower speed. The defects were causing a significant loss in productivity through the loss of finished parts per shift and the additional cost of quality assurance to remove the defective parts. With the anticipation of these presses operating at higher rates and thus lower cycle times, this level of defects was not acceptable. The defects consisted of "silver" spots and streaks that were mostly near the gate area. The splay defects were examined using optical microscopy, and a photomicrograph is shown in Figure 11.36. The defect was internal to the thin-walled part and was not directly on the surface. It consisted of a single point or tail closer to the gate area with a large fan of bubbles farther from the gate. The tail appeared to contain an "unmelted" polymer fragment, and the material surrounding the bubble fan appeared to be different than the material in the rest of the package, possibly a more viscous component or a different light diffraction caused by air entrapment. A photomicrograph of a single bubble is shown in Figure 11.37. It was unclear from the microscopic analysis whether these were air bubbles (air entrapment) or incompletely melted polymer or both. Both scenarios indicated that the problem was related to the screw design. The defects did not occur when the process was operated at considerably slower rates. On occasion, very small specks of dark-colored degraded resin were observed in the defects. The cycle time for the process was consistent at 6.1 s.

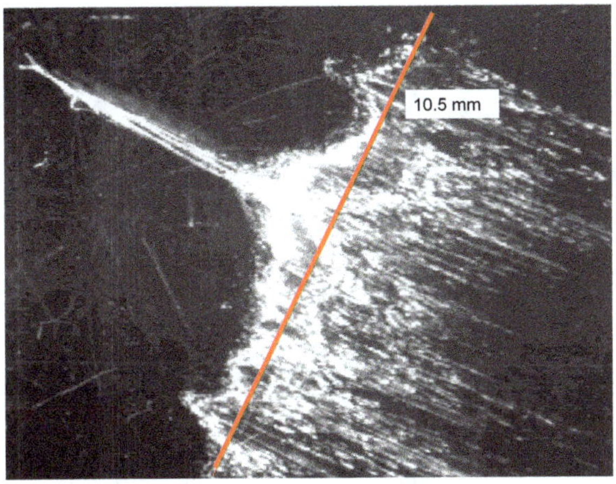

Figure 11.36 Photomicrograph of the silver-colored defect in a clear GPPS injection-molded packaging part. The flow direction was from the upper left to the lower right

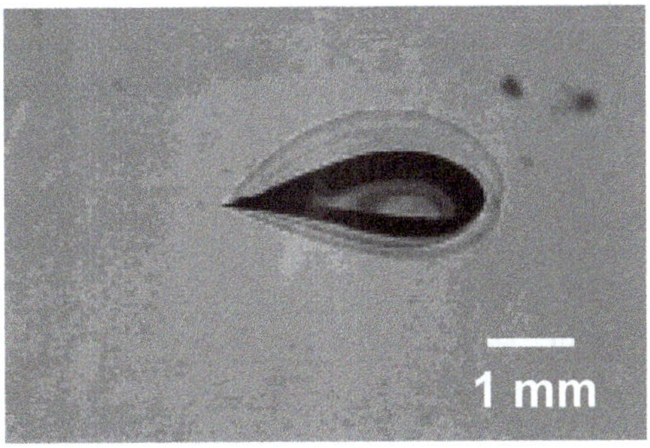

Figure 11.37 Photomicrograph of a single bubble defect. The resin surrounding the bubble appeared to be more viscous than the bulk material prior to solidification. The flow direction was from left to right

The injection-molding presses were 250 ton in size, and they were equipped with 63 mm diameter, 22 *L*/*D* plasticators. The screws were all identical and were conventional, single-flighted designs, and they had spiral dams positioned in the last

3.3 diameters of the transition sections. The specifications for the screws are provided in Table 11.15, and a schematic of a spiral dam is shown in Figure 11.38. These screws had a very low compression ratio of 2.4 and a compression rate of 0.0029. For GPPS resin and this application, a compression ratio near 3 and a compression rate of about 0.0035 is desired. The size of the flight radii in the metering section was relatively large compared to the depth of the channel. The specific rotational flow rate was calculated at 0.94 kg/(h·rpm). The resin had an MFR of 5.5 dg/min (200 °C, and 5 kg).

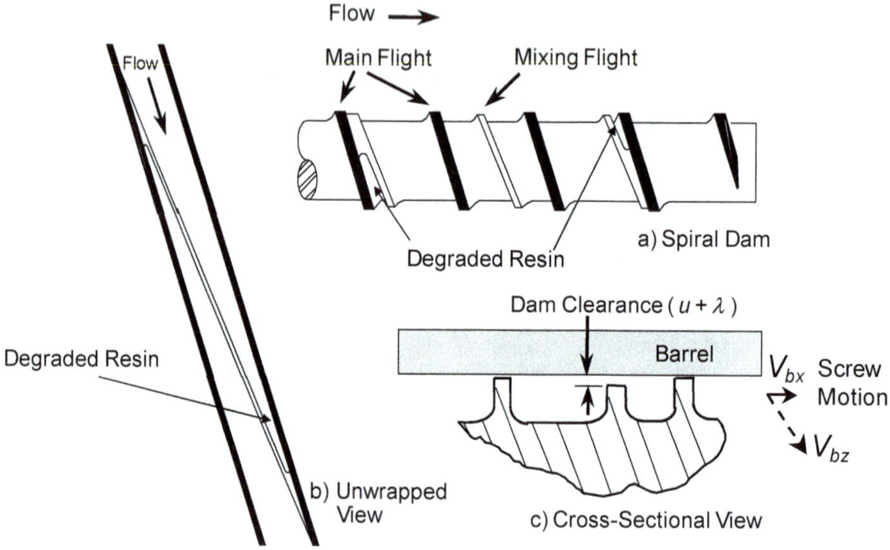

Figure 11.38 Schematic of a spiral dam: a) side view, b) unwrapped view, and c) a cross-sectional view perpendicular to the flight edge showing the clearance between the dam and the barrel wall. Degraded resin was observed at the pushing side of the channel just downstream of the entry to the spiral dam and at the trailing side of the channel just upstream of the exit to the spiral dam (courtesy of Jeff A. Myers of Robert Barr, Inc.)

Table 11.15 Screw Channel Dimensions for a 63 mm Diameter, 22 L/D Conventional Screw Running a GPPS Resin in an Injection-Molding Operation

	Depth, mm	Length, diameters	Notes
Feed section	8.26	11.7	
Transition section with a spiral dam		8	The spiral dam was 3.3 diameters in length and it was positioned at the end of the transition section
Meter section	3.43	2.3	

The lead length was 63 mm for the main flight of the screw. The main flight width and clearance were 7 and 0.06 mm, respectively, in all sections of the screw. The undercut of the spiral flight was 0.76 mm. The compression ratio was 2.4 and the compression rate was 0.0029. The ratio of the flight radii in the meter section to the meter depth was about 1.0. The specific rotational flow rate was calculated at 0.94 kg/(h·rpm).

The press chosen for the study was able to produce a 0.244 kg part and runner system with a plasticating time of 4.1 s, at a screw speed of 250 rpm and a tip pressure of 10 MPa. Using Equation 10.21, the screw was operating at a specific rate of 0.86 kg/(h·rpm). This specific rate was just slightly less than the specific rotational flow rate calculated at 0.94 kg/(h·rpm). The lower specific rate during operation was explained by the pressure gradient imposed in the metering channel of the screw during plastication as discussed next.

For the plastication process, the metering section of the screw must control the specific rate. If the metering section was not controlling the specific rate, then sections of the screw upstream of the metering section would control the rate, and some of the channels would be operating partially filled. For this application, partially filled channels would lead to the degradation of the resin. As the first diagnostic measure, the axial pressure profile for this screw and process was calculated to determine if the screw channels were operating full and thus under pressure. The axial pressure profile for this screw and process is shown in Figure 11.39. Based on the specific rate and the axial pressure profile, the screw was completely full of resin and operating under pressure, so the metering section was operating properly and thus controlling the specific rate of the process.

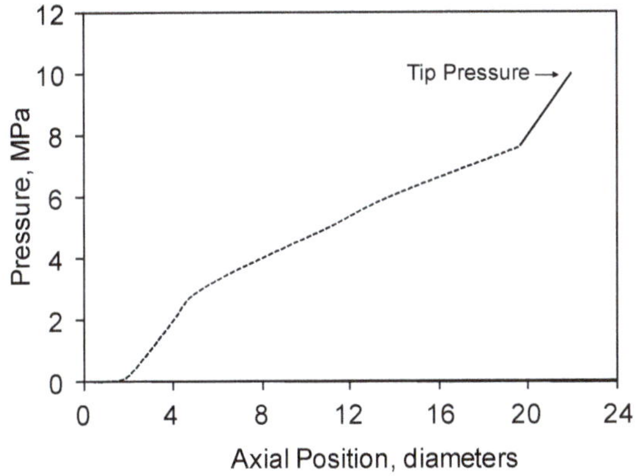

Figure 11.39 Axial pressure profile for a 63 mm diameter plasticator running GPPS resin at 215 kg/h and a screw speed of 250 rpm. The solid line in this figure was calculated using the methods described previously for metering sections, and the dashed line represents the expected pressure profile and was not calculated

Because the metering section of the screw was operating properly, it was then hypothesized that the screw was operating at too high of a screw speed and thus beyond its ability to melt resin. To test this hypothesis, process changes were made in an attempt to increase the discharge temperature to enhance melting. These attempts were unsuccessful. Process changes that were made included increasing the barrel temperature set points, increasing back pressure, and increasing the injection velocity. The only process variables that worked to minimize splay were lengthening the cycle time by increasing the dwell time in the barrel and decreasing the screw speed. Increasing the dwell time in the barrel was done to increase the temperature of the resin and remove unmelts. As previously stated, decreasing the screw speed to considerably lower levels eliminated the defect, but increased the cycle time. Increasing the cycle time was unacceptable to plant personnel.

The barrel temperatures were maintained at 250, 260, 270, and 270 °C for the zones starting at the feed and ending at the tip, respectively. These temperatures were higher than would be typically used. When the barrel temperatures were decreased, the scrap rate due to the defects increased. This information suggests that the screw was limited by its melting rate. It is well known that as the screw speed is increased, eventually the machine will be limited by its melting capacity, discharging solid polymer fragments with the injectate (or extrudate).

A screw was removed from one of the presses and examined. A large amount of dark-brown resin deposits was observed at the pushing side of the channel just downstream of the entry to the spiral dam. Similar deposits were observed at the trailing side of the channel just upstream of the exit to the spiral dam. The locations for these deposits are shown in Figure 11.38. These deposits were likely the source for the very small specks of dark-colored degraded resin that were observed in the defects. On further examination, the channels with the degraded resin were too deep at these locations, causing regions that were essentially stagnant.

In order to eliminate the splay problem, the compression ratio of the screw was increased from the original ratio of 2.4 to 3.0. The higher compression ratio should allow entrained air between the pellets to escape through the hopper and not be entrained with the injectate, and it should increase the melting capacity of the screw by increasing the pressure in the transition section [29]. This modification was made by increasing the feed channel depth from 8.26 to 10.3 mm by removing small amounts of metal from the feed channel. In order to maintain a constant compression rate on the transition section, deepening the feed section to 10.3 mm also decreased the feed section length by about 3.3 diameters and increased the length of the transition section by the same length. A summary of the channel dimensions for the modified screw is provided in Table 11.16 and shown schematically in Figure 11.40. The higher compression ratio should allow entrained air between the pellets to escape through the hopper and not be entrained with the injectate.

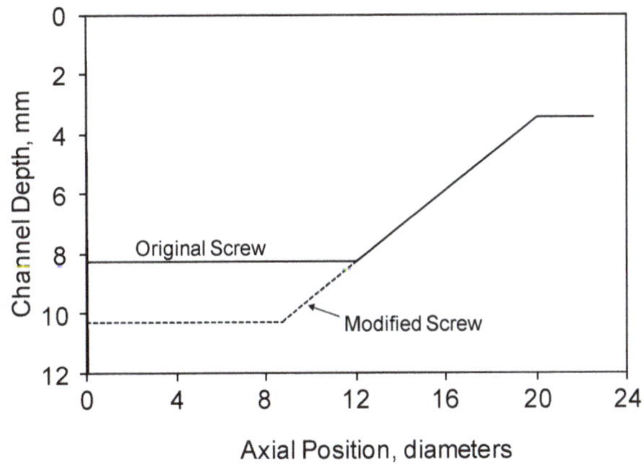

Figure 11.40 Channel depths for the original screw and the modified screw

Table 11.16 Modified Screw Channel Dimensions for a 63 mm Diameter, 22 *L*/*D* Conventional Screw Running a GPPS Resin in an Injection-Molding Operation

	Depth, mm	Length, diameters	Notes
Feed section	10.3	8.4	
Transition section with a spiral dam		11.3	The spiral dam was 3.3 diameters in length and was positioned at the end of the transition section
Meter section	3.43	2.3	

The lead length was 63 mm for the main flight of the screw. The main flight width and clearance were 7 and 0.06 mm, respectively, in all sections of the screw. The undercut of the spiral flight was 0.25 mm. The compression ratio was 3.0 and the compression rate was 0.0029. The ratio of the flight radii in the meter section to the meter depth was about 1.0.

The entrance and exit regions in the spiral dam were also modified to eliminate the stagnant sections of the channel. The modification is shown in Figure 11.41. This modification allowed a relatively small amount of resin to flow into the smaller channel at the entry such that stagnation of the resin cannot occur. A similar modification was made at the exit to allow a small amount of resin to flow out of the smaller channel into the main flow channel. To eliminate the unmelted particles or the particles that appeared to be more viscous because they were at a lower temperature, the clearance to the spiral dam was decreased from 0.76 to 0.25 mm. Since the meter channel depth was unchanged, the specific rotational flow rate for the modified screw was unchanged at 0.94 kg/(h·rpm).

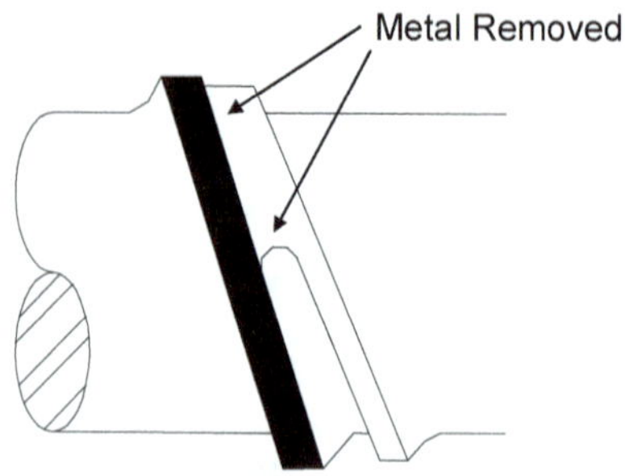

Figure 11.41 Location at the entry of the spiral dam where a small amount of metal was removed. Metal was also removed (not shown) at the exit of the spiral dam

The modified screw was placed back into the injection-molding press and evaluated for performance. The barrel temperatures were maintained at 245, 255, 260, and 260 °C for the feed zone through the last barrel zone, respectively. This temperature setting was lower than that used for the original screw. The screw was rotated at a speed of 235 rpm, and the back pressure was set so that the pressure at the tip was 10 MPa. The 0.244 kg part and runner system was plasticated in 4.2 s for a specific rate of 0.89 kg/(h·rpm). All parts produced were completely free of the splay defect. The modifications were able to eliminate the bubbles and the unmelted material.

The screw modifications allowed the cooling time to be decreased by 0.5 s as shown in Table 11.17. This decrease was due to the injectate having a lower temperature. For the original screw, the barrel temperatures had to be increased to 270 °C in the metering zone to increase the melting capacity of the screw. With the modified screw, the higher compression ratio and the smaller clearance on the spiral dam increased the melting capacity and allowed lower barrel temperatures (260 °C) and thus a lower injectate temperature. Since the cooling step was the rate-limiting step of the process, a decrease in cooling time resulted in an improved cycle time, as shown in Table 11.17. The modified screw decreased the cycle time from 6.1 to 5.6 s for a cycle time improvement of 8%. The plasticating time increased for the modified screw, but because the plasticating operation was not the rate-limiting step, the slightly longer plasticating times did not affect the cycle time.

Table 11.17 Key Cycle Times for the Molding Machine Using the Original Screw and the Modified Screw

	Original Screw	Modified Screw
Plasticating time, s	4.1	4.2
Cooling time, s	3.3	2.8
Cycle time, s	6.1	5.6

The modifications to the entry and the exit of the spiral dam section were successful in eliminating the stagnant portions of the section. That is, no dark-colored degraded resin specks were observed in the molded parts.

The injection-molding process is very complex and requires tuning for all sections of the process including the plasticating screw. The part defects presented here could not be eliminated without severely affecting the process economics, that is, by decreasing the screw speed and thus increasing the cycle time. Simple modifications to the screw, however, allowed the elimination of the defects and an 8% decrease in the cycle time. Moreover, with the elimination of the defective parts, the number of parts produced per shift increased by 14%. The modifications were performed very quickly and at a very low cost.

A systematic approach allowed the troubleshooting process to focus on the melting capacity of the screw as the root cause of the defects. The melting capacity was increased by increasing the compression ratio of the screw. The higher compression ratio caused the pressure in the melting section to increase. Higher pressures are known to increase the melting rate of GPPS resins. The higher compression ratio also increased the ability of the screw to force air that is entrained between the pellets back out through the hopper. Any solid polymer fragments that flowed downstream were then dispersed by the smaller clearance for the mixing flight in the spiral mixer. The stagnation regions that caused the resin to degrade were eliminated by allowing resin to flow through the passages at the entry and exit of the spiral dam.

11.12.5 The Injection-Molding Problem at Saturn

During the startup of Saturn Corporation's Spring Hill, TN, plant in September of 1990, a serious splay problem was encountered for the injection-molding of door panels from a PC/ABS resin [30]. The splay on the surface of the door panels created parts with unacceptable appearances after the painting process. The part rejection rate was higher than 25%, and it was high enough to nearly shut down the entire plant. Numerous teams were formed to determine the root cause of the splay. After a detailed analysis was performed, it was determined that the plasticating screw in the injection molder was not operating properly. The screw had a very low compression ratio of 2.25. A ratio of about 3 would have been more suitable for this resin and process. The low compression ratio caused the early part of the metering channel section to operate partially filled, leading to regions that were stagnant and allowed the resin to degrade. The splay was created by the volatile components from the degradation of the resin.

Two screws were designed with Energy Transfer (ET) sections from Robert Barr, Inc. [31], and the screws were labeled by Dow as ET-1 and ET-2. The original screw and both ET screw channel specifications are proprietary. The ET screw technology will be described in detail in Section 14.2.2. The ET-1 screw was made with a compression ratio of 3.4. The ET-2 screw was made as a backup in case the ET-1 screw did not work. The ET-2 screw had a compression ratio of 2.9. The ET-1 screw worked perfectly and virtually eliminated the splay from the part. This screw was able to operate this way because the design was streamlined and did not allow resin to become stagnant in the channels. Stagnant resin will have essentially infinite residence time at the processing temperature, allowing the material to degrade and cause defects in the molded parts. The ET-2 screw was also tested, but its compression ratio was not high enough to completely eliminate the splay defect for PC/ABS resin. Later the ET-2 screw was tested with a PC/polyester resin [32] and was shown to be an excellent design for eliminating splay in injection-molded parts.

11.13 Gels Caused by a Poorly Designed Transfer Line

A blown film line was experiencing a high level of gels in the product. The line had a configuration that positioned the extruder to the side of the die. The transfer line between the extruder and the die was 1.5 m in length and 31.8 mm in diameter (the existing design). The line was running 30 kg/h of a LDPE resin with a melt index of 1.9 dg/min (190 °C, 2.16 kg). The extruder was originally suspected of creating the degradation, but the investigation indicated that the extruder was operating properly and not causing the resin to degrade.

To investigate if the transfer line design was the root cause of the gels, numerical simulations were conducted to understand the flow field and the resin residence time inside of the transfer line. The simulation results are plotted in Figure 11.42 and Figure 11.43.

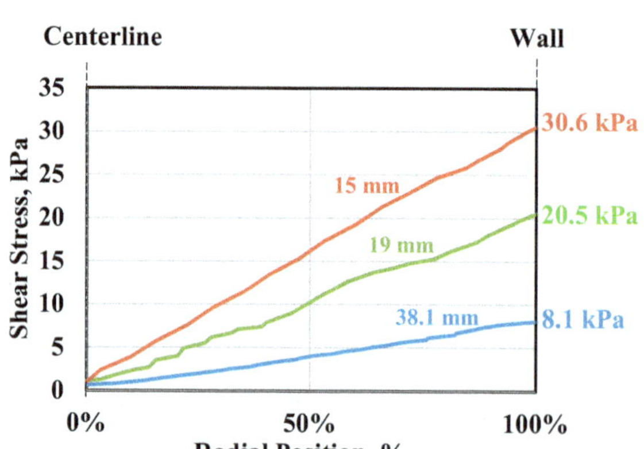

Figure 11.42 Shear stress comparison of the polymer from the centerline to the wall of the transfer line for the existing and proposed designs

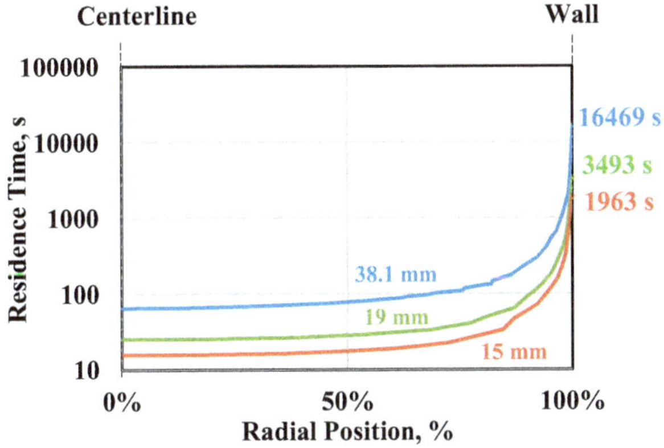

Figure 11.43 Residence time comparison of the polymer from the centerline to the wall of the transfer line for the existing and proposed designs

As shown in Figure 11.42, with the existing design, the wall shear stress was calculated to be 8.1 kPa. It was substantially lower than the recommended shear stress level of 20 kPa, which leads to very slow resin flow near the wall of the transfer line, poor surface renewal, and long residence times. The resin adjacent to the wall has a residence time as long as 16,500 s, as shown in Figure 11.43, which is going to lead to severe degradation of the LDPE resin.

In order to mitigate this issue, two improved designs were proposed. In the proposed designs, the transfer line diameter was reduced to 19 mm and 15 mm. Simulations were carried out to validate these designs. As shown in Figure 11.42, the wall shear stress increased to 20.5 kPa for the 19 mm transfer line and to 30.4 kPa for the 15 mm transfer line, which falls on the two ends of the recommended shear stress interval of 20 to 30 kPa [35]. As a result, the residence time at the wall and discharge was decreased significantly. The residence time at the wall was calculated to be 2000 s for the 15 mm transfer line, and 3500 s for the 19 mm transfer line, which should not cause polymer degradation or gels. Moreover, the high stress level will renew the resin at the wall.

Besides the residence time, the other factor that needs to be considered is the increase of the pressure change across the transfer line as a result of the smaller diameter in the proposed designs. As the transfer line was reduced from 31.8 to 19 mm, the pressure change was increased from 1.5 to 6.4 MPa. As the transfer line diameter was further reduced to 15 mm, the pressure change was increased to 12.3 MPa, which is considered to be excessively high for many applications. Because of this concern, the diameter of the transfer line was finalized at 19 mm. It yields 20.5 kPa wall shear stress, a residence time of 3500 s, and a pressure change of 6.4 MPa, which effectively mitigates the gel issue by reducing the residence time and producing a high stress at the wall, and without too high a pressure change required for flow.

11.14 The Incumbent Resin Effect

The incumbent resin effect starts off by running an extruder for extended time with the same resin (incumbent resin). Here the extrudate and PE film appear acceptable with only a few gels and black specks in the film product. These gels and black specks are generated in the stagnant regions of the screw. The majority of the degraded resin, however, is attached to the screw and is stable; i. e., not separating from the screw at a rate high enough to alarm quality specialists in the plant. Next, as a short trial or prototype run, the incumbent resin is switched with a challenger resin. Even though the challenger resin may be similar, it will likely process slightly differently than the incumbent resin. The differences could be subtle changes in the rheology, melting behavior, and/or solids conveying behavior. This slight difference in processing is often enough to cause the old degraded incumbent material that is adhering to the screw to separate from the screw and contaminate the extrudate. The old degradation will typically exit the die in about 5 minutes after the switch. The initial gel discharge is sometimes viewed as a gel shower, and then a high gel level will continue during the short trial run. In many cases, plant personnel will unknowingly blame the high gel level on the last change - in this case the switch to the challenger resin. That is, the challenger resin is incorrectly identified as the root cause for the high gel level. In severe cases, the trial is stopped and the challenger resin is eliminated as an acceptable resin for the application. The incumbent resin continues as the preferred resin. The root cause for the gels, however, is a poorly designed extruder screw and not the challenger resin. This is the incumbent resin effect.

A necessary condition for the incumbent resin effect is minor screw design flaws. These flaws include those presented earlier in this chapter. The degradation level is manageable at steady-state conditions for the incumbent resin. However, slight upsets that occur by introducing a challenger resin can cause the degraded material to separate from the screw at a faster rate. If the challenger resin was processed for an extended time, likely the same degradation level would eventually occur as observed for standard production with the incumbent resin. Typically, the challenger resin is extruded for only short trials, and is thus incorrectly identified as the root cause for the higher degradation level. If the screw is properly designed, the incumbent resin effect will not occur.

There are two technical solutions to mitigate the incumbent resin effect. The first and long-term solution is to design a new screw that does not contain regions where the resin flow will stagnate. The second and short-term technical solution is to remove and clean the existing screw before the challenger resin is introduced to the extruder. For this solution, the extruder may operate without producing degradation products for several hours. After this induction period, the degradation products that are formed may be stable and attached to the screw, causing only a low and manageable gel level. This gel level would be essentially equivalent to the level produced by the incumbent resin.

Nomenclature

d_w	diameter of the wires used to construct a filtering screen
d'_w	diameter of the wire in mm for Equation 11.1
d_{wp}	wire diameter for the warp wire of a Dutch weave filtering screen
d_{wt}	wire diameter for the weft wire of a Dutch weave filtering screen
d_o	average opening for flow for a mesh filtering screen
d'_o	width of the opening in the screen in mm for Equation 11.1
D_b	barrel diameter
D_s	diameter of the filtering screen surface
f_c	clearance between the mixing flight of a screw mixer and the barrel wall
H	channel depth
K	consistency factor for a power-law model
L	thickness of a breaker plate or the length of a transfer line
m	power law consistency index at a fixed temperature or mass rate of an extrusion process
M	mesh number for a screen in wires per inch
n	power law index
N	screw speed
N_h	number of holes in a breaker plate
Q	rate through a breaker plate or volumetric rate through a transfer line
R	radius of the holes in a breaker plate or the radius of a transfer line pipe
R_1	pushing flight radius
R_2	trailing flight radius
V_a	air rate that is being expelled from the extruder
V_N	recommended nitrogen flow rate for inerting the feed
W	mass flow rate through a filtering screen
$\dot{\gamma}$	shear rate

$\dot{\gamma}_w$ shear rate at the wall

ΔP_{screen} pressure change (or pressure drop) across a filtering screen

ΔP_{bp} pressure change across a breaker plate

η shear viscosity

ρ melt density

ρ_b bulk density of the feedstock in the hopper

ρ_s solid density of the resin

τ shear stress

τ_w shear stress at the wall

References

1. Smith, D., Stoughton, P., Morgan, C., and Hovis, G., "Auxiliary Systems," Chapter 5 in "The SPE Guide on Extrusion Technology and Troubleshooting," Vlachopoulos, J. and Wagner, J.R. (Eds.), Society of Plastics Engineers, Newtown, CT (2001)
2. Pahl, M.H., "Fundamental Principles of the Filtration of Non-Newtonian Fluids," Chapter 1 in "Filtration of Polymer Melts," English translation by Welling, M.S., VDI-Verlag GmbH, Dusseldorf (1981)
3. Carley, J.F. and Smith, W.C., "Design and Operation of Screen Packs," *SPE ANTEC Tech. Papers*, **21**, 594 (1975)
4. Todd, D.B., "Determining Pressure Drop in Extrusion," *Plast. Compd.*, **17**, 23 (1994)
5. Tadmor, Z. and Gogos, C.G., "Principles in Polymer Processing," 2nd ed., John Wiley & Sons, Hoboken, NJ (2006)
6. Butler, T.I., "Gel Troubleshooting," Chapter 19 in "Film Extrusion Manual," Butler, T.I. (Ed.), TAPPI Press, Atlanta, GA (2005)
7. Scheirs, J., "Compositional and Failure Analysis of Polymers," Wiley, New York, (2000)
8. Kodjie, S., Guerra, S., and Savargaonkar, N., "Analytical Techniques for Characterizing Defects and Morphology of Polymer Films," Microscopy and Microanalysis Proceedings, 715 (2012)
9. Garcia-Meitin, E.I., Bar, G., Blackson, J., and Reuschle, D., "High Resolution Polymer Imaging Using Scanning Transmission Electron Microscopy," Microscopy and Microanalysis, **14** (Suppl. 2), 1380 (2008)
10. Blackson, J., Garcia-Meitin, E., and Darus, M., "High Resolution Scanning Electron Microscopy Examination of Polymer Morphology," Microscopy and Microanalysis, **13** (Suppl. 02), 1062 (2007)
11. Hemsley, D.A., "Applied Polymer Light Microscopy," Elsevier Science (1989)

12. Ramanathan, R., Shanker, R., Rehg, T., Jons, S., Headley, S., and Schrenk, W.J., "Wave Pattern Instability in Multilayer Coextrusion - An Experimental Investigation," *SPE ANTEC Tech. Papers*, **42**, 224 (1996)
13. Tzoganakis, C. and Perdikoulias, J., "Interfacial Instabilities in Coextrusion Flows of Low-Density Polyethylenes: Experimental Studies," *Polym. Eng. Sci.*, **40**, 1056 (2000)
14. Grob, C. and Wortberg, J., "Optimizing Purging Time at Product Change in Blown Film Extrusion by Selective Modification of Process Parameters," *SPE ANTEC Tech. Papers*, **50**, 261 (2004)
15. Gilmor, C., Balke, S.T., Calidonio, F., and Rom-Roginski, A., "In-Line Color Monitoring of Polymers During Extrusion Using a Charge Coupled Device Spectrometer: Color Changeovers and Residence Time Distributions," *Polym. Eng. Sci.*, **43**, 356 (2003)
16. "Recommended Dimensional Guideline for Single Screws," The Society of the Plastics Industry, Inc.
17. Chen, L. and Lindt, J.T., "Three-Dimensional Flow Effects on Residence Time Distribution in Screw Extruders," *SPE ANTEC Tech. Papers*, **42**, 170 (1996)
18. Conner, J.H. and Bigio, D.I., "A Numerical Analysis of the Mixing Performance of a Single Screw Extruder," *SPE ANTEC Tech. Papers*, **38**, 1328 (1992)
19. Spalding, M.A., Dooley, J., Hyun, K.S., and Strand, S.R., "Three-Dimensional Numerical Analysis of a Single-Screw Extruder," *SPE ANTEC Tech. Papers*, **39**, 1533 (1993)
20. Spalding, M.A., Dooley, J., and Hyun, K.S., "The Effect of Flight Radii Size on the Performance of Single-Screw Extruders," *SPE ANTEC Tech. Papers*, **45**, 190 (1999)
21. Hyun, K.S., Spalding, M.A., and Powers, J.R., "Elimination of a Restriction at the Entrance of Barrier Flighted Extruder Screw Sections," *SPE ANTEC Tech. Papers*, **41**, 293 (1995)
22. Spalding, M.A., "Metering Channel Flows and Troubleshooting Single-Screw Extruders," *SPE ANTEC Tech. Papers*, **50**, 329 (2004)
23. Kamal, M.R., Isayev, A.I., and Liu, S-J., "Injection Molding, Fundamentals and Applications," Hanser Publications, Munich (2009)
24. Salamon, B.A., Koppi, K.A., and Little, J., "Halo Surface Defects on Injection-Molded Parts," *SPE ANTEC Tech. Papers*, **44**, 515 (1998)
25. Edwards, S.A. and Choudhury, N.R., "Variations in Surface Gloss on Rubber-Modified Thermoplastics: Relation to Morphological and Rheological Behavior," *Polym. Eng. Sci.*, **44**, 96 (2004)
26. Bryce, D.M., "Thermoplastic Troubleshooting for Injection Molders," Society of Plastics Engineers, Brookfield, CT (1991)
27. Spalding, M.A, Nelb, R.G., Patterson, J., Youngson, C., and Zawisza, M.J., "Processing Ignition-Resistant Polystyrene Resins in Injection-Molding Machines," *SPE ANTEC Tech. Papers*, **47**, 476 (2001)
28. Benkreira, H. and Brittin, R.N., "The Optimization of Masterbatch Formulations for Use in Single-Screw Machines," *Int. Polym. Process.*, **9**, 205 (1994)

29. Spalding, M. A., Hyun, K. S., and Cohen, B. R., "An Experimental Study of the Frictional and Viscous Forces Responsible for Solids Conveying and Melting in Single-Screw Plasticating Extruders," *SPE ANTEC Tech. Papers*, **43**, 202 (1997)
30. Kirkland, C., "Damage Control Saturn Style," *Injection Molding Mag.*, September (1994)
31. Chung, C. I. and Barr, R. A., "Energy Efficient Extruder Screw," U. S. Patent 4,405,239 (1983)
32. Ogando, J., "Screw Design Cures Splay Problems," *Plast. Technol.*, November (1995)
33. Spalding, M. A., Garcia-Meitin, E. I., and Kodjie, S. L., "Troubleshooting Gels," Chapter 5.10 in "TAPPI Film Manual," Global Editor Macnamara, J. F., Third Edition, TAPPI Press, Peachtree Corners, GA, 2020
34. Spalding, M. A., Gou, Q., Sun, X., and Shi, Q., "The Incumbent Resin Effect for Single-Screw Extrusion of Polyethylene Resin: The Effect of Resin Changeover on Gels in the Product," *J. Plast. Film Sheeting*, **34**, 381 (2018)
35. Perdikoulias, J., Compuplast, personal communication, 2018
36. Sun, X., Dutta, A., and Spalding, M. A., "Optimization of Transfer Lines for Mitigating Resin Degradation," *SPE-ANTEC Tech. Papers*, **65** (2019)
37. Fay, J. J. and King III, R. E., "Degradation and Stabilization of Polyethylene," Chapter 25 of "Handbook of Industrial Polyethylene and Technology", edited by Spalding, M. A. and Chatterjee, A. M., Wiley - Scrivener Publishing, Hoboken, NJ, 2017
38. Polychronopoulos, N. D. and Vlachopoulos, J., "Computer Flow Simulation of Moffatt Eddies in Single Screw Extrusion," *Int. Polym. Process.*, **33**, 662 (2018)

12 Flow Surging

Flow surging is defined as the oscillatory change in the rate of the extruder while maintaining constant set point conditions. Flow surging can originate from many different sources including improper solids conveying, melting instabilities, flow restrictions, and improper control algorithms [1–5]. Surging in most cases results in lower production rates, higher scrap rates, higher resin consumption, material degradation, and higher labor costs. In mild cases, flow surging will cause plant personnel to set the product at the low end of the dimension setting at the low rate portion of the surge. At the high rate portion of the surge, the dimensions of the product will be oversized. Oversized products will use more resin than necessary, adding cost to the product and decreasing the profitability of the plant. Obviously, a process that is very steady has the capability of minimizing resin usage and thus maximizing the profitability of the process. For a profile process where the dimension of the cross section is critical to downstream assembly processes, the extreme ends of the rate surges may result in a product that is outside of the specification, and some profiles will need to be scrapped or recycled. In this chapter numerous case studies along with diagnostic methods are presented for processes that flow surge.

The additional cost of producing products from a line that is flow surging can be substantial. If the flow surge is not too large and the line can be operated, the instability of mass flow at the die can cost the converter from 5 to 15% added costs in resins. The higher resin costs are incurred because the dimensions of the articles are larger or thicker than needed. For example, if a line is producing sheet for a downstream thermoforming process and it is operating unstably, then some thermoformed parts will have an acceptable mass while others will have a higher mass, costing the plant more in resin. Often the instability occurs only at a high rate while at lower rates the process is stable. In this case the plant may miss shipment dates since the line can only be operated at a fraction of its capacity, or the plant may incur higher labor costs because the line will need to operate over weekends. In severe cases, flow surging can cause the line to be incapable of producing product at any rate. Thus, in order to produce product at the lowest possible cost, the line must be operating stably so that the rate and product quality are maximized.

Processes that flow surge will often cause a higher level of degradation products to occur in the extrudate. For these cases, the unsteady nature of the flows in the screw channels will tend to break off small levels of degradation products adhering to the screw. The degradation products could occur at the flight radii and regions with long residence times, and they may not contaminate the extrudate under normal conditions. But the unsteady-state nature of the flow surge will tend to break them away from the screw surface.

12.1 An Overview of the Common Causes for Flow Surging

Improper process temperatures and poor temperature controls are common root causes for flow surging. For example, solids conveying depends on a balance of the forwarding forces at the barrel wall and the pushing flight and the retarding forces at the screw surface. These forces depend mainly on the geometry of the channel and are directly proportional to the coefficient of dynamic friction for temperatures less than the melting (or devitrification) temperature and on viscous forces for higher temperatures [6]. Since the coefficient depends on temperature, pressure, and velocity [7], surface temperature changes for the barrel and screw in the feeding section will strongly affect the performance of the extruder. If the surface temperatures become too different from the optimal values, flow surging and loss of specific rate will occur. If the solids-conveying section of the extruder is controlling rate, not the metering section as designed, then a portion of the screw channel between the sections will be partially filled at the low-rate swing of the cycle and most often will be completely filled at the high-rate region of the cycle.

Improper design and operation of the melting section of the screw can lead to extrusion instabilities. For example, solid bed breakup [3] can cause solids to migrate downstream. These solids can wedge into other sections of the screw and cause the extruder to flow surge [2, 4] or cause the extrudate to have periodic changes in temperature. Periodic changes in discharge temperature will cause some level of flow surging at the die [8].

12.1.1 Relationship Between Discharge Pressure and Rate at the Die

Dies are shaping devices that operate at a rate that is proportional to the upstream pressure. Thus, if the pressure to the die is not constant then a variable rate will occur at the die opening, causing the dimensions of the product to vary. Rate surges at the die can be estimated from the pressure surges using the following equations for flow through a cylindrical restriction (or die) for a power law fluid [4]:

$$\Delta Q = (Q_1 - Q_2) / Q_1 = 1 - (1 - \Delta P)^{1/n} \tag{12.1}$$

$$\Delta P = (P_1 - P_2) / P_1 \tag{12.2}$$

$$\text{or} \quad \frac{Q_1}{Q_2} = \left(\frac{P_1}{P_2}\right)^{1/n} \tag{12.3}$$

where n is the power law index, Q_1 and P_1 are the rate and discharge pressure at condition 1, and Q_2 and P_2 are the rate and pressure at condition 2. The pressure at the die lip is assumed to be zero. For example, a 5% variation in the discharge pressure (ΔP = 0.05) for a polymer with a power law index of 0.3 will cause a 16% change in the instantaneous rate (ΔQ = 0.16). An instantaneous rate change of this magnitude is unacceptable for most processes. The flow relationship with pressure is much more complicated than this for a commercial die, but the trend is the same.

12.2 Troubleshooting Flow Surging Processes

The analysis and troubleshooting of a process that is flow surging can be a difficult task, especially when the line is required to run production. The analysis can often be complicated by the operation of equipment downstream from the die. For example, if a pulling system is not operating at a constant speed then variations in velocity can cause the product to vary in dimension even though the extruder is operating stably. Worn components on a calendering roll stack can cause the speed of the rolls to vary or cause the gap between the rolls to change during a revolution. Both conditions will cause the product to change dimensions in the downstream direction. Unit operations downstream from the die must be checked to determine if they are the root cause of the product variation. The troubleshooter must be diligent to set a hypothesis and then test the hypothesis. If some problem other than the root cause is fixed, then the process will continue to flow surge.

The standard array of diagnostic equipment is required for the troubleshooting of a process that is flow surging. These tools include screw measuring devices, pyrometers, and devices to calibrate sensors in the process. These devices are discussed in Chapter 10. Often it is very difficult to impossible to determine a cause and effect relationship from process displays that are attached to typical extrusion lines. However, a portable data acquisition system that is capable of collecting process data as a function of time is highly useful in determining the cause and effect relationships between process parameters. In the cases presented here, the extrusion line was either equipped with a data acquisition system or a temporary acquisition system was connected to the machine during the trial.

12.3 Barrel Zone and Screw Temperature Control

Improper selection of process temperatures, poor temperature control, and inoperative temperature control devices are common causes for flow surging. As stated earlier, temperatures for the metal surfaces in the solids-conveying zone must be within a specific range for an application. This temperature range will depend on the resin, equipment design, placement of the temperature control sensor, and rate. Thermocouple placement on extruders is not standard, and thus they can be positioned at different axial positions for the zones and at different depths into the barrel wall. Because of these extruder and process differences, barrel temperatures typically need to be optimized for the machine and application. Optimization of barrel temperatures was presented in Section 10.9.

Equipment devices that are not functioning properly can cause a process to flow surge. For example, the feed casing of the extruder is typically cooled with water such that the outside temperature of the casing is about 50 °C or less. If the cooling water flow is turned off or is not flowing at a high enough rate, then the temperature of the inside wall of the casing will become too hot to convey solids into the machine. As a general rule for most resins, the outside temperature of the feed casing will be too hot to touch if the inside wall becomes too hot to convey solids, that is, at temperatures higher than 50 °C. At high casing temperatures, the rate-limiting step of the process is the solids conveying of resin from the casing to the barrel and not the metering channel of the screw. Thus, the specific rate will decrease and flow surging is very likely to occur. For specialty PE resins with very low solid densities, the temperature of the feed casing may need to be less than 35 °C. High temperatures on the feed casing can also cause the resin to bridge over the feed opening such that pellet flow to the extruder is severely or completely restricted.

Flow surging can occur if the temperature of the screw becomes too high in the solids-conveying section. In general, the temperature of the screw in this section needs to be less than the T_g for amorphous resins or less than the melting temperature for semicrystalline resins. Small-diameter screws will typically operate at feed zone screw temperatures that are low enough without the need for special cooling. For screws 150 mm in diameter and higher, the temperature of the screw, however, can become too hot for optimal solids conveying. In these cases, the temperature of the screw can be decreased by flowing water into the screw using a rotary union and piping assembly, as shown in Figure 12.1. Cool process water flows through the union and into a pipe that extends up to within 10 cm of the end of the cooling hole. The water then flows back out of the screw through a section of cast pipe. The cast pipe is attached by threads to the screw shank and rotary union. The length of the cooling hole and the flow rate of water are used to maintain the screw temperature in an optimal range. In general, the cooling hole is drilled into the screw up to the end of the feed section. Two case studies are presented that show flow surging processes that had poor temperature cooling on the feed section of the screw.

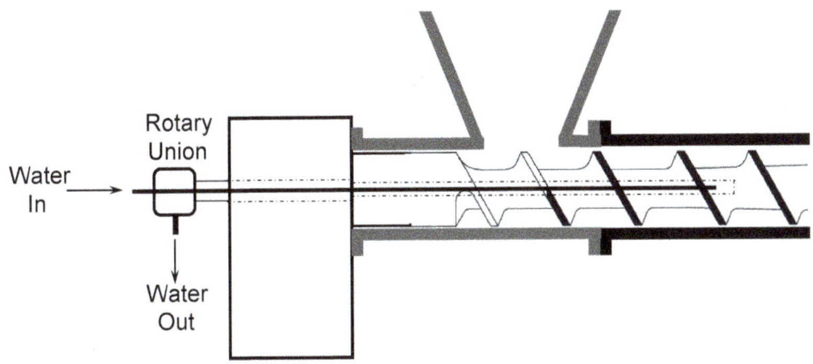

Figure 12.1 Diagram showing a rotary union piping assembly for cooling the feed section of a screw

Two-zone temperature control of the screw has been utilized to mitigate process instabilities in the solids-conveying zone and carbonaceous material buildup on the screw root in the melting zone for polyvinylidene chloride (PVDC) resins [9]. Two-zone screw temperature control can also be used to control the temperature of the solids-conveying zone and energy removal in the metering zone. The control device is similar to that shown in Figure 12.1 except that a second rotary union is required for the second fluid and a sealing device [10] is needed to isolate the cooling fluids.

12.3.1 Water- and Air-Cooled Barrel Zones

Heating and cooling of the barrel zones is typically done using modules that are equipped with electrical heaters and either water or air cooling. These modules are then clamped onto the outside of the barrel. Water cooling has the capability of removing more energy from the process, and it is well suited for extruders larger than 150 mm in diameter where the cooling demand is high, that is, where water flows to the modules for 10% or more of the time. If lower levels of cooling are required, however, water cooling can create temperature oscillations in the zone. For example when the zone becomes too hot, the controller will open the solenoid valve to the water flow line for the shortest possible duration. If this minimum amount of water flow is too large, then the cooling on the zone can be too much, causing the temperature of the zone to undershoot the set point temperature [11]. The control scheme will cause the zone temperature to oscillate. Variation in temperature for the barrel zones can affect the rate and discharge temperature. The oscillations can be mitigated by installing metering or needle valves in the water flow lines to reduce the water flow rate to the module. An in-line water filter is typically installed in the cooling line so that the needle valves do not get plugged with particulates.

Air-cooled zone modules do not have the ability to remove as much energy as do water-cooled units. For processes that only require a low level of cooling, air-cooled units will provide a more stable control of the temperature. Recent innovations in air cooling using high-flow fan systems [12] have allowed the replacement of some water-cooled systems with less costly and lower maintenance air-cooled systems [11].

12.4 Rotation- and Geometry-Induced Pressure Oscillations

Pressure transducers that are positioned in the barrel can be extremely useful for troubleshooting a process. Common positions include midway into the melting section and at the entry to the metering section. For two-stage screws, positioning of a transducer at the entry to the second-stage metering section provides information on the degree of fill of the stage and provides knowledge on the likelihood of vent flow. The pressures measured from these transducers provide three types of information: (1) the average pressure in the channel, (2) the pressure variation in the angular direction due to the rotating screw, and (3) the stability of the process by comparing the pressure oscillations during several screw rotations. The pressure in the angular direction is composed of two pressure components: (1) a pressure component in the downstream direction, $\partial P / \partial z$, and (2) the cross-channel pressure gradient, $\partial P / \partial x$. The shape of the angular pressure profile depends on the magnitudes of the components. In order to measure the pressures during rotation, high-speed data acquisition equipment is required. For example, a screw that is rotating at a speed of 60 rpm will require a data acquisition frequency of at least

20 Hz, providing 20 pressure measurements per rotation. Typical pressure measurements for transducers positioned in melting sections and metering sections that are filled with molten resin are shown in Figure 12.2.

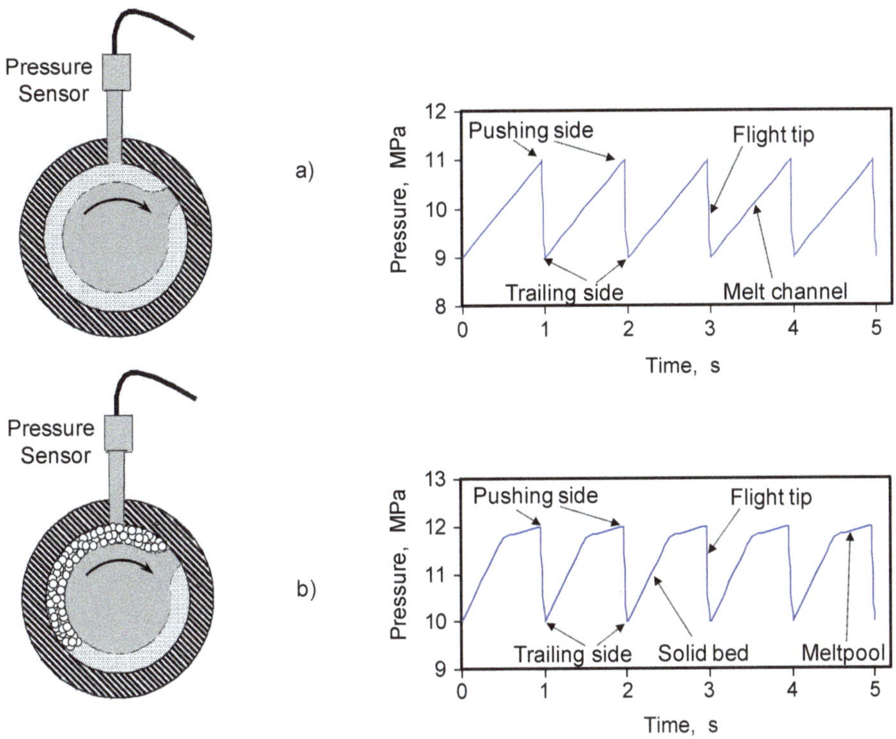

Figure 12.2 Typical pressure measurements for transducers positioned in the barrel for a screw speed of 60 rpm and a positive downstream pressure gradient ($\partial P / \partial z > 0$): a) for a transducer positioned in a metering section where the resin is completely molten, and b) for a transducer positioned in a single-flighted melting section

The pressure profile shown in Figure 12.2(a) is for a constant-depth metering channel that is completely filled with molten resin, a screw speed of 60 rpm, and a positive downstream pressure gradient ($\partial P / \partial z > 0$); five rotations are shown. The pressure is the highest at the pushing side of the channel and the lowest at the trailing side of the channel. The pressure typically increases nearly linearly with rotation from the trailing side of the channel to the pushing side. As the flight tip passes underneath the transducer, the pressure decreases quickly to that of the trailing side of the channel. Figure 12.2(b) shows a similar pressure profile with rotation in a conventional melting section. For this case, the solid bed extends across about 50% of the channel. The pressure profile is similar to that for the metering channel case except that the pressure gradient in the region over the solid bed is higher than that for the melt pool. The width of the molten resin can be estimated by the time fraction that the transducer spends over the melt pool and solid bed.

The pressure profiles with rotation shown in Figure 12.2 are ideal. In practice the pressure profiles contain a level of measurement error and unsteady-state behavior. Pressure in an actual channel operating at a screw speed of 30 rpm for an ABS resin is shown in Figure 12.3.

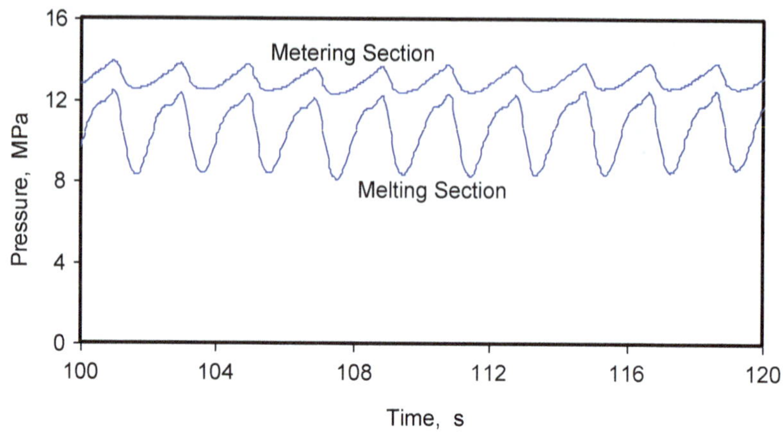

Figure 12.3 Measured pressure profiles with rotation for a 63.5 mm diameter extruder running an ABS resin at 30 rpm, a conventional single-flighted screw, and with a positive downstream pressure gradient in the metering section ($\partial P / \partial z > 0$)

As shown in Figure 12.3 for the metering channel, the highest pressure is at the pushing side of the channel and the lowest is at the trailing side of channel. The angular pressure profile in the melting section was typical and very similar to the ideal profile shown in Figure 12.2 because properly operating melting sections have positive pressure gradients in the downstream direction. The data in Figure 12.3 clearly shows that a level of measurement noise and unsteady-state activity is occurring in the process.

12.5 Gear Pump Control

Gear pumps are often positioned between the extruder and the die, and they provide several processing advantages. These advantages include the mitigation of pressure and flow surges from the extruder, a decrease in the discharge temperature by generating the pressure for the die by the pump instead of the extruder, and by decreasing the discharge pressure via the pump, a capacity increase is possible [13]. For gear pump assisted extrusion, the extruder control algorithms are set to maintain a constant pressure to the inlet side of the pump. The pump is operated at a constant rotational speed, and thus it delivers molten polymer at a very steady and controlled rate. A schematic of a gear pump assisted extrusion process is shown in Figure 12.4. If the pressure to the inlet of the pump is less than the set point value, then the control system will increase the screw speed of the extruder. Conversely, if the inlet pressure is too high, the control system will decrease the screw speed. Thus, processes that use a gear pump downstream of an extruder can show large variations in the screw speed in an attempt to compensate for an extruder that is flow surging.

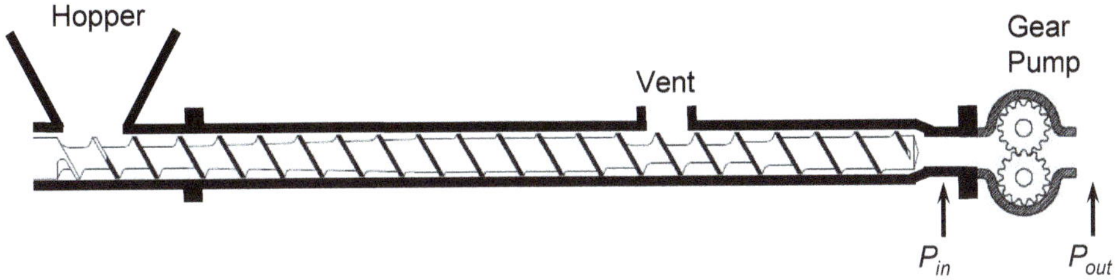

Figure 12.4 Schematic of a two-stage extruder with a downstream gear pump

A poor control algorithm for the pump can cause some variation in the extruder screw speed, causing large variations in the inlet pressure to the pump. This type of control-induced surging can occur even though the process as designed is inherently stable. To determine if the control algorithm is inducing the surging, the screw speed of the extruder should be operated in a manual mode and at a constant speed. If the controller is inducing the surging, placing the process in manual-control mode will stabilize the process. Transient process data were collected for an extruder with a downstream gear pump, as shown in Figure 12.4. For this case, the control algorithm was controlling the speed of the screw such that the inlet pressure to the pump was maintained at 8 MPa. Although the variation in screw speed was not excessive at 67±1.5 rpm, the variation in motor current seemed quite high at 540±90 A. At about 16 minutes into the run, the extruder was switched from automatic to manual screw control; the screw speed was held constant at 67 rpm. As shown by the data in Figure 12.5, the motor current variation was unchanged, indicating that the screw speed control algorithm was not inducing the variation in the motor current. During the period that the screw speed was held constant, the pressure to the inlet of the pump slowly increased, as shown in Figure 12.6. This pressure was increasing because the screw was operating at a speed that delivered a rate slightly higher than that needed by the pump. When the control was placed back into automatic mode, the screw speed was decreased initially to compensate for the higher than desired inlet pressure. This type of analysis is recommended when minor levels of flow surging are observed with a process where the screw speed is controlled from the inlet pressure of a gear pump.

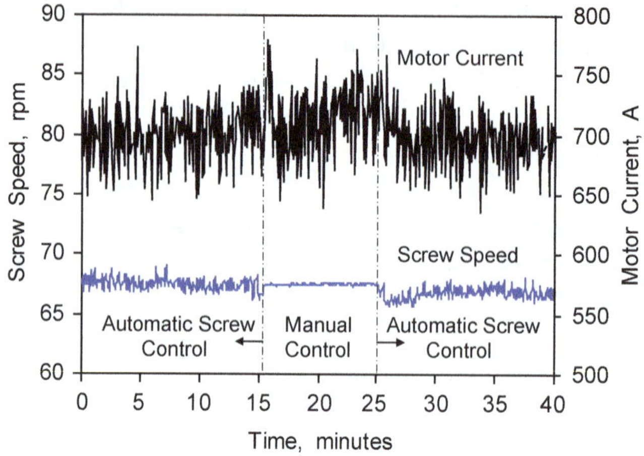

Figure 12.5 An extrusion process with a downstream gear pump with the screw operating in inlet pressure control and followed by the screw in manual operation (constant screw speed). The large level of variation in the motor current during constant screw speed control suggests that the extruder process is unstable, and the control algorithm is not the root cause for the variation in the motor current

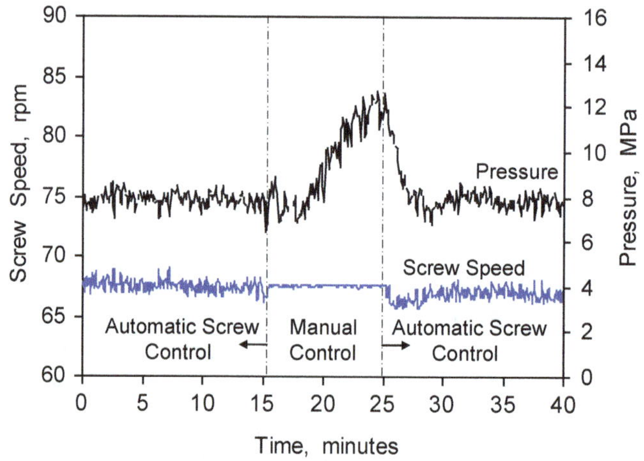

Figure 12.6 Pressure at the inlet to the gear pump for the data presented in Figure 12.5. The pressure increased during manual control because the flow rate of the extruder was slightly higher than the rate of the pump

12.6 Solids Blocking the Flow Path

Compacted solid polymer fragments can block and restrict the flow in a process. In order for this to occur, two defects typically exist in the process. The first defect causes the compacted solid to fragment and flow downstream in the screw channels. The second defect is a restriction in the channel where the fragments are trapped and accumulated. As the restriction builds, the local pressure just upstream of the restriction will increase while the pressure downstream will decrease. As the downstream pressure decreases, the pressure and rate at the discharge of the extruder will also decrease. The local and high pressure just upstream of the restriction will cause the melting rate of the fragments to increase, temporarily clearing the blockage [2]. When the blockage is removed the rate of the process returns to normal until the next solid fragment blocks the restricted region. Repeated blocking and clearing of the restricted region creates the flow surging.

To eliminate surging due to solid blockages, the troubleshooter must eliminate the defect that caused the solid bed to break up and must also mitigate the restriction in the downstream section of the screw. It is preferred to correct both defects to permanently eliminate surging from the process.

12.7 Case Studies for Extrusion Processes That Flow Surge

Numerous case studies are presented in the next sections that show some common flow surging problems. In these case studies, the problem is presented in a manner that the troubleshooter would encounter during a trial or information-gathering session. Incomplete data and erroneous data are often presented to the troubleshooter. These data were not included here because including them may mislead the reader. The troubleshooter, however, must be able to separate the facts of the process from misleading perceptions. In each case study, the modifications required to fix the process are detailed along with supporting fundamental information. In all cases, the rate of the process was limited and the cost to manufacture was high.

12.7.1 Poor Barrel Zone Temperature Control

A 203.2 mm diameter plasticating extruder was running GPPS resin and discharging to a specialty downstream process. Like most processes, the downstream equipment required a nearly steady supply of molten polymer. For this case, the barrel zones were electrically heated and water cooled. The barrel zone temperature is shown in Figure 12.7. The barrel heater for this zone was only used during startup. Once the extruder was operating, the energy dissipation from the screw to the resin was more than enough to keep the section hot. In fact, in order to maintain the zone at the set point temperature of 215 °C, the extruder was operated with a very small amount of barrel cooling. At the low-temperature portions of the cycle, both heating and cooling were off. The small amount of excess energy dissipated in the screw channel was causing the barrel temperature to increase slightly with time. When the temperature exceeded 220 °C, the control algorithm took action and opened the solenoid valve on the water line upstream of the heating and cooling barrel jacket, as shown in Figure 12.8(a). The controller opened the solenoid valve for the minimum amount of time, sending a short burst of water to the zone. The water would flash evaporate in the unit and then quickly cool the barrel to about 210 °C. Since the solenoid was opened for the shortest possible amount of time, the level of cooling that was utilized was the minimum. It was very obvious that the level of cooling water to this barrel zone was too high for this process. The barrel temperature oscillations shown in Figure 12.7 were enough to cause a small variation in the product dimensions. Although the variations in the product dimensions were acceptable, the variations did reduce the profitability of the process by causing too much resin to be used in the final product.

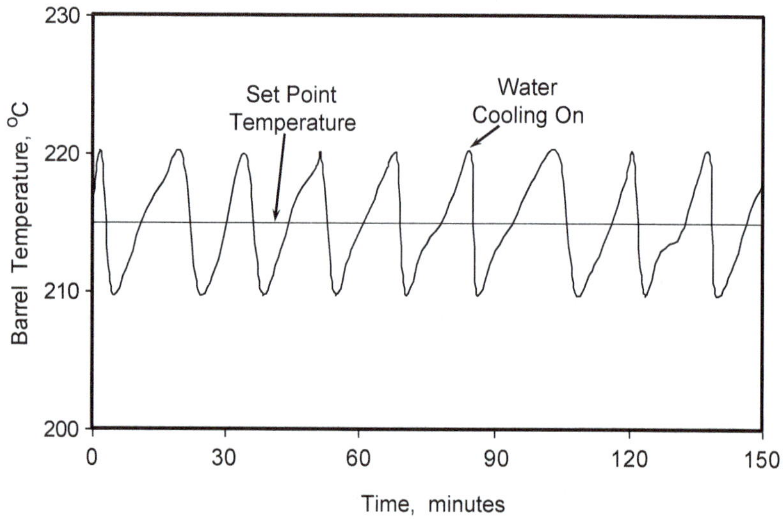

Figure 12.7 Barrel temperature data for a 203.2 mm diameter extruder running GPPS resin and with water cooling on the barrel heating and cooling. This extruder was configured with a water-cooling capability that was too high for the process

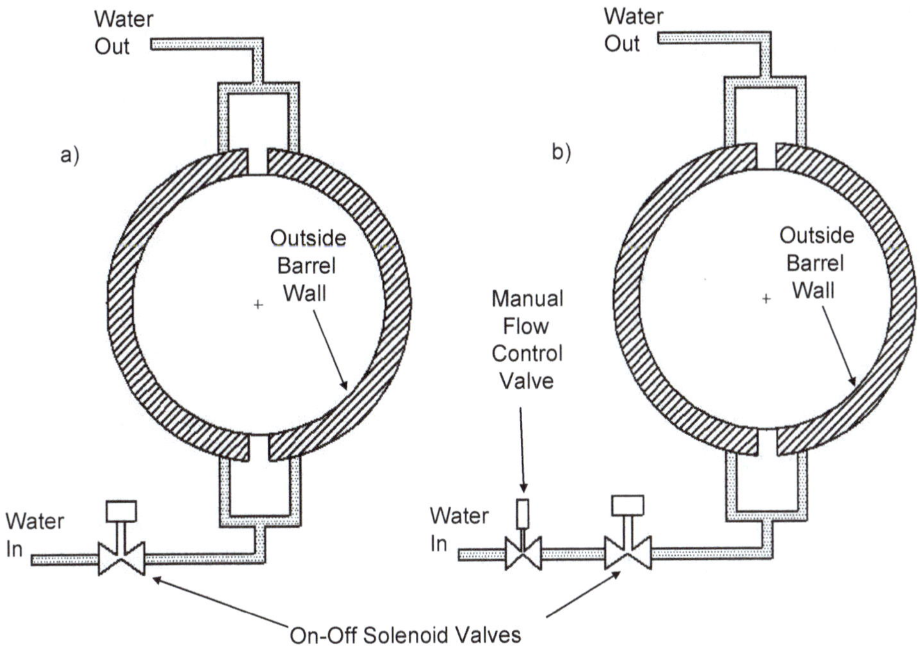

Figure 12.8 Heating and cooling system on the barrel: a) schematic of the original configuration that created the temperature oscillations in Figure 12.7 and b) a better configuration that minimized the temperature oscillations

In order to reduce the cooling level to the barrel zone, a metering valve was placed in the water line upstream of the solenoid valve as shown in Figure 12.8(b). Now when the controller opens the solenoid valve, a much lower quantity of water and thus cooling is available to the barrel zone. Prior to this modification, the barrel temperatures oscillated ±10 °C about the set point temperature. After the modification, the temperature oscillations were reduced to about ±3 °C, and the profitability of the process was improved due to the minimization of resin consumption.

This temperature control problem occurred due to the implementation of a high-performance-type screw. The original screw was fabricated with a relatively shallow metering channel. The shallow channel had a low specific rate and also dissipated a relatively high level of energy. The excess energy was easily removed through the barrel wall with the water cooling using the configuration shown in Figure 12.8(a). That is, the solenoid valve was in the open position enough to maintain cooling while not causing the barrel temperature to undershoot the set point temperature. The high-performance screw, however, was designed with a deeper metering section, had a considerably higher specific rate, and dissipated less energy. For this screw, less excess energy needed to be conducted through the barrel wall. Since the cooling system was designed for a process with a high heat flux through the barrel, the temperature became very oscillatory when the energy flux was reduced when the high-performance screw was implemented.

This simple case study shows the importance of verifying the control algorithms before proceeding with a troubleshooting trial. Before any testing or equipment modifications are performed, it is extremely important to have a deep understanding of the process and have all process controls and sensors in acceptable operation. If the sensors and controls are not functioning properly, then the troubleshooter may modify the wrong section of the process and obtain little to no improvement in the process.

12.7.2 Optimization of Barrel Temperatures for Improved Solids Conveying

Numerous complaints were logged by a single processor from several different manufacturing plants on flow surging and reduced rates for a specialty resin. The flow surging caused unacceptable variations in the final product. In all cases small-diameter extruders were used, but the operating conditions reported were different in the plants. In several of the plants, there were some extruders that did not flow surge, yet the design of these machines appeared to be identical to those that experienced flow surging. It was not apparent why some of the extruders were operating well while others were surging.

An extrusion trial was performed at the processor's plant using a 38.1 mm diameter production extruder, a proprietary screw design, and resin that had previously exhibited flow surging and reduced rate. The extruder was equipped with three barrel zone heaters with control thermocouples (labeled T1, T2, and T3) and two pressure sensors. One pressure sensor was located in the midsection (zone 2) of the barrel (P2) and the other at the end of the barrel near the tip of the screw (P3). Both transducers were positioned over the top of the screw such that a pressure variation due to screw rotation would be observed.

During the trial, process data were collected from each sensor at a frequency of once every 10 seconds using a portable data acquisition system. For barrel zone temperatures of 150, 163, and 174 °C for zones T1 through T3, respectively, the extruder was operating stably and at rates that were consistent with numerical simulations, and it was producing a high-quality product. Process data for steady-state operation are shown in Figure 12.9 for a screw speed of 50 rpm. As indicated by this figure, the barrel zone temperatures were steady and only small variations occurred for the P2 and P3 pressure sensors. Slight pressure variations were expected for this extruder because the sensors were positioned in the barrel and were measuring pressure in different regions of the channel as the screw rotated. The pressure patterns are not periodic like those in Figure 12.3 due to the screw speed and acquisition rate used. For this case, pressure samples were collected every 8.3 rotations. A faster data collection rate would have shown a periodic oscillation of the pressure. These data indicate that conditions exist for the stable processing of the resin.

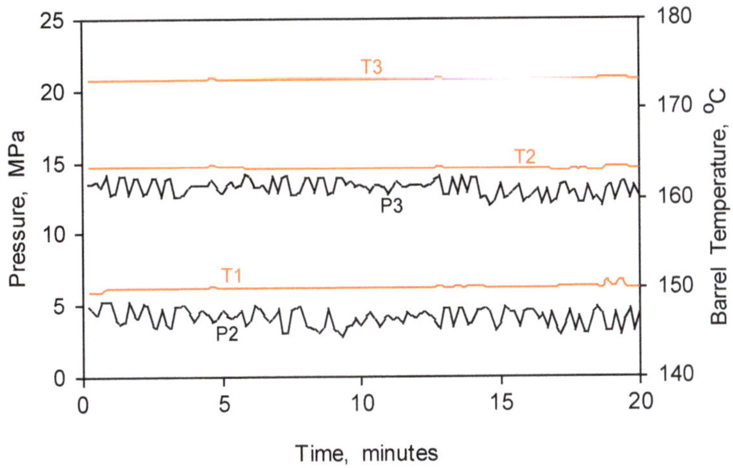

Figure 12.9 Barrel pressures and temperatures for the 38.1 mm diameter extruder operating stably. The temperature profiles are in red while the pressures are black

For a second experiment, the extruder was operated at barrel set point temperatures of 160, 163, and 174 °C for zones T1 through T3; the zone T1 temperature was increased by 10 °C, and zones T2 and T3 temperatures were unchanged. This increase in the T1 temperature caused the extruder to flow surge and decreased the rate by about 20%. The process data for the unstable conditions are shown in Figure 12.10. As indicated by this figure, the pressure for the mid-barrel pressure sensor, P2, was zero during the low pressure swing of the cycle, indicating that this portion of the channel was operating partially filled (or starved). Later in the experiment, the pressure sensor responses were checked when the pressure was known to be zero in the channels. The pressure was measured by the sensors at 1.4 MPa when the pressure was actually zero, explaining the offset pressure at the bottom of the pressure cycle in Figure 12.10. Numerical calculations and a Maddock solidification experiment confirmed that the midsection of the extruder was operating partially filled. Thus, a small 10 °C increase in the first barrel zone temperature was enough to cause the extruder to go from operating as a stable process producing high-quality product to one that was unstable with reduced rates and having a product with unacceptable product dimensions. Numerous other experiments had shown that the first barrel zone temperature needed for stable extrusion depended on the screw speed and the temperature of the feed resin. Moreover, the processor indicated that flow surging was experienced for some extruders at zone T1 barrel temperatures as low as 148 °C.

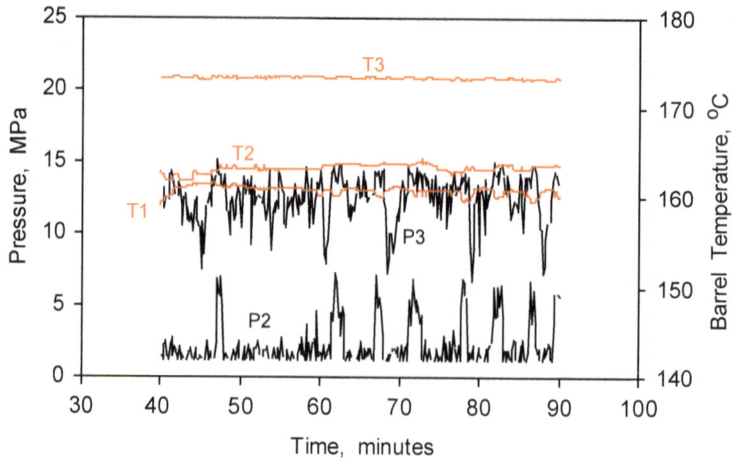

Figure 12.10 Barrel pressures and temperatures for the 38.1 mm diameter extruder operating at a zone 1 barrel temperature condition that caused the extruder to flow surge. The temperature profiles are in red while the pressures are black

Based on the data collected, the Maddock solidification experiment, and the numerical calculations, the problem was diagnosed as poor solids conveying from improper temperatures in the section. Slight differences between extruders, such as the axial and radial position of the zone T1 thermocouple, barrel zone controller tuning, screw geometry variations, and thermocouple accuracy likely caused conditions such that some of the extruders flow surged while others did not. These minor variations could influence the temperature of the inside barrel wall of the solids-conveying section. Moreover, different rate requirements for different products required that the extruder be operated at different screw speeds, which further complicated the solids-conveying problem. The problem could have been avoided if plant personnel had optimized the barrel temperatures for each extruder using the technique described in Section 10.9.

12.7.3 Flow Surging Due to High Temperatures in the Feed Section of the Screw

A severe and random flow surging problem limited the production rate for a large-diameter, two-stage, vented extruder. If it were not for a gear pump positioned between the extruder and die, this extrusion line would not have been operable. The surging did, however, limit the output of the line to about 70% of its potential rate. The maximum potential rate is the rate that the extruder can run at high screw speeds and with proper operation. The extruder was 203.2 mm in diameter and had a 40 *L/D* barrel. A schematic for the extruder and gear pump arrangement is shown in Figure 12.11, and the screw channel dimensions are provided in Table

12.1. The specific rotational flow rate for the first-stage metering section was calculated at 20.0 kg/(h·rpm). The extruder was fed a mixture of fresh HIPS resin with 30 to 60% recycled ground sheet from a downstream thermoforming process. The level of recycle affected the bulk density of the feedstock entering the extruder. The HIPS resin had an MFR of 3.9 dg/min (230 °C, 5.0 kg). The screw was single-flighted and typical of what is used for HIPS resins. Screw temperature control was accomplished by flowing cooling water through a rotary union into and out of a hole cut into the feed end of the screw as shown in Figure 12.1. This hole extended 3.8 diameters into the feed section. Pressure sensors were positioned in the barrel wall at the end of the first-stage transition section (P1), at the end of the first-stage metering section just before the vent (P2), and at the discharge. Additional pressure sensors were positioned at the discharge of the extruder and at the inlet (suction side) to the gear pump. A screen filtering system was positioned between these pressure sensors as shown in Figure 12.11. A commercial control scheme adjusted the screw speed to maintain a constant pressure of 9 MPa to the inlet of the gear pump. The gear pump was operated at constant speed in order to maintain a constant flow rate of material to the die.

Table 12.1 Screw Channel Dimensions for a 203.2 mm Diameter Two-Stage Vented Screw Running HIPS Resin

	Depth, mm	Length, diameters	Notes
Feed section	28.6	7	The compression ratio was 4.0 and the compression rate was 0.0032
First-stage transition		10	
First-stage meter	7.1	8	
Vent section	31.9	4.5	The pump ratio was 1.7
Second-stage transition		3.5	
Second-stage meter	12.3	6	

Lead length, flight width, and flight clearance were 203.2, 23.9, and 0.20 mm, respectively, in all sections of the screw. A 28.7 mm diameter screw cooling hole was drilled in the shank end of the screw, and it extended 3.8 diameters into the feed section. The first 2.5 diameters of the screw were inside a water-cooled feed casing. The specific rotational rate of the first-stage metering section was calculated at 20.0 kg/(h·rpm).

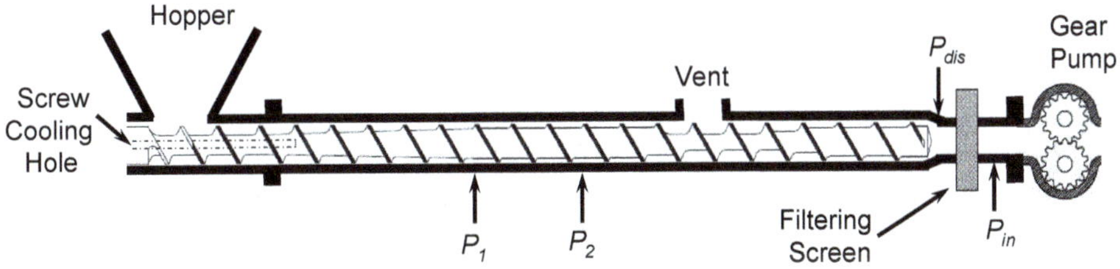

Figure 12.11 Schematic of the 203.2 mm diameter extrusion process for HIPS resin

In order to diagnose the problem, a data acquisition system was temporarily connected to the extrusion panel. All available sensor outputs were connected in parallel with the acquisition system. Electronic data were collected at a frequency of once every 9 s. Steady-state operation of the extruder is shown by the first 400 minutes in Figure 12.12, Figure 12.13, and Figure 12.14. The data for these figures were from the same production run. The extruder was running at 2250 kg/h and a screw speed of 99 rpm for a specific rate of 22.7 kg/(h·rpm). This specific rate is about 14% higher than the specific rotational flow rate calculated for the first-stage metering section, indicating that a negative pressure profile exists in the section. The negative pressure gradient is expected for a first-stage metering section of a vented screw that is operating properly; that is, the first-stage metering section was full of resin. To maintain the stability, the extruder screw speed was reduced such that the extruder was operating at about 70% of its potential maximum rate. That is, at screw speeds higher than 99 rpm the extruder was more likely to transition from a stable to an unstable operation. The barrel pressure at the end of the first-stage transition section, P1, had variations of about ±3 MPa about the average pressure. This pressure variation was considerably higher than expected and suggests that the extruder, although running stably, was on the verge of unstable operation. Some of the variation was due to the movement of the flight tip past the sensor. Barrel zone temperatures tracked the set point values and were stable.

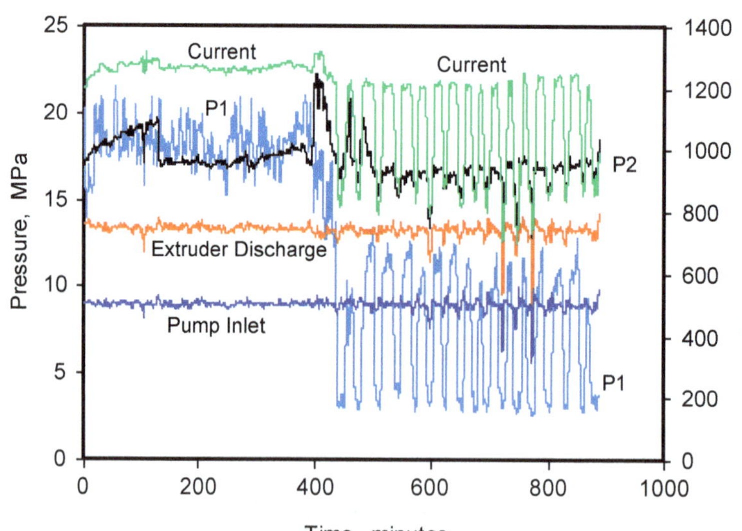

Figure 12.12 Barrel, discharge, and pump inlet pressures and motor current for stable and unstable extrusion for a large-diameter extruder running HIPS resin

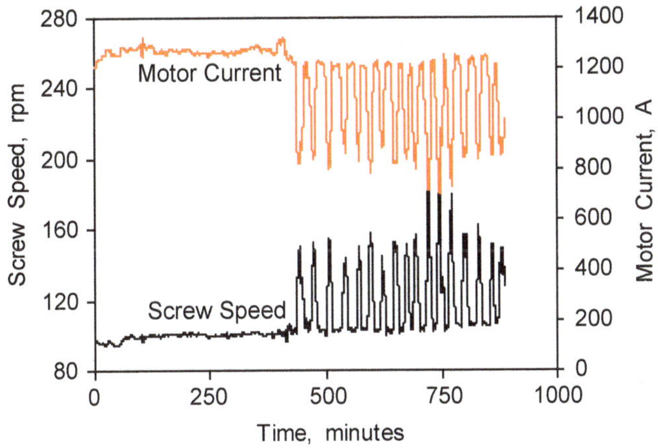

Figure 12.13 Screw speed and motor current for a large-diameter extruder running stably and unstably

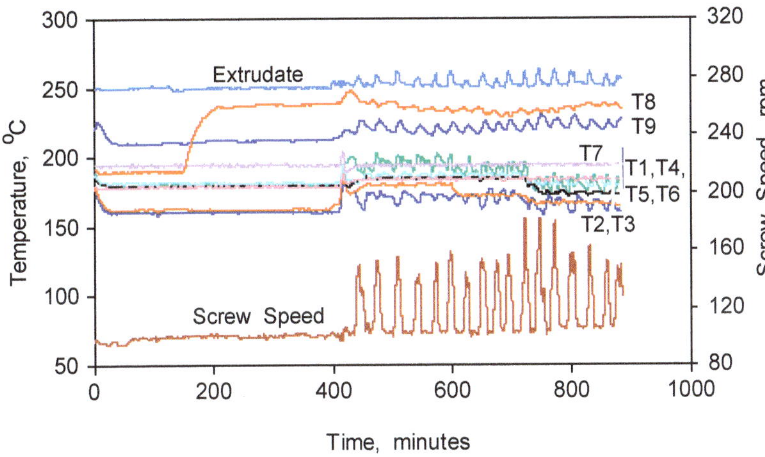

Figure 12.14 Screw speed, extrudate temperature, and barrel zone temperatures for a large-diameter extruder running stably and unstably

At about 410 minutes into the run, the extruder started to operate unstably, as indicated in Figure 12.12, Figure 12.13, and Figure 12.14. The processing change that caused the extruder to go from a stable operation to an unstable one was not known, but it could have been due to minor changes in the bulk density of the feedstock or cooling water fluctuations to the screw. As indicated by these figures, the event started when the P1 pressure decreased slightly, causing the rate and the P2 pressure to decrease. This decreased pressure transmitted down the extrusion system, eventually decreasing the pressure at the inlet to the gear pump. To correct for the lower pressure, the controller on the gear pump increased the speed of the screw from 99 rpm to about 160 rpm. Next the P1 pressure increased due to the higher screw speed and higher flow rate, as indicated in Figure 12.12. As the pressure increased at the gear pump inlet, the gear pump controller decreased the screw speed back to about 100 rpm, causing the extruder to flow surge. Flow surging caused the screw speed controller to oscillate about once every 25 minutes. As indicated in Figure 12.12, the screw speed controller was able to provide a relatively stable pressure to the pump inlet, allowing the process to run at reduced rates. The barrel zone temperatures, as indicated in Figure 12.14, were extremely oscillatory.

As indicated in Figure 12.12, the P1 pressure was considerably lower during the period of unstable operation. This result indicates that the cause of the problem originated in the first stage of the screw before the first-stage metering section. At a screw speed of 160 rpm, the extruder was still operating at a rate of 2250 kg/h, but the specific rate decreased to 14 kg/(h·rpm). This specific rate is considerably less than the specific rotational flow rate of 20 kg/(h·rpm), indicating that the first-stage metering section was operating improperly and only partially filled. The most likely reason for a partially filled or starved metering section was poor solids conveying from the feed section to the transition section. Poor solids conveying was likely due to improper temperature control of the metal surfaces in the feed section of the extruder and screw. Barrel feed zone heaters, controllers, and the feed casing were examined and determined to be operating properly at set point temperatures typically used for HIPS resin. Based on this information, the investigation was focused on the temperature control of the screw.

It was hypothesized that the screw temperature in the feed section was too hot to convey solids effectively to downstream sections of the screw. To test this hypothesis, the effect of internal screw cooling was determined during a period when the extruder was operating stably. For this period, cooling water was flowing to the screw-cooling device, and the extruder was operating stably and properly at a rate of 2360 kg/h and a screw speed of about 104 rpm. The metal surface temperatures of the pipes used to flow water into and out of the screw were measured at 29 and 37 °C, respectively. At about 28 minutes into the run, the cooling water flow to the screw was turned off, as indicated in Figure 12.15 and Figure 12.16. At about 30 minutes, the pressure at the end of the first-stage transition section, P1, started to decrease as shown in Figure 12.16, indicating that solids conveying was significantly reduced. Like before, the reduced solids flow caused the downstream pressures to decrease and ultimately to cause the extruder to flow surge. At about 36 minutes into the run, cooling water flow was turned on, and within about four minutes the extruder operation became stable, as indicated in Figure 12.15 and Figure 12.16. The surface temperature of the pipe for water flow out of the screw was measured at 81 °C just after the cooling water was turned on, a temperature change of 44 °C. As will be presented later in this section, solids conveying of HIPS resin becomes difficult or unstable at screw temperatures of about 150 °C and higher. The temperature of the screw surface was unknown, but it likely increased by at least 44 °C and possibly approached 150 °C.

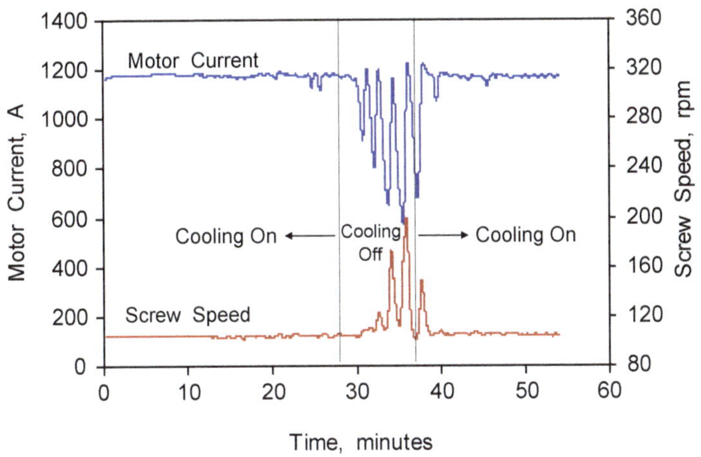

Figure 12.15 Screw speed and motor current for the screw cooling experiment

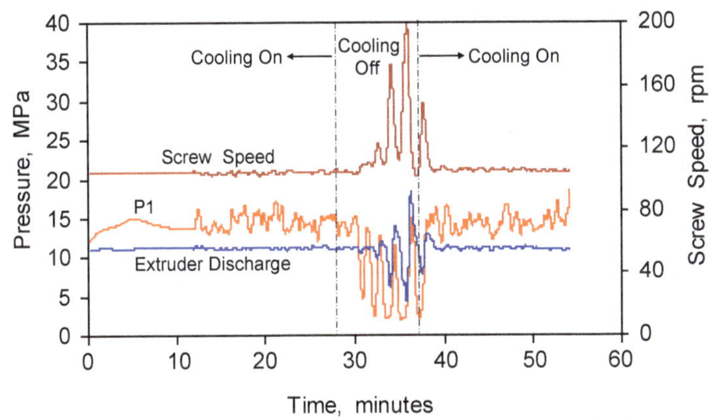

Figure 12.16 Screw speed, pressure at the entry to the first-stage meter (P1), and discharge pressure for the screw cooling experiment

Based on the above data, the cause of the extrusion instability was identified as high temperatures on the screw surfaces of the feed section. These high surface temperatures caused the coefficients of dynamic friction to increase, increasing the retarding forces on the solids at the screw surface. Since solids conveying depends on a combination of forwarding forces at the barrel wall and pushing flight and retarding forces at the screw root and trailing flight, an increase in the retarding forces will cause a reduction in the solids-conveying rate. The instability appeared to be random due to the complicated interactions of cooling water flow rate and temperature and due to changes in bulk density of the feedstock.

Several technical solutions were considered to increase the cooling level to the feed section of the screw, including increased water flow and the use of chilled water. The best technical solution and quickest to implement was to increase the length of the cooling hole in the screw. The length of the cooling hole was increased from 3.8 diameters into the flighted section to 7 diameters up to the end of the feed section. After the screw modification, the extruder has not experienced instabilities of this type, and the rate has increased to 100% of its maximum potential rate.

Cooling on the screw and feed casing are often limited by the water pressure at the supply and discharge sides. That is, if the water pressure on the discharge header is nearly the same as that of the supply side, then the water flow rate will be very low due to the lack of a pressure driving force. Thus, if the driving pressure for water flow is not available then adequate cooling to the screw and casing may not exist. A simple way to test if the cooling water flow is acceptable is to disconnect the discharge water line from the header and either flow this water to a drain or the parking lot using a temporary hose. The discharge water flow should be high and the temperature should be warm to the touch. A permanent arrangement might consist of a water pump and a rotameter in-line upstream of the rotary union attached to the screw.

To aid in the understanding of this solids-conveying problem, the coefficient of dynamic friction was measured for the resin as a function of temperature and sliding velocity at a pressure of 0.7 MPa. The equipment used to make the measurement is described in Section 4.3.1 and is shown in Figure 4.11. Since the coefficient of dynamic friction is only defined for solid-state processes, the friction values are reported here as stress at the interface because the stress can be described from ambient temperatures up to processing temperatures. The shear stress at the interface for HIPS resin is shown in Figure 12.17 at a pressure of 0.7 MPa. As indicated by this figure, the shear stress was nearly constant from ambient temperature up to about 110 °C, increased to a maximum stress near 150 °C, and then decreased as the temperature was increased further. Optimal performance of the solids-conveying section for this resin would be such that the forwarding forces are maximized with metal surface temperatures near 150 °C where the stress is a maximum, and the retarding forces minimized with metal surface temperatures of 110 °C or lower. Thus, optimal solids conveying for HIPS resin would occur with a feed zone barrel inner surface temperature near 150 °C and a screw surface temperature in the feed section no higher than 110 °C. In practice, screw temperatures less than 90 or 100 °C are preferred such that melting of the resin does not happen if an emergency shutdown should occur. For the solid-state temperature region, the shear stress at the interface can be converted to the coefficient of dynamic friction by the following:

$$f = \tau / P \tag{12.4}$$

where f is the coefficient of dynamic friction, τ is the shear stress at the polymer-metal interface, and P is the pressure (0.7 MPa in this case).

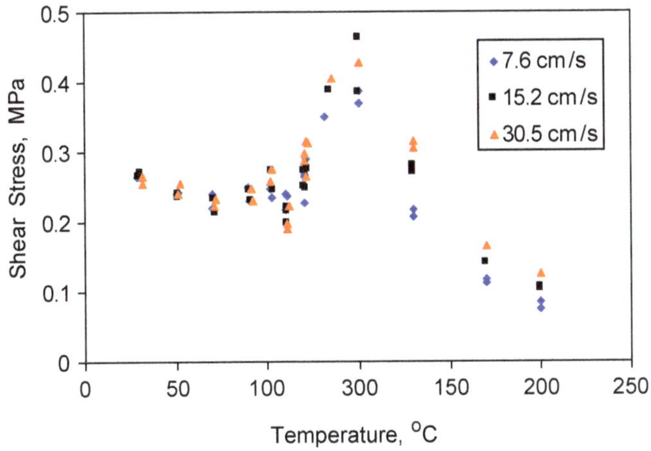

Figure 12.17 Shear stress between HIPS resin and a metal surface at a pressure of 0.7 MPa and as a function of temperature and sliding velocity

12.7.4 Flow Surging Due to High Temperatures in the Feed Casing

The extruder described in Figure 12.11 on a different occasion started to flow surge but with a slightly different frequency, as shown in Figure 12.18 and Figure 12.19. As indicated in these figures, there were short time periods when the discharge pressure and screw speed were stable and the motor current was high. During these periods, the extruder was operating well but at a reduced production rate. During periods of unstable operation, the motor current decreased by about 20%, the screw speed increased, and the discharge pressure became extremely oscillatory. Like the previous case, as the motor current decreased solids conveying decreased, causing the controller to increase the speed of the screw.

During the trial, the feed casing to the extruder had an outside surface temperature of about 80 °C. Although not measured, the inside cylinder wall of the feed casing for the first 1.5 diameters downstream of the feed opening was considerably hotter. These higher temperatures were caused by a combination of frictional heating of the solids on the wall and by conduction from the first heated zone of the barrel. It is estimated that temperatures as high as 170 °C occurred in the feed casing. As presented in Section 12.7.3, optimal solids conveying will occur when the stress at the polymer-metal interface at the barrel is a maximum, and for HIPS resin this surface temperature is near 150 °C. Surface temperatures higher than 150 °C in the feed section will reduce conveying and lead to starving of the screw channels and ultimately flow surging. Corrosion inside of the cooling channels of the feed casing prevented the flow of cooling water. Cleaning the cooling channels and adding a larger cooling water recirculation pump reduced the temperature of the feed casing and eliminated the flow surging problem.

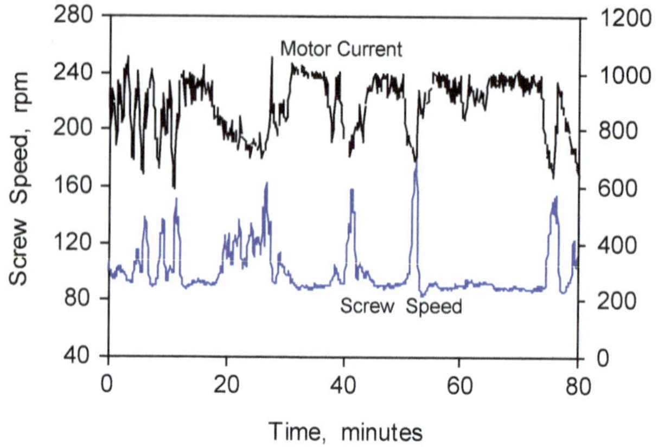

Figure 12.18 Screw speed and motor current for a large-diameter extruder with a feed casing that was too hot

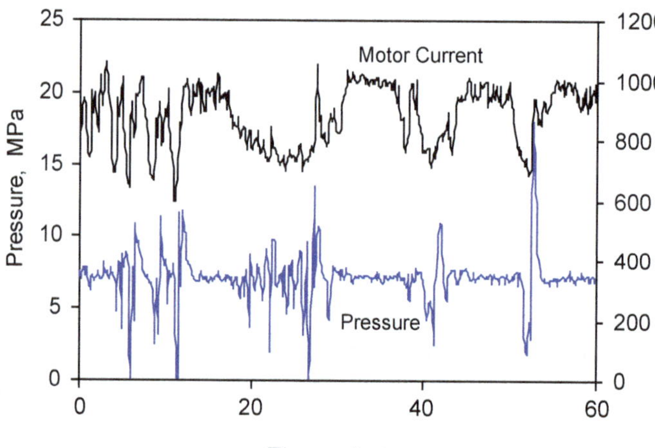

Figure 12.19 Discharge pressure and motor current for a large-diameter extruder with a feed casing that was too hot

This example and the example in Section 12.7.3 show the effect of improper surface temperatures in the feed section of a plasticating extruder. Acceptable solids conveying will cause the melting and metering sections to be full of resin and under pressure; the first-stage metering section must control the rate. If the solids-conveying section is rate controlling as in these two examples, then downstream sections of the screw channel will be starved, leading to low rates, material degradation, and flow surging.

12.7.5 Flow Surging Due to a Poorly Designed Barrier Entry for GPPS Resin

A precision profile extrusion line was producing parts that were machine assembled in a downstream process. In order for the assembly process to have a high yield and to be profitable, the dimensions of the profile had to be within a very narrow tolerance. If the profile was too large, the parts would not attach properly, and if the profile was too small the part would break during assembly. For this line, about 20% of the parts had to be scrapped because they did not have the proper dimensions. Moreover, if the part broke during the mechanical assembly process, the machine would often need to be momentarily shut down to remove the broken part. The part dimensions became more variable at high screw speeds, and there was no indication of resin degradation from the process. The high scrap rate, higher labor costs, and the higher level of quality control costs were reducing the profitability of the process. Plant personnel were unsure of the root cause for the problem, but it was believed to originate from the extruder. The plant was extruding a 1.5 dg/min (200 °C, 5 kg) GPPS resin. The extruder used for this process was a 63.5 mm diameter, single-stage machine with a 25 *L/D* barrel. The channel dimensions for the screw are provided in Table 12.2. The specific rotational flow rate for this screw was calculated at 1.41 kg/(h·rpm).

Table 12.2 Screw Channel Dimensions for a 63.5 mm Diameter Screw Running GPPS Resin for a Profile Extrusion Line

	Depth, mm	Length, diameters	Notes
Feed section	12.2	7	The barrier flight undercut was 0.25 mm
Barrier-flighted transition		12	
First meter section	5.1	2	
Maddock-style mixer		2	The compression ratio was 2.4 and the compression rate was 0.0039
Second meter section	5.1	2	

Lead length, flight width, and flight clearance were 63.5, 6, and 0.06 mm, respectively, for the feed and metering sections of the screw. The lead length was increased to 88.9 mm in the barrier section. The mixing flight of the Maddock-style mixer was undercut 0.64 mm from the main flight. The first 2.5 diameters of the screw were inside a water-cooled feed casing. The specific rotational rate of the metering section was calculated at 1.41 kg/(h·rpm).

A data acquisition system was attached to the control panel of the extruder in parallel with the existing controllers. Barrel temperatures, discharge pressure, motor current, and several downstream sensors were recorded at a frequency of once per second. The sensors downstream from the extruder were showing that the downstream part of the process was not contributing a significant level of variation to the profile dimensions. The rate was measured at 53 kg/h for a screw speed of

43 rpm, for a specific rate of 1.23 kg/(h·rpm). The discharge temperature was measured at 240 °C using a handheld pyrometer. The discharge pressure and the motor current for this process are shown in Figure 12.20. Variations in the barrel zone and die temperatures were very small and normal, and they likely did not contribute to the flow surging problem. The screw speed was very steady.

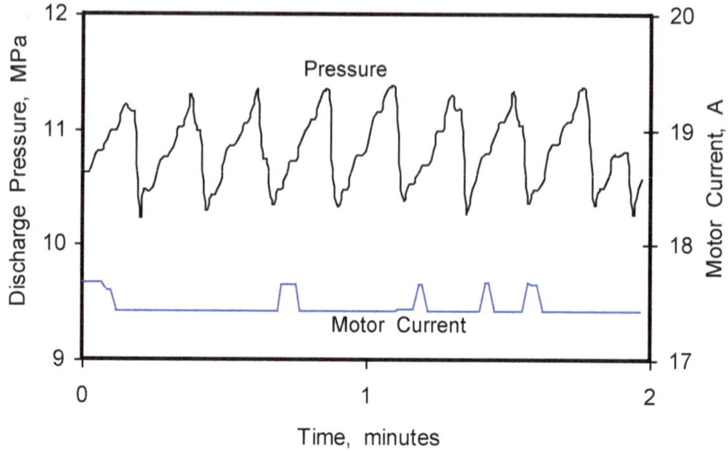

Figure 12.20 Discharge pressure and motor current for a 63.5 mm diameter extruder running GPPS resin at 53 kg/h and a screw speed of 43 rpm

As shown in Figure 12.20, the discharge pressure was periodic and had a range of about 10.2 to 11.4 MPa or about 6% about the average value, a pressure oscillation that is considered too large for this precision profile extrusion process. Thus, the discharge pressure oscillated over a range of about 1.2 MPa at a frequency of about four pressure surges per minute. The pressure would start out at the low value and would gradually increase in a linear manner until the maximum pressure was obtained, and then the pressure would abruptly decrease to the lower value. This type of pressure surge is typical of a process that is thermally driven or controlled by a blockage. In this case a blockage or restriction occurred at the entry to the barrier section. The motor current also tracked the pressure surge, as shown in Figure 12.20. In general, the motor current would increase as the machine either become full of resin or as the pressure upstream of the blockage became high. Not all pressure oscillations, however, showed a corresponding increase in motor current. It was not known why the motor current did not follow the exact trend, but it was likely due to the relatively small increase in the motor current during a cycle and electronic data filtering performed at the control panel.

The measured pressure surge and the time-averaged measured rate of 53 kg/h suggest that the rate cycled between about 44 kg/h at the low pressure point and 62 kg/h at the high pressure point via Equation 12.1 and Equation 12.2. Thus, the difference in rate between the low and high pressure points could be as much as

18 kg/h. Since the geometry of the downstream equipment and the die are much more complicated than the geometry assumed during the development of these equations, the rate change may be less than this. The exercise, however, indicates that the pressure and rate surge are the root cause for the dimensional change in the profile part. For a well-designed process, the pressure fluctuation should be on the order of the measurement error of the transducer. Transducers typically have measurement errors that are between 0.25% and 1.0% of the full scale. Thus, for a 35 MPa full-range transducer with a 0.5% full-scale error level, the measurement error and fluctuation are expected to be about 0.18 MPa.

The pressure surge did not seem to affect the appearance and temperature of the discharge as indicated by observation and by the temperature probe immersed in the flow stream just downstream from the extruder. The transient data from this temperature probe is shown in Figure 12.21. As indicated by this figure, the average discharge temperature as measured by the probe was 218 °C. Since this temperature measurement was highly influenced by the temperature of the metal surrounding the thermocouple [14], it was decided to measure the temperature of the discharge directly with a handheld measurement probe. This handheld temperature was measured at 240 °C, and it is considered to be the actual temperature of the discharge. This example clearly shows how a temperature sensor positioned in a transfer line can provide an erroneous measurement. For this case, the transfer line pipe where the sensor was positioned was maintained at a temperature of 215 °C. Even though this temperature was reported to measure the discharge temperature, it was actually measuring a temperature that was influenced mostly by the transfer line pipe and less by the discharge resin.

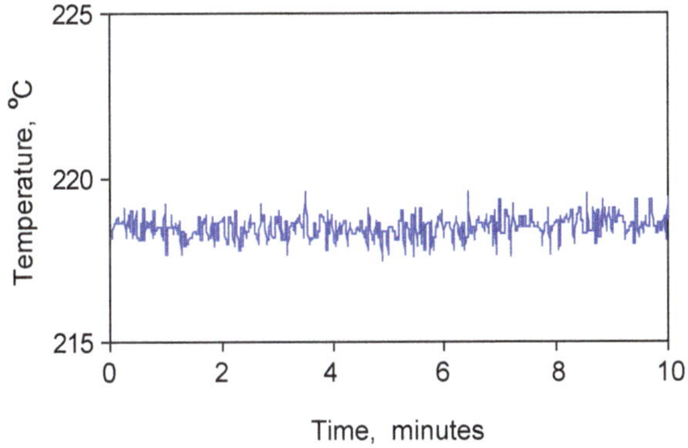

Figure 12.21 Discharge temperature as measured through a transfer line for a 63.5 mm diameter extruder running GPPS resin at 53 kg/h and a screw speed of 43 rpm. The actual discharge temperature was measured using a handheld probe at 240 °C

Rotational flow and pressure flow calculations for the screw and process data indicate that a positive pressure gradient of about 1.3 MPa/turn existed in the metering section. The 3.5 diameters of metering section length provide pressure generation of 4.6 MPa. For an average discharge pressure of 10.7 MPa and assuming that the Maddock-style mixer is operating without a pressure gradient, the pressure at the entry to the first metering section is estimated at 6 MPa. This meter section entry pressure is lower than expected, and it is highly influenced by the oscillating flow from the upstream barrier section. If this entry pressure were considerably higher, then the surging level would have been reduced considerably. A pressure at the entry to the first meter section of 15 to 20 MPa is normal for a properly operating extruder for this type of application.

The root cause for the flow surging was a poorly designed entry region of the barrier section and a low compression ratio. This entry section caused a restriction of the flow with the introduction of the barrier flight. This restriction could be mitigated by modifying the entry to the barrier section as presented in Section 11.10.1, and the process would likely operate stably. For the process here, a high-performance screw without a barrier-melting section was specified for this resin and application. A high-performance screw with the dimensions shown in Table 12.3 was fabricated and installed. This screw reduced the variation in the profile geometry considerably, and it increased the profitability of the plant. The flow surging was essentially eliminated, reducing the discharge pressure variation to that of the measurement error of the pressure sensor (0.18 MPa).

Table 12.3 Screw Channel Dimensions for a 63.5 mm Diameter High-Performance Screw Running GPPS Resin for a Profile Extrusion Line

	Depth, mm	Length, diameters	Notes
Feed section	12.2	6	The compression ratio was 3.0 and the compression rate was 0.0038
Melting section		12	
Average depth of the high-performance meter section	4.06	7	

Lead length, flight width, and flight clearance were 76.2, 6, and 0.06 mm, respectively, for all sections of the screw. The first 2.5 diameters of the screw were inside a water-cooled feed casing.

12.7.6 Solid Blockage at the Entry of a Spiral Mixer

An extrusion thermoforming operation was used to make large parts. In order for the process to be profitable, the process must be able to produce consistent, high-quality parts that have high strength with minimum resin consumption. Thus, the thermoforming process must be supplied with sheet stock that has a consistent thickness in both the cross direction and the machine direction. For this case, a 200 cm wide die was adjusted to provide a constant thickness extrudate in the cross direction, and the extruder must provide a nearly constant rate to maintain a constant thickness in the machine direction. The extrudate sheet was finished and cooled on a three calendar roll stack. The sheet was stored and then thermoformed into the desired end product. A 152.4 mm diameter, 36 *L/D* two-stage, vented extruder was used to plasticate an ABS resin. The extruder typically operated at a rate of 1180 kg/h and a screw speed of 75 rpm for a specific rate of 15.7 kg/(h·rpm). The screw channel dimensions are given in Table 12.4. The specific rotational flow rate for the first-stage metering section was calculated at 12.2 kg/(h·rpm). Thus, the extruder was operating at a rate that was about 1.3 times the specific rotational rate of the first-stage meter. The pump ratio was relatively high at 1.7 and the second-stage metering section was long at 7 diameters. The specific rotational flow rate for the second-stage metering section was calculated at 19.9 kg/(h·rpm). The extruder was equipped with 17 barrel zone temperature controllers, two pressure sensors, a motor current sensor, and numerous downstream equipment sensors. One pressure sensor was located in the transfer line to measure the discharge pressure, and the other was located in the barrel wall about 10 diameters from the start of the screw. This transducer in the barrel wall was monitored manually. The ABS resin had an MFR of 2 dg/min (230 °C, 3.8 kg). The recycle resin sheet from the extrusion-forming process was ground and added back to the extruder at rates between 0 and 50% to the fresh resin. Since the sheet was solid and the ground particles were about the same size as the pellets, the bulk density of the recycle material was only slightly less than that for the pellets. The resin was dried at 85 °C for 2 h prior to extrusion.

Table 12.4 Screw Channel Dimensions for a 152.4 mm Diameter Two-Stage Vented Screw Running ABS Resin

	Depth, mm	Length, diameters	Notes
Feed section	19.1	7	The compression ratio was 2.5 and the compression rate was 0.0021
First-stage transition		11	
First-stage meter	7.62	3	
Vent section	25.2	5.5	The pump ratio was 1.7
Second-stage transition		3	
Second-stage meter	12.7	7	
Lead length, flight width, and flight clearance were 152.4, 15, and 0.15 mm, respectively, in all sections of the screw. A spiral dam with a clearance of 2.03 mm was in the last 3.5 diameters of the first-stage transition section. The first 2.5 diameters of the screw were inside a water-cooled feed casing. The specific rotational flow rate for the first-stage metering section was calculated at 12.2 kg/(h·rpm).			

The extruder would operate for several hours to days at steady state, and then for no apparent reason it would flow surge for several hours. After a period of time, the extruder would return to a steady-state operation and would remain there until the cycle repeated. Problem diagnosis was impossible without transient process data. Moreover, molten resin would frequently flow out the vent, especially during times of unsteady-state operation.

A portable electronic data acquisition system was transported to the plant site and connected to the extruder panel. All available instrument outputs were connected in parallel with the acquisition system. Data collected included barrel zone temperatures, pressure in the transfer line, and motor current. A sample of the zone temperatures (labeled 1 to 17), discharge pressure, and motor current for steady-state operation are shown in Figure 12.22 and Figure 12.23. These data were collected for a screw speed of 75 rpm, a discharge pressure and temperature of 7.5 MPa and 250 °C, and a rate of 1180 kg/h. As previously mentioned, the specific rate of 15.7 kg/(h·rpm) was about 1.3 times higher than the specific rotational flow rate for the first-stage metering section. This result is typical for the first stage since a negative axial pressure gradient existed in the channel. That is, the pressure at the entry to the metering section was estimated at 22 MPa while the discharge of the section into the vent area had a pressure of zero. Numerical simulations indicated that the point of fill in the second stage was about 8 diameters downstream from the vent opening; only 6 diameters of the 7 diameter long second-stage metering section was needed to generate the 7.5 MPa discharge pressure. These data are consistent with the observation that vent flow did not occur during periods of steady operation. Large amounts of steady-state data needed to be collected since it could not be predicted when the extruder would go into an unstable operation.

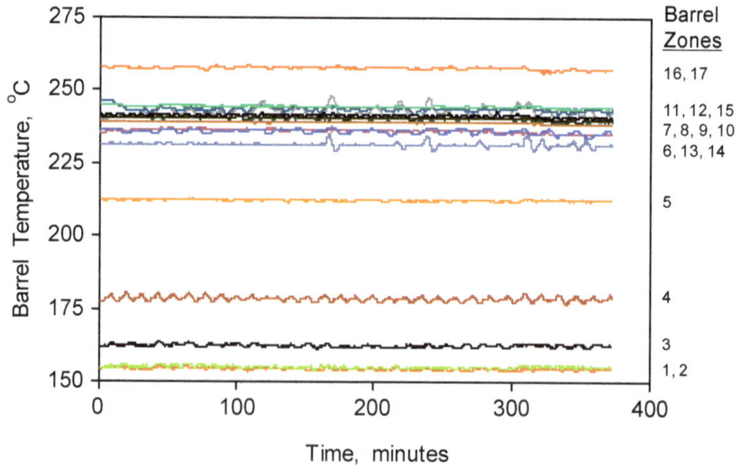

Figure 12.22 Barrel zone temperatures for stable ABS resin extrusion

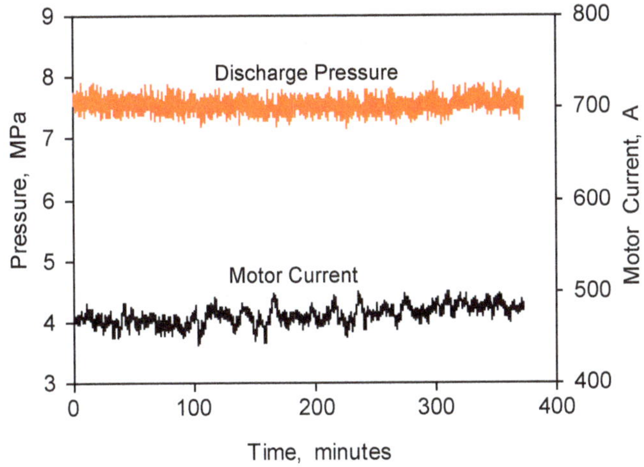

Figure 12.23 Discharge pressure and motor current for stable ABS resin extrusion

In the course of the trial, the extruder went into several periods of unsteady-state operation. The motor current, extruder discharge pressure, and some of the barrel zone temperatures for unstable extrusion are shown in Figure 12.24 and Figure 12.25. Some of the barrel zone temperatures were removed from Figure 12.24 because they did not fluctuate with time. As shown by these figures, the extruder was operating stably up to about 35 minutes on the graph. At about 40 minutes into the run, the discharge pressure decreased, the motor current increased, and some of the barrel zone temperatures became erratic. During this period, the machine was too unstable to produce sheet for the thermoforming process. The vent port plugged with resin about once every 15 minutes.

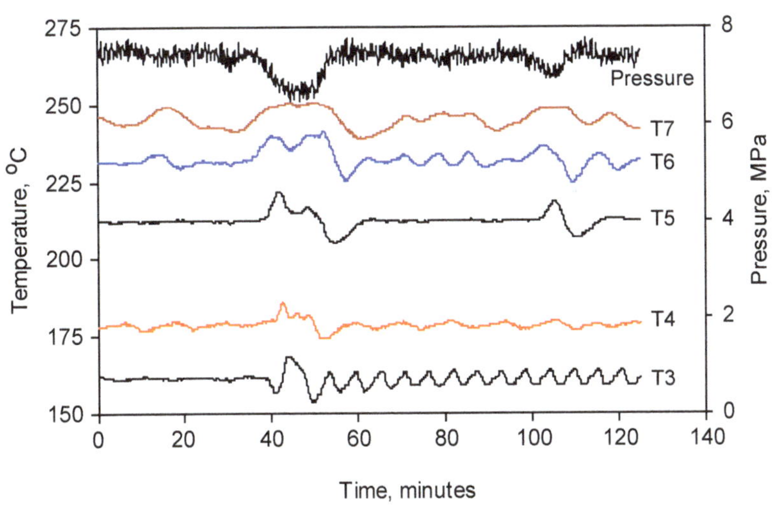

Figure 12.24 Select barrel zone temperatures and discharge pressures for unstable ABS resin extrusion

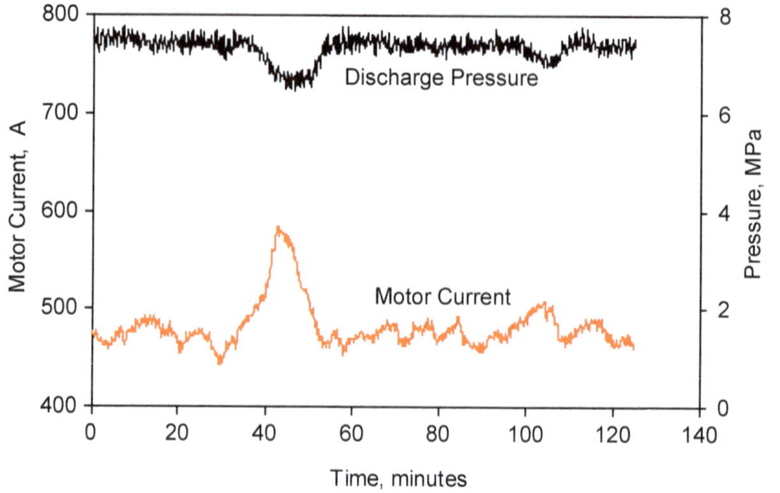

Figure 12.25 Discharge pressure and motor current for unstable ABS resin extrusion

For normal extrusion processes, the discharge pressure is directly proportional to the energy required to extrude the material. In other words, as the extrusion flow rate increases, the pressure and the motor current also increase. For this case, however, the opposite trend was observed. As the motor current increased, the transfer line pressure and flow rate decreased. It was hypothesized from the transient data that a breakup of the solid bed had occurred. Fragments of this bed flowed downstream and wedged into a section of the screw containing the spiral dam. This wedged material required additional energy to melt it. The additional energy was directly proportional to the increase in the motor current, and the blockage was evident by the decrease in the discharge pressure and flow rate. During this surge, the pressure transducer in the barrel just upstream of the spiral dam increased from 2 MPa up to 17 MPa. As the pressure increased at the entry to the spiral dam, the higher pressure caused the melting rate to increase [2, 15, 16].

The data were also complicated by the temperature variations of the first seven zones of the extruder because the amount of mechanically dissipated energy changed dramatically with every pressure surge. The blockage occurred at about barrel zone 7. As indicated in Figure 12.24, the temperatures for the zones upstream from zone 7, especially zone 5, fluctuated with the discharge pressure. For example, as the discharge pressure decreased, the blockage was occurring and the pressure upstream of the spiral dam increased. The increase in this upstream pressure caused additional energy to be dissipated in the early sections of the screw. Some of the dissipated energy escaped through the barrel, increasing the temperature of the zone, and some of the heat was used to melt the resin. The cycling of the blockage and melting caused the energy dissipation in the early sections of the screw to be variable.

As a result of this troubleshooting, the design of the spiral dam was modified to eliminate the random flow surging phenomena. For this modification, the length of the spiral dam section was increased from 3.5 diameters to 7 diameters. Thus, the entry of the section was moved upstream from 14.5 to 11 diameters from the start. The section length was increased in order to maintain the integrity of the solid bed much like a standard barrier section and thus minimize or eliminate solid bed breakup. The undercut of the barrier flight for the first diameter of the section was blended from the channel depth to the undercut depth of 2.03 mm. That is, the flight land was flush with the root of the screw at the entrance and smoothly transitioned to the final undercut of 2.03 mm after 1 diameter. This entry modification was very similar to those used for the case studies presented in Sections 11.10.1 and 11.10.3. Starting the dam upstream and the entry modification should mitigate the restriction that caused the original blockage. In order to mitigate the vent flow problem, the depth of the vent section was increased from 25.2 to 27.9 mm. The final screw modification that eliminated the problem and produced high-quality sheet is summarized in Table 12.5. After the modification to the screw was made, the random flow surging problem was eliminated, and the line has since produced high-quality sheet. This problem was so complicated that it could not have been diagnosed and solved without transient process data from all the process sensors collected by the acquisition system.

Table 12.5 Improved Screw Channel Dimensions for a 152.4 mm Diameter Two-Stage Vented Screw Running ABS Resin

	Depth, mm	Length, diameters	Notes
Feed section	19.1	7	The compression ratio was 2.5 and the compression rate was 0.0021
First-stage transition		11	
First-stage meter	7.62	3	
Vent section	27.9	5.5	
Second-stage transition		3	The pump ratio was 1.7
Second-stage meter	12.7	7	

Lead length, flight width, and flight clearance were 152.4, 15, and 0.15 mm, respectively, in all sections of the screw. A spiral barrier flight with a clearance of 2.03 mm was in the last 7 diameters of the first-stage transition section. The first 2.5 diameters of the screw were inside a water-cooled feed casing.

The optimal compression ratio for ABS screws is typically around 2.5. This compression ratio is lower than that used to extrude PS, LDPE, and other resins. If the compression ratio becomes greater than this, the solids-conveying rate can become too high, causing downstream problems such as incomplete melting, poor stability, and flow of resin out of the vent. The high solids-conveying rates are due to the stress at the interface for the resin, as shown in Figure 12.26. As shown in this figure, the stress at the optimal conditions of the screw (low velocity and temperatures between 70 and 100 °C) is at a minimum and causes the retarding forces at the screw to be small. The stress is at a maximum at the optimal conditions at the barrel wall at temperatures between 150 and 160 °C. The difference in stress between the optimal screw and barrel conditions is about 0.30 MPa, and it is considerably higher than that obtainable with other resins. As a comparison, the difference in optimal stress levels between the barrel and screw for HIPS resin is only about 0.20 MPa, as shown in Figure 12.17. This higher stress difference is the reason that ABS resins convey solids better than HIPS resins and other styrenic-based materials.

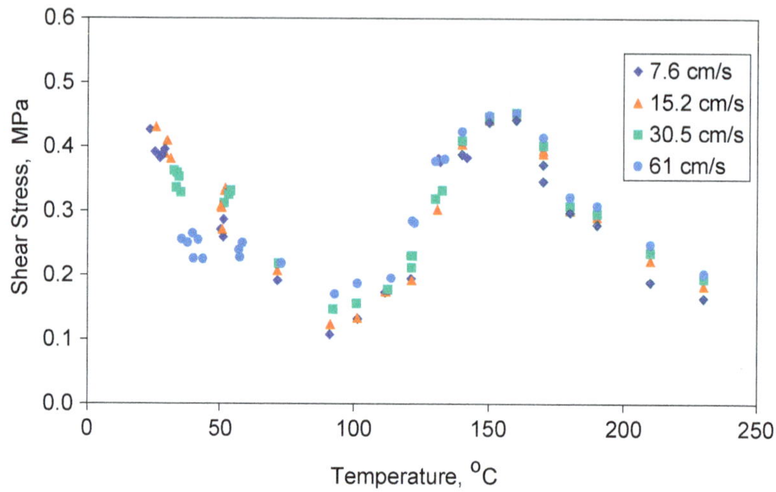

Figure 12.26 Shear stress at the interface for ABS resin at 0.7 MPa as a function of velocity and temperature

The difference in solids-conveying rates is quite significant when HIPS resins are run on lower compression ratio screws designed and optimized for ABS resins. For example, HIPS resins typically extrude at specific rates less than those for ABS resins on the same extruder and screw [17, 18]. Since the melt densities for these resins are essentially the same at the processing temperatures, the higher specific rate for ABS resins is attributed to a higher solids-conveying rate that produces a higher pressure at the entry to the metering section.

12.7.7 Flow Surging Caused by a Worn Feed Casing and a New Barrel

With continued use of a single-screw, plasticating extruder, the feed casing, barrel, and screw experience wear, increasing the clearance between the flights and the cylinder wall. These components all wear at different rates, with the screw flights wearing the fastest, the barrel wearing at intermediate rates, and the feed casing wearing the slowest. When the clearance becomes too large between the screw flights and barrel, the screw is generally replaced first. The effect of a large flight clearance on plasticating extrusion performance is well documented [1, 19, 20]. Effects include reduced rates due to material flow over the flight tips, high discharge temperatures due to reduced heat transfer rates, reduced melting rates, and the possibility of flow surging. With a new screw, the clearances are reduced to nearly that of the original design. Added clearance will, however, occur due to wear on the barrel surface. The barrel is eventually replaced after extended wear. Typical clearance between the flight land and the barrel wall is about 1/1000 the diameter of the barrel; for example, for a 200 mm diameter extruder the clearance is 0.20 mm. When the clearance increases to three to four times this amount due to wear, then either the screw or barrel is replaced. Wear also occurs in the feed casing but generally at a reduced rate due to the relatively low pressures in the section. Feed casings can be fabricated with hard, bimetallic liners, but they are generally made from cast steel. The normal clearance between the flight lands and the feed casing vary between extruder manufacturers, but they are in general about 0.003 times the diameter; the clearance is obtained by undercutting the flight lands. With extended use, however, the feed casing does show significant and detrimental levels of wear. The feed casing, however, is seldom replaced. Moreover, very little information on feed casing wear is available.

After several replacements of screws and barrels, it is possible to create a condition where a worn feed casing has a diameter that is significantly larger than originally specified and also larger than the barrel. For this condition, the clearance between the screw and barrel can be normal, but the clearance between the screw and feed casing can be up to ten to fifteen times larger than originally specified. Moreover, the transition between the feed casing and the barrel flange can create a ledge, as indicated in Figure 12.27. This ledge has the potential to create a restriction for solids conveying, disrupting flow and reducing solids-conveying rates. Moreover, the increased clearance in the casing may not convey solids to the same extent as a screw with a normal clearance.

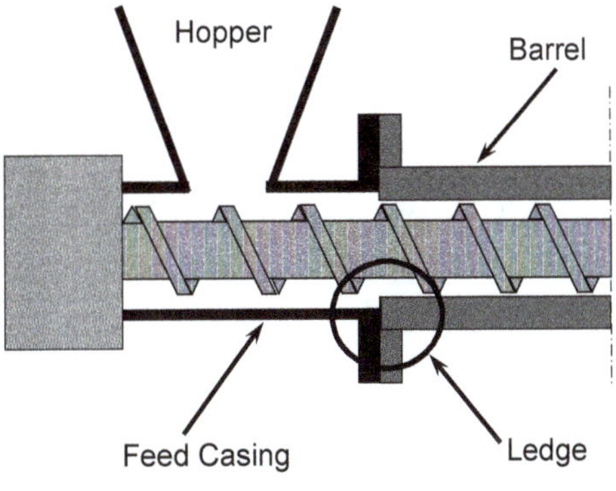

Figure 12.27 The solids-conveying section of a single-screw, plasticating extruder with a worn feed casing. The schematic is not to scale, and it shows a ledge formed between the oversized (worn) feed casing and the flange of a new barrel

A 114.3 mm diameter, smooth-bore, 24 *L/D* commercial extruder was experiencing severe and random motor current oscillations while operating at a reduced rate. The extruder started to flow surge immediately after the installation of a new barrel and screw. The screw geometry was very standard and acceptable for flood-fed, plasticating extrusion of GPPS resins. The GPPS resin had an MFR of 1.5 dg/min (200 °C, 5 kg). The bulk density for the pellets at ambient temperature and pressure was 0.61 g/cm^3. These commercial pellets were strand cut with an average diameter and length of 2.6 and 4 mm, respectively. Low-density recycle material in the form of ground sheet was added up to about 40 weight percent. With 40% recycle material, the feedstock to the extruder had an ambient density of about 0.40 g/cm^3.

The screw channel dimensions are provided in Table 12.6. The extruder was operating at the maximum screw speed, and it was the rate-limiting step for the line. Higher rates were expected, but unattainable. Calculations indicated that the extruder was operating at a rate of 360 kg/h at a screw speed of 150 rpm, a specific rate of 2.4 kg/(h·rpm) and a rate that was about 60% of the expected rate. Both the barrel and screw were new and thus had clearance specifications that were standard for this size machine. The specific rotational flow rate for the screw was calculated at 4.6 kg/(h·rpm). The extruder had adequate barrel temperature control, cooling water flow to the feed casing, and screw cooling via a rotary union. In order to diagnose the problem, a data acquisition system was temporarily connected to the extrusion panel. The motor current and discharge pressure, indicating unstable operation, are shown in Figure 12.28. As indicated in this figure, the motor current ranged from about 120 to 200 A, a level of variation that is abnormal for this size machine and operation. During this time period, however, the line was operating with a relatively stable discharge pressure, producing prime product, although at a rate that was less than desired. It was hypothesized that solids conveying was limiting the rate of the extruder.

Table 12.6 Screw Channel Dimensions for a 114.3 mm Diameter Two-Stage Vented Screw Running GPPS Resin

	Depth, mm	Length, diameters	Notes
Feed section	25.4	5.5	The compression ratio was 5.0 and the compression rate was 0.0071
First-stage transition		7.5	
First meter	5.1	3.3	
Blister	1.78 (clearance)	0.7	The blister clearance is the distance between the barrel wall and the top of the blister
Decompression	12.7 to 6.4	2	
Second-stage meter	6.4	4	
Distributive mixer		1	
Lead length, flight width, and flight clearance were 114.3, 11.1, and 0.12 mm, respectively, in all sections of the screw. A 25.4 mm diameter screw cooling hole was drilled in the shank end of the screw, and it extended 4.5 diameters into the feed section. The first 2.5 diameters of the screw were inside a water-cooled feed casing. The maximum screw speed was 150 rpm. The specific rotational flow rate for the screw was calculated at 4.6 kg/(h·rpm).			

The compression rate and compression ratio were considerably higher than what would be typically used for a 100% pellet feedstock. For pellet feedstocks, the target compression rate and compression ratio for PS resins are 0.0035 and 3.5, respectively. These target values are based on PS resin pellet feedstocks with a bulk density of 0.61 g/cm^3 at ambient conditions. Since this process is using a mixture containing 40% of a low-density recycle material, the bulk density of the feedstock, 0.40 g/cm^3, is considerably less than that for a pellet feedstock. The compression ratio and compression rate must be adjusted to allow the feeding of the lower density feedstock using Equations 6.3 and 6.4. Based on the bulk density data here, the compression ratio and compression rate needed for the process were 5.3 and 0.0053, respectively. As shown by the screw dimensions in Table 12.6, the compression ratio is in line with the guideline, and the compression rate is slightly high. Although the compression rate is higher than desired, it is likely in an acceptable range for the process.

The decompression section in the screw downstream from the blister was where a liquid plasticizing additive was injected into the extruder. The large channel depth in this section was used to accept the additional volume and to control pressure. The sections downstream from the first meter were not the focus of this case study and will not be discussed further.

Many standard diagnostic and troubleshooting techniques were tried, including temperature optimization for the feed section of the barrel and changes to the cooling for the feed casing and the screw, all without success. Feedstocks containing both 100% pellets and mixtures with up to 40% of a low-density recycle material were studied, all resulting in essentially the same reduced rate and with severe motor current oscillations. Due to the nature of the downstream processing equipment, flow surging at the die did not occur. A different downstream equipment arrangement, however, would likely have experienced unacceptable levels of flow surging at the die.

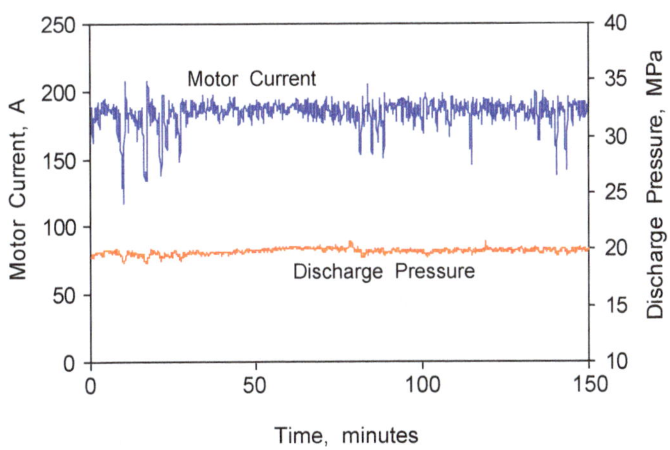

Figure 12.28 Motor current and discharge pressure for a 114.5 mm diameter GPPS resin extrusion process with unstable solids conveying

Upon further investigation, it was determined that the feed casing was worn and oversized in diameter by 5 mm, resulting in a 2.5 mm wide annular ledge between the feed casing and the new barrel flange, as indicated in Figure 12.27. This worn casing also caused the clearance between the flight lands and the casing to be 2.8 mm, a clearance that was about eight times higher than normal. It was hypothesized that this ledge was causing a restriction in the flow of solids from the feed casing to the barrel. Moreover, the increased clearance in the feed casing may have caused the solids-conveying ability in this section to decrease due to the flow of solids over the flight tips. For this case, the clearance was large enough to allow some of the pellets and recycle material to pass over the top of the flights, reducing the ability to convey resin out of the feed casing. As previously stated, the average pellet diameter was 2.5 mm, a diameter that was smaller than the flight clearance in the feed casing. Prior to the installation of the new screw and barrel, the old worn barrel and feed casing were such that flow surging and reduced rates did not occur, because the ledge did not exist. Due to commercial priorities, further testing and analysis on this extruder were not possible.

Due to the high cost of replacing the feed casing and the lack of direct evidence that the ledge and large flight clearance were the root cause of the reduced rate, an alternative experiment needed to be developed that simulated the process. Without these data plant personnel were unwilling to replace expensive components and incur downtime based on indirect evidence.

The effect of a worn feed casing was simulated on a 63.5 mm diameter solids-conveying device [21–24]. The device was constructed from a 63.5 mm diameter extruder, as described in Section 5.3.2. The standard barrel was removed from the extruder, and it was replaced with a short barrel, providing for a total L/D of 4.5. A schematic of the device is shown in Figure 12.29. The feed hopper casing had an effective barrel length of 1.9 diameters and an inside diameter of 63.5 mm, and it was water cooled. The removable barrel was 2.6 diameters in length, smooth bore, and it had an inside diameter of 63.5 mm. The feed opening was 1.5 diameters in length; the enclosed solids-conveying length by the barrel and casing was 3 diameters. The barrel temperature was controlled using a single-zone, 2000 W heater. The actual temperature of the barrel was measured using four strategically placed thermocouples at different axial positions.

Since fabricating several feed casings with different diameters is expensive, a less costly modification was developed. For this modification, a ledge simulation plate was positioned between the feed casing and the barrel, as shown in Figure 12.29. Two plates were fabricated from 12.7 mm thick steel plate, resulting in annular ledges of either 1.0 or 2.0 mm. To simulate the clearance that would occur for a worn casing, the flights of the screw for the first 1.9 diameters were undercut, up to the end of the ledge plate. The screw clearances and the ledge size are shown in Table 12.7.

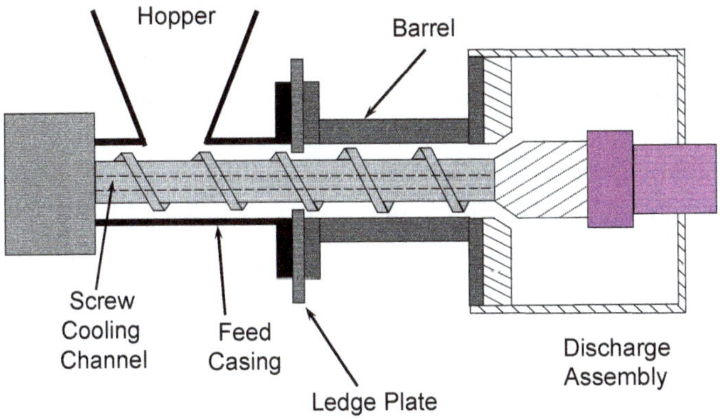

Figure 12.29 Schematic of the laboratory solids-conveying device with the ledge plate installed

Table 12.7 Ledge Sizes and Flight Clearances in the Feed Casing for the 63.5 mm Diameter Laboratory Solids-Conveying Device

Ledge Plate	Ledge Size, mm	Screw Clearance in the Feed Casing, mm
None	0.0	0.08
Plate 1	1.0	1.1
Plate 2	2.0	2.1

Three screws were used for this study. All screws were identical except for the flight clearance in the feed casing. As previously stated and as indicated in Table 12.7, the flight clearance was increased by undercutting the flight lands. All screws were square-pitched and had flight widths perpendicular to the flights of 6.4 mm. The screws had constant screw root diameters of 38.0 mm. This root diameter provided a constant channel depth of 12.7 mm for the screw that was not undercut. The shanks were bored for internal screw temperature control. The temperature of the screw was controlled by flowing pressurized heated water into the shank end of the screw using a rotary union assembly.

The discharge assembly could be used to apply pressure at the discharge. But since the focus of this study was the ledge and feed casing, the annular gap for material flow between the discharge end of the screw and the discharge assembly was fixed at the fully open position, providing zero pressure at the discharge. The feed casing temperature was maintained at about 30 °C using cooling water. The screw temperature was controlled at 50 °C, and the barrel temperature was controlled at 140 °C. All data were measured using flood feeding and at a screw speed of 50 rpm.

The first series of experiments were performed using 100% GPPS pellets. The solids-conveying rate using no ledge plate (that is, a standard feed casing) was

113 kg/h, resulting in a specific solids-conveying rate of 2.27 kg/(h·rpm). Thus for a new and properly designed extruder, the feed casing and first zone of the barrel are expected to convey a maximum of about 2.27 kg/(h·rpm). This size extruder, however, would likely melt and meter GPPS at a rate of about 0.9 kg/(h·rpm). Obviously, the meter section of the screw would be rate controlling as desired, and the solids-conveying section could maintain the rate necessary to keep the extruder filled. For this case, the pressure developed at the end of the solids-conveying section will balance the flow rate such that the solids-conveying rate is identical to the metering rate. When the ledge plates were used, the rate decreased dramatically, as shown in Table 12.8. For example, with the addition of the 1.0 and 2.0 mm ledges the rate decreased to 84 and 81 kg/h, respectively, for a maximum rate reduction of 28%.

Table 12.8 Solids-Conveying Rates for Commercial GPPS Pellets at 50 rpm

Ledge Size	Screw Clearance in the Feed Casing, mm	Rate, kg/h	Specific Rate, kg/(h·rpm)
None	0.08	113	2.27
1 mm (Plate 1)	1.1	84	1.68
2 mm (Plate 2)	2.1	81	1.62

The experiments were repeated using a mixture of 60% commercial pellets and 40% of the low-density recycle material. This mixture would approximate the mixture used in the commercial extrusion case. As indicated by the data with no ledge in Table 12.9, the solids-conveying rate was about 29 kg/h, resulting in a specific rate of 0.58 kg/(h·rpm). This rate was expected based on the reduction of the density alone. For this case, both the bulk density and solids-conveying rate decreased by 74%. When the 2.0 mm ledge plate was used, the rate decreased to 11 kg/h, but after a period of about four minutes the material would not convey; the rate had slowly reduced to zero. As solids conveying stopped, the energy added by the rotating screw was converted to heat, and this heat caused the material to partially melt. The data indicated that a combination of the 2.0 mm ledge and 2.1 mm flight clearance was incapable of conveying this material. As previously discussed, the 2.0 mm ledge caused a restriction to flow while the larger flight clearance may reduce the ability to convey material out of the feed casing.

Table 12.9 Solids-Conveying Rates for a Mixture of 60% GPPS Pellets and 40% of a Low-Density GPPS Recycle at 50 rpm

Ledge Size	Screw Clearance in the Feed Casing, mm	Rate, kg/h	Specific Rate, kg/(h·rpm)
None	0.08	29	0.58
1 mm (Plate 1)	1.1	8	0.16
2 mm (Plate 2)	2.1	0	0

The above experiments were performed using commercial GPPS pellets. These pellets had an average diameter of 2.6 mm and thus were not able to pass between the flight clearance of the solid-conveying device, even with the 2.0 mm ledge; the clearance in the feed casing was 2.1 mm. For the commercial operation, however, the clearance was 2.8 mm, enough clearance to allow some pellets to pass over the flight clearance. In order to determine the effect of clearance with pellet size, GPPS pellets were extruded and repelletized, producing strand-cut pellets with an average diameter of 1.6 mm and a length of 5 mm. These pellets were small enough to pass through the flight clearance using the 2.0 mm ledge, but too large to pass using the 1.0 mm ledge. The bulk density of the small-diameter pellets was 0.58 g/cm^3. This bulk density was only slightly less than the density of 0.62 g/cm^3 for the commercial pellets.

A similar trend in the solids-conveying rate was observed for the small-diameter pellets, as indicated by the data in Table 12.10. For no ledge, the rate was 127 kg/h, a rate that was slightly higher than that for the commercial pellets, and for the 2.0 mm ledge, the rate was very similar to that of the commercial pellets. As previously stated, the small-diameter pellets were small enough to pass over the flight clearance for the 2.0 mm ledge. But since the rate for the 2.0 mm ledge was essentially the same for both sizes of pellets, the restriction of the ledge is considered a more important factor than the loss of solids conveying due to pellets passing through the flight clearance.

Table 12.10 Solids-Conveying Rates for the Small-Diameter GPPS Pellets at 50 rpm

Ledge Size	Screw Clearance in the Feed Casing, mm	Rate, kg/h	Specific Rate, kg/(h·rpm)
None	0.08	127	2.54
1 mm (Plate 1)	1.1	91	1.82
2 mm (Plate 2)	2.1	83	1.66

The experiments were repeated with a mixture of 60% small-diameter pellets and 40% low-density GPPS recycle material. The bulk density for this feedstock was measured at 0.10 g/cm^3, a bulk density that was about 40% less than that for the commercial pellet-low-density recycle blend. This relatively large difference in density was attributed to the variability of the recycle material density. As indicated by the data in Table 12.11, the rate with no ledge was 20 kg/h, a rate that was about 30% less than that for the commercial pellet blend. Like before, the rate difference is primarily due to the differences between the feedstock bulk densities. When the ledge plates were positioned in the equipment, the solids-conveying rate was about 75% of the original rate. This rate decrease is very similar to the rate decrease that was experienced with the 114 mm diameter commercial extruder. Recall that the commercial extruder was operating at the maximum screw speed and at a rate that was only about 60% of the expected rate.

Table 12.11 Solids-Conveying Rates for a Mixture of 60% Small-Diameter GPPS Pellets and 40% of a Low-Density GPPS Recycle at 50 rpm

Ledge Size	Screw Clearance in the Feed Casing, mm	Rate, kg/h	Specific Rate, kg/(h·rpm)
None	0.08	20	0.40
1 mm (Plate 1)	1.1	14	0.28
2 mm (Plate 2)	2.1	15	0.30

The data clearly indicate that the ledge has the strongest influence on the solids-conveying rate. The effect on rate reduction of allowing pellets to pass over the flight clearance, as demonstrated using the small-diameter pellets, was not nearly as important. Since the ledge is the major factor, a simple technical solution was developed and evaluated using the solids-conveying device. For this solution, the 2.0 mm ledge plate was tapered such that an abrupt ledge was not present, as indicated in Figure 12.30(a). The solids-conveying rate data for the tapered plate along with the rates with no ledge plate are shown in Table 12.12. As indicated by the rates, all materials except the blend of small pellets with recycle conveyed at a rate that was nearly the same as the experiment with no plate. These data clearly indicate that the ledge is the cause of the restriction and that a tapered entry can mitigate its effects.

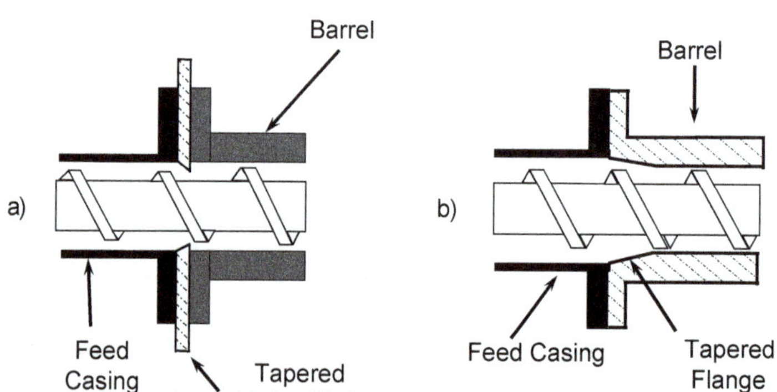

Figure 12.30 Modification to remove the ledge for both a) the solids-conveying device and b) a commercial extruder

Table 12.12 Solids-Conveying Rates for the 2.0 mm Tapered Ledge and with No Ledge at 50 rpm

Material	Rate for the Tapered Ledge Plate, kg/h	Rate with No Ledge Plate, kg/h
Commercial pellets	126	113
Small-diameter pellets	117	127
60% Commercial pellets and 40% recycle	28	29
60% Small pellets and 40% recycle	14	20

The experimental data clearly indicate the effect of a worn feed casing. With extensive wear, the oversized diameter of the feed casing creates a ledge between it and the flange of the new barrel. The ledge was shown to restrict the solids-conveying rate of material from the feed casing to the barrel, reducing the rate by 25 to 35% depending on the GPPS feedstock. In one case, the larger of the two restrictions (2.0 mm ledge) completely restricted the flow, reducing the solids-conveying rate to zero. These data are consistent with the data obtained from the commercial extrusion line. The 2.5 mm ledge on the commercial extruder reduced the rate of the machine and caused unstable oscillations in the motor current. Moreover, the ledge restriction will cause the pressure in the casing to be considerably higher, increasing the wear rate for the casing.

The data also indicated that material flow over the flight lands is not a significant factor in reducing the rate. The abrupt ledge created by the smaller flange diameter of the barrel, however, is the controlling factor. By tapering the barrel flange from the feed casing to the barrel diameter, as shown in Figure 12.30(b), the effect of the ledge can be almost completely eliminated. Instead of having an abrupt ledge that impedes flow, a force normal to the tapered surface likely pushed material into the channel and allowed flow to occur unimpeded.

Numerical analysis predicts that the commercial extruder should be able to convey the mixture containing 60% commercial pellets and 40% recycle at a maximum specific rate of 6 kg/(h·rpm) for zero pressure at the end of the feed casing. The measured specific rate, however, was 2.4 kg/(h·rpm), a specific rate that is about 60% less than the calculated maximum. Thus, the ledge for the commercial extruder restricted the flow more than that experienced with the lab device.

The commercial extruder was modified by tapering the entrance to the new barrel such that a ledge did not occur between the feed casing and the barrel. As predicted by the experimental research, the commercial line was able to operate stably and at a high specific rate of 4.2 kg/(h·rpm), and at a screw speed of 150 rpm the plant was operating at a rate of 630 kg/h. Plant personnel ordered a new feed casing and planned to install the casing at the next planned shutdown.

The inside diameter of the feed casing should be measured each time a worn barrel is removed and replaced with a new barrel. The problem described here could have been diagnosed more quickly had the measurements been made.

12.7.8 Flow Surging for a PC Resin Extrusion Process

A processor was extruding a PC resin (6 dg/min, 300 °C, 1.2 kg) into thin sheet using a 203.2 mm diameter, two-stage, vented extruder with a barrel *L*/*D* of 37. The processor was experiencing flow surging at high rates. This instability was limiting the rate of the process and the ability to produce new products. A gear pump was positioned between the extruder discharge and the die, and the process was very similar to that shown in Figure 12.11. The screw channel dimensions are proprietary, but the educational value of the problem is not limited by the nondisclosure of this information. The metering sections of the first and second stages of the screw had specific rotational flow rates of 28 and 54 kg/(h·rpm), respectively. Based on the screw channel dimensions, the compression ratio was 2.46, the compression rate was 0.0026, and the pump ratio was 1.90. The dimensions and design parameters were within the range for acceptable extrusion of PC resins. A cooling hole extended from the back of the screw up to 3.5 diameters into the feed section. This position was about midway into the solids-conveying section. Based on plant experience, the cooling hole was not used. Prior to the trial it was suspected that the lack of cooling on the screw was the root cause of the flow surging problem. A screw cooling system was connected to the extruder, but water was not flowing to the device.

A portable electronic data acquisition system was transported to the plant site and connected to the extruder panel. All available instrument outputs from the panel were connected in parallel with the acquisition system. Data collected included barrel zone temperatures, screw speed, motor current, pressure at the entry to the pump, transfer line temperature, and gear pump temperature. Process data were collected at a frequency of once every five seconds.

At the start of the trial, plant personnel wanted to operate the line with the water cooling full on. Next, the extruder and line were started up and operated at a rate of 1720 kg/h and a screw speed of 48 rpm. The line quickly operated in a stable manner, as shown by the time interval of 0 to 25 minutes in Figure 12.31 and Figure 12.32. All barrel zone temperatures were stable and tracking the set point temperatures. The line was stable for a longer period, but for clarity the data were omitted here. At about 25 minutes into Figure 12.31, the cooling water to the shank of the screw was turned off. Soon after the change the extruder went into a period of instability, as shown in Figure 12.31 and Figure 12.32 at about 30 minutes into the run. At 30 minutes, the inlet pressure to the gear pump started to decrease and then oscillate as shown in Figure 12.31. In order to maintain the inlet pressure near 5 MPa, the process control increased the screw speed. As shown in Figure 12.31, the screw speed increased up to about 61 rpm, reducing the specific rate of the screw from 36 to 28 kg/(h·rpm). During the start of the instability, the barrel zone temperatures for the first three zones decreased substantially (labeled T1, T2, and T3), as shown by the data in Figure 12.32. All other barrel zone temperatures were in control and tracking the set point temperatures. This decrease in barrel temperatures in the feed zone and early melting section indicates that the pressure in this section was reduced substantially. The reduced pressure and thus specific rate were indicated by the removal of a constant heat flux. When the flux was removed by the reduced pressure, the zone temperatures decreased. The heat flux due to this reduced pressure is described below.

$$F = fPV \tag{12.5}$$

where f is the coefficient of dynamic friction, P is the pressure, V is the relative sliding velocity between the PC solid material and the barrel wall, and F is the energy flux and has units of J/(m^2·s). Early in the unstable process, the coefficient of friction and the sliding velocity would be essentially constant. Thus, the only way that the energy flux could change is by a decrease in the channel pressure. A decrease in channel pressure is also indicative of a decrease in specific rate. This specific rate decrease was observed as previously stated. Moreover, the reduction in the energy flux out through the barrel wall was also obvious by the reduction in the motor current, as shown in Figure 12.31. The sheet line could be maintained operational at these conditions, but small surface flaws were observed in the sheet. These flaws were undesirable but acceptable for the product being produced.

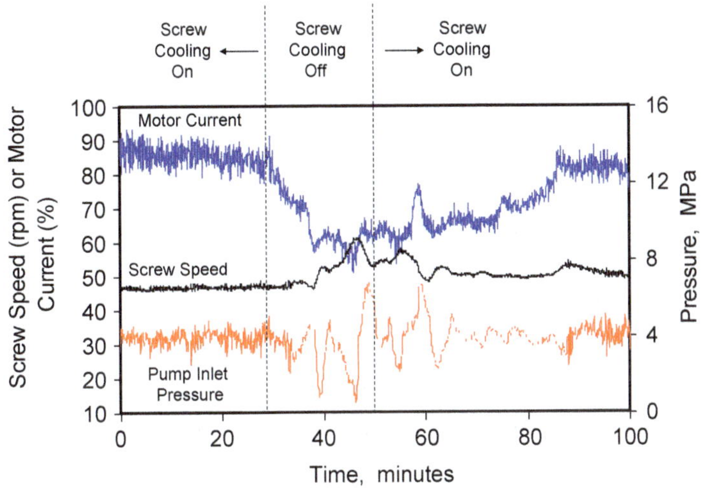

Figure 12.31 Operation of the extruder at a constant rate of 1720 kg/h with and without screw cooling for a PC resin

At about 45 minutes into the run of Figure 12.31 and Figure 12.32, the cooling water flow to the shank of the screw was turned on. Within 15 minutes the pump inlet pressure started to stabilize, and within 35 minutes the extruder was essentially operating in a stable mode, as shown in Figure 12.31. As the extruder returned to a stable operation, the screw speed returned to a normal speed of 50 rpm for a specific rate of 34 kg/(h·rpm), a specific rate that was slightly lower than that at the start of the run. The motor current increased during this period to about 80% load, a load level that was slightly lower than the 90% load for operation with higher levels of recycle resin. The barrel temperatures for zones 1, 2, and 3 increased with the improved stability, as shown in Figure 12.32.

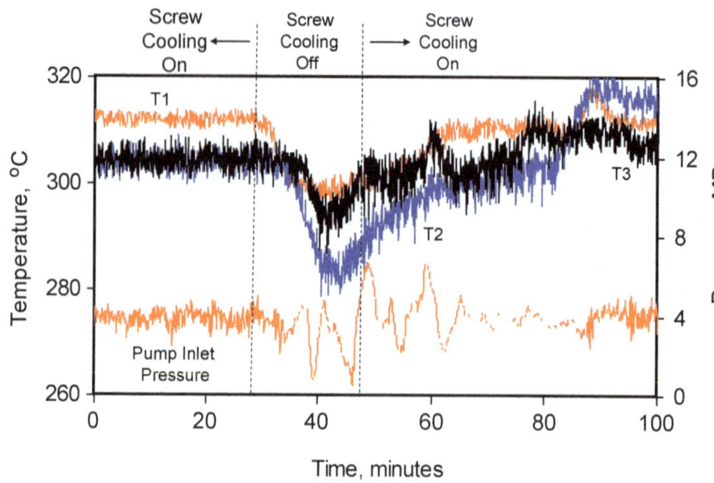

Figure 12.32 Select barrel temperatures for operation of the extruder at a constant rate of 1720 kg/h with and without screw cooling for a PC resin

To verify that the cooling water was the root cause for the flow surging, the cooling water flow was turned off and on several times. As expected, the stability of the extruder was acceptable when the cooling water was on and unacceptable when it was off. The line was operated for about a month with the cooling water on, and the line has operated stably. Thus, the level of cooling for this system was acceptable and major modifications to the extrusion process were not required.

The data clearly indicate that a high temperature on the feed section of the screw was the root cause for the flow surging observed on the line. This case was nearly identical to that for the HIPS resin extrusion problem of Section 12.7.3. For the HIPS resin case, additional cooling was needed and the length of the cooling hole was increased. For the PC resin case here, the cooling hole length was acceptable. For the HIPS resin case, the screw temperature when the process became unstable was about 150 °C. For PC resin, the screw temperature was unknown because the shank of the screw was not visible during operation.

The stress at the interface was measured as a function of temperature and sliding velocity for the resin using the equipment shown in Figure 4.11, and the data are shown in Figure 12.33. The stress curve had two maximums: the first peak was at the T_g of the resin at 150 °C, and the second peak occurred at a temperature of about 240 °C. In order to maximize solids conveying while maintaining a viable process, the optimal forwarding forces would occur at a barrel surface temperature near 240 °C, and the retarding forces at the screw surfaces would be minimized at temperatures less than about 120 °C. In order to maintain the high rate of this line and the inside barrel wall at a temperature near 240 °C, the first zone of the extruder needed to be maintained at a temperature of 310 °C.

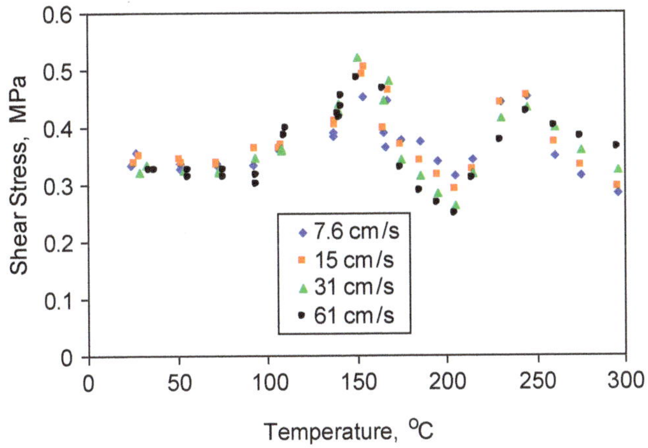

Figure 12.33 Shear stress between PC resin (6 dg/min, 300 °C, 1.2 kg) and a metal surface at a pressure of 0.7 MPa and as a function of temperature and sliding velocity [25]

Nomenclature

f	coefficient of dynamic friction
F	energy flux produced via frictional dissipation at a polymer-metal interface
n	power law index for a power law fluid
P	pressure
Q	mass flow rate
V	relative sliding velocity between a resin and metal surface
x	independent position variable for the cross-channel direction
z	independent position variable for the downstream channel direction
τ	shear stress at the polymer metal interface

References

1. Tadmor, Z. and Klein, I., "Engineering Principles of Plasticating Extrusion," Van Nostrand Reinhold Co., New York (1970)
2. Klein, I., "The Melting Factor in Extruder Performance," *SPE J.*, **28**, 47 (1972)
3. Fenner, R.T., Cox, A.P.D., and Isherwood, D.P., "Surging in Screw Extruders," *SPE ANTEC Tech. Papers*, **24**, 494 (1978)
4. Hyun, K.S. and Spalding, M.A., "Use of Process Data Obtained from a Data Acquisition System for Optimizing and Debugging Extrusion Processes," *Adv. Polym. Tech.*, **15**, 29 (1996)
5. Hyun, K.S., Spalding, M.A., and Powers, J., "Elimination of a Restriction at the Entrance of Barrier Flighted Extruder Screw Sections," *SPE ANTEC Tech. Papers*, **41**, 293 (1995)
6. Chung, C.I., "New Ideas About Solids Conveying in Screw Extruders," *SPE J.*, **26**, 32 (1970)
7. Spalding, M.A., Kirkpatrick, D.E., and Hyun, K.S., "Coefficients of Dynamic Friction for Low Density Polyethylene," *Polym. Eng. Sci.*, **33**, 427 (1993)
8. Thompson, J.R., "Effect of Head Pressure Variation on Stability: An Experimental Verification," *SPE ANTEC Tech. Papers*, **40**, 142 (1994)
9. Jenkins, S.R., Powers, J.R., Hyun, K.S., and Naumovitz, J.A., "High Rate Extrusion of PVDC Polymers," *J. Plast. Film Sheeting*, **6**, 90 (1990)
10. St. Louis, H.E. and Trumbull, W.A., "Adjustable Plug Seal Mechanism for Heat Transfer Zone Separation in Hollow Screws for Extrusion Machines," U.S. Patent 4,355,905 (1982)
11. Womer, T.M., Smith, W.S., and Wheeler, R.P., "Comparison of Two Different Cooling Methods for Extrusion Processes," *SPE ANTEC Tech. Papers*, **52**, 796 (2006)
12. Steward, E. and Kramer, W., "Air Versus Water Cooled Single-Screw Extruders," *SPE ANTEC Tech. Papers*, **49**, 87 (2003)
13. Pham, H.T. and Hyun, K.S., "Gear Pump Assisted Extrusion," *Polym. Eng. Sci.*, **32**, 488 (1992)
14. McCullough, T.W. and Spalding, M.A., "Predicting Actual Temperature Distributions in a Polymer Stream Using an Adjustable-Depth, Exposed-Tip Thermocouple Assembly," *SPE ANTEC Tech. Papers*, **42**, 412 (1996)
15. Mount, III, E.M. and Chung, C.I., "Melting Behavior of Solid Polymers on a Metal Surface at Processing Conditions," *Polym. Eng. Sci.*, **18**, 711 (1978)
16. Spalding, M.A., Hyun, K.S., and Cohen, B.R., "An Experimental Study of the Frictional and Viscous Forces Responsible for Solids Conveying and Melting in Single-Screw Plasticating Extruders," *SPE ANTEC Tech. Papers*, **43**, 202 (1997)
17. Powers, J.D., Spalding, M.A., and Hyun, K.S., "Extruding HIPS on ABS Screws," *SPE ANTEC Tech. Papers*, **40**, 151 (1994)

18. Altinkaynak, A., Gupta, M., Spalding, M.A., and Crabtree, S.L., "An Experimental Study on Shear Stress Characteristics of Polymers in Plasticating Single-Screw Extruders," *Polym. Eng. Sci.*, **49**, 471 (2009)

19. Barr, R.A. and Chung, C.I. "Effects of Radial Screw Clearance on Extruder Performance," *SPE J.*, **23**, 71 (1966)

20. Chung, C.I., "Extrusion of Polymers, Theory and Practice," Hanser Gardner Publications, Inc., Munich (2000)

21. Gogos, C.G., Zafar, M., Sebastian, D., Todd, D.B., Hyun, K.S., and Spalding, M.A., "A Device to Study Solids Compaction and Conveying in Single Screw Extrusion," *SPE ANTEC Tech. Papers*, **40**, 288 (1994)

22. Hyun, K.S., Spalding, M.A., and Hinton, C.E., "Theoretical and Experimental Analysis of Solids Conveying in Single-Screw Extruders," *SPE ANTEC Tech. Papers*, **42**, 199 (1996)

23. Spalding, M.A., Hyun, K.S., and Hoffmann, R., "An Experimental Investigation of Solids Conveying in Smooth and Grooved Barrel Single-Screw Plasticating Extruders," *SPE ANTEC Tech. Papers*, **44**, 136 (1998)

24. Spalding, M.A, Hughes, K.R., Sugden, J.L., Wagner, P.A., and Hyun, K.S., "The Effect of a Worn Feed Casing on Plasticating Extrusion," *SPE ANTEC Tech. Papers*, **47**, 180 (2001)

25. Altinkaynak, A., Gupta, M., Spalding, M.A., and Crabtree, S.L., "Shear Stress and Melting Flux for Polycarbonate Resins," *SPE ANTEC Tech. Papers*, **54**, 1405 (2008)

13 Rate-Limited Extrusion Processes

Minimizing the cost to produce a product can be achieved through many different means including down-gauging, down-engineering by resin replacement, improving the operation time or reliability of the equipment, minimizing scrap or off-grade product by eliminating a defect, process intensification, and increasing the rate. This section will present troubleshooting methods aimed at increasing the processing rate while maintaining quality. In general, increasing the rate of a line will decrease production costs by spreading the fixed costs over more product, by improving labor efficiencies, and by delaying the installation of new equipment to meet demand.

In order to increase the rate, it is imperative that the troubleshooter identify the rate-limiting step of the process. Obviously, if new technology is installed to improve a section of the process that is not rate limiting, then no rate gain and probably no economic gain will be obtained. Typically, the rate-limiting section of a process will involve the most expensive piece of equipment, or it will be a cooling or shaping process. For example, a tandem foam sheet extrusion line, if designed properly, will be rate limited by the secondary cooling extruder. The process is shown in Figure 13.1 and discussed in Section 15.4. The cooling extruder removes excess energy (cooling) from the molten resin so that a high-quality foam product can be produced at the die, and it is typically the most expensive single machine for a line. Other processes that are ideally rate limited by cooling include sheet lines, injection-molding machines, blow-molding machines, cast film lines, blown film lines, and thermoformers. Cooling resin at high rates will always be hampered by the relatively low thermal conductivity of resins compared to other materials such as metals. For example, the thermal conductivity of steel is about 400 times higher than that for common plastics. Shaping processes are often rate limited by melt fracture and draw resonance, which create unacceptable surface and geometric defects in the product.

Often an improvement in one section of a line or plant will cause the rate-limiting section to shift to either an upstream or downstream process. These types of projects often have considerable value associated with them because of the multiple bottlenecks that exist.

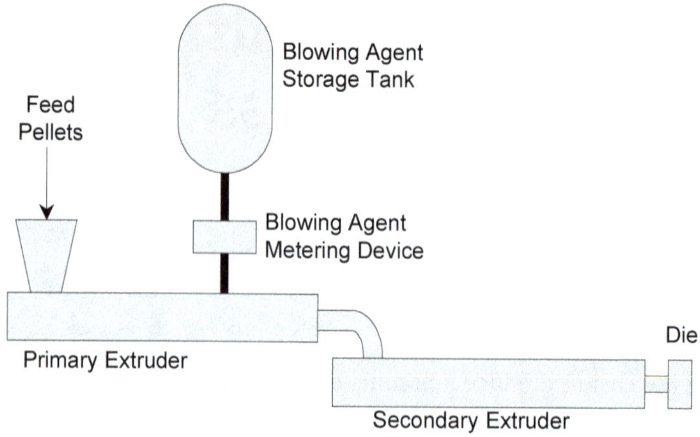

Figure 13.1 Schematic of a tandem extrusion process for foams. The secondary extruder or cooling extruder will be the rate-controlling step for a properly designed process

Extrusion processes are often rate limited by motor power or torque, discharge temperature, or the melting capacity of the screw. Other root causes associated with the design of the screw can limit rates as shown in previous sections. The problems, however, are typically associated with other defects such as flow surging or resin degradation. Chapters 11 and 12 discuss process defects associated with resin degradation and flow surging, respectively. Rate limitations due to inadequate motor power and torque are common problems for commercial plants. Two case studies are presented in the next sections that show rate limitations due to the lack of torque and motor power.

All extrusion processes will eventually become rate limited by the melting capacity of the screw if the screw is rotated at a high enough speed. That is, as the speed of the screw is increased, a maximum rate will be obtained where the extruder is discharging a fully molten stream. At a slightly higher screw speed, solid polymer fragments will start to discharge in the extrudate. This melting rate limitation was previously outlined by Chung [1] and Mount [2, 3] and was presented in Chapters 6 and 8. As expected, simple single-flighted conventional screws have the lowest rate for an acceptable discharge quality, that is, the most rate-limited design for melting. Barrier screw designs have improved performance while high-performance screw designs typically have the highest rates with an acceptable discharge quality; these have the least rate-limited designs. High-performance screws are discussed in Chapter 14. The melting performance of all designs can be improved by adding dispersive-type mixers near the end of the metering sections of the screws. As expected, the dispersive qualities of high-performance screw sections and mixers will trap, melt, and disperse solid polymer fragments.

Extrusion processes can be rate limited by the maximum allowable temperature of the discharge. For thermally sensitive resins, the extrudate will be required to be less than a maximum temperature typically specified by the resin manufacturer. Downstream equipment processes can also limit the maximum temperature of the extrudate. These downstream processes include blown film, cast film, profile extrusion, and sheet extrusion just to name a few. For most extrusion processes, the temperature of the extrudate increases as the rate is increased. In these cases, the maximum rate of the process is obtained by operating with the extrudate at the maximum operating temperature. At a slightly higher rate, the discharge temperature exceeds the critical temperature, and defects will start to occur in the product due to resin degradation or an unstable process.

Screw and barrel wear can reduce the performance of the extruder by causing the specific rate to decrease and the discharge temperature to increase. Screw wear is discussed in Section 13.3 and Section 13.4.1. Although extremely rare, wear in the feed casing can lead to a rate reduction and process instabilities. This case study is presented in Section 12.7.7.

13.1 Vent Flow for Multiple-Stage Extruders

Two-stage and multiple-stage vented extruders are commonly used to remove volatile components from molten resin streams prior to downstream processing. The vent ports can be open to the atmosphere, or they can be attached to elaborate vacuum systems. For very specialized systems, stripping agents such as water, nitrogen, and carbon dioxide can be added upstream of the vent, mixed into the resin stream, and then devolatilized in the vent area. This technique can be employed to remove difficult components or components at a higher rate from the resin.

The first-stage metering section (smooth bore extruder) must be the rate-controlling step for a multiple-stage extrusion process. If the second stage (or third stage) is rate controlling, then the second stage will not be able to convey away the resin as fast as the first stage is delivering. The excess resin will be forced out through the vent opening. This vent flow can cause safety issues and increase production costs. If a vacuum system is used, vent flow will force the line to shut down due to polymer solidifying in the piping system and condensers, plugging the system and making it inoperable. The pump ratio (see Section 10.8.2) must be specified such that the second stage is not the rate-limiting step. The pump ratio is the ratio of the rotational flow rate of the second stage to that of the first stage. For a screw with a constant lead length, the pump ratio is simply the ratio of the second-stage metering channel depth to that of the first-stage meter. Pump ratios typically range from 1.3 to 2.0, depending on the process and the rheology of the resin.

Most resins will expand slightly due to elastic forces and expanding volatile gases when they flow from a pressurized and confined region just upstream of the vent opening to the zero pressure and partially filled channels underneath the vent opening. This expansion causes the resin to move from the confines of the barrel cylinder into the vent port. The resin must be tucked back into the barrel cylinder, or it will flow out through the vent. The resin is tucked back into the barrel by installing a vent flow diverter in the vent port. Photographs of a diverter are shown in Figure 13.2, and a schematic is shown in Figure 13.3. The device allows the resin to expand at the leading edge of the diverter at the opening into a recessed area, allowing volatile gases to escape. Just downstream from the opening, the recessed area is slowly tapered flush to the inside barrel wall. This taper tucks the resin back into the barrel cylinder.

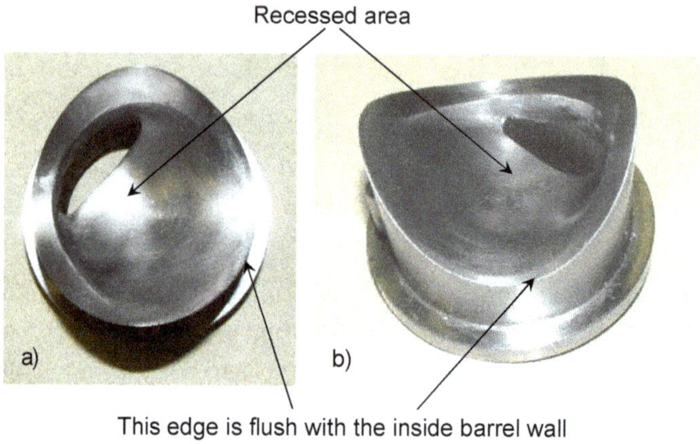

Figure 13.2 Photographs of a vent flow diverter for a 63.5 mm diameter extruder: a) top view showing the surface that is exposed to the inside of the extruder, and b) oblique view. The recessed areas that allow for the expansion of the resin are shown by the arrows. The outside edge is flush with the inside barrel wall when installed in the extruder. The opening is positioned in the upstream region in the vent while the tapered recess is in the downstream region

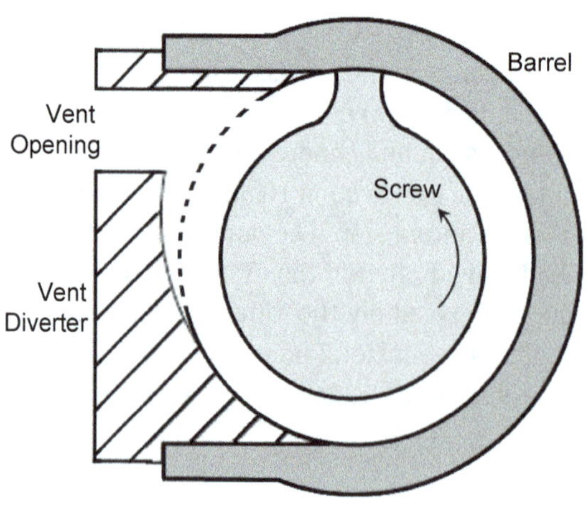

Figure 13.3 Schematic of a vent flow diverter in a two-stage vented extruder

Since both screw design and the vent diverter can cause resin to flow from the vent, the troubleshooter must evaluate both scenarios to determine the root cause. Typically, the calculation of the rotational flow and pressures in the channel using the methods in Section 1.4 should be performed first to determine if pressure exists in the channels at the vent opening. If the calculations indicate that pressure exists at the vent opening, then resin flow out the vent is likely to occur. If the analysis indicates that pressure occurs at the vent, then several solutions exist, such as changing the equipment downstream from the extruder so that the pressure at the discharge is decreased. The discharge pressure can often be reduced by changing to coarser screens in screen filtering systems. More expensive solutions include adding a gear pump [4] or fabricating a new screw with optimized channel depths in the first- and second-stage metering sections. An example of a vent flow problem is presented in Section 13.4.6. If the calculations indicate that the pressure in the channel underneath the vent opening is zero, then the investigation should focus on the operation of the vent diverter. The diverter could have been damaged or installed backwards. Most equipment suppliers, however, design the diverter such that it can be installed in the vent opening only one way.

13.2 Screw Wear

Screws and barrels wear over their service life with screws designed to wear at a higher rate than barrels. Screw wear was discussed previously in Section 10.4 and screw refurbishing in Section 10.11. The signs of wear are typically first observed by a slow reduction in the maximum rate at which a process can be operated. In many cases, the rate records for a line will show a gradual decrease in productivity over several years of operation. When the screw is removed for inspection, the flight clearance is determined to be several times more than the design clearance. For example, the design clearance for a 60 mm diameter screw is 0.06 mm in the metering section. After numerous years of service, the clearance can be as high as 0.4 mm. When the clearance becomes this large, excessive material will flow across the flight tip and cause the specific rate to decrease. Moreover, the heat transfer coefficient at the barrel wall will decrease significantly when excessive wear occurs. The heat transfer coefficient and its dependence on the clearance were discussed in Section 7.7.4. For the 60 mm diameter screw the heat transfer coefficient is expected to decrease by a factor of 2.5 using the Jepson model [5]. Those processes that require the transfer of energy through the barrel wall will be highly impacted by reduced heat transfer coefficients due to wear. For special cases where a plasticating extruder is starve-fed, the extruder can show process instabilities for screws that are highly worn.

For well-designed processes, operations downstream of the extruder (or plasticator) are the rate-limiting steps. For example, sheet extrusion lines are typically rate limited by the ability of the calender roll stacks to remove energy from the sheet. Thus, the screw speed and the rate of the extruder are increased until the cooling capacity of the roll stack is at a maximum. As the screw wears with time, the specific rate of the screw will decrease, and the screw speed will be increased to maintain rate and discharge pressure [6]. If the extruder is operating near the maximum screw speed, wear will eventually cause a reduction in rate because the decrease in specific rate cannot be compensated for by an increase in screw speed. Screw wear for an injection-molding machine can cause the plasticating time to increase [7]. If the plasticating time increases too much, the machine can switch from a cooling-limited process in the tooling to a plasticator rate-limited process, increasing the cycle time of the process and the cost to mold the part.

The amount of flight wear that is acceptable to a process depends on the economic operation of the process. That is, when the wear becomes large enough to affect negatively the profitability of the line, then the screw should be replaced. A spare screw should be available to minimize downtime during the screw change process. The worn screw should then be sent to the screw manufacturer to be rebuilt. Typically a screw can be rebuilt at the most three times. Rebuilding more times will cause the flights to become brittle due to the welding process, the flights to become narrower, and the flight radii to become smaller.

Wear can occur at all places in the screw and barrel, but the wear rates tend to be higher at locations with high levels of solids and high pressures. For example, if the solids-conveying capacity of the feed section is considerably higher than the rate of melting in the transition section, solids can become wedged at the end of the transition section. The wedging can create local pressures that are high enough to create a side force that pushes the screw against the barrel [8]. The contact between the screw and barrel causes galling, leading to the adhesive failure of the hard facing. Photographs of this type of failure are shown in Figure 13.4. Glass-filled engineering resins such as SMA resins can erode into the pushing flight at the end of the solids-conveying channel and the first parts of the transition section. The erosive wear can completely remove the flight in the affected area.

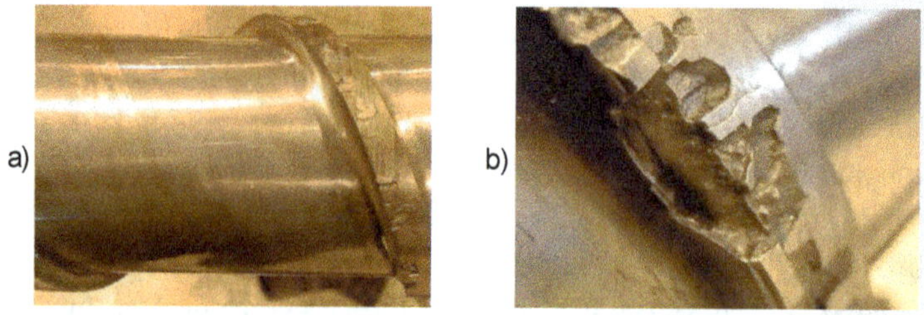

Figure 13.4 Flight damage due to solids wedging at the end of the transition section

13.3 High-Performance and Barrier Screws for Improved Rates

Extrusion processes that are rate limited by high discharge temperatures or solid polymer fragments in the extrudate can often be increased in rate through the use of advanced screw designs. These advanced designs allow the screw to operate at a higher specific rate, maintain a high-quality discharge, and decrease the temperature of the extrudate. Moreover, the small clearances between the barrier flight and the barrel wall and the dispersive mixing features of high-performance screws will trap and disperse solid polymer fragments. Barrier mixing sections were discussed in Chapter 6 while high-performance screws are presented in Chapter 14.

13.4 Case Studies That Were Rate Limited

Seven commercial case studies are presented in the next sections that describe rate-limited processes and the strategies and solutions to increase extrusion rates. Additional case studies that had limited rates and additional defects were presented in Chapters 11 and 12.

13.4.1 Rate Limitation Due to a Worn Screw

A specialty blown film product line was rate limited by the discharge temperature of the extrudate at 130 kg/h. At this rate, the discharge temperature from the 88.9 mm diameter extruder was 245 °C. At higher rates, the discharge temperature increased to even higher levels such that the bubble process could not be maintained. The operators decreased the barrel temperature setting over the metering zone to 190 °C, but the discharge temperature would not decrease. The proprietary screw design appeared acceptable for this process. When the process was numerically simulated, the discharge temperature of the process was estimated at 220 °C. Here the process based on the screw schematic could not be simulated to provide the measured discharge temperature of 245 °C.

The screw was removed from the extruder and inspected to determine if the screw was in specification as defined by the schematic. The measured geometries were consistent with the schematic, but the diameter at the flights was measured at 88.15 mm, indicating that the flight clearance was 0.38 mm. The original flight clearance was specified at 0.09 mm, and thus the screw had worn such that the flight clearance was more than four times the specified value. The screw was placed back into the extruder and production was continued at the reduced rate.

An identical screw was ordered with the correct flight clearance. As soon as the screw was delivered, the screw was installed in the extruder. Soon after startup, the extruder was discharging at a temperature near 220 °C at the base rate of 130 kg/h. Next the rate was slowly increased to 180 kg/h. The discharge temperature increased to about 224 °C. Higher rates are likely possible. The worn screw was sent back to the screw manufacturer for refurbishment. This screw was then stored as the spare screw.

This case study clearly shows that a worn screw will cause the discharge temperature to increase due to material flow over the flights and will decrease the heat transfer coefficient, limiting the transfer of energy through the barrel wall. With the restored flight clearance, the process was able to increase the rate by about 35%, and the quality of the product was improved.

13.4.2 Rate Limitation Due to Solid Polymer Fragments in the Extrudate

Thermoplastic polyurethane (TPU) resins are specialty resins used in a broad spectrum of applications including belting, hoses, cables, films, and tubing. They can be injection molded, blow molded, and extruded. For these processes, the resins are typically plasticated using single-screw smooth bore machines. In order to be competitive, the plasticator must be able to supply molten resin at very high rates and high qualities. That is, the extrudate must have a uniform temperature, be delivered to the die at constant pressure, be free of solid TPU fragments, and contain no degradation products. High-quality TPU extrudates are relatively easy to obtain at moderate rates, but high-quality extrudates can be very difficult to obtain at high rates.

The extruder was 114.3 mm in diameter and had a 32 length-to-diameter (L/D) barrel [9]. It was equipped with a 150 kW motor and a gear reduction set that provided a maximum screw speed of 105 rpm. The first 2.5 diameters consisted of a water-cooled feed throat and casing. The original screw was a very old-style screw designed for polyethylene resins. The screw was a conventional, single-flighted screw with a compression ratio of 2.7. The channel dimensions of the screw are shown in Table 13.1. In the middle of the metering section were three rings of pins, that is, three pin mixers. A schematic of this type of pin mixer is shown in Figure 8.25. It was not clear why the pin mixers were designed into this screw. As discussed in Section 8.5.4, this type of mixer is a distributive-type mixing device. A dispersive mixer, however, is required for this application such that the solid polymer fragments are trapped and then melted. The melt density of the TPU resin at

230 °C was 1.04 g/cm³, and the specific rotational flow rate was calculated at 5.0 kg/(h·rpm).

Table 13.1 Screw Channel Dimensions for the 114.3 mm Diameter Screw for the Specialty TPU Resin Extrusion Process. This Screw was Rate Limited by an Optical Defect

	Depth, mm	Length, diameters	Notes
Feed section	13.7	7	Lead length was 114.3 mm in all sections
Transition		13	
Metering section	5.1	12	

Flight width and flight clearance were 13.5 and 0.12 mm, respectively, in all sections of the screw. The extruder had a smooth-bore feed section. Three rings of pin mixers were evenly positioned in the middle portion of the metering section of the screw. All pins in a single ring had the same axial position as the type shown by Figure 8.25. The specific rotational flow rate for the metering section was calculated at 5.0 kg/(h·rpm). The compression ratio was 2.7.

Although the full history of this screw is unknown, the mixing pins appeared to have been added to the screw after the screw failed to mix and homogenize the extrudate. Pin mixers have been retrofitted to screws for many years as a way to improve the distributive mixing for an extruder. Here the pin mixers would distribute the solid polymer fragments evenly in the extrudate, but the mixers lacked the ability to trap and disperse them. Pin mixers were clearly the wrong type of mixer to use for this application. Instead, a dispersive mixer such as a Maddock-style mixer would be preferred; see Section 8.5.1.

The extruder typically operated at a rate of 204 kg/h at a screw speed of 32 rpm for a specific rate of 6.38 kg/(h·rpm). The extrudate temperature was difficult to measure since it was enclosed in a transfer line [10], but it was reported at 228 °C. The specific rate during operation was considerably higher than the calculated rate due just to rotation of 5.0 kg/(h·rpm). A negative pressure gradient ($\partial P/\partial z$) must exist to cause the flow rate to be higher than the calculated rotational rate. For this case, the discharge pressure of the process was 18 MPa, and the pressure gradient was calculated at about −2 MPa/turn. The axial pressure profile for this process is shown in Figure 13.5. The solid line was calculated while the dotted line was estimated, as outlined in Section 9.2. The calculation indicated that the pressure at the entry to the first metering section needs to be 42 MPa in order to maintain the negative pressure gradient and the measured discharge pressure. This pressure at the entry to the metering section is considered to be very high for this application. The high pressure was later verified by laboratory extrusions. The laboratory extrusions are beyond the scope of this case study.

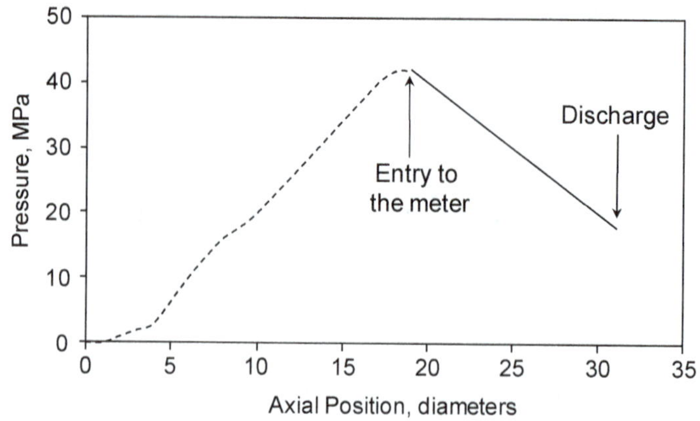

Figure 13.5 Axial pressure profile for the original screw operating at 204 kg/h, 32 rpm, and a discharge pressure of 18 MPa. The dotted line was estimated and the solid line was calculated

As previously stated, at rates higher than about 230 kg/h the extrudate would contain solid polymer fragments, and they would appear as gels in the final product. It was hypothesized that the high pressure at the inlet of the metering section was due to high solids-conveying rates. That is, at the screw speed of operation the solids-conveying process was conveying at a rate higher than the melting rate, pushing solids downstream. The pin mixer would break large fragments into small fragments and distribute them evenly in the extrudate. This hypothesis directs the design to a screw with a lower level of solids conveying, a higher melting capacity, and the ability to trap and disperse solid polymer fragments. Hot-stage microscopy confirmed that the gel defects were created from solid polymer fragments discharged with the molten TPU resin. That is, they fully melted at the expected stage temperature and did not re-form during cooling.

In order to design a screw that is capable of high rates and homogenous extrudates, the stress at a sliding interface and the melting flux were measured using a Screw Simulator (Section 4.3.1) by forcing (normal force) a slab of solid resin onto a rotating roll and then measuring the drag force (frictional or viscous force) on the sample. The stress at the interface at velocities of 18 and 36 cm/s and a pressure of 1.4 MPa is shown by Figure 13.6. At low temperatures, the coefficient of dynamic friction is simply the stress divided by the pressure used during the measurement. As shown in Figure 13.6, the coefficient of dynamic friction is about 0.8 between 50 and 100 °C. This coefficient is relatively high, causing a high level of energy to be dissipated at the interface, and it will promote a very high level of solids conveying in the feed section.

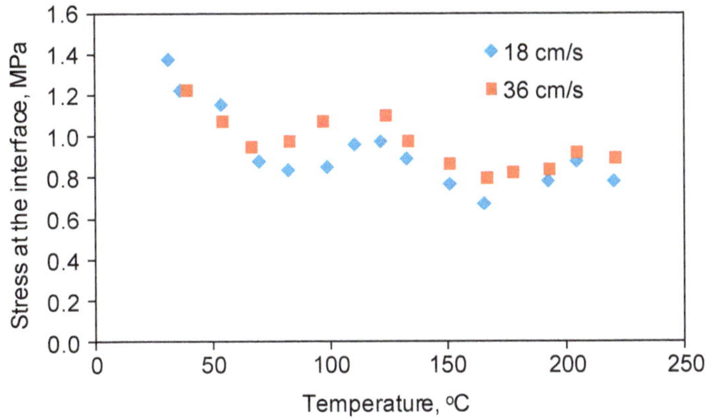

Figure 13.6 Stress at a sliding interface for TPU resin at a pressure of 1.4 MPa

The Screw Simulator was also used to measure the melting flux of TPU resins at conditions similar to those expected in the melting section of the screw. At these conditions, the melting flux was measured at 0.2 kg/($m^2 \cdot$ s). This melting flux is relatively low. For example, the melting fluxes [11] for linear low-density polyethylene (LLDPE) and low-density polyethylene (LDPE) resins are 0.3 and 0.4 kg/($m^2 \cdot$ s), respectively. The relatively low melting flux for TPU resins is a contributor to the solid polymer fragments in the extrudate.

The stress at the interface and melting flux data indicate that the solids-conveying forces will be high while the melting flux will be low. Thus, the new screw design must reduce the solids-conveying forces and increase the area available for melting. To achieve higher melting capacities and the ability to trap solid polymer fragments, a new screw was designed with a long barrier melting section and a Maddock-style mixer. A long barrier section was used to provide more area for melting and to maintain the integrity of the solid bed. This combination is known to provide higher melting capacities. Moreover, the screw was designed with a lower compression ratio of 2.5 in order to reduce the solids-conveying rate relative to the metering section. The screw was designed with five sections as shown in Table 13.2. The lead length was 114.3 mm in the feed and metering sections while the lead length in the barrier section was 140 mm. The undercut clearance on the barrier flight and mixer flights were 0.25 and 0.50 mm, respectively. The clearance on the barrier flight was very small and about a third that of a normal screw. This small clearance was designed to trap and disperse small solid fragments flowing between the barrier flight land and the barrel wall. Any solid fragments exiting the barrier section, likely from the end of the solids channel, would be trapped and dispersed in the Maddock mixer. The specific rotational flow rate for this screw for TPU resins was calculated at 7.4 kg/(h·rpm), a value that is about 50% higher than the original screw.

Table 13.2 High-Rate Screw Channel Dimensions for the 114.3 mm Diameter Extruder for the Specialty TPU Resin Extrusion Process

	Depth, mm	Length, diameters	Notes
Feed section	19.0	7	
Barrier section		15.3	0.25 mm barrier flight undercut
First meter	7.6	4.7	
Maddock-style mixer		2	0.50 mm undercut on the mixing flight
Second meter	7.6	3.0	

The lead length was 140 mm for the main flight of the barrier section and 114.3 mm for all other sections of the screw. The main flight width and clearance were 11 and 0.12 mm, respectively, in all sections of the screw. The barrier flight was undercut from the main flight by 0.25 mm. The first 2.5 diameters of the screw were inside a water-cooled smooth-bore feed casing. The compression ratio was 2.5. The specific rotational flow rate for the metering section was calculated at 7.4 kg/(h·rpm).

The original screw was removed from the extruder and the new screw was installed. Essentially the same barrel temperature settings were used initially. At these conditions, the extruder was operating near the maximum motor load. To mitigate the high motor loads, all barrel zone temperatures were increased by about 10 °C. This allowed the TPU resins to be processed at high rates. After several days of use, the performance of the screw was measured for several TPU resins. The extruder and new screw produced high-quality extrudates at a rate of 440 kg/h and a screw speed of 45 rpm for a specific rate of 9.8 kg/(h·rpm). The discharge temperature was reported as 242 °C, a temperature that was about 14 °C higher than that produced by the original screw. Although the rate was higher, the increase in extrudate temperature caused the discharge pressure to decrease to 9 MPa.

The extrudates were homogenous and free of gels for all products and rates tested using the new screw. Rates for several products were more than double the maximum rate for the original screw. The discharge temperatures were between 13 and 18 °C higher for the new screw. These higher temperatures were caused by a combination of the small clearances for the barrier flight and the Maddock-style mixer, the higher screw speeds, and the higher barrel zone temperature settings. Solid polymer fragments were not observed in any extrudate.

As discussed above, the new screw was operating at a specific rate higher than the calculated rotational rate, indicating that a negative pressure gradient existed in the metering channel. For this case, the discharge pressure of the process was 9 MPa, and the pressure gradient was calculated at about −0.9 MPa/turn. The axial pressure profile for this process is shown in Figure 13.7. The calculation indicated that the pressure at the entry to the first metering section needed to be 18 MPa in order to maintain the negative pressure gradient and the measured discharge pressure. This pressure is considered to be very normal and acceptable for this application.

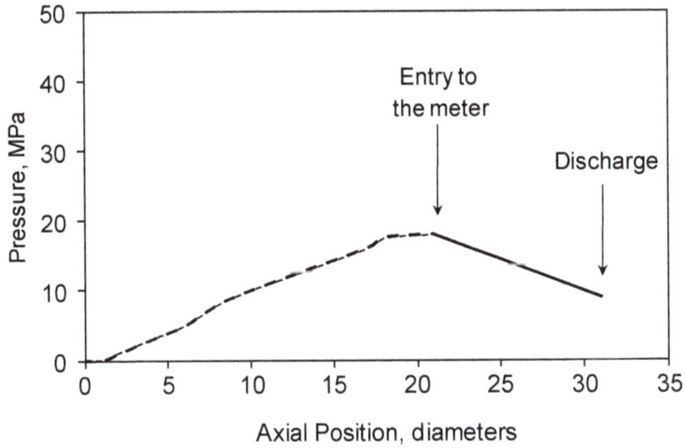

Figure 13.7 Axial pressure profile for the new screw operating at 440 kg/h, 45 rpm, and a discharge pressure of 9 MPa. The dotted line was estimated and the solid line was calculated

The combination of a lower compression ratio, a long barrier section with a very small barrier flight clearance, a Maddock-style mixer with a small clearance, and deeper metering channels allowed the TPU resins to extrude at twice the rate and provide high-quality extrudates that were free of solid polymer fragments. Here the reduced compression ratio reduced the solids-conveying rate such that the pressure at the end of the barrier melting section was considerably lower. The tight clearance on the barrier flight was successful at mitigating solid fragments. Removing the distributive pin mixers and replacing them with a dispersive Maddock-style mixer provided a trap that melted any solid fragments that managed to flow out of the solids channel of the barrier section. The deeper metering section allowed a higher specific rate.

At a maximum rate of 440 kg/h, the extruder was rate limited by motor torque and not by the quality of the extrudate. This extruder was fabricated with a gearbox and sheave ratio such that the maximum screw speed was 105 rpm. Since the process was operating at a screw speed of 45 rpm, the sheave ratio could be changed such that the maximum screw speed would be about 75 rpm. This might allow the process to operate at rates up to 700 kg/h. Changing the sheave ratio will increase the torque to the inlet side of the gearbox. Care must be taken to make sure that the additional torque does not harm the gearbox.

13.4.3 Rate Limited by the Discharge Temperature for a Pelletizing Extruder

Many compounding processes are configured with a twin-rotor mixer or twin-screw extruder to perform the compounding followed by a single-screw extruder to pressurize the material for pelletization. A typical process is shown in Figure 13.8.

In order to maintain a profitable process, the compounding operation must operate at very high rates. For most processes, the rate should be limited by the available torque on the compounding machine or by quality limitations from the compounder. That is, the feeders, single-screw extruder, and pelletizer should not be the rate-limiting steps. For this process, the compounding machine provides a homogenous compound, but it is typically not able to pressurize the compound for pelletization. Pressurization is provided by the single-screw extruder.

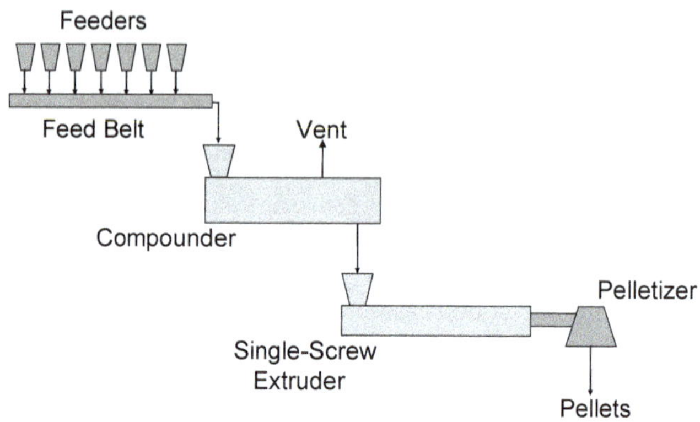

Figure 13.8 Schematic for a compounding process with a single-screw pumping extruder

The compounder is typically positioned on the second floor of the plant, and the discharge from it is dropped by gravity to the single-screw extruder positioned on the first floor. The feed throat opening of the single-screw extruder can be enlarged to not impede the flow of resin to the extruder. The screw is rotating at a speed such that the hopper is always empty. That is, the extruder is operating in a starve-fed mode such that the speed of the screw is not controlling the rate of the process. Instead the feeders upstream of the compounding machine are controlling the rate. Since the extruder is starve-fed and there is a pressure at the discharge, a positive pressure gradient $\partial P/\partial z$ must exist in the screw channel.

The single-screw flow equations used for this analysis were described by Equation 7.58. The equation is shown below for clarity of this section.

$$Q_m = \frac{p\rho_m V_{bz} WHF_d F_c}{2} - \frac{p\rho_m WH^3 F_p}{12\eta}\left[\frac{\partial P}{\partial z}\right]$$

As discussed numerous times, the first term on the right-hand side of the equation is the flow induced by the rotation of the screw. The second term on the right-hand side is the flow component induced by the pressure gradient in the channel. Since a positive pressure gradient exists in the screw, the mass flow rate in the channel (Q_m) will always be less than the rotational flow rate. Moreover, the rotational flow rate increases essentially linearly with increasing channel depth (H) while the pressure flow component increases to H^3; increasing the pressure flow component decreases the overall rate. Thus, an optimized channel depth exists for a specific resin and discharge pressure. If a screw is not designed with this optimization considered, then rate limitations for the compounding line may occur due to the single-screw extruder. A case study that demonstrates this optimization is presented next.

A compounding and pelletization line was commissioned to produce a new specialty product based on a PE resin feedstock and a flame retardant chemical system. For this process, the components for the formulation were compounded in an upstream continuous process, and then the material was dropped via gravity into a 381 mm diameter, melt-fed, 9.5 *L/D* single-screw extruder. The material was then discharged from the extruder and flowed through a screen pack system to a pelletizing system. Because of the decomposition of the flame retardant chemical, the material could not exceed a temperature of about 180 °C. During a preliminary trial, the process could be operated at a rate of 1490 kg/h at a screw speed of 15 rpm for a specific rate of 99 kg/(h·rpm). The discharge temperature could not be measured accurately due to the enclosed system and the inaccuracy of measuring the temperature through the transfer line wall [10]. Based on numerical simulations of the process, the discharge temperature was estimated at 182 °C. The pelletized product was of high quality with essentially no degradation of the flame retardant chemical. When the rate was increased to levels higher than 1490 kg/h, some of the flame retardant chemical decomposed and created off-specification product. In order for this process to be economically viable, the process needed to operate at a rate higher than 3500 kg/h.

The original screw used for this process was single flighted and was designed with three sections, as shown in Table 13.3. The lead length was 635 mm in the feed and transition sections and 381 mm in the metering section. The specific rotational flow rate for the metering section was calculated at 500 kg/(h·rpm). This calculated specific rate was about five times higher than the measured specific rate during the trial.

Table 13.3 Screw Channel Dimensions for the 381 mm Diameter Screw for the PE Flame Retardant Pelletization Process

	Depth, mm	Length, diameters	Notes
Feed section	108	2.9	Lead length was 635 mm
Transition		1.5	
Metering section	44	5.1	Lead length was 381 mm

Flight width and flight clearance were 38 and 0.38 mm, respectively, in all sections of the screw. The specific rotational flow rate for the metering section was calculated at 500 kg/(h·rpm).

Because the processing specific rate was extremely lower than the calculated specific rotational flow rate, it was hypothesized that the metering section of the screw was too deep for this resin. Although the resin produced here was extremely viscous, this screw was likely designed for a resin with a much higher shear viscosity. The hypothesis was made because the material entering the screw channel is at zero pressure, and it is then increased to the discharge pressure of 15 MPa. Thus, the axial pressure gradient in the channel is positive. This positive pressure gradient will cause the specific rate to decrease to a level less than the specific rotational flow rate. For a screw with a positive pressure gradient, increasing the channel depth of the metering section will cause the rotational flow rate to increase directly proportional to the depth. The rate reduction due to the positive pressure gradient, however, will reduce at a rate proportional to the cube of the channel depth. Because of these dependencies on channel depth, an optimal channel depth exists for a particular resin at a specified set of conditions.

The process was simulated using the existing (original) screw at a rate of 1490 kg/h and a screw speed of 15 rpm. The measured discharge pressure and estimated discharge temperature at the plant were 15 MPa and 180 °C, respectively. The simulated axial pressure and temperature profiles are shown in Figure 13.9. For this process, the molten resin from the compounder was conveyed forward in the partially filled channels in the single-screw machine for the first 5.5 diameters. At 5.5 diameters from the start, the channel became filled and started to generate pressure; that is, it reached the fill position. The pressure continued to increase to the discharge pressure of 15 MPa. The resin was discharged from the compounder at 160 °C, which is the feedstock temperature for the single-screw extruder. The simulation is predicting that the discharge temperature is 182 °C, a value that is consistent with the manufacturing observations; that is, the flame retardant chemical was not degraded.

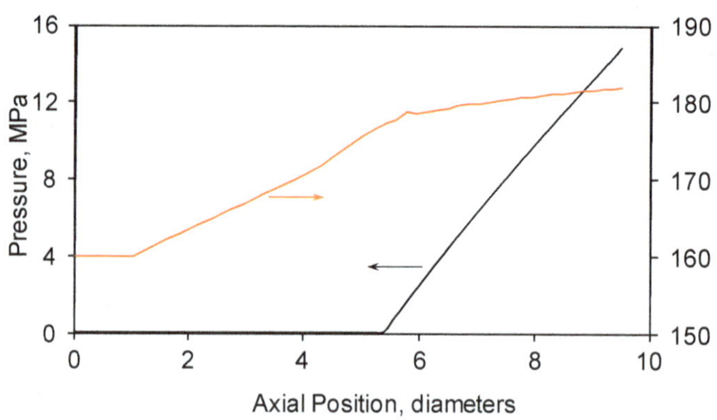

Figure 13.9 Simulated axial pressure and temperature profiles for the original pumping screw at a maximum rate of 1590 kg/h and a screw speed of 15 rpm

To test the hypothesis that the metering channel is too deep to run at a rate of 3860 kg/h without the degradation of the flame retardant chemical, simulations were performed as a function of the metering channel depth. Since the rate will be increased from 1490 to 3860 kg/h, the pressure required to operate the pelletizing die at the higher rate must be estimated. The die pressure at a rate of 3860 kg/h was estimated from the base rate data at 21 MPa. For these simulations, the screw section lengths were maintained and the compression ratio was 2.2, the feed depth divided by the meter depth. The metering channel depth was varied between 22 and 40 mm. The lead length was set at 457 mm. The screw speed was allowed to vary in order to maintain the rate at 3860 kg/h, and the fill position was fixed at 4.4 diameters from the screw start. This fill position was selected since it was at the start of the metering section. The specific rate for the process as a function of channel depth is shown in Figure 13.10. As shown in this figure, the optimal channel depth is about 30 mm where the specific rate is at a maximum. At this depth, the specific rate was calculated at 227 kg/(h·rpm) while the specific rotational rate was 410 kg/(h·rpm). The process is expected to operate at 3860 kg/h at a screw speed of 17 rpm. This channel depth is the best balance between the rotational flow rate and the reduction in the flow due to the positive pressure gradient in the channel. Moreover, the parabolic nature of this rate curve is due to the dominant rotational flow relationship at low channel depths and the dominant pressure flow relationship to the cube of the channel depth for deep channels. As previously stated, this analysis is very specific to the resin rheology and the discharge pressure. For a 30 mm deep channel, the discharge temperature was estimated at about 179 °C and is acceptable for this process.

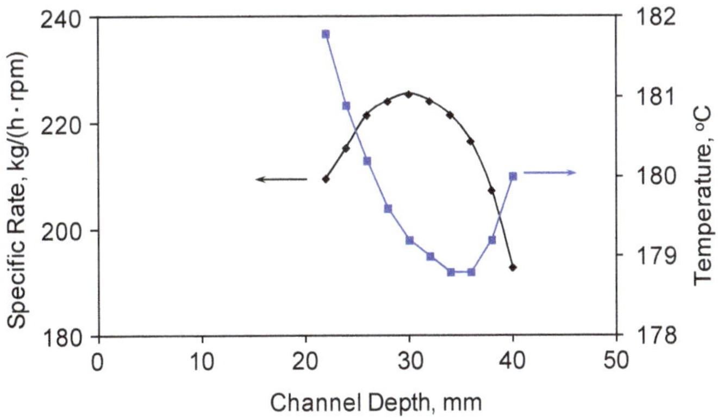

Figure 13.10 Simulated specific rates as a function of metering channel depth at a rate of 3860 kg/h and a discharge pressure of 21 MPa

The lead length in the metering section of the new screw was set at 457 mm, a value considerably higher than the 381 mm lead length specified for the original screw. The lead length was increased because it allows the specific rate to increase with a much lower dependence on the positive pressure gradient in the channel. For simplicity in design, all sections of the new screw were designed with the same lead length. The dimensions of the new screw are shown in Table 13.4.

Table 13.4 Screw Channel Dimensions for the 381 mm Diameter Screw for the PE Flame Retardant Pelletization Process

	Depth, mm	Length, diameters	Notes
Feed section	65	2.9	
Transition		1.5	Lead length was 457 mm
Metering section	30	5.1	

Flight width and flight clearance were 38 and 0.38 mm, respectively, in all sections of the screw.
The specific rotational flow rate for the metering section was calculated at 410 kg/(h·rpm).

The simulated axial pressure and temperature for the screw with a 30 mm deep metering channel (Table 13.4) are shown in Figure 13.11. As shown by this figure, the new screw meets the commercial targets of 3860 kg/h, a discharge pressure of 21 MPa, and a discharge temperature near 180 °C. The simulation was performed at a fill position of 4.4 diameters from the start of the screw (at the entry to the metering section), a screw speed of 17 rpm, and a specific rate of 227 kg/(h·rpm). If the screw speed is decreased slightly such that the fill position is at 3.5 diameters (just downstream of the start of the transition section), the screw could operate at a specific rate as high as 249 kg/(h·rpm) at 3860 kg/h. The motor and gearbox were such that sufficient torque was available to operate the new screw. Based on the simulation results, plant personnel decided to fabricate a new pumping screw based on a metering channel depth of 30 mm, a compression ratio of 2.2, a lead length of 457 mm, and section lengths equivalent to those of the original screw.

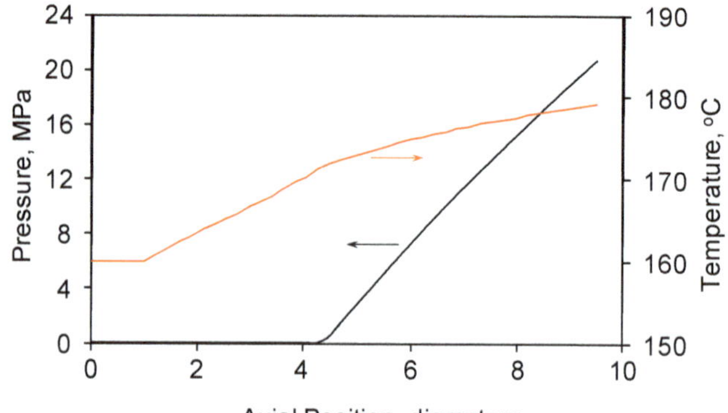

Figure 13.11 Simulated axial pressure and temperature profiles for the new pumping screw at a rate of 3860 kg/h and a screw speed of 17 rpm

The new pumping screw was fabricated, installed, and tested. The performance data for this screw is shown in Table 13.5. At the start of the trial, the process was operated at about 50% of the expected maximum rate. The extruder was operating at a rate of 1820 kg/h and a screw speed of 7.5 rpm for a specific rate of 243 kg/(h·rpm) and a discharge pressure of 17 MPa. Numerical simulations predicted that the discharge temperature was 175 °C. The rate of the process was then slowly increased to 3600 kg/h. At this rate, the extruder was operating with a screw speed of 15 rpm for a specific rate of 240 kg/(h·rpm). The compounded pellets were in specification and no degradation of the flame retardant chemical was observed. The 3600 kg/h rate was likely the highest rate possible with the current equipment setup. That is, the pressure at the discharge was the maximum allowed. Increased rates are likely possible with modifications to the screen pack and the die to reduce the pressure at the discharge of the extruder.

Table 13.5 Plant-Measured Extruder Performance with the New Pumping Screw

Rate, kg/h	Screw Speed, Rpm	Specific Rate, kg/(h·rpm)	Discharge Pressure, MPa	Estimated Discharge Temperature, °C
1820	7.5	243	17	175
3600	15	240	22	180

This case study is an excellent example of a screw that was originally designed to process a very viscous resin. This screw likely performed well for those compounds. Product and market changes, however, have occurred that required the line to compound materials that have a lower viscosity and are temperature sensitive. For the current compound, the extremely deep metering channel of the original screw caused the specific rate of the process to be very low at 99 kg/(h·rpm) due to the pressure gradient in the channel. The very low specific rate caused the discharge temperature to increase and thus limited the maximum rate of the process to 1490 kg/h. The new screw was designed with a shallower metering channel and a larger lead length, allowing the process to operate at a specific rate of 240 kg/(h·rpm) and a maximum rate of 3600 kg/h. The new screw allowed the rate to be increased by a factor of 2.4 and decreased the compounding costs for the product.

Melt pumping extruders used in compounding lines all operate with positive pressure gradients. This gradient occurs because the extruder is starve-fed, causing the pressure just downstream from the hopper opening to be zero and the high discharge pressure required to operate the pelletization die. From the design equations, this positive pressure gradient reduces the overall rate of the machine. Moreover, as the discharge pressure is increased the rate will decrease. For an existing process, the screw designer has essentially only two parameters to adjust, and these parameters are the metering channel depth and the lead length of the screw.

As demonstrated in the case study, an optimal channel depth exists where the specific rate is at a maximum. Designing the screw at this maximum specific rate is typically the desire. The parabolic shape of the specific rate curve shown in Figure 13.10 was caused by the rotational rate increasing linearly with increasing channel depth; see Equation 7.58. This increase in rotational rate causes the overall rate to increase. The pressure flow increases with the cube of the channel depth, and since the pressure gradient is positive, the pressure flow term decreases the overall rate. Other nonlinear contributions come from F_d, F_p, and F_c.

The lead length of the screw can be increased slightly to obtain an increase in specific rate. When the lead length is increased, the terms V_{bz} and W are increased, causing the rotational rate to increase. The magnitude of the pressure flow term increases linearly with W, and thus the contribution from pressure flow will not be affected as much as changing the channel depth. In general, the lead length can be between 1 to 1.3 times the diameter of the screw. As the lead length is increased, however, the total helical length (z) to generate pressure decreases for a fixed axial length.

Decreasing the discharge pressure required to operate the screen pack and the die will allow a rate increase. In this case, decreasing the discharge pressure will cause the pressure gradient in the channels to decrease. As mentioned previously, decreasing the pressure gradient will reduce the magnitude of the pressure-induced flow term of Equation 7.58. Reducing the restriction caused by the screen pack and die will cause the discharge pressure to decrease while maintaining the rate. For example, if the number of screens used can be reduced or if the mesh size can be decreased, then the restriction will be reduced, allowing higher rates. Or a new pelletizing die plate can be fabricated that has more holes or shorter lands.

The compounding line presented here was dedicated to producing only a single product. In most cases, a compounding line will be required to produce a range of products with different shear rheologies. If a range of compounds are produced on the line, the designer must consider all resins in the design of the screw. Often the screw is optimized for the resin that has the highest demand, but it still has the capability to produce the other resins at high but slightly less optimal rates.

For this type of problem, the specific rate of operation is designed to increase with the optimized screw. When the specific rate increases, the torque required to turn the screw at the same rate will increase. If the new screw is built and torque is not available to turn it, then the design process will fail and the rate increase will not be achieved. The designer must perform an overall energy balance to make sure that torque is available; for example, see Section 9.2.

13.4.4 Large Diameter Extruder Running PS Resin

A 203.2 mm diameter smooth-bore extruder running a PS resin experienced low rates when extruding to a complicated downstream sheet line. The extrudate flowed from the extruder to a screen pack system and then to a sheet die. The line was able to produce acceptable product but at reduced rates. The reduced rates increased the cost of manufacturing the product and caused the plant to miss shipment dates to its customers. At a screw speed of 110 rpm, the rate was measured at 1750 kg/h for a specific rate of 15.9 kg/(h·rpm). Under all processing conditions, the line operated stably without indications of contamination from degraded resin. The screw for this process was new and was recently installed. The dimensions for this two-stage, vented screw are summarized in Table 13.6. The specific rotational rate for the first-stage metering section was calculated at 20.1 kg/(h·rpm). If the machine was operating properly, the pressure gradient in the first-stage metering section must be negative because of a positive and high pressure at the entry and zero pressure at the start of the vent opening. Thus, the extruder should be operating at a specific rate higher than 20.1 kg/(h·rpm). The material extruded was a PS resin with an MFR of 1.5 dg/min (200 °C, 5 kg).

Table 13.6 Screw Channel Dimensions for a 203.2 mm Diameter Two-Stage Vented Screw Running PS Resin

	Depth, mm	Length, diameters	Notes
Feed section	30.5	5	The compression ratio was 4.12 and the compression rate was 0.0068
First-stage transition		5	
First-stage meter	7.37	7.5	
Vent section	30.5	4.5	
Second-stage transition		2	Pump ratio was 1.9
Second-stage meter	14.2	8	

Lead length, flight width, and flight clearance were 203.2, 20, and 0.20 mm, respectively, in all sections of the screw. A 38 mm diameter screw cooling hole was drilled in the shank end of the screw, and it extended 4 diameters into the feed section. The first 2.5 diameters of the screw were inside a water-cooled smooth-bore feed casing. The motor was 450 kW and had a base speed of 1750 rpm. The gearbox reduction was 17.33:1 for a maximum screw speed of 101 rpm at the base motor speed. The motor could be over speeded to provide higher screw speeds.

Detailed data collection and process analysis indicated several key findings related to the process. The barrel zone temperatures were normal for the second stage but were unusual for the first stage. The barrel temperatures in the first stage were 163, 151, 166, 174, 181, 185, and 185 °C for zones 1 through 7, respectively. The first two zones covered the feed section and the third and fourth zones covered the transition section of the screw. The temperatures in the first four zones were relatively low for PS resin extrusion. These temperatures were increased to maximum values of 190, 200, 210, and 210 °C for zones 1 through 4, respectively. The specific rate at a screw speed of 110 rpm, however, decreased to 14.8 kg/(h·rpm). Decreasing the barrel zone temperatures in the first three zones was not possible since these zones were operating with full cooling and at temperatures 20 to 30 °C higher than the set point temperatures. The temperature data suggests that a high pressure exists in the channel at zones 2 and 3. The high pressure causes the dissipation of a large amount of energy at the surface of the barrel wall. The energy dissipation for solids rubbing on the barrel wall is as follows:

$$F = fPV \tag{13.1}$$

where F is the energy flux (units of $J/(m^2 \cdot s)$), f is the coefficient of dynamic friction, P is the local pressure, and V is the relative velocity between the moving solids and the barrel wall. The process was consuming about 560 J/g and about 71% of the maximum motor current.

The dimensions of the screw were slightly out of the normal operating range. Before the compression rate and ratio can be compared to standard values, the standard values needed to be corrected for the lower bulk density of the feedstock. About 20% recycle edge trim with a bulk density of 0.35 g/cm^3 was added with the pellet feedstock with a bulk density of 0.64 g/cm^3. The resulting bulk density of the blend was measured at 0.58 g/cm^3. If the feedstock was 100% pellets with a bulk density of 0.64 g/cm^3, the optimal compression ratio and rate are about 3.0 and 0.0035, respectively. The compression rate and ratio corrected for operation with a feedstock bulk density of 0.58 g/cm^3 are 3.3 and 0.0040, respectively, using Equations 6.3 and 6.4. This calculation indicates that this screw has a compression ratio and rate that are too high for this application. The higher ratio and rate will create a mismatch between the feeding rate, melting rate, and pumping ability of the screw sections. That is, the feed section is too deep and is overfeeding the screw, and the transition section is too short to melt the resin. A large level of cool solids from the feed section is unable to deform to the quickly decreasing depth channel of the transition section. This creates a restriction at the entry to the melting section and results in a high channel pressure upstream of the restriction. The high pressure at the barrel wall creates the high energy flux that is removed in barrel zones 2 and 3. The observed behavior from the process is consistent with overfeeding and the lack of melting capacity.

To correct the problem, the screw was recut such that the solids-conveying and transition sections were longer. For this recut, small amounts of metal were removed between diameters 5 and 13.5. As discussed in Section 10.11, removing metal in this manner is relatively easy and inexpensive to perform. A summary of the modified screw geometry is provided in Table 13.7, and a schematic is shown in Figure 13.12. Lengthening these sections does not change the compression ratio. The only way to decrease the compression ratio effectively on this screw was to increase the depth of the first stage. Increasing the depth of the first stage was not done because the motor and gearbox may not be able to supply enough torque to the screw. Increasing the length of the transition section caused the compression rate to decrease from 0.0068 to 0.0046, a rate that is much closer to the optimal compression rate of 0.0040 for this process. The feed section was lengthened to allow more time for the resin to increase in temperature such that it may deform easier when it is compressed in the transition section. Since the metering channel depth was unchanged, the calculated rotational flow rate is the same at 20.1 kg/(h·rpm).

Table 13.7 Screw Channel Dimensions for the Modified 203.2 mm Diameter Two-Stage Vented Screw

	Depth, mm	Length, diameters	Notes
Feed section	30.5	6	The compression ratio was 4.12 and the compression rate was 0.0046
First-stage transition		7.5	
First-stage meter	7.37	4	
Vent section	30.5	4.5	
Second-stage transition		2	The pump ratio was 1.9
Second-stage meter	14.2	8	

Lead length, flight width, and flight clearance were 203.2, 20, and 0.20 mm, respectively, in all sections of the screw. A 38 mm diameter screw cooling hole was drilled in the shank end of the screw, and it extended 4 diameters into the feed section.

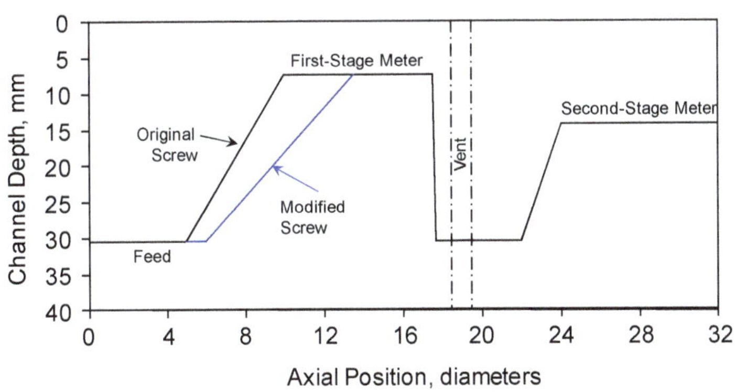

Figure 13.12 Schematic of the channel depths for the original and modified screws for the PS resin case study

The screw was modified and installed into the extruder. The extruder was then started up at essentially the same barrel temperature settings that were used for the original screw. After several days of operation, the extruder with the modified screw and modified conditions was operating at a rate of 2500 kg/h at a screw speed of 105 rpm for a specific rate of 23.8 kg/(h·rpm) and discharging at an acceptable temperature of 250 °C. The modifications made to the screw allowed higher specific rates by increasing the melting capacity of the screw. Moreover, the extruder was operating with barrel temperatures for zones 1, 2, and 3 of 195, 205, and 215 °C, temperatures that are normal for this application. All barrel temperatures were in control, indicating that the high pressure that was occurring in the original screw had been mitigated. Moreover, the specific rate was about 17% higher than the calculated rotational specific rate of 20.3 kg/(h·rpm), a specific rate that is expected for a properly operating two-stage, vented extruder.

13.4.5 Rate Limited by Discharge Temperature and Torque for Starch Extrusion

A process to make a starch-filled compound was rate limited by the discharge temperature and the available torque from the motor. The feedstock contained starch, water, and a resin binder. The components were preblended and then added to the hopper. The hopper was equipped with a cramming screw, and the screw was operated such that the extruder screw was starve-fed; the crammer speed was controlling the rate of the process, not the extruder screw speed. The extruder was used to compound, bring the formulation to a minimum temperature to denature or gelatinize the starch, and then pressurize the material for pelletization. The pelletization process must occur at a temperature low enough to not cause the water to vaporize and cause the pellets to turn into very low bulk density product; some regions of the pellets would foam due to the presence of water and a high discharge temperature. Low bulk density pellets caused processing problems in downstream processing steps, and thus it was important that the pellets were not partially foamed. The process was rate limited to 200 kg/h using a 114.3 mm diameter, 30 *L/D* smooth-bore single-screw extruder.

Two extruder screws were used to increase the rate of the process. Both screws were Double Wave-style screws [12, 13]. The channel dimensions and operation of Double Wave screws are presented in Section 14.2.1. For plasticating processes, solids that enter the Double Wave section will be forced over the relatively high shear regions of the secondary flight and channel peaks, providing additional energy that can be used to promote melting. In general, the Double Wave section will have between 4 and 10 peaks. A more detailed analysis of Double Wave screws will be presented in Section 14.2.1. The first Double Wave screw for this study was built with a barrier melting section and was not designed for use with starch feedstocks. A summary of the channel dimensions is shown in Table 13.8. The specific rotational flow rate was calculated at 18 kg/(h·rpm) based on the distance-averaged depth of the Double Wave section, a double-flighted system, and a melt density of 1.40 g/cm^3.

The power train on the extruder was also limiting the rate of the process. The extruder was driven by a 190 kW motor with a maximum speed of 1750 rpm and a maximum current of 262 A. The motor was directly coupled to a gearbox with a ratio of 13.91, providing a maximum screw speed of 125 rpm. The maximum torque available to the screw from the drive system was 14,200 Nm. As will be shown later, this gearbox-motor combination did not provide a high enough torque to the screw for this process.

Table 13.8 Screw 1 Channel Dimensions for a 114.3 mm Diameter, 30 *L*/*D* Barrier Double Wave Screw for Running a Starch-Resin Formulation

	Depth, mm	Length, diameters	Notes
Feed section	28	7	
Barrier section		14	1.3 mm barrier flight undercut
Double Wave meter	12.7	9	Entry depth of 11.4 mm

The lead length was 114.3 mm for the feed section and the main flight of the barrier section. The lead length in the Double Wave section was 152 mm. The main flight width and clearance were 18 and 0.11 mm, respectively, in the feed and barrier sections. The primary flight width and secondary flight width for the Double Wave section were 16 and 3.2 mm, respectively. The first 2.5 diameters of the screw were inside a water-cooled smooth-bore feed casing. The compression ratio was 2.2 and the compression rate was 0.0031. The Double Wave section was designed with six peaks. The peak clearances and the valley depths were 4.6 and 13.5 mm, respectively. The distance-averaged depth was calculated at 12.7 mm. A 23 mm diameter hole was bored down the length of the screw. The screw was constructed from 4140 steel, and it was capable of a maximum torque of 15,900 Nm.

Screw 1 was capable of a maximum rate of about 200 kg/h at a screw speed of 23 rpm, for a specific rate of 8.7 kg/(h·rpm). This specific rate was considerably less than the calculated specific rotational flow rate of 18 kg/(h·rpm). Since the rate of the process was controlled by the crammer feeder in the hopper, higher rates might be possible by increasing the feed rate of the crammer. For this process, the barrel zone temperature settings over the Double Wave section were set at 90 °C and were thus removing a small amount of energy from the starch formulation. Energy removal was necessary in order to keep the water in the formulation from foaming the pellets. At these conditions, the motor was consuming 170 A, and the specific energy was calculated at 392 J/g. The extruder was operating stably and the product was in specification. When the rate was increased slightly by increasing the crammer speed, the extruder became unstable with the motor current oscillating between 180 and 200 A. The pellet quality also deteriorated as some foaming of the formulation was occurring, causing the bulk density of the pellets to decrease. Although not proven, it was believed that the entry to the barrier section was causing a restriction in the process, limiting the rate and causing the process to become unstable.

To test the hypothesis, a second Double Wave screw was tested. This screw was labeled Screw 2 and was very similar to Screw 1 except that it did not have a barrier melting section. Instead, the melting section was a conventional single-flighted section. A summary of the screw geometry is provided in Table 13.9. The specific rotational flow rate for the Double Wave section was calculated at 19 kg/(h·rpm). This screw was expected to dissipate less energy because the average depth of the Double Wave channel was slightly deeper than Screw 1. Screw 2 was also expected to be more stable because there is no restriction at the entry to the melting section.

Table 13.9 Screw 2 Channel Dimensions for a 114.3 mm Diameter, 30 L/D Double Wave Screw for Running a Starch-Resin Formulation

	Depth, mm	Length, diameters	Notes
Feed section	20.3	8	
Transition section		7	
Double Wave meter	13.7	15	Entry depth of 7.1 mm

The lead length was 114.3 mm for the feed section and the main flight of the barrier section. The lead length in the Double Wave section was 152 mm. The main flight width and clearance were 18 and 0.11 mm, respectively, in the feed and barrier sections. The primary flight width and secondary flight width for the Double Wave section were 16 and 3.2 mm, respectively. The first 2.5 diameters of the screw were inside a water-cooled smooth-bore feed casing. The compression ratio was 1.5 and the compression rate was 0.0049. The Double Wave section was designed with 10 peaks. The peak clearances and the valley depths were 5.1 and 14.5 mm, respectively. The distance-averaged depth was calculated at 13.7 mm. A 23 mm diameter hole was bored down the length of the screw. The screw was constructed from 4140 steel, and it was capable of a maximum torque of 32,000 Nm.

The extruder with Screw 2 operated stably at a rate of 200 kg/h and a screw speed of 22 rpm, for a specific rate of 9.1 kg/(h·rpm). Like before, the crammer feeder was the rate-controlling step, causing the upstream channels in the screw to operate partially filled. The motor was consuming 150 A and was inputting 341 J/g to the formulation. The barrel set point temperatures were essentially the same as those used for Screw 1. The pellets produced were prime product. The rate of the extruder was slowly increased by increasing both the crammer speed and the screw speed. At a screw speed of 27 rpm the extruder was capable of stably producing a prime product at a rate of 230 kg/h. The extruder was operating with a motor current of 170 A and inputting 415 J/g to the formulation. At rates higher than this the starch formulation exiting the extruder was fully gelatinized but was exiting at too high of a temperature, causing the water to foam the product and thus reducing the bulk density of the pellets.

Since energy dissipation is strongly affected by the specific rate of the process, the extruder was operated at specific rates higher than 9.1 kg/(h·rpm). For this test, the rate of the extruder was increased to 260 kg/h at a screw speed of 22 rpm, for a specific rate of 11.8 kg/(h·rpm). The motor was consuming 215 A and inputting to the formulation a specific energy of 390 J/g. The process was operating stably and the pellets were prime product. From these experiments, it was clear that about 400 J/g was required to fully gelatinize the starch while not causing the water in the pellets to foam. The additional energy added by the motor would need to be removed in the last barrel temperature control zones. At these conditions, the motor was operating at 82% of its full motor load. Higher rates will require that the process be operated at a higher specific rate, requiring additional torque to the screw.

In order to provide a higher torque to the screw, a belt sheave system was installed between the motor and the gearbox. This sheave system had a reduction ratio of 1.95, reducing the maximum screw speed from 125 to 65 rpm. Sheaving the system allowed the maximum torque to the screw to be increased from 14,200 to 27,400 Nm. The manufacturer of the gearbox was contacted prior to the change to guarantee that the gearbox could accept the higher torque. Screw 2 can deliver a maximum torque of 32,000 Nm, providing a safety factor of about 1.2. Although this safety factor is very low for any application, the screw was still used for the process with the knowledge that a screw break due to high torque was possible.

With Screw 2 and the belt sheave system, the extruder was evaluated for maximum rate. The rate was slowly increased to a maximum rate of 360 kg/h at a screw speed of 26 rpm for a specific rate of 13.8 kg/(h·rpm). The extruder was operating stably and the pellets were in specification. The motor was operating with a current of 190 A for a specific energy input of 530 J/g. This energy input level corresponds to a torque on the screw of about 20,000 Nm, a torque level that could not be obtained without the installation of the belt sheave system. Although the energy input from the screw was considerably higher than the required 400 J/g, the excess energy was easily removed through the barrel wall in the last three zones of the extruder. A combination of screw optimization and modifications to the drive train allowed a rate increase from 200 to 360 kg/h, a rate increase of 80%.

13.4.6 Vent Flow for a Two-Stage Screw Running a Low Bulk Density PS Feedstock

During the production of a PS sheet product, the edge trim and off-specification sheet had to be pelletized before they could be added back into the process. The bulk density of this material was too low to economically add back into the main extruder. Instead, the feedstock was added into a reclaim pelletizing extruder so that the material would have essentially the same bulk density as the original pellet feedstock. During reclaim processing, the extruder had a very high rate of material exiting through the vent port of the extruder. Plant personnel had attempted to fix it in the past but had accepted the flow as normal for the process. The material exiting the vent was difficult to handle and was reducing the profitability of the plant.

The two-stage, vented extruder was 203.2 mm in diameter and had an L/D ratio of 31. At a screw speed of 37 rpm, the extruder was pelletizing at a rate of 410 kg/h, and the flow rate out through the vent opening was 51 kg/h. The extrudate was discharged from the extruder at a pressure of 4 MPa and an estimated temperature of 220 °C. The vent opening was at the 12 o'clock position. Numerous barrel temperature profiles were tested to determine if the venting rate could be reduced or eliminated. These condition changes, however, did not mitigate the vent flow. Two possible root causes were identified for the problem: (1) the first- and second-stage metering channel geometries were not balanced properly, causing pressure to occur in the channel underneath the vent opening, and (2) the vent diverter was not working properly. A description of the screw used in the extruder is provided in Table 13.10. The specific rotational flow rates for the first and second stages were 14.1 and 22.8 kg/(h·rpm) for a pump ratio of 1.62. The PS resin used had an MFR of 2 dg/min (200 °C, 5 kg).

Table 13.10 Screw Channel Dimensions for a Two-Stage, Vented Reclaim Extruder with a Diameter of 203.2 mm Running PS Resin

	Depth, mm	Length, diameters	Notes
Feed section	34	7	The compression ratio was 7 and the compression rate was 0.0039
First-stage transition		11	
First-stage meter	4.8	3	
Blister		0.3	Clearance was 1.4 mm
Vent section	21.6	3	
Second-stage transition		2	Pump ratio was 1.65
Second-stage meter	7.9	5	
Lead length, flight width, and flight clearance were 203.2, 20, and 0.20 mm, respectively, in all sections of the screw. The vent opening was 203.2 mm in diameter and positioned 23 diameters from the start of the screw. The extruder had a smooth-bore feed section.			

To test the hypothesis that the screw channels were not designed properly such that a pressure exists in the channels underneath the vent opening, the process was numerically simulated. The simulation of the first stage was performed at a rate of 461 kg/h while the second stage was at 410 kg/h. The estimated axial pressures for the process are shown in Figure 13.13.

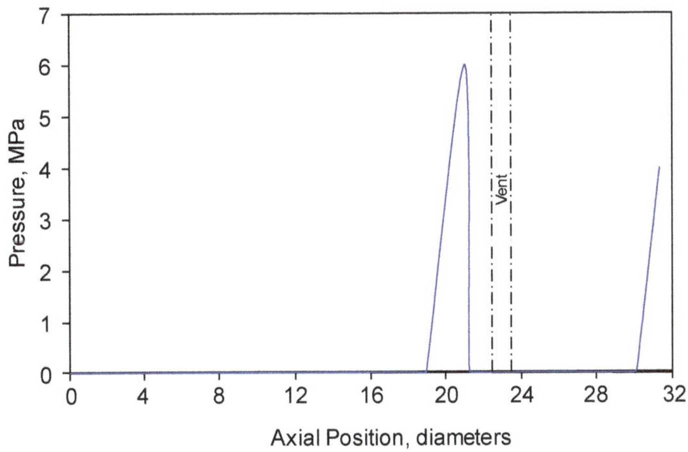

Figure 13.13 Simulated axial pressure profile for the 203.2 mm diameter reclaim extruder discharging PS resin at a rate of 410 kg/h at a screw speed of 37 rpm

As shown in Figure 13.13, the pressure underneath the vent opening was zero, indicating that the root cause of the vent flow was not caused by pressure. Only about 1 diameter of the second-stage metering section was completely filled with resin to generate the pressure for the pelletizing die. In the first-stage metering section, the fill position was about 2 diameters upstream of the blister section. The pressure generated by these 2 diameters was balanced by the pressure change required to flow resin over the blister. The pressure simulation, however, did indicate that the extruder and screw design are highly rate limited by the bulk density and the solids-conveying process. This conclusion is based on the relatively low pressures upstream of the blister section. If solids-conveying rates could be improved, the pressures in the first stage would increase and the fill position would be moved upstream towards the hopper. This type of process and pressure profile is typical for extruders that process low bulk density feedstocks. A crammer-type hopper could have been used to stuff the feedstock into the screw.

The axial pressure simulation suggested that the root cause of the vent flow was an improperly operating vent flow diverter. Testing this hypothesis, however, is more difficult since it requires a visual inspection of the device. The extruder was shut down and the diverter was removed from the vent port. On visual examination, the diverter was in poor condition with numerous indentations from hammer hits, and it was obvious that the diverter was modified from the original design. It appeared that metal was removed from the region in contact with the polymer and where the diverter flange connected to the vent port flange. The modifications were such that if the polymer expanded at the upstream side of the flow path the diverter may not be able to tuck the material back into the barrel cylinder. Based on this analysis, it was hypothesized that the altered vent flow diverter was the root cause of the flow problem. A new diverter was designed and built.

The new diverter was built according to the description presented in Section 13.1, and it was installed in the extruder. Immediately upon startup of the machine, resin flow from the vent did not occur. Thus, resin that expanded outside of the barrel cylinder at the entry of the vent opening was tucked back into the barrel cylinder. Based on the cost savings for the elimination of the vent flow and the installed cost of the diverter, the payback time was estimated at 18 days.

13.4.7 Increasing the Rate of a Large Part Blow-Molding Process

Blow-molding processes [14,] consists of five main operations: plastication of the resin, formation of the parison, inflation of the parison, solidification of the part, and removal of the part from the tooling. The best process economics will occur with a part optimized for weight and a minimum cycle time. In order to have a minimum cycle time, the cooling operation must be the rate-limiting step. For the case study here, a large blow-molded part required a very long cycle time to produce. The rate-controlling step of this process was the plastication of the resin. That is, after the part was ejected, the process had to wait for enough molten resin to accumulate before the next parison was formed. When the screw speed was increased to increase the rate, the extrudate temperature become too high and the parison could not be consistently formed. The extruder and screw design used were not capable of plasticating the resin at a high enough rate and at the proper temperature to allow shorter cycle times.

The parts produced were made using a PC/ABS resin blend. The extruder was capable of plasticating enough resin for a 13 kg parison at a screw speed of 12 rpm in 300 s. Thus, the extruder was operating at a specific rate of 13.0 kg/(h·rpm). If the line was limited by the cooling of the part in the mold, the cycle time could be decreased to less than 100 s. Instead, the line had to wait an additional 200 s for the extruder to produce enough material for the next parison. At these conditions, the discharge pressure was about 5.3 MPa. After a steady-state process was achieved, a parison was ejected and dropped onto the floor. The temperature of the discharge was measured with a handheld thermocouple probe at 235 °C. The process was capable of producing acceptable parts at these conditions with a cycle time of 300 s. Barrel temperatures were studied, and the highest specific rate for the process was obtained at barrel temperatures of 200, 210, 210, and 210 °C, respectively, for the feed zone through the last barrel zone. At higher screw speeds, the cycle time could be decreased but the parison temperature was too high to produce consistent high-quality parts. At the high temperatures, the viscosity of the resin was too low and the weight of the parison caused it to neck in near the die and cause dimensional problems in the finished part.

The screw used for this process was reported to be designed for a PP resin. The screw was 152.4 mm in diameter, was nominally 25 *L/D* in length, had a smooth-bore feed section, and was constructed with two gear mixing sections. The gear mixer style was essentially identical to that shown in Figure 8.27. The channel depths and section lengths are shown in Table 13.11. The calculated specific rotational flow rate for this screw for the PC/ABS resin was 13.2 kg/(h·rpm). The melt density of the PC/ABS resin at 235 °C was measured at 1.04 g/cm^3. The screw had a very short transition section and a very high compression rate of 0.0092. Moreover, the length of the gear mixing section was excessive, and the lead length of the screw was too low.

Table 13.11 Screw Channel Dimensions for a 152.4 mm Diameter Extruder for a Large-Part Blow-Molding Process Running a PC/ABS Blended Resin

	Depth, mm	Length, diameters	Notes
Feed section	25	5	The compression ratio was 2.8 and the compression rate was 0.0092
Transition		3	
First meter	9.1	8.5	
First gear mixer	9.1	1	
Second meter	9.1	3.8	
Second gear mixer	11.8	3.2	
Lead length, flight width, and flight clearance were 131, 15, and 0.15 mm, respectively, in all sections of the screw. The specific rotational flow rate for the screw running PC/ABS resin was calculated at 13.2 kg/(h·rpm). The extruder had a smooth-bore feed section.			

Plant data suggested that the root cause for the high discharge temperatures at high screw speeds was the energy dissipation rate of the screw. To verify this root cause, the metering sections and mixing sections of the screw were numerically simulated at a screw speed of 12 rpm and a rate of 156 kg/h, that is, at a plasticating rate that corresponds to a 13.0 kg parison and a 300 s plasticating time. The simulated axial pressure and temperature profiles are shown in Figure 13.14. The portion of the pressure profile where the gradient is positive corresponds to the flighted sections while the negative pressure gradient regions correspond to the gear mixer sections. The pressure and temperature at the entry to the first metering section were determined to be 7 MPa and 210 °C, respectively. The discharge temperature was simulated at 234 °C, a value near the experimentally measured discharge temperature of 235 °C. At higher simulated screw speeds, the discharge temperature increased to 240 °C and higher. The simulations were consistent with the experimental observations, and they also indicated that the screw was dissipating too much energy.

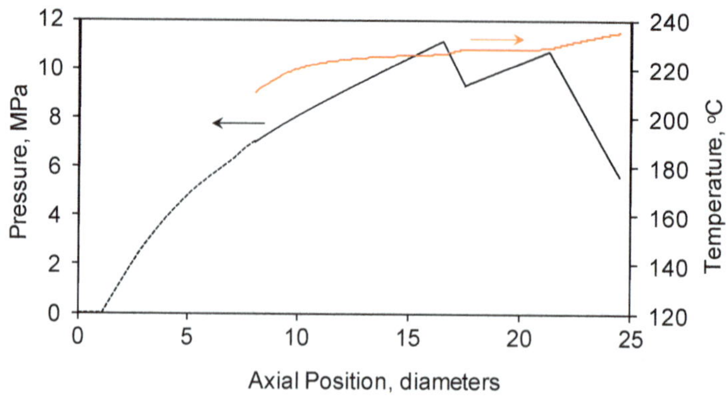

Figure 13.14 Simulated axial pressure and temperature profile for the original blow-molding screw at a rate of 156 kg/h and a screw speed of 12 rpm, for a specific rate of 13.0 kg/(h·rpm)

In order to increase the rate while not increasing the discharge temperature, the specific rate of the process must be increased. Increasing the specific rate, however, may cause solid polymer fragments to discharge into the parison. A high-performance screw design such as an Energy Transfer (ET) screw or a Double Wave screw is ideal for this application since they have the ability to operate at high specific rates while trapping and dispersing solid polymer fragments. ET and Double Wave screw designs are discussed in Section 14.2. For this application, an ET screw was designed and fabricated. The screw was designed with an ET section with a distance-averaged channel depth of 10.4 mm, a channel depth that was deeper than the original screw. Moreover, the lead length was set at 184 mm. The combination of increased channel depth and lead length resulted in a calculated specific rotational flow rate of 17.9 kg/(h·rpm). This specific rate is about 35% higher than that of the original screw, and thus it will provide an extrudate that is lower in temperature at the same rate. The channel peaks and undercuts in the ET flights were set at 4.1 mm to trap and disperse solid polymer fragments. A gear mixer was placed on the tip of the screw to mitigate thermal gradients. The channel dimensions of the ET screw are provided in Table 13.12. The target rate of the ET screw was 410 kg/h at a screw speed of 24 rpm and a discharge temperature less than 235 °C. At an extrusion rate of 410 kg/h the cycle time of the process would be decreased to about 115 s.

Table 13.12 ET Screw Channel Dimensions for a 152.4 mm Diameter Extruder for a Large-Part Blow-Molding Process Running a PC/ABS Blended Resin

	Depth, mm	Length, diameters	Notes
Feed section	26.7	6	The compression ratio was 2.8 and the compression rate was 0.0057
Transition		6.7	
ET section	10.4*	9.9	
Flighted meter	10.4	0.9	Double flighted
Gear mixer	10.4	1	
Lead length, flight width, and flight clearance were 184, 15, and 0.15 mm, respectively, in all sections of the screw. The specific rotational flow rate for the screw running PC/ABS resin was calculated at 17.9 kg/(h·rpm).			

*The distance-averaged depth of the ET section was calculated at 10.4 mm. The deepest and shallowest portions of the ET channel were 15.2 and 4.1 mm, respectively.

The ET section and downstream channel sections were numerically simulated to determine if the screw could meet the target performance. The simulation was performed at a rate of 410 kg/h using the same barrel temperatures for the original screw and process and at an entry pressure and temperature of 7 MPa and 210 °C, respectively. The axial pressure and temperature profiles are shown in Figure 13.15. The simulation indicates that 410 kg/h can be obtained at a screw speed of 24 rpm for a specific rate of 17.1 kg/(h·rpm) with a discharge temperature of 234 °C. Based on this simulation, the ET screw was fabricated.

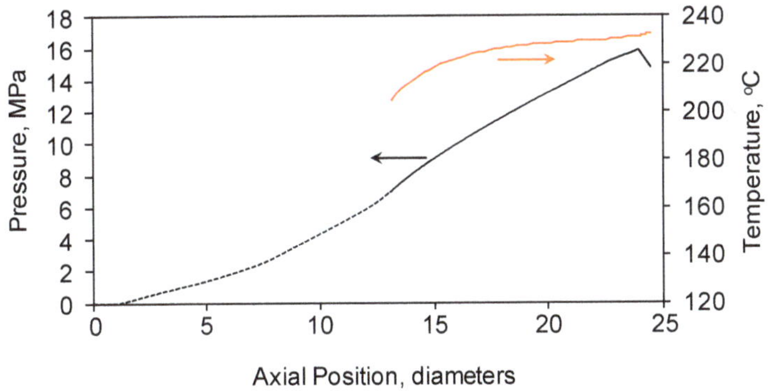

Figure 13.15 Simulated axial pressure and temperature profile for the high-rate ET screw at a rate of 156 kg/h and a screw speed of 24 rpm

The new ET screw was installed and tested. At essentially identical barrel temperature settings presented previously, the extruder was operating at 410 kg/h at a screw speed of 24 rpm for a specific rate of 17.0 kg/(h·rpm). At these conditions, the extrudate temperature was measured at 239 °C and the cycle time was 115 s. This temperature was slightly higher than the target temperature of 235 °C, but at this higher operating temperature the parison was acceptable, and essentially all parts produced were in specification. Increasing the metering channel depth and the lead length while utilizing a high-performance screw design allowed the specific rate to increase from 13.0 to 17.0 kg/(h·rpm) and reduced the cycle time from 300 to 115 s for a productivity improvement increase of 2.6 times.

Nomenclature

f	coefficient of dynamic friction
F	energy flux
F_c	rotational flow correction factor
F_d	shape factor for rotational flow
F_p	shape factor for pressure flow
H	local distance from the screw root to the barrel
p	number of flight starts
P	pressure
Q_m	mass flow rate
V	relative velocity between the moving solids and the barrel wall
V_{bz}	z component of velocity of the screw flight at the barrel wall
W	average width of the channel
z	independent variable in the down-channel direction (or helical direction)
η	shear viscosity of the polymer at the average channel temperature and average shear rate
ρ_m	melt density of the fluid

References

1. Chung, C.I., "On the Scale-Up of Plasticating Extruder Screws," *Polym. Eng. Sci.*, **24**, 626 (1984)
2. Mount, E.M., III, "A Review of Melting Mechanisms in Single-Screw Extrusion," *SPE ANTEC Tech. Papers*, **51**, 304 (2005)
3. Mount, E.M., III, "Evaluation of the Melting Capacity of Three Single Screws," *SPE ANTEC Tech. Papers*, **53**, 400 (2007)

4. Pham, H.T. and Hyun, K.S., "Gear Pump Assisted Extrusion," *Polym. Eng. Sci.*, **32**, 488 (1992)

5. Jepson, C.H., "Future Extrusion Studies," *Ind. Eng. Chem.*, **45**, 992 (1953)

6. Klein, I., "Predicting the Effect of Screw Wear on the Performance of Plasticating Extruders," *Polym. Eng. Sci.*, **15**, 444 (1975)

7. Potente, H., Effen, N., and Liu, J., "Throughput Calculation with Considerations of Leakage Flow on Injection Molding Plasticizing Units," *Polym. Eng. Sci.*, **36**, 1557 (1996)

8. Frankland, J., "Wedging Can Cause Serious Screw Wear," *Plast. Technol.*, **57** (2), 21 (2012)

9. Hyun, K.S. and Spalding, M.A., "Troubleshooting TPU Resin Extrusion Rate Limitations Due to Solids in the Discharge," *SPE ANTEC Tech. Papers*, **58**, (2012)

10. McCullough, T.W. and Spalding, M.A., "Predicting Actual Temperature Distributions in a Polymer Stream Using an Adjustable-Depth, Exposed-Tip Thermocouple Assembly," *SPE ANTEC Tech. Papers*, **42**, 412 (1996)

11. Altinkaynak, A., Gupta, M., Spalding, M.A., and Crabtree, S.L., "An Experimental Study on Shear Stress Characteristics of Polymers in Plasticating Single-Screw Extruders," *Polym. Eng. Sci.*, **49**, 471 (2009)

12. Kruder, G.A., "Extrusion Apparatus and Method," U.S. Patent 4,173,417 (1979)

13. Fan, P., Vlachopoulos, J., Smith, N., and Sheth, H., "Computer Simulation of Melt Flow in Wave Screws," *SPE ANTEC Tech. Papers*, **44**, 97 (1998)

14. Usman, M. and Siddiqui, A.S., "Blow Molding of Polyethylene," Chapter 16 of "Handbook of Industrial Polyethylene and Technology", edited by Spalding, M.A. and Chatterjee, A.M., Wiley - Scrivener Publishing, Hoboken, NJ, 2017

14 Barrier and High-Performance Screws

Increasing the rate of an extrusion line is a very common method of increasing the capacity of a plant and decreasing the costs to produce a product. A plant will typically operate the line at a rate where either the extruder or the downstream equipment are rate limiting. Thus, plant personnel will increase the screw speed until the rate limitation occurs. Several limitations can occur with the extruder, including: the screw speed is at its maximum value, the motor is operating at maximum torque (motor current), the discharge temperature becomes too high, or solid polymer fragments are observed in the discharge. Limitations with the motor and gearbox were discussed in Section 10.6. High discharge temperatures and solid polymer fragments in the discharge are common for extruders equipped with conventional single-flighted screws. Early screw designers would mitigate the solid fragments by positioning dispersive mixers in the metering section, trapping and melting the solids before they are discharged. Often the designer would increase the depth of the metering section in order to increase the specific rate and decrease the extrudate temperature. This approach worked well to control the extrudate temperature at low to moderate screw speeds. But when the screw was rotated at high speeds, the extruder was limited by the melting capacity, discharging solid polymer fragments in the extrudate. As expected the melting capacity of the screw was unchanged since the length and functionality of the screw was unchanged.

The first approach to mitigating the breakup of the solids bed and to increasing the melting capacity of a screw was the invention of the barrier melting section. A schematic of a screw with a barrier melting section is shown in Figure 14.1. This device maintained the solid bed intact and in the solids channel, allowing higher rates and lower discharge temperatures. Maillefer patented a barrier-type screw design in Europe in 1962 [1] while Lacher [2] was the first in the United States in 1966. This technology was the first of its kind to use the concept of separating the solid bed from the melt pool as it progressed down the transition section of the screw. Lacher's patent used the concept of an advancing lead for the barrier flight. That is, the barrier flight started at the pushing side of the channel and advanced to the trailing side at the end of the transition section. The barrier flight was undercut from the main flight to allow the flow of molten resin from the melt film into the melt channel. Numerous barrier screw patents followed Lacher's patent. These included barrier screw variations that were patented by Geyer [3], Dray and Lawrence [4], Barr [5], Hsu [6], Kim [7], and others. Although Geyer's barrier design patent was issued in 1968, it had a priority date of April 5, 1961, and thus was before Lacher (priority date of February 26, 1962). The design variations include the location of the barrier flight, the depth of the solids and melt channels, and the entry and exit geometries of the section.

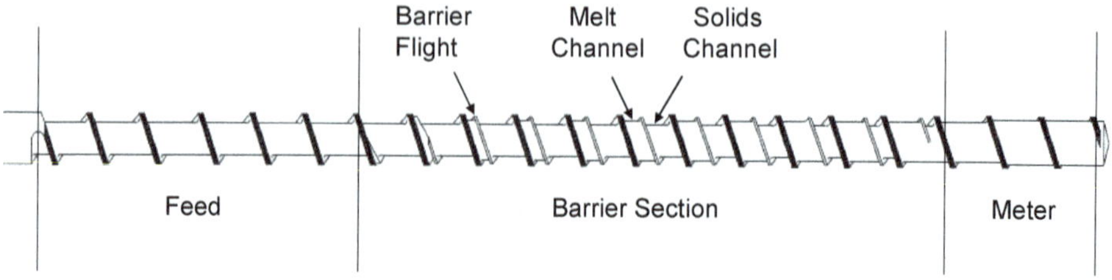

Figure 14.1 Schematic for a barrier melting section (courtesy of Jeff A. Myers of Robert Barr, Inc.). The schematic is for a Barr-III barrier screw

In the 1970s and early 1980s, a new class of screws was developed and implemented that allowed extruders to operate at higher rates before the extrudate temperature became excessive and before solid polymer fragments were discharged with the extrudate. These screws are known in the industry as high-performance screws. For most types, styles, and brands, the metering section of a conventional screw is replaced with the high-performance section, as shown by Figure 14.2 for an Energy Transfer™ (ET, Trademark of Robert Barr, Inc.) screw [8]. High-performance sections all work by forcing the material to flow repetitively through regions with relatively small clearances. As the material passes through these clearances, large solid polymer fragments will become trapped, partially melted, and broken into smaller solid polymer fragments. Small solid fragments that enter the device will be subjected to high elongational and shear stress fields that will aid in their melting. The lands for these small clearances tend to be relatively short in the flow direction while the rest of the channels are relatively deep. The deeper main channels allow the device to maintain high specific rates and low discharge temperatures, while the regions with the small clearances provide a level of dispersive mixing for completing the melting process.

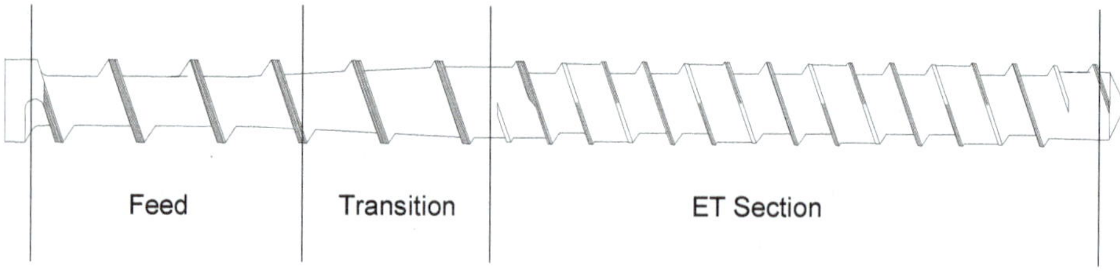

Figure 14.2 Schematic for an Energy Transfer (ET) screw (courtesy of Jeff A. Myers of Robert Barr, Inc.). The standard and conventional metering section was replaced by a high-performance ET section

The high-performance screws presented here and by others all have the objective of completing the melting process as early as possible in the device. As shown previously, complete melting is the first key to successful mixing in single-screw plasticating processes. High-performance screw sections are typically designed with multiple regions where the material must pass briefly through small channels, trapping and melting any solids present in the flow. Since the residence times in these small channels are very short, viscous dissipative heating of the resin is minimal, and thus temperature increases due to the devices are in general not large. But since the main channels for these devices are typically deeper than in a conventional metering section, the specific rates are higher and the discharge temperatures are typically lower than those for conventional screw designs. The data presented here were based on single-stage screws. For two-stage, vented screws, the high-performance section is positioned in the first-stage meter and the second stage is of a conventional design. Placing the high-performance section in the first stage ensures that the material entering the vent region is completely molten, allowing for maximum volatile removal.

This chapter will present and discuss the operations and designs for several barrier and high-performance screws that are commercially available. Numerous small companies that manufacture screws have recognized the performance of these screws and have manufactured look-a-like screws based on the technologies presented here and the descriptions in expired patents. These manufacturers will not be discussed here.

14.1 Barrier Screws

Barrier screws have been used commercially since they were first developed in the 1960s. Since then, the design of the section has varied in many ways, all with the goal of increasing the melting capacity of the screw and decreasing the extrudate temperature. As stated many times, if the melting capacity of the screw can be increased, the screw can be operated at higher screw speeds and thus higher rates. Typically, a well-designed barrier screw can provide an increase in the rate of about 20 to 30% over a well-designed conventional screw. The design features of a barrier section include the location or position downstream where the barrier section starts, the entry geometry, the lead length of the main flight and the barrier flight, the barrier flight undercut from the main flight, the length of the section, widths and depths of the channels, and the geometry of the exit of the section. For most designs, the barrier flight undercut is constant for the entire length of the section. As a general rule, the undercut is typically about 0.01 times the diameter of screw. Undercuts that are smaller than this rule, however, are often used. These design features can be difficult to specify. This section will describe the common commercially available devices and their operation. The melting mechanism for barrier sections is rather complex and the in-depth description was presented previously in Section 6.3.2.

The melting capacity for a barrier section is higher than that of a conventional screw due to two main reasons. First, the solids channel geometry is set such that the area available for melting in the melt film between the solid bed and the barrel wall (zone C) is maximized and adjusted for the fraction of the resin melted. Second, the barrier flight position is used to maintain the integrity of the solid bed. That is, it mitigates the breakup of the solid bed. During breakup, solid bed fragments can move downstream and get exposed to a lower level of dissipative melting and will possibly discharge with the extrudate. The location of the barrier flight and the tapered depth of the solids channel in the downstream direction mitigate the breakup of the bed and maximize the melting capacity of the screw.

A cross-sectional view of a typical barrier section is shown in Figure 14.3. This view is for channels that have been unwound from the screw and laid flat, and the flow direction is out of the plane of the page and towards the reader. For this design, the solid bed is forced to flow down the solids channel where the channel depth becomes shallower and the channel width becomes narrower with the downstream position. As the bulk of the material is melted in the film located between the solid bed and barrel wall, the motion of the screw forces the newly molten resin to flow across the barrier flight and into the melt-conveying channel. At the location of this cross section, the melt channel is deeper than the solids channel. The width of the melt-conveying channel increases as the barrier flight moves closer to the pushing side of the main channel, allowing for the addition of more molten material. The barrier flight is undercut from the primary flight to allow the flow of resin from the solids-conveying channel to the melt-conveying channel. The barrier flight is typically undercut from the primary flight by 0.3 to 1.5 mm depending on the diameter of the screw and the application. The actual clearance for flow is the total of the barrier flight undercut and the clearance of the primary flight. Barrier undercuts near 0.3 mm tend to be effective at mitigating small unmixed gels in PE resin streams because they have the ability to apply a higher level of shear stress to the flow over the barrier. The shear stress level can be estimated using Equation 14.1.

$$\tau = \frac{\pi N \eta \left(D_b - 2\lambda - 2u\right)}{(\lambda + u)} \tag{14.1}$$

where τ is the shear stress, N is the screw rotation speed in revolutions/s, D is the diameter, η is the shear viscosity, λ is the primary flight clearance, and u is the barrier flight undercut from the primary flight. Decreasing the barrier undercut clearance, however, will cause the discharge temperature to increase. Barrier clearances that are extremely small can cause a melt pool to collect on the pushing side of the barrier flight, leading to solid bed breakup in the solids-conveying channel [10]. If unmixed gels and solid polymer fragments are not a concern for the process and application, undercut clearances between 1.0 and 1.5 mm can be used. As expected, barrier flight clearances at this high level will cause the extrudate temperature and shear stress over the flight to be less than for a screw with a smaller undercut.

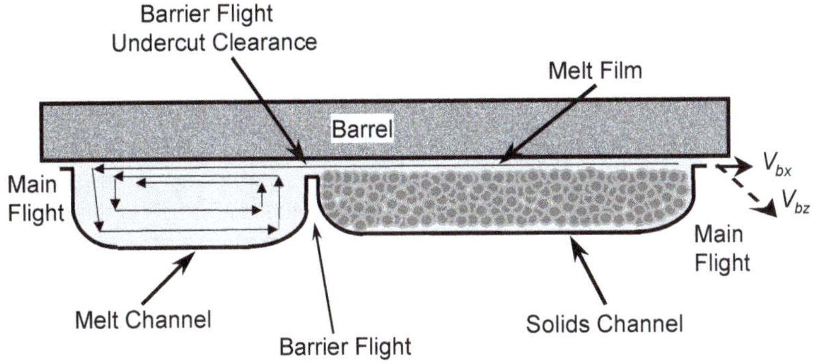

Figure 14.3 Cross-sectional schematic of a barrier melting section in the unwound channel configuration

The design of the entire barrier screw is equally complex. For example, a 114.3 mm diameter screw with a barrier melting section is shown in Figure 14.4. For this example, the lead length in the feed and meter sections is 114 mm, and thus the compression ratio is simply the feed channel depth divided by the meter channel depth; here the compression ratio is 2.8. At the end of the feed section, the barrier flight starts near the pushing side of the channel, and the barrier flight ends near the trailing side of the start of the metering section. The introduction of the barrier flight can restrict the flow of the solid bed by forcing the width of the solids channel to decrease suddenly. Many designers have increased the lead length in the barrier section to 1.2 to 1.5 times the lead length in the feed section to mitigate this restriction. For the design shown in Figure 14.4, the lead length of the primary flight in the barrier section was increased to 172 mm. The cross-sectional area perpendicular to the flight edge decreases with the downstream direction in the solids-conveying channel by decreasing the channel depth and decreasing the channel width. The melt channel cross-sectional area increases by increasing the width of the channel. For this screw, the melt channel depth was constant. A different barrier section design could have kept the widths of the solids- and melt-conveying channels the same in the downstream direction but with the solids channel depth decreasing and the melt channel depth increasing. The barrier undercut was 1.2 mm beyond the primary flight tips.

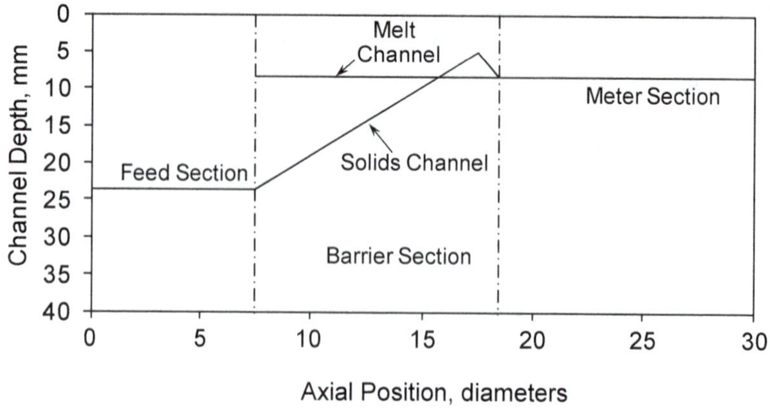

Figure 14.4 Channel depths for a 114.3 mm diameter screw with a barrier melting section. The lead length in the feed and meter sections is 114 mm; the lead length of the main flight in the barrier section is 172 mm. The barrier flight starts at the pushing side of the channel and ends near the trailing side. The barrier flight clearance is 1.2 mm

The design of the entry and exit geometries of the section is extremely important to maintaining rate, proper operation of the metering section, and not creating stagnant regions that can cause resin degradation. For example, if the feed section is too long and the addition of the barrier flight at the entry to the section is restrictive, then the restriction has the ability to control the specific rate of the process rather than the metering section. The restriction occurs because the upstream solid bed is highly compacted and is unable to rearrange and convey at a high rate around the entry where the downstream solids channel is narrower than the upstream channel. If the restriction is severe enough, the upstream channel of the meter section will operate partially filled and at zero pressure. During startup of a new screw, the process might operate at a specific rate that is about 50% of the expected specific rate calculated based on the metering channel. Soon after startup, the extrudate will start to show evidence of resin degradation due to the partially filled channels. The amount of time required for the first degradation products to appear will depend on the resin. For example, LLDPE resins may show crosslinked gels in less than 30 minutes while LDPE resins could take more than 12 days. Several examples that show this restriction problem and its elimination were presented in Chapter 11.

The evolution of the barrier section entry region is clearly shown by comparing the Barr-II and Barr-III screws in Figure 14.5. These screws are designed with a primary flight that has a lead length that is constant for the entire screw. The barrier flight is positioned such that the lead length is identical to the primary flight lead length, creating a melt channel that has a width that is about 25 to 30% of the total channel width. The depth of the solids channel decreases in the downstream direction while the melt channel depth increases. The entry region of the Barr-II barrier section has three features: (1) the width of the melt flow channel at the entry is the same as the entire section, (2) the melt channel depth starts at a depth that is at the solids-conveying channel depth and then decreases in depth over about 1 to 2 diameters, and (3) the barrier flight is removed for the first 1 to 2 diameters. A schematic for the entry section is shown in Figure 14.5(b). The very large initial depth of the melt channel and the removal of the barrier flight are designed to mitigate the restriction at the entry to the barrier section such that the metering section controls the specific rate. If the melt channel had been designed with a very shallow depth that would be reflective of the very small level of molten polymer at this location, then the entry section would have been highly restrictive and possibly the rate-controlling feature of the screw. The entry region for the Barr-III barrier section evolved from the Barr-II as a less restricted entry, as shown in Figure 14.5(d). This entry is similar to the Barr-II except that the width of the melt channel is tapered from the pushing flight to the full width of the melt channel where the barrier flight starts. These designs permit a small level of solids to enter the melt-conveying channel.

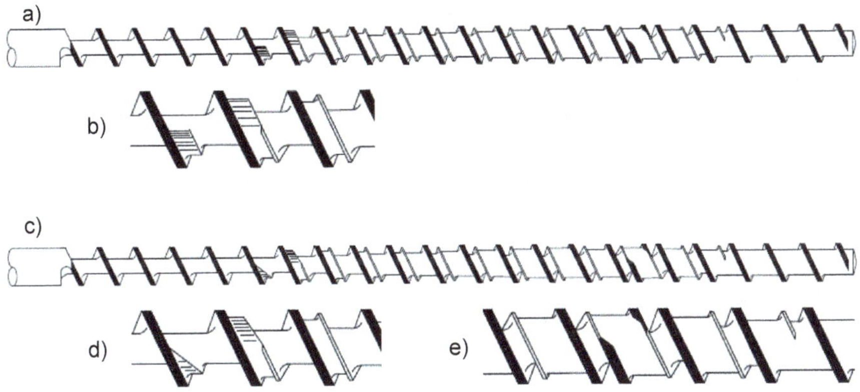

Figure 14.5 Schematics for the Barr-II and Barr-III barrier screw technologies: a) full-length schematic of a Barr-II barrier screw, b) expanded view of the barrier entry section of a Barr-II screw, c) full-length schematic of a Barr-III barrier screw, d) expanded view of the barrier entry section of a Barr-III screw, and e) expanded view of the crossover flow rearrangement section at the exit of Barr barrier screws (Schematics courtesy of Jeff Myers of Robert Barr, Inc.)

The exit region of the solids-conveying channel must be streamlined and not restrictive. For example, if a high level of solid material is trapped at the exit of a restrictive solids channel, then the local pressure at that point can become relatively high, creating degradation of the resin at the root of the screw channel. For resins such as LLDPE, HIPS, ABS, TPU, and many others, a carbonaceous deposit or coating can occur on the screw root in the region. The deposits will eventually break off, creating black specks in the extrudate. Moreover, if the exit channel geometry is not streamlined, then a region can exist where the resin can have a very long residence time, leading to the degradation of the resin. Often the best way to mitigate these defects is to relieve the restriction by removing metal from the screw at the exit channel, allowing a small amount of solid polymer fragments to exit and flow downstream. Maddock-style mixers are often positioned downstream of barrier sections to trap and disperse these solid polymer fragments. A well-designed barrier screw will operate at higher rates with a lower discharge temperature while not degrading the resin. Steward [11] points out that a poorly designed barrier screw can have a performance that is considerably less than that of a properly designed conventional screw.

The exit region of a barrier section can have a transition or melt-rearrangement section, as shown in Figure 14.5(e). For this case, the primary and barrier flights are switched such that the upstream barrier flight now becomes the downstream primary flight. This causes the melt channel to be between the barrier flight and the trailing flight downstream from the interchange. This technique can mitigate a stagnation region at the end of the solids channel and may trap and melt some solid polymer fragments.

A series of high-performance barrier screws are marketed by Davis Standard under the label DSB screws, that is, Davis Standard Barrier screws. The original barrier screw design was patented by Wheeler in 1982 [12], has the label of DSB-I, and is widely used in polyethylene blown film processes. The screw is typically designed with a square pitch in the feed and metering sections, and the main flight of the barrier section will have a lead length that is about 1.5 times the pitch of the feed section. Like other barrier screw designs, the increase in lead length mitigates the restriction at the entry of the section where the barrier flight is introduced. The barrier flight starts at the pushing flight and is positioned such that the width of the melt channel increases in the downstream direction. The width increases relatively quickly at the start of the section and more slowly towards the end. The depth of the melt channel increases in the downstream direction while the solids channel decreases. The DSB-II and DSB-III screws are proprietary designs and incorporate design modifications to the DSB-I screw.

The DSB-IV and DSB-V screws [13, 14] were designed to improve the melting and mixing processes in the barrier section of the screw. For both screws, the barrier flight starts at the pushing side of the channel much like that of a DSB-I screw. Several turns downstream from the barrier section entry a third flight is introduced in the solids-conveying channel. In the latter stages of the section, bumps are positioned in the solids channels such that the solid bed is repeatedly compressed and then relaxed. The deformation to the bed is reported to improve the melting and mixing processes and thus improve the quality of the extrudate [15, 16]. The DSB-VI screw [17] was developed for grooved barrel extruders and was designed for high melting rates. The screw is similar to the DSB-I design except that the feed section is double flighted and the barrier section has two DSB-I helical channels that operate independently. For this design, two barrier flights are introduced at the entry of the barrier section at the pushing sides of the two primary flights. Like the DSB-I, the barrier flights are undercut from the primary flights to allow molten resin to flow into the melt-conveying channel.

14.2 Wave Dispersion Screws

The first wave-dispersion-type screw was developed and patented by Kruder in 1975 [18], and the device was trademarked as the "Wave" screw. Numerous other wave dispersion screws were developed later based on Kruder's design. The term wave dispersion screw refers to screws with metering sections that have two or more channels with a flight between them that is selectively undercut to allow the dispersion of solid polymer fragments and molten resin. Several commercially available screws utilize this type of technology and are discussed in this section. These screws include Double Wave screws, Energy Transfer screws, Variable Barrier Energy Transfer screws, DM2 screws, and Fusion screws.

14.2.1 Double Wave Screw

The first wave dispersion screw was developed by Kruder in the early 1970s and was called the Double Wave screw [19, 20]. Double Wave screws are constructed by replacing the metering section of a conventional single-flighted screw with a Double Wave section. The Double Wave section consists of two channels, labeled A and B, that oscillate in depth as shown by Figure 14.6 and Figure 14.7. The Double Wave screw design is very complex and can be built with four to 14 channel peaks and valleys. The main flight in the section is built with a tight barrel clearance (0.001 times the diameter) to minimize the flow of resin over the flight and to maintain heat transfer between the resin and the barrel wall. The lead length of the feed and transition sections are typically set to the diameter of the screw; that is, they are square pitched. The main flight and secondary flight of the Double Wave section are typically designed with longer lead lengths, nearly 1.4 times the diameter. The secondary flight is undercut the entire length of the section. This undercut provides the dispersive mixing that traps and melts solid polymer fragments. That is, as material flows down a channel towards a peak, some of the material that cannot flow through the peak opening is forced to flow across the secondary flight. The material that flows across the secondary flight is collected in the opposite channel. When the opposite channel depth starts to decrease as flow approaches the next peak, the material is again subjected to a dispersive mixing process before it collects in the original channel. This dispersive mixing process typical occurs between 4 and 14 times before the material is discharged. Triple Wave screws with three alternating channels also exist.

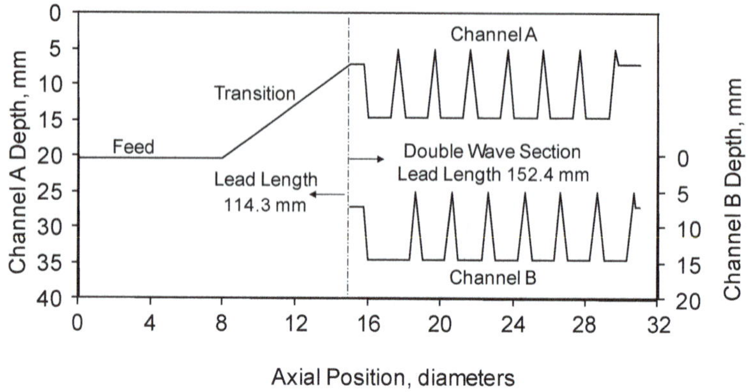

Figure 14.6 Typical channel depths for a 114.3 mm diameter Double Wave screw. The channel depths in the peaks of the Double Wave section and the undercut on the secondary flight were 5.0 mm

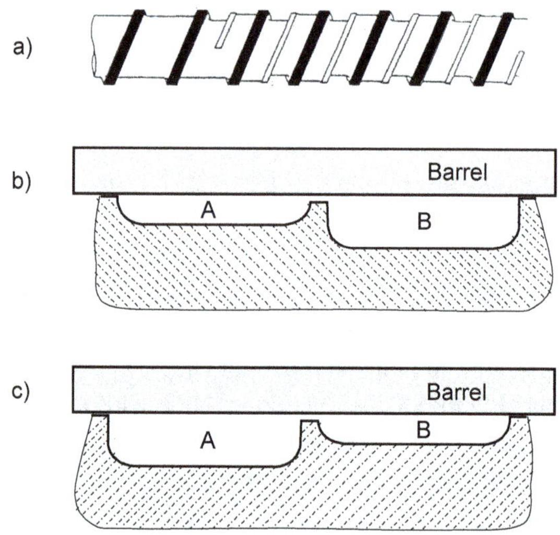

Figure 14.7 Schematics of a Double Wave screw channel: a) schematic of a Double Wave section. The unfilled flight is the secondary flight, and it is undercut with respect to the filled primary flight. b) and c) are cross-sectional views perpendicular to the flight. Section c) is about 1 diameter downstream from section b). The flow direction is out from the page surface

The performance of Double Wave screws was provided earlier by Kruder and others [20–22]. For these trials, the extrusion process was capable of higher rates and lower discharge temperatures than well designed conventional single-flighted screws. Three-dimensional flow simulations for a Double Wave and a Triple Wave section were performed by Fan [23] and Perdikoulias [24]. These simulations showed the complex nature of the flows, including the acceleration of the flow as it passes from the shallow channel over the secondary flight, and then the deceleration of the flow as it collects in the deep channel. As expected, the Triple Wave screw had a much higher complexity for the flows and thus provided a higher level of both dispersive and distributive mixing.

14.2.2 Energy Transfer Screws

The Energy Transfer (ET) screw was developed and patented by Chung and Barr in 1983 [8]. This screw has been used in all types of extrusion processes, blow molding, injection molding, and tandem cooling extruders. Moreover, an ET screw was the key to eliminating a severe splay problem that occurred at the Saturn Corporation's Spring Hill, TN, plant [25] as discussed in Sections 1.2.1 and 11.12.5. Like the Double Wave screw, this screw is constructed by replacing the metering section with a high-performance ET section. Although the ET section appears similar to the Double Wave section, the flows of the ET section are considerably more complex. The ET section [26] consists of two channels as shown in Figure 14.8. One channel is positioned one-half turn downstream from the other, and both channels are of equal width. In order to create cross-channel mixing, the depths of both channels are periodic and out of phase. An undercut in the flight to the rear of channel A starts when that channel decreases in depth, permitting flow from the channel to the channel behind it. The undercut ends when the depth of the channel in front of that flight starts to increase. Thus, the switching in the undercut position acts like a valve and a dispersive mixer. This is the main difference between the ET and Double Wave sections. Most ET screws are designed with three regions where the channels become shallow, creating peaks in the channel profile. The peak clearance and flight undercuts are typically the same for all three peaks. These peaks and flight undercuts create cross-channel mixing and disperse and melt solid polymer fragments entering the section. Some of the larger solids will remain as solids after passing through the first peak region but will be reduced in size. The increased surface area of the smaller fragments will have a higher melting rate though energy conduction while being transported to the next peak region. Repetitive flow through the three peak regions produces extrudates with improved temperature homogenization, lower temperatures, lower levels of pressure and temperature fluctuations, and increased rates as compared to conventional screws. As a comparison and as shown in Section 8.4.1, conventional screws operating at high rates will sometimes permit solids to flow close to the screw tip or into the vented area of the second stage, producing extrudates with higher volatile levels, large temperature gradients, and with transfer line pressures and temperatures that vary with time. Due to the complex geometry of the ET section, dynamic pressure fluctuations as measured with a transducer positioned in the barrel wall and over the ET section can be relatively high, indicating complicated cross-channel flows [27, 28]. These pressure fluctuations were considerably larger than those for a single-flighted metering section (see Section 12.3).

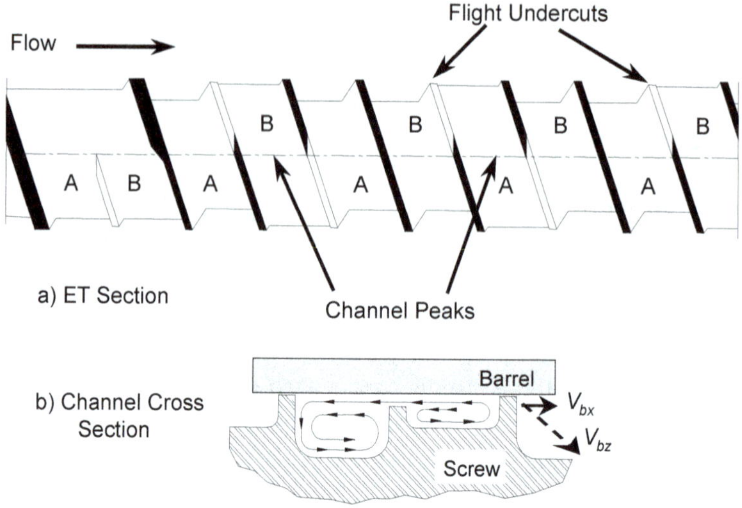

Figure 14.8 Schematic of an ET section: a) the section is used in place of a metering section, and it has two channels labeled A and B. The unfilled flight in the ET section is undercut with respect to the filled flight. b) cross-sectional view perpendicular to the flight tips showing the two channels and a flight undercut between them. The arrows show the direction of the flows perpendicular to the flight (courtesy of Jeff A. Myers of Robert Barr, Inc.)

The geometry of the ET section is very complex and has several features that can be adjusted for optimal processing performance. For example, the channel depths for a 114.3 mm diameter ET screw designed for use with a polyolefin resin is shown by Figure 14.9. For this screw, the lead length was 140 mm for the entire length of the screw. Lead lengths, however, can range from 0.8 to 1.25 times the diameter, with lead lengths of 0.8 times the diameter occurring mostly on screws for injection-molding plasticators with a short *L*/*D* barrel length. The position and depths of the A and B channels are apparent in this diagram. The depth of the deep channel of the ET section or "valley depth" is set to minimize the level of viscous energy dissipation and thus is set to control the extrudate temperature. The depth of the channel at the shallow point of the ET section, or the "peak depth," is set to provide the level of dispersive mixing required by the process. The length of the ET section and thus the length between the peaks and the valleys are set such that the target discharge pressure can be produced. Often a 1 to 3 diameter long single-flighted metering section is positioned downstream from the ET section to help mitigate pressure fluctuations. Like all metering sections, the ET section must be set such that it controls the specific rate for the process. The rotational flow rate can be estimated for the ET section by using Equation 1.13 with a double-flighted channel and the distance-averaged depth of the ET section. ET and Variable Barrier Energy Transfer (VBET) sections are often used with barrier-flighted melting sections.

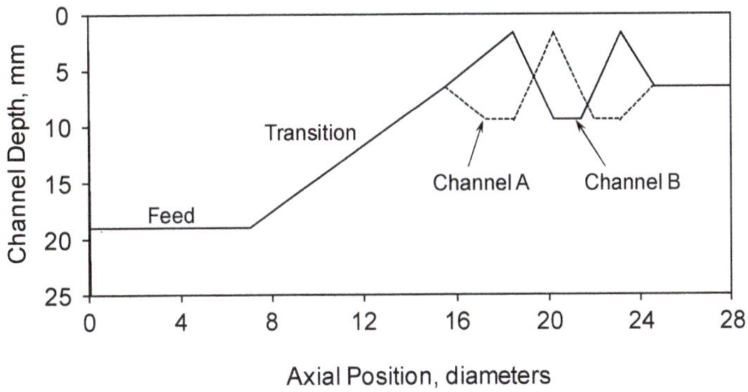

Figure 14.9 Typical channel depths for a 114.3 mm diameter ET screw. The channel depths in the peaks of the ET section and the undercut on the flight were 3.3 mm. The lead length was 140 mm

Several different designs are available at the entry to the ET and VBET sections depending on the application.

A Maddock solidification experiment was performed using a 21 *L/D*, 63.5 mm diameter extruder with an ET screw [29]. The extrusion equipment was described in detail in Section 8.4.1. The layout of the screw was similar to that shown in Figure 14.9. The screw had a lead length of 76.2 mm and a primary flight clearance of 0.08 mm. It had an 8 diameter long feed section that was 10.9 mm deep, a 6 diameter long transition section, and a 7 diameter long ET section. The ET channel depths were 3.18 mm at the entrance and exit of the section, and within the section they had peak and valley depths of 1.45 mm and 6.35 mm, respectively. The undercut of the flight in the ET section was 1.40 mm. Changes in the channel depths are obvious from the cross-sectional views shown by Figure 14.10. For this experiment, the main feedstock was ABS resin with 2% of TiO_2 compounded into the pellets. A small amount of black-colored ABS pellets were added to the feedstock at a letdown ratio of 220:1. The mostly white pellets are clearly visible in the solid bed, and regions that are molten and well mixed at the moment that screw rotation was stopped are an even black color.

As expected, the cross-sectional views for the ET screw were very similar to those for a single-flighted screw at axial lengths less than about 14 diameters for the sections upstream of the ET section, as shown by Figure 14.10. A comparable solidification experiment for a conventional single-flighted screw was shown previously in Figure 8.19. At axial distances of 14.4 diameters and greater (downstream), the channels of the ET section are clearly visible; that is, two channels are connected by an undercut flight. The view at 14.4 diameters was only 0.4 diameters into the ET section; much of the resin was still only softened, with most of the dark molten resin accumulating at the pushing side of channel A. This molten material originated from the melt pool of the upstream transition section. Breakup of the solid bed is evident at 15 diameters, where there was striated molten resin across the entire channel, with no trace of a solid bed. It is clear that this was solid bed breakup and not the completion of the melting and mixing processes, because unmixed and likely solid white resin was evident in downstream views. At 15.6 diameters, the view indicated a segregated dark melt pool in channel A, and most of channel B contained white resin, similar to the view at 14.4 diameters. For this view at 15.6 diameters, channel B was shallow, so the resin there experienced additional viscous dissipation. In addition, the white resin in channel B could flow over the flight undercut and into the adjoining and deeper channel A where heat transfer and mixing could occur between the portions of resin that were previously in different channels. This flow caused the white resin to occupy the center of channel A one turn downstream at 16.8 diameters. Here, channel A was no longer linked to its downstream channel B neighbor by a flight undercut. However, downstream at 17.4 diameters, the flight undercut reappeared and connected the now shallow channel A to the upstream and deepening channel B neighbor. Because channel A was shallow at 17.4 diameters, many of the remaining fragments of white resin were trapped, melted, and dispersed. Furthermore, the undercut allowed resin to flow into the downstream channel B, providing another level of dispersive mixing and melting. The flight undercut ended at 18 diameters. These trends continued at downstream positions, with the flight undercut reappearing, channel B becoming shallower, channel A becoming deeper, and the level of unmixed fragments of white resin becoming smaller due to the dispersive melting and mixing process.

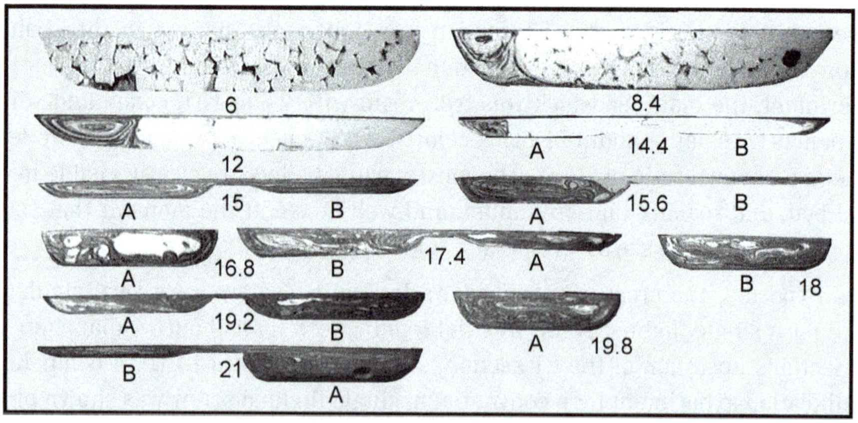

Figure 14.10 Cross-sectional views for a Maddock solidification experiment for the ET screw at a letdown ratio of 220:1 white-pigmented ABS resin to a black color concentrate [29]. The A and B channels are labeled along with the axial positions in diameters. The views were for a screw speed of 66 rpm and a rate of 70 kg/h using a 63.5 mm diameter extruder

Extrudate samples from a melting-mixing experiment were also collected for the ET screw at screw speeds ranging from 30 to 150 rpm using the experimental method described in Section 8.4.1. The ET screw used was the 63.5 mm diameter screw described above. The cross-sectional views of the extrudate samples are shown by Figure 14.11 as a function of screw speed. At a screw speed of 30 rpm, a few white streaks were evident in the extrudate samples, likely due to inadequate mixing at the lower screw speed. When the screw speed was increased to 60 rpm, the extrudate was free of defects. The same result was achieved when the screw speed was increased to 90 rpm. As the screw speed was further increased to 120 rpm, subtle spiral patterns became obvious and solid resin fragments began to appear in the extrudate samples. When the screw speed was increased to 150 rpm, the amount and size of the solid resin fragments in the extrudate increased further, as shown in Figure 14.11. The highest possible rate while providing an acceptable quality discharge for this ET screw will likely occur at about 90 rpm with a rate of about 85 kg/h. This rate is about 20% higher than the comparable single-flighted conventional screw described in Section 8.4.1 and as shown by the extrudate samples in Figure 8.18, showing the enhanced mixing and melting abilities of this high-performance design. Processes operating at high rates should always consider a high-performance design in order to mitigate mixing problems due to exceeding the melting capacity of the screw.

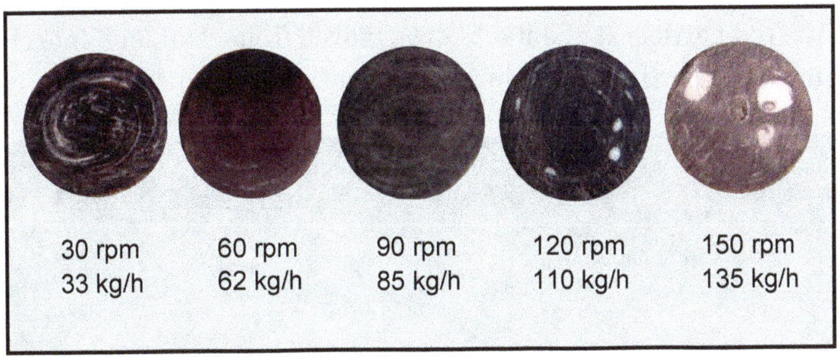

Figure 14.11 Cross-sectional views of extrudate samples at a letdown ratio of 100:1 of a white-pigmented ABS resin with a black color concentrate for an ET screw [30]. The samples were produced at screw speeds ranging from 30 to 150 rpm. The large round white regions for the cross sections at screw speeds of 120 rpm and higher are solid particles discharged with the extrudate

The melting and mixing results shown in Figure 14.11 are highly dependent upon the geometry of the screw and especially the geometry of the ET section. For example, if the ET peak clearances were decreased from the 1.4 mm used for the above trial to 0.7 mm, then the level of solid fragments at high screw speeds would decrease. The discharge temperature, however, would increase due to a higher level of energy dissipation for flow over the ET peaks and flight undercuts. The challenge for the screw designer is to set the channel peaks and valleys such as to provide a high-quality extrudate at the target rate and discharge temperature.

The same ET section of the screw used for the Maddock solidification and cross-sectional views was numerically simulated using an LDPE resin at a screw speed of 60 rpm [28, 31]. These simulations were performed using a three-dimensional finite element technique, and they clearly show the complex flows and temperature fields that exist in the channels. Complex flows are more easily viewed using particle trajectories. For example, Figure 14.12(a) shows the three-dimensional trajectories [28] of two particles released in the flow path of a conventional single-flighted channel at the same radial positions with one particle slightly more downstream than the other. These particles were in a location where the cross-channel flow was essentially zero. As shown in Figure 14.12(a), the trajectories are very repetitive helical motions. Starting the particles in different positions at the entry will cause the residence times and the trajectories to change. The trajectories, however, are still helical in nature. In contrast, the trajectories of two particles released into an ET section are shown in Figure 14.12(b). As before, the particles were released at the same radial position and into channel A and channel B at the inlet. As the particles moved with the flow for the first 2.5 turns, they stayed in the channels that they were released in, and they moved at velocities that depended on the depth of the channel. After about 2.5 turns, however, the particle released closest to the inlet corner (the green particle in Figure 14.12[b]) traveled from channel A over an undercut region and into channel B to the rear. This same particle later moved across a second undercut at about 4 turns and flowed back into channel A. Moreover, the red particle at about 4.5 turns moved from channel B into channel A. Consequently, both particles exited from channel A.

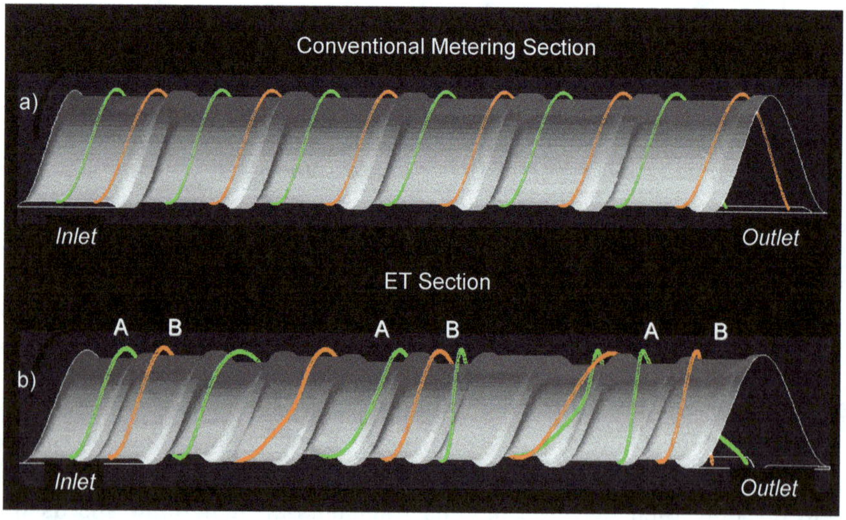

Figure 14.12 Simulated three-dimensional particle paths [28]: a) conventional single-flighted screw metering section, and b) an ET section

Similar particle releases are shown for views perpendicular to the screw axes by Figure 14.13. The particles in these views are spaced evenly with time, and thus more space between the particles indicates higher velocities. The spacing, however, is complicated by the projection of the three-dimensional flow onto a two-dimensional surface. Like the three-dimensional view of the conventional metering channel of Figure 14.12(a), the flows as viewed from the side are also helical in nature as shown by Figure 14.13(a). The particle spacing for this view is relatively even, indicating that the velocities for these trajectories are nearly constant. The particle trajectories for the ET section are considerably more complex, as shown by Figure 14.13(b), and the velocity variations are apparent from the compressed and extended regions of the paths for the particles. These flow paths were responsible for the improved thermal homogenization and mixing that occurs in the ET screw section.

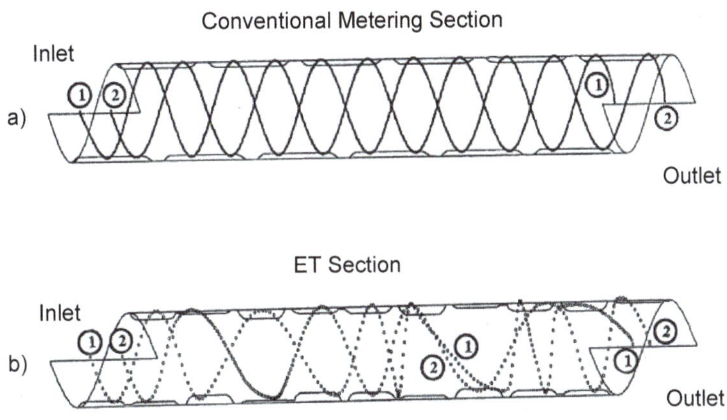

Figure 14.13 Two-dimensional projection of particle trajectories [28]: a) a conventional metering section, and b) an ET section

14.2.3 Variable Barrier Energy Transfer Screws

The Variable Barrier Energy Transfer™ (VBET, Trademark of Robert Barr, Inc.) screw is an enhanced design based on the ET screw [32, 33]. The original VBET design was a replacement design for blown film applications where a barrier section was followed by an ET section. For this application, the VBET section replaces some of the barrier section and all of the ET section. In most applications, however, the VBET section replaces the metering channel. In general, there are more peaks and undercuts in a VBET screw than in a standard ET screw. As previously discussed, ET screws have three peaks or channels with shallow regions, and all peak depths are identical. A VBET screw, however, will have three or more peaks, and the peaks become more restrictive as flow moves downstream.

Many of the geometric features of a VBET screw are similar to those for an ET screw. The VBET screw, however, has three additional enhanced features that can be adjusted for optimal processing performance. The channel depths for a 114.3 mm diameter VBET screw designed for use with a specific polyolefin resin is shown by Figure 14.14. Like the ET screw, the lead length for this VBET screw was 140 mm for the entire length of the screw. Lead lengths can range from 0.8 to 1.25 times the diameter, with lead lengths of 0.8 times the diameter occurring mostly on screws for injection-molding plasticators with a short *L/D* barrel length. The additional enhancements include a larger number of peaks, variable peak clearance, and an increasing length between the peak and the valley in the downstream direction. For example, the screw shown by Figure 14.14 had a first peak depth of 2.54 mm, and the distance between this peak and the next valley was 2.3 diameters. The last and fourth peak depth was 1.0 mm, and the distance between this peak and the upstream valley was 3.2 diameters. Like the ET design, the valley depth is set to minimize the level of viscous energy dissipation and thus is set to control the extrudate temperature while the peak depths are set to provide dispersive mixing. As the peak depth decreases with successive peaks in the downstream direction, the peak-to-valley distance is increased to minimize the pressure gradient across the undercut flight. This technique provides a higher level of dispersive mixing than a similar ET screw since the final peak depth is smaller than those for the ET screw, providing a higher shear rate and likely higher shear stress in the flow over the flight undercut. The larger peak depths upstream work well for dispersing large solid bed fragments into smaller solid fragments, while the downstream peaks add a higher level of dispersive mixing and mitigate the smaller solid fragments in the extrudate.

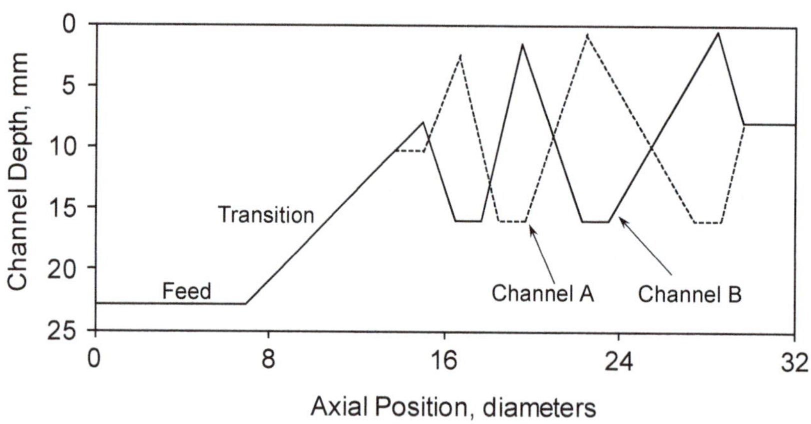

Figure 14.14 Channel depths for a 114.4 mm diameter VBET screw

Like the ET screw, a typical VBET screw will be followed by a 1 to 3 diameter long single-flighted metering section to help mitigate pressure fluctuations. The rotational flow rate for the VBET section can be estimated by standard calculation methods using a double-flighted section and the distance-averaged depth of the section.

A 63.5 mm diameter VBET screw was designed and built that was comparable to the ET screw presented previously [30] and in Section 14.2.2. However, unlike the ET screw, the undercut clearances were gradually decreased from 2.3 mm at the entrance of the section to 1 mm at the discharge end of the screw for a total of five peaks. In contrast, the ET screw had 3 peaks, and the undercut clearance was a constant 1.40 mm. The specific rotational flow rate for ABS resin was calculated at 1.0 kg/(h·rpm) for the VBET screw, which was slightly higher than that for the ET screw. A Maddock solidification experiment and a melting-mixing experiment were performed on this screw.

A Maddock solidification experiment was performed using ABS resin at a screw speed of 60 rpm [30]. Like the solidification experiment for the ET screw of Section 14.2.2, the feedstock resin contained 2% TiO_2 compounded into the pellets, and a black-colored masterbatch was added at a letdown ratio of about 220:1. The cross-sectional views of the resin solidified in the channels are shown by Figure 14.15. The white pellets and unmixed resin are visible in the views as white regions. The regions that are well mixed are an evenly colored dark or black color.

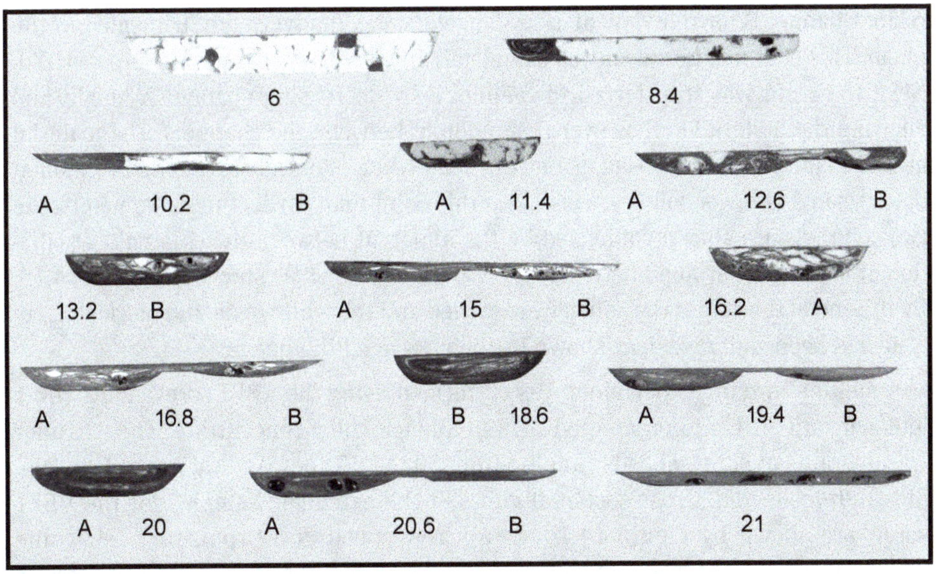

Figure 14.15 Cross-sectional views for a Maddock solidification experiment for a 63.5 mm diameter VBET screw running ABS resin at a screw speed of 60 rpm [30]. The A and B channels are labeled along with the axial positions in diameters

As shown in Figure 14.10 and Figure 14.15, the cross-sectional views at axial distances less than 10 diameters were similar for both ET and VBET screws. For example, the views at 6 diameters were composed of a compacted bed of discrete pellets. The views at 8.4 diameters were in the transition section, where the pellets compacted, softened, and began to melt. For the ET screw, the compaction process had just begun at 8.4 diameters, which was only 0.4 diameters into the transition section, so discrete pellets were still visible. For the VBET screw, melting was further along at 8.4 diameters, which was 3.4 diameters into the transition section. Fewer individual pellets were distinguishable, and the melt pool was more mixed with finer circular striations.

At axial distances of 10 diameters and greater, the views were different, as the VBET section on the VBET screw started at this point, while the ET section for the ET screw started at 14 diameters. The view at 10.2 diameters for the VBET screw in Figure 14.15 shows that individual pellets were no longer distinguishable and the melt pool had increased in width. This view was also at the start of the secondary flight, which can be seen as a depression on the middle of the lower edge of the view. From this point forward, the secondary flight started and separated the channel into the A and B channels. At 11.4 diameters, channel B depth decreased and forced molten resin and solid bed material into the A channel, as can be seen from the white and dark regions in the view. As the solid bed and melt advanced to 12.6 diameters, the material flowed over the secondary flight into channel B. The white domains are regions of the broken-up solid bed that are surrounded by regions of black molten resin. Channel B for the view at 12.6 diameters was inadvertently truncated in the photo. This channel, however, was completely filled with resin and was pressurized. Next, the resin was transferred to channel B by decreasing channel A height and allowing the material to flow over the undercut between the channels. The material at 13.2 diameters is composed of a mix of solid white material surrounded by molten black resin, but fewer solids are visible at this point than earlier. At 15 diameters, the two channels are shown connected by the undercut between the channels. Oscillation of material continued between the A and B channels as shown in Figure 14.15. By the end of the screw, the solids were melted and the white resin and black concentrate had been well mixed, as shown by the view at 21 diameters.

A melting-mixing experiment was performed using the VBET screw and a 100:1 letdown ratio of white pigmented ABS to a black color concentrate. The extruder was operated at screw speeds ranging from 30 to 150 rpm, and the extrudate samples were collected. Cross-sectional views of the extrudate samples for the VBET screw are shown by Figure 14.16. At a screw speed of 30 rpm, very few white streaks were evident in the extrudate strand samples. When the screw speed was increased to 60 and 90 rpm, the extrudate changed little. As the screw speed was further increased to 120 rpm, spiral patterns became obvious. There were no solid polymer fragments, however, in the extrudate samples. Some solid fragments, however, were evident in the extrudate for the ET screw at this speed, as shown in Figure 14.11. When the screw speed was increased to 150 rpm, a few solid resin fragments in the extrudate began to appear as shown by Figure 14.16. The specific rate was nearly constant at 1.0 kg/(h·rpm) over the range of the screw speed. These results clearly indicate that this VBET screw has improved melting and mixing capacity, as there were fewer unmelted fragments in the extrudate at higher screw speeds and at higher specific rates than that for the ET screw.

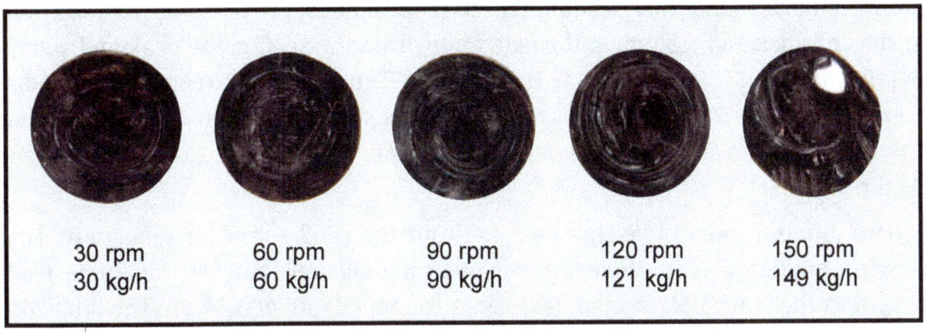

Figure 14.16 Extrudate samples as a function of screw speed for a 100:1 letdown ratio of white ABS (TiO_2) to black ABS resin using a VBET screw [30]

The data presented here clearly show the advanced melting and mixing capability of the VBET screw section. As shown by the extrudate samples, the VBET screw can be designed to allow higher screw speeds and rates before solids are discharged to the extrudate. For the cases presented here, the ET screw can be operated at a maximum rate of about 85 kg/h at a screw speed of 90 rpm while the VBET screw can run at about 121 kg/h at 120 rpm. Thus, the maximum rate for the VBET screw was about 40% higher than that for the ET screw, while maintaining a high melt quality. The additional two peaks and the strategic tightening of the undercut clearance with each successive peak created the additional mixing and performance of the screw. In comparison, the conventional screw was only able to run at a maximum rate at a quality discharge of 70 kg/h at 75 rpm. All screws, including high-performance designs, will discharge solid material with the extrudate if the screw is rotated fast enough. Additional performance data for VBET screws are in the references [30, 33].

14.2.4 Distributive Melt Mixing Screws

Distributive melt mixing (DM2™, Trademark of Glycon Corporation) screws from Glycon Corporation are based on the ET screw patent and thus are constructed as described by Section 14.2.2 and as shown in Figure 14.8. Variations in the entry and exit sections occur, however, between ET and DM2 designs. Like the ET screw, the DM2 screw is capable of operating at high rates and low discharge temperatures [34].

A mixing experiment was performed using a 63.5 mm diameter DM2 screw as outlined in Section 8.4.1. The DM2 main screw was 18.5 diameters long and was fitted with a 2.5 diameter long removable tip. The main section of the screw had a lead length of 70 mm and a primary flight clearance of 0.07 mm. It had a 5 diameter long feed section that was 10.9 mm deep, a 5 diameter transition section, and an 8.5 diameter long DM2 section. The DM2 section channel depths were 3.0 mm at the entrance and 3.3 mm at the exit, and within the section the peak and valley depths were 1.27 and 6.86 mm, respectively. The flights between the channels were undercut to 1.27 mm at strategic locations so that flow could occur between the channels. The specific rotational flow rate was calculated at 1.0 kg/(h·rpm) for ABS resin.

Two different removable tips were built for the DM2 screw for this study. The first tip studied was a single-flighted metering section that had the same lead length as the main DM2 section and had a channel depth of 3.3 mm. The other tip was an Eagle™ (trademark of Reiloy Westland Corporation) mixer [35], as shown by Figure 14.17. The Eagle mixing tip was constructed with two spiral in-flow channels and two spiral out-flow channels. Some material flows from the in-flow channel across a mixing flight and then into the out-flow channel. The lead length and the undercut clearance for the mixing flights were 170 mm and 1.27 mm, respectively. These mixing flights provided a level of dispersive mixing. Several bypass channels were designed into the mixing flight to allow some material to pass directly from the in-flow flute to the out-flow flute. These channels were 11 mm wide in the helical direction, and they provided a level of distributive mixing.

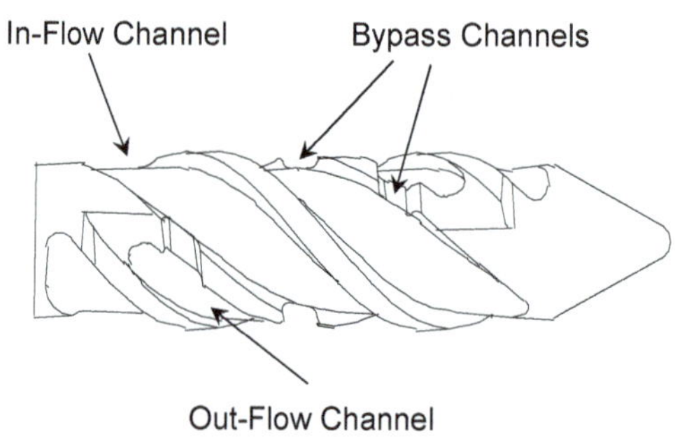

Figure 14.17 Schematic of an Eagle mixing tip [34, 35] (courtesy of Dave Larson of Westland Corporation)

A melting-mixing experiment was performed at screw speeds between 30 and 180 rpm, and the extrudate cross-sectional views are shown by Figure 14.18. As shown by this figure, the views for the DM2 screw with the metering tip indicated that the extrudates were fairly well mixed at screw speeds up to about 120 rpm. At 150 rpm, all resin was molten but small and possibly unacceptable partially mixed regions existed. At a screw speed of 180 rpm, a small and unacceptable level of solid polymer fragments was discharged, as indicated by the white-colored material. By comparing these cross-sectional views with those for the conventional screw shown in Figure 8.18, the enhanced mixing and melting abilities of the DM2 screw are obvious. The maximum rate that this screw could operate at before discharging solid polymer fragments was about 150 rpm for a rate of 131 kg/h. The cross-sectional views for the DM2 screw with the Eagle mixing tip are also shown by Figure 14.18. For this screw and mixer combination, the extrudates were all completely molten and well mixed. Thus, the DM2-Eagle combination has the capability of operating at screw speeds exceeding 180 rpm and at rates higher than 163 kg/h. As expected, the melting and mixing performance of the DM2 screw with the Eagle mixing tip exceeds that of just the DM2 screw with the metering tip, allowing for the production of high-quality products at higher rates.

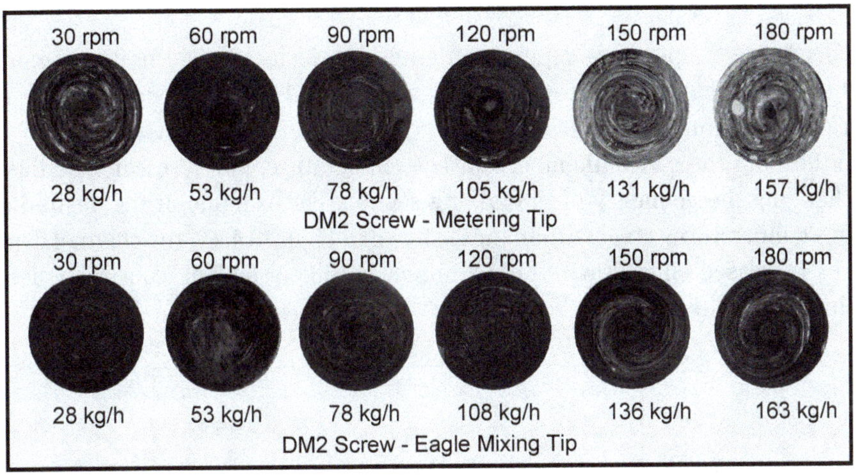

Figure 14.18 Cross-sectional views of extrudate samples for a DM2 screw with either a single-flighted metering tip or an Eagle mixing tip for a melting-mixing experiment

For injection-molding processes that are rate limited by the plasticating step, replacing the screw with a high-performance screw can reduce the cycle time of the process. For example, if a molding operation required a 0.5 kg shot size to fill the tool and runner system, the plasticating times for the screw combinations presented here could be calculated using the highest screw speed and rate that will produce an acceptable quality discharge. For the screw combinations here, the highest screw speeds that deliver an acceptable discharge were 50 rpm for the conventional screw (Section 8.4.1), 120 rpm for the DM2 screw with the metering tip, and 180 rpm for the DM2 screw with the Eagle mixing tip. At these maximum screw speeds, the instantaneous rates and the plasticating times for a 0.5 kg shot size are shown by Table 14.1. As shown by this table, the plasticating time required for the conventional screw is about 38 s. This time period would likely be longer than the cooling step of the process; this process is rate limited by the plasticating step. For the DM2 screw with the metering tip, the plasticating time decreases to 17 s, and for the DM2 screw with the Eagle mixing tip, the cycle time decreases to 11 s. For an 11 s plasticating time, the process may be rate limited by the cooling step and thus operating at the maximum production rate and at the lowest manufacturing cost.

Table 14.1 Maximum Operating Screw Speeds and Rates That Provide an Acceptable Quality Discharge for an Injection-Molding Operation

Screw	Maximum Operating Screw Speed, rpm	Continuous Rate, kg/h	Time to Plasticate 0.5 kg, s
Conventional screw	50	47	38
DM2 screw with the metering tip	120	105	17
DM2 screw with the Eagle mixing tip	180	163	11

The discharge temperature depends on numerous factors including the geometry of the high-performance screw and the screw speed for the operation. The discharge temperatures for the DM2 screw combinations were considerably higher than those for the conventional screw. The temperature, however, can be adjusted by changing the geometry of the mixing section. For example, if the desired discharge temperatures are less than those shown in Figure 14.19, the channel depths can be increased slightly with only a minor and likely insignificant reduction in mixing performance.

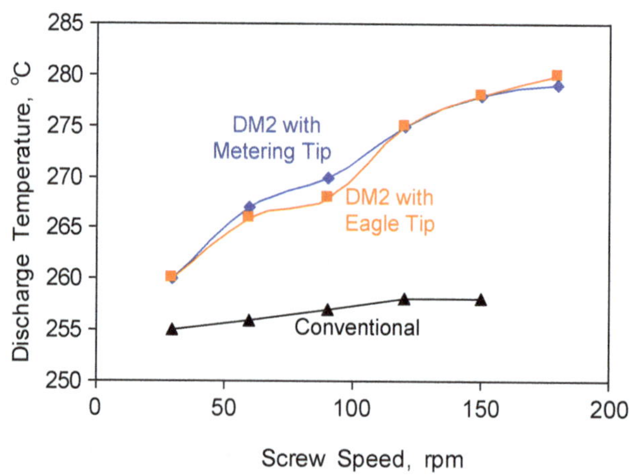

Figure 14.19 Discharge temperature as a function of screw speed for the conventional screw (Section 8.4.1), the DM2 screw with a metering tip, and the DM2 screw with an Eagle mixing tip

The mixing tip clearly provided an additional level of mixing as shown by Figure 14.18. This additional mixing is highly desirable at high screw speeds or color mixing operations for difficult-to-color concentrate mixing systems. The improved mixing is obtained without a reduction in rate or an increase in discharge temperature. Moreover, due to long lead length of the flights and the deep channels on the mixing tip, a slight increase in rate is obtained.

14.2.5 Fusion Screws

The success and breakthrough innovation of Kruder's wave screw was further enhanced by coupling the wave technology with an upstream barrier section and a material reorientation section. The reorientation section was positioned between the barrier section and the wave section. This spin-off technology was patented by Womer, Buck, and Hudak [36] in 2004. Other improvements were patented later [37, 38].

The Fusion™ (trademark of Xaloy) screw uses an Efficient™ (trademark of Xaloy) Barrier [4] section followed by a simple reorientation section and a modified Double Wave section. The Efficient Barrier design typically uses a lead length that is about 1.2 to 1.3 times the lead length of the feed section. The barrier flight starts at the pushing flight and transverses to a position that is about one-third the width between the primary flights after about 2 diameters downstream from the start of the section. After the first 2 diameters, the lead length of the barrier flight and primary flight are identical so as to provide constant widths for the melt and solids channels. The barrier flight is removed for the first 2 diameters of the section. The combination of the starting location, the position of the melting and solids channels for the first two diameters, the lead length, and the removal of the flight all mitigate the restriction at the entry of the section. The melt channel starts out shallow and increases in depth with the downstream position. The reorientation section is typically just a diameter length of a single-flighted metering channel. The wave section of the screw has a lead length that is the same as that used in the feed section. As discussed in Section 14.2.1, Kruder's wave screw design used a lead length that was longer than the lead length in the feed and transition sections. The channel depths for the wave section of the Fusion screw are also different since the channel valleys (deepest) are not maintained at a constant depth. Instead, once the valley's depth is obtained in the wave, the channel depth immediately starts to decrease. A schematic of the wave section of a Fusion screw is shown in Figure 14.20, and a schematic showing channel depths for a 144.3 mm diameter Fusion screw is shown in Figure 14.21.

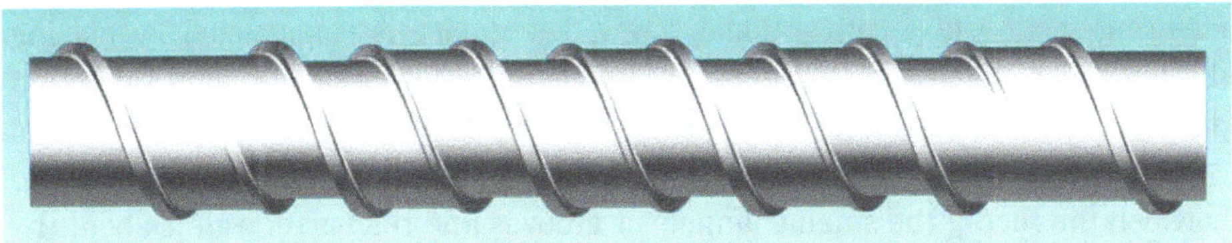

Figure 14.20 Schematic of a wave section of a Fusion high-performance screw (courtesy of Timothy W. Womer of Xaloy Incorporated)

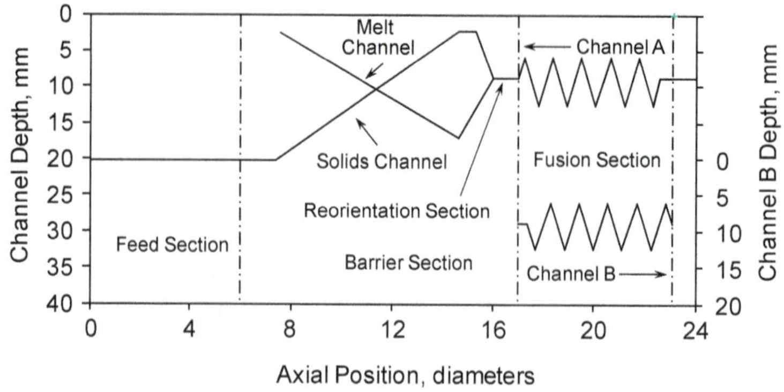

Figure 14.21 Schematic of a 114.3 mm diameter Fusion screw with a barrier melting section. The lead length of the screw was 131 mm for all sections of the screw except for the barrier melting section. The lead length in the barrier section was 159 mm. The undercut clearance on the barrier flight and the secondary flight in the Fusion sections was 1.3 and 2.5 mm, respectively

14.3 Other High-Performance Screw Designs

Numerous other high-performance screws are available commercially that are not based on wavy channels. Instead they are designed with flow paths that disrupt the normal flow that would occur in a conventional flighted metering section. The most common screws are the Stratablend™ (trademark of Xaloy) designs and the Unimix™ (trademark of Glycon Corporation) screw. These screws will be discussed in the next sections.

14.3.1 Stratablend Screws

The Stratablend and Stratablend II screws are also high-performance designs that are capable of high rates while discharging high-quality extrudates. The Stratablend screw is designed by positioning a Stratablend mixing section [39] in the metering section of a conventional screw. The device is constructed by placing three smaller advancing channels (or grooves) in the root of the main flight and parallel to the main flight, as shown by Figure 14.22. The base depth of the channel is generally very shallow relative to a standard metering section. The distance between the root of the small channels or grooves and the barrel wall is about the same depth as a standard metering channel for the same rate. The small channels are typically a half a screw turn in length. These series of small channels provide multiple regions where solid fragments can be trapped, melted, and dispersed into the main flow of the resin. Experiments were performed on a 63.5 mm diameter extruder using a Stratablend design and were presented previously [40]. A Maddock solidification experiment was performed, and the cross-sectional views are shown in Figure 14.23.

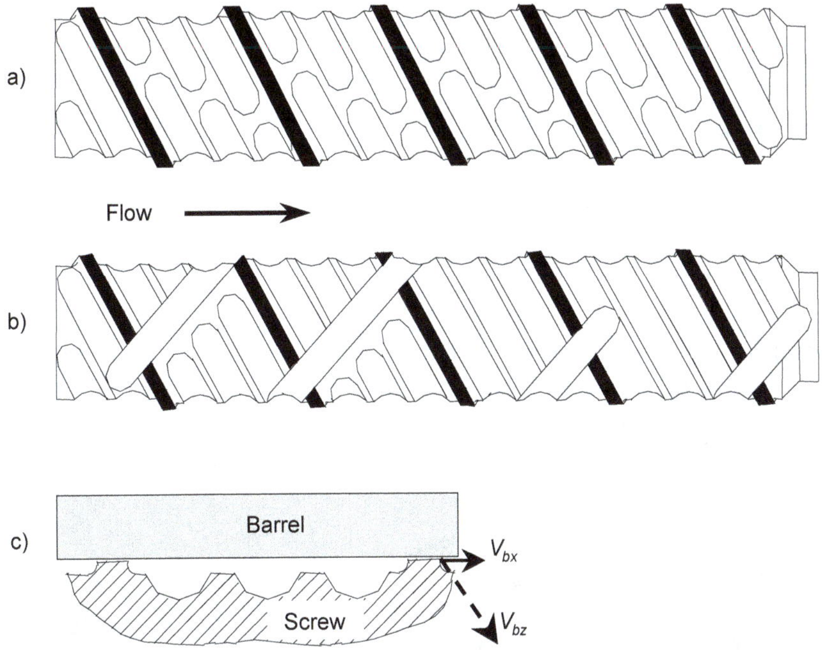

Figure 14.22 Schematics for Stratablend mixing sections; a) Stratablend schematic, b) Stratablend II schematic, and c) cross-sectional view perpendicular to the flight tips showing the three small channels or grooves and the flight undercuts between them (courtesy of Timothy W. Womer of Xaloy Incorporated)

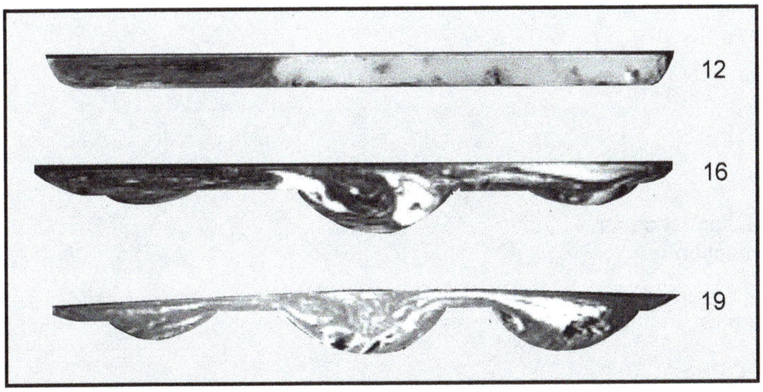

Figure 14.23 Photographs of cross-sectional views of resin removed from a Maddock solidification experiment for a Stratablend screw. The labels are the axial positions of the views in diameters

Like the Stratablend screw, the Stratablend II mixing screw [41] is constructed by placing a Stratablend II mixing section in the metering section of the screw. The Stratablend II section is similar to the earlier design in that three advancing grooves are positioned between and parallel to the flight edges. The enhancement in mixing and melting performance is obtained by the addition of traversing grooves that allow flows in the cross-channel direction, as shown by Figure 14.22(b).

14.3.2 Unimix Screws

Like the previously discussed high-performance screws, the Unimix [42] screw is designed by replacing all or part of the metering section with a Unimix mixing section. The Unimix section is designed with three channels positioned between the main flights, as shown by Figure 14.24. All channels vary (oscillate) in depth with respect to the downstream direction. The center channel is the deepest for most sections of the mixer, and the depth can be as large as twice that of the meter section channel that it replaced. At strategic locations, the center channel depth will decrease to a minimum depth for a very short axial length. At this location all

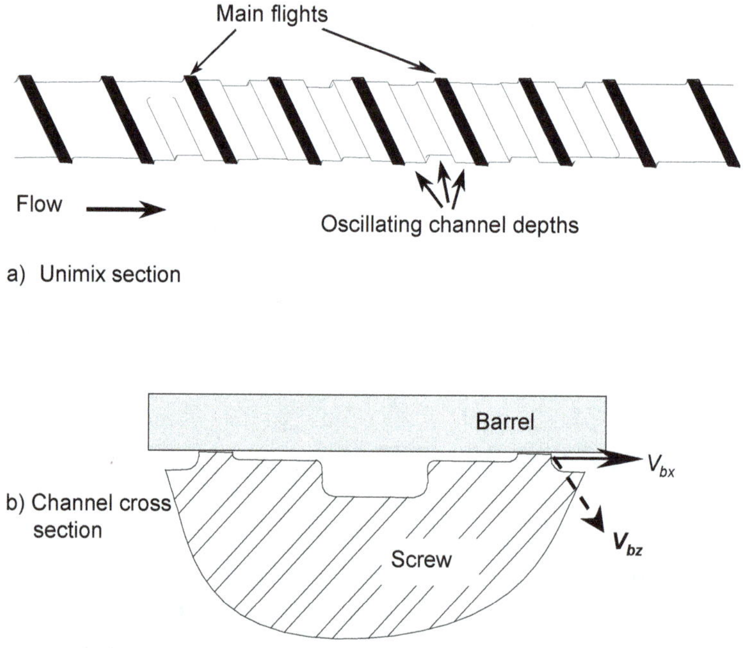

Figure 14.24 Schematic of a Unimix section: a) the section is used in place of a metering section, and it has three channels with oscillating depths, and b) cross-sectional view perpendicular to the flight tips showing channels at a local position (courtesy of Jeffrey A. Kuhman of Glycon Corporation)

three channels have identical depths, providing a mechanism to trap and melt solid polymer fragments. The channels near the pushing and trailing flight edges typically vary to a lesser extent, and these channels are shallower than the center channel. A typical Unimix screw is designed with at least three of these melting traps where the channel depths are the same and at a minimum. For a 63.5 mm diameter screw, the minimum channel depth of the Unimix section is on the order of 1 mm.

14.4 Calculation of the Specific Rotation Rate

The specific rotation rate for a high-performance screw section can be estimated using standard methods, and scaling can be performed using the techniques described in Chapter 9. The channel depth used for the calculation is the axial distance-averaged depth of the channels. For example, the specific rotation rate of an ET section would base the calculation on a double-flighted section, the lead length of the section, and the axial distance-averaged depth of both channels for the entire length of the section. For example, a 63.5 mm diameter ET screw with a lead length of 76 mm has a distance-averaged depth of 4.31 mm. This average depth was based on peak depths, valleys, and entry and exit depths of 1.45, 6.35, 3.18, and 3.18 mm, respectively. Next, the specific rotational rate is calculated for a double-flighted channel, an average depth of 4.31 mm, and for LDPE resin at 0.89 kg/(h·rpm).

The technique has been used to estimate successfully the rotational and the pressure flow rates in high-performance screw channels for Double Wave, ET, VBET, DM2, and Fusion screws. Since these screws all use a double-flighted geometry in the section, the aspect ratio (h/W) of the channels can become relatively large. For example, the ET case above had a calculated h/W of about 0.14. As discussed in Section 7.5.3, rotational and pressure flows cannot be calculated using the pseudo-Newtonian method for channels with aspect ratios greater than about 0.04, especially for highly shear-thinning resin rheologies such as an LDPE resin. The pseudo-Newtonian method would estimate the specific rotation rate at 1.04 kg/(h·rpm), a value that is too high by about 15%.

The compression ratio for a high-performance screw is typically based on the axial distance-averaged depth, while the compression rate is based on the channel depth at the entry to the section. The screw described above has a feed section with a depth of 10.9 mm, and thus the compression ratio is about 2.5.

Nomenclature

D	screw diameter
h	meter channel depth
L	barrel length
N	screw rotation speed in revolutions/s
u	barrier flight undercut from the primary flight
W	width of the channel at the barrel wall
λ	primary flight clearance
η	shear viscosity
τ	shear stress between the barrier flight tip of a barrier screw and the barrel wall

References

1. Maillefer, C., Swiss Patent 363,149 (1962)
2. Lacher, F.K., "Extruder," U.S. Patent 3,271,819 (1966)
3. Geyer, P., "Method and Apparatus for Refining and Separating Plastic Materials," U.S. Patent 3,375,549 (1968)
4. Dray, R.F. and Lawrence, L., "Apparatus for Extruding Plastic Material," U.S. Patent 3,650,652 (1972)
5. Barr, R.A., "Extruder, or Extruder-Like Melting Apparatus," U.S. Patent 3,698,541 (1972)
6. Hsu, J.S., "Extruder Screw," U.S. Patent 3,858,856 (1975)
7. Kim, H.T., "Apparatus for Extruding Polymeric Material," U.S. Patent 3,867,079 (1975)
8. Chung, C.I. and Barr, R.A., "Energy Efficient Extruder Screw," U.S. Patent 4,405,239 (1983)
9. Somers, S.A., Spalding, M.A., Dooley, J., and Hyun, K.S., "An Experimental Study of the Flows in an Energy Transfer Screw," *SPE ANTEC Tech. Papers*, **48**, 307 (2002)
10. Han, C.D., Lee, K.Y., and Wheeler, N.C., "A Study on the Performance of Barrier-Screw Extruders," *Polym. Eng. Sci.*, **31**, 831 (1991)
11. Steward, E., "Barrier Screws, Their History and Their Function," *SPE ANTEC Tech. Papers*, **48**, 69 (2002)
12. Wheeler, Jr., N.C., "Extruder Screw," U.S. Patent 4,341,474 (1982)

13. Thompson, M. R. and Christiano, J. P., "Extruder Screw," U. S. Patent 6,176,606 (2001)
14. Christiano, J. P. and Thompson, M. R., "Extruder Screw," U. S. Patent 6,179,461 (2001)
15. Christiano, J. P. and Thompson, M. R., "A New Barrier Screw Design Utilizing Solid Bed Deformation with Forced Melt Removal," *SPE ANTEC Tech. Papers*, **46**, 78 (2000)
16. Slusarz, K. R., Thompson, M. R., and Christiano, J. P., "An Experimental Comparison of the Improved Mixing Obtained from a New Barrier Screw Design," *SPE ANTEC Tech. Papers*, **47**, 262 (2001)
17. Christiano, J. P. and Thompson, M. R., "Extruder Screw Having Multi-Channeled Barrier Section," U. S. Patent 6,139,179 (2000)
18. Kruder, G. A., "Extruder Screws," U. S. Patent 3,870,284 (1975)
19. Kruder, G. A., "Extrusion Apparatus and Method," U. S. Patent 4,173,417 (1979)
20. Kruder, G. A., "Ultra High Output Extrusion on a Commercial Scale," *SPE ANTEC Tech. Papers*, **30**, 726 (1972)
21. Kruder, G. A. and Nichols, R. J., "New Applications of Wave Screw Designs for Ultra High Output Extrusion," *SPE ANTEC Tech. Papers*, **31**, 137 (1973)
22. Kruder, G. A. and Calland, W. N., "Fundamentals of Wave Screw Extrusion," *SPE ANTEC Tech. Papers*, **36**, 74 (1990)
23. Fan, P., Vlachopoulos, J., Smith, N., and Sheth, H., "Computer Simulation of Melt Flow in Wave Screws," *SPE ANTEC Tech. Papers*, **44**, 97 (1998)
24. Perdikoulias, J., Svabik, J., and Womer, T., "3D FEM Analysis of a Wave Type Screw Channel," *SPE ANTEC Tech. Papers*, **55**, 1180 (2009)
25. Kirkland, C., "Damage Control Saturn Style," *Injection Molding Mag.*, September (1994)
26. Chung, C. I., and Barr, R. A., "A Novel Energy Efficient Screw Design," *SPE ANTEC Tech. Papers*, **29**, 168 (1983)
27. Hyun, K. S. and Spalding, M. A., "Use of Process Data Obtained from a Data Acquisition System for Optimizing and Debugging Extrusion Processes," *Adv. Polym. Tech.*, **15**, 29 (1996)
28. Somers, S. A., Spalding, M. A., Dooley, J., and Hyun, K. S., "Numerical Analysis of the Thermal Mixing Effects of an Energy Transfer (ET) Screw Section," *SPE ANTEC Tech. Papers*, **41**, 222 (1995)
29. Somers, S. A., Spalding, M. A., Dooley, J., and Hyun, K. S., "An Experimental Study of the Flows in an Energy Transfer Screw," *SPE ANTEC Tech. Papers*, **48**, 307 (2002)
30. Hogan, T. A., Spalding, M. A., Kim, E. K., Barr, R. A., and Myers, J. A., "Performance Analysis of a Variable Barrier Energy Transfer Screw," *SPE ANTEC Tech. Papers*, **49**, 92 (2003)
31. Plumley, T. A., Spalding, M. A., Dooley, J., and Hyun, K. S., "Three Dimensional Numerical Analysis of an Energy Transfer (ET) Screw Section of a Single-Screw Extruder," *SPE ANTEC Tech. Papers*, **40**, 332 (1994)
32. Barr, R. A., "Extruder Screw with Improved Energy Efficient Melting," U. S. Patent 6,599,004 (2003)

33. Myers, J.A. and Barr, R.A., "Improved Screw Design for Maximum Conductive Melting," *SPE ANTEC Tech. Papers*, **48**, 154 (2002)

34. Spalding, M.A., Kuhman, J.A., Larson, D., Kuhman, J., and Prettyman, H.L., "Performance of a Distributive Melt Mixing Screw with an Advanced Mixing Tip," *SPE ANTEC Tech. Papers*, **50**, 599 (2004)

35. Davis, M.E. and Hook, J.W., "Extruder Mixing Screw," U.S. Patent 5,215,764 (1993)

36. Womer, T.W., Buck, E.J., and Hudak, Jr., B.J., "Apparatus for Plasticating Thermoplastics," U.S. Patent 6,672,753 (2004)

37. Womer, T.W., Buck, E.J., and Hudak, Jr., B.J., "Plasticating Screw and Apparatus," U.S. Patent 7,014,353 (2006)

38. Womer, T.W. and Smith, W., "Apparatus for Plasticating Thermoplastic Resin Including Polypropylene," U.S. Patent 7,156,550 (2007)

39. Frankland, Jr., J.D., "Extrusion Screw," U.S. Patent 4,639,143 (1987)

40. Somers, S.A., Spalding, M.A., Hughes, K.R., and Frankland, J.D., "Performance of a Stratablend Mixing Screw for Single-Screw Extrusion," *SPE ANTEC Tech. Papers*, **44**, 272 (1998)

41. Womer, T.W. and Lepore, B.L., "Apparatus for Plasticating Thermoplastics," U.S. Patent 6,488,399 (2002)

42. Myers, J.A., "Feedscrew for Injection Molding and Extrusion," U.S. Patent 5,342,125 (1994)

15 Melt-Fed Extruders

Many large-scale processes utilize melt-fed extruders to complete an operation prior to pelletization of the product. As the name implies, the feed to these extruders is a molten polymer feedstock. In general, an upstream process either plasticates the resin or provides it as a high viscosity fluid such as from a reactor system. These extruders typically provide pumping, devolatilization, and the incorporation of additives via mixing sections. Other operations are possible. For example, machines have been made that have dispersive mixing devices designed into the screws that allow for the shear refinement of LDPE resins [1]. Shear refining of LDPE resins was discussed in Section 8.10. Because melt-fed processes do not require solids conveying and melting operations, the screw designs are considerably different than plasticating screws.

This chapter will present the three most common uses for melt-fed extruders: (1) pumping of resin from a compounding operation to a pelletizer, (2) large-diameter extruders used for finishing sections in resin production plants, and (3) secondary cooling extruders for foam sheet lines.

15.1 Simulation Methods

The design of most melt-fed screws is straightforward using numerical simulation techniques. As will be discussed in the next sections, often the aspect ratio of the channels (*H/W*) is relatively high due to a combination of deep channels and multiple flight starts. As presented in Chapter 7, channels with high aspect ratios cannot be accurately simulated using the simple pseudo-Newtonian calculation method. If this method is used, the estimation of the specific rotational flow rate can be in error by as much as 20% and the pressure gradients can be in error by as much as a factor of seven [2–4]. Instead, these simulations must be performed using a three-dimensional numerical technique such as a finite element method or a finite difference method. Calculation of the channel flows using a correction technique for the pseudo-Newtonian method as outlined in Section 7.5.3 provides an acceptable estimate of the rotational and pressure flow components for these processes.

15.2 Compounding Processes

Most corotating, intermeshing twin-screw compounding extruders are capable of generating sufficient pressure at the discharge of the machine to support a pelletization process at moderate rates. These pressures are typically in the 6 to 15 MPa range. At high rates, a twin-screw machine may not be able to generate adequate pressure. Other compounding machines such as continuous mixers and twin-screw extruders with extremely deep channels typically cannot generate adequate pressure to operate a die. Deep channels in twin-screw machines occur when the ratio of the outside diameter (OD) to the inside diameter (ID) of the screw is in the high range of 1.8 (OD/ID ratio). These high-range OD/ID machines are designed to discharge at low temperatures or to mitigate the agglomeration of solid additives in the feedstock. For these machines, pressurization and pumping of the molten resin is often performed using either a gear pump or a melt-fed single-screw extruder. A common compounding process using a single-screw pressurization extruder is shown in Figure 15.1. For this process, the feed components are metered at the correct rate onto a moving feed belt. These components are then gravity-fed into the feed port of the continuous mixer where they are melted and compounded. The material discharging from the continuous mixer is then gravity-fed into the feed hopper of a single-screw extruder where the material is pressurized and pumped to an underwater pelletizer. Next the wet pellets are dried in a spin drier, small dust particles and large-size pellets are removed using a classifier, and the pellets are then stored in bins or packaged. The melt-fed extruder is only used here to pump the compounded product to the pelletizer. A complete description of compounding processes can be obtained elsewhere [5, 6].

The melt-fed extruder is typically designed with a relatively short barrel length of 6 to 12 L/D. Since the resin is molten when it enters the extruder, the motor size is relatively small compared to a plasticating extruder with the same barrel diameter. The extruders are typically starve-fed. That is, the extruder is operating at a screw speed that is higher than required, and thus the rate is not controlled by the screw speed. Instead, the rate of the overall process is controlled by the feedstock addition rate to the compounding machine.

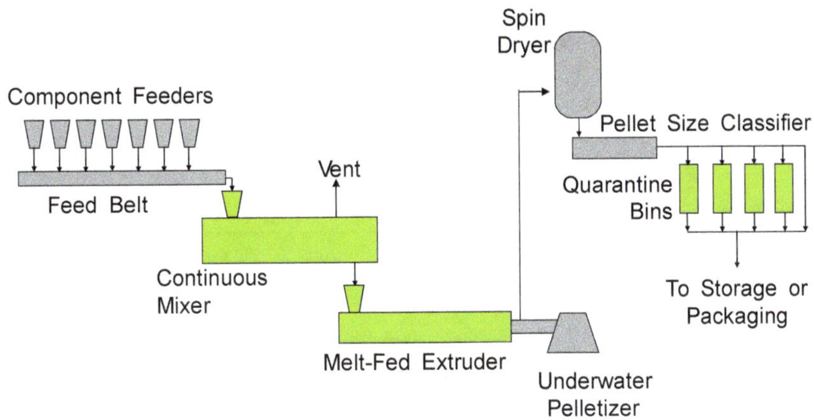

Figure 15.1 Schematic of a compounding process using a continuous mixer as the compounding machine and a melt-fed single-screw extruder to pressurize and pump the product stream to the underwater pelletizing system

Since the only purpose of these extruders is to pump and pressurize the resin, the design of the screw is relatively simple. Often the screw is single-flighted with a lead length equal to the barrel diameter. The length of the feed section is typically about one diameter longer than the feed opening while the transition section can be as short as two diameters. The length and depth of the metering section is the key to the successful operation of the extruder. The channel should be deep enough that excessive energy dissipation does not occur, while not being too deep such that pressure cannot be developed for the downstream process. An example of a metering channel design that was too deep was discussed in Section 13.4.3. In summary, the length and depth of the metering section must be specified so as to provide an acceptable level of pressurization and pumping to the pelletizer while not overheating the resin.

A typical screw design for compounding an ABS resin product is shown in Figure 15.2. This screw was 304.8 mm in diameter, had a length of 7 *L*/*D* and a compression ratio of 2.1. The lead length and flight width were 304.8 and 19 mm, respectively. This screw typically operated at 5500 kg/h at a screw speed of 70 rpm. Compression ratios near 2 are very common for these designs. Mixing devices are typically not needed and thus are not used on these types of processes.

The operation of the screw described in Figure 15.2 is shown by the simulated axial pressure and temperature profiles shown in Figure 15.3. The simulation was performed at a rate of 5500 kg/h and at a screw speed of 70 rpm for an ABS resin with an MFR of 7.5 dg/min (230 °C, 3.8 kg). As shown in this figure, the pressure was zero for the first 3 diameters of the screw. In this section, the channels were operating partially filled with the screw operating in a starve-fed mode. At 3 diameters from the start, the channel becomes full and pressure in the channel starts to increase. The pressure increase is nearly linear for this process. The temperature increased by about 13 °C.

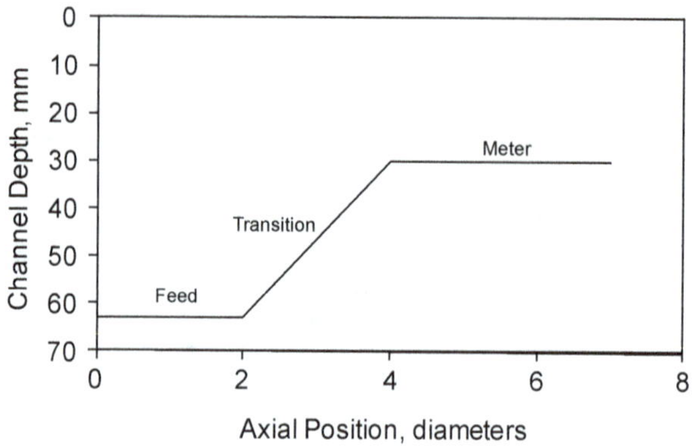

Figure 15.2 Schematic showing the channel depths for a 304.8 mm diameter melt-fed extruder designed to pump an ABS resin

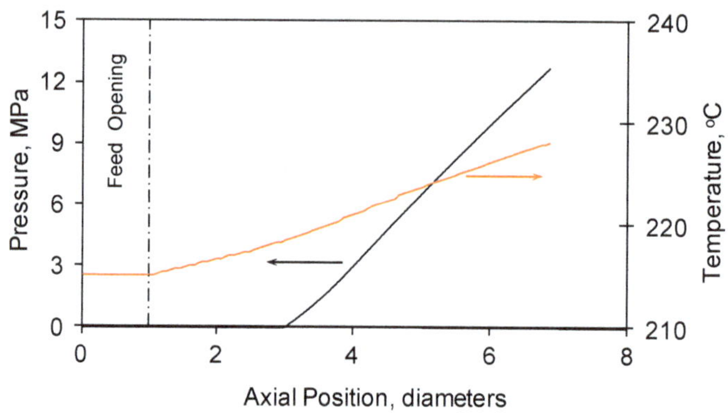

Figure 15.3 Simulated axial pressure and temperature for a melt-fed extruder running a compounded ABS resin. The process is operating at a rate of 5500 kg/h and a screw speed of 70 rpm

Ideally, the depth of the channels would be optimized for the shear viscosity of the resin. In most compounding plants, the line will typically be used for many products with a range of shear viscosities. In these cases, compromises to the designs must be performed such that all compounds can be produced at acceptable rates and qualities.

15.2.1 Common Problems for Melt-Fed Extruders on Compounding Lines

Several common problems can occur for melt-fed extruders. These include rate limitations due to improper screw design, high discharge temperatures, and long purging times between product changes. These problems can be mitigated through screw design. In Section 13.4.3, a case study was presented where the rate was limited by high discharge temperatures and the inability of the screw to generate enough pressure to operate the pelletizing die. For this case, the depth of the metering channel was too deep for the resin being processed. When a new screw was fabricated with a metering channel that was optimized for the viscosity of the resin, the rate of the process was increased by a factor of 2.4. If the metering channel depth is too shallow for the resin, then the rate will be less than optimal and high discharge temperatures will occur.

Most compounding lines are not dedicated to a single product or resin color. Instead, these lines run multiple products and colors. Product changes will always have a period of time where the process is producing off-specification product due to the carryover of the old product into the new product. The most economical process will occur when the amount of off-specification product is minimized and thus the purging time is minimized. Most sections of a process will purge the old resin very quickly. If stagnant regions exist, however, in the compounder, melt-fed extruder, transfer line, or die, then very long purge times are likely to occur. Simple compounding screws are most likely to have stagnant regions at the radii of the flights. As discussed in Section 11.7, screws with small flight radii can create stagnant regions at the corners of the channel due to Moffat eddies. Moffat eddies are discussed in Section 8.3.2. Melt-fed screws should always be fabricated with flight radii that are at least as large as the depth of the channel.

15.3 Large-Diameter Pumping Extruders

Large-diameter, single-screw melt-fed extruders are often used in the pelletization unit operation of polymer production plants for the manufacturing of LDPE [26], PS, and PB resins, and specialty copolymers of polyethylene including EVA, EMA, EEA, and EAA resins [7, 8]. For example, LDPE resin is produced by the high-pressure reaction of ethylene [9, 10]. The product stream from the reactor is under very high pressure (81–276 MPa) and contains about 22 to 35% LDPE in essentially unreacted ethylene [7]. Most of the ethylene is removed from the stream in a high-pressure separator (or flash tank), and it is recycled to the reactors. The high-pressure separator is operated at a pressure near 26 MPa [11]. The effluent from the high-pressure separator is then sent to a low-pressure separator where all but trace amounts of ethylene are removed and recycled. The effluent from the low-pressure separator will contain between 1000 and 2000 ppm of dissolved and entrained ethylene in the resin [10, 11], and this stream is fed to the extruder. The level of dissolved ethylene depends on the operating pressure and temperature of the low-pressure separator. The operating conditions for the separator depend on the equipment design, rates, resin, and the technologies used by the resin manufacturer. Typically the low-pressure separator will be operated at an absolute pressure of 0.04 to 0.14 MPa and at a temperature near 250 °C [11, 12]. The low-pressure separator is positioned over the top of the extruder feed port, and typically the separator has a liquid level that is about 0.5 to 1 m above the feed opening. A slide valve can be positioned between the low-pressure separator and the extruder. The valve is useful for maintenance purposes to shut off flow to the extruder or for use as a rate-controlling device. The feed pipe and slide valve are designed such that the extruder will be flood-fed when the valve is completely open. A schematic of a typical extrusion operation is shown in Figure 15.4.

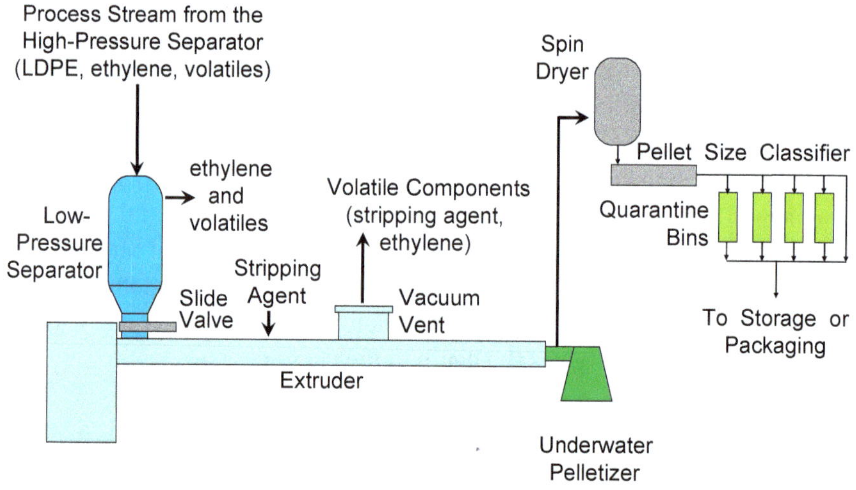

Figure 15.4 Schematic for a melt-fed devolatilizing extruder for an LDPE resin production plant

These extruders typically provide additional unit operations beyond just pressurization and pumping to a pelletization system. The extruder can also be used to devolatilize ethylene, to compound additives, and to shear refine the resin. A devolatilization process can be designed with either a single vent or with multiple vents. Since the ethylene content is at very low levels, a strong vacuum is applied to the vent port opening of the extruder. Typically, about 70 to 90% of the volatile components can be removed in a single vent stage. If additional devolatilization is necessary, more vents can be placed downstream of the first vent. Moreover, a stripping agent can be added upstream of the vent. The stripping agent can be distributed into the resin using a gear-type mixing section. Common stripping agents include nitrogen and water [13]. The stripping agents are then removed in a downstream vent along with the volatile components. A detailed analysis of polymer devolatilization in extruders was presented previously [14-16].

Additives such as stabilizers [27], slip agents [28], and processing aids [29] can be added to the melt stream using a small single-screw extruder that is attached to the melt-fed extruder. For this operation, the additive components are precompounded into a pelletized masterbatch, and these pellets are plasticated and pumped by the small extruder. The molten masterbatch is then injected through a port in the barrel wall of the melt-fed extruder in a convenient location downstream from the low-pressure separator. The components can be distributed into the resin flow using a gear-type mixer. If the additives have a high vapor pressure at the processing conditions, then they would need to be added downstream of the last vent.

Shear refining of LDPE resin is a practice where the resin is placed in a high-stress field and some of the polymer chains are temporarily reoriented, causing the resin to have better optical properties. This process is a physical change and is reversible; the relaxation time is relatively long. It is a well-known process for improving the haze and gloss properties for LDPE blown-film-grade resins [1, 18-20]. The shear refining process was described in Section 8.10. Shear refinement in a melt-fed extruder is typically performed using a spiral-fluted mixer where the resin must pass across a very small clearance passage, as shown in Figure 15.5 and Figure 15.6. For example, a 500 mm diameter extruder with a shear refining Maillefer mixer would have an undercut for the refining (mixing) flight from the main flights of 0.4 to 0.8 mm and a total clearance between the refining flight and the barrel wall of 0.8 to 1.2 mm.

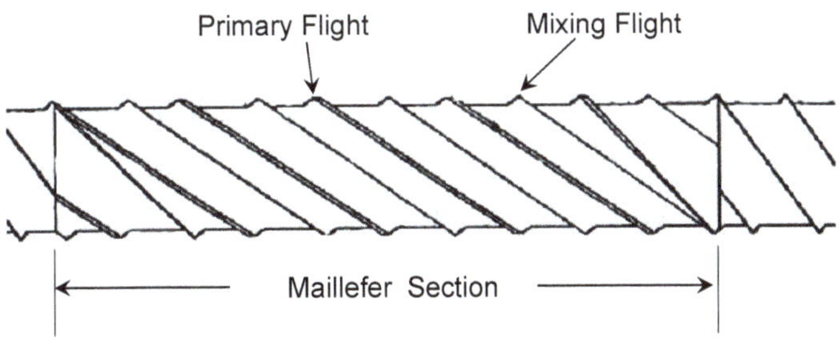

Figure 15.5 Schematic for a Maillefer fluted mixing section [21] for the shear refinement of LDPE resin

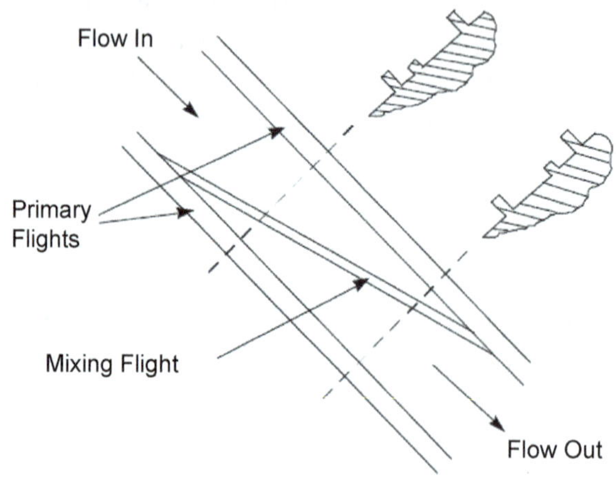

Figure 15.6 Flow path for a single flute set of a Maillefer mixing section

Screw design for these large machines is very complex, and their manufacture is very costly and has a long delivery time. Very large screws can take more than six months to fabricate and only a few manufacturers have adequately sized machining equipment. Moreover, since the extruder is a necessary component of a resin production plant, the screw must be able to process the entire product mix when it is installed. If the screw is poorly designed and built, the entire plant will not be able to operate and must be kept shut down until the screw is modified and made operational.

The rates for resin manufacturing plants have a wide range, and consequently large-diameter, melt-fed extruders have been built with a wide range of diameters and lengths. Specialty copolymer resin plants with relatively low production rates would use a smaller extruder with a diameter of 250 to 300 mm. High-pressure LDPE resin plants, however, are operated at rates of 25,000 kg/h and higher and will require an extruder with a 500 mm diameter barrel or larger. Typical extrusion rates and motor sizes as a function of diameter and barrel length are shown in Table 15.1. Currently, the largest extruder built and operated is a machine designed and built by Berstorff at 800 mm in diameter, and it is capable of rates as high as 67,000 kg/h. A photograph of this extruder is shown in Figure 15.7. The length of the extruder will depend on the number of unit operations that need to be performed. For example, a 300 mm diameter extruder with an L/D of 12 would be used primarily for pressurization and pumping to a pelletizer. Machines of the same diameter but with one and two vents are designed with longer barrel lengths of 30 and 40 L/D, respectively. Operations such as the incorporation of additives and stripping agents will also require additional barrel length.

Table 15.1 Typical Extruder Specifications for Melt-Fed Extruders Used in the Manufacturing of LDPE Resins (Some of the Data Courtesy of Peter Hunziker of Berstorff USA). The Data Are for Resins with an MI of 2 dg/min (190 °C, 2.13 kg)

Barrel Diameter, mm	Barrel Length, L/D	Rate, kg/h	Motor Size, kW	Specific Rate, kg/(h·rpm)
250	18	4200	500	40
300	12	8000	670	70
300	18	7000	750	70
300	26 (1 vent)	6000	750	70
300	41 (2 vents)	5000	950	70
400	18	13,000	1400	160
500	18	22,000	2450	260
500	24	18,000	3000	260
600	18	34,000	3700	490
700	18	46,000	5100	760
800	12	67,000	5500	1000
800	18	61,000	6800	1000

Figure 15.7 Photograph (top view with the pelletizer in the foreground) of a 800 mm diameter KE-800 extruder with an L/D of 18 manufactured by Berstorff (courtesy of Peter Hunziker of Berstorff USA)

As previously mentioned, screw designs for these machines can be very complicated due to the need for numerous unit operations. For example, a 300 mm diameter, two-stage vented screw that has the capability to use stream stripping by the injection of water is shown in Figure 15.8. For this screw, the feed channel is 2 diameters in length, ending about 1 diameter downstream from the edge of the feed port opening. Typically, the feed port opening is 1 to 1.5 diameters in axial length. The transition section on the screw is relatively short at 2 to 4 diameters. The first-stage metering section is about 2 to 6 diameters in length and has a typical depth between 12 and 20 mm. The next section downstream from the metering section has two functions: (1) adjustment of the specific rate using two removable blisters, and (2) a gear mixer between the blisters that is used to distribute the stripping agent. A schematic of this section is shown in Figure 15.9. The second blister is also used to disperse the stripping agent into the resin. After the second blister, the material flows into the deep vented area. The depth of this channel is typically about the same depth as the feed channel. The vent openings for these machines are very long in order to maximize the level of volatiles removed. For the screw shown in Figure 15.8, the vent opening is 3 diameters in axial length. The second-stage transition section is short at between 2 and 4 diameters. Pressurization and pumping to the pelletizer is performed in the second-stage metering section. The second-stage metering section is typically between 15 and 24 mm deep and is about 8 to 12 diameters in length. The lead length is normally equal to the diameter of the extruder except for the vented channel section. In the vented channel, the lead length is typically increased to 1.5 to 2 times the diameter of the barrel. Larger diameter screws are often multiflighted. For this 300 mm screw, all sections were designed with two flights except for the vent channel with six flights. The larger number of flights and the larger lead length in the vent section aid in the generation of surface area for the removal of volatile components. The aspect ratio (H/W) of the second-stage metering section can be relatively high ($H/W \geq 0.16$) due to the deep channels and the two flight starts. This high ratio is very common for melt-fed extruders, and it can create difficulties in screw design and numerical simulation. As previously discussed in Section 15.1 and Section 7.5.1, the pseudo-Newtonian calculation method cannot be used due to calculation errors in the rotational flow and pressure gradient. Instead a three-dimensional numerical method or a correction technique for the pseudo-Newtonian method must be used for these calculations; see Sections 7.5.3 and 7.5.4.

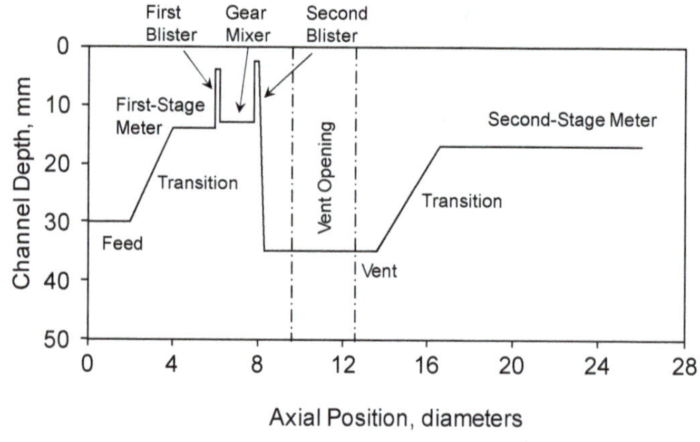

Figure 15.8 Channel depths and section placements for a 300 mm diameter, two-stage vented extruder

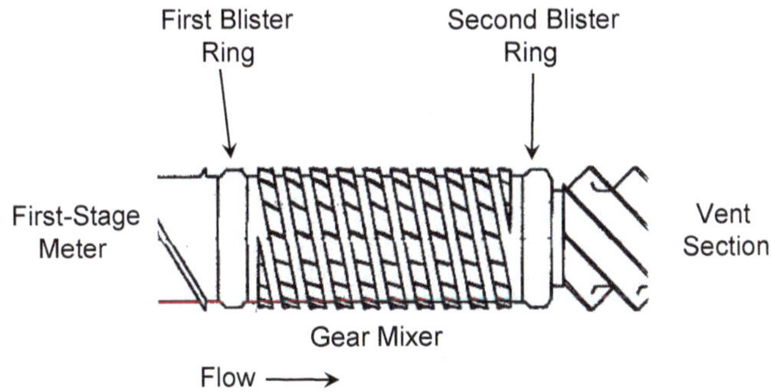

Figure 15.9 Schematic for a blister and gear mixing section. The blisters are used to adjust the specific rate of the process, and the gear mixer and second blister are used to distribute and disperse the stripping agent into the resin

Early design and simulation of large-diameter, melt-fed extruders were described by Fenner [17]. A numerical simulation of the axial pressure and temperature for a screw similar to that shown in Figure 15.8 is shown in Figure 15.10. This simulation was performed using a three-dimensional method using a finite difference approach. The process starts with an LDPE resin (2 dg/min, 2.16 kg, 190 °C) in the low-pressure separator at a pressure of 0.04 MPa (gauge) and a temperature of 230 °C. The pressure increases in the flighted section of the channel up to an axial position of 6 diameters. At 6 diameters, the pressure decreases initially due to the first blister, decreases at a lower rate in the gear mixer section, and then decreases to zero pressure at the end of the second blister. The balance between the pressure-generation ability of the flighted section and the pressure consumption of the blisters and gear mixer will set the specific rate of the process. This balance will be discussed fully later. Channels between an axial position of 8 and about 15 diameters are partially filled. Since these channels are partially filled, they are at the operation pressure of the vent. If the fill position (15 diameters) occurred underneath the vent, then flow of resin into the vent would occur. The vent opening needs to be sealed to exclude oxygen from the atmosphere and must have the capability to remove volatile components. In the second stage, the pressure starts to increase from the location where the channel first becomes fully filled, that is, at the fill position (15 diameters). The pressure increases to match the pressure required to operate the die at the production rate.

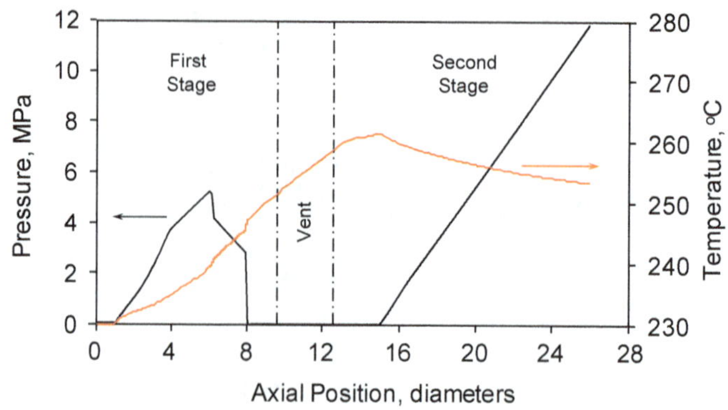

Figure 15.10 Simulated axial pressure and temperature for a 300 mm diameter screw with channel geometry that is similar to that shown in Figure 15.8. The simulation was performed at 5000 kg/h and a screw speed of 95 rpm for a 2 dg/min (2.16 kg, 190 °C) MI LDPE resin

The blister and gear mixer section used for the screw described in Figure 15.8 is shown in Figure 15.9. As discussed previously, this section is used to adjust the specific rate and the distribution and dispersion of the stripping agent into the resin. The stripping agent is injected through the barrel wall and just downstream from the first blister. The clearance between the first blister and the barrel wall is relatively large, but it does create a seal such that the stripping agent does not flow upstream. The stripping agent is distributed in the molten resin stream by the gear mixer. The second blister clearance is smaller than the first blister clearance, and it is designed to disperse the stripping agent and to adjust the specific rate of the screw. For modern screws, the blisters are constructed such that they can be changed quickly on-site at the resin manufacturing plant. The blisters are made in symmetrical halves and are bolted onto the screw. A spare set of blisters with zero clearance is typically maintained at the plant. If operations at the plant change, such as the production of a new product with an MI (Melt Index) outside of the current operating range, then a new second blister may be required to optimize the rate or to prevent flow of resin into the vent. The new clearance on the blisters is set via numerical simulation. The blister halves are then mounted on a cylindrical shaft and are machined to the desired clearance. Next, the screw is removed from the extruder, and the old blister is removed and replaced with the new blister. This technique minimizes the downtime for the plant.

The specific rate of a two-stage vented extruder and screw is set by the balance between the first-stage flighted section and the blister-mixer section for a flood-fed operation. The resin enters from the low-pressure separator at a pressure just above zero. The pressure increases in the flighted section, and then the pressure decreases as the resin flows across the blister-mixer section. As the resin flows into the vent channel, the pressure is zero. Thus, the specific rate is determined by a balance between the ability of the flighted section to generate pressure and of the blister-mixer section to consume pressure. The second stage must be able to pump the resin while maintaining the location for complete channel fill downstream from the vent. If the specific rate is too high for the second stage, then the location for complete channel fill will occur in the vent region, causing resin to be forced into the vent dome and vent pipes. When the vent becomes plugged with resin, the extruder and plant must be shut down so that the vent can be cleaned. The specific rate as a function of the second blister clearance for a 300 mm diameter extruder running a resin with an MI of 2 dg/min (190 °C, 2.16 kg) is shown in Figure 15.11(a). The geometry of the screw used for this calculation is similar to that shown in Figure 15.8. As shown by Figure 15.11(a), the specific rate as determined by the first stage is relatively low for small clearances on the second blister. As the clearance is increased the specific rate increases from 56 to 74 kg/(h·rpm). At specific rates less than about 70 kg/(h·rpm), the second stage is able to pump and pressurize the resin to an acceptable level for operating the pelletizing die. At specific rates greater than 70.5 kg/(h·rpm), the second stage cannot pump the resin and vent flow is predicted. For this case and in order to maximize the rate for the plant while not creating resin flow at the vent, the ideal second blister clearance is about 4.0 to 4.5 mm.

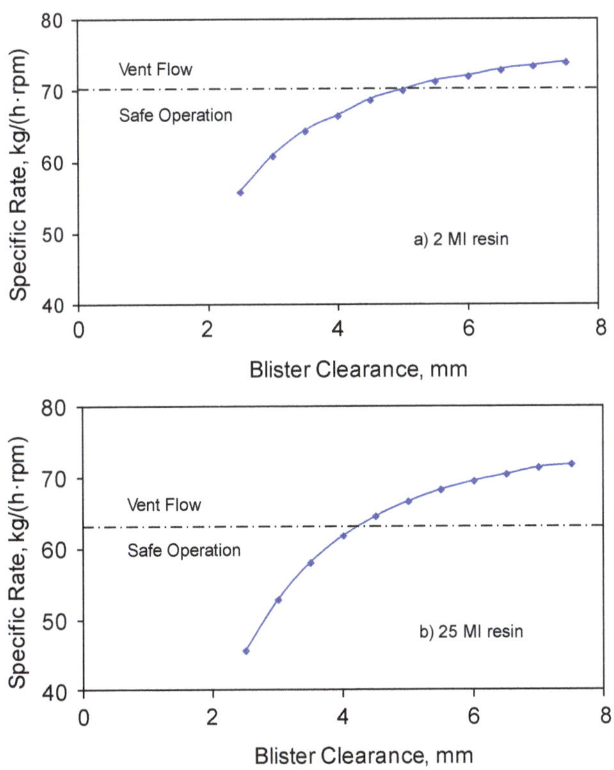

Figure 15.11 Simulated specific rate as a function of the second blister clearance for a 300 mm diameter extruder operating at constant rate: a) for a 2 MI LDPE resin, and b) for a 25 MI LDPE resin

The same analysis was performed using the identical screw design and an LDPE resin with a 25 dg/min (190 °C, 2.16 kg) MI, as shown in Figure 15.11(b). For this case, the specific rate was 5 to 20% less than that for the 2 MI resin shown in Figure 15.11(a). Moreover, the highest specific rate possible before resin flows into the vent was calculated at 63 kg/(h·rpm), a value that was about 10% less than that for the 2 MI resin. For the 25 MI resin, the optimal clearance for the second blister is about 3.0 to 3.8 mm, providing the highest specific rate while not forcing resin to flow into the vent dome.

Most resin manufacturing plants produce resins with a range of molecular weights. Thus, a diagram such as Figure 15.11 must be calculated for all major resins, especially those resins at the low and high end of the MI range. In most cases and as shown in Figure 15.11, resins with a higher MI will extrude at a lower specific rate for the same second blister clearance. Typically the second blister clearance will be set by the resin with the highest MI. All other resins that need to be produced will then run at reduced specific rates and possible reduced rates. If the extruder has enough power and can operate at the maximum screw speed, resins with MIs in the lower region of the product range may be rate limited by the extruder screw speed.

A visco-seal must be designed into the shank of the screw to prevent resin from flowing out of the barrel and onto the floor. These seals are relatively simple and consist of a small channel that is spiral cut into the shank and in the same direction as the main flights of the screw, as shown in Figure 15.12. The pressure in the feed channel causes the resin to flow out of the extruder through the annular region between the shank of the screw and the barrel wall and also down the channels of the visco-seal. The rotational flow for the rotating spiral channels pumps the resin back into the feed channel. A more complete description of visco-seals was provided in Section 10.4.8. A poorly designed or malfunctioning visco-seal can cause several severe problems. If the seal is not pumping enough resin back into the extruder, then molten resin will slowly flow out of the extruder and into the back housing of the extruder and then to the floor, eventually solidifying and causing a cleaning issue. If the seal is not filled adequately with molten polymer, small amounts of oxygen from the atmosphere can enter the extrusion system. Oxygen can cause LDPE resins to crosslink at extrusion conditions. These gels will eventually contaminate the finished product. Under specific conditions, some ethylene that is dissolved in the feed resin can come out of solution and exist as gas bubbles in the feed channel. If the visco-seal is not working properly, some ethylene gas can escape to the atmosphere, creating safety issues. Modern machines with properly designed seals will not allow ethylene to leak out of the seal.

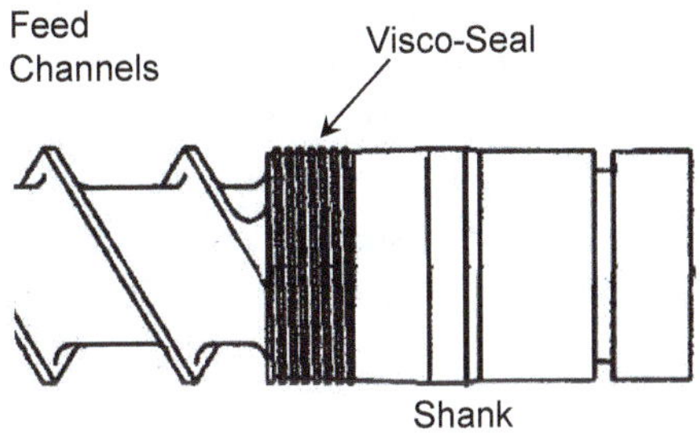

Figure 15.12 Schematic of a visco-seal on the shank of a melt-fed extruder screw

Minor resin flow out through the visco-seal can sometimes be mitigated or eliminated by increasing the flow of cooling water to the cooling channels on the barrel side of the visco-seal. Increasing the cooling level may be able to cool the resin in the visco-seal, causing the viscosity in the channels to increase. An increase in viscosity will reduce the level of flow in the annulus due to the pressure gradient.

15.3.1 Loss of Rate Due to Poor Material Conveyance in the Feed Section

Large-diameter, melt-fed extruders are almost always flood-fed as described above. As such they will typically operate at a specific rate that is very predictable based on numerical simulations. The extruder will operate at these rates in a very stable manner. That is, the motor current and discharge pressure are essentially constant with time. An improperly designed extruder or a malfunctioning extruder can, however, operate at much lower specific rates and in a very unstable manner. This mode of operation is often referred to as slipping because the feed section of the screw is not conveying resin at a high enough rate to maintain the extruder at the designed specific rate. Slipping occurs when some of the dissolved ethylene comes out of solution at the barrel wall of the extruder. In this case the inside barrel wall temperature is typically too hot. The ethylene gas immediately develops a foam consistency. In order for the resin to move forward at the expected rates, the material must have a no-slip boundary condition at the barrel wall, and the resin must be able to transfer a relatively high level of stress to the barrel wall. But since the material near the wall is foamed, the shear strength of the resin is reduced and the material slips at the barrel wall. When the material slips at the wall the melt-conveying rate is reduced. Since the no-slip and then slip behavior are very erratic, the extrusion process tends to be unstable, creating fluctuations in the discharge pressure and motor current. Pressure changes at the discharge of the extruder will cause variations in the sizes of the pellets, decreasing the quality of the pelletized product.

The slipping problem occurs due to process changes and the solubility of ethylene in the LDPE resin. In order to understand the fundamentals of the process, it is important to understand the solubility of ethylene in LDPE resin. Maloney and Prausnitz [22] studied the solubility of several solutes in LDPE resin, including ethylene. They measured Henry's constant as a function of temperature. Henry's law was shown to be acceptable for predicting the concentration of ethylene in LDPE resin as follows:

$$w = \frac{yP}{H_c} \tag{15.1}$$

where P is the absolute pressure, y is the mole fraction of ethylene in the gas phase, H_c is Henry's constant, and w is the weight fraction of dissolved ethylene in LDPE resin at equilibrium. For the low-pressure separator, the gas above the LDPE resin is essentially ethylene, and thus the value of y is 1. Henry's constant at several temperatures is shown in Table 15.2. At the conditions in the low-pressure separator, the level of dissolved ethylene is shown in Figure 15.13. More sophisticated models for estimating the level of dissolved ethylene in LDPE resin have been developed [11, 12].

The conditions in the low-pressure separator are typically in the range of 0.04 to 0.14 MPa absolute pressure and at a temperature near 250 °C [11, 12]. At these conditions, the LDPE resin in the low-pressure separator is expected to contain about 750 ppm of dissolved ethylene at equilibrium. Another 250 to 750 ppm of ethylene can also be entrained as a gas or at a supersaturated condition in the LDPE. As the resin flows from the low-pressure separator into the first diameter of the extruder, the pressure is expected to be nearly the same since the flow pipes are not restrictive. If the temperature of the inside wall of the feed casing of the extruder is less than 250 °C, then the ethylene will remain in solution, foaming will not happen, and the extruder will operate as expected. If the feed casing temperature, however, is at 260 °C, the resin at the wall will be saturated with 700 ppm ethylene, and the excess ethylene will form small bubbles in the resin along with the entrained ethylene. Although the level of ethylene released as a gas is small on a mass basis, it is enough to cause slipping at the wall of the feed casing. The only method to eliminate slipping is to maintain the feed casing temperature at a level that is less than the resin temperature in the low-pressure separator. Because it is difficult to measure the inside wall temperature of the extruder, the feed casing should be maintained at 30 to 40 °C less than the temperature of the resin in the low-pressure separator.

Table 15.2 Henry's Constant for LDPE Resin and Dissolved Ethylene [22]

Temperature, °C	Henry's Constant, MPa
150	157
200	185
250	208
300	229

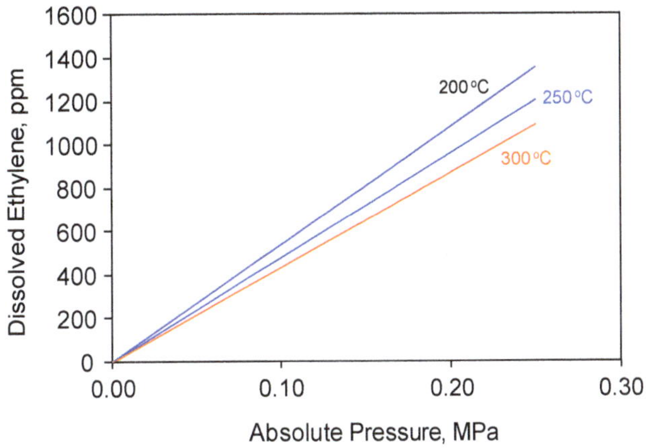

Figure 15.13 Dissolved ethylene in LDPE at saturation as a function of the absolute pressure and temperature [22]

15.3.2 Operation of the Slide Valve

Ideally, the slide valve between the low-pressure separator and the extruder is maintained in the fully open position during standard operation, allowing the screw to be flood-fed. This operation works well when the resin being processed is within the MI operating range for the screw. As market conditions change, resin plants are often required to manufacture materials outside of the design MI range. For example, if a two-stage vented extruder has a screw that was designed for resins with an MI range of 2 to 20 dg/min, and plant personnel are asked to manufacture a resin with an MI of 100 dg/min, the extruder may not process the resin acceptably. For this case and flood feeding, the extruder is likely to have resin flowing into the vent because the first stage can deliver more resin than the second stage can pump. One practical solution to this problem is to operate the slide valve in a partially open manner such that it controls the rate of the extrusion process. This will allow the extruder to operate at a lower specific rate and allow operation without the flow of resin into the vent.

Operation of the slide valve in a partially open position, however, can create additional process issues. For this mode of operation, the rate of the process depends on the pressure gradient between the extruder and the low-pressure separator and on the level of the restriction caused by the valve position. As expected, the valve position is fixed, but the pressure in the low-pressure separator can be variable. If the pressure in the separator increases suddenly, then the instantaneous flow to the extruder will increase. If the pressure change is not too high, the extruder will be able to process the additional resin. If the pressure change is too high, however, then the specific rate will increase dramatically, causing the flow of resin into the vent. Moreover, if the process is operating near the maximum motor current and a pressure increase in the low-pressure separator occurs, it can increase the specific rate of the process and cause additional need for motor power. Since the motor is already near its operating limit, the pressure surge may cause the shutdown of the process due to high motor current.

15.3.3 Nitrogen Inerting on Vent Domes

Secondary casings are always designed around the outside of seals that are in contact with polymer and vacuums. Seals around the vacuum domes of vents are the primary concern. These secondary casings are continuously purged with nitrogen, so if a small leak occurs in the vacuum seal, nitrogen instead of oxygen is introduced into the process. If oxygen were allowed to leak through a seal, crosslinked gels may occur, contaminating the pelletized product. If a high level of crosslinked gels is present in the pelletized product, oxygen intrusion into the vent domes is a likely root cause.

15.4 Secondary Extruders for Tandem Foam Sheet Lines

Foamed polymer products are used in a variety of applications, including low-density containers such as egg cartons and insulation boards. The products can be made from a relatively broad range of polymers using a tandem foam sheet line. For this process, the pellets and possibly recycle resin are fed to a single-screw plasticating extruder where the resin is melted and mixed with a physical blowing agent. For PS resins, the temperature of the extrudate is about 230 °C and considerably too hot to produce a high-quality foam product. Next the resin and blowing agent mixture is conveyed using a short transfer line to a secondary extruder, as shown in Figure 15.14. Often the line is configured with a gear pump positioned between the extruders. The secondary extruder (or cooling extruder) is used to cool the resin mixture down to the foaming temperature. For PS resins, the foaming temperature is about 140 °C. For foam sheet, the extrudate from the secondary extruder is forced through an annular die and then shaped and cooled over a mandrel. The foam annulus is then slit with a knife blade and laid flat so the sheet can be wound onto a roll. The sheet is aged to allow the diffusion of the blowing agent from the cells. If the sheet is thermoformed in a subsequent operation, the edge trim scrap will have a very low bulk density, and typically it must be pelletized before it is recycled back to the primary extruder. A good overview of the PS foam sheet process was presented previously [23].

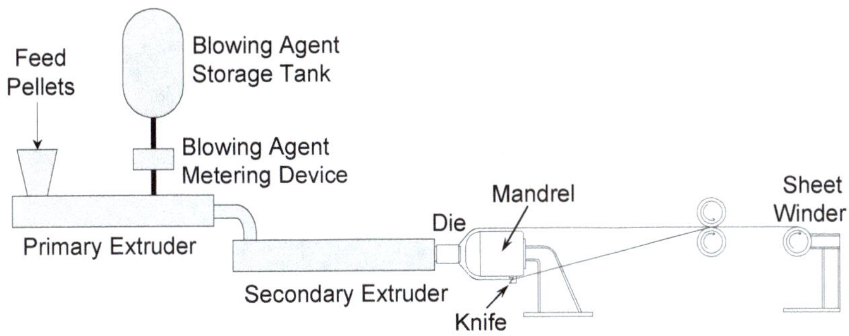

Figure 15.14 Schematic of a typical tandem foam sheet line

The secondary extruder is typically larger in diameter than the primary machine, and it rotates at a lower screw speed. For example, a midsized tandem line would be constructed using a 114.3 mm diameter primary extruder and a 152.4 mm secondary extruder. The primary extruder would operate at a normal screw speed (50 to 70 rpm) and could generate the pressure required to run the downstream equipment. The secondary extruder would operate at a lower screw speed (10 to 15 rpm) so that the entry and exit pressure for the extruder are about the same. Cooling water flow to the barrel jacket coolers for the secondary extruder removes the excess energy in the resin dissipated by the primary extruder and also the energy dissipated by the screw in the secondary machine. For a properly designed line, the performance of the secondary extruder determines the overall rate of the process.

The maximum rate for operation depends on several key design options. The primary extruder must be able to plasticate and discharge at a proper rate and with a minimal discharge temperature. The secondary extruder must be designed with the proper specific rotation rate for the screw, the screw must provide a high level of mixing capability, and the barrel must be equipped with efficient barrel coolers. The specific rotation rate and the mixing capability of the secondary screw are designed to minimize the amount of energy dissipated by the screw and to mitigate thermal gradients in the melt. These two design features must be balanced. For example, the level of energy dissipated by the cooling screw is minimized by designing the channels very deep and by operating at a very high specific rotational rate. The deep channels, however, will create a high level of thermal gradients because the excess energy is conducted out through the barrel. Due to this energy conduction and the motion of the fluid due to screw rotation, the temperature of the melt in the channel will be the lowest at the metal surfaces and the highest at an internal region, as shown in Figure 15.15. If the melt is discharged with a high level of thermal gradients, the quality of the foam sheet will be poor and not uniform. The thermal gradients are eliminated by specialized mixing techniques.

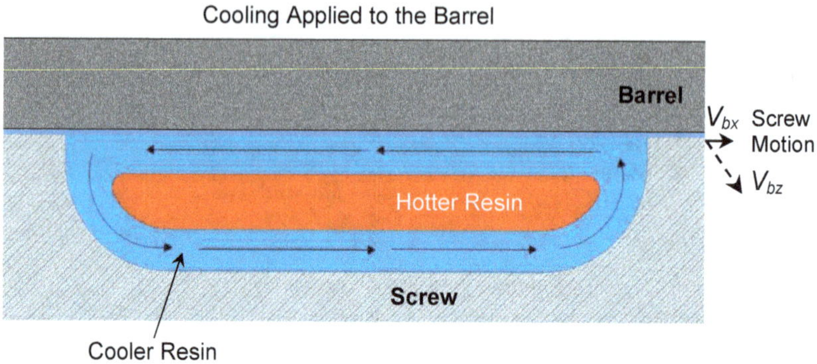

Figure 15.15 Schematic of the flow lines and thermal gradients in a deep channel of a cooling screw

Energy transported out through the barrel wall of the secondary extruder is extremely important for maximizing the rate of the process. The best barrel designs are those that have the water cooling coils integral to the barrel. Typically, the outside surface of the barrel is grooved to accept a spiral-wound cooling pipe. The pipe is then welded to the outside of the barrel. Several cooling zones are positioned on the barrel. Less effective cooling occurs if the cooling jackets are strapped onto the outside surface of the barrel wall. These zones must contain heating elements to enable the startup of the process. Heat transfer is improved by designing the screw with multiple flights such that the inside barrel wall is wiped as often as possible.

Design of the primary extruder is also a key to the operation of the line at maximum rates. For example, if the primary screw is discharging the resin-blowing agent mixture at excessive temperatures, the extra energy load must be removed by the secondary extruder. Designing the primary extruder for lower discharge temperatures reduces the energy removal load on the secondary machine.

The screw geometry for the secondary extruder is very critical for the economical operation of the line and for the quality of the foam sheet. For example, if the process operates at half of the design rate, then the fixed costs of the process will be spread over a reduced amount of product, increasing the unit price. Moreover, if thermal gradients occur in the extrudate from the secondary extruder or if the discharge temperature is too high for foaming, the quality of the sheet will be unacceptable. The design of conventional screws from the original equipment manufacturers is typically proprietary. In general, the screws are multi-flighted with lead lengths between 1.5 to 2.5 times the diameter, deep channels, and strategically positioned slots in the flights. A conventional cooling extruder screw is shown in Figure 15.16. The slots are made by removing a portion of the flight down to the root of the screw. Long lead lengths are used to accommodate multiple flights and to increase the specific rotation rate. Deep channels are also used to increase the specific rotation rate. Slots in the flights periodically allow some resin to pass to the channel behind it, causing a disruption in the flow and the mitigation of the thermal gradients shown in Figure 15.15. The slots are positioned on the flights such that all sections of the barrel wall are wiped by an advancing flight. These slots, however, will cause the local heat transfer coefficient to decrease because the local area at the barrel wall is wiped less often in the vicinity of a slot.

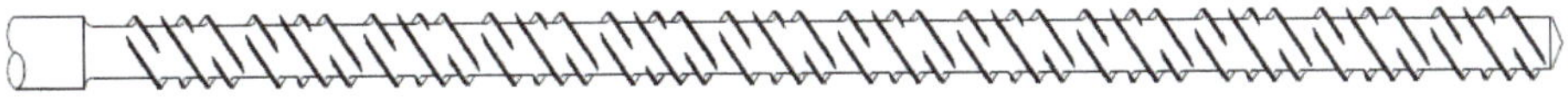

Figure 15.16 Schematic of a conventional cooling screw. A viscous seal is required for this screw, but it was omitted from the drawing (courtesy of Jeff A. Myers of Robert Barr, Inc.)

The screw is typically operated at the specific rotation rate such that the pressure gradient in the axial direction is near zero. The barrel cooling zone temperatures are set such that the energy is removed quickly at the start of the screw and less at the discharge. Thus, the bulk temperature of the resin-blowing agent mixture decreases quickly early in the process and approaches an asymptotic value at the discharge, that is, the foaming temperature. As the temperature of the mixture decreases in the downstream direction, the viscosity of the mixture increases, and a higher amount of energy is dissipated locally by the rotation of the screw. This viscosity effect is the reason for the asymptotic nature of the temperature profile.

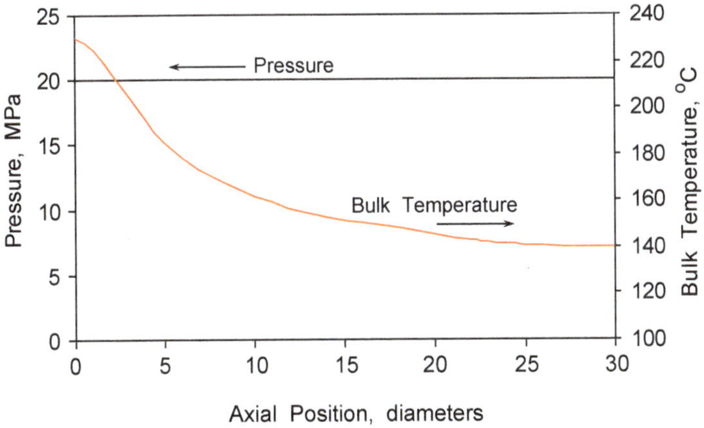

Figure 15.17 Typical axial pressure and temperature profiles for a secondary cooling extruder running a PS-blowing agent mixture

A typical pressure and temperature profile for a secondary cooling extruder is shown in Figure 15.17 for a PS resin.

The resin-blowing agent mixture from the primary extruder enters the secondary extruder at relatively high pressures, typically about 20 MPa. In order to eliminate leakage through the shank end of the screw, a specialized viscous seal is fabricated into the shank. The geometry of the channels is essentially identical to that used on large-diameter melt-fed extruders as described in Figure 15.12 except that the length of the seal can be 2 to 3 diameters in length. The extra length is needed to compensate for the high pressure of the feedstock entering the secondary extruder. The actual length needed will depend on the viscosity of the feedstock, the clearance between the shank of the screw and the barrel wall, and the dimensions of the viscous seal. Here the pressure gradient between the feed pressure and the outside chamber causes resin to flow out through the annulus created between the shank and the barrel wall. The channels cut into the shank convey the resin back into the feed area via rotational flow. The design is such that these flows are identical, and thus flow out through the back of the extruder does not occur. Although many design variations are used, a typical viscous seal design for a 152.4 mm diameter cooling screw is shown in Table 15.3. A cooling hole is drilled into the shank end up to the end of the viscous seal as shown in Figure 10.8. Cooling water is forced through a rotary union to the shank of the screw and then back out through the rotary union. Cooling the shank decreases the amount of material flow through the gap between the barrel and shank by increasing the viscosity of the resin in the region. Water cooling is also applied to the barrel side of the shank.

Table 15.3 Typical Viscous Seal Channel Dimensions for a 152.4 mm Diameter Secondary Cooling Extruder

	Dimension
Number of flight starts	4
Lead length, mm	75
Channel depth, mm	1.5
Length, diameters	3
Flight width perpendicular to the edge, mm	9

15.4.1 High-Performance Cooling Screws

High-performance cooling screws can operate at higher rates and higher qualities than conventional multichannel, slotted screws. These screws are designed to operate at a specific rotation rate that is 20 to 50% higher than a conventional cooling screw. At higher specific rates, the high-performance screw will rotate at a slower speed for the same rate and thus dissipate less energy. The high level of thermal gradients as shown in Figure 15.15 is eliminated using advanced mixing processes. The two most widely used high-performance screws are the four-channel Energy Transfer™ (ET) screw from Robert Barr, Inc. and the Turbo-Screws™ (TS) technology from Plastic Engineering Associates Licensing, Inc. The design and general operation of these screws will be presented in the next sections.

Thermal gradients that exist within the channels of the screw can be apparent in the foam sheet as regions with different densities, and they can often be observed with thermal imaging cameras. For example, Figure 15.18(a) is a thermal image of PS resin foam moving across the cooling mandrel at a rate that exceeds the cooling capacity of the secondary extruder with a conventional screw. The patterns of the flights are clearly visible as regions that are cooler than the bulk temperature of the sheet. In this case the bulk temperature of the sheet was 75 °C. Visual examination of the sheet showed imperfections due to different densities in the regions of the flight pattern. A high-performance screw was placed in the secondary extruder and then operated at the same rate, and the thermal image is shown in Figure 15.18(b). As shown in this figure, the screw pattern is not visible, and the bulk temperature of the sheet was reduced to 60 °C. The sheet had a consistent appearance and was acceptable for downstream thermoforming operations. Figure 15.18 clearly shows the performance advantages of a high-performance cooling screw design over a conventional multichannel, slotted screw design.

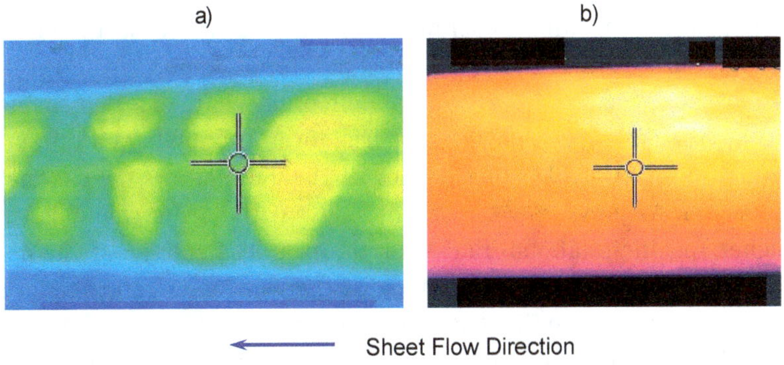

Figure 15.18 Thermal images of PS foam sheet over the mandrel: a) image is for a conventional cooling screw operating beyond the designed rate, causing thermal gradients in the foam in a pattern of the screw flights (average temperature was 75 °C), and b) an image of foam produced using a high-performance cooling screw at the same rate (average temperature was 60 °C)

The actual rate improvement obtained from a high-performance cooling screw will depend on equipment factors, including the design of the conventional cooling screw, the barrel cooler designs, and the availability of torque from the motor. Since high-performance cooling screws operate at higher specific rates, the screws will rotate slower for the same rate. The lower rotation rates will require a higher level of torque from the motor. If a tandem line is retrofitted with a high-performance cooling screw, the designer must make sure that enough torque is available to operate at higher specific rates. Often torque availability limits the maximum rate of a retrofitted operation.

15.4.1.1 Four-Channel Energy Transfer Cooling Screws

Four-channel Energy Transfer (ET) cooling screws can be used to increase the rate and sheet quality of foamed products. The screw is constructed much like a plasticating ET screw as described in Section 14.2.2, except that a second pair of channels runs parallel with the A and B channels. These new channels are labeled C and D with the A and C channels having the same depth at any axial location. Likewise, the B and D channels are designed at the same depth at any axial location. A schematic of a four-channel ET cooling screw is shown by Figure 15.19, and the channel depths for a 154.2 mm diameter screw are shown in Figure 15.20. A typical screw will be flightless directly underneath the feed transfer line from the primary extruder. Pressure is used to flow the resin into the first diameter of the screw. A flightless feed area allows less restriction at the entry area and mitigates pressure surges due to a passing flight. The next 1 to 2 diameters are double flighted, and both channels have identical and constant depths. At the end of the double-flighted section two additional flights start, creating the entry to the ET section of the cooling screw. The four channels created are consecutively labeled A, B, C, and D. Like a plasticating ET screw, undercuts periodically occur in the flights, allowing material to flow to the channel behind it. The undercuts on the pushing side of each channel occur only when the depth of the channel is decreasing in the downstream direction. The flow of material over the undercut mitigates the high-temperature regions shown in Figure 15.15, and it allows the channels to be designed deeper. As previously discussed, screws with deeper channels will operate at higher specific rotational rates and dissipate less energy. At the discharge end of the screw, two of the flights end such that the screw returns to a double-flighted section. The double-flighted section has a constant depth and is designed to stabilize the flow. The lead length is typically two times the diameter in order to accommodate all four channels and to maintain a high specific rate.

Figure 15.19 Schematic of a four-channel ET cooling screw (courtesy of Jeff A. Myers of Robert Barr, Inc.). The flights that are shaded have a normal clearance with the barrel wall while the flights that are not shaded are undercut

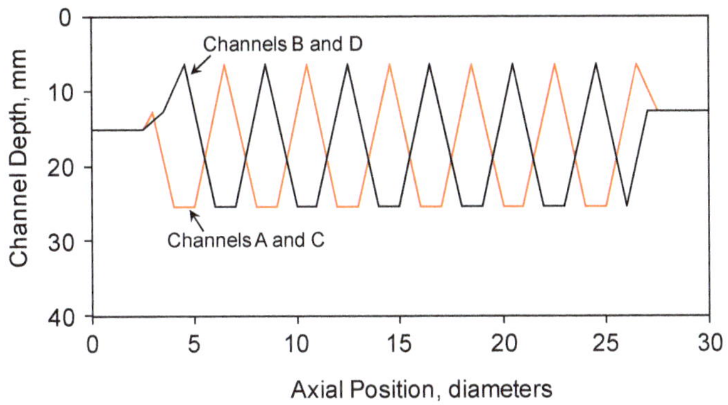

Figure 15.20 Channel depths for a 152.4 mm diameter four-channel ET cooling screw with an *L/D* of 30. The lead length of all sections of the screw was 305 mm. Channels A and C are the same depth at any axial position. Channels B and D are also identical in depth

The periodic undercut on the flights of the channel and the oscillating depth of the channels allow the screw channels to be designed very deep. Undercutting the flights, however, will reduce the heat transfer coefficient because the local barrel area will not be wiped as often. The large degree of thermal homogenization due to the flow over the flight tends to compensate for the loss in heat transfer.

15.4.1.2 The Turbo-Screws Technology for a Cooling Extruder

The Turbo-Screws (TS) cooling screw is designed with multiple flights and with very deep and constant-depth channels [24, 25]. A schematic of a TS cooling screw is shown in Figure 15.21. Like the ET screw, the TS cooling screw dissipates minimal amounts of energy to the resin via the deep channel design. The thermal gradients created by the deep channels, however, are eliminated using circulation channels cut into the side of the flight. These circulation channels allow some resin near the pushing flight to flow to the channel behind it. The circulation channels are designed to provide elongational flow. Elongational flows occur because the opening of the circulation channel on the pushing side of the channel is relatively large and then tapers to a smaller exit opening on the trailing side of the channel to the rear. The elongational flows provide a level of mixing to the melt, and the flows between the channels reduce the thermal gradients, allowing higher rates and higher quality foams.

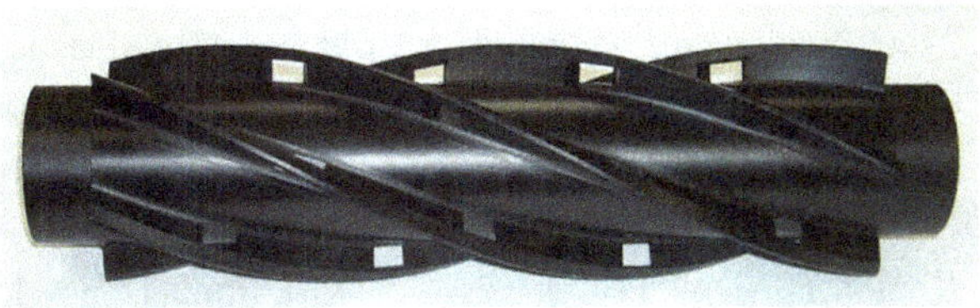

Figure 15.21 Photograph of a TS cooling screw model segment showing the multiple, deep flights and the circulation channels positioned in the flights (courtesy of James D. Fogarty of Plastic Engineering Associates Licensing, Inc.)

A unique feature of the TS cooling screw is that the circulation channels allow flow to the channel behind it without decreasing the level of wiping of the barrel by the flights. That is, all flights are designed with a tight clearance between the flight tip and the barrel wall, allowing for maximum heat transfer between the melt and the barrel wall.

Nomenclature

- H channel depth
- H_c Henry's constant
- P absolute pressure for Henry's law
- w weight fraction of dissolved ethylene in LDPE resin at equilibrium (or other reactant-polymer system)
- W average width of the channel perpendicular to the flight
- y mole fraction of ethylene (or other reactant) in the gas phase

References

1. Chung, C.I., Powell, T.M., and Werling, C.L., "Shear Refining of Branched Low Density Polyethylene," *SPE ANTEC Tech. Papers*, **44**, 76 (1998)
2. Spalding, M.A., Campbell, G.A., Carlson, F., and Nazrisdoust, K., "The Effect of Barrel Diameter and Rotation on the Numerical Simulation of Single-Screw Extruders," *SPE ANTEC Tech. Papers*, **52**, 792 (2006)
3. Spalding, M.A. and Campbell, G.A., "The Accuracy of Standard Drag Flow and Pressure Gradient Calculations for Single-Screw Extruders," *SPE ANTEC Tech. Papers*, **54**, 262 (2008)
4. Campbell, G.A., Spalding, M.A., and Carlson, F., "Prediction of Screw Temperature Rise in Single-Screw Pump Extruders," *SPE ANTEC Tech. Papers*, **54**, 267 (2008)
5. Todd, D.B., "Introduction to Compounding," Chapter 1 in "Plastics Compounding, Equipment and Processing," Todd, D.B. (Ed.), Hanser Publishers, Munich (1998)
6. White, J.L. and Kim, E.K., "Twin Screw Extrusion, Technology and Principles," 2nd ed., Hanser Publishers, Munich (2010)
7. Maraschin, N., "Polyethylene, Low Density," in "Kirk-Othmer Encyclopedia of Chemical Technology," John Wiley & Sons, Inc., New York, (2005)

8. Platzer, N., "Design of Continuous and Batch Polyethylene, PVC, and Styrene Resins," *Ind. Eng. Chem.*, **62**, 6 (1970)

9. Golding, B., "Polymers and Resins, Their Chemistry and Chemical Engineering," Van Nostrand Company, Inc., Princeton, NJ (1959)

10. Bokis, C.P., Ramanathan, S., Franjione, J., Buchelli, A., Call, M.L., and Brown, A.L., "Physical Properties, Reactor Modeling, and Polymerization Kinetics in the Low-Density Polyethylene Tubular Reactor Process," *Ind. Eng. Chem. Res.*, **41**, 1017 (2002)

11. Buchelli, A., Call, M.L., Brown, A.L., Bokis, C.P., Ramanathan, S., and Franjione, J., "Nonequilibrium Behavior in Ethylene/Polyethylene Flash Separators," *Ind. Eng. Chem. Res.*, **43**, 1768 (2004)

12. Costa, G.M.N., Guerrieri, Y., Kislansky, S., Pessoa, F.L.P., Vieira de Melo, S.A.B., and Embirucu, M., "Simulation of Flash Separation in Polyethylene Industrial Processing: Comparison of SRK and SL Equations of State," *Ind. Eng. Chem. Res.*, **48**, 8613 (2009)

13. Biesenberger, J.A. and Sebastian, D.H., "Principles of Polymerization Engineering," John Wiley and Sons, New York (1983)

14. Biesenberger, J.A., "Devolatilization of Polymers: Fundamentals, Equipment, Applications," Hanser Publishers, Munich (1983)

15. Biesenberger, J.A., Dey, S.K., and Brizzolara, J., "Devolatilization of Polymer Melts: Machine Geometry and Scale Factors," *Polym. Eng. Sci.*, **30**, 1493 (1990)

16. Yang, C.-T., Smith, T.G., Bigio, D.I., and Anolick, C., "A Model for Foam Devolatilization in an Extruder," *Ind. Eng. Chem. Res.*, **37**, 1464 (1998)

17. Fenner, R.T., "The Design of Large Hot Melt Extruders," *Polymer*, **16**, 298 (1975)

18. Baker, W.E., Rudin, A., Schreiber, H.E., and El-Kindi, M., "The Effect of Processing on Rheological and Molecular Characteristics of a Low Density Polyethylene," *Polym. Eng. Sci.*, **33**, 377 (1993)

19. Huck, N.D. and Clegg, P.L., "The Effect of Extrusion Variables on the Fundamental Properties of Tubular Polyethylene Film," *SPE Trans.*, 121 (July 1961)

20. Rokudai, M., Mihara, S., and Fujiki, T., "Influence of Shearing History on the Rheological Properties and Processability of Branched Polymers. II. Optical Properties of Low-Density Polyethylene Blown Films," *J. Appl. Polym. Sci.*, **23**, 3289 (1979)

21. Maillefer, C., "Screw for Extrusion Apparatus," U.S. Patent 3,358,327 (1967)

22. Maloney, D.P. and Prausnitz, J.M., "Solubilities of Ethylene and Other Organic Solutes in Liquid, Low-Density Polyethylene in the Region 124 to 300 °C," *AIChE J.*, **22**, 1, 74 (1976)

23. Skochdopole, R.E. and Welsh, G.C., "Polystyrene Foams," in "Styrene Polymers," Moore, E.R. (Ed.), 2nd ed., John Wiley & Sons, Inc., New York (1989)

24. Fogarty, J., "Thermoplastic Foam Extrusion Screw with Circulation Channels," U.S. Patent 6,015,227 (2000)

25. Fogarty, J., Fogarty, D., and Rios, A., "Turbo-Screws™, New Screw Design for Foam Extrusion," *SPE ANTEC Tech. Papers*, **47**, 167 (2001)

26. Burdett, I. D. and Eisinger, R. S., " Ethylene Polymerization Processes and Manufacture of Polyethylene," Chapter 3 of "Handbook of Industrial Polyethylene and Technology", edited by Spalding, M. A. and Chatterjee, A. M., Wiley - Scrivener Publishing, Hoboken, NJ, 2017.

27. Fay, J.J. and King, III, R. E., " Degradation and Stablization of Polyethylene," Chapter 25 of "Handbook of Industrial Polyethylene and Technology", edited by Spalding, M. A. and Chatterjee, A. M., Wiley - Scrivener Publishing, Hoboken, NJ, 2017.

28. Gray, J. and Breuer, T., "Slip Agents," Chapter 28 of "Handbook of Industrial Polyethylene and Technology", edited by Spalding, M. A. and Chatterjee, A. M., Wiley - Scrivener Publishing, Hoboken, NJ, 2017.

29. Seiler, D. A., Beaume, F., Devisme, S., and Pomante, J. A., "Fluorinated Polymer Processing Aids for Polyethylene," Chapter 33 of "Handbook of Industrial Polyethylene and Technology", edited by Spalding, M. A. and Chatterjee, A. M., Wiley - Scrivener Publishing, Hoboken, NJ, 2017.

Appendix A1 Polymer Abbreviation Definitions

This text and others use abbreviations for most resin types. The common resin definitions are presented below.

ABS	acrylonitrile-butadiene-styrene terpolymer
aPS	atactic polystyrene
CPE	chlorinated polyethylene
EAA	polymers containing ethylene and acrylic acid
EEA	polymers containing ethylene and ethyl acetate
EMA	polymers containing ethylene and methyl acrylate
EVA	polymers containing ethylene and vinyl acetate
GPPS	general purpose polystyrene or "crystal clear" polystyrene
HDPE	high density polyethylene
HIPS	high impact polystyrene
LDPE	low density polyethylene
LLDPE	linear low density polyethylene
MDPE	medium density polyethylene
OBC	olefinic block copolymers
PA	polyamide
PA 6	polyamide made from caprolactam
PA 6,6	polyamide made from hexamethylenediamine and adipic acid
PAN	polyacrylonitrile
PB	polybutylene
PBT	polybutylene terephthalate
PC	polycarbonate
PC/ABS	a physical blend of polycarbonate and acrylonitrile-butadiene-styrene terpolymer
PE	polyethylene
PEEK	polyetheretherketone

PEN	polyethylene naphthalate
PEO	polyethylene oxide
PET	polyethylene terephthalate
PIB	polyisobutylene
PLA	polylactic acid
PMMA	poly(methyl methacrylate)
PP	polypropylene
PPO	polyphenylene oxide
PS	polystyrene
PTFE	polytetrafluoroethylene
PU	polyurethane
PVC	polyvinyl chloride
PVDC	polyvinylidene chloride
PVDF	polyvinylidene fluoride
SAN	styrene-acrylonitrile copolymer
SBR	styrene-butadiene rubber
SBS	styrene-butadiene-styrene block copolymer
SIS	styrene-isoprene-styrene block copolymer
SMA	styrene maleic anhydride copolymer
sPS	syndiotactic polystyrene
TPE	thermoplastic elastomer
TPU	thermoplastic polyurethane
ULDPE	ultralow density polyethylene

Appendix A3 Rheological Calculations for a Capillary Rheometer and for a Cone and Plate Rheometer

This appendix contains the detailed development for the equations that are presented in Chapter 3. Full understanding of these developments is not required for detailed analysis and troubleshooting of the extrusion process. They are presented here for those who desire a deep understanding of the mathematics involved with the screw rotation analysis. Some of the equations and figures are duplicated in this appendix for clarity. The nomenclature used here is consistent with that used earlier. The reader is directed to Chapter 3 for nomenclature.

A3.1 Capillary Rheometer

This section will present the derivations for a capillary rheometer. A schematic of the device is shown in Figure A3.1. The main discussion is presented in Section 3.6.1. The assumptions for the derivations are as follows:

1. Flow is only in the z direction.
2. The fluid is incompressible.
3. A symmetric plane exists normal to the z direction.
4. The capillary is long and thus the end effects can be neglected.
5. The stress tensor is symetric.
6. Pressure gradient in the z direction is constant.
7. Stress is finite at all values of r.

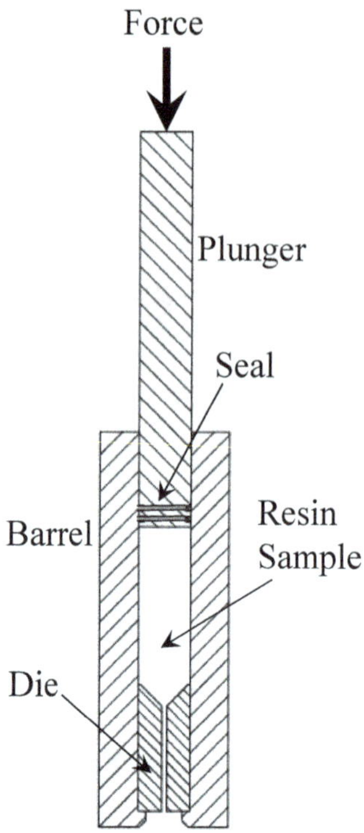

Figure A3.1 Schematic of a capillary rheometer

The capillary in the die has a diameter of $2R$ and a length in the z direction (axial) of L. The pressure gradient is considered to be constant over the length of the capillary. Since the z direction velocity is not a function of z and it is not a function of θ, the z component of the cylindrical coordinate equation of motion reduces to an ordinary differential equation [1]:

$$\frac{dP}{dz} = \frac{\mu}{r}\frac{d}{dr}\left(r\frac{dV_z}{dr}\right) \tag{A3.1}$$

At the entry to the capillary:

$$z = 0 \quad P = P_0 \tag{A3.2}$$

At the discharge end:

$$z = L \quad P = P_L \tag{A3.3}$$

At the center and wall of the capillary:

$$r = 0 \quad \frac{dV_z}{dr} = 0 \quad \text{center of the capillary} \tag{A3.4}$$

$$r = R \quad V_z = 0 \quad \text{wall of the capillary} \tag{A3.5}$$

The resulting pressure and velocity can be easily determined:

$$P = -\frac{P_0 - P_L}{L} z + P_0 \tag{A3.6}$$

$$V_z(r) = \frac{(P_0 - P_L)R^2}{4\mu L}\left(1 - \left(\frac{r}{R}\right)^2\right) \tag{A3.7}$$

It follows that the volumetric flow rate Q can be calculated as follows:

$$Q = \int_0^R \int_0^{2\pi} V_z(r) r d\theta dr \tag{A3.8}$$

$$Q = \frac{\pi(P_0 - P_L)R^4}{8\mu L} \tag{A3.9}$$

It follows that:

$$V_z(r) = \frac{2Q}{\mu R^2}\left(1 - \left(\frac{r}{R}\right)^2\right) \tag{A3.10}$$

From the nonzero terms of the equation of motion:

$$-\frac{dP(z)}{dz} = \frac{1}{r}\frac{d}{dr}\left(r\tau_{rz}(r)\right) \tag{A3.11}$$

Using the same pressure boundary conditions:

$$\tau_{rz} = -\frac{(P_0 - P_L)r}{2L} + \frac{C_1}{r} \tag{A3.12}$$

C_1 must be 0 if the shear stress is to be finite at all r values.

$$\tau_{rz} = -\frac{(P_0 - P_L)r}{2L} = \tau_R \frac{r}{R} \tag{A3.13}$$

Where:

$$\tau_{rz}\big|_{r=R} = \tau_R = -\frac{(P_0 - P_L)R}{2L} \tag{A3.14}$$

Now:

$$\dot{\gamma} = -\frac{dV_z(r)}{dr} = \frac{4Q}{\pi R^3}\frac{r}{R} \tag{A3.15}$$

$$\dot{\gamma}_R = \frac{4Q}{\pi R^3} \tag{A3.16}$$

$$\eta = \frac{-\tau_{21}}{\dot{\gamma}_0} = \frac{-\tau_R}{\dot{\gamma}_R} = \frac{(P_0 - P_L)R}{2L}\left(\frac{\pi R^3}{4Q}\right) = \mu \tag{A3.17}$$

Where μ is referred to as the Newtonian viscosity. The apparent shear rate is often used when dealing with a non-Newtonian polymer:

$$\dot{\gamma}_a = \frac{4Q}{\pi R^3} = \frac{1}{\mu}\frac{(P_0 - P_L)R}{2L} \tag{A3.18}$$

As pointed out by Morrison [1], there is nothing apparent about $\dot{\gamma}_a$; that is, this is the shear rate that is obtained when it is assumed that the polymer behaves like a Newtonian fluid.

$$\dot{\gamma}_a = \frac{1}{\mu}\tau_R \tag{A3.19}$$

Now for a power law fluid the viscosity μ is replaced by the shear rate dependent viscosity η.

$$\eta = m\dot{\gamma}^{n-1} \tag{A3.20}$$

$$\frac{dP}{dz} = -\frac{m}{r}\frac{d}{dr}\left(r\left(-\frac{dV_z}{dr}\right)^n\right) \tag{A3.21}$$

Again:

$$P = -\frac{P_0 - P_L}{L}z + P_0 \tag{A3.22}$$

And the velocity is determined by integration:

$$V_z = \frac{(P_0 - P_L)^{\frac{1}{n}} R^{\frac{1}{n}+1}}{2mL}\left(\frac{1}{\frac{1}{n}+1}\right)\left(1 - \left(\frac{r}{R}\right)^{\frac{1}{n}+1}\right) \tag{A3.23}$$

Finding the shear rate by differentiation at the wall:

$$\dot{\gamma}_w = -\frac{dV_z}{dr}\bigg|_{r=R} = \frac{4Q}{\pi R^3}\left(\frac{1/n+3}{4}\right) = \dot{\gamma}_a\left(\frac{3n+1}{4n}\right) = \left(\frac{\tau_R}{m}\right)^{\frac{1}{n}} \tag{A3.24}$$

A3.2 Cone and Plate Rheometer

This section will present the derivations for a cone and plate rheometer. A schematic of the device is shown in Figure A3.2. The main discussion is presented in Section 3.6.2.

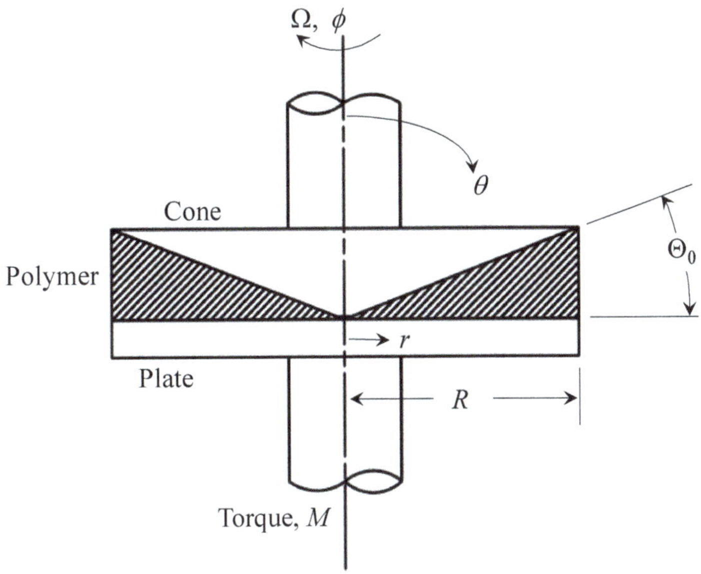

Figure A3.2 Schematic for a cone and plate rheometer

The velocity in vector format for the above cone and plate rheometer is:

$$\underline{v} = \begin{pmatrix} 0 \\ 0 \\ v_\varphi \end{pmatrix}_{r\theta\varphi} \tag{A3.25}$$

We now assume that we have simple shear flow in the φ direction and that the gradient is in the $(-r\theta)$ direction. From continuity, $\partial v_\varphi / \partial \varphi = 0$, which leads to:

$$v_\varphi = C_1(-r\theta) + C_2 \tag{A3.26}$$

For boundary conditions:

$$v_\varphi = 0 \qquad \theta = \frac{\pi}{2} \tag{A3.27}$$

$$v_\varphi = r\Omega \qquad \theta = \frac{\pi}{2} - \Theta_0 \tag{A3.28}$$

This leads to:

$$v_\varphi = \frac{r\Omega}{\Theta_0}\left(\frac{\pi}{2}-\theta\right) \tag{A3.29}$$

Using the parameters in the above figure, the deformation rate tensor is defined as follows:

$$\underline{\underline{\dot{\gamma}}} = \begin{pmatrix} 0 & 0 & r\dfrac{\partial}{\partial r}\left(\dfrac{v_\varphi}{r}\right) \\ 0 & 0 & \dfrac{\sin\theta}{r}\dfrac{\partial}{\partial\theta}\left(\dfrac{v_\varphi}{\sin\theta}\right) \\ r\dfrac{\partial}{\partial r}\left(\dfrac{v_\varphi}{r}\right) & \dfrac{\sin\theta}{r}\dfrac{\partial}{\partial\theta}\left(\dfrac{v_\varphi}{\sin\theta}\right) & 0 \end{pmatrix}_{r\theta\varphi} = \begin{pmatrix} 0 & 0 & 0 \\ 0 & 0 & \dot{\gamma}_{\theta\varphi} \\ 0 & \dot{\gamma}_{\theta\varphi} & 0 \end{pmatrix}_{r\theta\varphi} \tag{A3.30}$$

$$\dot{\gamma}_{\theta\varphi} = \frac{\sin\theta}{r}\frac{\partial}{\partial\theta}\left(\frac{v_\varphi}{\sin\theta}\right) = \frac{1}{r}\frac{\partial v_\varphi}{\partial\theta} = -\frac{\Omega}{\Theta_0} \tag{A3.31}$$

Thus:

$$\underline{\underline{\dot{\gamma}}} = \begin{pmatrix} 0 & 0 & 0 \\ 0 & 0 & -\dfrac{\Omega}{\Theta_0} \\ 0 & -\dfrac{\Omega}{\Theta_0} & 0 \end{pmatrix}_{r\theta\varphi} \tag{A3.32}$$

It follows that:

$$\dot{\gamma} = \left|\underline{\underline{\dot{\gamma}}}\right| = \frac{\Omega}{\Theta_0} \tag{A3.33}$$

The strain for the cone and plate rheometer is as follows:

$$\gamma(0,t) = \int_0^t \dot{\gamma}(t')\,dt' = \int_0^t \frac{\Omega}{\Theta_0}\,dt' = \frac{\Omega t}{\Theta_0} \tag{A3.34}$$

Now it follows that:

$$\dot{\gamma}_0 = \frac{1}{r}\frac{\partial v_\varphi}{\partial(-\theta)} = \frac{\Omega}{\Theta_0} = \dot{\gamma} \tag{A3.35}$$

For shear flow we utilize the τ_{21} component of the stress tensor and in this coordinate system it follows that:

$$\tau_{21} = -\tau_{\theta\varphi} \tag{A3.36}$$

Therefore the viscosity is defined as:

$$\eta = \frac{-\tau_{21}}{\dot{\gamma}_0} = \frac{-\tau_{\theta\varphi}}{\dot{\gamma}} \tag{A3.37}$$

Now the stress must be evaluated from the torque on the bottom plate, M:

$$M = \int_A (stress)(lever\ arm)dA = \int_0^{2\pi}\int_0^R \left(\tau_{\theta\varphi}\bigg|_{\theta=\frac{\pi}{2}}\right)(r)(rd\varphi dr) \tag{A3.38}$$

Now the shear stress is constant as is the shear rate throughout the polymer in the gap, and this allows the shear stress to be removed from the integrals as follows:

$$M = \frac{2}{3}\pi R^3 \left(\tau_{\theta\varphi}\bigg|_{\theta=\frac{\pi}{2}}\right) \tag{A3.39}$$

Solving for the shear stress and using the definition of viscosity one obtains:

$$\eta = \frac{-\tau_{21}}{\dot{\gamma}_0} = \frac{-\tau_{\theta\varphi}}{\dot{\gamma}} = \frac{3M\Theta_0}{2\pi R^3\Omega} \tag{A3.40}$$

References

1. Morrison, F.A., Table C.6, "Understanding Rheology," Oxford University Press, New York (2001)

Appendix A4 Shear Stress at a Sliding Interface and Melting Fluxes for Select Resins

As presented in Chapters 5 and 6, solids conveying and melting are two key processes in plasticating single-screw extruders. In order for the extruder to operate properly, the solids conveying section must have a rate high enough to supply material to the transition and metering sections of the screw (smooth bore machines). If the solids conveying rate is less than the designed rate for the metering section, channels downstream of the solids conveying section will operate starved, leading to reduced rates, high discharge temperatures, degraded resin, and flow surging. Moreover, the melting section must be long enough and capable of completely melting the resin. If the extruder is operated at rates that exceed the melting capacity of the screw, then solid resin fragments will discharge with the extrudate, creating defects in the finished product. This appendix contains shear stress at a sliding interface data and melting fluxes for select resins [1-6]. These data were measured with the Screw Simulator at Dow. The Screw Simulator is described in Section 4.3.1.

A4.1 Shear Stress at a Sliding Interface for Select Resins

The coefficient of dynamic friction is, in general, a function of temperature, pressure, and the relative velocity of the resin with respect to the metal surface on which the polymer is sliding. Forces acting on the solid bed in a single-screw extruder are directly proportional to the coefficient of dynamic friction for temperatures less than the melting or devitrification temperature of the polymer. Once the temperature on the polymer-metal interface exceeds the melting or devitrification temperature, these forces are controlled by a viscous mechanism. Both types of forces, however, are proportional to the shear stress at a sliding interface. The shear stress for a polymer bed sliding on a metal surface is related to the coefficient of dynamic friction as follows:

$$\tau = fP \tag{A4.1}$$

where τ is the shear stress at a sliding interface, f is the coefficient of dynamic friction, and P is the pressure at the interface. Shear stress at a sliding interface allows the low temperature stresses from a frictional mechanism to be plotted with the high temperature stresses from a viscous mechanism.

Data for shear stresses at a sliding interface are presented in several sections within the main chapters of this book. The data are used to show the nature of the stress and provide insight into troubleshooting case studies. The resins and sections where the stresses are presented in the chapters are listed in Table A4.1.

Table A4.1 Data for Shear Stresses at a Sliding Interface for Select Resins in this Book

Resin	Section
PET	4.3.1
HIPS, 3.9 dg/min (230 °C, 5.0 kg)	12.7.3
ABS, 2 dg/min (230 °C, 3.8 kg)	12.7.6
PC, 6 dg/min (300 °C, 1.2 kg)	12.7.8
TPU	13.4.2
LDPE, 2.0 dg/min (190 °C, 2.16 kg)	A5.6

Stresses at a sliding interface for select resins are shown in Figures A4.1 through A4.8. All resins were commercially available when tested, and they contained only low levels of standard additives.

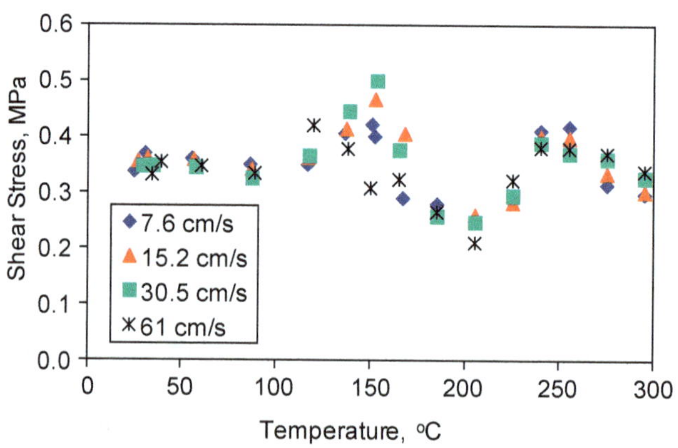

Figure A4.1 Shear stress at a sliding interface for a PC resin with an MFR of 3 dg/min (300 °C, 1.2 kg) as a function of temperature and sliding velocity at a pressure of 0.7 MPa

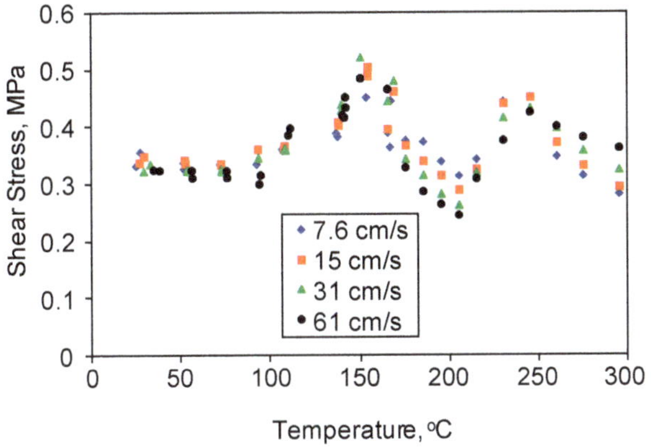

Figure A4.2 Shear stress at a sliding interface for a PC resin with an MFR of 6 dg/min (300 °C, 1.2 kg) as a function of temperature and sliding velocity at a pressure of 0.7 MPa

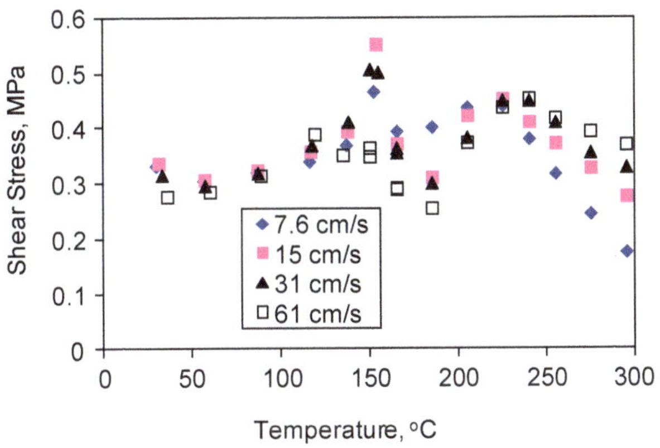

Figure A4.3 Shear stress at a sliding interface for a PC resin with an MFR of 22 dg/min (300 °C, 1.2 kg) as a function of temperature and sliding velocity at a pressure of 0.7 MPa

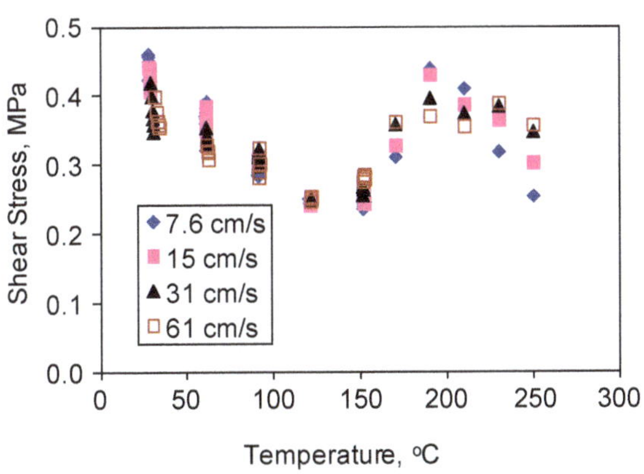

Figure A4.4 Shear stress at a sliding interface for a PC/ABS resin blend with an MFR of 1.7 dg/min (260 °C, 3.8 kg) as a function of temperature and sliding velocity at a pressure of 0.7 MPa. The melting flux for this resin is shown in Figure 4.13

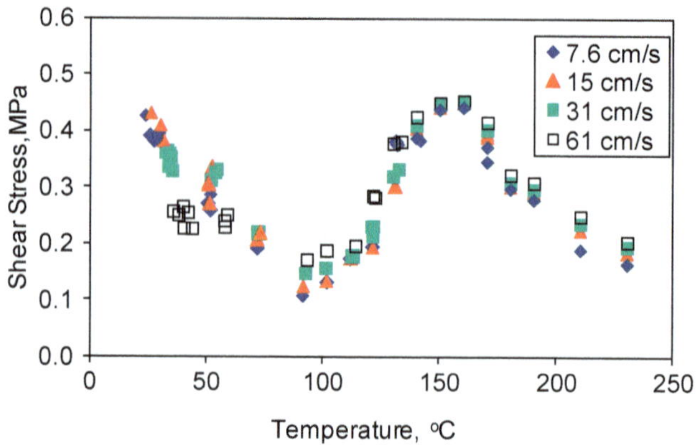

Figure A4.5 Shear stress at a sliding interface for an ABS resin with an MFR of 2.5 dg/min (230 °C, 3.8 kg) as a function of temperature and sliding velocity at a pressure of 0.7 MPa

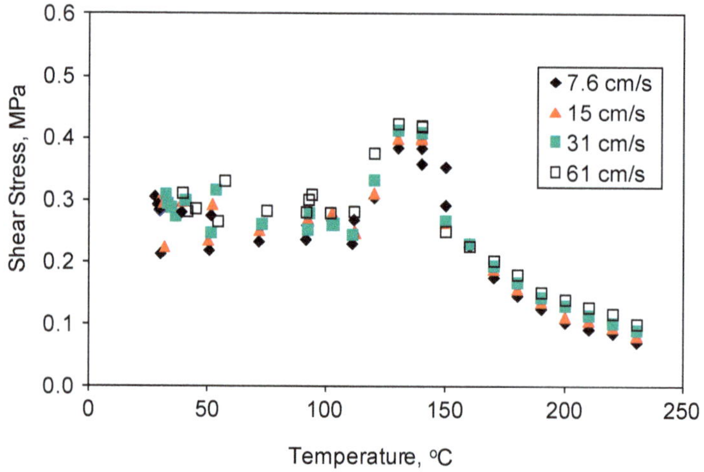

Figure A4.6 Shear stress at a sliding interface for a HIPS resin with an MFR of 2.8 dg/min (200 °C, 5 kg) as a function of temperature and sliding velocity at a pressure of 0.7 MPa

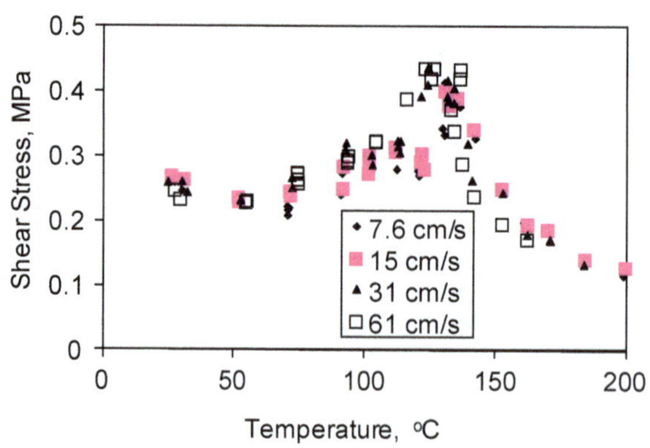

Figure A4.7 Shear stress at a sliding interface for an IRPS resin with an MFR of 9.3 dg/min (200 °C, 5 kg) as a function of temperature and sliding velocity at a pressure of 0.7 MPa

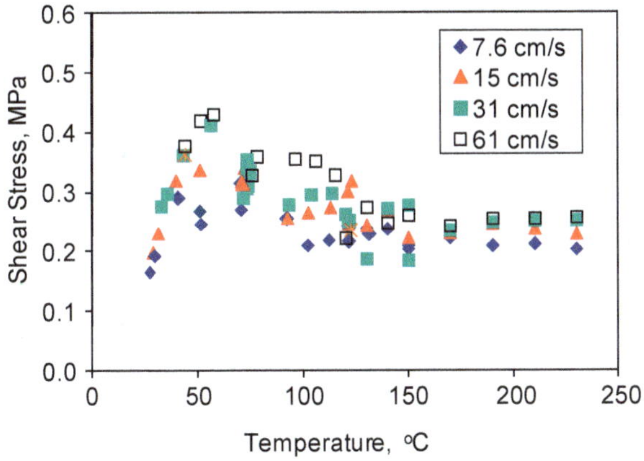

Figure A4.8 Shear stress at a sliding interface for an LLDPE resin with an MI of 1.0 dg/min (190 °C, 2.16 kg) as a function of temperature and sliding velocity at a pressure of 0.7 MPa

A4.2 Melting Fluxes for Select Resins

The melting flux has units of kg/(m^2·s) and represents the rate of phase change for a solid resin bed to a molten resin. This flux is important to understand in order to obtain high rates and extrudates that are free of solid resin particle fragments. As presented in Section 6.3, the primary melting process occurs at the interface of the solid bed and the thin melt film located between the solid bed and the barrel wall. Thus, the area on the barrel wall is important for melting since this is where the largest portion of melting is occurring. This area multiplied by the melting flux determines the melting rate of the process. Melting fluxes are presented in several sections within the main chapters of the book. The data are used to show the nature of the stress and provide insight into troubleshooting case studies. The resins and sections where the melting fluxes are presented in the chapters are listed in Table A4.2.

Table A4.2 Melting Flux Data for Select Resins in this Book

Resin	Section
PC/ABS, 1.7 dg/min (260 °C, 3.8 kg)	4.4
PP, LLDPE, LDPE resin comparative	4.4
ABS, 2 dg/min (230 °C, 3.8 kg)	12.7.6
PC, 6 dg/min (300 °C, 1.2 kg)	12.7.8

Melting fluxes for select resins are shown in Figures A4.9 through A4.16. Like the resins in Section A4.1, these resins were commercially available when tested, and they contained only low levels of standard additives.

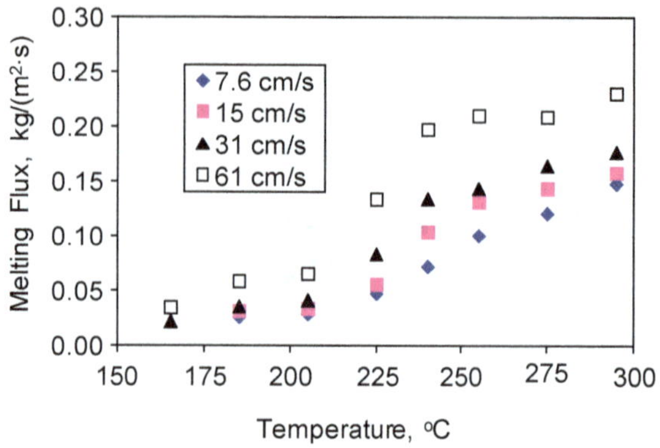

Figure A4.9 Melting flux for a PC resin with an MFR of 3 dg/min (300 °C, 1.2 kg) as a function of temperature and sliding velocity at a pressure of 0.7 MPa

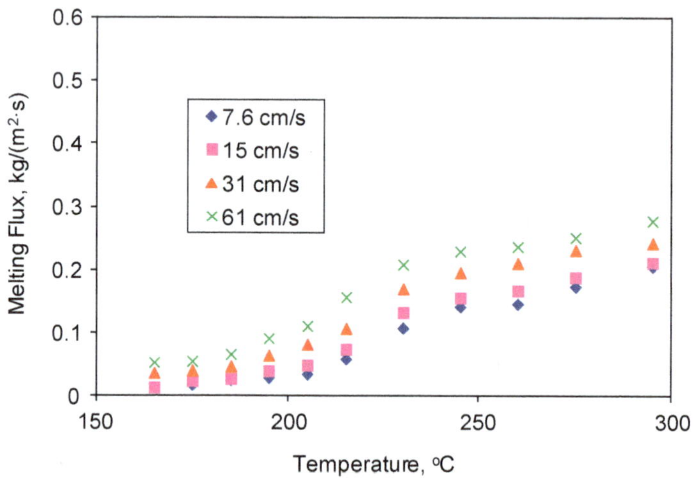

Figure A4.10 Melting flux for a PC resin with an MFR of 6 dg/min (300 °C, 1.2 kg) as a function of temperature and sliding velocity at a pressure of 0.7 MPa

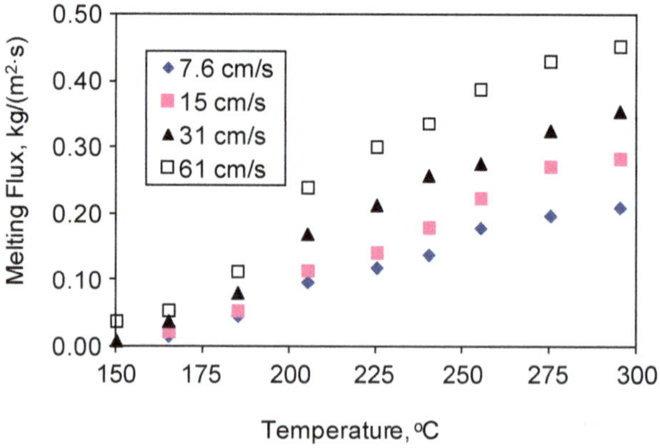

Figure A4.11 Melting flux for a PC resin with an MFR of 22 dg/min (300 °C, 1.2 kg) as a function of temperature and sliding velocity at a pressure of 0.7 MPa

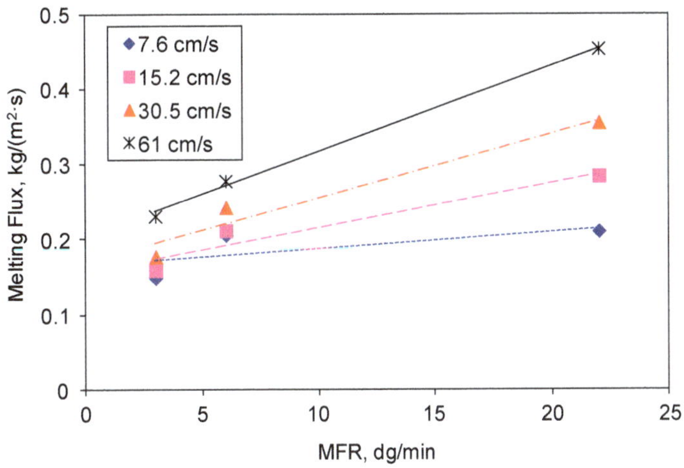

Figure A4.12 Melting fluxes for PC resins at 300 °C as a function of MFR and sliding velocity at a pressure of 0.7 MPa

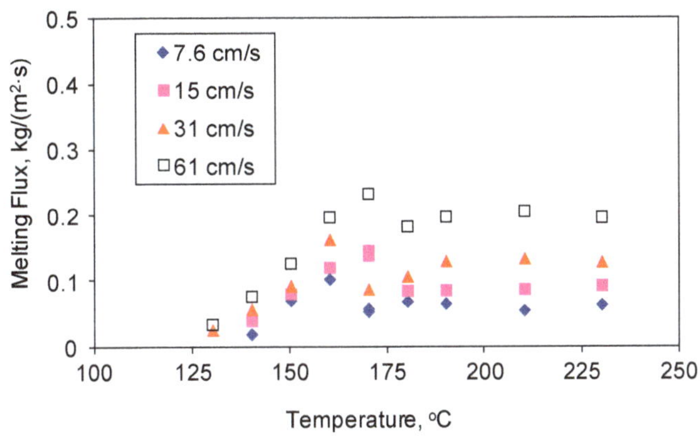

Figure A4.13 Melting flux for an ABS resin with an MFR of 2.5 dg/min (230 °C, 3.8 kg) as a function of temperature and sliding velocity at a pressure of 0.7 MPa

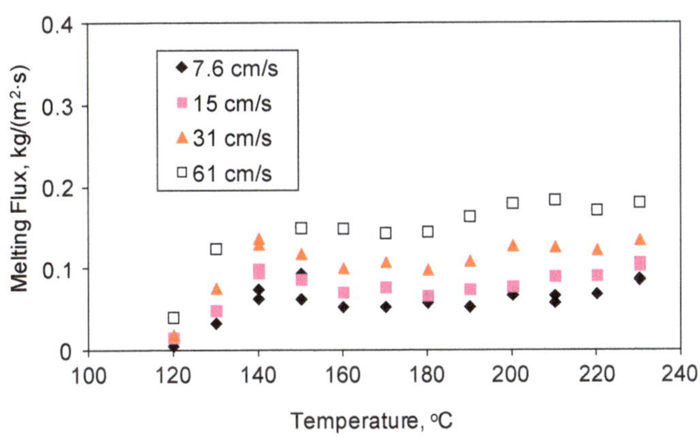

Figure A4.14 Melting flux for a HIPS resin with an MFR of 2.8 dg/min (200 °C, 5 kg) as a function of temperature and sliding velocity at a pressure of 0.7 MPa

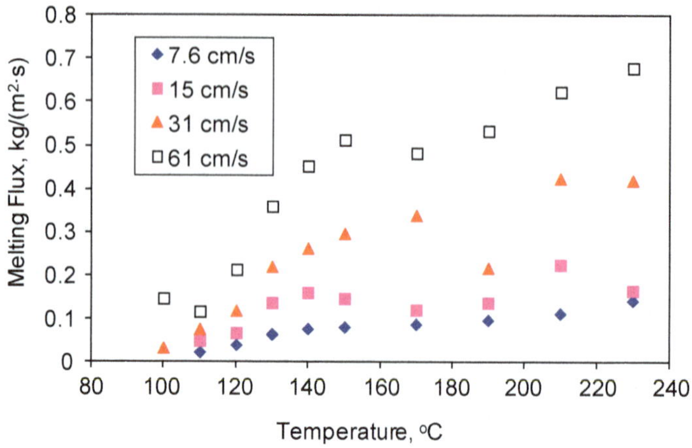

Figure A4.15 Melting flux for an LDPE resin with an MI of 2.0 dg/min (190 °C, 2.16 kg) as a function of temperature and sliding velocity at a pressure of 0.7 MPa. This resin was identical to that used in the solids conveying study in Section A5.6

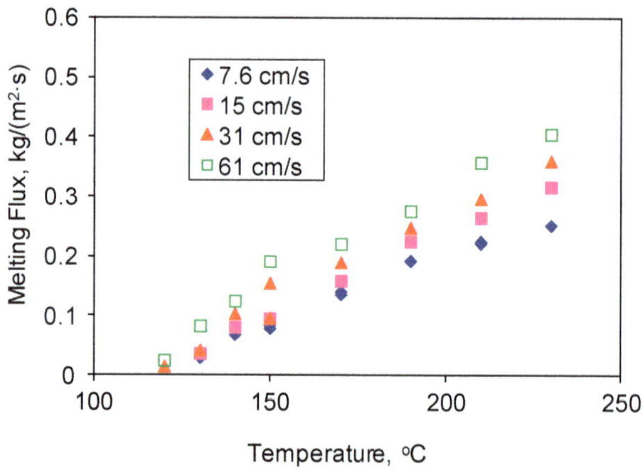

Figure A4.16 Melting flux for an LLDPE resin with an MI of 1.0 dg/min (190 °C, 2.16 kg) as a function of temperature and sliding velocity at a pressure of 0.7 MPa

References

1. Spalding, M.A., Kirkpatrick, D.E., and Hyun, K.S., "Coefficients of Dynamic Friction for Low Density Polyethylene," *Polym. Eng. Sci.*, **33**, 423 (1993)
2. Spalding, M.A., Hyun, K.S., Jenkins, S.R., and Kirkpatrick, D.E., "Coefficients of Dynamic Friction and the Mechanical Melting Mechanism for Vinylidene Copolymers," *Polym. Eng. Sci.*, **35**, 1907 (1995)

3. Spalding, M. A., Hyun, K. S., and Cohen, B. R., "An Experimental Study of the Frictional and Viscous Forces Responsible for Solids Conveying and Melting in Single-Screw Plasticating Extruders," *SPE ANTEC Tech. Papers*, **43**, 202 (1997)

4. Spalding, M. A, Nelb, R. G., Patterson, J., Youngson, C., and Zawisza, M. J., "Processing Ignition-Resistant Polystyrene Resins In Injection Molding Machines," *SPE ANTEC Tech. Papers*, **47**, 476 (2001)

5. Altınkaynak, A., Gupta, M, Spalding, M. A., and Crabtree, S. L., "Shear Stress and Melting Flux for Polycarbonate Resins," *SPE ANTEC Tech. Papers*, **54**, 1405 (2008)

6. Altinkaynak, A., Gupta, M, Spalding, M. A., and Crabtree, S. L., "An Experimental Study on Shear Stress Characteristics of Polymers in Plasticating Single-Screw Extruders," *Polym. Eng. Sci.*, **49**, 471 (2009)

Appendix A5 Solids Conveying Model Derivations and the Complete LDPE Solids Conveying Data Set

This appendix contains the development for three of the more recent solids conveying models that describe the processes that occur in the feed section of the screw. The model developments include the Campbell-Dontula model [1], the Hyun-Spalding model [2–4], and the model byYamamuro et al. [5, 6]. These models and processes are discussed in detail in Chapter 5. The equations are derived using screw rotation. A full understanding of these developments is not required for detailed analysis and troubleshooting of the extrusion process. Some of the equations and figures are duplicated in this appendix for clarity. The nomenclature used here is consistent with that used earlier, and the reader is directed to Chapter 5 for the nomenclature. The complete Dow solids conveying data set for LDPE resins is also provided.

A5.1 Channel Dimensions, Assumptions, and Basic Force Balances

All models are dependent on the differential slab in the z direction, as shown in Figure A5.1. This figure shows the dimensions at the barrel wall, at the screw core, and the average. Like the nomenclature in other chapters including Chapter 5, the dimension at the screw core will have a subscript of c, dimensions at the barrel wall will have a subscript of b, and the average values when they occur will not be subscripted. The dimensions are shown here for clarity. These key dimensions will be removed from the force balance diagrams to follow.

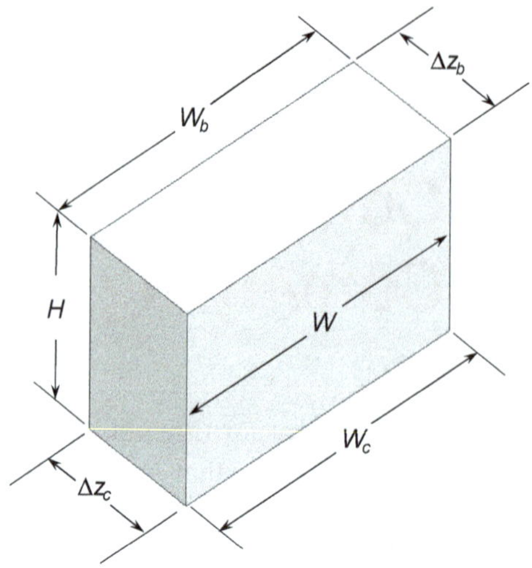

Figure A5.1 Dimensions for the slab in the *z* direction used for the force balances in the next sections. A complete description of the dimensions are provided in Section 7.1

Campbell et al. and Yamamuro et al. assume a form for F_N (or F^*) and then determine the solids conveying angle φ. Hyun and Spalding calculated φ using a force balance and two torque balances. Campbell and Dontula and Yamamuro et al. did not use a torque balance in their analysis. The following is a list of assumptions from all the models discussed. Additional assumptions will be added as needed. Assumptions that do not have keys were used by all authors:

1. The polymer pellets in the slab behave as a continuum.
2. Plug flow exists (the velocity of the polymer is constant in time and space) and the plug contacts all sides of the channel.
3. The channel depth, *H*, is constant.
4. The flight clearance (between the flight and the barrel) is neglected.
5. Gravitational forces are neglected.
6. The density of the plug is constant (Campbell model [1] and Yamamuro model [5, 6]).
7. The pressure is a function of the down-channel direction only.
8. The coefficients of friction, f_b and f_s, are independent of pressure, but may be functions of temperature (Campbell model [1] and Yamamuro model [5, 6]).
9. The coefficients of friction, f_b and f_s, are dependent on temperature, pressure, and velocity (Hyun-Spalding model [2]).

Figure A5.2 is a generalized force diagram of the screw rotation forces and Figure A5.3 illustrates an elemental slice of the plug with the forces labeled. Note that not all the forces or angles shown on the diagram in Figure A5.2 are inclusive of all of the models. The major difference between barrel and screw rotation assumptions are in the literature barrel assumption F^* is evaluated and in the screw rotation models F_N is evaluated. The force diagrams are thus very similar.

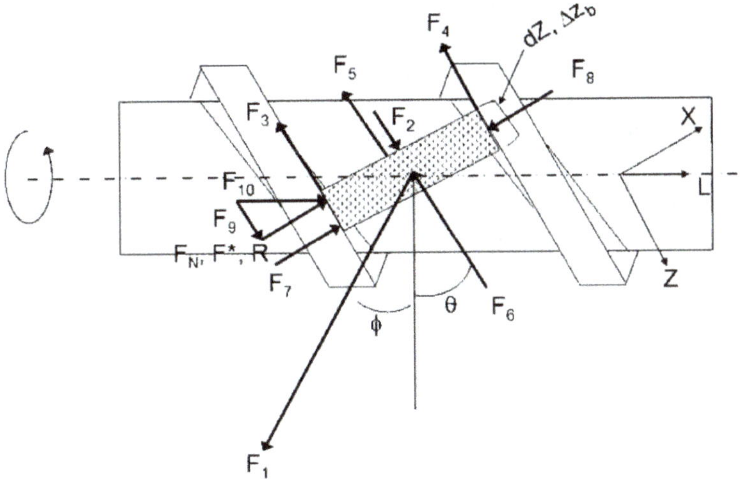

Figure A5.2 Slab and force diagram for a rotating screw and stationary barrel [1]

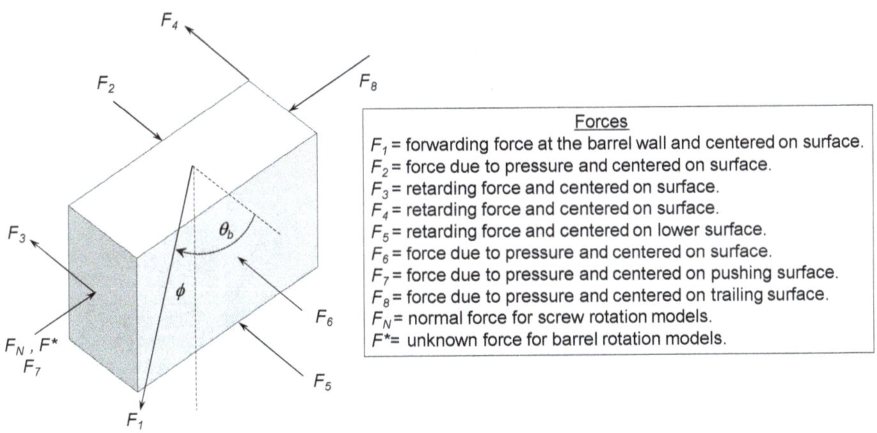

Figure A5.3 Typical force diagram for rotating screws and barrels

■ A5.2 Campbell-Dontula Model

Campbell and Dontula [1] revisited the original solids conveying problem. They proposed that the flights of the screw cause the movement of the solids down the screw, and that the polymer pellets behaved more like an elastic fluid than a solid polymer plug. Therefore, a torque balance would not be applicable. The following is a list of the additional assumptions they made:

1. The screw is rotated and not the barrel.
2. The polymer bed behaves as an elastic fluid.
3. The previously unresolved force, F_N (F^* in previous models) is assumed to be proportional to the frictional force between the polymer bed and the barrel.
4. The temperature down the barrel is assumed to be constant.

The Campbell-Dontula model [1] is fundamentally different from the previous models because of assumptions 10 to 12. Instead of solids conveying resulting from the friction between the polymer plug-barrel surfaces and retarded by the screw; the solid plug motion is due to the action of the screw flights pushing the polymer forward. Regardless of the new assumptions made, Campbell's original model over predicted conveying rates at exit pressures above atmospheric when compared to experimental data [3]. This is due to the assumption that the lateral stress ratio was 1.0 because at the time acceptable solids conveying data was not available.

A5.2.1 Modified Campbell-Dontula Model

The modified Campbell-Dontula solids conveying model was based on screw rotation and a normal force at the pushing flight that was directly proportional to the frictional force between the solid bed and the barrel wall. The force balance on the slab is provided in Figure A5.4

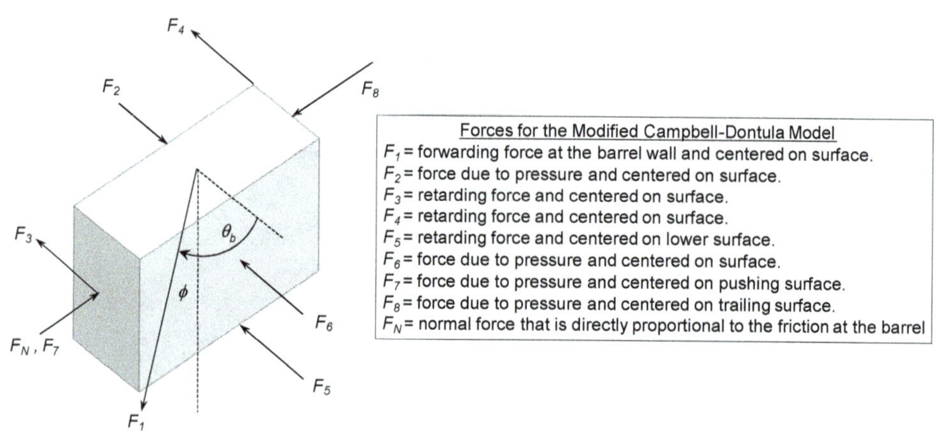

Figure A5.4 Force diagram for modified Campbell-Dontula model

The modified Campbell-Dontula model used a lateral stress ratio K_{xz} [5, 6] that is defined below:

$$K_{xz} = \frac{\sigma_x}{\sigma_z} \quad \text{(A5.1)}$$

Unfortunately, K_{xz} cannot be easily determined analytically, and is often either estimated using methods from the field of soil mechanics (from the normal stress ratio, K_o), or simply adjusted to a value which produces good fits for the experimental data. The forwarding force F_1 at the barrel wall is positioned in the plane of the barrel surface. This force is adjusted using K_{xz} as found in Equation A5.1 and results in Equation A5.2. This is because σ_z is the local pressure in the channel. Referring to Figure A5.4, the forces are defined as follow:

$$F_1 = f_b K_{xz} P W_b dz_b \tag{A5.2}$$

$$F_p = (F_6 - F_2) = HWdP \tag{A5.3}$$

$$F_7 = f_b K_{xz} PHdz \tag{A5.4}$$

$$F_8 = f_b K_{xz} PHdz \tag{A5.5}$$

$$F_N = f_b K_{xz} P W_b dz_b \tag{A5.6}$$

$$F_3 = (F_7 + F_N) f_H \tag{A5.7}$$

$$F_4 = f_H F_8 \tag{A5.8}$$

$$F_5 = f_s K_{xz} P W_c dz_c \tag{A5.9}$$

Following the procedure in Campbell and Dontula [1], the static force balance is:

$$0 = -F_1 \sin\varphi - F_p \sin\theta + F_N \cos\theta_b - (F_7 f_H \sin\theta - F_N f_H \sin\theta_b)$$
$$- F_4 \sin\theta - F_5 \sin\theta_c \tag{A5.10}$$

Remembering that:

$$\sin\theta_b = \frac{dz}{dz_b}\sin\theta \quad \text{and} \quad \sin\theta_b = \frac{dz_c}{dz_b}\sin\theta_c \tag{A5.11}$$

Substitution leads to the following differential equation:

$$0 = (-f_b W_b \sin\varphi + f_b W_b \cos\theta_b - f_H H \sin\theta_b - f_N f_b W_b \sin\theta_b - f_H H \sin\theta_b$$
$$- f_s W_c \sin\theta_b) dz_b \tag{A5.12}$$
$$-\frac{HW \sin\theta}{K_{xz}} \frac{dP}{P}$$

Upon integration and rearrangement the screw rotation design Equation A5.13 is obtained.

$$\sin(\varphi)=\cos(\theta_b)-\left(2\frac{Hf_H}{W_b f_b}+f_H+\frac{W_c f_s}{W_b f_b}\right)\sin(\theta_b)-\frac{HW\sin\theta}{K_{xz}W_b f_b Z_b}\ln\left(\frac{P_z}{P_o}\right) \quad \text{(A5.13)}$$

When this screw rotation equation with the variable K_{xz} is used then the equation provides a credible solution for fitting literature solids conveying data at all pressures.

A5.3 Hyun-Spalding Model

Hyun and Spalding [2] sought to refine the solids conveying problem and derive a model which predicted accurately the behavior of several different types of polymers at various extruder operating conditions. Their model was similar to historic models discussed in Chapter 5, but had a few deviations which included a force acting at the pushing flight at an angle (with normal and tangential components) that satisfies two torque balances [2]. They also included a lateral stress ratio compensation. The inclusion of a lateral stress ratio was done unidirectionally to describe the anisotropic pressure distribution at the flight tip (a plane perpendicular to the flight tip), however more complicated and multi-directional ratios could have been used. Assumptions 6 and 8 are not used here. Instead, the bulk density is allowed to vary with temperature and pressure, and the friction factors were allowed to vary with temperature, pressure, and velocity. As a result, the solid bed velocity was allowed to vary in order to satisfy continuity relationships. Figure A5.5 illustrates the changes made to Figure A5.3.

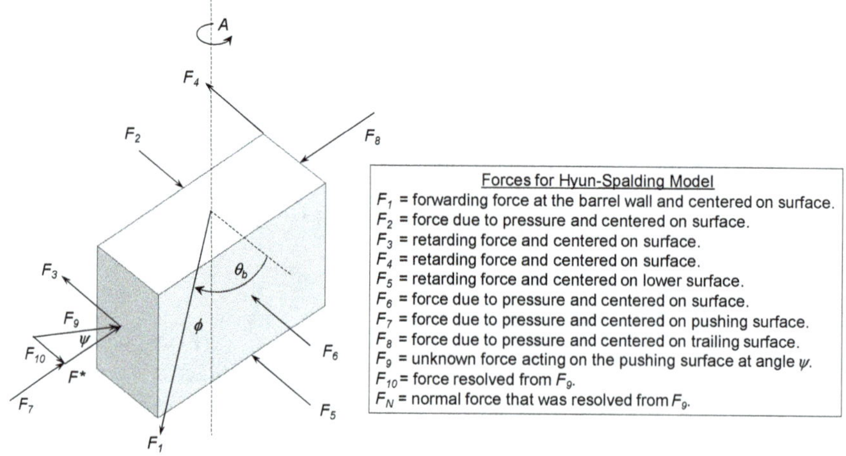

Figure A5.5 Force diagram for Hyun-Spalding model

Here, F_9 plays a similar role to that of the unknown force, F_N, in Campbell's models. F_9 is resolved into forces F_{10} and F^*, and acts at an angle Ψ to F_7 (the flight edge). These forces are in a plane parallel to the screw root surface. The new static force balance yields:

$$0 = -F_1 \sin\varphi - F_2 \sin\theta_s - (F_3 + F_4 - F_{10})\sin\theta - (F_2 - F_6)\sin\theta + F^* \cos\theta \quad \text{(A5.14)}$$

However, the third term of Equation A5.14 can be simplified using a torque balance about the axis A of Figure A5.5:

$$-F_3 \frac{W}{2} + F_4 \frac{W}{2} + F_{10} \frac{W}{2} = 0 \quad \text{(A5.15)}$$

Coupling Equation A5.15 with forces F_3, F_4, and F_{10}, as defined using the screw core friction factor, yields the following:

$$F_3 + F_4 - F_{10} = 2 f_s F_7 \quad \text{(A5.16)}$$

The remaining forces can be defined using pressures, coefficients, and surface areas:

$$F_1 = K_{xz} f_b P W_b \Delta z_b \quad \text{(A5.17)}$$

$$F_2 = K_{xz} f_s P W_c \Delta z_b \frac{\sin\theta_b}{\sin\theta_c} \quad \text{(A5.18)}$$

$$F_7 = K_{xz} P H \Delta z_b \frac{\sin\theta_b}{\sin\theta} \quad \text{(A5.19)}$$

where P is defined as the pressure at the center of the slab of polymer.

Performing a balance using the above forces and a torque balance gives the following:

$$0 = -K_{xz} f_b P W_b \Delta z_b \sin\varphi - K_{xz} f_s P W_s \Delta z_b \frac{\sin\theta_b}{\sin\theta_c} \sin\theta_c$$

$$-2K f_s P H \Delta z_b \frac{\sin\theta_b}{\sin\theta} \sin\theta \quad \text{(A5.20)}$$

$$-\Delta P H W \sin\theta + F^* \cos\theta$$

$$\frac{D_b}{2} F_1 \cos\varphi - \frac{D_c}{2} F_2 \cos\theta_c - \frac{D}{2}(F_3 + F_4 - F_{10})\cos\theta \quad \text{(A5.21)}$$

$$-\frac{D}{2}(F_5 - F_6)\cos\theta - \frac{D}{2} F^* \sin\theta = 0$$

Like other models, substitution and simplification of Equations A5.20 and A5.21 eliminates F^* and gives an equation for determining φ. Due to the lack of assumption 7 and the inclusion of a lateral stress ratio (K_{xz}), the following is left in differential form in terms of P:

$$\cos\varphi = K\sin\varphi + 2\frac{Hf_s}{W_b f_b}\sin\theta_b\left[K + \frac{D}{D_b}\cot\theta\right] + \frac{W_c f_s}{W_b f_b}\sin\theta_b\left[K + \frac{D_c}{D_b}\cot\theta_c\right] \tag{A5.22}$$

$$+\frac{WH}{K_{xz}PW_b f_b}\sin\theta\left[C + \frac{D}{D_b}\cot\theta\right]\frac{\partial P}{\partial z}$$

where K is defined using screw geometry as follows:

$$K = \frac{D}{D_b}\tan\theta \tag{A5.23}$$

A5.4 Yamamuro-Penumadu-Campbell Model

Yamamuro, Penumadu, and Campbell [5] developed another model with a force R of the same magnitude as F_1 but acting at a skew angle α, which is an adjustable parameter. The form of the analysis is similar to that by Tadmor and Klein [7] and Campbell and Dontula [1] in nature. Figure A5.6 illustrates forces for the model and the changes made to Figure A5.3. The following is an extended list of assumptions they made:

1. The relationship between the normal stress, σ_x, acting on the barrel and screw and the down-channel pressure, P, is constant and cross-anisotropic ($\sigma_x = \sigma_y$) [6].
2. The line of action of the reaction force, R, is at an angle ($\alpha + \varphi + \theta_b$) measured from the flight [6].

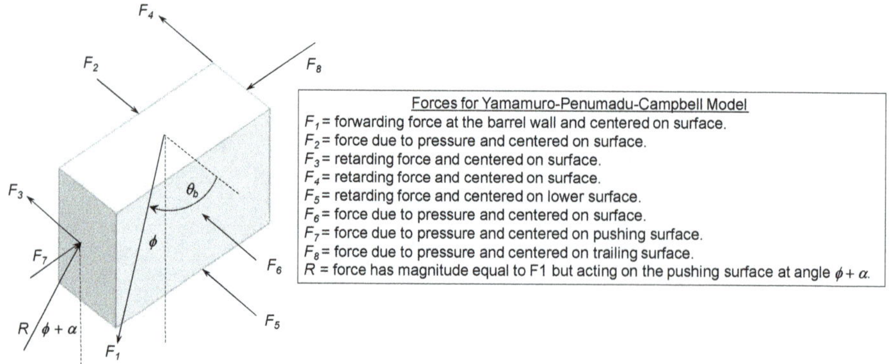

Figure A5.6 Force diagram for the Yamamuro-Penumadu-Campbell model

To incorporate stress effects into the solution for φ, the following forces must be redefined from their original definitions:

$$F_1 = K_{xy} f_b P W_b \partial z_b \quad \text{(A5.24)}$$

$$F_p \equiv (F_6 - F_2) = HW\partial P \quad \text{(A5.25)}$$

$$F_7 = F_8 = K_{xz} PH\partial z \quad \text{(A5.26)}$$

$$R = f_b K_{xz} P W_b \partial z \quad \text{(A5.27)}$$

$$F_3 = \left[F_7 + R\sin(\alpha + \varphi + \theta_b\right] f_H \quad \text{(A5.28)}$$

Forces F_1 and R are considered to have the same magnitude, but R is skewed by α to the pushing flight. That is, R acts at the pushing flight at an angle of ($\varphi + \alpha$) from the vertical plane of the flight edge. This definition of R was chosen simply because it is reasonable to expect that the magnitude and direction of R varies with the magnitude of the down-channel pressure and barrel friction [6]. Applying a force balance in the axial direction, simplifying, and integrating yields:

$$\sin\varphi = \sin(\alpha + \varphi) - 2\frac{f_H H}{f_b W_b}\sin\theta_b - f_H \sin(\alpha + \varphi + \theta_b)\sin\theta_b - \frac{f_c W_c}{f_b W_b}\sin\theta_b$$

$$-\frac{WH\sin\theta}{K_{xz} Z_b W_b f_b}\ln\frac{P_z}{P_o} \quad \text{(A5.29)}$$

Unfortunately, K_{xz} cannot be easily determined analytically, and is often either estimated using methods from the field of soil mechanics (from the normal stress ratio, K_o), or simply adjusted to a value which produces good fits for experimental data. The same can be said for the skew angle, α.

Yamamuro's paper also demonstrates the performance of their model versus experimental data using estimated values of both K_{xz} and α. The model fits virtually all the data accurately, with the exception of the data at a low barrel temperature.

All of the solids solution equation results in either a sin or cos of the angle ϕ. To be able to use equation A5.29 to calculate the solid flow rate Q, we need the tangent of ϕ:

$$\tan\varphi = \frac{\sin\varphi}{\sqrt{1 - \sin^2\varphi}} \quad \text{(A5.30)}$$

$$\tan\varphi = \frac{\sqrt{1 - \cos^2\varphi}}{\cos\varphi} \quad \text{(A5.31)}$$

A5.5 Campbell-Spalding Model

It is interesting that the model by Yamamuro et al. fits the data quite well; however, it has an extra parameter, the angle of attack α of the force R that limits the model's ability to predict for new materials. Reexamining the model of Campbell and Dontula the following change in the model yields a model more responsive to the process parameters.

The force F_N is altered to act at an angle related to φ.

$$F_N = f_b K_{xz} P W_b dz_b \tag{A5.32}$$

Is defined to operate at an angle similar to R but in terms of φ:

$$F_N = f_b K_{xz} P W_b \sin\left(\frac{\pi}{2} + \varphi + \theta\right) dz_b \tag{A5.33}$$

Substituting in the force balance of Campbell et al. and integrating:

$$\sin\varphi = \cos\theta_b \sin\left(\frac{\pi}{2} + \varphi + \theta\right) - \left(2\frac{Hf_H}{W_b f_b} + f_H \sin\left(\frac{\pi}{2} + \varphi + \theta\right) + \frac{W_s f_s}{W_b f_b}\right)\sin\theta_b - \frac{HW\sin\theta}{K_{xz} W_b f_b Z_b}\ln\left(\frac{P_z}{P_o}\right) \tag{A5.34}$$

This model does not have a free parameter.

A5.6 The Complete Dow Solids Conveying Data Set

The complete solids conveying data set for a 2.0 MI (190 °C, 2.16 kg) LDPE resin with a solid density of 0.92 g/cm^3 are presented in this section. The data were collected for shallow and deep screws, as a function of discharge temperature, barrel temperature, screw temperature, and screw speed. The solids conveying device was described in Section 5.3.2. The data points shown are an average of at least 3 measurements. The solids conveying data are presented in Figures A5.7 through A5.18. The geometries of the screws with the small flight radii used for these experiments are provided in Table A5.1.

Table A5.1 Channel Dimensions for the Shallow and Deep Screws for the 63.5 mm Diameter, 4.5 L/D Solids Conveying Device at Dow

	Shallow Screw	Deep Screw
Channel depth, mm	8.89	11.1
Lead length, mm	63.5	63.5
Flight width, mm	6.4	6.4
Pushing flight radius, mm	6	16
Trailing flight radius, mm	6	16
L/D	4.5	4.5
Feed opening length, diameters	1.5	1.5
Active length for solids conveying, diameters	3.0	3.0

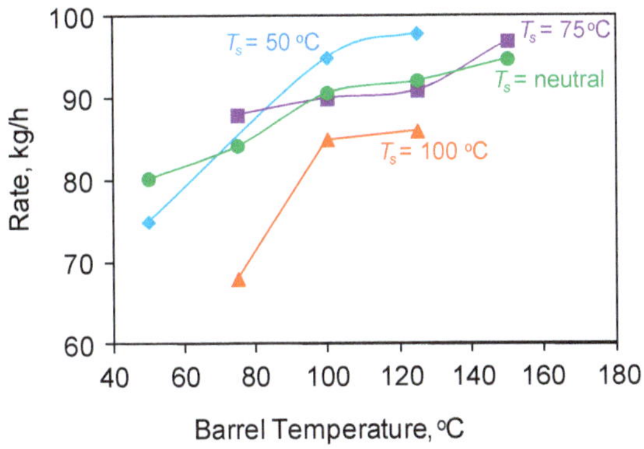

Figure A5.7 Solids conveying rate as a function of barrel and screw temperature for the shallow screw (8.89 mm) at 50 rpm and zero discharge pressure

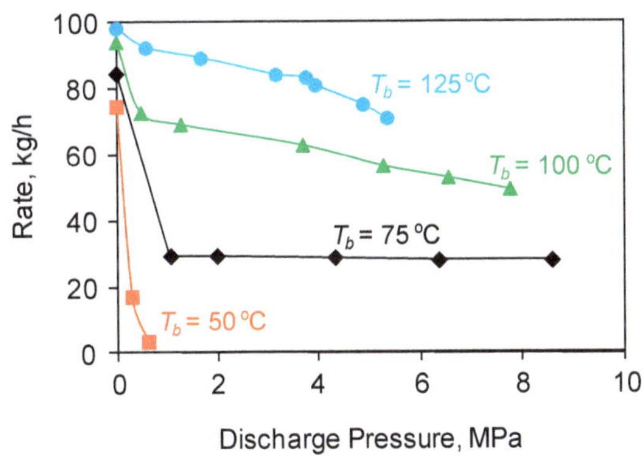

Figure A5.8 Solids conveying rate as a function of discharge pressure and barrel temperature for the shallow (8.89 mm) screw. The screw was maintained at a temperature of 50 °C and 50 rpm

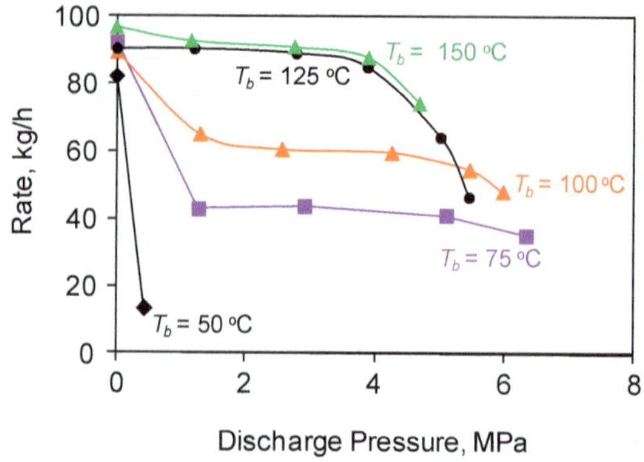

Figure A5.9 Solids conveying rate as a function of discharge pressure and barrel temperature for the shallow (8.89 mm) screw. The screw was maintained at a temperature of 75 °C and 50 rpm

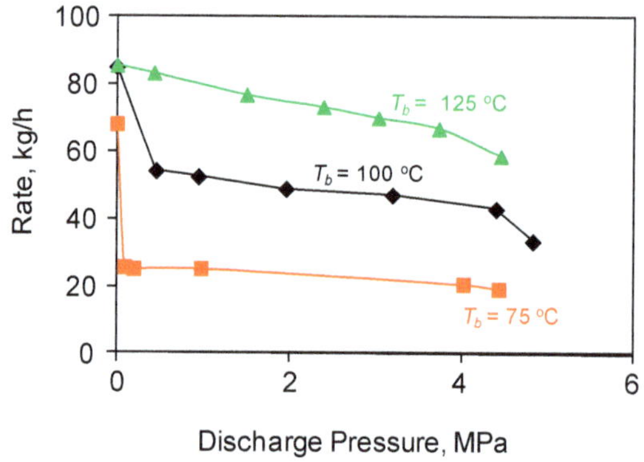

Figure A5.10 Solids conveying rate as a function of discharge pressure and barrel temperature for the shallow (8.89 mm) screw. The screw was maintained at a temperature of 100 °C and 50 rpm

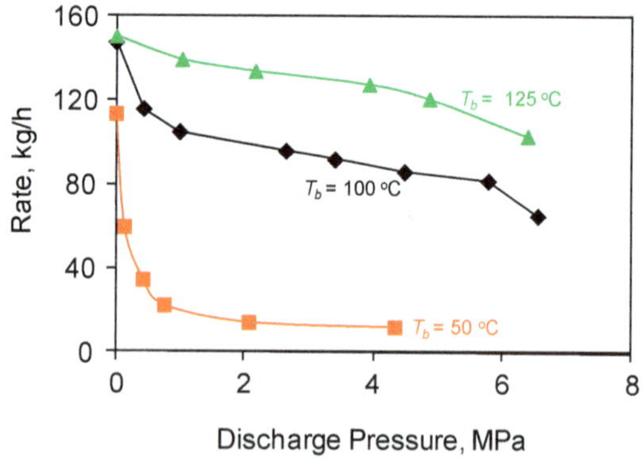

Figure A5.11 Solids conveying rate as a function of discharge pressure and barrel temperature for the shallow (8.89 mm) screw. The screw was maintained at a temperature of 50 °C and 80 rpm

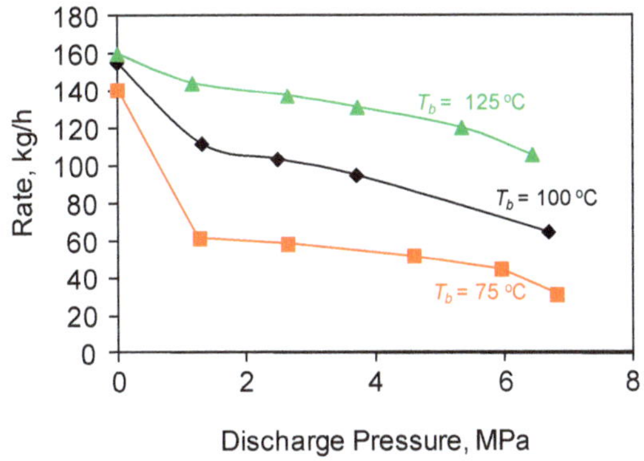

Figure A5.12 Solids conveying rate as a function of discharge pressure and barrel temperature for the shallow (8.89 mm) screw. The screw was maintained at a temperature of 75 °C and 80 rpm

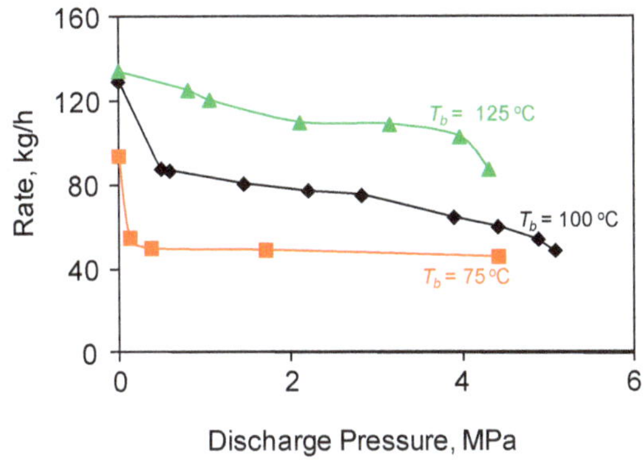

Figure A5.13 Solids conveying rate as a function of discharge pressure and barrel temperature for the shallow (8.89 mm) screw. The screw was maintained at a temperature of 100 °C and 80 rpm

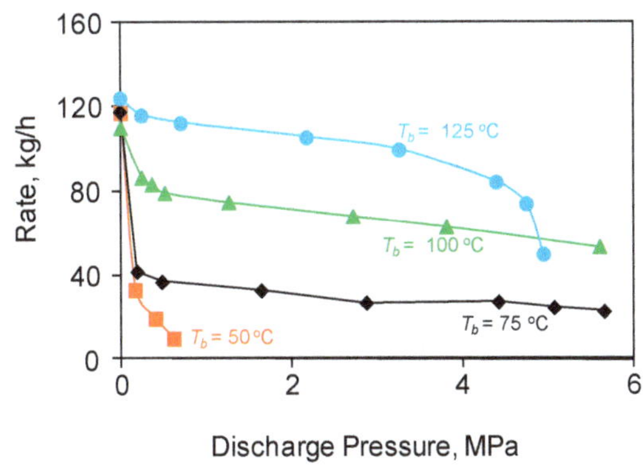

Figure A5.14 Solids conveying rate as a function of discharge pressure and barrel temperature for the deep (11.1 mm) screw. The screw was maintained at a temperature of 50 °C and 50 rpm

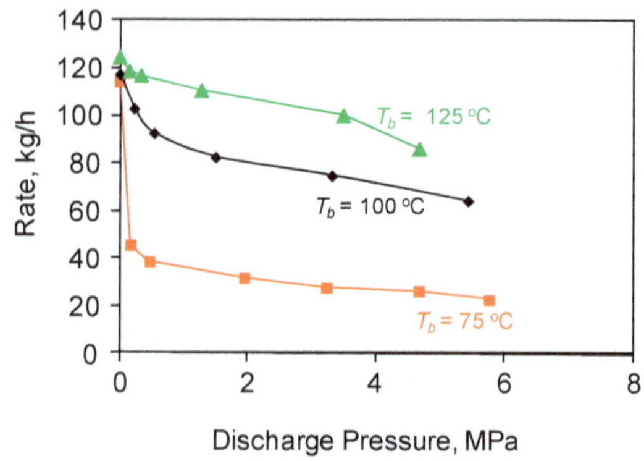

Figure A5.15 Solids conveying rate as a function of discharge pressure and barrel temperature for the deep (11.1 mm) screw. The screw was maintained at a temperature of 75 °C and 50 rpm

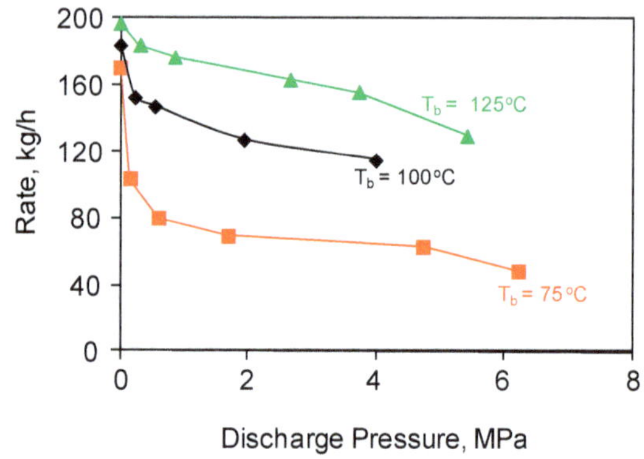

Figure A5.16 Solids conveying rate as a function of discharge pressure and barrel temperature for the deep (11.1 mm) screw. The screw was maintained at a temperature of 75 °C and 80 rpm.

The stress at the sliding interface was measured using a Screw Simulator as outlined in Section 4.3.1 for the LDPE resin and is shown in Figure A5.17 at a pressure of 0.7 MPa and shown in Figure A5.18 at a pressure of 3.4 MPa. At temperatures of less than about 110 °C, the stress at the interface is due to a frictional force mechanism. At temperatures higher than 110 °C, the forces are from a viscous mechanism. The coefficients of friction were reported as stresses so that the data from both mechanisms can be reported on the same figure. The coefficient of friction is related to the stress at the interface τ as follows:

$$\tau = fP \tag{A5.35}$$

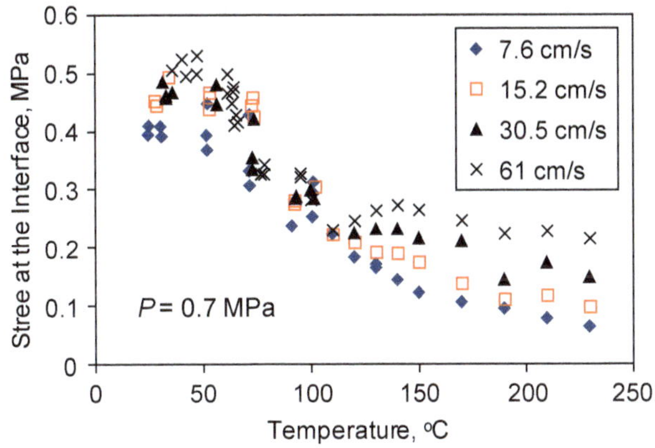

Figure A5.17 Stress at the interface for LDPE resin as a function of temperature and sliding velocity and at a pressure of 0.7 MPa

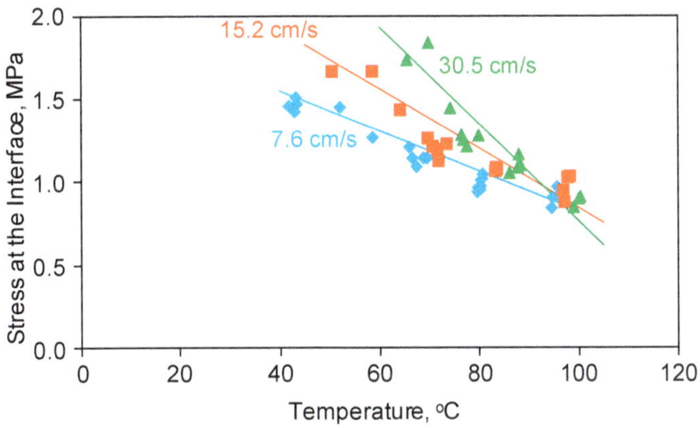

Figure A5.18 Stress at the interface for LDPE resin as a function of temperature and sliding velocity and at a pressure of 3.4 MPa

References

1. Campbell, G.A. and Dontula, N., "Solids Transport in Extruders," *Int. Polym. Process.*, **10**, 30 (1995)
2. Hyun, K.S. and Spalding, M.A., "A New Model for Solids Conveying In Single-Screw Plasticating Extruders," *SPE ANTEC Tech. Papers*, **43**, 211 (1997)
3. Hyun, K.S., Spalding, M.A., and Hinton, C.E., "Theoretical and Experimental Analysis of Solids Conveying in Single-Screw Extruders," *SPE ANTEC Tech. Papers*, **42**, 199 (1996)
4. Spalding, M.A., Hyun, K.S., and Hughes, K.R., "Stress Distribution in Solid Polymer Compacts," *SPE ANTEC Tech. Papers*, **42**, 191 (1996)

5. Yamamuro, J. A., Penumadu, D., and Campbell, G. A., "Modeling Solids Conveying in Polymer Extruders," *Int. Polym. Process.*, **13**, 3 (1998)

6. Penumadu, D., Yamamuro, J., Abrantes, A. E., and Campbell, G. A., "Stress-Strain behavior of Polymer Pellets," *SPE ANTEC Tech. Papers*, **43**, 224 (1997)

7. Tadmor, Z. and Klein, I., "Engineering Principles of Plasticating Extrusion," Van Nostrand Reinhold Co., New York, 1970

Appendix A6 Melting Rate Model Development

This appendix contains the developments for the melting rates and energy balances that describe the melting processes in the transition section of the screw. These balances and processes are discussed in detail in Chapter 6. The equations are derived using screw rotation. A full understanding of these developments is not required for detailed analysis and troubleshooting of the extrusion process. Some of the equations and figures are duplicated in this appendix for clarity. The nomenclature used here is consistent with that used earlier; the reader is directed to Chapter 6 for the nomenclature.

A6.1 Derivation of the Melting Performance Equations for a Conventional Channel

This analysis starts with the assumption that melting occurs in all four melt films that surround the solid bed. The initial analysis will be carried out for Film C in Figure A6.1. The film is located between the barrel and the solid bed interface. This analysis describes the viscous energy dissipation in the film and the energy conduction from the barrel wall and how they relate to the melting flux at the solid bed-melt interface.

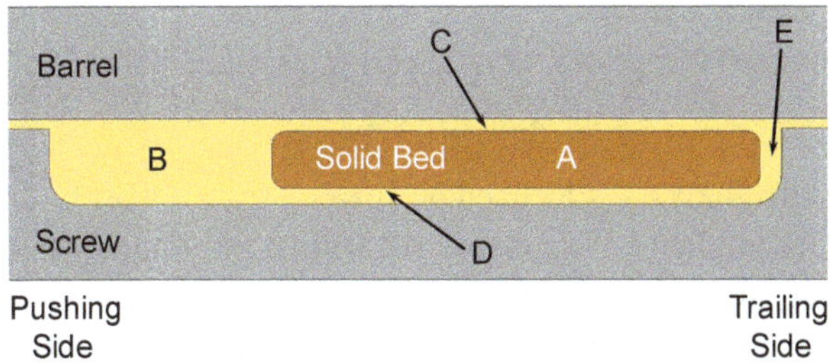

Figure A6.1 Schematic for the zones of the melting model: Zone A is the solid bed, Zone B is the melt pool, Zone C is the melt film located between the solid bed and the barrel wall, Zone D is the melt film between the solid bed and the screw root, and Zone E is the melt film between the solid bed and the trailing flight. The cream color represents molten resin

The hot barrel provides the initial energy to develop a melt film due to Q_{cond}, which is limited by the thermal conductivity of the molten resin (k_m), and driven by the temperature difference ($T_b - T_m$) between the barrel temperature (T_b) and the melting point of the resin (T_m). Energy dissipation in the film, Q_{dis}, is due to viscous dissipation ($\mu\dot{\gamma}^2$). The melting rate is then assumed to be due to the transfer of all of this energy to the solid bed surface, resulting in the energy term Q_{melt}. The energy to melt the solid bed in the negative y direction is shown schematically in Figure A6.2. The analysis used in this development closely follows the initial part of the theoretical development for melting in Tadmor and Klein [1]. This analysis deviates from that analysis in two major elements: 1) this analysis focuses on the velocity of all four solid surfaces as they recede toward the cold center of the solid bed, and 2) this analysis calculates the melting energy in all four of the films surrounding the solid bed, one at the barrel solid interface with a thickness of δ_C, and three films located between the screw surface and the solid bed with thicknesses of δ_D, δ_E, and δ_B. The Tadmor and Klein model uses barrel rotation. Screw rotation is used for the derivations here. The melting rates at the interfaces of these last three films are typically dominated by the energy transferred from the screw rather than from viscous dissipation. This is due to the rather low shear rate developed because of the screw's relative motion with the solid bed. Moreover, the screw temperature in the transition section is considerably higher than the melting or devitrification temperature; see Section 10.10. The goal of this analysis is to develop relationships that predict the rate that the solid bed disappears from the four surfaces shown in Figure A6.1. We demonstrate that the relative velocity of the solid bed surface close to the barrel and the relative velocities in the three films between the screw and the solid bed can be calculated in a straight forward manner.

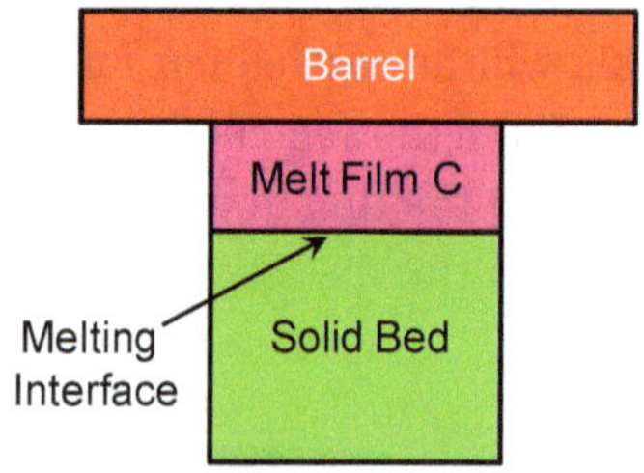

Conduction $Q_{cond} = k_m(T_b - T_m)/\delta_C$

Dissipation $Q_{dis} = (\mu\dot{\gamma}^2)\delta_C$

Melting Energy $Q_{melt} = Q_{cond} + Q_{dis}$

Figure A6.2 Schematic for the two sources of energy (as a flux) for melting resin at the solid bed-melt Film C interface

The first part of the analysis is focused on the energy transfer to the solid bed and what assumptions might be reasonable regarding the temperature profile in the solid bed. For this analysis the barrel and solid interface will be addressed. It is desired that an infinite bed assumption can be justified. Once this assumption is justified, the heat transfer analysis for the melting is quite straightforward.

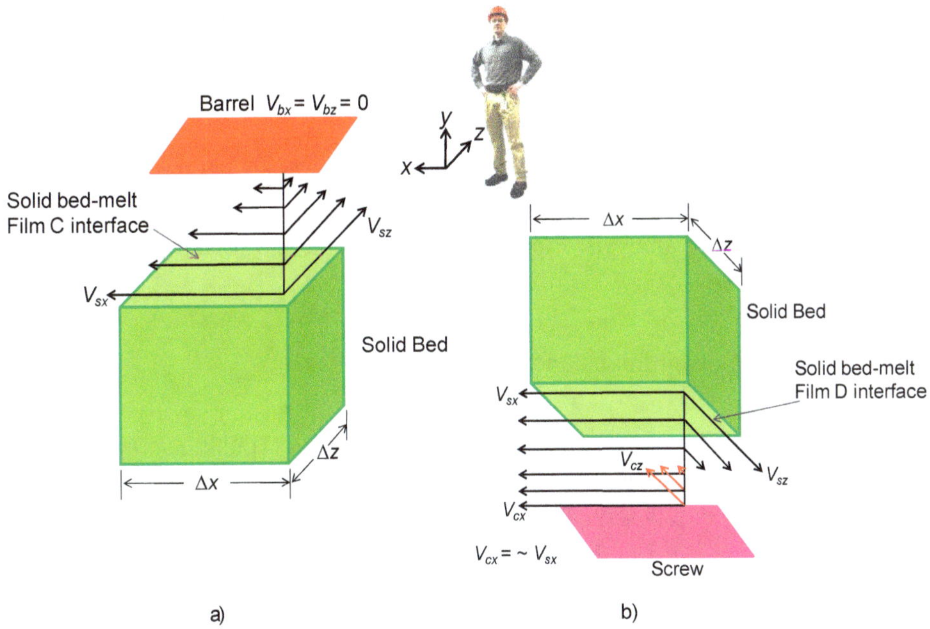

Figure A6.3 Melting differential element for a rotating screw: a) melt film in Zone C (barrel velocity is zero), and b) melt film in Zone D with the screw moving with a velocity of V_{cz} and V_{cx}. The red colored vectors are moving out of the plane of the page towards the upstream direction while the black vectors in the *z* direction are moving into the page and in the downstream direction

Screw rotation analysis is used here so the barrel velocity is 0. For Film C, as diagrammed in Figure A6.3(a), two velocities are relevant: V_{sz}, the velocity of the solid bed in the down channel direction, and V_{sx}, the velocity of the solid bed in the *x* direction due to the screw rotation.

$$V_{sx} = V_b \sin\theta_b = \pi N D_b \sin\theta_b \tag{A6.1}$$

The resulting vectorial velocity that produces the dissipation in Film C can be calculated as follows:

$$V_{j,C} = \sqrt{V_{sz}^2 + V_{sx}^2} \tag{A6.2}$$

The velocity of the solid bed V_{sz} is determined using a mass balance and the dimensions of the channel at the entry to the transition section:

$$V_{sz} = \frac{Q_m}{\rho_s W H_1} \tag{A6.3}$$

where Q_m is the mass rate for the process, ρ_s is the solid density of the bed, W is the average channel width, and H_1 is the channel depth at the start of the transition section. The generalized energy equation for Film C (see Equation A, Table 2.8 in

Tadmor and Klein [1]) reduces to the following after eliminating nonsignificant terms:

$$k_m \frac{d^2T}{dy^2} + \mu\left(\frac{dv_j}{dy}\right)^2 = 0 \quad \text{for Film C} \tag{A6.4}$$

where k_m is the thermal conductivity of the melt, T is temperature, and y is the depth direction of the analysis. From Equation A, Table 2.6 in Tadmor and Klein [1], with $x = j$:

$$\frac{d^2v_j}{dy^2} = 0 \quad \text{momentum balance for Film C} \tag{A6.5}$$

Following the historical approach from Tadmor and Klein [1] using screw rotation mechanics, the boundary condition at the top of the solid bed ($y = 0$) and at the barrel wall ($y = \delta_C$) are as follows:

$$y = 0 \quad v_j = V_{j,C} \quad \text{at the solid bed-melt Film C interface} \tag{A6.6}$$

$$y = \delta_C \quad v_j = 0 \quad \text{at the barrel wall} \tag{A6.7}$$

Integrating Equation A6.5 twice,

$$v_j = C_1 y + C_2 \tag{A6.8}$$

After applying the boundary conditions:

$$v_j = -\frac{V_{j,C}\, y}{\delta_C} + V_{j,C} \tag{A6.9}$$

Therefore:

$$\frac{dv_j}{dy} = -\frac{V_{j,C}}{\delta_C} \quad \text{shear rate for Film C} \tag{A6.10}$$

which leads to the energy balance Equation A6.11 for melt Film C:

$$k_m \frac{d^2T}{dy^2} + \mu\left(\frac{V_{j,C}}{\delta}\right)^2 = 0 \quad \text{energy balance for Film C} \tag{A6.11}$$

Integrating:

$$T = -\frac{\mu}{k_m}\left(\frac{V_{j,C}}{\delta}\right)^2 \frac{y^2}{2} + C_1 y + C_2 \tag{A6.12}$$

$$y = 0 \quad T = T_m \quad \text{at the solid bed-melt Film C interface} \tag{A6.13}$$

$$y = \delta_C \quad T = T_b \quad \text{at the barrel wall} \tag{A6.14}$$

Applying the boundary conditions and simplifying:

$$\frac{T - T_m}{T_b - T_m} = \frac{\mu}{2k_m} \frac{V_{j,C}^2}{(T_b - T_m)} \frac{y}{\delta_C}\left(1 - \frac{y}{\delta_C}\right) + \frac{y}{\delta_C}$$

temperature profile in Film C (A6.15)

Equation A6.15 represents the temperature profile in Film C. The temperature appears to decrease as the square of the distance (y^2) away from the barrel surface; assumes that $T_b > T_m$. To visualize this see Figure 5.29 in Tadmor and Klein [1]. Now the heat flow per unit area, or the heat flux into the interface, is obtained by differentiating Equation A6.15 and multiplying by k_m as follows:

$$-\left[q_y\right]_{y=0} = k_m \left(\frac{dT}{dy}\right)_{y=0} = \frac{k_m}{\delta_C}(T_b - T_m) + \frac{\mu V_{j,C}^2}{2\delta_C}$$

energy flux into the solid bed at the interface (A6.16)

Assuming one-dimensional heat transfer is the mode of the solid bed heating due to the heating of the film by conduction and dissipation, the temperature will only change in the y direction. The same assumption that was made by Tadmor and Klein will be made here; that the heat transfer model is a semi-infinite slab moving at a velocity $V_{sy,C}$ (melting velocity) with the boundary conditions $T(0) = T_m$ and $T(-\infty) = T_s$. This assumption is not strictly correct because it will also be proposed that the other four surfaces are melting. The major error will occur at the corners of the solid bed. $V_{sy,C}$ is the velocity of the solid bed surface adjacent to Film C as it moves toward the center of the solid bed in the y direction.

The energy equation for the solid with simplification and deleting negligible and irrelevant terms as before leads to:

$$k_s \frac{d^2T}{dy^2} = \rho_s C_v V_{sy,C} \frac{dT}{dy}$$

energy balance for the solid bed near the Film C interface (A6.17)

where ρ_s, C_v, and k_s, are the density, specific heat, and thermal conductivity of the solid bed, respectively. This equation can be integrated by substituting:

$$u = \frac{dT}{dy} \tag{A6.18}$$

$$\frac{du}{dy} = \frac{\rho_s C_v V_{sy,C}}{k_s} u \tag{A6.19}$$

$$u=\frac{dT}{dy}=C_1\exp\left(\frac{\rho_s C_v V_{sy,C}}{k_s}y\right) \tag{A6.20}$$

Integration on y yields:

$$T=C_1\frac{k_s}{\rho_s C_v V_{sy,C}}\exp\left(\frac{\rho_s C_v V_{sy,C}}{k_s}y\right)+C_2 \tag{A6.21}$$

Applying the appropriate boundary conditions to solve for C_1 and C_2 yields:

$$y=0 \quad T=T_m \quad \text{at the solid bed-melt Film C interface} \tag{A6.22}$$

$$y=-\infty \quad T=T_s \quad \text{deep in the solid bed} \tag{A6.23}$$

$$\frac{T-T_s}{T_m-T_s}=\exp\left(\frac{V_{sy,C}}{\alpha_s}y\right) \tag{A6.24}$$

where

$$\alpha_s=\frac{k_s}{\rho_s C_v} \quad \text{solid bed thermal diffusivity} \tag{A6.25}$$

As pointed out by Tadmor and Klein [1], the solid bed temperature decreases very rapidly from the melting surface, and thus, a reasonable first assumption is that it can be considered to be infinitely deep for heat transfer and melting. Given this assumption the heat flux away from melt film into the solid will be:

$$-\left[q_y\right]_{y=0}=k_s\left(\frac{dT}{dy}\right)_{y=0}=\rho_s C_v V_{sy,C}\left(T_m-T_s\right) \tag{A6.26}$$

The energy available for melting is calculated from an energy balance for the solid bed-Film C interface as follows:

$$\left[\frac{k_m}{\delta_c}\left(T_b-T_m\right)+\frac{\mu V_{cj}^2}{2\delta_c}\right]-\rho_s C_v V_{sy,C}\left(T_m-T_s\right)=V_{sy,C}\rho_s\lambda \tag{A6.27}$$

where λ is the heat of fusion of the polymer. This derivation is strictly only valid for semicrystalline polymers. Solving for $V_{sy,C}$ sets the melting velocity at the solid bed-melt film interface for Film C.

$$V_{sy,C}=-\left.\frac{dY}{dt}\right|_{y=\delta_c}=\frac{\left[\frac{k_m}{\delta_c}\left(T_b-T_m\right)+\frac{\mu V_{cj}^2}{2\delta_c}\right]}{\rho_s C_v\left(T_m-T_s\right)+\rho_s\lambda} \tag{A6.28}$$

where Y is the local thickness of the solid bed. This is the function for the rate of loss of the solid bed at Film C, as shown in Figure A6.1, and is in terms of the film thickness that at this point in the analysis is unknown. Again this analysis follows the classical analysis for calculation of the gap thickness that was originally developed in Tadmor and Klein [1] Equation 5-31.

A mass balance is developed for the down channel flow in Film C that will lead to an expression for the film thickness.

$$Q_{m,z} = \frac{V_{sz}}{2} \rho_m \left(X \delta_c \right) \Big|_z$$

rate of mass flow into the Film C increment at z (A6.29)

$$Q_{m,z+\Delta z} = \frac{V_{sz}}{2} \rho_m \left(X \delta_c \right) \Big|_{z+\Delta z}$$

rate of mass flow out of the Film C increment at $z + \Delta z$ (A6.30)

$$Q_{m,melt} = V_{sy,C} \rho_s \left(\overline{X} \Delta z \right)$$

rate of mass flow entering Film C through melting at the interface (A6.31)

$$Q_{m,pool} = \frac{V_{bx}}{2} \rho_m \left(\overline{\delta} \Delta z \right)$$

rate of mass flow leaving Film C to the melt pool (Film B) (A6.32)

where ρ_m is the melt density, X is the local width of the solid bed, $\overline{X}$ is the average solid bed width for the Δz increment, and $\overline{\delta}$ is the average film thickness for the increment. Since mass flow is incompressible and there is no reaction, a mass balance for the increment leads to the following differential equation.

$$\frac{V_{sz}}{2} \rho_m \frac{d \left(X \delta_c \right)}{dz} + V_{sy,C} \rho_s \overline{X} - \frac{V_{bx}}{2} \rho_m \overline{\delta} = 0 \quad \text{(A6.33)}$$

Tadmor and Klein demonstrated that the differential term, the first term in the above equation, was small so the equation reduces to a good approximation as follows:

$$V_{sy,C} \rho_s X = \frac{V_{bx}}{2} \rho_m \delta_c \quad \text{(A6.34)}$$

Substituting for $V_{sy,C}$ from above and solving:

$$\delta_C = \left\{ \frac{\left[2k_m \left(T_b - T_m \right) + \mu V_{j,C}^2 \right] X}{\rho_m V_{bx} \left[C_v \left(T_m - T_s \right) + \lambda \right]} \right\}^{1/2} \quad \text{(A6.35)}$$

A fitting factor can be used to force the Equation A6.35 to fit experimental data:

$$\delta_C = \left\{ \frac{\left[2k_m \left(T_b - T_m \right) + \mu V_{j,C}^2 \right] \frac{X}{\alpha}}{\rho_m V_{bx} \left[C_v \left(T_m - T_s \right) + \lambda \right]} \right\}^{1/2} \tag{A6.36}$$

The analysis for Film D is developed in a similar manner except the vectorial velocity creating the dissipation is different and the film thickness calculation is different, as shown in Figure A6.2(b). For Film D, the vectorial velocity gradient relative to the fixed barrel and rotating screw is defined as:

$$V_{j,D} = V_{sz} + \left| V_{cz} \right| \tag{A6.37}$$

$$\dot{\gamma}_D = \frac{V_{j,D}}{\delta_D} \quad \text{shear rate in Film D} \tag{A6.38}$$

$$V_{cz} = -\pi N D_c \cos\theta_c \tag{A6.39}$$

where V_{cz} is the velocity of the screw core in the z direction. Recall that the core of the screw is moving in the negative z direction. The mass and energy balances are developed in the same manner as for $V_{sy,C}$ except adjacent to Film D at the solid interface. It might be expected that the screw core motion relative to the solid bed motion would create a recirculation flow field. However, as explained in Chapter 6, the current assumption is that most of the flow leaks into melt pool B. This leakage flow is evident in Figure 6.35.

$$V_{sy,D} = -\left. \frac{dY}{dt} \right|_{y=\delta_D} = \frac{\left[\frac{k_m}{\delta_D} \left(T_{screw} - T_m \right) + \frac{\mu V_{j,D}^2}{2\delta_D} \right]}{\rho_s C_v \left(T_m - T_s \right) + \rho_s \lambda} \tag{A6.40}$$

The definition for δ_D must be developed. The definition requires dimensional consistency; the gap cannot be calculated as for δ_C. The dimensional analysis in the y direction involves δ_D, δ_C, and the remaining solid bed at any position z down the channel. The following equation was developed to meet these criteria.

$$\delta_D = H(z) - Y - \delta_c \tag{A6.41}$$

where $H(z)$ is the local channel depth at position z. The two differential equations for the velocity of the melting surfaces are solved simultaneously. These equations cannot be solved at this point in the analysis because the equations for the X bed width must be determined to evaluate the loss of solid in the y direction. This is necessary because δ_C is a function of the X bed width in the cross channel direction.

The analysis is extended to the gap next to the trailing flight. The expression for δ_E is analogous to the expression for δ_C. The difference comes from the different flow mechanism in this gap. The flow is assumed to be a recirculating flow similar to the cross channel (x direction) flow in the pumping section. This is because the flight drags fluid from the pumping section toward the solid conveying zone and the flow is reversed at the beginning of the fluid Film E. It is unlikely that all of the fluid is dragged to that very small gap at or near the initiation of melting. The velocity equation, assuming that $x = 0$ at the solid bed surface, as a first approximation at any local z is thus:

$$v_z = x_E \frac{V_{j,E}}{\delta_E}\left[2 - \left(\frac{3x_E}{\delta_E}\right)\right] \tag{A6.42}$$

The dissipation in Film E due to this flow is therefore,

$$\dot{\omega} = 4\mu\left(\frac{V_{j,E}}{\delta_E}\right)^2 \tag{A6.43}$$

It should be pointed out at this point that this is a first approximation for this analysis and it is realized that the process in this gap is most likely more complicated than the analysis presented here. It might be expected that all of the heat generated will not enter the solid bed to induce melting. Some of the heat generated will likely increase the fluid temperature in Film E as the fluid flows toward the pumping zone of the screw. Given these simplified first order assumptions, it follows using the same reasoning as used in Equation A6.28 that the velocity in the x direction for solid bed loss is:

$$V_{sx,E} = -\left.\frac{dX}{dt}\right|_{y=\delta_E} = \frac{\left[\frac{k_m}{\delta_E}\left(T_b - T_m\right) + \frac{4\mu V_{j,E}^2}{2\delta_E}\right]}{\rho_s C_v\left(T_m - T_s\right) + \rho_s \lambda} \tag{A6.44}$$

Notice that the dissipation term is four times as large as the one in Equation A6.28. The barrel temperature is used here rather than the temperature of the trailing flight. In general, the temperature of the trailing flight is unknown. Derezinski [2] has shown, however, that the temperature of the flight is close to the local barrel temperature. The vectorial velocity $V_{j,E}$ in Film E is calculated as follows:

$$V_{j,E} = V_{sz} + \left|\overline{V}_{fz}\right| \tag{A6.45}$$

where $\overline{V}_{fz}$ is the average velocity of the flight in the z direction. As discussed in Chapter 7, the velocity of the flight in the z direction is towards the hopper and thus is in the negative z direction. The magnitude is the lowest at the screw root and the highest at the flight tip.

$$\overline{V}_{fz} = \left(\frac{V_{cz} + V_{bz}}{2}\right) \tag{A6.46}$$

$$V_{bz} = -\pi N D_b \cos\theta_b \tag{A6.47}$$

where V_{bz} is the velocity of the flight tips in the z direction. The average velocity of the flight in the z direction is the same for both the leading and trailing flights. It is assumed that the flights will drag material into the films between the solid bed and the flights, and the same amount of material, plus what has melted will leave the film at the end of the melting zone. It is also assumed that the material entering and leaving by the flight drag mechanism will be at the melt temperature. Therefore, the gap width is assumed to be a function of the velocity of the solid bed and the flight in the z direction. This is the simplest approach that follows the approach developed by Tadmor and Klein [1] for the melting at the barrel interface. Further research may lead to a more comprehensive set of assumptions.

A mass balance is developed for the y direction channel flow in Film E that will lead to an expression for the film thickness that causes most of the solid bed loss in the cross channel (x direction).

$$Q_{m,y} = V_{Ey}\rho_m \left(Z\delta_E\right)\Big|_y$$

rate of mass flow into the Film E increment at y (A6.48)

$$Q_{m,y+\Delta y} = \frac{V_{Ey}}{2}\rho_m \left(Z\delta_E\right)\Big|_{y+\Delta y}$$

rate of mass flow out of the Film E increment at $y + \Delta y$ (A6.49)

$$Q_{m,melt} = V_{sx,E}\rho_s \left(\overline{Z}\Delta y\right)$$

rate of mass flow entering Film E through melting at the interface (A6.50)

where V_{Ey} is the average velocity in Film E in the y direction. This velocity is not known and will be eliminated later in the derivation. The rate of mass flow leaving Film E to the metering section, $Q_{m,meter}$ at the end of melting (at $z = Z$):

$$Q_{m,meter} = V_{Ez}\rho_m \left(\frac{2\overline{\delta}_E}{3}\Delta y\right)$$

recirculation flow, it is modeled using average out flow in gap (A6.51)

where ρ_m is the melt density, $\overline{Z}$ is the average solid bed length, V_{Ez} is the average velocity in Film E in the z direction between the side surface of the bed and 2/3 of the way into the film, and $2\,\overline{\delta}_E/3$ is the average film for outflow remembering that the velocity goes to 0 as the film is crossed in the x direction and then reverses direction. Since mass flow is incompressible and there is no reaction, a mass balance leads to the following differential equation:

$$V_{Ey}\rho_m \frac{d(Z\delta_E)}{dy} + V_{sx,E}\rho_s \overline{Z} - \frac{2V_{Ez}}{3}\rho_m \overline{\delta}_E = 0 \tag{A6.52}$$

Tadmor and Klein demonstrated that the differential term for the balance on top of the bed, the first term in the above equation, was small: the same assumption is made here so the equation reduces to a good approximation as follows assuming that $\overline{\delta}$ is equal to δ_E; this appears to a reasonable assumption since the melting rate would be small compared to the flight driven fluid convection:

$$V_{sx,E}\rho_s Z = \frac{2V_{Ez}}{3}\rho_m \delta_E \tag{A6.53}$$

Substituting for $V_{sx,E}$ from above and solving:

$$\delta_E = \left\{ \frac{\left[3k_m\left(T_b - T_m\right) + 6\mu V_{j,E}^2\right]Z}{\rho_m V_{Ez}\left[C_v\left(T_m - T_s\right) + \lambda\right]} \right\}^{1/2} \tag{A6.54}$$

where Z is the solid bed length in the down channel direction from point of the initiation of melting. The first order evaluation of V_{Ez} is calculated as follows:

$$V_{Ez} = \frac{\int_0^{\frac{2}{3}\delta_E} x_E \frac{V_{j,E}}{\delta_E}\left[2 - \left(\frac{3x_E}{\delta_E}\right)\right]dx}{\frac{2}{3}\delta_E} = \frac{\frac{V_{j,E}}{\delta_E}\left[x_E^{\,2} - \left(\frac{x_E^{\,3}}{\delta_E}\right)\right]\Bigg|_0^{\frac{2}{3}\delta_E}}{\frac{2}{3}\delta_E} \tag{A6.55}$$

$$V_{Ez} = \frac{2}{9}V_{j,E} = 0.22V_{j,E} \tag{A6.56}$$

This is the velocity that is to be used in Equation A6.56. The film width at the pushing flight is calculated in a similar manner as the film thickness at the trailing flight surface.

$$\delta_B = W(z) - X - \delta_E \tag{A6.57}$$

This is then related to the velocity of the bed surface in the negative x direction.

$$V_{sx,B} = -\frac{dX}{dt}\bigg|_{y=\delta_B} = \frac{\left[\frac{k_m}{\delta_B}\left(T_b - T_m\right) + \frac{4\mu V_{j,B}^2}{2\delta_B}\right]}{\rho_s C_v\left(T_m - T_s\right) + \rho_s\lambda} \tag{A6.58}$$

$$V_{j,B} = V_{sz} + \left|\overline{V}_{fz}\right| \tag{A6.59}$$

Thus the evaluation of the shape of the solid bed requires the simultaneous solution of four differential equations that are functions of time. For the solution to be relevant to the melting process, it must be related to the down channel distance from the initiation of melting and that is calculated from:

$$\frac{dZ}{dt} = V_{sz} \tag{A6.60}$$

All of the pertinent variables are now differential functions of the time parameter. These are stiff equations, however, that can be solved using an appropriate stiff differential equation solver.

A6.2 Effect of Static Pressure on Melting

It is well know that melting occurs more rapidly if there is a positive pressure at the end of the transition zone. If Film B is considered to be large and if the pressure in the melt Film C is essentially zero at the trailing flight, a first order approximation of this pressure effect can be achieved by adding a pressure dissipation term to the rate of material loss in the y direction of the solid surface adjacent to the Film C:

$$V_{sy,C,P} = -\left.\frac{dY}{dt}\right|_{y=\delta_c} = \frac{\left[\frac{k_m}{\delta_c}\left(T_b - T_m\right) + \frac{\mu V_{j,C}^2}{2\delta_c}\right] + \left[\frac{V_{j,B}\delta_c \left|\Delta P_x(z)\right|}{2X}\right]}{\rho_s C_v \left(T_m - T_s\right) + \rho_s \lambda} \tag{A6.61}$$

where $\Delta P_x(z)$ is the pressure change across (x direction) the solid bed of width X, and as expected it is a function of the downstream position z.

References

1. Tadmor, Z. and Klein, I., "Engineering Principles of Plasticating Extrusion," Van Nostrand Reinhold Company, New York, 1970
2. Derezinski, S.J., "Heat Transfer in Extruder Screws," *SPE ANTEC Tech. Papers*, **46**, 83 (2000)

Appendix A7 Flow and Energy Equation Development for the Metering Channel

This appendix contains the detailed development for the velocity and energy dissipation equations that are presented in Chapter 7. Full understanding of these developments is not required for detailed analysis and troubleshooting of the extrusion process. They are presented here for those who desire a deep understanding of the mathematics involved with the screw rotation analysis. Some of the equations and figures are duplicated in this appendix for clarity. The nomenclature used here is consistent with that used earlier. The reader is directed to Chapters 1 and 7 for nomenclature. Flow rates can be easily calculated from the velocity equations.

A7.1 Transformed Frame Flow Analysis

Traditionally the fluid mechanics of the extrusion process are summarized by the simple plate model illustrated in Figure A7.1 and as described in Section 7.4. The motion of the screw is unchanged, but the reference frame has been moved to transform the problem to a fixed boundary problem for the observer. The flow in the rectangular channel is reduced into the *x*-direction flow across the channel and the *z*-direction flow down the channel.

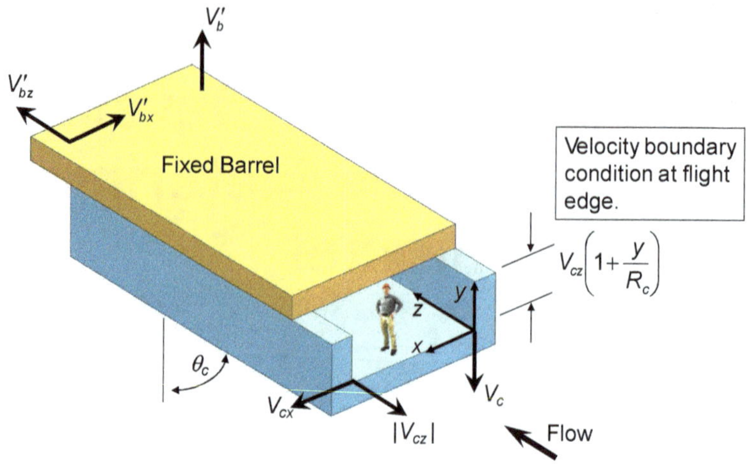

Figure A7.1 Transformed (Lagrangian) frame for the analysis of extruder fluid flow. Here the reference frame is positioned on the bottom of the screw channel. The observer on the frame would see the barrel move with the component velocities of V'_{bx} and V'_{bz}

The velocity at the flight tips is provided by Equation A7.1 and the components by Equations A7.2 and A7.3 as follows:

$$V_b = \pi N D_b \tag{A7.1}$$

$$V_{bx} = \pi N D_b \sin\theta_b \tag{A7.2}$$

$$V_{bz} = -\pi N D_b \cos\theta_b \tag{A7.3}$$

Additional assumptions were made in order to derive a two-dimensional analytical model to calculate the flow rate and viscous dissipation:

1. The fluid is an incompressible Newtonian fluid.
2. The flow is steady state and fully developed in the down-channel direction.
3. The flow is laminar.
4. There is no slip at the wall.
5. The polymer is so viscous that the inertial force can be neglected.
6. The gravitational force is also negligible.
7. The flow is isothermal.

The simplified governing equations of continuity and motion are as follows:

$$0 = \left[\frac{\partial V_x}{\partial x} + \frac{\partial V_y}{\partial y}\right] \tag{A7.4}$$

$$\frac{\partial P}{\partial x}\frac{1}{\mu} = \left[\frac{\partial^2 V_x}{\partial x^2} + \frac{\partial^2 V_x}{\partial y^2}\right] \tag{A7.5}$$

$$\frac{\partial P}{\partial y}\frac{1}{\mu}=\left[\frac{\partial^2 V_y}{\partial x^2}+\frac{\partial^2 V_y}{\partial y^2}\right] \tag{A7.6}$$

$$\frac{\partial P}{\partial z}\frac{1}{\mu}=\left[\frac{\partial^2 V_z}{\partial x^2}+\frac{\partial^2 V_z}{\partial y^2}\right] \tag{A7.7}$$

The principle of superposition is used to break the complicated flow into the com ponent velocities. These component velocities will be derived in the next sections.

A7.1.1 *x*-Directional Flow

The *x*-direction velocity component is perpendicular to the flight edge. It is assumed that the flow in the *x* direction is fully developed, forcing $\partial V_x/\partial x$ to be zero. Moreover, from the equation of continuity (Equation A7.4), both $\partial V_y/\partial y$ and V_y are zero for this simplified analysis. The *x*-component equation of motion (Equation A7.5) reduces to:

$$\frac{\partial P}{\partial x}\frac{1}{\mu}=\left[\frac{\partial^2 V_x}{\partial y^2}\right] \tag{A7.8}$$

Equation A7.8 is solved with the boundary conditions below for the transformed reference frame shown in Figure A7.1. Here the flight clearance is ignored.

$$y=0 \quad V_x=0 \qquad \text{screw core} \tag{A7.9}$$

$$y=H \quad V_x=-V_{bx} \qquad \text{flight tips} \tag{A7.10}$$

$$\int_0^H V_x dy=0 \qquad \text{mass balance in the } x \text{ direction} \tag{A7.11}$$

Equation A7.8 is integrated twice and the resulting equation is solved using the boundary conditions of Equations A7.9 and A7.10. The resulting equation for V_x is as follows:

$$V_x=-V_{bx}\frac{y}{H}-\frac{1}{2\mu}y(H-y)\frac{\partial P}{\partial x} \tag{A7.12}$$

The pressure term is eliminated using the mass balance from Equation 7.12. Upon integration the velocity function for the *x* direction is as follows:

$$V_{xL}=V_{cx}\left(1+\frac{H}{R_c}\right)\frac{y}{H}\left(2-3\frac{y}{H}\right)=V_{bx}\frac{y}{H}\left(2-3\frac{y}{H}\right) \tag{A7.13}$$

Equation A7.13 is the cross-channel flow in the transformed (Lagrangian) frame and concludes the derivation of Equation 7.18. Equation A7.13 also applies to a physical device where the barrel is actually rotated. Transforming Equation A7.13 to the laboratory (Eulerian) reference frame as follows for a physical device where the screw is rotated:

$$V_{xE} = V_{cx}\left(1+\frac{H}{R_c}\right)\frac{y}{H}\left(2-3\frac{y}{H}\right)+V_{cx}\left(1+\frac{y}{R_c}\right) \tag{A7.13a}$$

A7.1.2 *z*-Directional Flow

The *z*-direction velocity is in the downstream direction and parallel to the flight edges and described by Equation A7.7. The *z*-direction flow is considered to be the combination of pure pressure flow and pure drag flow. The boundary conditions for Equation A7.7 are as follows for barrel rotation:

$$y = 0 \quad V_z = 0 \quad \text{screw core} \tag{A7.14}$$

$$y = H \quad V_z = V_{bz} \quad \text{barrel surface} \tag{A7.15}$$

$$x = 0 \quad V_x = 0 \quad \text{pushing flight} \tag{A7.16}$$

$$x = W \quad V_x = 0 \quad \text{trailing flight} \tag{A7.17}$$

By using the principle of superposition and separation of variables, the solution to Equation A7.7 is provided by Equations A7.18 and A7.19 [pages 190 to 199 in reference 1]. For barrel-rotation-driven flow in the *z* direction for the transformed reference frame, the drag flow and pressure flow components are as follows:

$$V_{dzL} = V_{cz}\left(1+\frac{H}{R_c}\right)\frac{4}{\pi}\sum_{i=1,3,5\ldots}^{\infty}\frac{\sinh\left(i\pi\frac{y}{W}\right)\sin\left(i\pi\frac{x}{W}\right)}{i\sinh\left(i\pi\frac{H}{W}\right)}$$

$$= \frac{4V_{bz}}{\pi}\sum_{i=1,3,5\ldots}^{\infty}\frac{\sinh\left(i\pi\frac{y}{W}\right)\sin\left(i\pi\frac{x}{W}\right)}{i\sinh\left(i\pi\frac{H}{W}\right)} \tag{A7.18}$$

$$V_{pz} = \frac{H^2}{2\mu}\frac{\partial P}{\partial z}\left[\left(\frac{y}{H}\right)^2-\frac{y}{H}+\frac{8}{\pi^3}\sum_{i=1,3,5\ldots}^{\infty}\frac{\cosh\left(i\pi\frac{W}{H}\left(\frac{x}{W}-\frac{1}{2}\right)\right)\sin\left(i\pi\frac{y}{H}\right)}{i^3\cosh\left(i\pi\frac{W}{2H}\right)}\right] \tag{A7.19}$$

Equation A7.18 is identical to Equation 7.20 and Equation A7.19 is identical to Equation 7.23. These equations will be derived next.

A7.1.2.1 z-Directional Flow: Generalized Equation for Screw, Barrel, and Core Rotations

Using the boundary functions defined in Figure A7.2 with the z-direction flow coming out of the page, Equation A7.18 can be developed as follows from a general solution where either or both of the surfaces at the screw root or at the surface at the barrel interface is translated in the z direction. Thus, for barrel rotation in the transformed frame, V_b is nonzero and V_c is zero ($N_c = 0$). For screw core rotation, V_c is nonzero and V_b is zero ($N_b = 0$). If in the special case where both the barrel and screw are moving, then V_b and V_c are nonzero. The solution involves Fourier analysis, and as a result, the task is to convert the designated equation into a form where the solution can be just written down [2]. Fourier analysis requires that at least one of the four boundaries have 0 as a boundary condition. The following analysis is developed for either or both of the x-direction surfaces moving out of the page.

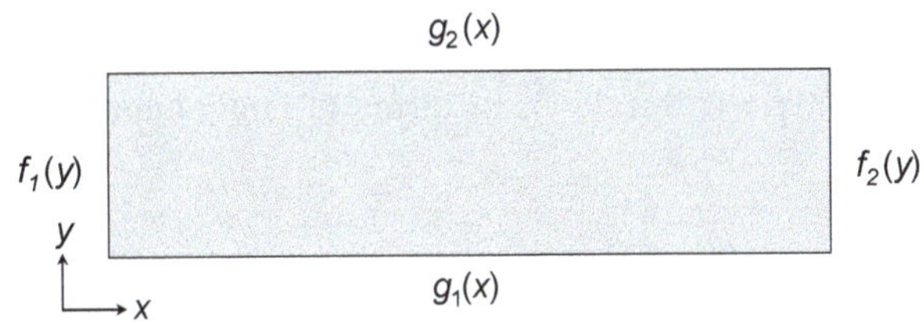

Figure A7.2 Schematic for the generalized boundary functions for flow in the z direction (out of the page)

Starting with Equation A7.7, the following boundary conditions are defined for generalized screw, barrel, or core rotations:

$$V_z(x, 0) = V_{cz} \qquad \text{screw core} \tag{A7.20}$$

$$V_z(x, H) = V_{bz} \qquad \text{barrel surface} \tag{A7.21}$$

$$V_z(0, y) = V_H(y) \qquad \text{pushing flight} \tag{A7.22}$$

$$V_z(W, y) = V_H(y) \qquad \text{trailing flight} \tag{A7.23}$$

$$V_H(y) = V_{cz}(1+y/R_c) \tag{A7.24}$$

$$V_{cz} = \pi N_c D_c \cos(\theta_c) \quad V_{bz} = \pi N_b D_b \cos(\theta_b) \tag{A7.25}$$

where N_c and N_b are the rotation rates of the screw core and barrel, respectively, in revolutions/s.

For the Fourier analysis solution, assume:

$$V_z(x,y) = X(x)Y(y) \tag{A7.26}$$

$$\frac{\partial V_z}{\partial x} = X'(x)Y(y) \qquad \frac{\partial^2 V_z}{\partial^2 x} = X''(x)Y(y) \tag{A7.27}$$

$$\frac{\partial V_z}{\partial y} = X(x)Y'(y) \qquad \frac{\partial^2 V_z}{\partial^2 y} = X(x)Y''(y) \tag{A7.28}$$

Substituting into Equation A7.7:

$$X''(x)Y(y)+X(x)Y''(y)=0 \tag{A7.29}$$

Rearranging and recognizing that the function must equal a constant because the two resulting terms cannot be functions of the other:

$$\frac{X''(x)}{X(x)}=-\frac{Y''(y)}{Y(y)}=-\lambda^2 \tag{A7.30}$$

λ^2 = constant < 0, thus:

$$X''(x)+\lambda^2 X(x)=0 \tag{A7.31}$$

$$Y''(y)-\lambda^2 Y(y)=0 \tag{A7.32}$$

These equations satisfy the Fourier analysis for Cartesian coordinates, and the solution can be directly written down in terms of trigonometric functions as follows:

$$X(x)=A\sinh(\lambda x)+B\cosh(\lambda x) \tag{A7.33}$$

$$Y(y)=C\sin(\lambda y)+D\cos(\lambda y) \tag{A7.34}$$

Let u be the velocity in the z direction. Define $g_1(x)$ as the velocity at $y = 0$ and $g_2(x)$ as the velocity at $y = H$:

$$V_z(x,y)=u=\sum_{i=1}^{\infty}\sin(\mu_i x)\left[\frac{A_i\sinh(\mu_i y)+B_i\sinh((a-y)\mu_i)}{\sinh(\mu_i a)}\right] \tag{A7.35}$$

where

$$\mu_i=\frac{i\pi}{b} \tag{A7.36}$$

$$A_i=\frac{2}{b}\int_0^b g_2(x)\sin(\mu_i x)\,dx \tag{A7.37}$$

$$B_i=\frac{2}{b}\int_0^b g_1(x)\sin(\mu_i x)\,dx \tag{A7.38}$$

If $g_1(x)$ equals the screw root velocity, $g_1(x) = V_{cz}$, then:

$$B_i=\frac{2}{b}\int_0^b V_{cz}\sin(\mu_i x)\,dx \tag{A7.39}$$

$$B_i=\frac{2}{\mu_i b}V_{cz}\left[-\cos(\mu_i x)\right]_0^b \tag{A7.40}$$

$$B_i = \frac{-2}{\mu_i b} V_{cz} \left[\cos(i\pi) - \cos(0)\right] \tag{A7.41}$$

$$B_i = \frac{-2}{\mu_i b} V_{cz} \left[(-1)^i - 1\right] \tag{A7.42}$$

$$B_i = \frac{-2}{i\pi} V_{cz} \left[(-1)^i - 1\right] \tag{A7.43}$$

If $g_1(x)$ equals the screw root velocity V_{cz} and $g_2(x) = V_{bz} = V_{cz}(1+H/R_c)$, then:

$$A_i = \frac{2}{b} \int_0^b V_{bz} \sin(\mu_i x) dx \tag{A7.44}$$

$$A_i = \frac{2}{\mu_i b} V_{bz} \left[-\cos(\mu_i x)\right]_0^b \tag{A7.45}$$

$$A_i = \frac{-2}{\mu_i b} V_{bz} \left[\cos(i\pi) - \cos(0)\right] \tag{A7.46}$$

$$A_i = \frac{-2}{\mu_i b} V_{bz} \left[(-1)^i - 1\right] \tag{A7.47}$$

$$A_i = \frac{-2}{i\pi} V_{bz} \left[(-1)^i - 1\right] \tag{A7.48}$$

Substituting and recognizing that u represents $V_z(x, y)$:

$$V_z(x,y) = \frac{4}{\pi} \sum_{i=1,3,5,\ldots}^{\infty} \left[\frac{V_{bz} \sinh\left(i\pi \frac{y}{\mathbf{b}}\right) + V_{cz} \sinh\left(i\pi \frac{(a-y)}{\mathbf{b}}\right)}{i \sinh\left(i\pi \frac{a}{\mathbf{b}}\right)} \right] \sin\left(i\pi \frac{x}{\mathbf{b}}\right) \tag{A7.49}$$

where $a = H$ and $b = W$ for a typical screw from Figure A7.1.

$$V_z(x,y) = \frac{4}{\pi} \sum_{i=1,3,5,\ldots}^{\infty} \left[\frac{V_{bz} \sinh\left(i\pi \frac{y}{W}\right) + V_{cz} \sinh\left(i\pi \frac{(H-y)}{W}\right)}{i \sinh\left(i\pi \frac{H}{W}\right)} \right] \sin\left(i\pi \frac{x}{W}\right) \tag{A7.50}$$

Equation A7.50 is the generalized velocity in the z direction due to any combination of barrel rotation and screw core rotation in the transformed reference frame.

A7.1.2.2 *z*-Directional Flow for Barrel Rotation

For the barrel rotation case where $N_c = 0$, the term $V_{cz} = 0$ and the second term in the numerator of Equation A7.50 drops out. The *z*-direction velocity for barrel rotation in the transformed reference frame is as follows:

$$V_{dzL}(x,y) = \frac{4V_{bz}}{\pi}\sum_{i=1,3,5,...}^{\infty}\left[\frac{\sinh\left(i\pi\frac{y}{W}\right)\sin\left(i\pi\frac{x}{W}\right)}{i\sinh\left(i\pi\frac{H}{W}\right)}\right]$$

$$= V_{cz}\left(1+\frac{H}{R_c}\right)\frac{4}{\pi}\sum_{i=1,3,5,...}^{\infty}\left[\frac{\sinh\left(i\pi\frac{y}{W}\right)\sin\left(i\pi\frac{x}{W}\right)}{i\sinh\left(i\pi\frac{H}{W}\right)}\right] \tag{A7.51}$$

Recognizing that

$$V_{cz}\left(1+\frac{H}{R_c}\right) = V_{bz} \tag{A7.52}$$

the *z*-directional velocity Equation A7.51 derived for barrel rotation is identical to Equations A7.18 and 7.20.

A7.1.2.3 *z*-Directional Flow for Screw Rotation

The solution for screw rotation is obtained by moving Equation A7.51 from the transformed frame back to the laboratory (Eulerian) frame:

$$V_{dzE}(x,y) = V_{cz}\left(1+\frac{H}{R_c}\right)\frac{4}{\pi}\sum_{i=1,3,5,...}^{\infty}\left[\frac{\sinh\left(i\pi\frac{y}{W}\right)\sin\left(i\pi\frac{x}{W}\right)}{i\sinh\left(i\pi\frac{H}{W}\right)}\right] - V_{cz}\left(1+\frac{y}{R_c}\right) \tag{A7.53}$$

A7.1.2.4 Flow Rate Calculation

Since flow is frame indifferent, then the transformed flow rate calculation for rate can be calculated using the above Equation A7.51. The flow rate is obtained by integrating the *z*-direction velocity of the area as follows:

$$Q_d = \int_0^W\int_0^H V_{dzL}\,dx\,dy \tag{A7.54}$$

Defining a new set of variables:

$$\chi = \frac{x}{W} \quad W d\chi = dx \quad \varepsilon = \frac{y}{H} \quad H d\varepsilon = dy \quad h = \frac{H}{W} \tag{A7.55}$$

It follows:

$$Q_d = V_{bz} WH \frac{4}{\pi} \int_0^1 \int_0^1 \sum_{i=1,3,5,\ldots}^{\infty} \left[\frac{\sinh(i\pi h\varepsilon)}{i \sinh(i\pi h)} \right] \sin(i\pi\chi) d\chi d\varepsilon \tag{A7.56}$$

$$Q_d = V_{bz} WH \frac{4}{\pi} \sum_{i=1,3,5,\ldots}^{\infty} \left[\frac{\cosh(i\pi h\varepsilon)}{i \sinh(i\pi h) i\pi \, i\pi h} \right] \left[-\cos(i\pi\chi) \right] \Big|_0^1 \Big|_0^1 \tag{A7.57}$$

$$Q_d = V_{bz} WH \frac{4}{\pi^3} \frac{W}{H} \sum_{i=1,3,5,\ldots}^{\infty} \left[\frac{1}{i^3} \frac{(\cosh(i\pi h) - 1)(-\cos(i\pi) + \cos(0))}{\sinh(i\pi h)} \right] \tag{A7.58}$$

Using the trigonometric identity:

$$\tanh\left(\frac{i\pi h}{2}\right) = \frac{\cosh(i\pi h) - 1}{\sinh(i\pi h)} \tag{A7.59}$$

Therefore:

$$\begin{aligned} Q_d &= V_{bz} WH \frac{4}{\pi^3} \frac{W}{H} \sum_{i=1,3,5\ldots}^{\infty} \frac{1}{i^3} \tanh\left(\frac{i\pi h}{2}\right)(1+1) \\ &= V_{bz} WH \frac{8}{\pi^3} \frac{W}{H} \sum_{i=1,3,5\ldots}^{\infty} \frac{1}{i^3} \tanh\left(\frac{i\pi h}{2}\right) \end{aligned} \tag{A7.60}$$

$$Q_d = \frac{V_{bz} WHF_d}{2} \tag{A7.61}$$

with F_d defined as

$$F_d = \frac{16W}{\pi^3 H} \sum_{i=1,3,5\ldots}^{\infty} \frac{1}{i^3} \tanh\left(\frac{i\pi H}{2W}\right) \tag{A7.62}$$

A7.1.2.5 *z*-Directional Flow for Core Rotation

The development of the *z*-directional flow due to the motion of the screw core with a stationary barrel and stationary helix starts from Equation A7.50. As discussed in Section A7.1.2.1, Equation A7.50 was developed for generalized motion of the top and bottom surfaces. For the core rotation case here, N_b and V_{bz} are 0, and the first term in the numerator for Equation A7.50 drops out. The z-directional velocity for core rotation in the transformed (Lagrangian) reference frame is as follows:

$$V_{czL}(x,y)=\frac{4V_{cz}}{\pi}\sum_{i=1,3,5,\ldots}^{\infty}\left[\frac{\sinh\left(i\pi\frac{(H-y)}{W}\right)\sin\left(i\pi\frac{x}{W}\right)}{i\sinh\left(i\pi\frac{H}{W}\right)}\right] \tag{A7.63}$$

Transforming Equation A7.63 into the laboratory (Eulerian) reference frame is as follows:

$$V_{czE}(x,y)=\frac{4V_{cz}}{\pi}\sum_{i=1,3,5,\ldots}^{\infty}\left[\frac{\sinh\left(i\pi\frac{(H-y)}{W}\right)\sin\left(i\pi\frac{x}{W}\right)}{i\sinh\left(i\pi\frac{H}{W}\right)}\right]-V_{cz}\left(1+\frac{y}{R_c}\right) \tag{A7.64}$$

A7.1.3 *z*-Directional Flow for Helix Rotation with a Stationary Screw Core and Barrel

The velocity in the z direction due to the motion of just the helix is derived here. Here the core and barrel are stationary. For the rotation of the flights out of the page for Figure A7.2, the above analysis is modified for the following boundary conditions:

$V_z(x, 0) = 0$	screw core	(A7.65)
$V_z(x, H) = 0$	barrel surface	(A7.66)
$V_z(0, y) = V_H(y)$	pushing flight	(A7.67)
$V_z(W, y) = V_H(y)$	trailing flight	(A7.68)
$V_H(y) = V_{cz}(1+y/R_c)$		(A7.69)
$V_{cz} = \pi N_c D_c \cos(\theta_c)$		(A7.70)

Again assume:

$$V_z(x,y)=X(x)Y(y) \tag{A7.71}$$

$$\frac{\partial V_z}{\partial x}=X'(x)Y(y) \qquad \frac{\partial^2 V_z}{\partial^2 x}=X''(x)Y(y) \tag{A7.72}$$

$$\frac{\partial V_z}{\partial y}=X(x)Y'(y) \qquad \frac{\partial^2 V_z}{\partial^2 y}=X(x)Y''(y) \tag{A7.73}$$

Substituting into Equation A7.7:

$$X''(x)Y(y)+X(x)Y''(y)=0 \tag{A7.74}$$

Rearranging and recognizing that the function must equal a constant because the two resulting terms cannot be functions of the other:

$$\frac{X''(x)}{X(x)} = -\frac{Y''(y)}{Y(y)} = -\lambda^2 \tag{A7.75}$$

λ^2 = constant < 0, thus:

$$X''(x) + \lambda^2 X(x) = 0 \tag{A7.76}$$

$$Y''(y) - \lambda^2 Y(y) = 0 \tag{A7.77}$$

These equations satisfy the Fourier analysis for Cartesian coordinates and the solution can be directly written down in terms of trigonometric functions as follows:

$$X(x) = A\sinh(\lambda x) + B\cosh(\lambda x) \tag{A7.78}$$

$$Y(y) = C\sin(\lambda y) + D\cos(\lambda y) \tag{A7.79}$$

Let u be the velocity in the z direction. Define $f_1(y)$ as the velocity at $x = 0$ and $f_2(y)$ as the velocity at $x = W$, then:

$$V_z(x,y) = u = \sum_{i=1}^{\infty} \sin\lambda_i x \left[\frac{A_i \sinh\lambda_i y + B_i \sinh(a-y)\lambda_i}{\sinh\lambda_i a}\right] \tag{A7.80}$$

where

$$\lambda_i = \frac{i\pi}{b} \tag{A7.81}$$

$$A_i = \frac{2}{b}\int_0^b f_2(y)\sin(\lambda_i y)\,dy \tag{A7.82}$$

$$B_i = \frac{2}{b}\int_0^b f_1(y)\sin(\lambda_i y)\,dy \tag{A7.83}$$

If $f_1(y)$ equal $f_2(y)$:

$$f_1(y) = V_{cz}\left(1 + \frac{y}{R_c}\right) \tag{A7.84}$$

then

$$A_i = B_i = \frac{2V_{cz}}{b}\int_0^b (1)\sin(\lambda_i y)dy + \int_0^b \left(\frac{y}{R_c}\right)\sin(\lambda_i y)\,dy \tag{A7.85}$$

$$A_i = B_i = \frac{2}{1} V_{cz} \left[\left(-\frac{1}{\lambda_i} \cos(\lambda_i y) \right) + \frac{1}{R_c} \left(\frac{1}{\lambda_i^2} \sin(\lambda_i y) - \frac{y}{\lambda_i} \cos(\lambda_i y) \right) \right]_0^b \tag{A7.86}$$

$$A_i = B_i = \frac{2}{i\pi} V_{cz} \left[-\left((-1)^i - 1 \right) - \frac{H}{R_c} \left((-1)^i - 1 \right) \right] \tag{A7.87}$$

$$A_i = B_i = \frac{4V_{cz}}{i\pi} \left[1 + \frac{H}{R_c} \right] \tag{A7.88}$$

Substituting and recognizing that u represents $V_{zHL}(x,y)$, the velocity in the z direction is due to the motion of the flights (stationary core and barrel) in the transformed (Lagrangian) reference frame.

$$V_{zHL}(x,y) = \frac{4V_{cz}}{\pi} \left[1 + \frac{H}{R_c} \right] \sum_{i=1,3,5\ldots}^{\infty} \left[\frac{\sinh\left(i\pi \frac{x}{H} \right) + \sinh\left(i\pi \frac{(W-x)}{H} \right)}{i \sinh\left(i\pi \frac{W}{H} \right)} \right] \sin\left(i\pi \frac{y}{H} \right) \tag{A7.89}$$

Transforming A7.89 into the laboratory (Eulerian) reference frame is as follows:

$$V_{zHE}(x,y) = \frac{4V_{cz}}{\pi} \left[1 + \frac{H}{R_c} \right] \sum_{i=1,3,5\ldots}^{\infty} \left[\frac{\sinh\left(i\pi \frac{x}{H} \right) + \sinh\left(i\pi \frac{(W-x)}{H} \right)}{i \sinh\left(i\pi \frac{W}{H} \right)} \right]$$

$$\sin\left(i\pi \frac{y}{H} \right) - V_{cz} \left(1 + \frac{y}{R_c} \right) \tag{A7.90}$$

A7.1.4 *z*-Directional Flow Due to a Pressure Gradient

The *z*-direction velocity that is induced by a pressure gradient was first solved by Boussinesq [3]. Starting with Equation A7.7 the pressure flow Equation A7.19 can be developed as follows:

$$\frac{\partial P}{\partial z} \frac{1}{\mu} = \left[\frac{\partial^2 V_z}{\partial x^2} + \frac{\partial^2 V_z}{\partial y^2} \right] \tag{A7.91}$$

Redefining the variables:

$$\left[\frac{\partial^2 f}{\partial x^2}+\frac{\partial^2 f}{\partial y^2}+\frac{\partial^2 g}{\partial x^2}+\frac{\partial^2 g}{\partial y^2}\right]=\frac{\partial P}{\partial z}\frac{1}{\mu} \tag{A7.92}$$

$$\left[\frac{\partial^2 f}{\partial x^2}+\frac{\partial^2 f}{\partial y^2}\right]=\frac{\partial P}{\partial z}\frac{1}{\mu} \quad \text{pressure flow} \tag{A7.93}$$

$$\left[\frac{\partial^2 g}{\partial x^2}+\frac{\partial^2 g}{\partial y^2}\right]=0 \quad \text{drag flow} \tag{A7.94}$$

Boundary conditions are as follows:

Pressure Flow	Drag Flow		
$f(0, y) = 0$	$g(0, y) = 0$	pushing flight	(A7.95)
$f(W, y) = 0$	$g(W, y) = 0$	trailing flight	(A7.96)
$f(x, 0) = 0$	$g(x, 0) = 0$	screw core	(A7.97)
$f(x, H) = 0$	$g(x, H) = V_{bz}$	barrel surface	(A7.98)

The drag flow solution to this general approach was previously developed and shown by Eq A7.51. For pressure flow:

$$\text{Let } f(xy)=\alpha(x,y)+\beta(y) \tag{A7.99}$$

Then:

$$\left[\frac{\partial^2 \alpha}{\partial x^2}+\frac{\partial^2 \alpha}{\partial y^2}+\frac{\partial^2 \beta}{\partial y^2}\right]=\frac{\partial P}{\partial z}\frac{1}{\mu} \tag{A7.100}$$

Rearranging:

$$\left[\frac{\partial^2 \beta}{\partial y^2}\right]=\frac{\partial P}{\partial z}\frac{1}{\mu} \tag{A7.101}$$

$$\left[\frac{\partial^2 \alpha}{\partial x^2}+\frac{\partial^2 \alpha}{\partial y^2}\right]=0 \tag{A7.102}$$

The transformed boundary conditions are as follows:

$\alpha(0, y) = -\beta$	$\beta(0) = 0$	pushing flight	(A7.103)
$\alpha(W, y) = -\beta$	$\beta(0) = 0$	trailing flight	(A7.104)
$\alpha(x, 0) = 0$		screw core	(A7.105)
$\alpha(x, H) = 0$		barrel surface	(A7.106)

First Equation A7.101 is integrated twice as follows:

$$\left[\frac{\partial \beta}{\partial y}\right] = \frac{\partial P}{\partial z}\frac{1}{\mu}y + c_1 \qquad \text{(A7.107)}$$

$$\beta(y) = \frac{\partial P}{\partial z}\frac{1}{\mu}\frac{y^2}{2} + c_1 y + c_2 \qquad \text{(A7.108)}$$

For $\beta(0) = 0 = c_2$ (A7.109)

For $\beta(H) = 0$ $\quad 0 = \frac{\partial P}{\partial z}\frac{1}{\mu}\frac{H^2}{2} + c_1 H$ (A7.110)

$$c_1 = -\frac{\partial P}{\partial z}\frac{1}{\mu}\frac{H}{2} \qquad \text{(A7.111)}$$

$$\beta(y) = \frac{1}{\mu}\frac{y}{2}\frac{\partial P}{\partial z}(y + H) \qquad \text{(A7.112)}$$

Turning to the companion equation Equation A7.102:

$$\left[\frac{\partial^2 \alpha}{\partial x^2} + \frac{\partial^2 \alpha}{\partial y^2}\right] = 0 \qquad \text{(A7.102)}$$

$$\alpha(x, y) = X(x)Y(y) \qquad \text{(A7.113)}$$

$$X''(x)Y(y) + X(x)Y''(y) = 0 \qquad \text{(A7.114)}$$

Rearranging and recognizing that the function must equal a constant because the two resulting terms cannot be a function of the other:

$$\frac{X''(x)}{X(x)} = -\frac{Y''(y)}{Y(y)} = \lambda^2 \qquad \text{(A7.115)}$$

λ^2 = constant > 0, thus:

$$X''(x) - \lambda^2 X(x) = 0 \qquad \text{(A7.116)}$$

$$n^2 = \lambda^2 \qquad \text{(A7.117)}$$

$$n = \pm\lambda \qquad \text{(A7.118)}$$

$$X(x) = c_1 e^{\lambda x} + c_2 e^{-\lambda x} \qquad \text{(A7.119)}$$

$$X(x) = c_1\left(\cosh(\lambda x) + \sinh(\lambda x)\right) + c_2\left(\cosh(\lambda x) - \sinh(\lambda x)\right) \qquad \text{(A7.120)}$$

$$X(x) = (c_1 + c_2)(\cosh(\lambda x)) + (c_1 - c_2)(\sinh(\lambda x)) \quad \text{(A7.121)}$$

$$X(x) = A(\cosh(\lambda x)) + B(\sinh(\lambda x)) \quad \text{(A7.122)}$$

In the case of Y:

$$Y''(y) + \lambda^2 Y(y) = 0 \quad \text{(A7.123)}$$

Now:

$$m^2 = -\lambda^2 = \pm i\lambda \quad \text{(A7.124)}$$

$$Y(y) = c_1 e^{i\lambda x} + c_2 e^{-i\lambda x} \quad \text{(A7.125)}$$

Using the same technique as for $X(x)$:

$$Y(y) = C(\cos(\lambda y)) + D(\sin(\lambda x)) \quad \text{(A7.126)}$$

Boundary conditions are as follows:

$\alpha(x, 0) = 0$ it follows that $c = 0$

$\alpha(x, H) = 0$ then

$$D\sin(\lambda H) = 0 \qquad D \neq 0 \therefore \sin(\lambda H) = 0 \quad \text{(A7.127)}$$

where $\lambda H = i\pi$ where $i = 0, 1, 2, 3, \ldots$

$$\lambda = \frac{i\pi}{H} \quad \text{(A7.128)}$$

$$Y(y) = \sum_{i=1}^{\infty} D_i \sin\left(i\pi \frac{y}{H}\right) \quad \text{(A7.129)}$$

From: $\alpha(x, y) = X(x)Y(y)$ and $A_i = D_i A$ and $B_i = D_i B$

$$\alpha(x,y) = \sum_{i=1}^{\infty}\left(A_i \cosh(\lambda x)\sin\left(i\pi\frac{y}{H}\right) + B_i \sinh(\lambda x)\sin\left(i\pi\frac{y}{H}\right)\right) = \frac{y\partial P}{2\mu\partial z}(H - y) \quad \text{(A7.130)}$$

From:

$$\alpha(0, y) = -\beta(y) \quad \text{(A7.131)}$$

$$\sum_{i=1}^{\infty}\left(A_i \sin\left(i\pi\frac{y}{H}\right)\right) = -\beta(y) = \frac{y\partial P}{2\mu\partial z}(H - y) \quad \text{(A7.132)}$$

$$A_i = \frac{y\partial P}{2\mu\partial z}(H - y)\frac{1}{\sin\left(i\pi\frac{y}{H}\right)} \quad \text{(A7.133)}$$

$$\alpha(W,y) = -\beta(y) \tag{A7.134}$$

$$\sum_{i=1}^{\infty}\left(A_i\cosh(\lambda W)\sin\left(i\pi\frac{y}{H}\right)+B_i\sinh(\lambda W)\sin\left(i\pi\frac{y}{H}\right)\right)=\frac{y\partial P}{2\mu\partial z}(H-y) \tag{A7.135}$$

Multiply $\sum_{i=1}^{\infty}\left(A_i\sin\left(i\pi\frac{y}{H}\right)\right)=-\beta(y)=\frac{y\partial P}{2\mu\partial z}(H-y)$ by $\sin\left(i\pi\frac{y}{H}\right)$

The summations are replaced by integrations:

$$\int_0^H\left(\sin^2\left(i\pi\frac{y}{H}\right)\right)dy=\int_0^H\frac{\partial P}{2\mu\partial z}(yH-y)\left(\sin\left(i\pi\frac{y}{H}\right)\right)dy \tag{A7.136}$$

$$\int_0^H\frac{\partial P}{2\mu\partial z}(yH-y)\left(\sin\left(i\pi\frac{y}{H}\right)\right)dy=A_i\sinh\left(i\pi\frac{W}{H}\right)\left(1+B_i\cosh\left(i\pi\frac{W}{H}\right)\right)$$

$$\int_0^H\sin^2\left(i\pi\frac{y}{H}\right)dy \tag{A7.137}$$

Now solving the two previous equations for A_i and B_i:

$$B_i=\frac{4H^2\partial P}{i^3\pi^3\mu\partial z}\qquad i = 1, 3, 5, \ldots \tag{A7.138}$$

$$A_i=B_i\frac{1-\cosh\left(i\pi\frac{W}{H}\right)}{\sinh\left(i\pi\frac{W}{H}\right)}\qquad i = 1, 3, 5, \ldots \tag{A7.139}$$

Substituting into the definition of $\alpha(x,y)$ above and rearranging to get α as a function of x and y:

$$\alpha(x,y)=\frac{4H^2}{\pi^3}\left(\frac{1}{\mu}\frac{\partial P}{\partial z}\right)\sum_{i=1,3,5\ldots}^{\infty}\frac{\cosh\left(i\pi\frac{W}{H}\left(\frac{x}{H}-\frac{1}{2}\right)\right)}{i^3\cosh\left(i\pi\frac{W}{2H}\right)}\sin\left(i\pi\frac{y}{H}\right) \tag{A7.140}$$

The pressure-driven velocity $V(x, y)$ is the sum of $\alpha(x, y)$ and $\beta(y)$. Therefore the z-direction velocity as a function of x and y is as follows:

$$V_{pz}(x,y)=\frac{H^2}{2\mu}\frac{\partial P}{\partial z}\left\{\left(\frac{y}{H}\right)^2-\frac{y}{H}+\frac{8}{\pi^3}\sum_{i=1,3,5\ldots}^{\infty}\frac{\cosh\left(i\pi\frac{W}{H}\left(\frac{x}{W}-\frac{1}{2}\right)\right)}{i^3\cosh\left(i\pi\frac{W}{2H}\right)}\sin\left(i\pi\frac{y}{H}\right)\right\} \tag{A7.141}$$

Equation A7.141 is identical to Equations A7.19 and 7.23.

A7.2 Viscous Energy Dissipation for Screw Rotation

The viscous energy dissipation is calculated by multiplying the normal shear stress for the screw surface by the velocity of the screw surface integrated over the screw surface area. The method for calculation of dissipation for screw rotation is as follows:

$$\dot{\omega}_s = \iint_s n_i \tau_{i,j} v_j ds \tag{A7.142}$$

where $\dot{\omega}_s$ is the rate of work (or power) that is added to the system through surface s for screw rotation, n_i is the normal unit vector, τ is the shear stress tensor, and v is the velocity vector. $\dot{\omega}_s$ has components relating to the screw core, flights, and clearance for both rotational flow and for pressure flows. $\dot{\omega}_s$ is the rate of work (or power) inputted to the system via screw rotation, and the units are J/s. The shear stresses, velocities, and rate of work for the components for the length of the metering section are provided in the next sections. The calculations are shown for the generalized solution and for a channel where the H/W is less than 0.1. The generalized solution provides the best calculation of the energy dissipated and is thus the preferred method due to the availability of high-speed computers.

A7.2.1 Viscous Energy Dissipation for Screw Rotation: Generalized Solution

The viscous energy dissipation for screw rotation is calculated using the laboratory (Eulerian) frame velocities. The velocities were shown previously to be as follows.

For x-direction rotational flow:

$$V_{xE} = V_{cx}\left(1+\frac{H}{R_c}\right)\frac{y}{H}\left(2-3\frac{y}{H}\right)+V_{cx}\left(1+\frac{y}{R_c}\right) \tag{A7.143}$$

For z-direction rotational flow:

$$V_{dzE} = V_{cz}\left(1+\frac{H}{R_c}\right)\frac{4}{\pi}\sum_{i=1,3,5\ldots}^{\infty}\frac{\sinh\left(i\pi\frac{y}{W}\right)\sin\left(i\pi\frac{x}{W}\right)}{i\sinh\left(i\pi\frac{H}{W}\right)}-V_{cz}\left(1+\frac{y}{R_c}\right) \tag{A7.144}$$

For z-direction pressure flow:

$$V_{pz} = \frac{H^2}{2\mu}\frac{\partial P}{\partial z}\left[\left(\frac{y}{H}\right)^2 - \frac{y}{H} + \frac{8}{\pi^3}\sum_{i=1,3,5\ldots}^{\infty}\frac{\cosh\left(i\pi\frac{W}{H}\left(\frac{x}{W}-\frac{1}{2}\right)\right)\sin\left(i\pi\frac{y}{H}\right)}{i^3\cosh\left(i\pi\frac{W}{2H}\right)}\right] \quad \text{(A7.145)}$$

The viscous dissipation is obtained by multiplying the shear stress on the screw surface by the velocity on the screw surface and integrating over the screw surface area. As a consequence of screw rotation, the viscous dissipation in the extruder arises from seven components: (1) z-direction drag flow at the screw core, (2) x-direction drag flow at the core, (3) z-direction-driven flow at the flight, (4) x-direction-driven flow at the flight, (5) z-direction pressure flow at the screw core, (6) z-direction pressure flow at the screw flights, and (7) leakage flow over the flight lands. The principle of superposition is used, and the viscous dissipation is the combination of the seven components.

The rate of work to the material in the flight clearance, $\dot{\omega}_\lambda$, is the same for screw rotation and barrel rotation. The rate of work was previously provided in the literature [1] as follows:

$$\dot{\omega}_\lambda = \frac{p\mu_f \pi^2 N^2 D_b^2 e Z_b}{\lambda} \quad \text{(A7.146)}$$

The rate of work contribution from screw core rotation for driven flow in the z direction is calculated as follows:

$$V_{dzE} = V_{cz}\left(1+\frac{H}{R_c}\right)\frac{4}{\pi}\sum_{i=1,3,5\ldots}^{\infty}\frac{\sinh\left(i\pi\frac{y}{W}\right)\sin\left(i\pi\frac{x}{W}\right)}{i\sinh\left(i\pi\frac{H}{W}\right)} - V_{cz}\left(1+\frac{y}{R_c}\right) \quad \text{(A7.147)}$$

$$\tau_{yz}\,|_{y=0} = \mu\frac{\partial V_{dzE}}{\partial y}\,|_{y=0} = \mu V_{cz}\left(1+\frac{H}{R_c}\right)\frac{4}{\pi}\left[\sum_{i=1,3,5\ldots}^{\infty}\frac{\sin\left(i\pi\frac{x}{W}\right)}{\sinh\left(i\pi\frac{H}{W}\right)}\frac{\pi}{W}\right] - \frac{\mu V_{cz}}{R_c} \quad \text{(A7.148)}$$

$$V_{dzE}\big|_{y=0} = -V_{cz} \qquad \text{simple core motion in the } z \text{ direction} \quad \text{(A7.149)}$$

For viscous dissipation, the absolute value of $V_{dzE}\big|_{y=0}$ is used. The rate of work inputted by the core in the z direction for screw rotation, $\dot{\omega}_{core\_z}$, is as follows:

$$\dot{\omega}_{core\_z} = p\int_0^{W_c}\int_0^{Z_c}\left(\tau_{yz}\,|_{y=0}\right)\left(V_{dzE}\,|_{y=0}\right)dz\,dx \quad \text{(A7.150)}$$

$$\dot{\omega}_{core\_z} = p\mu Z_c V_{cz}^2 \left(1+\frac{H}{R_c}\right)\frac{4}{\pi}\left[\sum_{i=1,3,5...}^{\infty} \frac{-\cos\left(i\pi\frac{W_c}{W}\right)+1}{i\sinh\left(i\pi\frac{H}{W}\right)}\right] - \frac{p\mu V_{cz}^2 Z_c W_c}{R_c} \quad \text{(A7.151)}$$

The above equation is the generalized solution for the viscous dissipation due to the motion of the core in the z direction. The term W_c/W approaches 1 in the limit and this equation approaches the following:

$$\dot{\omega}_{core\_z} = p\mu Z_c V_{cz}^2 \left(1+\frac{H}{R_c}\right)\frac{8}{\pi}\left[\sum_{i=1,3,5...}^{\infty} \frac{1}{i\sinh\left(i\pi\frac{H}{W}\right)}\right] - \frac{p\mu V_{cz}^2 Z_c W_c}{R_c} \quad \text{(A7.152)}$$

The rate of work for flow in the x direction due to the motion of the core $\dot{\omega}_{core\_x}$ is as follows:

$$\dot{\omega}_{core\_x} = p\int_0^{W_c}\int_0^{Z_c}\left(\tau_{yx}\big|_{y=0}\right)\left(V_{xE}\big|_{y=0}\right)dxdz \quad \text{(A7.153)}$$

$$V_{xE}\big|_{y=0} = \left[V_{cx}\left(1+\frac{H}{R_c}\right)\frac{y}{H}\left(2-3\frac{y}{H}\right)+V_{cx}\left(1+\frac{y}{R_c}\right)\right]_{y=0} = V_{cx}$$

simple core rotation in the x direction (A7.154)

$$\tau_{yx}\big|_{y=0} = \mu\frac{\partial V_{xE}}{\partial y}\big|_{y=0} = \mu\left[\frac{V_{cx}}{H}\left(1+\frac{H}{R_c}\right)+\frac{V_{cx}}{R_c}\right] = \mu\left[\frac{2V_{cx}}{H}+\frac{3V_{cx}}{R_c}\right] \quad \text{(A7.155)}$$

$$\dot{\omega}_{core\_x} = p\mu\left(\frac{2V_{cx}^2}{H}+\frac{3V_{cx}^2}{R_c}\right)Z_c W_c \quad \text{(A7.156)}$$

The rate of work for flow in the z direction due to the motion of the flights $\dot{\omega}_{flight\_z}$ is as follows:

$$V_{dzE} = V_{cz}\left(1+\frac{H}{R_c}\right)\frac{4}{\pi}\sum_{i=1,3,5...}^{\infty}\frac{\sinh\left(i\pi\frac{y}{W}\right)\sin\left(i\pi\frac{x}{W}\right)}{i\sinh\left(i\pi\frac{H}{W}\right)} - V_{cz}\left(1+\frac{y}{R_c}\right) \quad \text{(A7.157)}$$

$$\tau_{xz}\big|_{x=0} = \mu\frac{\partial V_{dzE}}{\partial x}\big|_{x=0} = \mu V_{cz}\left(1+\frac{H}{R_c}\right)\frac{4}{\pi}\sum_{i=1,3,5...}^{\infty}\frac{\sinh\left(i\pi\frac{y}{W}\right)}{\sinh\left(i\pi\frac{H}{W}\right)}\frac{\pi}{W} \quad \text{(A7.158)}$$

$$V_{dxE}\big|_{x=0} = -V_{cz}\left(1+\frac{y}{R_c}\right) \quad \text{simple flight motion in the } z \text{ direction} \tag{A7.159}$$

$$\dot{\omega}_{flight\_z} = 2pZ_c \int_0^H \left(\tau_{xz}\big|_{x=0}\right) V_{cz}\left(1+\frac{y}{R_c}\right) dy \tag{A7.160}$$

$$\dot{\omega}_{flight\_z} = 2pZ_c \int_0^H \left(\mu V_{cz}\left(1+\frac{H}{R_c}\right)\frac{4}{\pi}\sum_{i=1,3,5\ldots}^{\infty} \frac{\sinh\left(i\pi\frac{y}{W}\right)}{\sinh\left(i\pi\frac{H}{W}\right)}\frac{\pi}{W}\right) V_{cz}\left(1+\frac{y}{R_c}\right) dy \tag{A7.161}$$

The 2 preceding the integral is due to the existence of two flight surfaces. As before, the absolute value of the velocity is used since dissipation is always a positive quantity. The integration shown in Equation A7.161 is complex and is performed numerically.

Since the velocity in the x direction is assumed not to be dependent on x, the dissipation component for the x-direction-driven flow at the flight is assumed to be zero.

The viscous dissipation that occurs due to pressure-induced flow is calculated in the same manner and it has two components due to the motion at the core and at the flights.

For pressure-induced flow in the z direction and the motion of the screw core:

$$V_{pz} = \frac{H^2}{2\mu}\frac{\partial P}{\partial z}\left[\left(\frac{y}{H}\right)^2 - \frac{y}{H} + \frac{8}{\pi^3}\sum_{i=1,3,5\ldots}^{\infty} \frac{\cosh\left(i\pi\frac{W}{H}\left(\frac{x}{W}-\frac{1}{2}\right)\right)\sin\left(i\pi\frac{y}{H}\right)}{i^3\cosh\left(i\pi\frac{W}{2H}\right)}\right] \tag{A7.162}$$

$$\frac{\partial V_{pz}}{\partial y} = \frac{H^2}{2\mu}\frac{\partial P}{\partial z}\left[\frac{2y}{H^2} - \frac{1}{H} + \frac{8}{\pi^3}\sum_{i=1,3,5\ldots}^{\infty} \frac{\cosh\left(i\pi\frac{W}{H}\left(\frac{x}{W}-\frac{1}{2}\right)\right)\cos\left(i\pi\frac{y}{H}\right)}{i^3\cosh\left(i\pi\frac{W}{2H}\right)}\frac{i\pi}{H}\right] \tag{A7.163}$$

$$\frac{\partial V_{pz}}{\partial y}\bigg|_{y=0} = \frac{H^2}{2\mu}\frac{\partial P}{\partial z}\left[-\frac{1}{H} + \frac{8}{\pi^3}\sum_{i=1,3,5\ldots}^{\infty} \frac{\cosh\left(i\pi\frac{W}{H}\left(\frac{x}{W}-\frac{1}{2}\right)\right)}{i^2\cosh\left(i\pi\frac{W}{2H}\right)}\frac{\pi}{H}\right] \tag{A7.164}$$

$$\tau_{yz}\big|_{y=0} = \mu\frac{\partial V_{pz}}{\partial y}\bigg|_{y=0} = \frac{H^2}{2}\left|\frac{\partial P}{\partial z}\right|\left(-\frac{1}{H} + \frac{8}{\pi^3}\sum_{i=1,3,5\ldots}^{\infty} \frac{\cosh\left[i\pi\frac{W}{H}\left(\frac{x}{W}-\frac{1}{2}\right)\right]}{i^2\cosh\left(i\pi\frac{W}{2H}\right)}\frac{\pi}{H}\right) \tag{A7.165}$$

Equation A7.165 is a negative stress. Since dissipation is always positive, the sign in Equation A7.165 will be changed. The rate of work due to pressure-induced flow in the z direction at the core is as follows:

$$\dot{\omega}_{core\_p} = \frac{pH^2 Z_c V_{cz}}{2}\left|\frac{\partial P}{\partial z}\right| \int_0^{W_c} \frac{1}{H} - \frac{8}{\pi^3}\sum_{i=1,3,5\ldots}^{\infty} \frac{\cosh\left[i\pi \frac{W}{H}\left(\frac{x}{W}-\frac{1}{2}\right)\right]}{i^2 \cosh\left(i\pi \frac{W}{2H}\right)} \frac{\pi}{H}\, dx \quad \text{(A7.166)}$$

$$\dot{\omega}_{core\_p} = \frac{pH^2 Z_c V_{cz}}{2}\left|\frac{\partial P}{\partial z}\right| \left[\frac{W_c}{H} - \frac{8}{\pi^3}\sum_{i=1,3,5\ldots}^{\infty} \frac{\sinh\left[i\pi \frac{W}{H}\left(\frac{W_c}{W}-\frac{1}{2}\right)\right] - \sinh\left[\frac{-i\pi W}{2H}\right]}{i^3 \cosh\left(i\pi \frac{W}{2H}\right)}\right] \quad \text{(A7.167)}$$

When $\frac{W_c}{W}$ approaches 1, then the above equation reduces to:

$$\dot{\omega}_{core\_p} = \frac{pH^2 Z_c V_{cz}}{2}\left|\frac{\partial P}{\partial z}\right| \left[\frac{W_c}{H} - \frac{8}{\pi^3}\sum_{i=1,3,5\ldots}^{\infty} \frac{\sinh\left[i\pi \frac{W}{2H}\right] - \sinh\left[\frac{-i\pi W}{2H}\right]}{i^3 \cosh\left(i\pi \frac{W}{2H}\right)}\right] \quad \text{(A7.168)}$$

$$\dot{\omega}_{core\_p} = \frac{pH^2 Z_c V_{cz}}{2}\left|\frac{\partial P}{\partial z}\right| \left[\frac{W_c}{H} - \frac{8}{\pi^3}\sum_{i=1,3,5\ldots}^{\infty} \frac{\sinh\left[i\pi \frac{W}{2H}\right] + \sinh\left[\frac{i\pi W}{2H}\right]}{i^3 \cosh\left(i\pi \frac{W}{2H}\right)}\right] \quad \text{(A7.169)}$$

$$\dot{\omega}_{core\_p} = \frac{pH^2 Z_c V_{cz}}{2}\left|\frac{\partial P}{\partial z}\right| \left(\frac{W_c}{H} - \frac{16}{\pi^3}\sum_{i=1,3,5\ldots}^{\infty} \frac{\tanh\left[i\pi \frac{W}{2H}\right]}{i^3}\right) \quad \text{(A7.170)}$$

For pressure-induced flow in the z direction and the motion of the screw flights:

$$\tau_{xz}\,|_{x=0} = \mu \frac{\partial V_{pz}}{\partial x}\,|_{x=0} = \frac{H^2}{2}\frac{\partial P}{\partial z}\frac{8}{\pi^3}\left(\sum_{i=1,3,5\ldots}^{\infty} \frac{\sinh\left(i\pi \frac{W}{2H}\right)\sin\left(i\pi \frac{y}{H}\right)}{i^2 \cosh\left(i\pi \frac{W}{2H}\right)} \frac{\pi}{H}\right) = \frac{H^2}{2}\frac{\partial P}{\partial z}\frac{8}{\pi^3}$$

$$\left(\sum_{i=1,3,5\ldots}^{\infty} \frac{\tanh\left(i\pi \frac{W}{2H}\right)\sin\left(i\pi \frac{y}{H}\right)}{i^2} \frac{\pi}{H}\right) \quad \text{(A7.171)}$$

To maintain the stress as a positive quantity:

$$\tau_{xz}\big|_{x=0} = \mu\frac{\partial V_{pz}}{\partial x}\big|_{x=0} = \frac{H^2}{2}\left|\frac{\partial P}{\partial z}\right|\frac{8}{\pi^3}\left(\sum_{i=1,3,5\ldots}^{\infty}\frac{\tanh\left(i\pi\frac{W}{2H}\right)\sin\left(i\pi\frac{y}{H}\right)}{i^2}\frac{\pi}{H}\right) \quad \text{(A7.172)}$$

$$\dot{\omega}_{flight\_p} = 2p\int_0^{Z_c}\int_0^H\left(\tau_{xz}\big|_{x=0}\right)V_{cz}\left(1+\frac{y}{H}\right)dydz = 2pZ_c\int_0^H\left(\frac{H^2}{2}\left|\frac{\partial P}{\partial z}\right|\frac{8}{\pi^3}\right.$$

$$\left.\left(\sum_{i=1,3,5\ldots}^{\infty}\frac{\tanh\left(i\pi\frac{W}{2H}\right)\sin\left(i\pi\frac{y}{H}\right)}{i^2}\frac{\pi}{H}\right)\right)V_{cz}\left(1+\frac{y}{H}\right)dy \quad \text{(A7.173)}$$

The integration shown in Equation A7.173 is complex and is performed numerically.

The heat transferred through the barrel wall over the channel, E_T, is given by Equation A7.174. A heat transfer expression could also be written for energy transferred through the barrel wall above the flight land to the molten resin.

$$E_T = pK_bW_bZ_b(T_m - T_b) \quad \text{(A7.174)}$$

$$\Delta E_T = pK_bW_b(T_m - T_b)\Delta z_b \quad \text{(A7.175)}$$

where K_b is the heat transfer coefficient at the barrel wall, T_m is the melt temperature of the resin in the channel, and T_b is the temperature of the barrel wall.

In order to use the dissipation equations in a control volume temperature calculation, the energy dissipation must be normalized to a unit length of Z_c for a single channel. The energy dissipations for unit length of Z_c with units of J/(m·s) for a single channel are as follows:

$$\dot{\omega}'_{core\_z} = \mu V_{cz}^2\left(1+\frac{H}{R_c}\right)\frac{8}{\pi}\left[\sum_{i=1,3,5\ldots}^{\infty}\frac{1}{i\sinh\left(i\pi\frac{H}{W}\right)}\right] - \frac{\mu V_{cz}^2 W_c}{R_c} \quad \text{(A7.176)}$$

$$\dot{\omega}'_{core\_x} = \mu\left(\frac{2V_{cx}^2}{H}+\frac{3V_{cx}^2}{R_c}\right)W_c \quad \text{(A7.177)}$$

$$\dot{\omega}'_{core\_p} = \frac{H^2 V_{cz}}{2} \left| \frac{\partial P}{\partial z} \right|_k \left(\frac{W_c}{H} - \frac{16}{\pi^3} \sum_{i=1,3,5...}^{\infty} \frac{\tanh\left[i\pi \frac{W}{2H} \right]}{i^3} \right) \tag{A7.178}$$

where $\partial P/\partial z$ is the local pressure gradient at the start of the increment at k. As before, the absolute value of the local pressure gradient is used to force the dissipation term to be positive.

$$\dot{\omega}'_{flight\_z} = 2\int_0^H \left(\mu V_{cz} \left(1 + \frac{H}{R_c}\right) \frac{4}{\pi} \sum_{i=1,3,5...}^{\infty} \frac{\sinh\left(i\pi \frac{y}{W} \right)}{\sinh\left(i\pi \frac{H}{W} \right)} \frac{\pi}{W} \right) V_{cz} \left(1 + \frac{y}{R_c}\right) dy \tag{A7.179}$$

$$\dot{\omega}'_{flight\_p} = 2\int_0^H \left(\frac{H^2}{2} \left| \frac{\partial P}{\partial z} \right|_k \frac{8}{\pi^3} \left(\sum_{i=1,3,5...}^{\infty} \frac{\tanh\left(i\pi \frac{W}{2H} \right) \sin\left(i\pi \frac{y}{H} \right)}{i^2} \frac{\pi}{H} \right) \right) V_{cz} \left(1 + \frac{y}{H}\right) dy \tag{A7.180}$$

$$\dot{\omega}'_{\lambda} = \frac{\mu_f \pi^2 N^2 D_b^2 e Z_b}{\lambda Z_c} \tag{A7.181}$$

The integrations shown in Equations A7.179 and A7.180 are complex and are performed numerically. The heat transferred through the barrel wall over the channel for unit length Z_c is E'_T, and it is calculated using Equation A7.182. A heat transfer expression could also be written for energy transferred from the molten polymer to the barrel wall above the flight land and transferred from the melt to the screw.

$$E'_T = \frac{K_b W_b Z_b}{Z_c} (T_m - T_b) \tag{A7.182}$$

The energy balance for the $k+1$ increment in the Δz_c direction is as follows:

$$T_{k+1} = T_k + \frac{\left(\dot{\omega}'_{core\_x} + \dot{\omega}'_{core\_z} + \dot{\omega}'_{flight\_p} + \dot{\omega}'_{flight\_z} + \dot{\omega}'_{core\_p} + \dot{\omega}'_{\lambda} - E'_T \right)}{Q_m C_v} p\Delta z_c \tag{A7.183}$$

A7.2.2 Viscous Energy Dissipation for Screw Rotation for Channels with Small Aspect Ratios ($H/W < 0.1$)

For channels with small aspect ratios, the complex summations can be eliminated from the velocity and dissipation equations. For the x component of flow at the core, the dissipation calculation is the same as that for the generalized solution as follows:

$$\dot{\omega}_{core\_x} = p\mu\left(\frac{2V_{cx}^2}{H} + \frac{3V_{cx}^2}{R_c}\right)Z_c W_c \tag{A7.184}$$

For the z component of flow due to core rotation:

$$V_{dzE} = \frac{y}{H}V_{cz}\left(1 + \frac{H}{R_c}\right) - V_{cz}\left(1 + \frac{y}{R_c}\right) \tag{A7.185}$$

$$V_{dzE}\big|_{y=0} = -V_{cz} \tag{A7.186}$$

$$\tau_{yz}\,|_{y=0} = \mu\frac{\partial V_{dzE}}{\partial y}\,|_{y=0} = \frac{\mu V_{cz}}{H}\left(1 + \frac{H}{R_c}\right) - \frac{\mu V_{cz}}{R_c} \tag{A7.187}$$

$$\dot{\omega}_{core\_z} = p\int_0^{W_c}\int_0^{Z_c}\left(\tau_{yz}\,|_{y=0}\right)\left(V_z\,|_{y=0}\right)dz\,dx \tag{A7.188}$$

$$\dot{\omega}_{core\_z} = \frac{p\mu W_c Z_c V_{cz}^2}{H}\left(1 + \frac{H}{R_c}\right) - \frac{p\mu W_c Z_c V_{cz}^2}{R_c} = \frac{p\mu W_c Z_c V_{cz}^2}{H} \tag{A7.189}$$

The rate of work for flow in the z direction due to the motion of the flights $\dot{\omega}_{flight\_z}$ is as follows:

$$V_{dzE} = \frac{y}{H}V_{cz}\left(1 + \frac{H}{R_c}\right) - V_{cz}\left(1 + \frac{y}{R_c}\right) \tag{A7.190}$$

$$V_{dzE}\big|_{x=0} = \frac{y}{H}V_{cz}\left(1 + \frac{H}{R_c}\right) - V_{cz}\left(1 + \frac{y}{R_c}\right) \tag{A7.191}$$

$$\tau_{xz}\,|_{x=0} = \mu\frac{\partial V_{dzE}}{\partial x}\,|_{x=0} = 0 \tag{A7.192}$$

$$\dot{\omega}_{flight\_z} = 2pZ_c\int_0^{H}\left(\tau_{xz}\,|_{y=0}\right)\left(\frac{y}{H}V_{cz}\left(1 + \frac{H}{R_c}\right) - V_{cz}\left(1 + \frac{y}{R_c}\right)\right)dy = 0 \tag{A7.193}$$

For pressure-induced flow in the z direction and the motion of the screw core:

$$V_{pz} = \frac{H^2}{2\mu}\frac{\partial P}{\partial z}\left[\left(\frac{y}{H}\right)^2 - \frac{y}{H}\right] \tag{A7.194}$$

$$\tau_{yz}\,|_{y=0} = \mu\frac{\partial V_{pz}}{\partial y}\,|_{y=0} = -\frac{H}{2}\frac{\partial P}{\partial z} \tag{A7.195}$$

$$\dot{\omega}_{core\_z} = p\int_0^{W_c}\int_0^{Z_c}\left(\tau_{yz}\,|_{y=0}\right)\left(V_{cz}\right)dzdx \tag{A7.196}$$

$$\dot{\omega}_{core\_p} = \frac{pHW_cZ_cV_{cz}}{2}\left|\frac{\partial P}{\partial z}\right| \tag{A7.197}$$

The absolute value of the pressure gradient is used to force the rate of work term to be positive.

For pressure-induced flow in the x direction and the motion of the screw core:

$$V_{pz} = \frac{H^2}{2\mu}\frac{\partial P}{\partial z}\left[\left(\frac{y}{H}\right)^2 - \frac{y}{H}\right] \tag{A7.198}$$

$$\tau_{xz}\,|_{x=0} = \mu\frac{\partial V_{pz}}{\partial x}\,|_{x=0} = 0 \tag{A7.199}$$

$$\dot{\omega}_{flight\_p} = 2p\int_0^{H}\int_0^{Z_c}\left(\tau_{yz}\,|_{y=0}\right)V_{cz}\left(1+\frac{y}{H}\right)dzdy = 0 \tag{A7.200}$$

The simplified method for calculation of the temperature profile using the control volume method and screw rotation is shown by Equation A7.201. The simplified calculation under-predicts the energy dissipation.

$$T_{k+1} = \tag{A7.201}$$

$$T_k + \frac{\left(\frac{2\mu V_{cx}^2W_c}{H} + \frac{3\mu V_{cx}^2W_c}{R_c} + \frac{\mu W_cV_{cz}^2}{H} + \frac{V_{cz}HW_c}{2}\left|\frac{\partial P}{\partial z}\right|_k + \frac{\pi^2\mu_fN^2D_b^2eZ_b}{\lambda Z_c} - \frac{K_bW_bZ_b}{Z_c}\left(T_k - T_0\right)\right)p\Delta z_c}{Q_mC_v}$$

■ A7.3 Viscous Energy Dissipation for Barrel Rotation

The viscous energy dissipation is calculated by multiplying the normal shear stress for the barrel surface by the velocity of the barrel surface integrated over the barrel surface area. The traditional method for calculation of dissipation for barrel rotation is as follows:

$$\dot{\omega}_b = \iint_s n_i \tau_{i,j} v_j ds \tag{A7.202}$$

where $\dot{\omega}$ is the rate of work (or power) that is added to the system through surface s for barrel rotation, n_i is the normal unit vector, τ is the shear stress tensor, and v is the velocity vector. The shear stresses, velocities, and surfaces for this integral for the length of the metering section are as follows:

$$\dot{\omega}_b = p\int_0^{W_b}\int_0^{Z_b}\left(\tau_{yz}\,|_{y=H}\right)\left(V_z\,|_{y=H}\right)dzdx + p\int_0^{W_b}\int_0^{Z_b}$$

$$\left(\tau_{yx}\,|_{y=H}\right)\left(V_x\,|_{y=H}\right)dzdx + p\dot{\omega}_{clear} \tag{A7.203}$$

A7.3.1 Viscous Energy Dissipation for Barrel Rotation: Generalized Solution

For the generalized solution for viscous energy dissipation, the Lagrangian component velocities are used for the analysis as follows.

The velocity of the barrel for the x and z directions is as follows:

$$V_x\Big|_{y=H} = -V_{bx} \tag{A7.204}$$

$$V_z\Big|_{y=H} = V_{bz} \tag{A7.205}$$

For the x component of flow and the stress at the barrel for the component:

$$V_{xL} = \frac{y}{H}\left(2 - 3\frac{y}{H}\right)V_{bx} \tag{A7.206}$$

$$\tau_{yx}\Big|_{y=H} = \mu\frac{\partial V_{xL}}{\partial y}\Big|_{y=H} = \frac{-4\mu V_{bx}}{H} \tag{A7.207}$$

For viscous dissipation the absolute values of the velocity and stress will be used. The rate of work added for barrel rotation and flow in the x direction:

$$\dot{\omega}_{b\_x} = p\int_0^{W_b}\int_0^{Z_b}\left(\tau_{yx}\,|_{y=H}\right)\left(V_x\,|_{y=H}\right)dzdx \tag{A7.208}$$

$$\dot{\omega}_{b\_x} = \frac{4p\mu V_{bx}^2 Z_b W_b}{H} \tag{A7.209}$$

The z-direction rotational flow and stress at the barrel is calculated as follows:

$$V_{dzL} = V_{cz}\left(1+\frac{H}{R_c}\right)\frac{4}{\pi}\sum_{i=1,3,5...}^{\infty}\frac{\sinh\left(i\pi\frac{y}{W}\right)\sin\left(i\pi\frac{x}{W}\right)}{i\sinh\left(i\pi\frac{H}{W}\right)}$$

$$=\frac{4V_{bz}}{\pi}\sum_{i=1,3,5...}^{\infty}\frac{\sinh\left(i\pi\frac{y}{W}\right)\sin\left(i\pi\frac{x}{W}\right)}{i\sinh\left(i\pi\frac{H}{W}\right)} \tag{A7.210}$$

$$\tau_{yz}\Big|_{y=H} = \mu\frac{\partial V_{dzL}}{\partial y}\Bigg|_{y=H} = \frac{4\mu V_{bz}}{\pi}\sum_{i=1,3,5...}^{\infty}\frac{\cosh\left(i\pi\frac{H}{W}\right)\sin\left(i\pi\frac{x}{W}\right)}{i\sinh\left(i\pi\frac{H}{W}\right)}\frac{i\pi}{W} \tag{A7.211}$$

$$\dot{\omega}_{b\_z} = p\int_0^{W_b}\int_0^{Z_b}\left(\tau_{yz}\big|_{y=H}\right)\left(V_z\big|_{y=H}\right)dz\,dx \tag{A7.212}$$

$$\dot{\omega}_{b\_z} = \frac{4p\mu Z_B V_{bz}^2}{\pi}\left[\sum_{i=1,3,5...}^{\infty}\frac{-\cosh\left(i\pi\frac{H}{W}\right)\cos\left(i\pi\frac{x}{W}\right)}{i\sinh\left(i\pi\frac{H}{W}\right)}\right]_{x=0}^{x=W_b} \tag{A7.213}$$

$$\dot{\omega}_{b\_z} = \frac{4p\mu Z_B V_{bz}^2}{\pi}\left[\sum_{i=1,3,5...}^{\infty}\frac{-\cosh\left(i\pi\frac{H}{W}\right)\cos\left(i\pi\frac{W_b}{W}\right)}{i\sinh\left(i\pi\frac{H}{W}\right)}-\frac{-\cosh\left(i\pi\frac{H}{W}\right)}{i\sinh\left(i\pi\frac{H}{W}\right)}\right] \tag{A7.214}$$

$$\dot{\omega}_{b\_z} = \frac{4p\mu Z_b V_{bz}^2}{\pi}\left[\sum_{i=1,3,5...}^{\infty}\frac{\cosh\left(i\pi\frac{H}{W}\right)}{i\sinh\left(i\pi\frac{H}{W}\right)}-\frac{\cosh\left(i\pi\frac{H}{W}\right)\cos\left(i\pi\frac{W_b}{W}\right)}{i\sinh\left(i\pi\frac{H}{W}\right)}\right] \tag{A7.215}$$

Equation A7.215 is the proper dissipation term for flow in the z direction due to barrel rotation. For relatively shallow channels the term $\frac{W_b}{W}$ approaches 1, so the above equation reduces to:

$$\dot{\omega}_{b\_z} = \frac{8p\mu Z_b V_{bz}^2}{\pi}\left[\sum_{i=1,3,5...}^{\infty}\frac{\cosh\left(i\pi\frac{H}{W}\right)}{i\sinh\left(i\pi\frac{H}{W}\right)}\right] \tag{A7.216}$$

For z-direction pressure flow:

$$V_{pz} = \frac{H^2}{2\mu}\frac{\partial P}{\partial z}\left[\left(\frac{y}{H}\right)^2 - \frac{y}{H} + \frac{8}{\pi^3}\sum_{i=1,3,5\ldots}^{\infty}\frac{\cosh\left(i\pi\frac{W}{H}\left(\frac{x}{W}-\frac{1}{2}\right)\right)\sin\left(i\pi\frac{y}{H}\right)}{i^3\cosh\left(i\pi\frac{W}{2H}\right)}\right] \tag{A7.217}$$

$$\tau_{yz}\,|_{y=H} = \mu\frac{\partial V_{pz}}{\partial y}\,|_{y=H} = \frac{H^2}{2}\frac{\partial P}{\partial z}\left(\frac{1}{H} - \frac{8}{\pi^3}\sum_{i=1,3,5\ldots}^{\infty}\frac{\cosh\left[i\pi\frac{W}{H}\left(\frac{x}{W}-\frac{1}{2}\right)\right]}{i^3\cosh\left(i\pi\frac{W}{2H}\right)}\frac{i\pi}{H}\right) \tag{A7.218}$$

$$\dot{\omega}_{barrel\_p} = p\int_0^{W_b}\int_0^{Z_b}\left(\tau_{yz}\,|_{y=H}\right)\left(V_z\,|_{y=H}\right)dzdx \tag{A7.219}$$

$$\dot{\omega}_{barrel\_p} = \frac{pH^2Z_bV_{bz}}{2}\left|\frac{\partial P}{\partial z}\right|\int_0^{W_b}\left(\frac{1}{H} - \frac{8}{\pi^3}\sum_{i=1,3,5\ldots}^{\infty}\frac{\cosh\left[i\pi\frac{W}{H}\left(\frac{x}{W}-\frac{1}{2}\right)\right]}{i^3\cosh\left(i\pi\frac{W}{2H}\right)}\frac{i\pi}{H}\right)dx \tag{A7.220}$$

$$\dot{\omega}_{barrel\_p} = \frac{pH^2Z_bV_{bz}}{2}\left|\frac{\partial P}{\partial z}\right|\left[\frac{W_b}{H} - \frac{8}{\pi^3}\sum_{i=1,3,5\ldots}^{\infty}\frac{\sinh\left[i\pi\frac{W}{H}\left(\frac{W_b}{W}-\frac{1}{2}\right)\right]}{i^3\cosh\left(i\pi\frac{W}{2H}\right)} - \frac{\sinh\left[-i\pi\frac{W}{2H}\right]}{i^3\cosh\left(i\pi\frac{W}{2H}\right)}\right] \tag{A7.221}$$

$$\dot{\omega}_{barrel\_p} = \frac{pH^2Z_bV_{bz}}{2}\left|\frac{\partial P}{\partial z}\right|\left[\frac{W_b}{H} - \frac{8}{\pi^3}\sum_{i=1,3,5\ldots}^{\infty}\frac{\sinh\left[i\pi\frac{W}{H}\left(\frac{W_b}{W}-\frac{1}{2}\right)\right]}{i^3\cosh\left(i\pi\frac{W}{2H}\right)} + \frac{\sinh\left[i\pi\frac{W}{2H}\right]}{i^3\cosh\left(i\pi\frac{W}{2H}\right)}\right] \tag{A7.222}$$

In order to use the dissipation equations in a control volume temperature calculation, the energy dissipation must be normalized to a unit length of Z_b. The energy dissipations for unit length of Z_b with units of J/(m·s) are as follows for a single channel:

$$\dot{\omega}'_{b\_x} = \frac{4\mu V_{bx}^2 W_b}{H} \tag{A7.223}$$

$$\dot{\omega}'_{b\_z} = \frac{4\mu V_{bz}^2}{\pi}\left[\sum_{i=1,3,5\ldots}^{\infty}\frac{\cosh\left(i\pi\frac{H}{W}\right)}{i\sinh\left(i\pi\frac{H}{W}\right)} - \frac{\cosh\left(i\pi\frac{H}{W}\right)\cos\left(i\pi\frac{W_b}{W}\right)}{i\sinh\left(i\pi\frac{H}{W}\right)}\right] \tag{A7.224}$$

$$\dot{\omega}'_{barrel\_p} = \frac{H^2 V_{bz}}{2}\left|\frac{\partial P}{\partial z}\right|\left[\frac{W_b}{H} - \frac{8}{\pi^3}\sum_{i=1,3,5...}^{\infty}\frac{\sinh\left[i\pi\frac{W}{H}\left(\frac{W_b}{W}-\frac{1}{2}\right)\right]}{i^3\cosh\left(i\pi\frac{W}{2H}\right)} + \frac{\sinh\left[i\pi\frac{W}{2H}\right]}{i^3\cosh\left(i\pi\frac{W}{2H}\right)}\right] \quad \text{(A7.225)}$$

$$\dot{\omega}'_{\lambda b} = \frac{\mu_f \pi^2 N^2 D_b^2 e}{\lambda} \quad \text{(A7.226)}$$

$$E'_{Tb} = K_b W_b (T_m - T_b) \quad \text{(A7.227)}$$

The energy balance for the $k+1$ increment in the Δz_b direction is as follows:

$$T_{k+1} = T_k + \frac{\left(\dot{\omega}'_{b\_x} + \dot{\omega}'_{b\_z} + \dot{\omega}'_{b\_p} + \dot{\omega}'_{\lambda b} - E'_{Tb}\right)}{Q_m C_v} p \Delta z_b \quad \text{(A7.228)}$$

A7.3.2 Viscous Energy Dissipation for Barrel Rotation for Channels with Small Aspect Ratios (*H*/*W* < 0.1)

For the x component of flow:

$$V_{xL} = \frac{y}{H}\left(2 - 3\frac{y}{H}\right)V_{bx} \quad \text{(A7.229)}$$

$$\tau_{yx}\Big|_{y=H} = \mu \frac{\partial V_{xL}}{\partial y}\Big|_{y=H} = \frac{-4\mu V_{bx}}{H} \quad \text{(A7.230)}$$

For the z component of flow (both drag and pressure flows):

$$V_{zL} = V_{bz}\frac{y}{H} - \left(\frac{H^2}{2\mu}\frac{\partial P}{\partial z}\right)\left(1 - \frac{y}{H}\right)\frac{y}{H} \quad \text{(A7.231)}$$

$$\tau_{yz}\Big|_{y=H} = \mu \frac{\partial V_{bz}}{\partial y}\Big|_{y=H} = \frac{\mu V_{bz}}{H} + \frac{H}{2}\left|\frac{\partial P}{\partial z}\right| \quad \text{(A7.232)}$$

Substituting the shear stresses and velocities and then integrating provides the rate of work for the channel as provided by Equation A7.233 for barrel rotation. The units of $\dot{\omega}_b$ are J/s.

$$\dot{\omega}_b = \frac{\mu V_{bz}^2 W_b Z_b}{H} + \frac{V_{bz} H W_b Z_b}{2}\left|\frac{\partial P}{\partial z}\right| + \frac{4\mu V_{bx}^2 W_b Z_b}{H} + \dot{\omega}_\lambda \quad \text{(A7.233)}$$

The incremental rate of work $\Delta\dot{\omega}_b$ inputted into an incremental channel length Δz_b is as follows:

$$\Delta\dot{\omega}_b = \left(\frac{\mu V_{bz}^2 W_b}{H} + \frac{V_{bz} H W_b}{2} \left| \frac{\partial P}{\partial z} \right| + \frac{4\mu V_{bx}^2 W_b}{H} + \frac{\mu_f \pi^2 N^2 D_b^2 e}{\lambda} \right) p \Delta z_b \tag{A7.234}$$

The energy balance for the $k+1$ increment in the Δz_b direction is as follows:

$$T_{k+1} = \tag{A7.235}$$

$$T_k + \frac{\left(\frac{\mu V_{bz}^2 W_b}{H} + \frac{V_{bz} H W_b}{2} \left| \frac{\partial P}{\partial z} \right|_k + \frac{4\mu V_{bx}^2 W_b}{H} + \frac{\mu_f \pi^2 N^2 D_b^2 e}{\lambda} - K_s W_b (T_m - T_b) \right) p \Delta z_b}{Q_m C_v}$$

References

1. Tadmor, Z. and Klein, I., "Engineering Principles of Plasticating Extrusion," Van Nostrand Reinhold Company, New York (1970)
2. Powers, D. L., "Boundary Value Problems and Partial Differential Equations," 6th ed., Elsevier, Amsterdam (2010)
3. Boussinesq, M. J., *J. Mathematique Pures et Appliquees*, serials 2.13, 377 (1868)

Author Index

G

H

I

J

K

T

U

V

W

Subject Index

A

B

C

D

O

P

Q

R

S

T

U

V

W

Z

www.ingramcontent.com/pod-product-compliance
Lightning Source LLC
LaVergne TN
LVHW082052011026
849162LV00020B/1629
* 9 7 8 1 5 6 9 9 0 7 8 4 9 *